Microbiology

Laboratory Theory and Application

2nd Edition

Michael J. Leboffe
San Diego City College

Burton E. Pierce
San Diego City College

Morton Publishing Company
925 W. Kenyon Ave., Unit 12
Englewood, Colorado 80110
http://www.morton-pub.com

Book Team

Publisher: Douglas N. Morton
David Ferguson Biology Editor
Production Manager: Joanne Saliger
Typography: Ash Street Typecrafters, Inc.
Copyediting: Carolyn Acheson
Cover Design: Bob Schram, Bookends, Inc.

ISBN 10: 0-89582-708-5
ISBN 13: 978-0-89582-708-1

10 9 8 7 6 5 4 3 2 1

Printed in the United States of America

Preface

This edition of *Microbiology Laboratory Theory and Application* has been four years in the making. That is, we started thinking about improvements almost immediately after the publication of the first edition. Global changes include replacing many photographs with better ones, reducing the number of organisms used throughout the book, and the addition of a new component called, "In This Exercise..." that provides an overview of what students can expect from each exercise.

Because the choice of appropriate educational organisms is limited, and because of the need to provide an optimal learning experience, species used in laboratory exercises are often duplicated in the photos. For this reason we have decided not to include organism names in this edition's photo captions. This has enabled us to use standard control organisms for each medium in both the photos *and* procedures without revealing specific results prior to performing tests.

Specific changes in each section are as follows.

- **Introduction**
  - Safety procedures have been updated.
  - The section on graphing has been rewritten.
- **Section 1 Fundamental Skills for the Microbiology Laboratory**
  - Instructions for common aseptic transfers have been incorporated into Exercise 1-2.
  - Less commonly used and more specialized transfers, including pipetting, have been moved to the appendices.
  - The streak plate (Exercise 1-3) and spread plate (Exercise 1-4) techniques have been moved into this section.
- **Section 2 Microbial Growth**
  - Many new photos of colony morphology have been added.
  - The thermal death exercise has been moved to Section 6, Quantitative Techniques.
  - The Kirby-Bauer exercise has been moved to the newly-created Section 7, Medical Microbiology.
  - Cultivation of Anaerobes has been split into two exercises: Exercise 2-8, Thioglycollate Broth, and Exercise 2-9, The Anaerobic Jar.
  - Exercise 2-15, Hand Scrubbing, has been rewritten.
- **Section 3 Microscopy and Staining**
  - Exercise 3-6, The Gram Stain, has been expanded to include more photomicrographs, including direct smears and commonly seen artifacts.
- **Section 4 Selective Media**
  - Most photos have been replaced.
  - The Streak Plate Method of Isolation has been moved to Section 1.
- **Section 5 Differential Tests**
  - Many photos have been replaced.
  - Where safety permits, procedures now include standard positive and negative control organisms.
- **Section 6 Quantitative Techniques**
  - Exercise 6-6, Thermal Death Time and Decimal Reduction Value (previously in Section 2) has been moved into this section and rewritten to eliminate organizational difficulties.
  - All applicable exercises have been updated and revised for understandability (and consistency with dilution formulas) using milliliter measurements. Alternative Exercises 6-1, 6-4, and 6-6 written for microliter measurements and digital pipettes are included in Appendix F.
- **Section 7 Medical Microbiology**
  - This is a new section derived from the old Environmental, Food, and Medical Microbiology.
  - Two unknowns exercises—one for enterics and the other for Gram-positive cocci—have been developed from the identification flowcharts found in Appendix A of the First Edition.
- **Section 8 Environmental and Food Microbiology**
  - EMB Agar, used in Exercise 8-1, Membrane Filter Technique, has been changed to Endo Agar.

- Exercise 8-2, MPN, has been rewritten and includes a new procedural diagram.
- Exercise 8-3, Bioluminescence, has been rewritten for solid medium to produce better results and increase student success.

Section 9 Microbial Genetics

- Exercise 9-1, Extraction of DNA from Bacterial Cells, has been rewritten.
- Exercise 9-2, Ultraviolet Radiation Damage and Repair, has been rewritten.

Section 10 Hematology and Serology

- New micrographs of blood cells have been used in Exercise 10-1, Differential Blood Cell Count.
- The graph showing lattice formation during precipitation reactions in Exercise 10-2 has been modified.
- New photographs of blood types have been included in Exercise 10-5, Blood Typing.

Section 11 Eukaryotic Microbes

- New micrographs and photographs of fungi have been added to Exercise 11-1, The Fungi—Common Yeasts and Molds.
- New photographs and micrographs of parasitic worms have been added to Exercise 11-3, Parasitic Helminths.

Appendices

- Appendix A now includes a greater diversity of metabolic pathways and summary tables of reactants and products.
- Appendix B covers less commonly used transfer methods in an attempt to make Exercise 1-2 a more manageable size.
- Appendices B and C treat the use of glass and digital pipettors, respectively.
- Appendix F contains alternate dilution schemes using microliter volumes for Exercises 6-1, 6-4, and 6-6.
- The recipes in Appendix G have been updated.

Our goal in this revision was to produce a user-friendly laboratory manual—one that facilitates learning and contributes to a positive experience for students and instructors alike. Time will tell if we have succeeded. As always, your comments are welcome.

Michael J. Leboffe
Burton E. Pierce
San Diego City College

Acknowledgments

Microbiology Laboratory Theory and Application, 2nd Edition, is truly an amalgamation of work comprising three editions of the *Photographic Atlas of Microbiology* and *Exercises for the Microbiology Laboratory* covering over a decade of writing. It would be impossible to thank everyone who has assisted us along the way, but be assured that these works would not be anywhere near what they are without your contributions. We remain indebted to all of you for your assistance.

Some individuals specifically helped us on this edition. We are indebted to Roberta Pettriess of Wichita State University for critically reviewing the first edition of the Lab Manual. We have incorporated many of her suggestions throughout. Our thanks also go to Dr. Melissa Scott and Allison Shearer of San Diego City College, Donna Mapston and Dr. Ellen Potter of Scripps Institute for Biological Studies, and Dr. Sandra Slivka of Miramar College for their helpful suggestions on improving the DNA extraction exercise.

We also have been the beneficiaries of suggestions from our students as they experience these activities from the student perspective. Two students, Julie Fitzpatrick and Kerrylea O'Brien contributed extra hours and valuable insights as we worked out some bugs from several protocols. In addition, Dr. David Singer of San Diego City College was a readily available resource for feedback from the professor's perspective. Marlene DeMers of San Diego State University also provided feedback and ideas for improvement of several exercises. Thanks to all of you.

Thanks to Bob Schram from Bookends, Inc. for the handsome cover and interior design. A special thanks to Joanne Saliger from Ash Street Typecrafters, Inc., who, for the last 11+ years has consistently taken our raw files and transformed them into the beautiful compositions that have been so successful. Thanks, Joanne, for your talent, your artistry, and your hard work.

And as always, thanks to the team at Morton Publishing—Doug Morton, President; Chrissy Morton DeMier, Business Manager; and David Ferguson, Biology Editor, for your continued belief in our projects and the loose reins on our creativity.

Finally, thanks to our wives, Karen and Michele, for your ability to recognize the difference between productive determination and obsession. Thanks to your support and encouragement throughout this lengthy project, we can now enjoy the fruits of our labor together.

Contents

Introduction

Safety and Laboratory Guidelines

Microbiology lab can be an interesting and exciting experience, but at the outset you should be aware of some potential hazards. Improper handling of chemicals, equipment and/or microbial cultures is dangerous and can result in injury or infection. Safety with lab equipment will be addressed when you first use that specific piece of equipment, as will specific examples of chemical safety. Our main concern here is to introduce you to safe handling and disposal of microbes.[1]

Because microorganisms present varying degrees of risk to laboratory personnel (students, technicians, and faculty), people outside the laboratory, and the environment, microbial cultures must be handled safely. The classification of microbes into four biosafety levels (BSLs) provides a set of minimum standards for laboratory practices, facilities, and equipment to be used when handling organisms at each level. These biosafety levels, defined in the U.S. Government publication, *Biosafety in Microbiological and Biomedical Laboratories*, are summarized below and in Table I-1. For complete information, readers are referred to the original document.

BSL-1: Organisms do not typically cause disease in healthy individuals and present a minimal threat to the environment and lab personnel. Standard microbiological practices are adequate. These microbes may be handled in the open, and no special containment equipment is required.

BSL-2: Organisms are commonly encountered in the community and present a moderate environmental and/or health hazard. These organisms are associated with human diseases of varying severity. Individuals may do laboratory work that is not especially prone to splashes or aerosol generation, using standard microbiological practices.

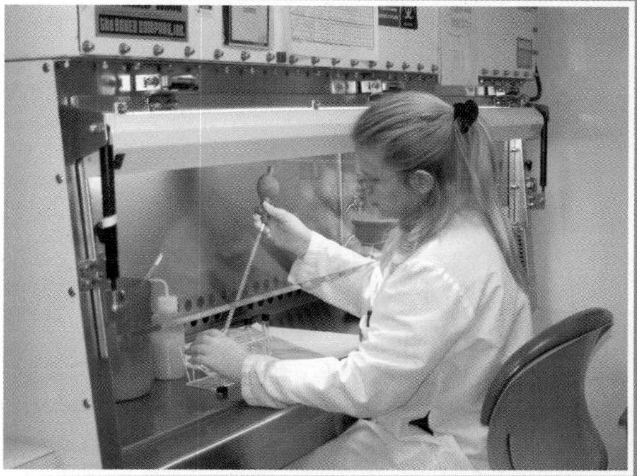

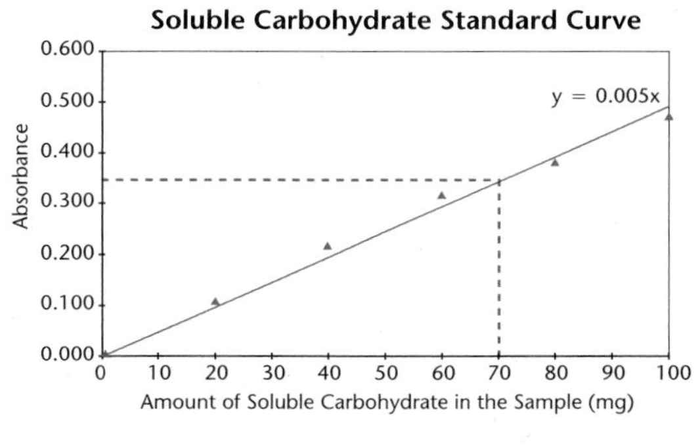

[1] Your instructor may augment or revise these guidelines to fit the conventions of your laboratory

BSL	Agents	Practices	Safety Equipment (Primary Barriers)	Facilities (Secondary Barriers)
1	Not known to consistently cause disease in healthy individuals	Standard microbiological practices	None required	Open benchtop sink required
2	Associated with human disease, hazard is through percutaneous injury, ingestion, exposure to mucous membrane	BSL-1 practice plus: ● Limited lab access ● Biohazard warning signs ● "Sharps" precautions ● Biosafety manual defining any needed waste decontamination or medical surveillance policies	Primary barriers = Class I or II BSCs or other physical containment devices used for all manipulations of agents that cause splashes or aerosols of infectious materials; Personal Protective Equipment (PPEs): laboratory coats, gloves, face protection, as needed	BSL-1 plus: Autoclave available
3	Indigenous or exotic agents with potential for aerosol transmission; disease may have serious or lethal consequences	BSL-2 practices plus: ● Controlled lab access ● Decontamination of all waste ● Decontamination of all lab clothing before laundering ● Baseline serum	Primary barriers = Class I or II BSCs or other physical containment devices used for all open manipulations of agents; PPEs: protective lab clothing, gloves, respiratory protection as needed	BSL-2 plus: ● Physical separation from access corridors ● Access to self-closing, double door ● Exhausted air not recirculated ● Negative airflow into laboratory
4	Dangerous/exotic agents that pose high risk of life-threatening disease, aerosol-transmitted lab infections; or related agents with unknown risk of transmission	BSL-3 practices plus: ● Clothing change before entering ● Shower on exit ● All material decontaminated on exit from facility	Primary barriers = All procedures conducted in Class III BSCs or Class I or II BSCs *in combination with* full-body, air-supplied, positive pressure personnel suit	BSL-3 plus: ● Separate building or isolated zone ● Dedicated supply and exhaust, vacuum, and decon systems ● Other requirements outlined in the text

Source: Reprinted from *Biosafety in Microbiological and Biomedical Laboratories*, 4th edition (Washington: U. S. Government Printing Office, 1999).

TABLE I-1 Summary of Recommended Biosafety Levels for Infectious Agents

BSL-3: Organisms are of local or exotic origin and are associated with respiratory transmission and serious or lethal diseases. Special ventilation systems are used to prevent aerosol transmission out of the laboratory, and access to the lab is restricted. Specially trained personnel handle microbes in a Class I or II biological safety cabinet (BSC), not on the open bench (see Figure I-1).

BSL-4: Organisms have a great potential for lethal infection. Inhalation of infectious aerosols, exposure to infectious droplets, and autoinoculation are of primary concern. The lab is isolated from other facilities, and access is strictly controlled. Ventilation and waste management are under rigid control to prevent release of the microbial agents to the environment. Specially trained personnel perform transfers in Class III BSCs. Class II BSCs may be used as long as personnel wear positive pressure, one-piece body suits with a life-support system.

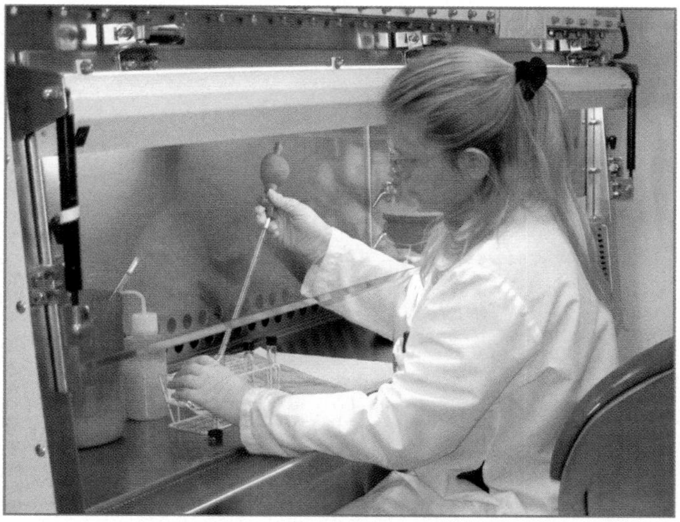

FIGURE I-1 Biological Safety Cabinet in a BSL-2 Laboratory
In this Class II BSC, air is drawn in from the room and is passed through a HEPA filter prior to release into the environment. This airflow pattern is designed to keep aerosolized microbes from escaping from the cabinet. The microbiologist is pipetting a culture. When the BSC is not in use at the end of the day, an ultraviolet light is turned on to sterilize the air and the work surface. (San Diego County Public Health Laboratory)

The microorganisms used in introductory microbiology courses depend on the institution, objectives of the course, and student preparation. Most introductory courses use organisms that may be handled at BSL-1 and BSL-2 levels so we have followed that practice in designing this set of exercises. Below are general safety rules to reduce the chance of injury or infection to you and to others, both inside and outside the laboratory. Although they represent a mixture of BSL-1 and BSL-2 guidelines, we believe that students should learn and practice the safest level of standards (relative to the organisms they are likely to encounter) at all times. Please follow these and any other safety guidelines required by your college.

Student Conduct

- To reduce the risk of infection, do not smoke, eat, drink, or bring food or drinks into the laboratory room—even if lab work is not being done at the time.
- Do not apply cosmetics or handle contact lenses in the laboratory.
- Wash your hands *thoroughly* with soap and water after handling living microbes and before leaving the laboratory each day. Also, wash your hands after removing gloves.
- Lab time is precious, so come to lab prepared for that day's work. Figuring out what to do as you go is likely to produce confusion and accidents.
- Do not remove any organisms or chemicals from the laboratory.
- Work carefully and methodically. Do not hurry through any laboratory procedure.

Basic Laboratory Safety

- Wear protective clothing (*i.e.*, a lab coat) in the laboratory when handling microbes. Remove the coat prior to leaving the lab and autoclave it regularly (Figure I-2).
- Do not wear sandals or open-toed shoes in the laboratory.
- Wear eye protection whenever you are heating chemicals, even if you wear glasses or contacts (Figure I-2).
- Turn off your Bunsen burner when it is not in use. In addition to being a fire and safety hazard, it is an unnecessary source of heat in the room.
- Tie back long hair, as it is a potential source of contamination as well as a likely target for fire.

FIGURE I-2 Safety First
This student is prepared to work safely with microorganisms. The lab area is uncluttered, tubes are upright in a test tube rack, and the flame is accessible but not in the way. The student is wearing a protective lab coat, gloves, and goggles, all of which are to be removed prior to leaving the laboratory. Not all procedures require gloves and eye protection. Your instructor will advise you as to the standards in your laboratory.

- If you are feeling ill, go home. A microbiology laboratory is not a safe place if you are ill.
- If you are pregnant, immune compromised, or are taking immunosuppressant drugs, please see the instructor. It may be in your best *long-term* interests to postpone taking this class.
- Wear disposable gloves while staining microbes and handling blood products—plasma, serum, antiserum, or whole blood (Figure I-2). Handling blood can be hazardous, even if you are wearing gloves. Consult your instructor before attempting to work with any blood products.
- Use an antiseptic (*e.g.*, Betadine) on your skin if it is exposed to a spill containing microorganisms. Your instructor will tell you which antiseptic you will be using.
- Never pipette by mouth. Always use mechanical pipettors (see Figure C-1, Appendix C).
- Dispose of broken glass in an appropriate "sharps" or broken glass container (Figure I-3).
- Use a fume hood to perform any work involving highly volatile chemicals or stains that need to be heated.
- Find the first-aid kit, and make a mental note of its location.

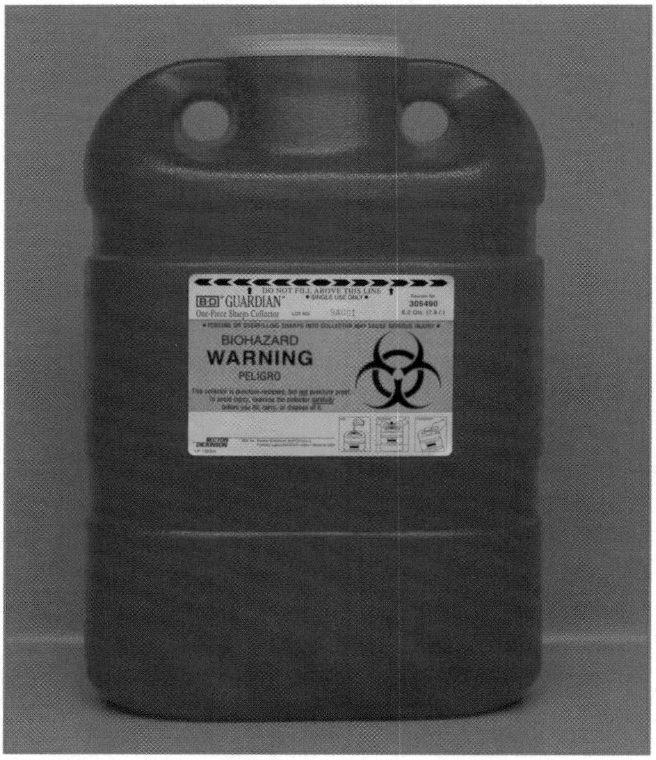

FIGURE I-3 SHARPS CONTAINER
Needles, glass, and other contaminated items that can penetrate the skin should be disposed of in a sharps container. Do not fill above the dashed black line.

- Find the fire blanket, shower, and fire extinguisher, note their locations, and develop a plan for how to access them in an emergency.
- Find the eye wash basin, learn how to operate it, and remember its location.

Reducing Contamination of Self, Others, Cultures, and the Environment

1. Wipe the desktop with a disinfectant (*e.g.*, Amphyl or 10% chlorine bleach) before *and* after each lab period. Never assume that the class before you disinfected the work area. An appropriate disinfectant will be supplied. Allow the disinfectant to evaporate; do not wipe it dry.
2. Never lay down culture tubes on the table; they always should remain upright in a tube holder (Figure I-2). Even solid media tubes contain moisture or condensation that may leak out and contaminate everything it contacts.
3. Cover any culture spills with paper towels. Soak the towels immediately with disinfectant, and allow them to stand for 20 minutes. Report the spill to your instructor. When you are finished, place the towels in the container designated for autoclaving.
4. Place all nonessential books and papers under the desk. A cluttered lab table is an invitation for an accident that may contaminate your expensive school supplies.
5. When pipetting microbial cultures, place a disinfectant-soaked towel on the work area. This reduces contamination and possible aerosols if a drop escapes from the pipette and hits the tabletop.

Disposing of Contaminated Materials

In most instances, the preferred method of decontaminating microbiological waste and reusable equipment is the autoclave (Figure I-4).

1. Dispose of plate cultures (if plastic Petri dishes are used) and other contaminated nonreusable items in the appropriate container, to be decontaminated when you are finished with them (Figure I-5). Petri dishes should be taped closed.
2. Remove all labels from tube cultures and other contaminated reusable items and place them in the container designated for autoclaving.
3. Dispose of all blood product samples, as well as disposable gloves in the container designated for autoclaving.
4. Place used microscope slides of bacteria in a container designated for autoclaving, or soak them in

FIGURE I-4 AN AUTOCLAVE
Media, cultures, and equipment to be sterilized are placed in the basket of the autoclave. Steam heat at a temperature of 121°C (produced at atmospheric pressure plus 15 psi) for 15 minutes is effective at killing even bacterial spores. Some items that cannot withstand the heat, or have irregular surfaces that prevent uniform contact with the steam, are sterilized by other means.

FIGURE I-5
AN AUTOCLAVE BAG
Nonreusable items (such as plastic Petri dishes) are placed in an autoclave bag for decontamination. Petri dishes should be taped closed. Do not overfill the bag.

disinfectant solution for at least 30 minutes before cleaning or discarding them. Follow your laboratory guidelines for disposing of glass.

5. Place contaminated broken glass in the container designated for autoclaving (Figure I-3). After decontamination, dispose of the glass in a "sharps" container or specialized broken glass container.

References

Barkley, W. Emmett, and John H. Richardson. 1994. Chapter 29 in *Methods for General and Molecular Bacteriology*, edited by Philipp Gerhardt, R. G. E. Murray, Willis A. Wood, and Noel R. Krieg. American Society for Microbiology, Washington, DC.

Collins, C. H., Patricia M. Lyne, and J. M. Grange. 1995. Chapters 1 and 4 in *Collins and Lyne's Microbiological Methods*, 7th ed. Butterworth-Heineman, Oxford.

Darlow, H. M. 1969. Chapter VI in *Methods in Microbiology,* Volume 1, edited by J. R. Norris and D. W. Ribbins. Academic Press, Ltd., London.

Fleming, Diane O., and Debra L. Hunt (Editors). 2000. *Laboratory Safety—Principles and Practices*, 3rd ed. American Society for Microbiology, Washington, DC.

Koneman, Elmer W., Stephen D. Allen, William M. Janda, Paul C. Schreckenberger, and Washington C. Winn, Jr. 1997. *Color Atlas and Textbook of Diagnostic Microbiology*, 5th ed. Lippincott-Raven Publishers, Philadelphia and New York.

Power, David A., and Peggy J. McCuen. 1988. Pages 2 and 3 in *Manual of BBL™ Products and Laboratory Procedures*, 6th ed. Becton Dickinson Microbiology Systems, Cockeysville, MD.

Richmond, Jonathan Y., and Robert W. McKinney (Editors). May 1999. U. S. Department of Health and Human Services, *Biosafety in Microbiological and Biomedical Laboratories,* 4th ed. U. S. Government Printing Office, Washington, DC.

A Word About Experimental Design

Like most sciences, microbiology has descriptive and experimental components. Here we are concerned with the latter. Science is a philosophical approach to finding answers to questions. In spite of what you may have been taught in grade school about *THE* "Scientific Method," science can approach problems in many ways, rather than in any *single* way. The nature of the problem, personality of the scientist, intellectual environment at the time, and good, old-fashioned luck all play a role in determining which approach is taken. Nevertheless, in experimental science, one component that is always present is a **control** (or controls).

A controlled experiment is one in which all **variables** except one—the **experimental variable**—are maintained without change. This is the only way the results can be considered reliable. By maintaining all variables except one, other potential sources of an observed event can be eliminated. Then (presumably), a **cause and effect relationship** between the event and the experimental variable can be established. If the event changes when the experimental variable changes, we provisionally link that variable and the event. Alternatively, if there is no observed change, we can eliminate the experimental variable from involvement with the event.

Throughout science experimentation—and this book—you will see the word *control.* Controls are an essential and integral part of all experiments. As you work your way through the exercises in this book, pay attention to all the ways controls are used to improve the reliability of the procedure and your confidence in the results.

Microbiological experimentation often involves tests that determine the ability of an organism to use or produce some chemical, or to determine the presence or absence of a specific organism in a sample. Ideally, a positive result in the test indicates that the microbe has the ability or is present in the sample, and a negative result indicates a lack of that ability or absence in the sample. The tests we run, however, have limitations and occasionally may give **false positive** or **false negative** results. An inability to detect small amounts of the chemical or organism in question would yield a false negative result and would be a failure in the **sensitivity** of the test (Figure I-6). An inability to discriminate between the chemical or organism in question and similar chemicals or organisms would yield a false positive result and be a failure in the **specificity** of the test (Figure I-6). Sensitivity and specificity can be quantified using the following equations:

$$\text{Sensitivity} = \frac{\text{True Positives}}{\text{True Positives} + \text{False Negatives}}$$

$$\text{Specificity} = \frac{\text{True Negatives}}{\text{True Negatives} + \text{False Positives}}$$

The closer sensitivity and specificity are to a value of one, the more useful the test is. As you perform the tests in this book, be mindful of each test's limitations, and be open to the possibility of false positive and false negative results.

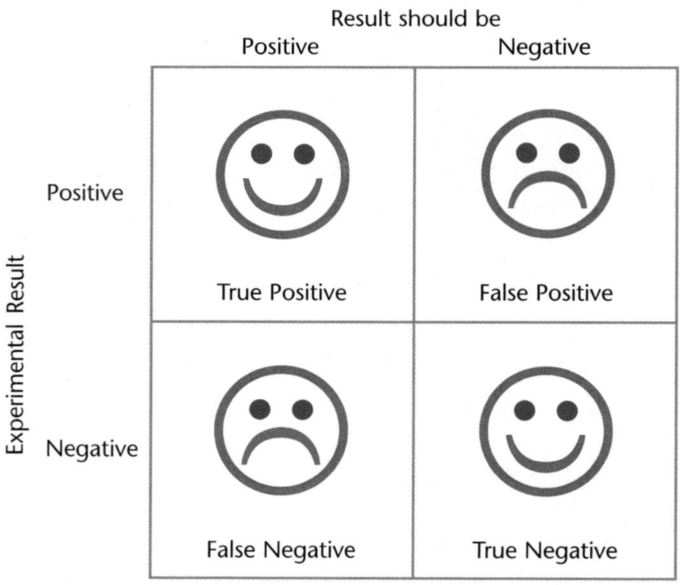

FIGURE I-6 LIMITATIONS OF EXPERIMENTAL TESTS
Ideally, tests should give a positive result for specimens that are positive, and a negative result for specimens that are negative. False positive and false negative results do occur, however, and these are attributed to a lack of specificity and a lack of sensitivity, respectively, of the test system.

References

Forbes, Betty A., Daniel F. Sahm, and Alice. S. Weissfeld. 1998. Chapter 5 in *Bailey and Scott's Diagnostic Microbiology*, 10th ed. Mosby-Year Book, St. Louis.

Lilienfeld, David E. and Paul D. Stolley. 1994. Page 118 in *Foundations of Epidemiology*, 3rd ed. Oxford University Press, New York.

Mausner, Judith S., and Shira Kramer. 1985. Pages 217–220 in *Epidemiology: An Introductory Text*, 2nd ed. W.B. Saunders Company, Philadelphia.

Data Presentation: Tables and Graphs

In microbiology, we perform experiments and collect data, but it is often difficult to know what the data mean without some method of organization. Tables and graphs allow us to summarize data in a way that makes interpretation easier.

Tables

A table is often used as a preliminary means of organizing data. As an example, Table I-2 shows the winning times for each male and female age division in a half-marathon race. Again, the aim of a table is to provide information to the reader. Notice the meaningful title, the column labels, and the appropriate measurement units. Without these, the reader cannot completely understand the table and your work will go unappreciated! Data tables are provided for you on the Data Sheets for each exercise in this book, but you may be required to fill-in certain components (units, labels, *etc.*) in addition to the data.

Male Runners		Female Runners	
Winner's Age (Years)	Winning Time (Minutes)	Winner's Age (Years)	Winning Time (Minutes)
15	73	15	88
25	67	24	82
31	67	30	82
35	71	39	84
40	71	42	85
52	78	50	109
62	95	62	108
70	123	70	126

TABLE I-2 Winning Half-Marathon Times By Sex and Age Division

Graphs

Table I-2 does give the information, but what it is telling us may not be entirely clear. It appears that the times increase as runners get older, but we have difficulty determining if this is truly a pattern. That is why data also are presented in graphic form at times; a graph usually shows the relationship between variables better than a table of numbers.

X-Y Scatter Plot

The type of graph you will be using in this manual is an "X–Y Scatter Plot," in which two variables are graphed against each other. Figure I-7 shows the same data as Table I-2, but in an X–Y Scatter Plot form.

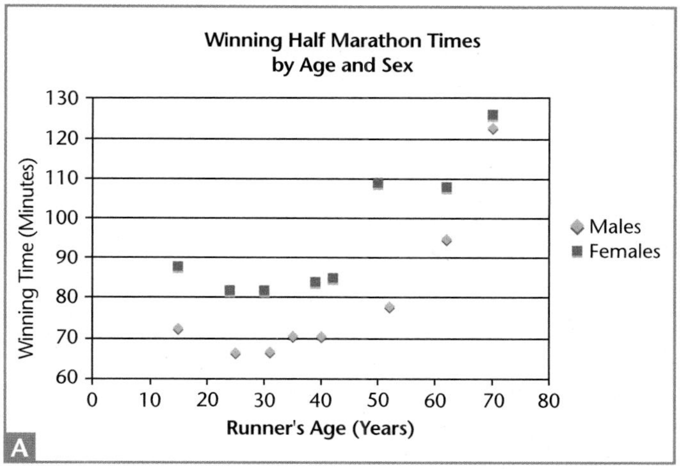

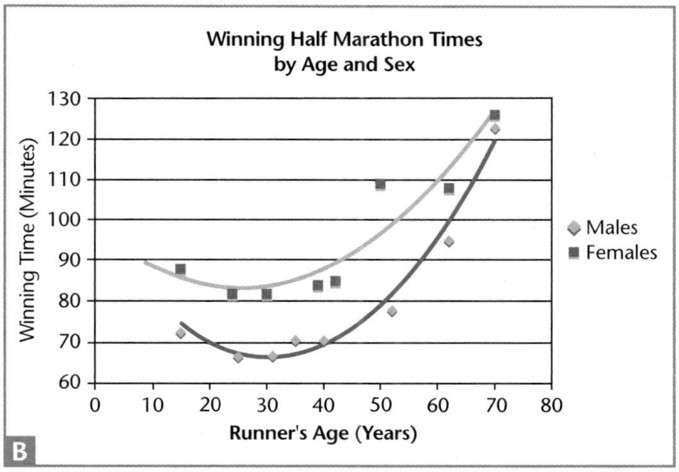

FIGURE I-7 Sample *X-Y* Scatter Plot
A graph often shows the relationship between variables better than a table of numbers. Examine this sample and identify the essential components of a quality graph (see text). (a) Presentation of data without a best-fit line is acceptable if there are not enough data points to justify illustrating a trend. (b) Shown here are the same data but with a trend line. Notice that the points do not fall directly on the line but, rather, that the line gives the general trend of the data. "Connecting the dots" is not appropriate.

Notice the following important features of the graph in Figure I-7:

- *Title:* The graph has a meaningful title—which should tell the reader what the graph is about.
- *Dependent and independent variables:* The graph is read from left to right. In our example, we might say for the male runners, "As runners get older, winning times get longer." *Winning time* depends on *age*, so winning time is the *dependent* variable and age is the *independent* variable. By convention, the independent variable is plotted on the x-axis and the dependent variable is plotted on the y-axis. (Age does *not* depend on the winning time.) By way of comparison, notice the consequence of plotting age on the y-axis and winning time on the x-axis: "As runners get slower, they get older"—which doesn't reflect the actual relationship between the variables.
- *Axis labels:* Each axis is labeled, including the appropriate units of measure. "Age" without units is meaningless. Does the scale represent months? years? centuries?
- *Axis scale:* The scale on each axis is uniform. The distance between marks on the axis is always the same and represents the same amount of that variable. (But increments on the x-axis don't have to equal those on the y-axis, as shown.) The size of each increment is up to the person making the graph and is dictated by the magnitude and range of the data. Most of the time, we choose a length for the axis that fills the available graphing space.
- *Axis range:* The scale for an axis does not have to begin with "0." Use a scale that best presents the data. In this case, the smallest y-value was 67 minutes, so the scale begins at 60 minutes.
- *Multiple Data Sets and the Legend:* The two data series (male and female times) are plotted on the same set of axes, but with different symbols that are defined in the legend at the right. The symbols shown differ in color *and* shape, but one of these is adequate.
- *Best-fit Line:* If a line is to be drawn at all, it should be an average line for the data points, not one that "connects the dots" (Figure I-7b). Notice that the points are not necessarily *on* the line. The purpose of a best-fit line is to illustrate the general trend of the data, not the specifics of the individual data points. (Be assured that most graphs in your textbooks where a smooth line is shown were experimentally determined and the lines are derived from points scattered around the line.) There is a mathematical formula that allows one to compute the slope and

y-intercept of the **trend line** if the relationship is linear or a **best-fit line** if the relationship is nonlinear (as in the half-marathon times example), but this is beyond our needs. For our purposes, a hand-drawn trend line that looks good is good enough. (If you use a computer graphing program, then it will produce the trend line without you doing any of the math—the best of all situations!)

Bar Graphs

Bar graphs are used to illustrate one variable. Using a bar graph to show the relationship between winning times and ages is inappropriate. Examine Figure 1-8a. Notice that the space each bar fills is meaningless; that is, the only important part of the bar is the top—which is the value used in the *X-Y* scatter plot. Appropriate use of a bar graph would be the distribution of student performance on an exam (Figure 1-8b). Notice that the space each bar fills has meaning. Each student in a particular group adds height to the bar.

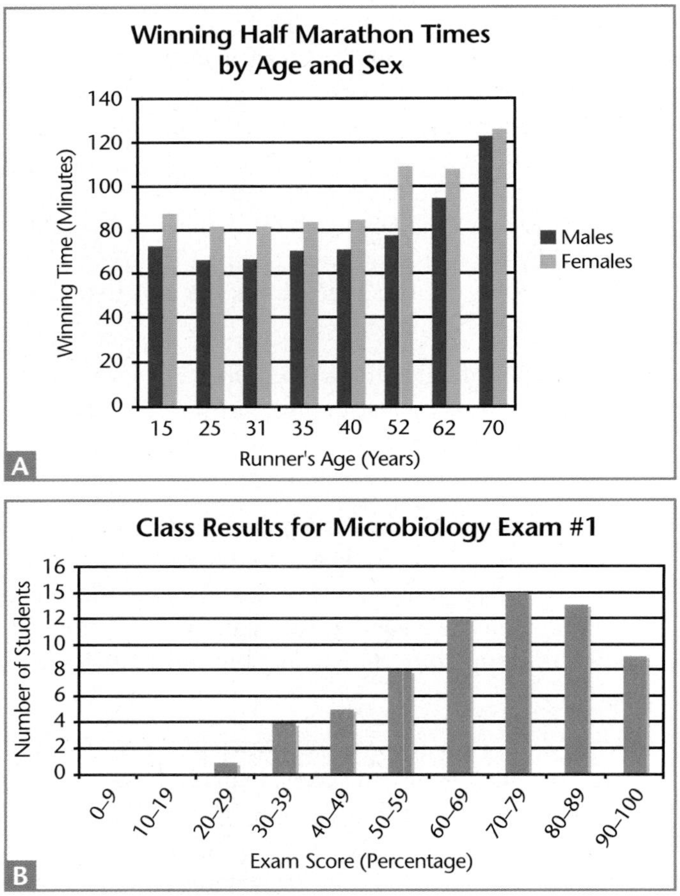

FIGURE 1-8 BAR GRAPHS
A bar graph is appropriate to present data involving a single variable. (a) Plotting the winning half marathon times from Table I-2 using a bar graph is inappropriate because the only meaningful point is at the top. (b) A bar graph is useful in presenting data of a single variable, such as the number of students earning a specific score on their microbiology exam.

There is no single correct way to produce a graph for a particular data set. Actually, most people working independently would graph the same data set in different ways (*e.g.,* different scales, colors, wording of the title and axis labels), but the essential components listed above would have to be there. You will be asked to graph some of the data you collect. Be sure your graphs tell a complete and clear story of what you've done.

Standard Curve

Besides making interpretation easier, graphs sometimes are used to establish an experimental value. Each point on the trend line represents a correlation between the *x* and *y* values (Figure I-9). If we know one value, we can read the other off the graph. We use this process with something called a "standard curve" or "calibration curve." To produce a standard curve, samples with a known amount of the independent variable (*e.g.,* soluble carbohydrate concentration) are subjected to the experiment. The resulting data (the *y* values—*e.g.,* absorbances) are plotted and a trend line is drawn.

Now a sample with an *unknown* amount of the independent variable can be subjected to the same experimental procedure to determine its *y* value. Once the *y* value is known, the corresponding *x* value is read directly off the graph to determine the unknown amount. (If the relationship is linear, the "average" line is described by the equation $y = mx + b$. Once the *y* value is determined experimentally, the *x* value can be calculated by substitution.)

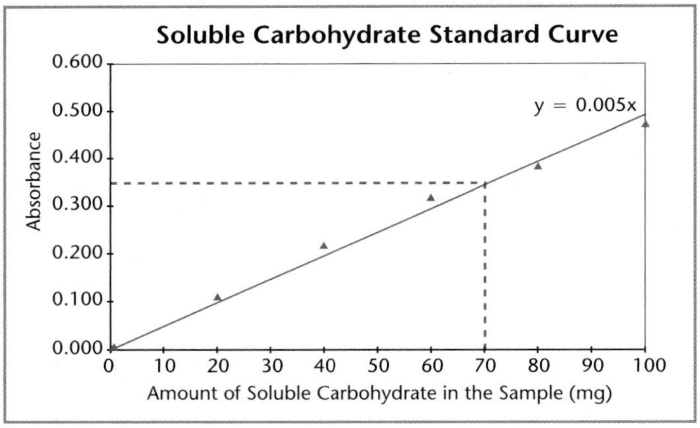

FIGURE I-9 A STANDARD CURVE
Standard curves are used to determine the value of some unknown in a sample. In this example, the absorbances of six solutions with known soluble carbohydrate concentrations were determined experimentally, plotted (blue triangles) and used to make a trend line (blue line). Absorbance of a sample with an unknown soluble carbohydrate concentration then was determined to be 0.35 by the same experimental method. The point where the *y* value (0.35) intersects the trend line gives the *x* value that corresponds to it (red dashed lines): 70 µg of soluble carbohydrate. The equation for the trend line ($y = 0.005x$) is also given and can be used to determine the *x* value by substituting 0.35 for *y*.

Fundamental Skills for the Microbiology Laboratory

SECTION 1

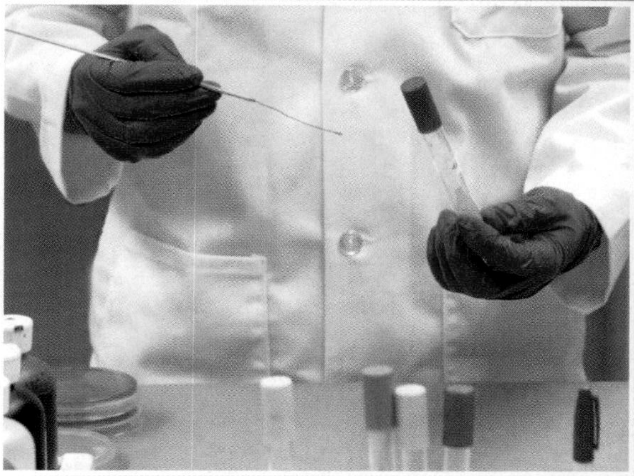

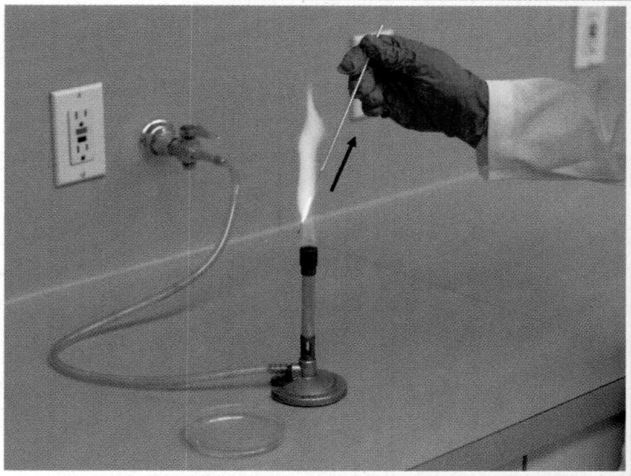

Bacterial and fungal **cultures** are grown and maintained on or in solid and liquid substances called **media**. Preparation of these media involves weighing ingredients, measuring liquid volumes, calculating proportions, handling basic laboratory glassware, and operating a pH meter and an autoclave. In Exercise 1-1 you will learn and practice these fundamental skills by preparing a couple of simple growth media. When you have completed the exercise, you will have the skills necessary to prepare almost any medium if given the recipe.

A second fundamental skill necessary for any microbiologist is the ability to transfer bacterial cells from one place to another without contaminating the original culture, the new medium, or the environment (including the microbiologist). This **aseptic** (sterile) transfer technique is required for virtually all procedures in which living microbes are handled, including isolations, staining, and differential testing. Exercises 1-2 through 1-4 present descriptions of common transfer and inoculation methods. Less frequently used methods are covered in Appendices B through D.

Basic Growth Media

To cultivate microbes, microbiologists use a variety of growth media. Although these media may be formulated from scratch, they more typically are produced by rehydrating commercially available powdered media. Media that are routinely encountered in the microbiology laboratory range from the widely used, general-purpose growth media, to the more specific selective and differential media used in identification of microbes. In Exercise 1-1 you will learn how to prepare simple general growth media.

Exercise 1-1

Nutrient Broth and Nutrient Agar Preparation

Theory Nutrient broth and nutrient agar are common media used for maintaining bacterial cultures. To be of practical use, they have to meet the diverse nutrient requirements of routinely cultivated bacteria. As such, they are formulated from sources that supply carbon and nitrogen in a variety of forms—amino acids, purines, pyrimidines, monosaccharides to polysaccharides, and various lipids. Generally, these are provided in digests of plant material (phytone) or animal material (peptone and others). Because the exact composition and amounts of carbon and nitrogen in these ingredients are unknown, general growth media are considered to be **undefined.**

In most classes (because of limited time), media are prepared by a laboratory technician. Still, it is instructive for novice microbiologists to at least gain exposure to what is involved in media preparation. Your instructor will provide specific instructions on how to execute this exercise using the equipment in your laboratory.

Application Microbiological growth media are prepared to cultivate microbes. These general growth media are used to maintain bacterial stock cultures.

In This Exercise . . .

You will prepare 1-liter batches of two general growth media: nutrient broth and nutrient agar. Over the course of the semester, a laboratory technician will probably do this for you, but it is good to gain firsthand appreciation for the work done behind the scenes!

Materials

Per Student Group

- 2-liter Erlenmeyer flasks
- three or four 500 mL Erlenmeyer flasks and covers (can be aluminum foil)
- stirring hotplate
- magnetic stir bars
- all ingredients listed below in the recipes (or commercially prepared dehydrated media)
- sterile Petri dishes
- test tubes (16 mm × 150 mm) and caps
- balance
- weighing paper or boats
- spatulas

Medium Recipes

Nutrient Broth

• beef extract	3.0 g
• peptone	5.0 g
• distilled or deionized water	1.0 L
pH 6.6–7.0 at 25°C	

Nutrient Agar

• beef extract	3.0 g
• peptone	5.0 g
• agar	15.0 g
• distilled or deionized water	1.0 L
pH 6.6–7.0 at 25°C	

Preparation of Medium

Day One

To minimize contamination while preparing media, clean the work surface, turn off all fans, and close any doors that might allow excessive air movements.

Nutrient Agar Tubes

1. Weigh the ingredients on a balance (Figure 1-1).

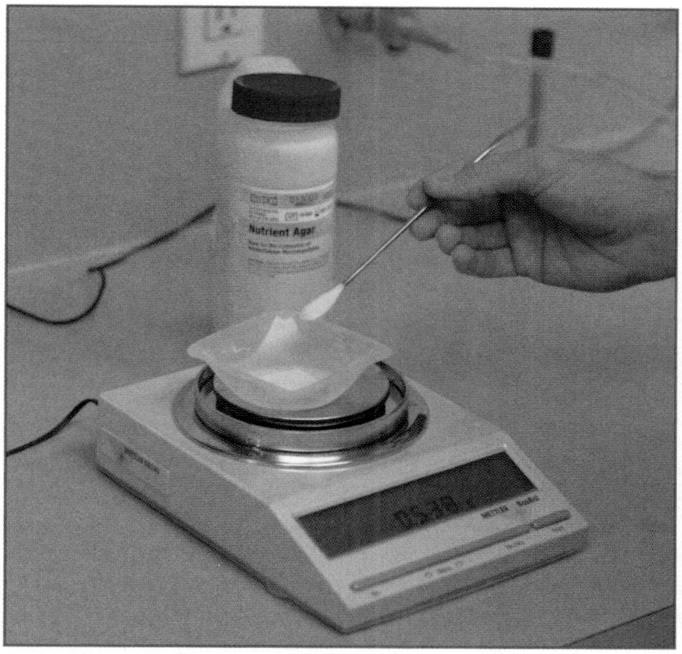

FIGURE 1-1 Weighing Medium Ingredients
Solid ingredients are weighed with an analytical balance. A spatula is used to transfer the powder to a tared weighing boat. Shown here is dehydrated nutrient agar, but the weighing process is the same for any powdered ingredient.

2. Suspend the ingredients in one liter of distilled or deionized water in the two-liter flask, mix well, and boil until fully dissolved (Figure 1-2).
3. Dispense 7 mL portions into test tubes and cap loosely (Figure 1-3). If your tubes are smaller than those listed in Materials, adjust the volume to fill 20% to 25% of the tube. Fill to approximately 50% for agar deeps.
4. Sterilize the medium by autoclaving for 15 minutes at 121°C (Figure 1-4).
5. After autoclaving, cool to room temperature with the tubes in an upright position for agar deep tubes. Cool with the tubes on an angle for agar slants (Figure 1-5).
6. Incubate the slants and/or deep tubes at 35 ± 2°C for 24 to 48 hours.

Nutrient Agar Plates

1. Weigh the ingredients on a balance (Figure 1-1).
2. Suspend the ingredients in one liter of distilled or deionized water in the two-liter flask, mix well, and boil until fully dissolved (Figure 1-2).
3. Divide into three or four 500 mL flasks for pouring. Smaller flasks are easier to handle when pouring plates. Don't forget to add a magnetic stir bar and to cover each flask before autoclaving.
4. Autoclave for 15 minutes at 121°C to sterilize the medium.
5. Remove the sterile agar from the autoclave, and allow it to cool to 50°C while you are stirring it on a hotplate.
6. Dispense approximately 20 mL into sterile Petri plates (Figure 1-6). *Be careful! The flask will still be*

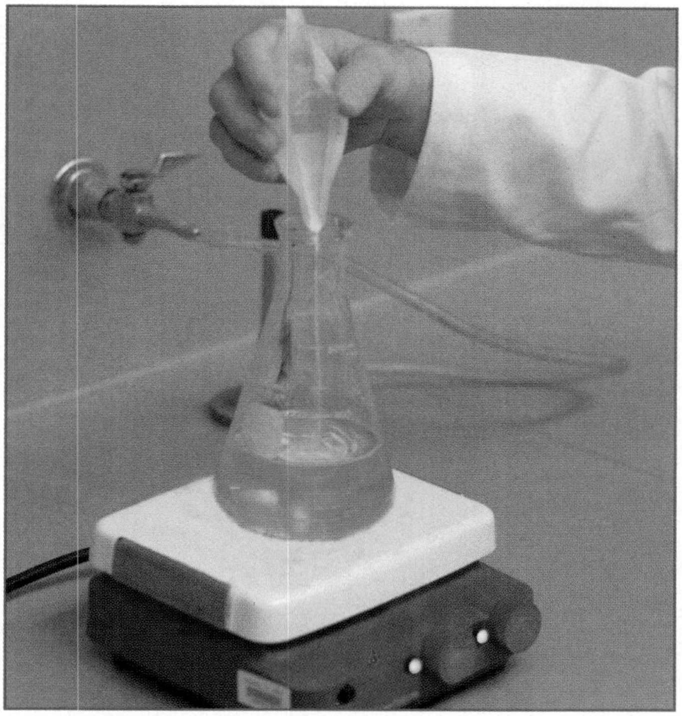

FIGURE 1-2 Mixing the Medium
The powder is added to a flask of distilled or deionized water on a hotplate. A magnetic stir bar mixes the medium as it is heated to dissolve the powder.

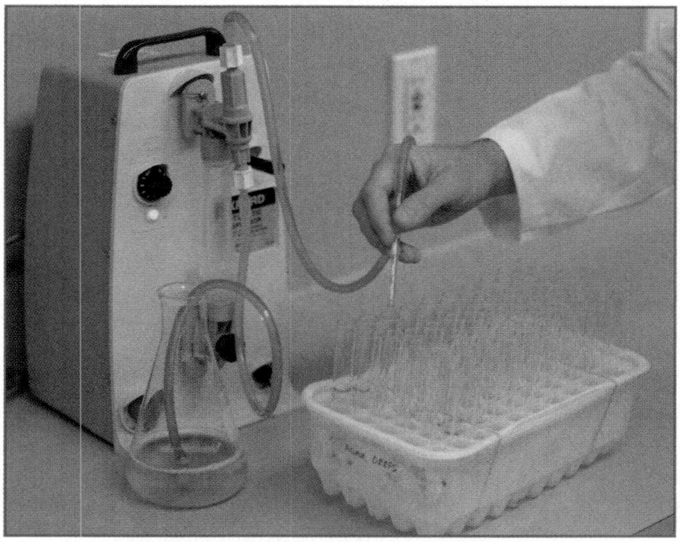

FIGURE 1-3 Dispensing the Medium into Tubes
An adjustable pump can be used to dispense the appropriate volume (usually 7–10 mL) into tubes. The tubes then are loosely capped.

FIGURE 1-4 AUTOCLAVING THE TUBED MEDIUM
The basket of tubes is sterilized for 15 minutes at 121°C in an autoclave. When finished, the tubes are cooled.

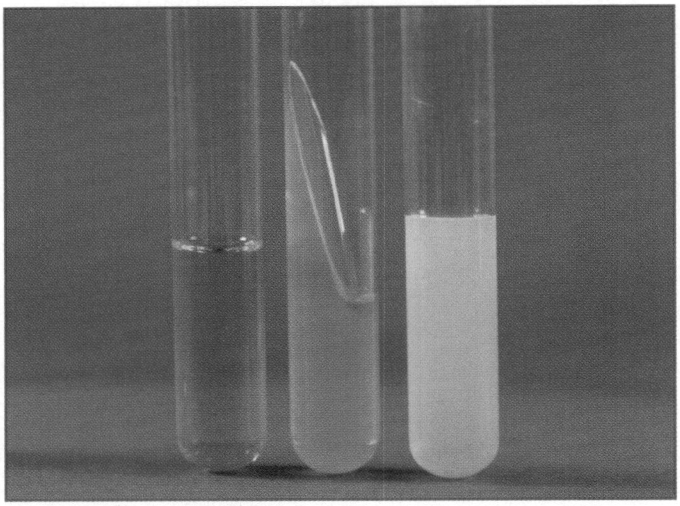

FIGURE 1-5 TUBED MEDIA
From left to right: a broth, an agar slant, and an agar deep tube. The solid media are liquid when they are removed from the autoclave. Agar deeps are allowed to cool and solidify in an upright position, whereas agar slants are cooled and solidified on an angle.

hot so wear an oven mitt. While you pour the agar, shield the Petri dish with its lid to reduce the chance of introducing airborne contaminants. If necessary, *gently* swirl each plate so the agar completely covers the bottom; do not swirl the agar up into the lid. Allow the agar to cool and solidify before moving the plates.

7. Store these plates on a counter top or in the incubator for 24 hours to allow them to dry prior to use.

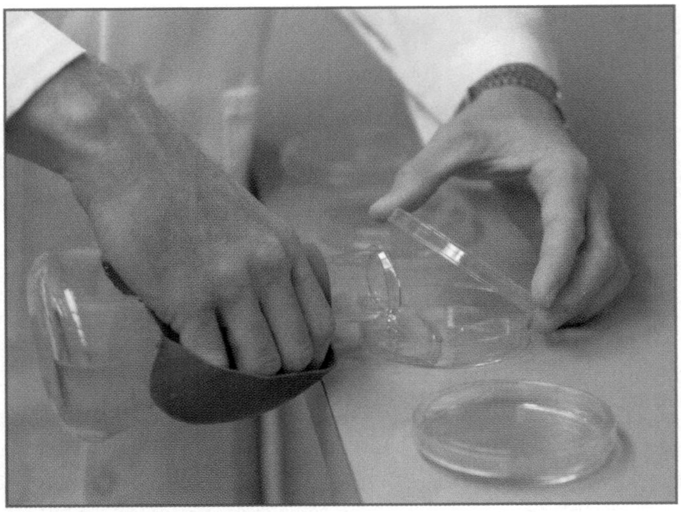

FIGURE 1-6 POURING AGAR PLATES
Agar plates are made by pouring sterilized medium into sterile Petri dishes. The lid is used as a shield to prevent airborne contamination. Once poured, the dish is gently swirled so the medium covers the base. Plates are then cooled and dried to eliminate condensation.

Nutrient Broth

1. Weigh the ingredients on a balance (Figure 1-1).
2. Suspend the ingredients in one liter of distilled or deionized water in the two-liter flask. Agitate and heat slightly (if necessary) to dissolve them completely (Figure 1-2).
3. Dispense 7 mL portions into test tubes and cap loosely (Figure 1-3). As with agar slants, if your tubes are smaller than those recommended in Materials, add enough broth to fill them approximately 20% to 25%.
4. Sterilize the medium by autoclaving for 15 minutes at 121°C (Figure 1-4).
5. Incubate the tubes at 35 ± 2°C for 24 to 48 hours.

Day Two

Record your observations on the Data Sheet.

Reference

Zimbro, Mary Jo and David A. Power. 2003. Pages 404–405 and 408 in *DIFCO™ & BBL™ Manual—Manual of Microbiological Culture Media.* Becton, Dickinson and Company, Sparks, MD.

Common Aseptic Transfers and Inoculation Methods

As a microbiology student, you will be required to transfer living microbes from one place to another **aseptically** (*i.e.*, without contamination of the culture, the sterile medium, or the surroundings). While you won't be expected to master all transfer methods right now, you will be expected to perform most of them over the course of the semester. Refer back to this section as needed.

To prevent contamination of the sample, inoculating instruments (Figure 1-7) must be sterilized prior to use. Inoculating loops and needles are sterilized immediately before use in an incinerator or Bunsen burner flame. The mouths of tubes or flasks containing cultures or media are also incinerated at the time of transfer. Instruments that are not conveniently or safely incinerated, such as Pasteur pipettes, cotton applicators, glass pipettes, and digital pipettor tips, are sterilized inside wrappers or containers by autoclaving prior to use.

Aseptic transfers are not difficult; however, a little preparation will help assure a safe and successful procedure. Before you begin, you will need to know where the sample is coming from, its destination, and the type of transfer instrument to be used. These exercises provide step-by-step descriptions of different transfer methods. In an effort to avoid too much repetition, skills that are basic to most transfers are described in detail once under "The Basics" and mentioned only briefly as they apply to transfers in the discussion of "Specific Transfer Methods." These are printed in regular type. New material in each specific transfer will be introduced in blue type. Certain less routine transfer methods are discussed in Appendices B through D.

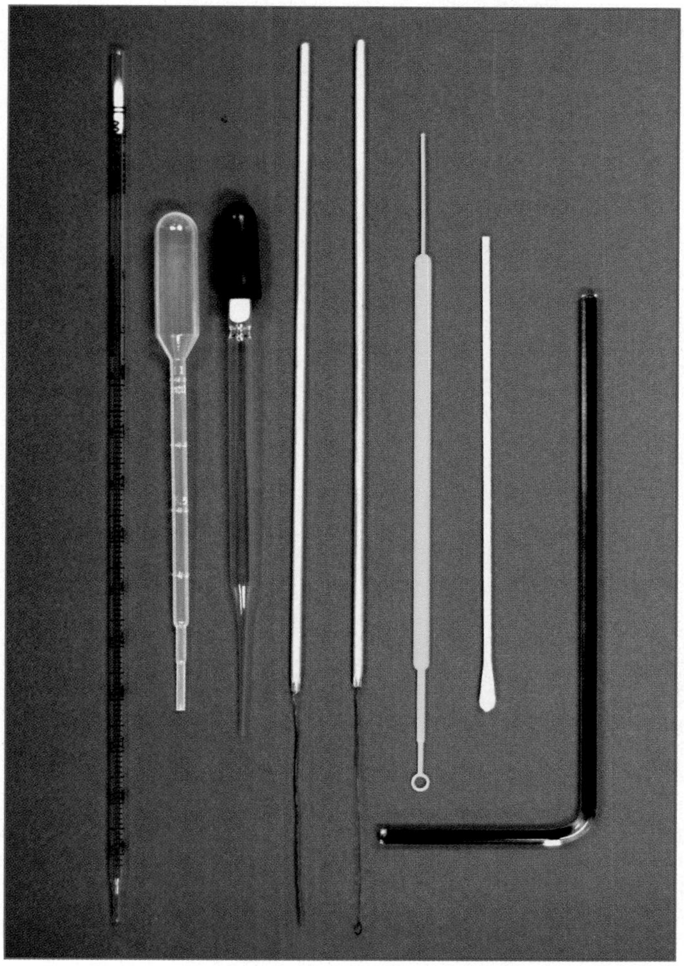

FIGURE 1-7 INOCULATING TOOLS
Any of several different instruments may be used to transfer a microbial sample, the choice of which depends on the sample source, its destination, and any special requirements imposed by the specific protocol. Shown here are several examples of transfer instruments. From left to right: serological pipette (see Appendix C), disposable transfer pipette, Pasteur pipette, inoculating needle, inoculating loop, disposable inoculating needle/loop, cotton swab (see Appendix B and Exercise 1-3), and glass spreading rod (see Exercise 1-4).

Exercise 1-2

Common Aseptic Transfers and Inoculation Methods

Application—The Basics

The following is a listing of general techniques and practices and is not presented as sequential.

- *Be organized.* Arrange all media in advance and clearly label them with your name, the date, the medium and the inoculum. Tubes are typically labeled with tape or paper held on with rubber bands; you may write directly on the base of Petri plates. Be sure not to place labels in such a way as to obscure your view of the inside of the tube or plate.
- *Take your time.* Work efficiently, but *do not hurry*. You are handling potentially dangerous microbes.
- *Place all media tubes in a test tube rack when not in use* whether they are or are not sterile. Tubes should never be laid on the table surface (Figure 1-8).
- *Hold the handle of an inoculating needle or loop like a pencil* in your dominant hand and relax (Figure 1-8)!
- *Adjust your Bunsen burner* so its flame has an inner and an outer cone (Figure 1-9).
- *Sterilize a loop/needle by incinerating it in the Bunsen burner flame* (Figure 1-10). Pass it through the tip of the flame's inner cone, holding it at an angle with the loop end pointing downward. Begin flaming about 2 cm up the handle, then proceed down the wire by pulling the loop backward through the flame until the entire wire has become uniformly orange-hot. Flaming in this direction limits aerosol production by allowing the tip to heat up more slowly than if it were thrust into the flame immediately.
- *Hold a culture tube in your nondominant hand* and move it, not the loop, as you transfer. This will minimize aerosol production from loop movement.

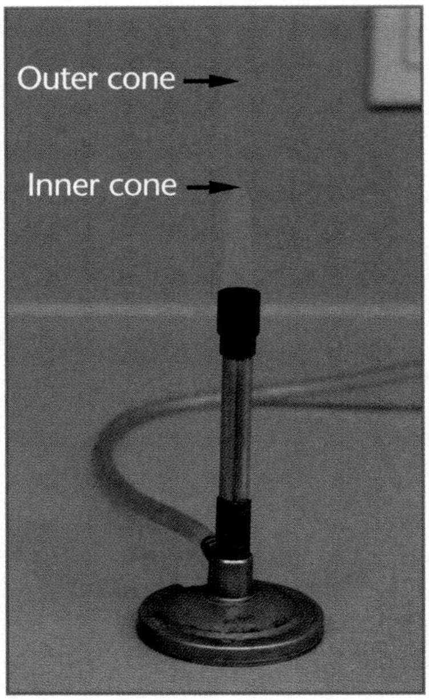

FIGURE 1-9 **Bunsen Burner Flame**
When properly adjusted, a Bunsen burner produces a flame with two cones. Sterilization of inoculating instruments is done in the hottest part of the flame—the tip of the inner cone. Heat-fixing bacterial smears on slides and incinerating the mouths of open glassware items may be done in the outer cone.

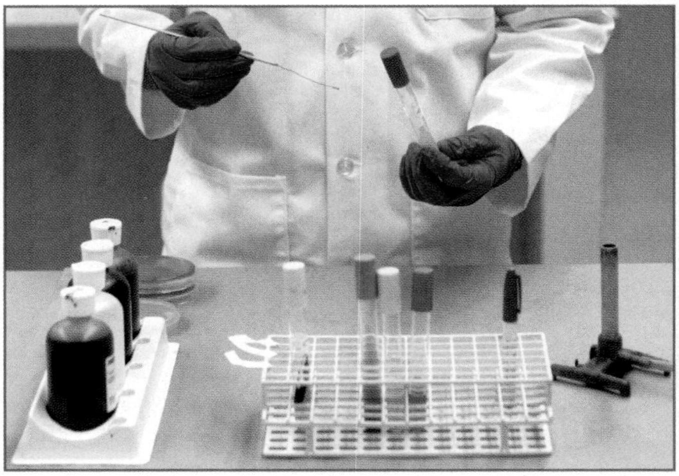

FIGURE 1-8 **Microbiologist at Work**
Materials are neatly positioned and not in the way. To prevent spills, culture tubes are stored upright in a test tube rack. They are never laid on the table. The microbiologist is relaxed and ready for work—in the photo here, holding the loop like a pencil, not gripping it like a dagger.

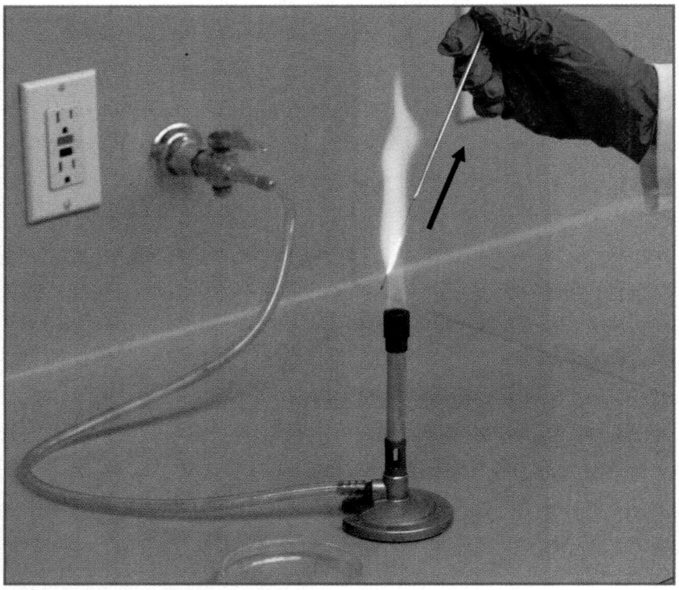

FIGURE 1-10 **Flaming Loop**
Incineration of an inoculating loop's wire is done by passing it through the tip of the flame's inner cone. Begin at the wire's base and continue to the end, making sure that all parts are heated to a uniform orange color. Allow the wire to cool before touching it or placing it on/in a culture. The former will burn you; the latter will cause aerosols of microorganisms.

- *Grasp the tube's cap with your little finger* and remove it by pulling the *tube* away from the cap. Hold the cap in your little finger during the transfer (Figure 1-11). (The cap should be loosened prior to transfer, especially if it's a screw-top cap.) When replacing the cap, move the tube back to the cap to keep your loop hand still. The replaced cap doesn't have to be on firmly at this time—just enough to cover the tube.
- *Flame tubes* by passing the open end through the Bunsen burner flame two or three times (Figure 1-12).

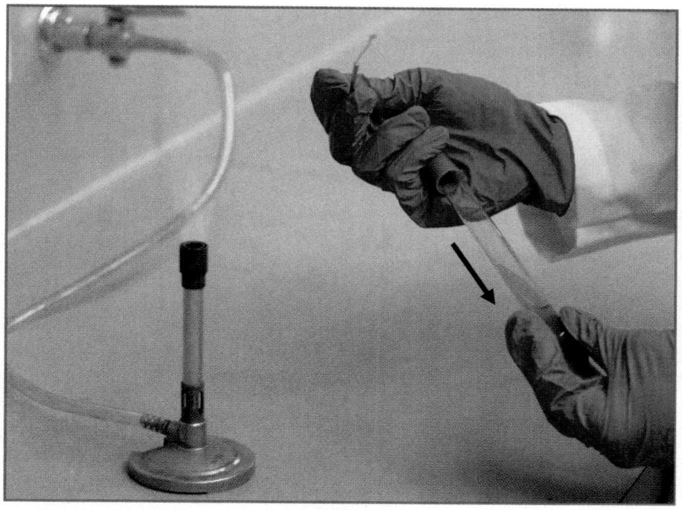

FIGURE 1-11 Removing the Tube Cap
The loop is held in the dominant hand and the tube in the other hand. Remove the tube's cap with your little finger of your loop hand by pulling the tube away with the other hand; keep your loop hand still. Hold the cap in your little finger during the transfer. When replacing the cap, move the tube back to the cap to keep your loop hand still. The replaced cap doesn't have to be on firmly at this time—just enough to cover the tube.

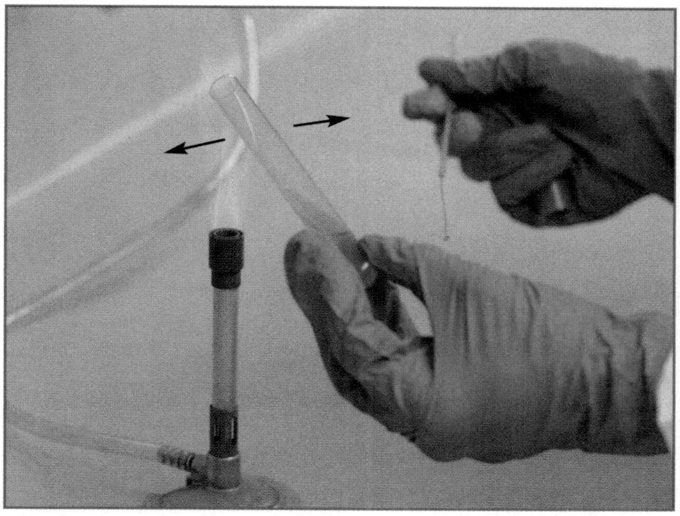

FIGURE 1-12 Flaming the Tube
The tube's mouth is passed quickly through the flame a couple of times to sterilize the lip of the tube's lip and the surrounding air. Notice that the tube's cap is held in the loop hand.

- *Hold open tubes at an angle* to minimize the chance of airborne contamination (Figure 1-13).
- *Suspend bacteria in a broth* with a vortex mixer prior to transfer (Figure 1-14). Be sure not to mix so vigorously that broth gets into the cap or that you

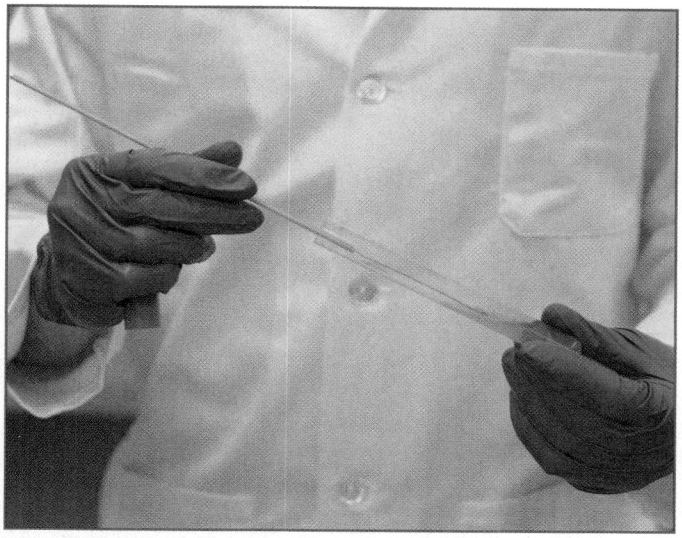

FIGURE 1-13 Holding the Tube at an Angle
The tube is held at an angle to minimize the chance that airborne microbes will drop into it. Notice that the tube's cap is held in the loop hand.

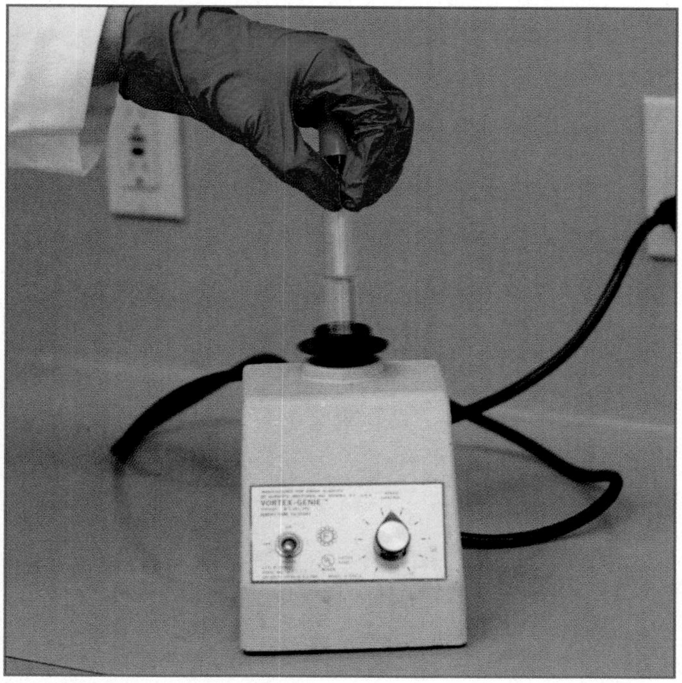

FIGURE 1-14 Vortex Mixer
Bacteria are suspended in a broth with a vortex mixer. The switch on the left has three positions: on (up), off (middle), and touch (down). The rubber boot is activated when touched only if the "touch" position is used; "on" means the boot is constantly vibrating. On the right is a variable speed knob. Caution must be used to prevent broth from getting into the cap or losing control of the tube and causing a spill.

lose control of the tube. Start slowly, then gently increase the speed until the tip of the vortex reaches the bottom of the tube. Alternatively, broth may be agitated by drumming your fingers along the length of the tube several times (Figure 1-15). Be careful not to splash the broth into the cap or lose control of the tube.

- *When opening a plate, use the lid as a shield* to minimize the chances of airborne contamination (Figure 1-16).

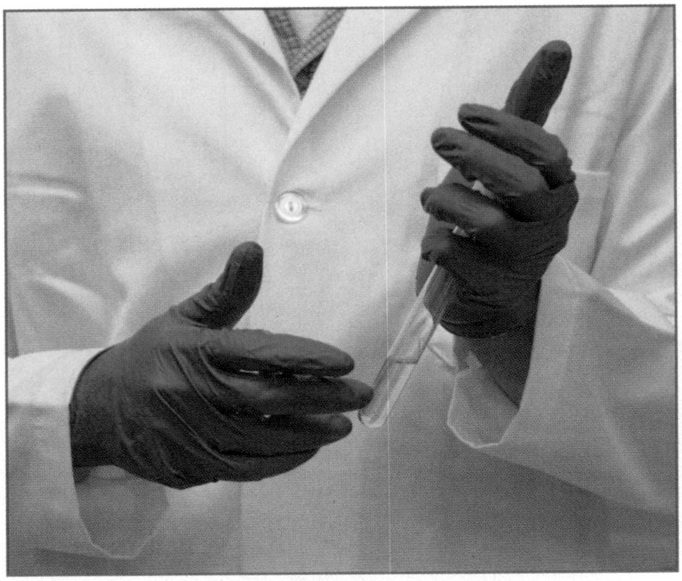

FIGURE 1-15 Mixing Broth by Hand
A broth culture always should be mixed prior to transfer. Tapping the tube with your fingers gets the job done safely and without special equipment.

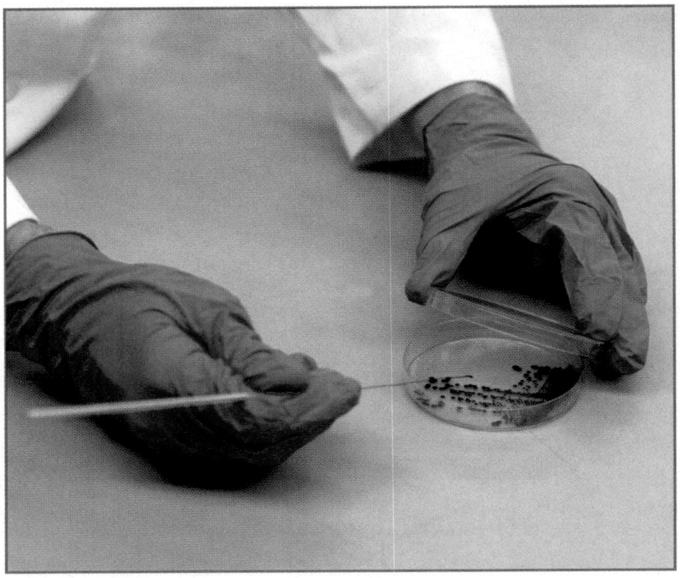

FIGURE 1-16 Using the Lid as a Shield
When transferring bacteria to or from a Petri dish, the lid is used to cover the surface of the agar to minimize airborne contamination.

Application—Specific Transfer Methods

Transfers occur in two basic stages:

1. obtaining the sample to be transferred, and
2. transferring to the sterile culture medium.

These two stages may be combined in various ways. The following descriptions are organized to reflect that flexibility. (Recall that steps in the transfer *not* covered in the Basics Section are printed in blue type.)

Transfers Using an Inoculating Loop or Needle

Inoculating loops and needles are the most commonly used instruments for transferring microbes between all media types—broths, slants, or plates can be the source, and any can be the destination. Because loops and needles are handled in the same way, we refer only to loops in the following instructions, for ease of reading.

Obtaining a Sample with an Inoculating Loop or Needle

- **From a broth**

1. Suspend bacteria in the broth with a vortex mixer (Figure 1-14) or by agitating the tube with your fingers (Figure 1-15).
2. Flame the loop (Figure 1-10).
3. Remove and hold the tube's cap with the little finger of your loop hand (Figure 1-11).
4. Flame the open end of the tube by passing it through a flame two or three times (Figure 1-12).
5. Hold the open tube at an angle to prevent airborne contamination (Figure 1-13).
6. Holding the loop hand still, move the tube up the wire until the tip is in the broth. Continuing to hold the loop hand still, *remove the tube* from the wire (Figure 1-17). There should be a film of broth in the loop (Figure 1-18). Be especially careful not to catch the loop tip on the tube lip. This springing action of the loop creates bacterial aerosols.
7. Flame the tube lip as before. Keep your loop hand still.
8. Keeping the loop hand still (remember, it has growth on it), move the tube to replace its cap.

What you do next depends on the medium to which you are transferring the growth. Please continue with the appropriate inoculation section.

- **From a Slant**

1. Flame the loop (Figure 1-10).

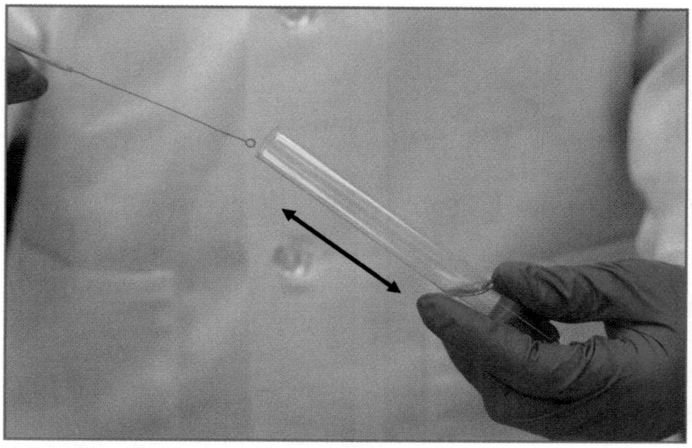

FIGURE 1-17 Loop In/Out of Broth
The open tube is held at an angle to minimize airborne contamination. When placing a loop into a broth tube or removing it, keep the loop hand still and move the tube. Be careful not to catch the loop on the tube's lip when removing it. This produces aerosols that can be dangerous or produce contamination.

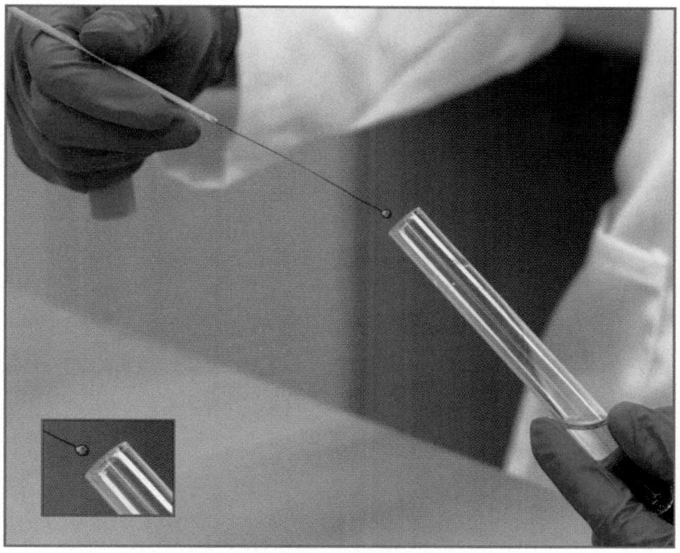

FIGURE 1-18 Removing the Loop from Broth
Notice the film of broth in the loop (see inset). Be careful not to catch the loop on the lip of the tube when removing it. This would produce aerosols that can be dangerous or produce contamination.

2. Remove and hold the culture tube's cap with the little finger of your loop hand (Figure 1-11).
3. Flame the open end of the tube by passing it through a flame two or three times (Figure 1-12).
4. With the agar surface facing upward, hold the open tube at an angle to prevent airborne contamination (Figure 1-13).
5. Holding the loop hand still, move the tube up the wire until the wire tip is over the desired growth (Figure 1-19). Touch the loop to the growth and obtain the smallest visible mass of bacteria. Then, holding the loop hand still, *remove the tube* from the wire. Be especially careful not to catch the loop tip on the tube lip. This springing action of the loop creates bacterial aerosols.

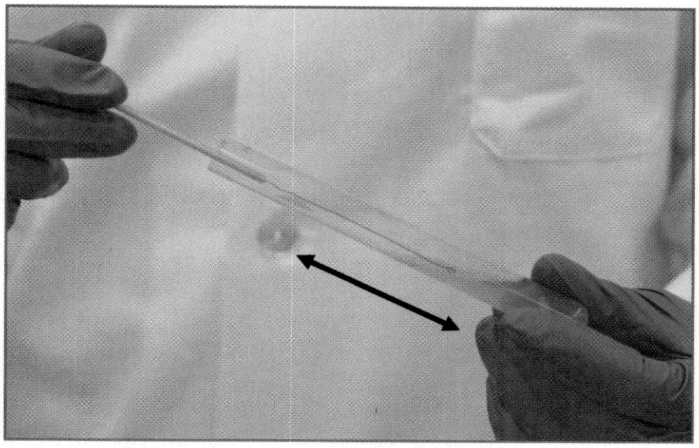

FIGURE 1-19 A Loop and an Agar Slant
When placing a loop into a slant tube or removing it, the loop hand is kept still while the tube is moved. Hold the tube so the agar is facing upward.

6. Flame the tube lip as before. Keep your loop hand still.
7. Keeping the loop hand still (remember—it has growth on it), move the tube to replace its cap.

What you do next depends on the medium to which you are transferring the growth. Please continue with the appropriate inoculation section.

● From an Agar Plate

1. Flame the loop (Figure 1-10).
2. Lift the lid of the agar plate, but continue to use it as a cover to prevent contamination from above (Figure 1-16).
3. Touch the loop to an uninoculated portion of the plate to cool it. (Placing a hot wire on growth may cause the growth to spatter and create aerosols.) Obtain a small amount of bacterial growth by gently touching a colony with the wire tip (Figure 1-16).
4. Carefully remove the loop from the plate and hold it still as you replace the lid.

What you do next depends on the medium to which you are transferring the growth. Please continue with the appropriate inoculation section.

Inoculating Media with an Inoculating Loop or Needle

● Fishtail Inoculation of Agar Slants

Agar slants generally are used for growing stock cultures that can be refrigerated after incubation and maintained for several weeks. Many differential media used in identification of microbes are also slants.

1. Remove the cap of the sterile medium with the little finger of your loop hand and hold it there (Figure 1-11).
2. Flame the tube by quickly passing it through the flame a couple of times. Keep your loop hand still (Figure 1-12).
3. Hold the open tube on an angle to minimize airborne contamination. Keep your loop hand still.
4. With the agar surface facing upward, carefully move the tube over the wire. Gently touch the loop to the agar surface near the base.
5. Beginning at the bottom of the exposed agar surface, drag the loop in a zigzag pattern as you withdraw the tube (Figure 1-20). Be careful not to cut the agar surface, and be especially careful not to catch the loop tip on the tube lip as you remove it. This springing action of the loop creates bacterial aerosols.
6. Flame the tube mouth as before. Keep your loop hand still.
7. Keeping the loop hand still (remember—it has growth on it), move the tube to replace its cap.
8. Sterilize the loop as before by incinerating it in the Bunsen burner flame. It is especially important to flame it from base to tip now because the loop has lots of bacteria on it.
9. Label the tube with your name, date, medium, and organism. Incubate at the appropriate temperature for the assigned time.

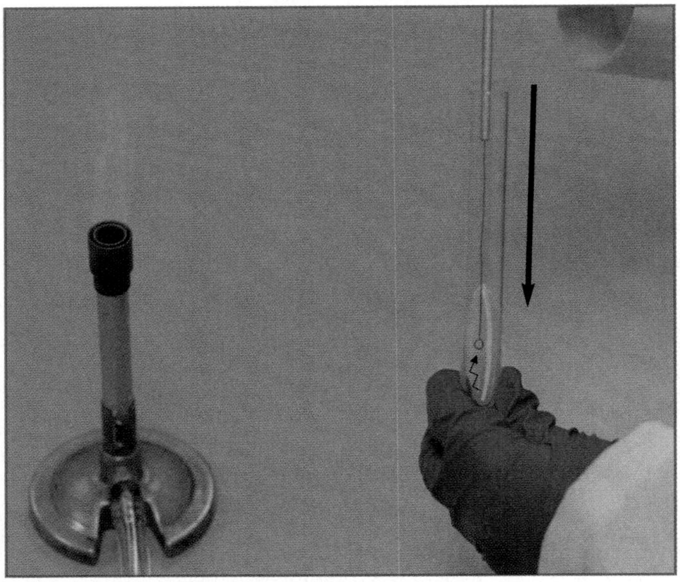

FIGURE 1-20 Fishtail Inoculation of a Slant
Begin at the base of the slant surface and gently move the loop back and forth as you withdraw the tube. Be careful not to cut the agar. After completing the transfer, sterilize the loop.

Inoculating Broth Tubes

Broth cultures are often used to grow cultures for use when fresh cultures or large numbers of cells are desired. Many differential media are also broths.

1. Remove the cap of the sterile medium with the little finger of your loop hand and hold it there (Figure 1-11).
2. Sterilize the tube by quickly passing it through the flame a couple of times. Keep your loop hand still (Figure 1-12).
3. Hold the open tube on an angle to minimize airborne contamination. Keep your loop hand still.
4. Carefully move the broth tube over the wire (Figure 1-21). Gently swirl the loop in the broth to dislodge microbes.
5. Withdraw the tube from over the loop. Before completely removing it, touch the loop tip to the glass to remove any excess broth (Figure 1-22). Then be especially careful not to catch the loop tip on the tube lip when withdrawing it. This springing action of the wire creates bacterial aerosols.
6. Flame the tube lip as before. Keep your loop hand still.
7. Keeping the loop hand still (remember—it has growth on it), move the tube to replace its cap.
8. Sterilize the loop as before by incinerating it in the Bunsen burner flame. It is especially important to flame it from base to tip now because the loop and wire have lots of bacteria on them.
9. Label the tube with your name, date, medium, and organism. Incubate at the appropriate temperature for the assigned time.

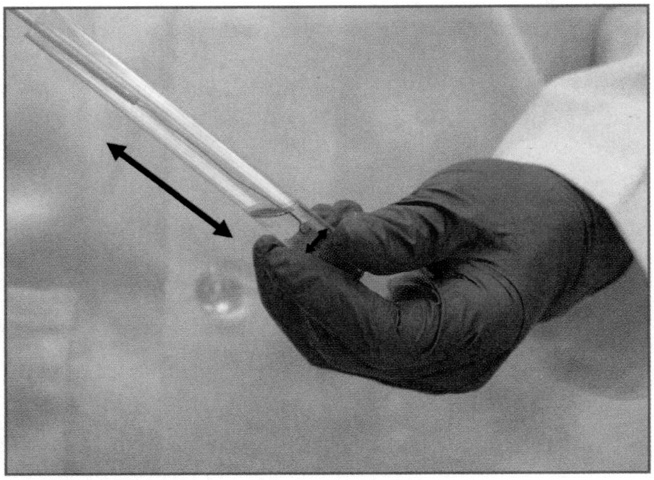

FIGURE 1-21 Inoculation of a Broth
When entering or leaving the tube, move the tube and keep the loop hand still. Gently swirl the loop in the broth to transfer the organisms.

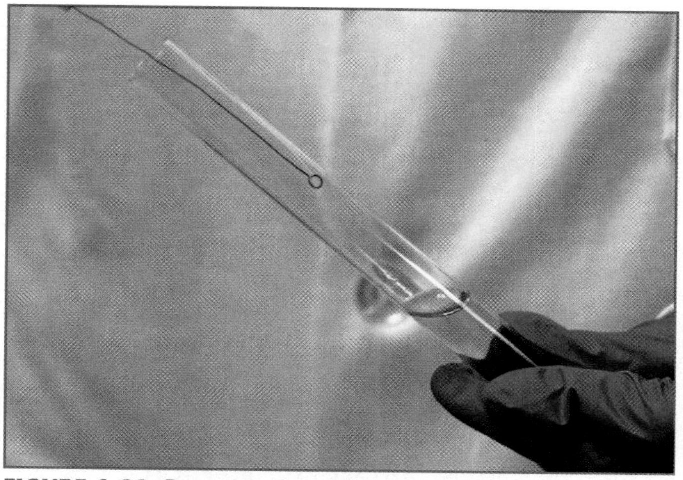

FIGURE 1-22 Removing Excess Broth from Loop
Before removing it from the tube, touch the loop to the glass to remove excess broth. Failing to do so will result in splattering and aerosols when sterilizing the loop in a flame.

In This Exercise . . .

You will perform some simple aseptic transfers: slant to slant and broth, broth to slant and broth, and plate to slant and broth. Master these, and you are well on your way to becoming a microbiologist!

Materials

Per Student Group

- inoculating loop (one per student)
- five nutrient broth tubes
- five nutrient agar slants
- marking pen and labels
- vortex mixer (optional)
- nutrient agar slant cultures of:
  - *Bacillus subtilis*
  - *Escherichia coli*
  - *Micrococcus luteus*
- nutrient broth culture of:
  - *Staphylococcus epidermidis*
- nutrient agar plate culture of:
  - *Kocuria rosea (Micrococcus roseus)*

Procedure

Lab One

1. To make the transfers listed, refer to the appropriate section in "Specific Transfer Methods" above.
   a. *B. subtilis* to slant and broth using an inoculating loop.
   b. *E. coli* to slant and broth using an inoculating loop.
   c. *M. luteus* to slant and broth using an inoculating loop.
   d. *S. epidermidis* to slant and broth using an inoculating loop.
   e. *K. rosea* (*M. roseus*) to slant and broth. (For each transfer choose a well-isolated colony and touch the center with the loop as in Figure 1-16).
2. Label all tubes clearly with your name, the organisms' names, and the date.
3. Incubate *M. luteus* and *M. roseus* at 25°C and the rest at 35 ± 2°C until next class.

Lab Two

1. Remove your cultures from the incubators and examine the growth. Record your observations on the Data Sheet.
2. Your instructor may ask you to save your cultures for later use. Otherwise, dispose of all materials in the appropriate autoclave containers.

References

Barkley, W. Emmett and John H. Richardson. 1994. Chapter 29 in *Methods for General and Molecular Bacteriology*. American Society for Microbiology, Washington, DC.

Claus, G. William. 1989. Chapter 2 in *Understanding Microbes—A Laboratory Textbook for Microbiology*. W. H. Freeman and Company, New York, NY.

Darlow, H. M. 1969. Chapter VI in *Methods in Microbiology*, Volume 1. Edited by J. R. Norris and D. W. Ribbins. Academic Press, Ltd., London.

Fleming, Diane O. 1995. Chapter 13 in *Laboratory Safety—Principles and Practices*, 2nd ed. Edited by Diane O. Fleming, John H. Richardson, Jerry J. Tulis, and Donald Vesley. American Society for Microbiology, Washington, DC.

Koneman, Elmer W., Stephen D. Allen, William M. Janda, Paul C. Schreckenberger and Washington C. Winn, Jr. 1997. Chapter 2 in *Color Atlas and Textbook of Diagnostic Microbiology*, 5th ed. Lippincott-Raven Publishers, Philadelphia.

Murray, Patrick R., Ellen Jo Baron, Michael A. Pfaller, Fred C. Tenover, and Robert H. Yolken. 1995. *Manual of Clinical Microbiology*, 6th ed. American Society for Microbiology, Washington, DC.

Power, David A. and Peggy J. McCuen. 1988. *Manual of BBL™ Products and Laboratory Procedures*, 6th Ed. Becton Dickinson Microbiology Systems, Cockeysville, MD.

Exercise 1-3

Streak Plate Methods of Isolation

Theory A microbial culture consisting of two or more species is said to be a **mixed culture**, whereas a **pure culture** contains only a single species. Obtaining isolation of individual species from a mixed sample is generally the first step in identifying an organism. A commonly used **isolation technique** is the **streak plate** (Figure 1-23).

In the streak plate method of isolation, a bacterial sample (always assumed to be a mixed culture) is streaked over the surface of a plated agar medium. During streaking, the cell density decreases, eventually leading to individual cells being deposited separately on the agar surface. Cells that have been sufficiently isolated will grow into **colonies** consisting only of the original cell type. Because some colonies form from individual cells and others from pairs, chains, or clusters of cells, the term **colony-forming unit** (**CFU**) is a more correct description of the colony origin.

Several patterns are used in streaking an agar plate, the choice of which depends on the source of inoculum and microbiologist's preference. Although streak patterns range from simple to more complex, all are designed to separate deposited cells (CFUs) on the agar surface so individual cells (CFUs) grow into isolated colonies. A quadrant streak is generally used with samples suspected of high cell density, whereas a simple zigzag pattern may be used for samples containing lower cell densities.

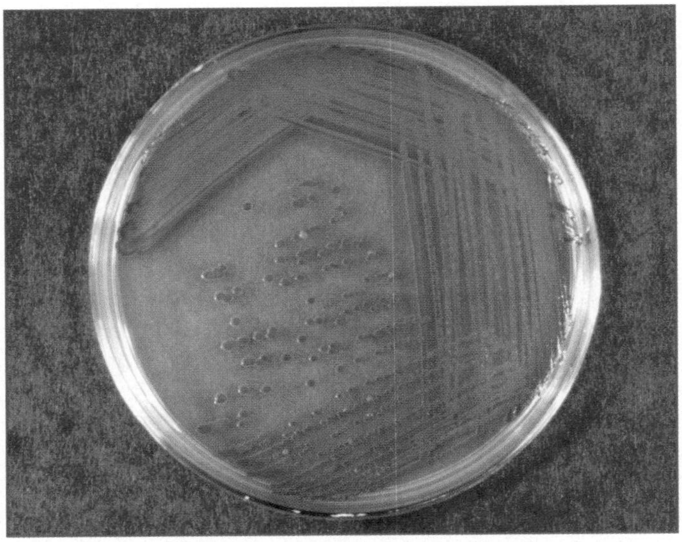

FIGURE 1-23 Streak Plate of *Serratia marcescens*
Note the decreasing density of growth in the four streak patterns. On this plate, isolation is first obtained in the fourth streak. Cells from an individual colony may be transferred to a sterile medium to start a pure culture.

Application The identification process of an unknown microbe relies on obtaining a pure culture of that organism. The streak plate method produces individual colonies on an agar plate. A portion of an isolated colony then may be transferred to a sterile medium to start a pure culture.

Following are descriptions of streak techniques. As in Exercise 1-2, basic skills are printed in the regular black type and new skills are printed in blue.

Inoculation of Agar Plates Using the Quadrant Streak Method

This inoculation pattern is usually performed as the initial streak for isolation of two or more bacterial species in a mixed culture with suspected high cell density.

1. Obtain the sample of mixed culture with a sterile loop.
2. Lift the lid of the sterile agar plate and use it as a shield to prevent airborne contamination.
3. Starting at the edge of the plate lightly drag the loop back and forth across the agar surface as shown in Figure 1-24a. Be careful not to cut the agar surface.
4. Remove the loop and replace the lid.
5. Sterilize your loop as before. It is especially important to flame it from base to tip now because the loop has lots of bacteria on it.
6. Let the loop cool for a few moments, then perform another streak with the sterile loop beginning at one end of the first streak pattern (Figure 1-24b).
7. Sterilize the loop, then repeat with a third streak beginning in the second streak (Figure 1-24c).
8. Sterilize the loop, then perform a fourth streak beginning in the third streak and extending into the middle of the plate. Be careful not to enter any streaks but the third (Figure 1-24d).
9. Sterilize the loop.
10. Label the plate's base with your name, date and organism(s) inoculated.
11. Incubate the plate in an inverted position for the assigned time at the appropriate temperature.

Zigzag Inoculation of Agar Plates Using a Cotton Swab

This inoculation pattern is usually performed when the sample does not have a high cell density and with pure cultures when isolation is not necessary.

1. Hold the swab comfortably in your dominant hand and lift the lid of the Petri dish with the other. Use

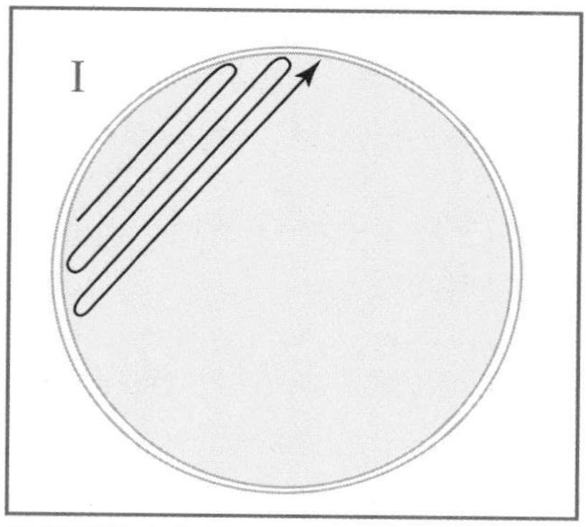

FIGURE 1-24A
Beginning the Streak Pattern
Streak the mixed culture back and forth in one quadrant of the agar plate. Use the lid as a shield and do not cut the agar with the loop. Flame the loop, then proceed.

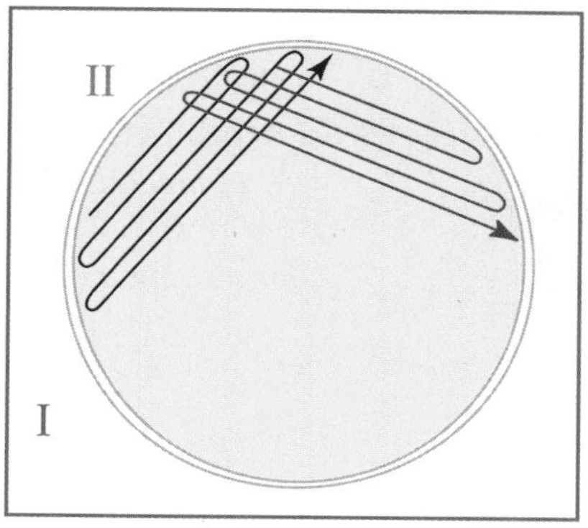

FIGURE 1-24B Streaking Again
Rotate the plate nearly 90° and touch the agar in an uninoculated region to cool the loop. Streak again using the same wrist motion. Flame the loop.

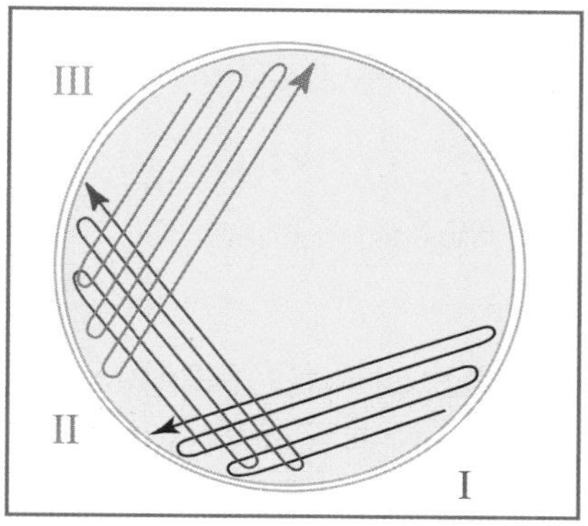

FIGURE 1-24C Streaking Yet Again
Rotate the plate nearly 90° and streak again using the same wrist motion. Be sure to cool the loop prior to streaking. Flame again.

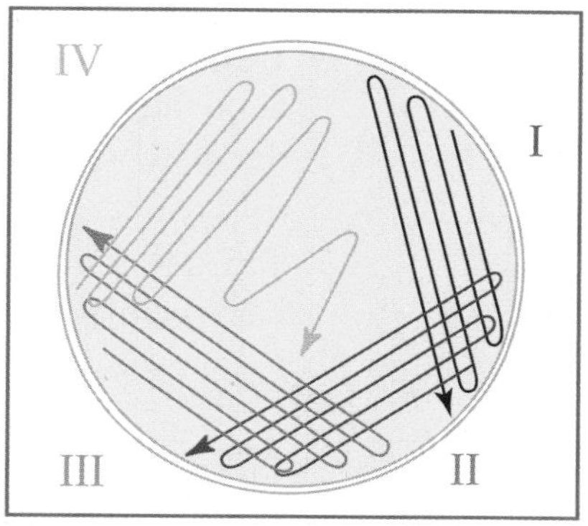

FIGURE 1-24D
Streaking Into the Center
After cooling the loop, streak one last time into the center of the plate. Flame the loop and incubate the plate in an inverted position for the assigned time at the appropriate temperature.

the lid as a shield to protect the agar from airborne contamination.

2. Lightly drag the cotton swab across the agar surface in a zigzag pattern. Be careful not to cut the agar surface (Figure 1-25).
3. Replace the lid.
4. Dispose of the swab according to your lab's practices.
5. Label the base of the plate with your name, date, and sample.
6. Incubate the plate in an inverted position for the assigned time at the appropriate temperature.

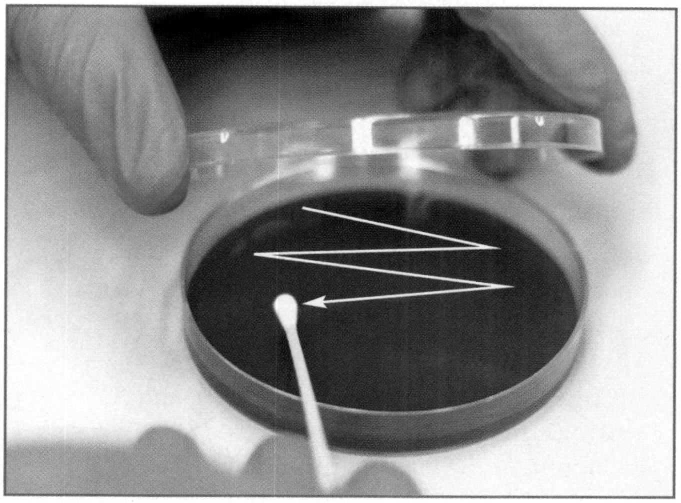

FIGURE 1-25 Zigzag Inoculation
Use the cotton swab to streak the agar surface to get isolated colonies after incubation. Be careful not to cut the agar. Properly dispose of the swab in a biohazard container.

Inoculation of Agar Plates with a Cotton Swab in Preparation for a Quadrant Streak Plate

This inoculation pattern is usually performed as the initial streak for isolation of two or more bacterial species in a mixed culture with suspected high cell density.

1. Hold the swab comfortably in your dominant hand and lift the lid of the Petri dish with the other. Use the lid as a shield to protect the agar from airborne contamination.
2. Lightly drag the cotton swab back and forth across the agar surface in one quadrant of the plate (Figure 1-26). This replaces the first streak as shown in Figure 1-24a.
3. Dispose of the swab according to your lab's practices.
4. Further streaking is performed with a loop as described in Figures 1-24b through 1-24d.
5. Label the plate's base with your name, date and sample.
6. Incubate the plate in an inverted position for the assigned time at the appropriate temperature.

In This Exercise . . .

You will learn how to isolate individual organisms from a mixed culture, the first step in producing a pure culture. Three related streaking techniques will be used, the choice of which is determined by the anticipated cell density of the sample.

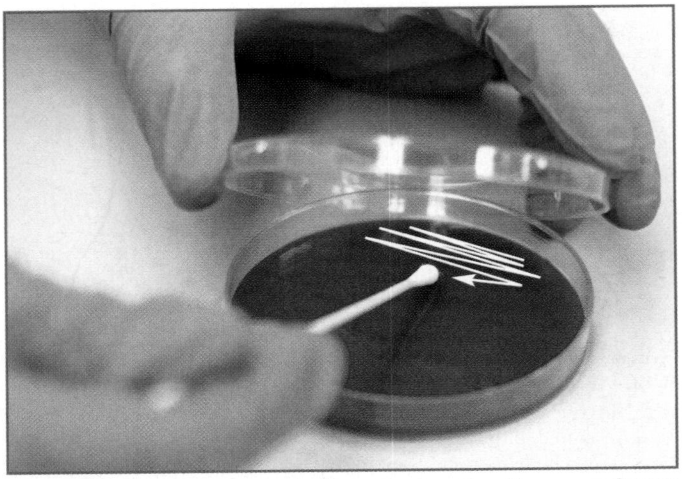

FIGURE 1-26 INOCULATION IN PREPARATION FOR A QUADRANT STREAK
If the sample is expected to have a high density of organisms, streak one edge of the plate with the swab. Then continue with the quadrant streak, using a loop. Be careful not to cut the agar. Properly dispose of the swab in a biohazard container.

Materials

Per Student Pair

- inoculating loop (one per student)
- four Trypticase Soy Agar (TSA) or Nutrient Agar (NA) plates
- two sterile cotton swabs in sterile distilled water
- four sterile transfer pipettes
- two sterile microtubes
- broth cultures of:
  - *Serratia marcescens*
  - *Chromobacterium violaceum*
  - *Staphylococcus aureus*
  - *Staphylococcus epidermidis*

Procedure

Lab One

1. Using a pencil, practice quadrant-streaking the plate on the Data Sheet before trying it with living bacteria. *Hint*: keep your wrist relaxed. Also, avoid digging into or cutting the agar, which ruins the plate and may create dangerous aerosols.
2. Transfer a few drops of *C. violaceum* and *S. marcescens* to a microtube and mix well. Transfer a loopful of the mixture to a sterile Trypticase Soy Agar or Nutrient Agar plate and follow the diagrams in Figure 1-24 to streak for isolation. Label the plate with your name, the date, and the organisms.
3. Repeat step one with a mixture of *S. aureus* and *S. epidermidis* on a second Trypticase Soy Agar or Nutrient Agar plate. Label the plate with your name, the date, and the organisms.
4. Use the cotton swab to sample an environmental source (see Appendix B), then do a simple zigzag streak on the third Trypticase Soy Agar or Nutrient Agar plate (Figure 1-25). Dispose of the swab in an autoclave container. Label the plate with your name, the date, and the sample source.
5. Use the second sterile cotton swab to sample the inside of one partner's cheek or along the gum line of the teeth. Use the swab to perform the first streak of a quadrant streak on the fourth plate (Figure 1-26). Properly dispose of the swab, and then complete the quadrant streak with your loop. Label the plate with your name, the date, and the sample source.
6. Tape the four plates together, invert them, and incubate them at $35 \pm 2°C$ for 24 to 48 hours.

Lab Two

1. After incubation, examine the plates for isolation.
2. Compare your streak plates with your lab partner's plates and critique each other's technique. Remember, a successful streak plate is one that has isolated colonies; the pattern doesn't have to be textbook quality—it's just that textbook quality provides you with a greater chance of getting isolation.

References

Collins, C. H. and Patricia M. Lyne. 1995. Chapter 6 in *Collins and Lyne's Microbiological Methods*, 7th ed. Butterworth-Heineman.

Delost, Maria Dannessa. 1997. Chapter 1 in *Introduction to Diagnostic Microbiology.* Mosby, Inc., St. Louis, MO.

Forbes, Betty A., Daniel F. Sahm, and Alice S. Weissfeld. 2002. Chapter 1 in *Bailey and Scott's Diagnostic Microbiology,* 11th ed. Mosby-Yearbook, St. Louis, MO.

Koneman, Elmer W., Stephen D. Allen, William M. Janda, Paul C. Schreckenberger and Washington C. Winn, Jr. 1997. Chapter 2 in *Color Atlas and Textbook of Diagnostic Microbiology*, 5th ed. J. B. Lippincott Company, Philadelphia, PA.

Power, David A. and Peggy J. McCuen. 1988. Pages 2 and 3 in *Manual of BBL™ Products and Laboratory Procedures,* 6th ed. Becton Dickinson Microbiology Systems, Cockeysville, MD.

Exercise 1-4

Spread Plate Method of Isolation

Theory The spread plate technique is a method of isolation in which a diluted microbial sample is deposited on an agar plate and spread uniformly across the surface with a glass rod. With a properly diluted sample, cells (CFUs) will be deposited far enough apart on the agar surface to grow into individual colonies.

Application After incubation, a portion of an isolated colony can be transferred to a sterile medium to begin a pure culture. The spread plate technique also has applications in quantitative microbiology (see Section Six).

Following is a description of the spread plate technique. As in the previous exercises, basic skills are printed in the regular black type and new skills are printed in blue.

Spread Plate Technique

1. Arrange the alcohol beaker, Bunsen burner, and agar plate as shown in Figure 1-27. This arrangement minimizes the chances of catching the alcohol on fire.
2. Lift the plate's lid and use it as a shield to protect from airborne contamination.
3. Using an appropriate pipette, deposit the designated inoculum volume on the agar surface. (Please see Appendices C and D for use of pipettes.) From this point, the remainder of steps should be completed within about 15 seconds to prevent the inoculum from soaking into the agar.
4. Properly dispose of the pipetting instrument used to inoculate the medium, because it is contaminated. Each lab has its own specific procedures and your instructor will advise you what to do.
5. Remove the glass spreading rod from the alcohol and pass it through the flame to ignite the alcohol (Figure 1-28). Remove the rod from the flame and allow the alcohol to burn off completely. Do not leave the rod in the flame; the combination of the alcohol and brief flaming are sufficient to sterilize it. *Be careful not to drop any flaming alcohol on the work surface. Be especially careful not to drop flaming alcohol back into the alcohol beaker.*
6. After the flame has gone out on the glass rod, lift the lid of the plate and use it as a shield from airborne contamination. Then touch the rod to the agar surface away from the inoculum to cool it.
7. To spread the inoculum, hold the plate lid with the base of your thumb and index finger and use the tip of your thumb and middle finger to rotate the base (Figure 1-29). At the same time, move the rod in a back-and-forth motion across the agar surface. After a couple of turns, do one last turn with the rod next to the plate's edge. Alternatively, place the plate on a rotating platform and spread the inoculum (Figure 1-30).
8. Remove the rod from the plate and replace the lid.
9. Return the rod to the alcohol in preparation for the next inoculation. There is no need to flame it again.
10. Label the plate base with your name, date, organism, and any other relevant information.

FIGURE 1-27 SPREAD PLATE SET-UP
The spread plate technique requires a Bunsen burner, a beaker with alcohol, a glass spreading rod, and the plate. Position these components in your work area as shown: isopropyl alcohol, flame, and plate. This arrangement reduces the chance of accidentally catching the alcohol on fire.

FIGURE 1-28 STERILIZING THE GLASS ROD
Remove the glass spreading rod from the alcohol and pass it through the flame to ignite the alcohol. Remove the rod from the flame and allow the alcohol to burn off completely. Do not leave the rod in the flame; the combination of the alcohol and brief flaming are sufficient to sterilize it. *Be careful not to drop any flaming alcohol on the work surface.*

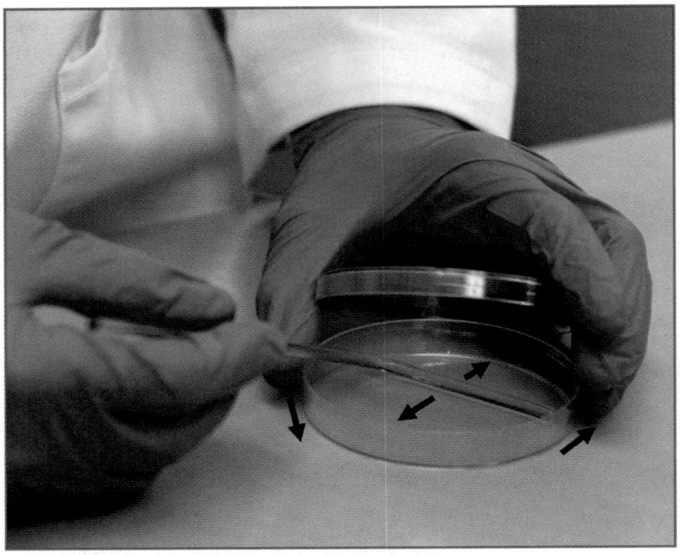

FIGURE 1-29 Spreading Plate Inoculum
After the flame has gone out on the rod, lift the lid of the plate and use it as a shield from airborne contamination. Then, touch the rod to the agar surface away from the inoculum in order to cool it. To spread the inoculum, hold the plate lid with the base of your thumb and index finger, and use the tip of your thumb and middle finger to rotate the base. At the same time, move the rod in a back-and-forth motion across the agar surface. After a couple of turns, do one last turn with the rod next to the plate's edge.

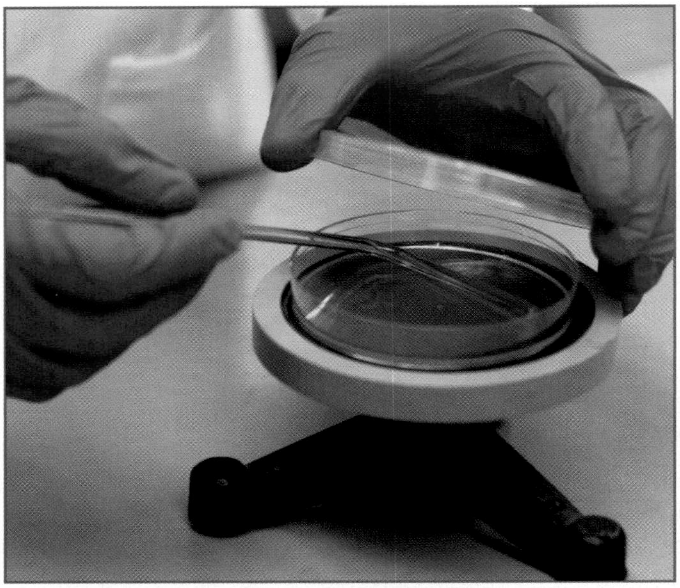

FIGURE 1-30 Inoculating with a Turntable
A turntable makes it easier to rotate the plate during the spread plate technique.

11. Incubate the plate in an inverted position at the appropriate temperature for the assigned time. (If you plated a large volume of inoculum (greater than 0.5 mL), wait a few minutes and allow it to soak in before inverting the plate.)

In This Exercise…

You will perform a spread plate inoculation. In the context of this exercise, it used as an isolation procedure, but it also is used in quantifying cell densities in broth samples.

Materials

Per Student Pair

- inoculating loop (each student)
- six sterile plastic transfer pipettes
- 500 mL beaker with 50 mL of isopropyl alcohol
- glass spreading rod
- Bunsen burner and striker
- four nutrient agar plates
- one sterile microtube
- four capped microtubes with about 1 mL sterile distilled or deionized water (dH_2O)
- vortex mixer (optional)
- broth cultures of:
  - *Escherichia coli*
  - *Serratia marcescens*

Procedure

Lab One

1. Using a different pipette for each, transfer a few drops of *E. coli* and *S. marcescens* to a microtube. Cap the tube and mix well with a vortex mixer. *Or* use the second pipette to mix well by gently drawing and dispensing the mixture in and out of the tube a couple of times. Do not spray the mixture!
2. Label the four microtubes containing sterile dH_2O "A," "B," "C," and "D."
3. Label the four nutrient agar plates "A," "B," "C," and "D."
4. Transfer a loopful of the mixture to Tube A and mix well with the loop or a vortex mixer. If using a vortex mixer, be sure to cap the tube.[1]
5. Transfer a loopful of the mixture in Tube A to Tube B and mix well with the loop or a vortex mixer.
6. Transfer a loopful of the mixture in Tube B to Tube C and mix well with the loop or a vortex mixer.
7. Transfer a loopful of the mixture in Tube C to Tube D and mix well with the loop or a vortex mixer.
8. Using a sterile transfer pipette, place a couple of

[1] Because this is not a quantitative procedure, it is not necessary to flame the loop between transfers.

drops of sample from Tube A on Plate A. Spread the inoculum with a glass rod as described above. Let the plate sit for a few minutes.

9. Repeat Step 8 for Tubes B, C, and D and Plates B, C, and D, respectively.
10. Tape the four plates together, invert them and incubate them at 25°C for 24 to 48 hours.

Lab Two

1. After incubation, examine the plates for isolation. *S. marcescens* produces reddish colonies, and *E. coli* produces buff-colored colonies.
2. Fill in the Data Sheet.

References

Clesceri, WEF, Chair; Arnold E. Greenberg, APHA; Andrew D. Eaton, AWWA ; and Mary Ann H. Franson. 1998. Pages 9–38 in *Standard Methods for the Examination of Water and Wastewater*, 20th edition. Joint publication of American Public Health Association, American Water Works Association and Water Environment Federation. APHA Publication Office, Washington, DC.

Downes, Frances Pouch, and Keith Ito. 2001. Page 57 in *Compendium of Methods for the Microbiological Examination of Foods*. American Public Health Association. Washington DC.

Gerhard, Philipp, R.G.E Murray, Willis A. Wood, and Noel R. Kreig. 1994. Pages 255–257 in *Methods for General and Molecular Bacteriology*. American Society for Microbiology, Washington, D.C.

Microbial Growth

SECTION 2

Microorganisms are extraordinarily diverse. Every species demonstrates a unique combination of characteristics, some of which can be easily observed. In this section we illustrate some of those characteristics and factors that affect them.

Allowing for variables, such as the growth medium used, as well as incubation time and temperature, much can be determined about an organism by simply looking at the colonies it produces. Distinguishing different growth patterns is an important skill—one that you can use as you progress through the semester. Note the growth characteristics of all the organisms provided for your laboratory exercises and jot them down or even draw sketches of them. When the time comes to identify your unknown species, you will find that your good record keeping is a valuable asset.

You will begin this section with an exercise intended to sensitize you to the vast microbial population living all around us. Then you will examine some microbial growth characteristics and cultivation methods. Next you will look at some environmental factors affecting microbial growth, including pH, oxygen, temperature, and osmotic pressure. Finally, you will examine the resistance and susceptibility of microorganisms to antiseptics and disinfectants.

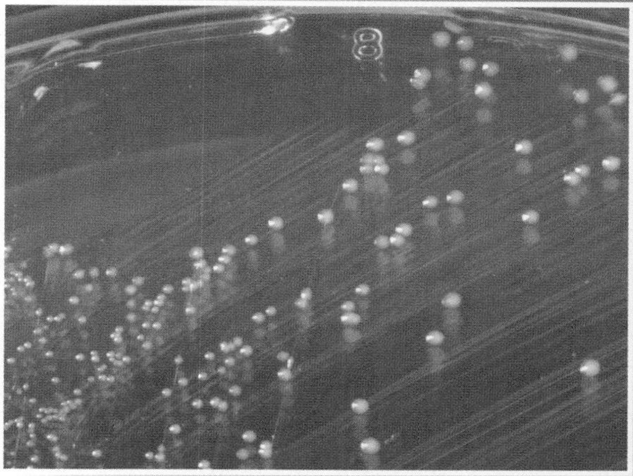

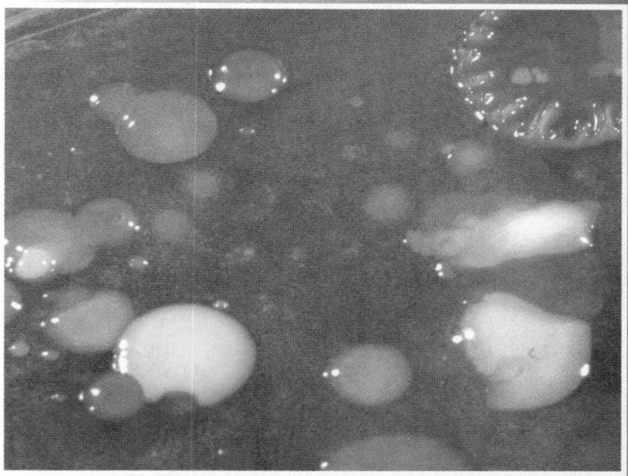

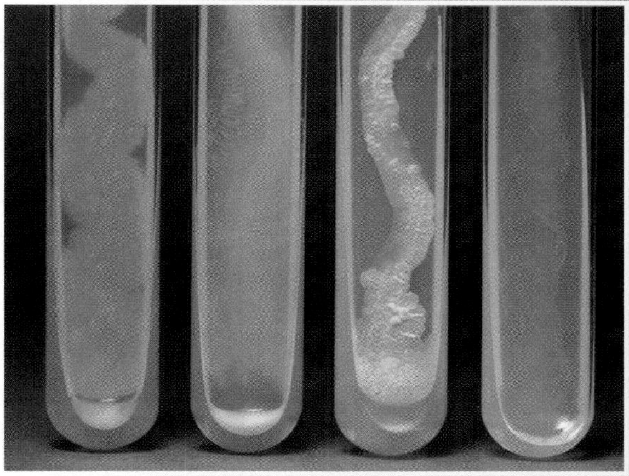

Diversity and Ubiquity of Microorganisms

Microorganisms are found everywhere that other forms of life exist. They can be isolated from soil, bodies of water, even from the air. As unwanted parasites or colonizers, some cause diseases or infections. Most, however, are harmless saprophytes; they simply live in, on, or around plants and animals and decompose dead organic matter. In so doing, they perform the essential function of nutrient recycling in ecosystems.

In this section, you will grow microorganisms from seemingly uninhabited sources. You then will learn to identify the various types of growth characteristics that these "invisible" cohabitants produce when they are cultivated in broth and on solid media.

Exercise 2-1

Ubiquity of Microorganisms

Theory In the literature on microorganisms, you will frequently encounter the phrase "ubiquitous in nature." This means that the organism being considered can be found just about everywhere. More specifically, the organism likely could be isolated from soil, water, plants, and animals (including humans). Although the word *ubiquitous* doesn't apply to every species, it does apply to many, and certainly to microorganisms as a group.

Many microorganisms are **free-living—**they do not reside on or in a specific plant or animal **host** and are not known to cause disease. They are **nonpathogenic**. Other microorganisms are **pathogens** and generally are associated with their host (or hosts). Even many of the **commensal** or **mutualistic** strains inhabiting our bodies are **opportunistic pathogens**. That is, they are capable of producing a disease state if introduced into a suitable part of the body. Any area including sites outside the host organism where a microbe resides and serves as a potential source of infection is called a **reservoir**.

Application This exercise is designed to demonstrate the ubiquitous nature of microorganisms and the ease with which they can be cultivated. (Interestingly enough, many more are uncultivatable at this time and we are hard-pressed to even know of their presence!)

In This Exercise...

Today, you will work in small groups to sample and culture several locations in your laboratory. Your instructor may have other locations outside of the lab to sample as well. It is wise to remember that even relatively "harmless" bacteria, when cultivated on a growth medium, are in sufficient numbers to constitute a health hazard. Treat them with care.

Materials

Per Student Group

- eight Nutrient Agar plates
- one sterile cotton swab

Procedure

Lab One

1. Number the plates 1 through 8.
2. Open plate number 1 and expose it to the air for 30 minutes or longer. Set it aside and out of the way of the other plates.
3. Use the cotton swab to sample your desk area and then streak plates 2 and 3, as in Figure 2-1. (Pressing very lightly, roll the swab on the agar as you streak it.)
4. Cough several times on the agar surface of plate 4.
5. Rub your hands together, and then touch the agar surface of plate 5 lightly with your fingertips. (A light touch is sufficient; touching too firmly will crack the agar.)
6. Remove the lid of plate 6 and vigorously scratch your head above it. (Keep this plate away from plate 1 to avoid cross-contamination.)
7. Leave plates 7 and 8 covered; do not open them.
8. Label the base of each plate with the date, type of exposure it has received, and your group's name.
9. Invert all plates and incubate them for 24 to 48 hours at the following temperatures:

   Plates 1, 2, and 8: 25°C

   Plates 3, 4, 5, 6 and 7: 37°C

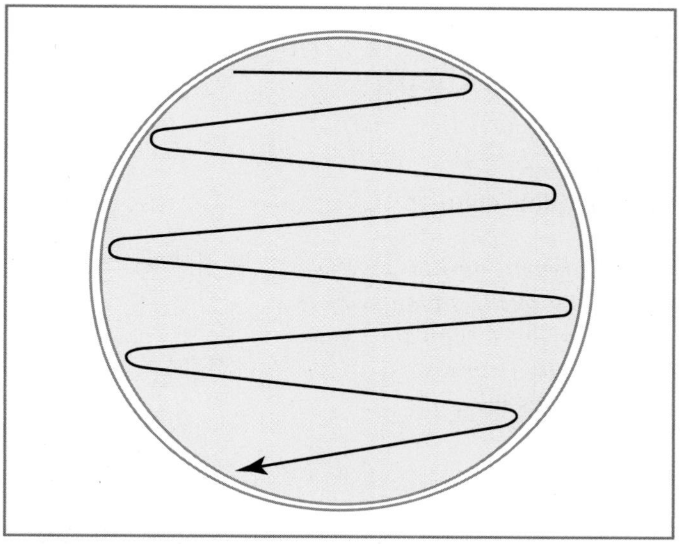

FIGURE 2-1 Simple Streak Pattern on Nutrient Agar
Roll the swab as you inoculate the plate. Do not press so hard that you cut the agar.

Lab Two

1. Using the plate diagrams on the Data Sheet, draw the growth patterns on each of your agar plates. Be sure to label them according to incubation time, temperature, and source of inoculum.
2. Combine these plates with those in Exercise 2-2, "Colony Morphology," to examine and compare the growth. (***Note:*** Remember that many microorganisms are opportunistic pathogens, so be sure to handle them carefully. Do not open plates containing fuzzy growth, as a fuzzy appearance suggests fungal growth with spores that can easily spread and contaminate the laboratory and other cultures. If you are in doubt, check with your instructor.)

References

Forbes, Betty A., Daniel F. Sahm, Alice S. Weissfeld. 2002. Chapter 10 in *Bailey & Scott's Diagnostic Microbiology,* 11th ed. Mosby, St. Louis.

Holt, John G. (Editor). 1994. *Bergey's Manual of Determinative Bacteriology,* 9th ed. Williams and Wilkins, Baltimore.

Koneman, Elmer W., Stephen D. Allen, William M. Janda, Paul C. Schreckenberger, and Washington C. Winn, Jr. 1997. *Color Atlas and Textbook of Diagnostic Microbiology,* 5th ed. J. B. Lippincott.

Varnam, Alan H., and Malcolm G. Evans. 2000. *Environmental Microbiology.* ASM Press, Washington, DC.

Exercise 2-2
Colony Morphology

Theory

When a single bacterial cell is deposited on a solid nutrient medium, it begins to divide. One cell makes two, two make four, four make eight . . . one million make two million, and so on. Eventually a visible mass of cells—a **colony**—appears where the original cell was deposited. Color, size, shape, and texture of microbial growth are determined by the genetic makeup of the organism, but also greatly influenced by environmental factors including nutrient availability, temperature, and incubation time.

The basic categories of colony morphology include colony shape, margin (edge), elevation, texture, and pigment production (color).

1. *Shape* may be described as **circular**, **irregular**, or **punctiform** (tiny, pinpoint).
2. The *margin* may be **entire** (smooth, with no irregularities), **undulate** (wavy), **lobate** (lobed), **filamentous**, or **rhizoid** (branched like roots).
3. *Elevations* include **flat, raised, convex, pulvinate** (very convex), and **umbonate** (raised in the center).
4. *Texture* may be **moist, mucoid**, or **dry**.
5. *Pigment production* (color), is another useful characteristic and may be combined with optical properties such as **opaque, translucent, shiny**, or **dull**.

Application

Recognizing different bacterial growth morphologies on agar plates is a useful step in the identification process. Once purity of a colony has been confirmed by an appropriate staining procedure, cells can be cultivated and maintained on sterile media for a variety of purposes.

In This Exercise...

Today you will be viewing colony characteristics on prepared streak plates. Colony characteristics may be viewed with the naked eye or with the assistance of a colony counter (Figure 2-2). Figures 2-3 through 2-28 show a variety of bacterial colony forms and characteristics. Where applicable, contrasting environmental factors are indicated.

Materials

Per Student Group

- colony counter (optional)
- metric ruler
- streak plate cultures of:
  - *Micrococcus luteus*
  - *Corynebacterium xerosis*
  - *Lactobacillus plantarum*
  - *Mycobacterium smegmatis*
  - *Bacillus subtilis*
  - *Proteus mirabilis*

Procedure

1. Using the terms in Figure 2-3, describe the colonies in today's cultures and the colonies on the plates from Exercise 2-1. Measure colony diameters (in mm) with a ruler and include them with your descriptions in the table on the Data Sheet. It may be helpful to use a colony counter (Figure 2-2).
2. Save the plates from today's exercise to compare with the slants and broths in Exercises 2-3 and 2-4. Discard all plates from Exercise 2-1 in an appropriate autoclave container.

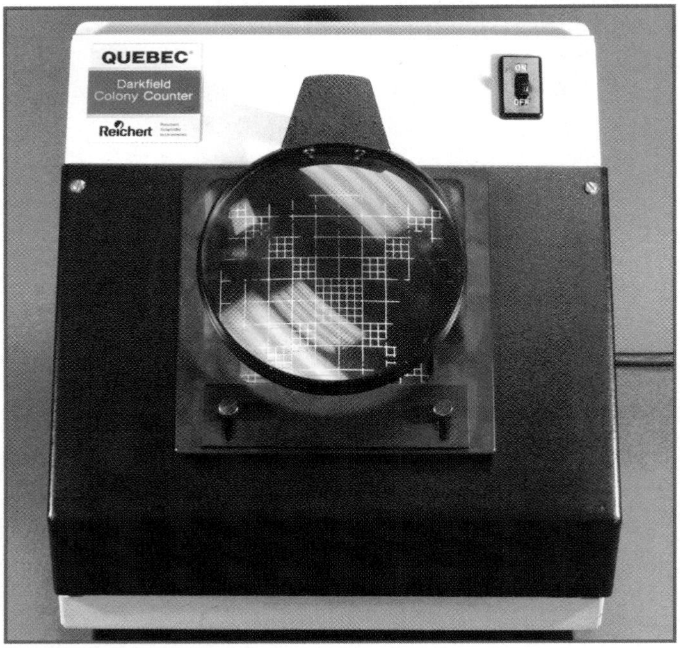

FIGURE 2-2 COLONY COUNTER

Subtle differences in colony shape and size can best be viewed on the colony counter. The **transmitted light** and magnifying glass allow observation of greater detail; however, colony color is best determined with **reflected light**. The grid in the background is a counting aid.

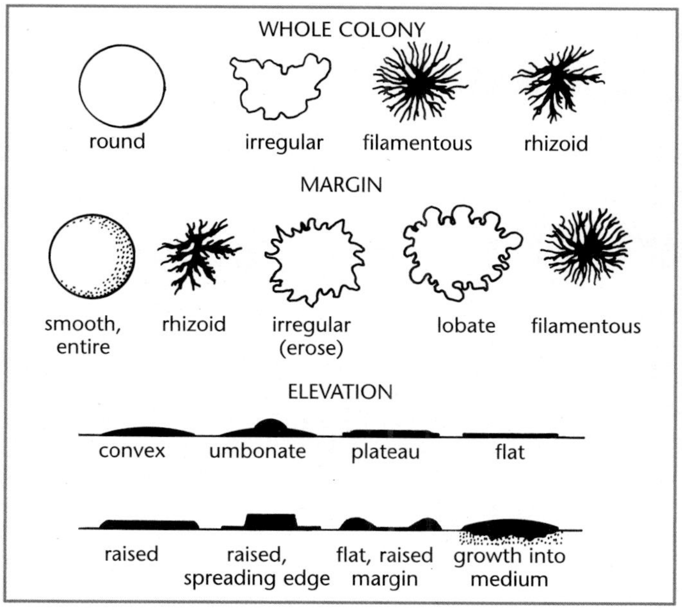

FIGURE 2-3 A SAMPLING OF BACTERIAL COLONY FEATURES
These terms are used to describe colonial morphology. Color, surface characteristics (dull or shiny), consistency (dry, butyrous-buttery, or moist) and optical properties (opaque or translucent), should also be included in colony morphology descriptions.

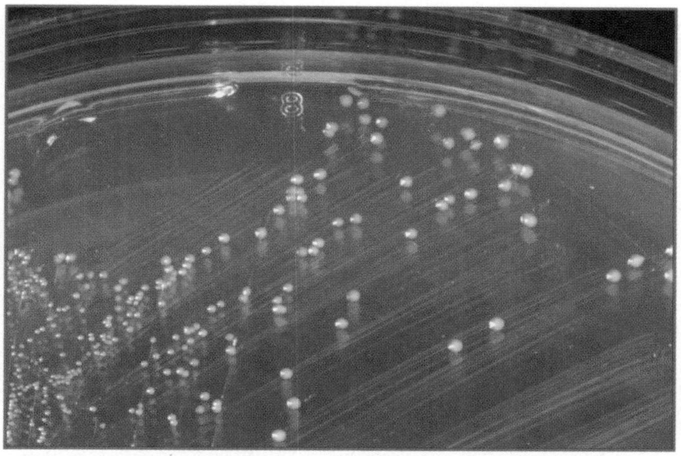

FIGURE 2-4 *ENTEROCOCCUS FAECIUM* GROWN ON NUTRIENT AGAR
The colonies are white, circular, convex, and have an entire margin. *E. faecium* (formerly known as *Streptococcus faecium*) is found in human and animal feces.

FIGURE 2-5 *STAPHYLOCOCCUS EPIDERMIDIS* GROWN ON SHEEP BLOOD AGAR
The colonies are white, raised, circular and entire. *S. epidermidis* is an opportunistic pathogen.

FIGURE 2-6 *CHROMOBACTERIUM VIOLACEUM* GROWN ON SHEEP BLOOD AGAR
C. violaceum produces shiny, purple, convex colonies. It is found in soil and water, and rarely produces infections in humans.

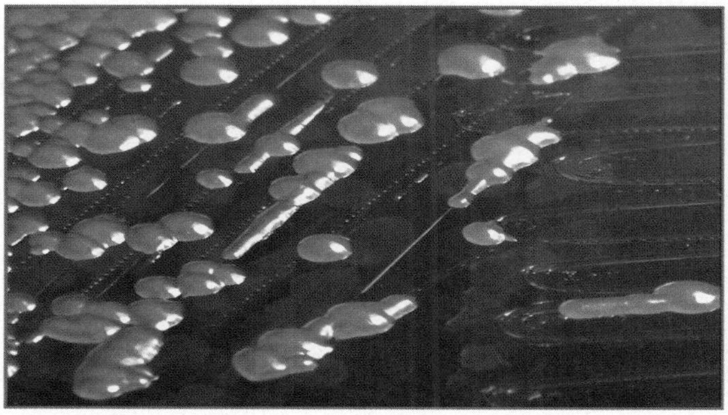

FIGURE 2-7 *PROVIDENCIA STUARTII* GROWN ON NUTRIENT AGAR
The colonies are shiny, buff, and convex. *P. stuartii* is a frequent isolate in urine samples obtained from hospitalized and catheterized patients.

FIGURE 2-8 *KLEBSIELLA PNEUMONIAE* GROWN ON NUTRIENT AGAR
The colonies are mucoid, raised, and shiny.

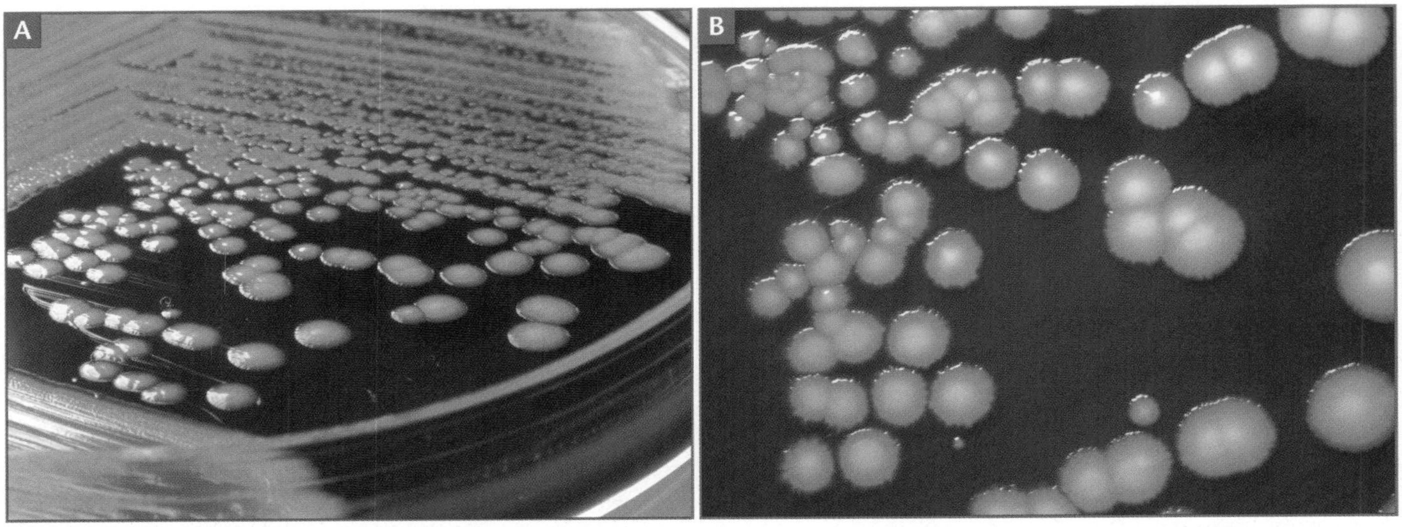

FIGURE 2-9 *Alcaligenes faecalis* Colonies on Sheep Blood Agar

The colonies of this opportunistic pathogen are umbonate with an opaque center and a spreading edge. (A) Side view: Note the raised center. (B) Close-up of the *A. faecalis* colonies showing spreading edge.

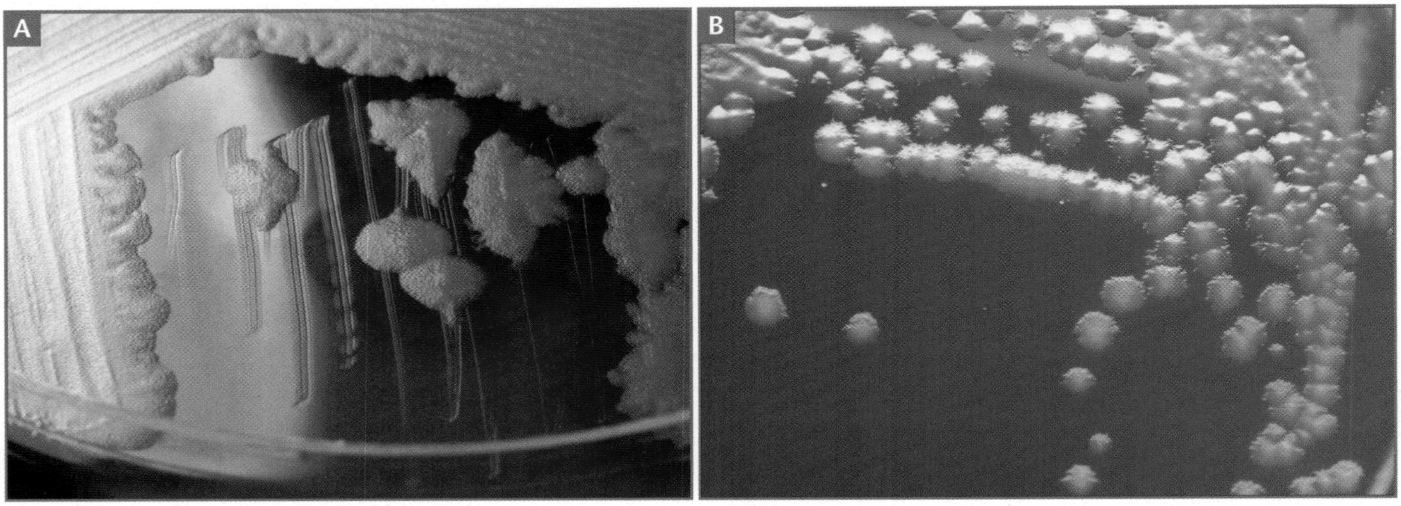

FIGURE 2-10 Comparison of Two *Bacillus* species.

The colonies are dry, dull, raised, rough-textured, and gray. (A) *Bacillus cereus* on Sheep Blood Agar. (B) *Bacillus anthracis* on Sheep Blood Agar.

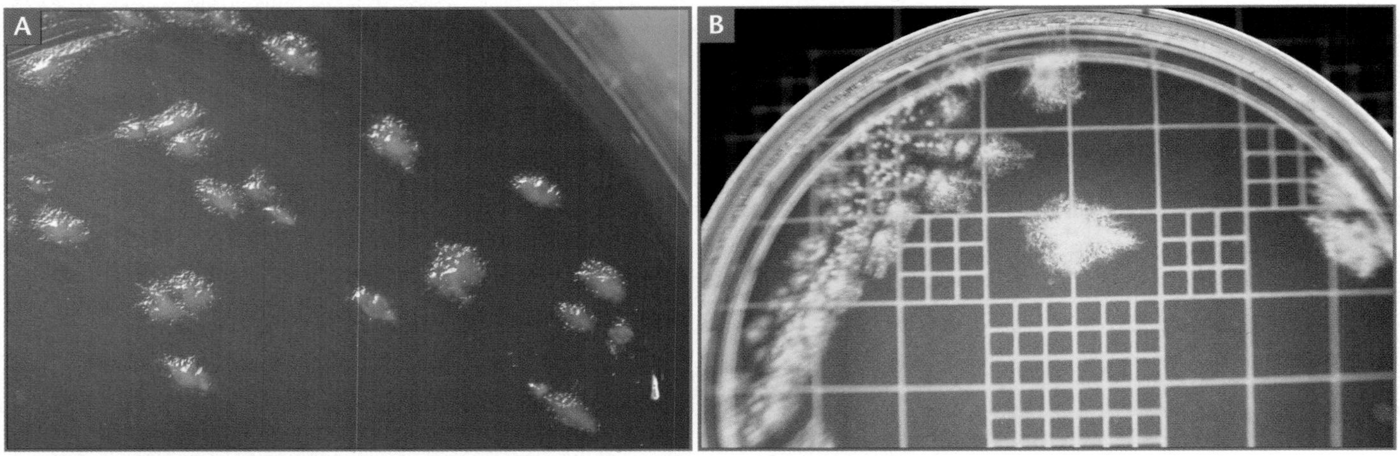

FIGURE 2-11 Comparison of *Clostridium sporogenes* Colonies Grown on Different Media

The colonies are irregular and rhizoid. *C. sporogenes* is found in soils worldwide. (A) *C. sporogenes* grown anaerobically on Sheep Blood Agar and viewed with reflected light. (B) *C. sporogenes* grown anaerobically on Nutrient Agar and viewed with transmitted light.

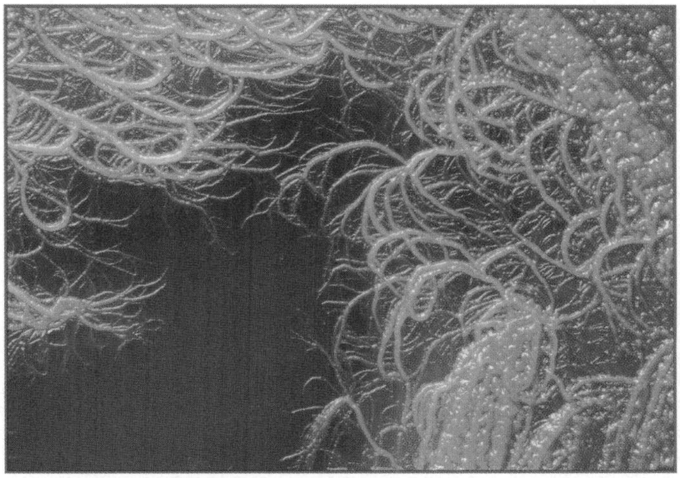

FIGURE 2-12 FILAMENTOUS GROWTH
This is an unknown soil organism grown on Sheep Blood Agar.

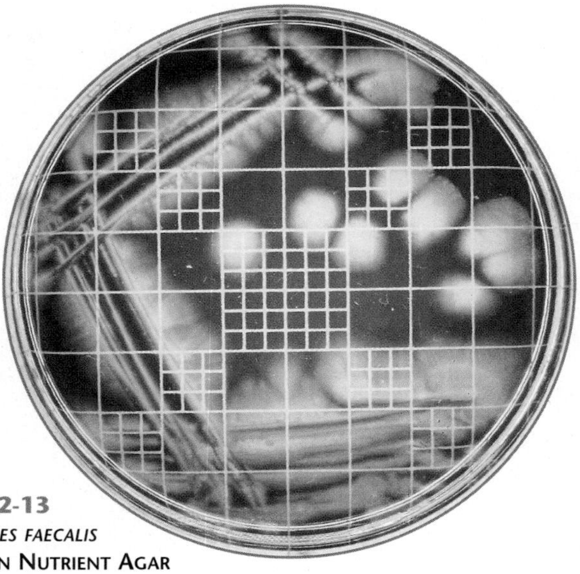

FIGURE 2-13
ALCALIGENES FAECALIS GROWN ON NUTRIENT AGAR
The growth demonstrates spreading attributable to motility and is translucent. Compare with the growth in Figure 2-9.

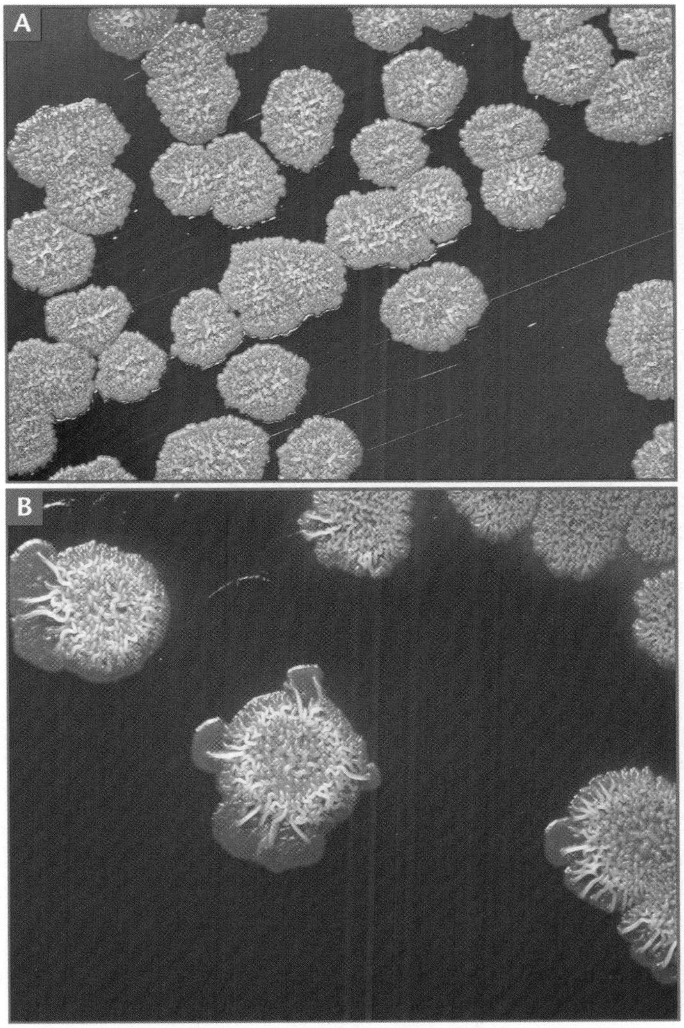

FIGURE 2-15 EFFECT OF AGE ON COLONY MORPHOLOGY
(A) Close-up of *Bacillus subtilis* on Sheep Blood Agar after 24 hours' growth. (B) Close-up of *Bacillus subtilis* on Sheep Blood Agar after 48 hours' growth. Note the wormlike appearance.

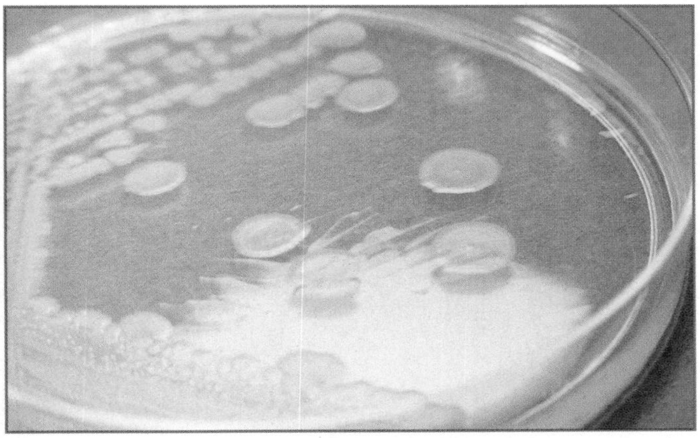

FIGURE 2-14 *BACILLUS SUBTILIS* COLONIES GROWN ON NUTRIENT AGAR VIEWED FROM THE SIDE
B. subtilis produces colonies with a raised margin and a dull surface. Compare with *B. cereus* and *B. anthracis* in Figure 2-10 and *B. subtilis* in Figure 2-15.

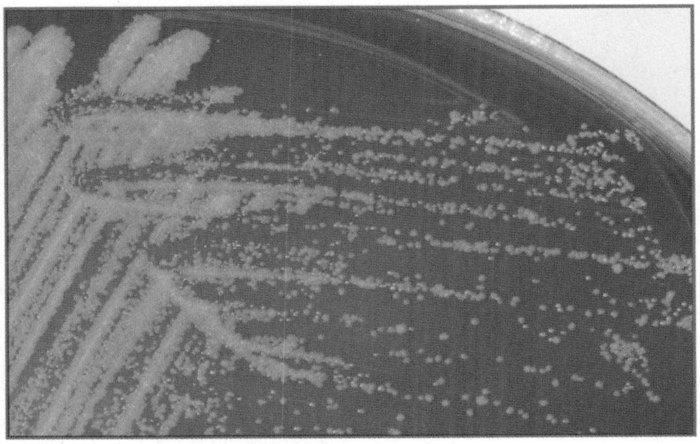

FIGURE 2-16 *MYCOBACTERIUM SMEGMATIS* GROWN ON SHEEP BLOOD AGAR
The colonies of this slow-growing relative of *M. tuberculosus* are punctiform.

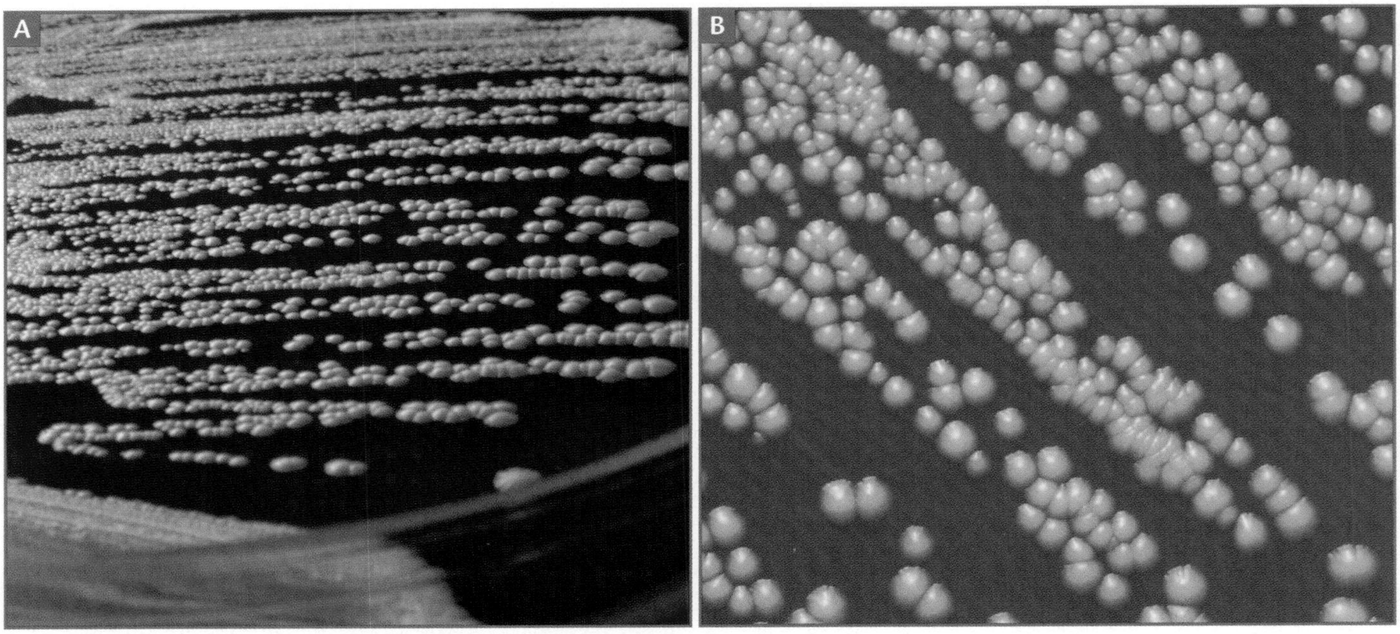

FIGURE 2-17 *Corynebacterium xerosis* Grown on Sheep Blood Agar
(A) As seen in this view from the side, the colonies are round, dull, buff, and convex. (B) Close-up of circular *C. xerosis* colonies. *C. xerosis* is rarely an opportunistic pathogen.

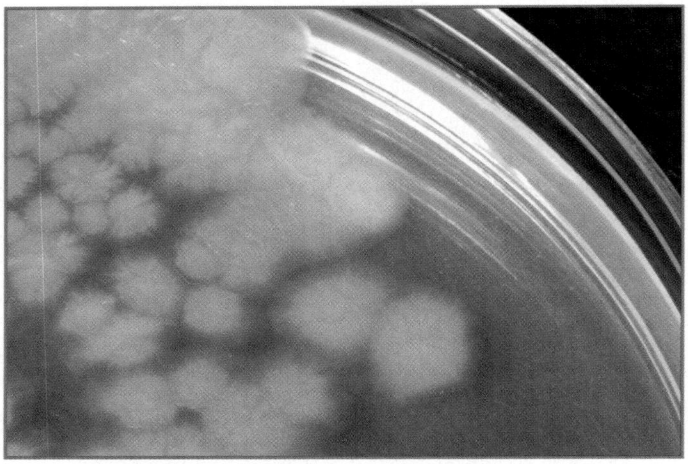

FIGURE 2-18 *Erwinia amylovora* Colonies Grown on Nutrient Agar
Note the irregular shape and spreading edges. *E. amylovora* is a plant pathogen.

FIGURE 2-19 Swarming Growth Pattern
Members of the genus *Proteus* will swarm at certain intervals and produce a pattern of concentric rings because of their motility. This photograph demonstrates the swarming behavior of *P. vulgaris* on DNase agar.

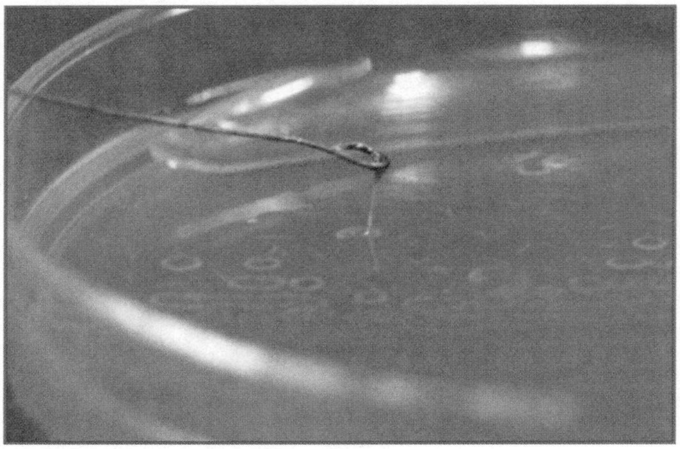

FIGURE 2-20 Mucoid Colonies
Pseudomonas aeruginosa grown on Endo agar illustrates a mucoid texture. *P. aeruginosa* is found in soil and water and can cause infections of burn patients.

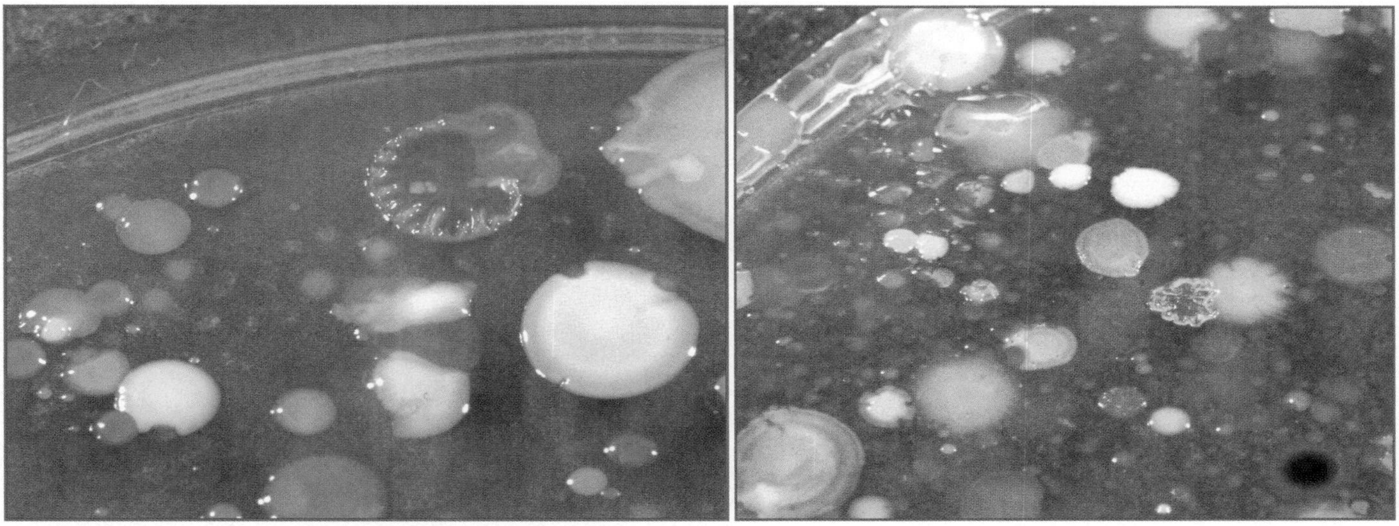

FIGURE 2-21 Two Mixed Soil Cultures on Nutrient Agar
These plates show the morphological diversity present in two soil samples.

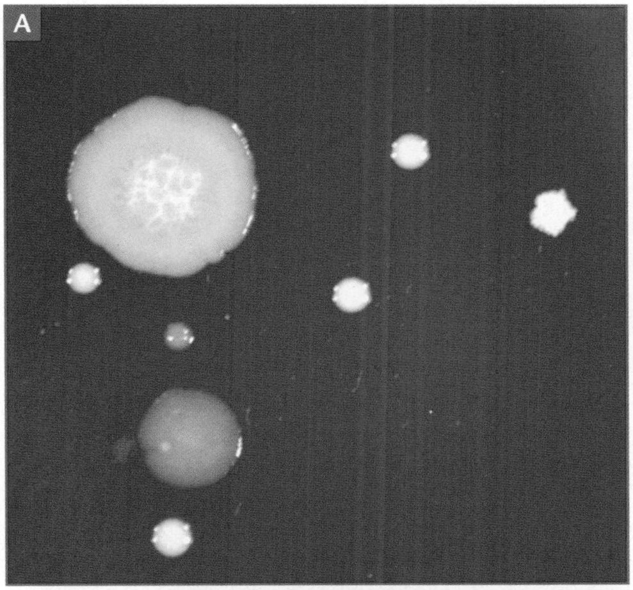

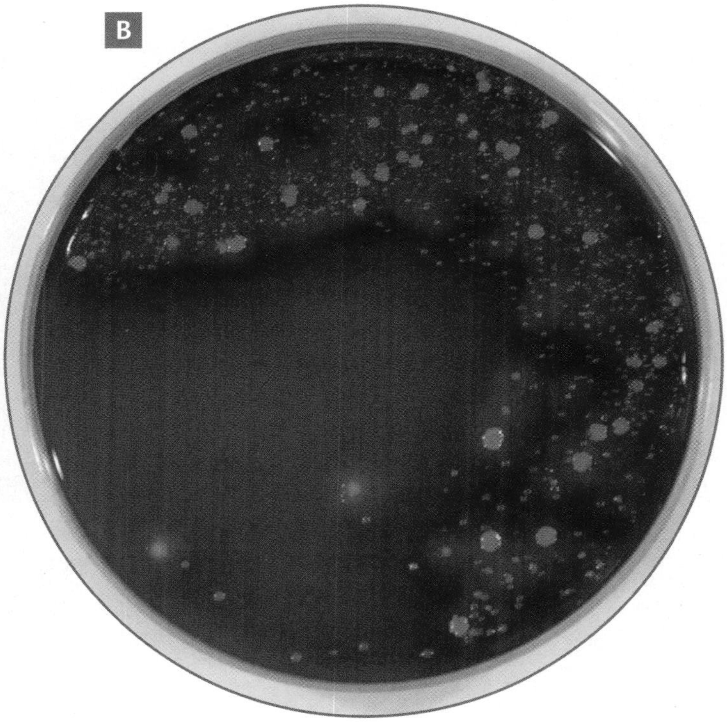

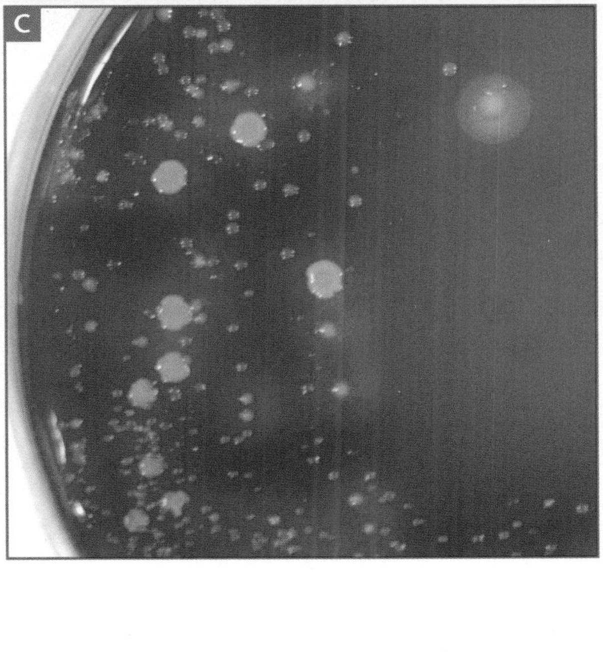

FIGURE 2-22 Two Throat Cultures on Sheep Blood Agar
(A) Several different species are growing on this plate. (B) Note the α-hemolysis (darkening of the agar) from most of the growth. (C) This is a close-up of the same plate as in (B). Note the weak β-hemolysis of the white colony. White growth with β-hemolysis is characteristic of *Staphylococcus aureus*.

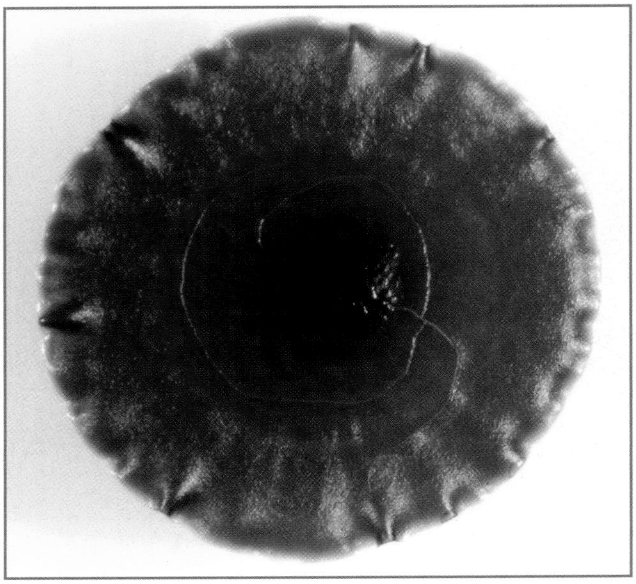

FIGURE 2-23 Chromobacterium violaceum Colony
This is a magnified *C. violaceum* colony (approximately X10) after one week of incubation on Trypticase Soy Agar. Compare this colony with those in Figure 2-6.

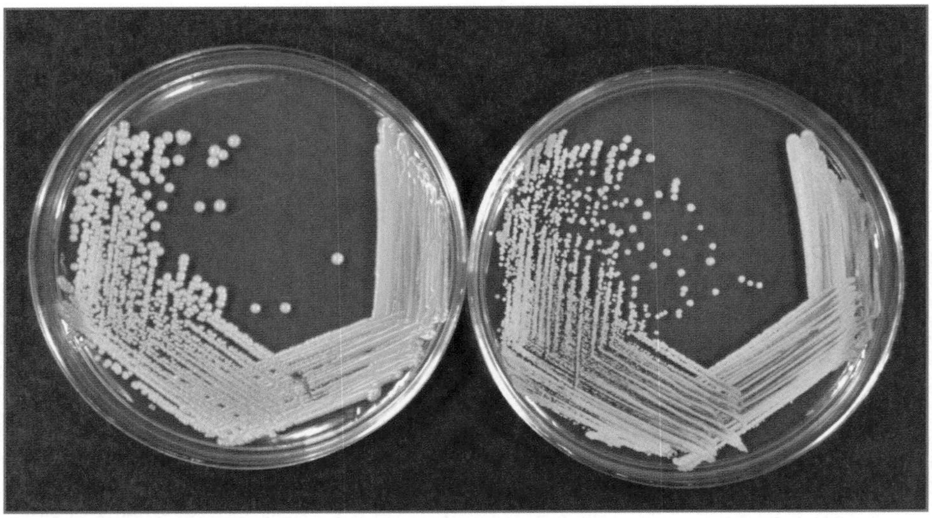

FIGURE 2-24 Pigment Production
Closely related species may look very different, as seen in these plates of *Micrococcus luteus* (left) and *Kocuria rosea* (formerly *Micrococcus roseus*, right).

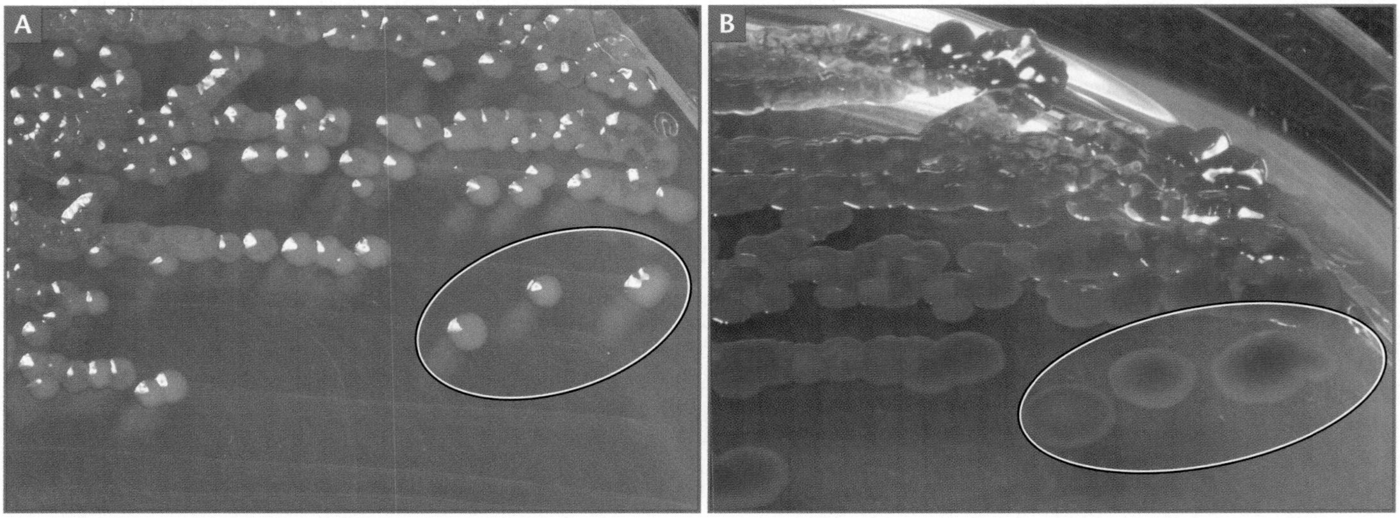

FIGURE 2-25 Influence of Age on Pigment Production
(A) *Serratia marcescens* grown on Sheep Blood Agar after 24 hours growth. (B) The same plate of *S. marcescens* after 48 hours growth. Note in particular the change in the three colonies in the lower right (encircled).

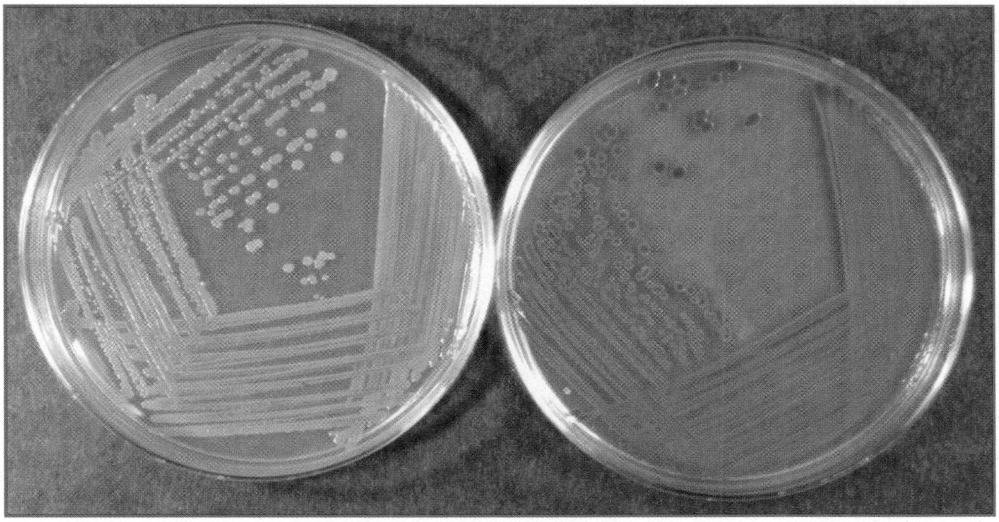

FIGURE 2-26 Influence of Temperature on Pigment Production
Pigment production also may be influenced by temperature. *Serratia marcescens* produces less orange pigment when grown at 37°C (left) than when grown at 25°C (right).

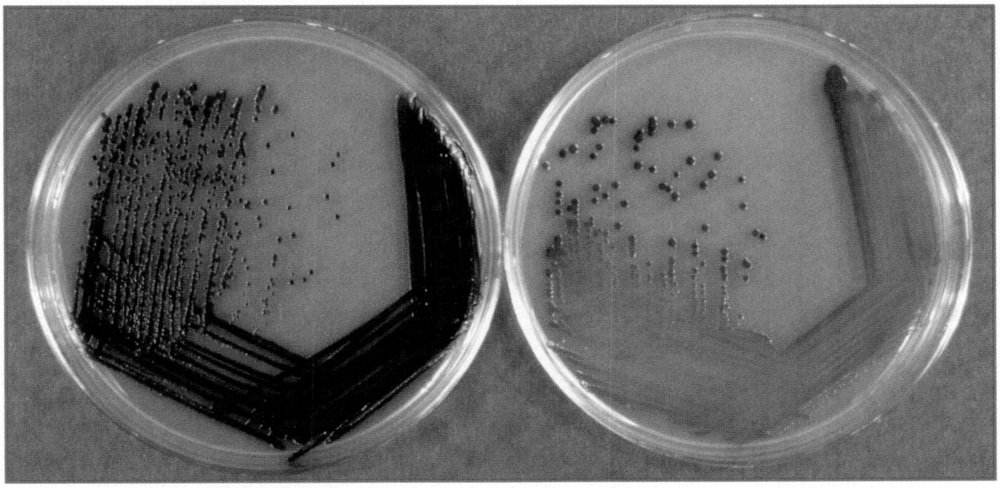

FIGURE 2-27 Influence of Nutrient Availability on Pigment Production
Pigment production may be influenced by environmental factors such as nutrient availability. *Chromobacterium violaceum* produces a much more intense purple pigment when grown on Trypticase Soy Agar (left) than when grown on Nutrient Agar (right), a less nutritious medium.

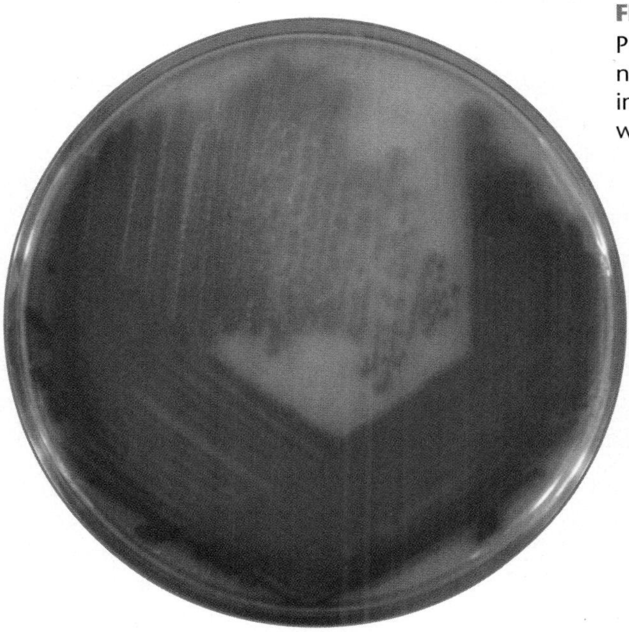

FIGURE 2-28 Diffusible Pigment
Here, *Pseudomonas* is growing on Trypticase Soy Agar. *P. aeruginosa* often produces a characteristic diffusible, blue-green pigment.

References

Claus, G. William. 1989. Chapter 14 in *Understanding Microbes—A Laboratory Textbook for Microbiology*. W. H. Freeman and Co., New York.

Collins, C. H., Patricia M. Lyne, and J. M. Grange. 1995. Chapter 6 in *Collins and Lyne's Microbiological Methods,* 7th ed. Butterworth-Heineman, Oxford, England.

Forbes, Betty A., Daniel F. Sahm, and Alice S. Weissfeld. 2002. Chapter 10 in *Bailey and Scott's Diagnostic Microbiology,* 11th ed. Mosby-Yearbook, St. Louis.

Koneman, Elmer W., Stephen D. Allen, William M. Janda, Paul C. Schreckenberger, and Washington C. Winn, Jr. 1997. Chapter 2 in *Color Atlas and Textbook of Diagnostic Microbiology,* 5th ed. J. B. Lippincott Co., Philadelphia.

Exercise 2-3

Growth Patterns on Slants

Theory

Agar slants are useful primarily as media for cultivation and maintenance of stock cultures. Organisms cultivated on slants do, however, display a variety of growth characteristics. We offer these more as an item of interest than one of diagnostic value.

Most of the organisms you will see in this class produce **filiform** growth (dense and opaque with a smooth edge). All of the organisms in Figure 2-29 are filiform and **pigmented.** Most species in the genus *Mycobacterium* produce **friable** (crusty) growth (Figure 2-30—*Mycobacterium phlei*). Some motile organisms produce growth with a **spreading edge** (Figure 2-30—*Alcaligenes faecalis*). Still others produce **translucent** or **transparent** growth (Figure 2-30—*Lactobacillus plantarum*).

Application

Although not definitive by themselves, growth characteristics on slants can provide useful information when attempting to identify an organism.

In This Exercise...

In the previous exercise, you examined growth on agar plates. Today you will be looking at the growth patterns of six organisms on prepared agar slants. Compare the growth with that on the plates from the previous exercise (Exercise 2-2). These slants will be used to inoculate the broths in Exercise 2-4. It might be useful to save the slants until after the broths have incubated for further comparison.

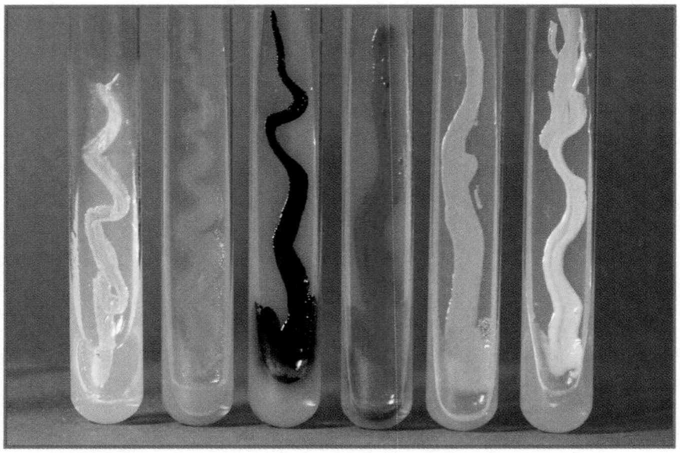

FIGURE 2-29 Pigment Production on Slants
From left to right, *Staphylococcus epidermidis* (white), *Pseudomonas aeruginosa* (green), *Chromobacterium violaceum* (violet), *Serratia marcescens* (red/orange), *Kocuria rosea* (rose), *Micrococcus luteus* (yellow).

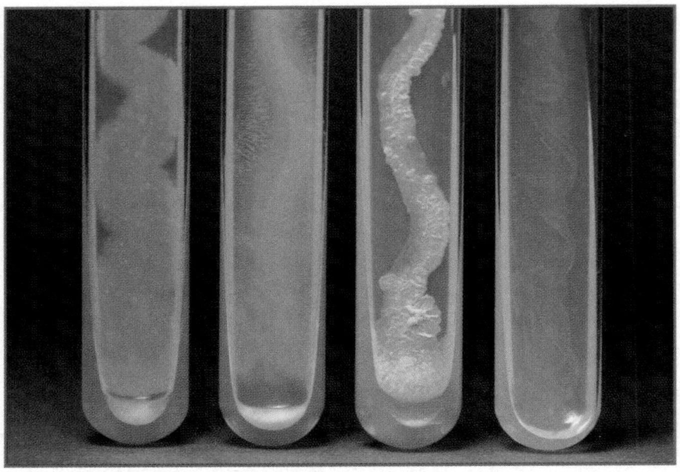

FIGURE 2-30 Growth Texture on Slants
From left to right, *Bacillus spp.* (flat, dry), *Alcaligenes faecalis* (spreading edge), *Mycobacterium phlei* (crusty/friable), *Lactobacillus plantarum* (transparent, barely visible).

Materials

Per Student Group

- five Nutrient Agar slants
- Tryptic Soy Agar or Brain Heart Infusion Agar Slant cultures of:
  - *Micrococcus luteus*
  - *Corynebacterium xerosis*
  - *Lactobacillus plantarum*
  - *Mycobacterium smegmatis*
  - *Bacillus subtilis*
  - *Proteus mirabilis*

Procedure

Lab One

1. Examine the tubes and describe the different growth patterns on the Data Sheet. Include a sketch of a representative portion of each.
2. Compare the slant growth to that on the agar plates from Exercise 2-2.
3. Use these cultures to inoculate the broths in Exercise 2-4. If you are not doing Exercise 2-4 as part of today's lab, refrigerate the slants until they are needed.

Reference

Claus, G. William. 1989. Chapter 17 in *Understanding Microbes—A Laboratory Textbook for Microbiology*. W. H. Freeman and Co., New York.

Exercise 2-4

Growth Patterns in Broth

Theory

Microorganisms cultivated in broth display a variety of growth characteristics. Some organisms float on top of the medium and produce a type of surface membrane called a **pellicle**. Others sink to the bottom as **sediment**. Some bacteria produce **uniform fine turbidity**, and others appear to clump in what is called **flocculent** growth. Refer to Figures 2-31 and 2-32.

Application

Bacterial genera—and frequently individual species within a genus—demonstrate characteristic growth patterns in broth that provide useful information when attempting to identify an organism.

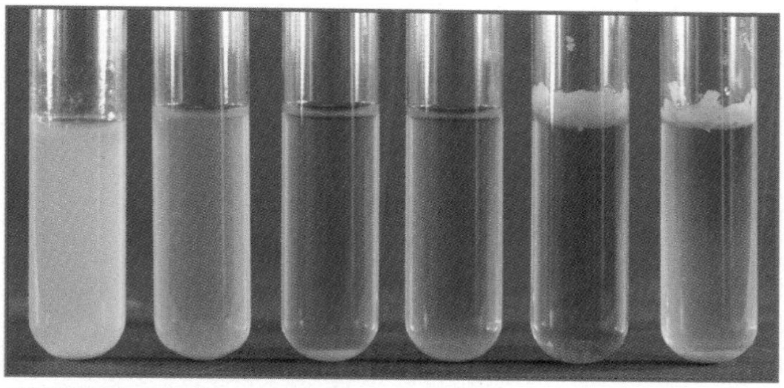

FIGURE 2-31 GROWTH PATTERNS IN BROTH
From left to right in pairs (of similar species): *Enterobacter aerogenes* and *Citrobacter diversus*—motile members of *Enterobacteriaceae* (uniform fine turbidity—UFT), *Enterococcus faecalis* and *Staphylococcus aureus*—nonmotile Gram-positive cocci (sediment), *Mycobacterium phlei* and *Mycobacterium smegmatis* (relatives of *Mycobacterium tuberculosis*)—nonmotile with a waxy cell wall (pellicle).

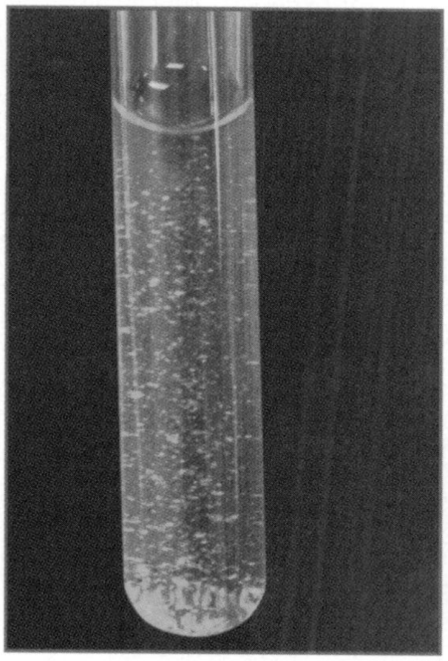

FIGURE 2-32
FLOCCULENCE IN BROTH
This is a *Streptococcus* species from a throat culture demonstrating flocculence in Todd-Hewitt Broth.

In This Exercise...

Today, you will be inoculating six Nutrient Broths with the slants from Exercise 2-3. You will then examine the growth characteristics and combine them with those seen on the plates and slants from Exercises 2-2 and 2-3.

Materials

Per Student Group

- seven Nutrient Broth tubes
- your slants from Exercise 2-3

Procedure

Lab One

1. Aseptically inoculate each Nutrient Broth with a slant saved from Exercise 2-3. For best results, use a very light inoculum. Leave the seventh broth uninoculated. It will be used as a control tube for comparison. Refer to Exercise 1-2 for help with transfers.
2. Label each tube with your name, the date, the medium, and the name of the organism.
3. Incubate the *M. luteus* broth and the uninoculated control at 25°C. Incubate all other broths at 35 ± 2°C for 24 to 48 hours.

Lab Two

1. Examine the tubes and describe the different growth patterns in the table on the Data Sheet.

References

Claus, G. William. 1989. Chapter 17 in *Understanding Microbes—A Laboratory Textbook for Microbiology*. W. H. Freeman and Co., New York.

Forbes, Betty A., Daniel F. Sahm, and Alice S. Weissfeld. 2002. Chapter 10 in *Bailey and Scott's Diagnostic Microbiology*, 11th ed. Mosby-Yearbook, St. Louis.

Environmental Factors Affecting Microbial Growth

Bacteria have limited control over their internal environments. Whereas many eukaryotes have evolved sophisticated internal control mechanisms, bacteria are almost completely dependent upon external factors to provide suitable conditions for their existence. Minor environmental changes can dramatically change a microorganism's ability to transport materials across the membrane, perform complex enzymatic reactions, and maintain critical cytoplasmic pressure.

One way to observe bacterial responses to environmental changes is to artificially manipulate external factors and measure growth rate. Understand that growth rate, as it is used in this manual, is synonymous with reproductive rate. Optimal growth conditions, as might be expected, result in faster growth and greater cell density (as evidenced by greater turbidity) than do less than optimal conditions.

In this series of laboratory exercises, you will examine the effects of oxygen, temperature, pH, osmotic pressure, and the availability of nutritional resources on bacterial growth rate. You also will learn some methods for cultivating anaerobic bacteria. When possible, you will attempt to classify organisms based on your results.

Exercise 2-5

Evaluation of Media

Theory Living things are composed of compounds from four biochemical families:

1. proteins,
2. carbohydrates,
3. lipids, and
4. nucleic acids.

Even though all organisms share this fundamental chemical composition, they differ greatly in their ability to make these molecules. Some are capable of making them out of the simple carbon compound, carbon dioxide (CO_2). These organisms, called **autotrophs,** require the least "assistance" from the environment to grow. The remaining organisms, called **heterotrophs,** require preformed organic compounds from the environment.

Some heterotrophs are metabolically flexible and require only a few simple organic compounds from which to make all their biochemicals. Others require a greater portion of their organic compounds from the environment. An organism that relies heavily on the environment to supply ready-made organic compounds is referred to as **fastidious.** Heterotrophs vary greatly in their requirements for these available compounds and, as such, range from highly fastidious to **nonfastidious.** Autotrophs are less fastidious than the most nonfastidious heterotrophs.

Successful cultivation of microbes in the laboratory requires an ability to satisfy their nutritional needs. The absence of a single required chemical resource prevents their growth. In general, the more fastidious the organism, the more ingredients a medium must have. **Undefined** media are composed of extracts from plant or animal sources and are very rich in nutrients. Even though the exact composition of the medium and the amount of each ingredient are unknown, undefined media are useful in growing the greatest variety of microbes. A **defined** medium is one in which the amount and identity of every ingredient is known. Defined media typically support a narrower range of organisms.

Application The ability of a microbiologist to cultivate microorganisms requires some knowledge of their metabolic needs. One quick way to make this determination is to transfer an organism to media containing different nutritional components and observe how well it grows. In this exercise you will evaluate the ability of four media—Nutrient Broth, Glucose Broth, Yeast Extract Broth, and Glucose Salts Broth—to support bacterial growth.

In This Exercise...

Today you will be comparing the density of growth between organisms in a variety of liquid media. It is important to make your inoculations as uniform as possible. Make sure your loop is fully closed so it will hold enough broth to produce a film across its opening (similar to a toy loop for blowing bubbles). Inoculate each medium with a single loopful of broth.

Materials

Per Student Group

- five tubes each of:
  - Nutrient Broth
  - Glucose Broth
  - Yeast Extract Broth
  - Glucose Salts Medium
- Nutrient Broth cultures of:
  - *Escherichia coli*
  - *Lactococcus lactis*
  - *Moraxella catarrhalis*
  - *Staphylococcus epidermidis*

Medium Recipes

Nutrient Broth

• Beef extract	3.0 g
• Peptone	5.0 g
• Distilled or deionized water	1.0 L

Glucose Broth

• Peptone	10.0 g
• Glucose	5.0 g
• NaCl	5.0 g
• Distilled or deionized water	1.0 L

Yeast Extract Broth

• NaCl	5.0 g
• Yeast extract	5.0 g
• Nutrient broth	1.0 L

Glucose Salts Medium

• Glucose	5.0 g
• NaCl	5.0 g
• $MgSO_4$	0.2 g
• $(NH_4)H_2PO_4$	1.0 g
• K_2HPO_4	1.0 g
• Distilled or deionized water	1.0 L

Procedure

Lab One

1. Inoculate each medium with a single loopful of each organism. Leave one of each tube uninoculated.
2. Incubate all tubes (including the uninoculated ones) at 35 ± 2°C for 24–48 hours.

Lab Two

1. Mix all tubes well and examine them for turbidity. Score relative amounts of growth using "0" for no growth, and "1," "2," and "3" for successively greater degrees of growth.
2. Record what you see. Do not expect one of each result just because there are four organisms.
3. Record your results on the Data Sheet.

References

Delost, Maria Dannessa. 1997. Page 144 in *Introduction to Diagnostic Microbiology. A Text and Workbook*. Mosby, St. Louis.

Zimbro, Mary Jo, and David A. Power, Editors. 2003. *Difco™ and BBL™ Manual—Manual of Microbiological Culture Media.* Becton Dickinson and Co., Sparks, MD.

Aerotolerance

Because of microorganisms' natural tendency to populate areas where conditions are most favorable, a growth medium that contains a gradient of oxygen concentrations allows visual determination of **aerotolerance.** Aerotolerance is the term used to indicate the ability or inability of an organism to grow in the presence of oxygen.

Most growth media are sterilized in the autoclave during preparation. Although this process removes the oxygen from the medium, it begins to diffuse back into the medium as it cools and solidifies. This process creates a gradient of oxygen concentrations in the medium, ranging from **aerobic** at the top to **anaerobic** at the bottom.

Obligate (strict) **aerobes** grow at the top where oxygen is most plentiful. **Facultative anaerobes** and **aerotolerant anaerobes**—organisms that either don't require oxygen or are not adversely affected by it—live throughout the medium. **Microaerophiles**—organisms that require a reduced level of oxygen—will be seen somewhere near the middle, and **obligate** (strict) **anaerobes** typically inhabit the lower half to two-thirds, depending on how far oxygen has diffused into the medium.

Obligate aerobes are organisms that respire aerobically and use oxygen as the final electron acceptor in an electron transport chain. Facultative anaerobes grow in the presence *or* absence of oxygen. When oxygen is available, they respire aerobically. When oxygen is not available, they either respire anaerobically (reducing sulfur or nitrate instead of oxygen) or ferment. Refer to Appendix A and Section Five for more information on anaerobic respiration and fermentation.

Aerotolerant anaerobes ferment or respire anaerobically even in the presence of free oxygen. Microaerophiles, as the name suggests, survive only in environments containing lower than atmospheric levels of oxygen. Some microaerophiles called **capnophiles** can survive only if carbon dioxide levels are elevated. Finally, obligate anaerobes are organisms for which even small amounts of oxygen are lethal and, thus, live only in oxygen-free environments.

Exercise 2-6

Agar Deep Stabs

Theory

Agar deep stabs are prepared with Tryptic Soy Agar (TSA) enriched with yeast extract to promote growth of a broad range of organisms. As stated in the introduction to this section, oxygen, which is removed from the medium during preparation and autoclaving, immediately begins to diffuse back into the medium as the agar cools and solidifies. This process creates an oxygen gradient in the medium, ranging from aerobic at the top to anaerobic at the bottom.

Agar deeps are prepared with 10 mL instead of the customary 7 mL used in slants. This extra depth ensures (if inoculated soon after preparation) that the bottom portion of the medium is anaerobic. The agar is stab-inoculated with an inoculating needle to introduce as little air as possible. The location of growth that develops indicates the organism's aerotolerance (Figure 2-33).

Application

This test is a good visual indicator of oxygen tolerance in microorganisms.

In This Exercise...

Today you will stab-inoculate three agar tubes to determine the aerotolerance categories of three bacteria.

Materials

Per Student Group

- four agar deep stab tubes
- inoculating needle
- fresh cultures of:
  - *Clostridium sporogenes*
  - *Pseudomonas aeruginosa*
  - *Staphylococcus aureus*

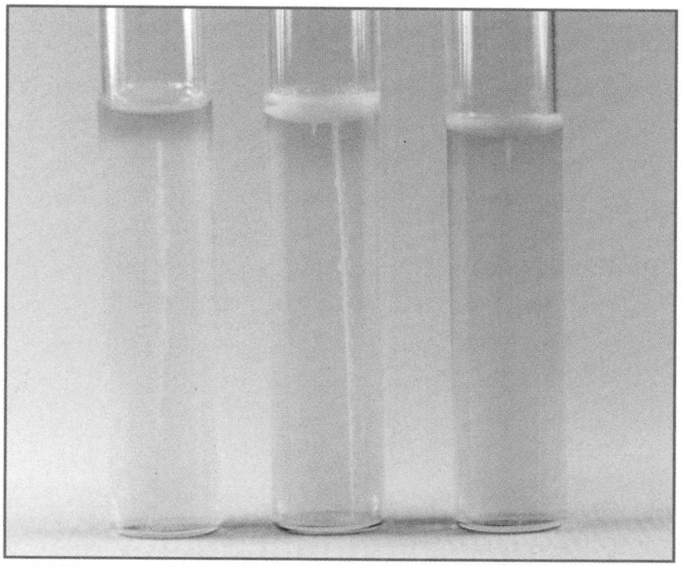

FIGURE 2-33 Agar Deep Stab Tubes
From left to right: strict anaerobe (possible microaerophile), facultative anaerobe, and strict aerobe.

Medium Recipe

Enriched TSA

Tryptone	15.0 g
Soytone	5.0 g
Sodium Chloride	5.0 g
Yeast Extract	5.0 g
Agar	15.0 g
Distilled or deionized water	1.0 L

Procedure

Lab One

1. Obtain 4 agar deep stab tubes.
2. With a heavy inoculum on your inoculating needle, carefully stab the agar tubes with the test organisms. Refer to Appendix B for the proper stab technique. (***Note:*** To minimize the introduction of unwanted oxygen to the medium, try to avoid lateral motion of the needle during inoculation.)
3. Stab the fourth tube with your sterile needle.
4. Label each tube with your name, the date, the medium, and the name of the organism.
5. Incubate the tubes at 35 ± 2°C for 24 to 48 hours.

Lab Two

1. Examine the tubes and determine the aerotolerance category of each organism.
2. Record your results on the Data Sheet.

References

Forbes, Betty A., Daniel F. Sahm, and Alice S. Weissfeld. 2002. Chapter 10 in *Bailey and Scott's Diagnostic Microbiology,* 11th ed. Mosby, St. Louis.

Koneman, Elmer W., Stephen D. Allen, William M. Janda, Paul C. Schreckenberger, and Washington C. Winn, Jr. 1997. Chapter 14 in *Color Atlas and Textbook of Diagnostic Microbiology*, 5th ed. J. B. Lippincott Co. Philadelphia.

Exercise 2-7
Agar Shakes

Theory

Agar shakes are prepared with enriched TSA, but they differ from agar deep stabs (Exercise 2-6) in that, after autoclaving, they are placed in a 50°C water bath until time for inoculation. Agar shakes are inoculated in liquid form, mixed gently to distribute the bacteria evenly throughout the medium, and allowed to solidify. Like agar deep stabs, the growth that develops in agar shakes indicates the aerotolerance of the organism (Figure 2-34).

Application

This test is a good visual indicator of oxygen tolerance in microorganisms.

In This Exercise...

Today you will inoculate three molten agar tubes with a transfer pipette. After inoculation, you will have to mix the liquefied agar well to get the proper distribution of cells. Do not actually shake the medium, as this will introduce excessive oxygen; simply roll it between your hands until the mixture is complete.

Materials

Per Student Group

- four agar shake tubes (liquefied and held in a 50°C water bath)
- three sterile transfer pipettes

FIGURE 2-34 Agar Shake Tubes
From left to right: strict anaerobe, facultative anaerobe, uninoculated control, and strict aerobe.

- fresh broth cultures of:
  - *Clostridium sporogenes*
  - *Pseudomonas aeruginosa*
  - *Staphylococcus aureus*

Medium Recipe

Enriched TSA

Tryptone	15.0 g
Soytone	5.0 g
Sodium Chloride	5.0 g
Yeast extract	5.0 g
Agar	15.0 g
Distilled or deionized water	1.0 L

Procedure

Lab One

1. Prepare four labels, one for each organism and one control. Because it is important to leave the liquid agar in the water bath until it is needed, place the labels on the tubes one at a time as you obtain them.
2. Remove one liquid agar from the water bath. (***Note:*** Check to see that the temperature of the water bath is 50°C. If the temperature is too high, it may kill the organism; if it is too low, it will solidify before you inoculate it.) Using a sterile pipette transfer a heavy inoculum (about 0.5 mL) to the liquefied agar. Immediately mix it thoroughly. Place the tube in a rack and allow it to solidify.
3. One at a time, do the same with the other three organisms. Thoroughly mix the control tube and allow it to solidify uninoculated.
4. When all agar tubes have solidified, incubate them at 35 ± 2°C for 24 to 48 hours.

Lab Two

1. After incubation, examine the tubes and determine the aerotolerance category of each organism.
2. Record your observations and interpretations on the Data Sheet.

References

Forbes, Betty A., Daniel F. Sahm, and Alice S. Weissfeld. 2002. Chapter 10 in *Bailey and Scott's Diagnostic Microbiology*, 11th ed. Mosby, St. Louis.

Koneman, Elmer W., Stephen D. Allen, William M. Janda, Paul C. Schreckenberger, and Washington C. Winn, Jr. 1997. Chapter 14 in *Color Atlas and Textbook of Diagnostic Microbiology*, 5th ed. J. B. Lippincott Co., Philadelphia.

Exercise 2-8

Fluid Thioglycollate Medium

Theory

Fluid Thioglycollate Medium is prepared as a basic medium (as used in this exercise) or with a variety of supplements, depending on the specific needs of the organisms being cultivated. As such, this medium is appropriate for a broad variety of aerobic and anaerobic, fastidious and nonfastidious organisms. It is particularly well adapted for cultivation of strict anaerobes and microaerophiles.

Key components of the medium are yeast extract, pancreatic digest of casein, dextrose, sodium thioglycollate, L-cystine, and resazurin. Yeast extract and pancreatic digest of casein provide nutrients; sodium thioglycollate and L-cystine reduce oxygen to water; and resazurin (pink when oxidized, colorless when reduced) acts as an indicator. A small amount of agar is included to slow oxygen diffusion.

Oxygen is removed from the medium during autoclaving but begins to diffuse back into the medium as the tubes cool to room temperature. This produces a gradient of concentrations from fully aerobic at the top to anaerobic at the bottom. Thus, fresh media will appear clear to straw-colored with a pink region at the top where the dye has become oxidized (Figure 2-35).

FIGURE 2-35 Aerobic Zone in Thioglycollate Broth
Note the pink region in the top (oxidized) portion of the broth resulting from the indicator resazurin. In the bottom (reduced) portion, the dye is colorless and the medium is its typical straw color.

FIGURE 2-36 Growth Patterns in Thioglycollate Medium
Growth patterns of a variety of organisms are shown in these Fluid Thioglycollate broths. Pictured from left to right are: obligate aerobe, facultative anaerobe, uninoculated control, microaerophile, and strict anaerobe. The colored region at the top of the control illustrates the oxidation/reduction status of the medium.

Figure 2-36 demonstrates some basic bacterial growth patterns in the medium as influenced by the oxygen gradient.

Application

Fluid Thioglycollate Medium is a liquid medium designed to promote growth of a wide variety of fastidious microorganisms. It can be used to grow microbes representing all levels of oxygen tolerance; however, it is generally associated with the cultivation of anaerobic and microaerophilic bacteria.

In This Exercise...

Today you will be inoculating Fluid Thioglycollate Medium to determine the aerotolerance categories of three bacteria.

Materials

Per Student Group

- four fluid Thioglycollate Medium tubes
- fresh cultures of:
  - *Pseudomonas aeruginosa*
  - *Clostridium sporogenes*
  - *Staphylococcus aureus*

Medium Recipe

Fluid Thioglycollate Medium

Yeast extract	5.0 g
Pancreatic digest of casein	15.0 g
Dextrose	5.5 g
Sodium chloride	2.5 g
Sodium thioglycollate	0.5 g
L-cystine	0.5 g
Agar	0.75 g
Resazurin	0.001 g
Distilled or deionized water	1.0 L

Procedure

Lab One

1. Obtain four Fluid Thioglycollate tubes and label them with your name, the date, medium, and organism.
2. Using your loop, inoculate three broths with the organisms provided. Do not inoculate the fourth tube as it will be your control.
3. Incubate the tubes at 35 ± 2°C for 24 to 48 hours.

Lab Two

1. Check the control tube for growth to assure sterility of the medium. Note any changes that may have occurred as a result of incubation, especially in the colored region at the surface.
2. Using the control as a comparison, examine and note the location of the growth in all tubes.
3. Enter your observations and interpretations in the chart provided on the Data Sheet.

References

Allen, Stephen D., Christopher L. Emery, and David M. Lyerly. 2003. Chapter 54 in *Manual of Clinical Microbiology*, 8th ed. Edited by Patrick R. Murray, Ellen Jo Baron, James H. Jorgensen, Michael A. Pfaller, and Robert H. Yolken. ASM Press, American Society for Microbiology, Washington, DC.

Forbes, Betty A., Daniel F. Sahm, and Alice S. Weissfeld. 2002. Chapter 10 in *Bailey and Scott's Diagnostic Microbiology*, 11th ed. Mosby, St. Louis.

Koneman, Elmer W., Stephen D. Allen, William M. Janda, Paul C. Schreckenberger and Washington C. Winn, Jr. 1997. Chapter 14 in *Color Atlas and Textbook of Diagnostic Microbiology*, 5th ed. J. B. Lippincott Co., Philadelphia.

Zimbro, Mary Jo and David A. Power, editors. 2003. *Difco™ and BBL™ Manual—Manual of Microbiological Culture Media*. Becton Dickinson and Company. Sparks, MD.

Exercise 2-9

Anaerobic Jar

Theory

The GasPak® Anaerobic System by BBL™ that is used for this exercise is a plastic jar in which to place inoculated agar plates. It includes a chemical gas generator packet containing sodium borohydride and sodium bicarbonate, and a paper strip containing methylene blue to indicate the presence of oxygen (Figure 2-37). Methylene blue is colorless when reduced and blue when oxidized. Also included in the gas generator packet is a small envelope of palladium to act as a catalyst.

After the inoculated plates are placed inside the jar, the gas generator packet is opened and placed inside the jar along with the methylene blue indicator strip. Water is added to the gas generator packet and the jar lid is immediately fastened down. The sodium borohydride and sodium bicarbonate in the packet react with the water to produce hydrogen and carbon dioxide gases. The palladium catalyzes a reaction between the hydrogen and free oxygen in the jar to produce water, as shown in Figure 2-38. Removal of free oxygen produces anaerobic conditions in the jar within approximately one hour, as evidenced by a white indicator strip and moisture on the inside of the jar.

FIGURE 2-37
THE ANAEROBIC JAR
Note the white methylene blue strip and the open packet, which has discharged H_2 and CO_2 gases. The palladium, contained in the packet, catalyzes the conversion of H_2 and O_2 as shown in Figure 2-38.

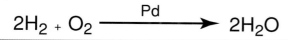

$$2H_2 + O_2 \xrightarrow{Pd} 2H_2O$$

FIGURE 2-38 Conversion of H_2 and O_2 to Water Using a Palladium Catalyst.

Figure 2-39 illustrates some typical growth patterns under aerobic and anaerobic conditions.

Application

This procedure provides a means of cultivating anaerobic and microaerophilic bacteria.

In This Exercise...

Today each group will inoculate two Nutrient Agar plates with three organisms. One of the plates will go into the jar, and the second plate will not. Both plates will be incubated at 35 ± 2°C for 24 to 48 hours. Following incubation, the growth on the two plates will be examined and compared.

Materials

Per Class

- one anaerobic jar with gas generator packet*

Per Group

- two Nutrient Agar plates
- fresh cultures of:
  - *Pseudomonas aeruginosa*
  - *Clostridium sporogenes*
  - *Staphylococcus aureus*

Procedure

Lab One

1. Obtain two Nutrient Agar plates. Using a marking pen, divide the bottom of each plate into three sectors.
2. Label each plate with your name, the date, and organism by sector.
3. Using your loop, inoculate the sectors of both plates with the organisms provided. (***Note***: Inoculate with single streaks about one centimeter long.) Tape the lids in place.
4. Place one plate in the anaerobic jar in an inverted position.
5. When all groups have placed their plates in the jar, discharge the packet as follows (or follow the instructions for your system):
   a. Stick the methylene blue strip on the wall of the jar.

* Available from Becton Dickinson Microbiology Systems, Sparks, MD http://www.bd.com

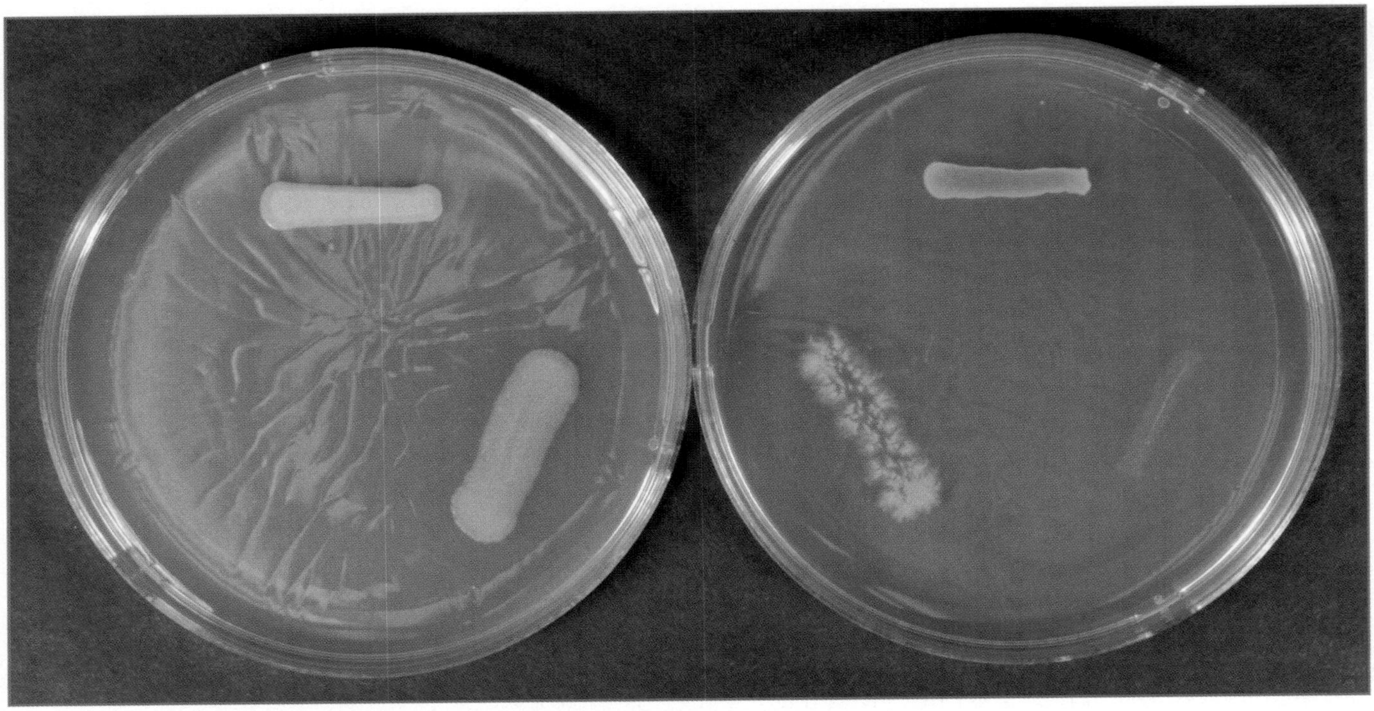

FIGURE 2-39 PLATES INCUBATED INSIDE AND OUTSIDE THE ANAEROBIC JAR
Both Nutrient Agar plates were spot-inoculated with a facultative anaerobe (top), strict aerobe (right), and strict anaerobe (left). The plate on the left was incubated outside the jar, and the plate on the right was incubated inside the jar. Note the relative amounts of growth of the three organisms.

b. Open the packet and add 10 mL of distilled water.
c. Place the open packet in the jar with the label facing inward.
d. Immediately close the jar.

6. Place the second plate and the anaerobic jar in the 35 ± 2°C incubator for 24 to 48 hours.

Lab Two

1. Examine and compare the growth on the plates. (*Note:* Density of growth will be the most useful basis for comparison.)
2. Record your results and interpretations in the chart provided on the Data Sheet.

References

Allen, Stephen D., Christopher L. Emery, and David M. Lyerly. 2003. Chapter 54 in *Manual of Clinical Microbiology,* 8th ed. Edited by Patrick R. Murray, Ellen Jo Baron, James H. Jorgensen, Michael A. Pfaller, and Robert H. Yolken. ASM Press, American Society for Microbiology, Washington, DC.

Forbes, Betty A., Daniel F. Sahm, and Alice S. Weissfeld. 2002. Chapter 10 in *Bailey and Scott's Diagnostic Microbiology,* 11th ed. Mosby, St. Louis.

Koneman, Elmer W., Stephen D. Allen, William M. Janda, Paul C. Schreckenberger and Washington C. Winn, Jr. 1997. Chapter 14 in *Color Atlas and Textbook of Diagnostic Microbiology,* 5th ed. J. B. Lippincott Co., Philadelphia.

Zimbro, Mary Jo and David A. Power, editors. 2003. *Difco™ and BBL™ Manual—Manual of Microbiological Culture Media.* Becton Dickinson and Company, Sparks, MD.

Exercise 2-10

The Effect of Temperature on Microbial Growth

Theory

Bacteria have been discovered living in habitats ranging from –10 degrees Celsius to more than 110 degrees Celsius. The temperature range of any single species, however, is a small portion of this vast range. As such, each species is characterized by a minimum, maximum and optimum temperature—collectively known as its **cardinal temperatures** (Figure 2-40). Minimum and maximum temperatures are, simply, the temperatures below and above which the organism will not survive. Optimum temperature is the temperature at which an organism shows the greatest growth over time—its highest growth rate.

Organisms that grow only below 20 degrees Celsius are called **psychrophiles**. These are common in ocean, Arctic, and Antarctic habitats where the temperature remains permanently cold with little or no fluctuation. Organisms adapted to cold habitats that fluctuate from about 0 degrees to above 30 degrees Celsius are called **psychrotrophs.** Bacteria adapted to temperatures between 15 degrees and 45 degrees Celsius are known as **mesophiles.**

Most bacterial residents in the human body, as well as numerous human pathogens, are mesophiles. **Thermophiles** are organisms adapted to temperatures above 40 degrees Celsius. They typically are found in composting organic material and in hot springs. Heat-adapted organisms that will not grow at temperatures below 40 degrees are called **obligate thermophiles;** and those that will grow below 40 degrees are known as **facultative thermophiles.** Bacteria isolated from hot ocean floor ridges living between 65 and 110 degrees Celsius are called **extreme thermophiles.** Extreme thermophiles grow best above 80 degrees Celsius. Figure 2-41 illustrates bacterial temperature ranges and classifications.

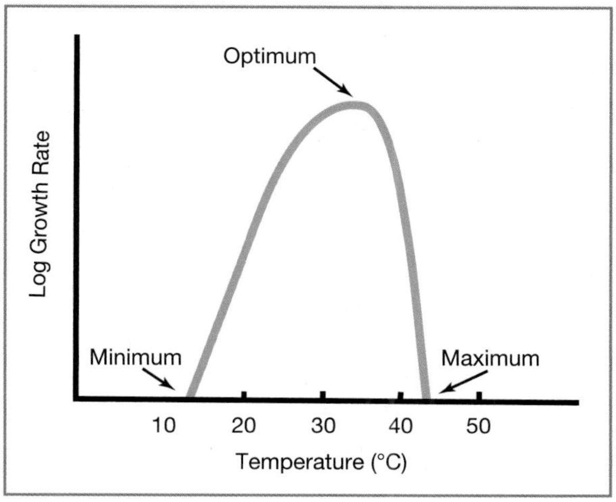

FIGURE 2-40 TYPICAL GROWTH RANGE OF A MESOPHILE
The "minimum" and "maximum" are temperatures beyond which no growth takes place. The "optimum" is the temperature at which growth rate is highest. After Prescott (1999).

Application

This procedure is used to determine the cardinal temperatures for a species.

In This Exercise...

Today you will examine the growth characteristics of four organisms at five different temperatures. In addition, you will observe the influence of temperature on pigment production.

Materials

Per Class

- Five incubating devices set at 10°C, 20°C, 30°C, 40°C, and 50°C. These devices may be any combination of the following:
  - refrigerator,
  - incubator,
  - hot water bath, or
  - cold water bath.
- spectrophotometers (optional)
- cuvettes (optional)

Per Student Group

- twenty-five sterile Nutrient Broths
- two Trypticase Soy Agar (TSA) plates

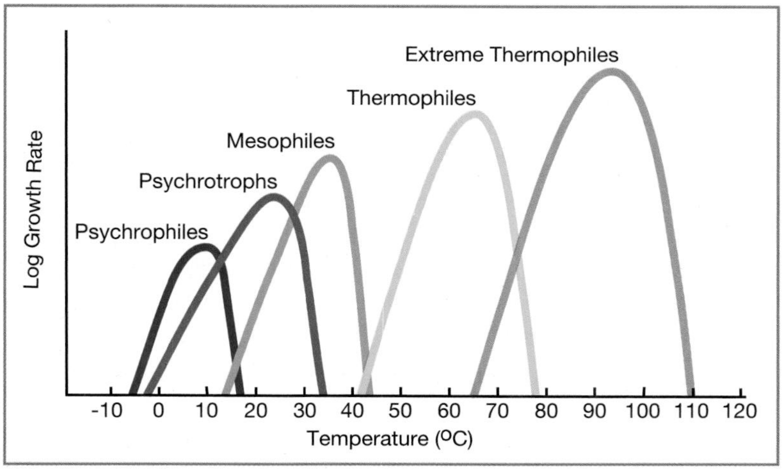

FIGURE 2-41 THERMAL CLASSIFICATIONS OF BACTERIA
After Prescott (1999).

- three sterile transfer pipettes
- three nonsterile transfer pipettes
- fresh nutrient broth cultures of:
    - *Escherichia coli*
    - *Serratia marcescens*
    - *Bacillus stearothermophilus*
    - *Pseudomonas fluorescens*

Procedure

Lab One

1. Obtain twenty-five Nutrient Broths—one broth for each organism at each temperature plus one control for each temperature. Label them accordingly. Also obtain two TSA plates and label them 20°C and 35°C, respectively.
2. Mix each culture thoroughly before the following transfers. Using a sterile pipette, transfer a *single drop* of each broth culture to its appropriate Nutrient Broth tube. (***Note:*** Because you will be comparing the amount of growth in the Nutrient Broth tubes, you must be sure to begin by transferring the same volume of culture to each one. Use the same pipette for all transfers with a single organism.)
3. Using a simple zigzag pattern (as in Exercise 1-3), inoculate each plate with *Serratia marcescens.*
4. Incubate all tubes in their appropriate temperatures for 24 to 48 hours. Incubate the plates in the 20°C and 35°C incubators.

Lab Two (without a spectrophotometer)

1. Clean the outside of all tubes with a tissue and place them in a test tube rack organized into groups *by organism.*
2. Gently shake each broth until uniform turbidity is achieved.
3. Examine the controls for turbidity. If no turbidity is present, place them in the rack with the test broths.
4. Compare all tubes in a group to each other and to a control. Rate each one as 0, 1, 2, or 3, according to its turbidity (0 is clear and 3 is very turbid). Record these in the chart on the Data Sheet.
5. Examine the plates incubated at different temperatures, compare the growth characteristics, and enter your results in the chart on the Data Sheet.
6. Using the data from the chart, determine the cardinal temperatures and classification of each of the four organisms.
7. On the graph paper provided, plot the data (numeric values versus temperature) for the four organisms.

Lab Two (using a spectrophotometer)

1. Completely dry all tubes and place them in a test tube rack organized into groups *by temperature.*
2. Shake each broth gently until uniform turbidity is achieved.
3. Set the spectrophotometer wavelength at 650 nm. If the spectrophotometer is digital, set it to "absorbance." (***Note:*** For instruction on spectrophotometer operation, refer to Appendix E.)
4. Blank the spectrophotometer with the control for the first temperature group. Take absorbance readings of all tubes in the group and enter the results in the chart provided on the Data Sheet. Turbidity (a qualitative measure) increases with higher cell density. Measuring absorbance—which increases proportionately with turbidity—allows us to quantify the differences in growth. (***Note:*** Your instructor may provide you with cuvettes for this portion of the exercise. Use a nonsterile transfer pipette to fill the cuvettes. After use, dispose of the cuvettes and transfer pipettes in an autoclave receptacle.)
5. Using the control for each group to blank the spectrophotometer, continue reading all of the broths and enter the results in the chart provided in the Data Sheet.
6. Examine the plates incubated at different temperatures, compare the growth characteristics, and enter your results in the table provided on the Data Sheet.
7. Using the data from the chart and the terms from the discussion, determine the cardinal temperatures and classification of each of the four test organisms.
8. On the graph paper provided, plot the data (absorbance versus temperature) of the four organisms.

References

Forbes, Betty A., Daniel F. Sahm, and Alice S. Weissfeld. 2002. Chapters 2 and 10 in *Bailey and Scott's Diagnostic Microbiology,* 11th ed. Mosby-Yearbook, St. Louis.

Holt, John G. (Editor). 1994. *Bergey's Manual of Determinative Bacteriology,* 9th ed. Williams and Wilkins, Baltimore.

Koneman, Elmer W., Stephen D. Allen, William M. Janda, Paul C. Schreckenberger, and Washington C. Winn, Jr. 1997. *Color Atlas and Textbook of Diagnostic Microbiology,* 5th ed. J.B. Lippincott Co., Philadelphia.

Moat, Albert G., John W. Foster, and Michael P. Spector. 2002. Pages 597-601 in *Microbial Physiology,* 4th ed. Wiley-Liss, New York.

Prescott, Lansing M., John P. Marley, and Donald A. Klein. 2005. Chapter 6 in *Microbiology,* 6th ed., WCB McGraw-Hill, Boston.

Varnam, Alan H. and Malcolm G. Evans. 2000. *Environmental Microbiology.* ASM Press, Washington, DC.

White, David. 2000. Pages 384-387 in *The Physiology and Biochemistry of Prokaryotes,* 2nd ed. Oxford University Press, New York.

Exercise 2-11

The Effect of pH on Bacterial Growth

Theory

The conventional means of expressing the concentration of hydrogen ions in a solution is pH. Values of pH range from 0 to 14 and actually are negative logarithms of the hydrogen ion concentration in moles per liter. Mathematically, the formula appears as:

$$pH = -\log [H^+]$$

Pure water contains 10^{-7} moles of hydrogen ions per liter and has a pH of 7. As hydrogen ions increase, the solution becomes more acidic and the pH decreases (see Table 2-1).

Bacteria live in habitats throughout the pH spectrum; however, the range of most individual species is small. Like temperature and salinity, pH tolerance is used as a means of classification. The three major classifications are

1. **acidophiles**, organisms adapted to grow well in environments below about pH 5.5,
2. **neutrophiles,** organisms that prefer pH levels between 5.5 and 8.5, and
3. **alkaliphiles,** organisms that live above pH 8.5.

Under normal circumstances, bacteria maintain a near-neutral internal environment regardless of their habitat; pH changes outside an organism's range may destroy necessary membrane potential (in the production of ATP) and damage vital enzymes beyond repair. This **denaturing** of cellular enzymes may be as minor as conformational changes in the proteins' tertiary structure, but usually is lethal to the cell.

In vitro acids from carbohydrate fermentation and alkaline products from protein metabolism are sufficient to disrupt enzyme integrity. This is why buffers made from weak acids such as hydrogen phosphate are added to bacteriological growth media. In solution, buffers are able to alternate between weak acid ($H_2PO_4^-$) and conjugate base (HPO_4^{2-}) to maintain H^+/OH^- equilibrium.

$$H^+ + HPO_4^{2-} \longrightarrow H_2PO_4^-$$

$$OH^- + H_2PO_4^- \longrightarrow HPO_4^{2-} + H_2O$$

Application

This procedure is used to determine the minimum, maximum, and optimum pH for growth of a bacterial species.

In This Exercise...

Today you will cultivate and observe the effects of pH on three organisms. Then you will classify them based on your results.

Acidity/ Alkalinity	pH	H+ Concentration In Moles/Liter	Common Examples	Bacterial Ranges
↑	0	10^0	Nitric acid	↑
	1	10^{-1}	Stomach acid	
	2	10^{-2}	Lemon juice	
	3	10^{-3}	Vinegar, cola	
	4	10^{-4}	Tomatoes, orange juice	
	5	10^{-5}	Black coffee	Acidophiles
Acidic	6	10^{-6}	Urine	
Neutral	7	10^{-7}	Pure water	Neutrophiles
Alkaline	8	10^{-8}	Sea water	
	9	10^{-9}	Baking soda	Alkaliphiles
	10	10^{-10}	Soap, milk of magnesia	
	11	10^{-11}	Ammonia	
	12	10^{-12}	Lime water [$Ca(OH)_2$]	
	13	10^{-13}	Household bleach	
↓	14	10^{-14}	Drain cleaner	↓

TABLE 2-1. pH Scale

Materials

Per Student Group

- five of each pH adjusted Nutrient Broth as follows: pH 2, pH 4, pH 6, pH 8, and pH 10
- spectrophotometer (optional)
- cuvettes (optional)
- fresh nutrient broth cultures of:
  - *Lactobacillus plantarum*
  - *Lactococcus lactis*
  - *Staphylococcus aureus*
  - *Alcaligenes faecalis*

Medium Recipe

pH-Adjusted Nutrient Broth

Beef extract	3.0 g
Peptone	5.0 g
Distilled or deionized water	1.0 L
NaOH or HCl as needed to adjust pH	

Procedure

Lab One

1. Obtain five tubes of each pH broth—one of each pH per organism plus one of each to be used as a control (25 tubes total). Label them accordingly.
2. Mix each culture thoroughly before the following transfers. Using a sterile pipette, transfer a *single drop* of each broth culture to its appropriate Nutrient Broth tube. (***Note***: Because you will be comparing the amount of growth in the Nutrient Broth tubes, you must be sure to begin by transferring the same volume of culture to each tube. Use the same pipette for all transfers with a single organism.)
3. Incubate all tubes at 35±2°C for 48 hours.

Lab Two (without a spectrophotometer)

1. Clean the outside of all tubes with a tissue and place them in a test tube rack organized into groups *by organism*.
2. Shake each broth gently until uniform turbidity is achieved.
3. Compare all tubes in a group to each other and to a control. Rate each one as 0, 1, 2, or 3 according to its turbidity (0 is clear and 3 is very turbid). Enter your observations in the chart on the Data Sheet. (***Note:*** Some color variability may exist between the different pH broths; therefore, base your conclusions solely on turbidity as compared to the control, not color.)
4. Using your data and that from Table 2-1, determine the range and classification of each of the four test organisms, and enter it on the Data Sheet.
5. On the graph paper provided, plot the data (numeric values versus pH) of the three organisms.

Lab Two (using a spectrophotometer)

1. Clean the outside of all tubes with a tissue and place them in a test tube rack organized into groups by *pH*.
2. Shake each broth gently until uniform turbidity is achieved. Continue to agitate the tubes as needed to keep solids from settling to the bottom.
3. Set the spectrophotometer wavelength at 650 nm. If the spectrophotometer is digital, set it to "absorbance." (For instruction on how to use a spectrophotometer, refer to Appendix E.)
4. Blank the spectrophotometer with the control for the first pH group. Take absorbance readings of all the tubes in the group and enter the results in the chart provided. Turbidity (a qualitative measure) increases with higher cell density. Measuring absorbance—which increases proportionately with turbidity—allows us to quantify the differences in growth. (***Note***: Your instructor may provide you with cuvettes for this portion of the exercise. Use a nonsterile transfer pipette to fill the cuvettes. After use, dispose of the cuvettes and transfer the pipettes into an autoclave receptacle.)
5. Using the control for each group to blank the spectrophotometer, continue reading all of the broths and enter the results in the table provided on the Data Sheet.
6. Using your data and that from Table 2-1, determine the range and classification of each of the three test organisms and enter it on the Data Sheet.
7. On the graph paper provided, plot the data (absorbance versus pH) of the four organisms.

References

Forbes, Betty A., Daniel F. Sahm, and Alice S. Weissfeld. 2002. Chapter 10 in *Bailey and Scott's Diagnostic Microbiology*, 11th ed. Mosby, St. Louis.

Koneman, Elmer W., Stephen D. Allen, William M. Janda, Paul C. Schreckenberger, and Washington C. Winn, Jr. 1997. *Color Atlas and Textbook of Diagnostic Microbiology*, 5th ed. J. B. Lippincott Co., Philadelphia.

Holt, John G. (Editor). 1994. *Bergey's Manual of Determinative Bacteriology*, 9th ed. Williams and Wilkins, Baltimore.

Varnam, Alan H. and Malcolm G. Evans. 2000. *Environmental Microbiology*. ASM Press, Washington, DC.

Exercise 2-12

The Effect of Osmotic Pressure on Microbial Growth

Theory

Water is essential to all forms of life. It is not only the principal component of cytoplasm, but also an essential source of electrons and hydrogen ions. Prokaryotes, like plants, require water to maintain cellular **turgor pressure**. Whereas eukaryotic animal cells burst with a constant influx of water, prokaryotes require it to prevent **plasmolysis**—shrinking of the cell membrane resulting in separation from the cell wall.

Many bacteria regulate turgor pressure by transporting in and maintaining a relatively high cytoplasmic potassium or sodium ion concentration, thereby creating a concentration gradient that promotes inward **diffusion** of water. For bacteria living in saline habitats, the job of maintaining turgor pressure is continuous because of the constant efflux of water.

Irrespective of a cell's efforts to control its internal environment, natural forces will cause water to move through its semipermeable membrane from an area of low **solute** concentration to an area of high solute concentration. In a solution in which solute concentration is low, water concentration is high, and *vice versa*. Therefore, water moves through a cell membrane from where its concentration is high to where its concentration is low. This process is called **osmosis**, and the force that controls it is called **osmotic pressure.**

Osmotic pressure is a quantifiable term and refers, specifically, to the ability of a solution to *pull water toward itself* through a semipermeable membrane. If a bacterial cell is placed into a solution that is **hyposmotic** (*i.e.* a solution having low osmotic pressure), there will be a *net* movement of water into the cell. If an organism is placed into a **hyperosmotic** solution, there will be a net movement of water out of the cell. For a bacterial cell in an **isosmotic** solution, water will tend to move in both directions equally; that is, there is no net movement (Figure 2-42).

Bacteria constitute a diverse group of organisms and, as such, have developed many adaptations for survival. Microorganisms tend to have a distinct range of salinities that are optimal for growth, with little or no survival outside that range. For example, some bacteria called **halophiles** grow optimally in NaCl concentrations of 3% or higher. **Extreme halophiles** are organisms with specialized cell membranes and enzymes that require salt concentrations from 15% up to about 25% and will not survive where salinity is lower. Except for a few **osmotolerant** bacteria, which will grow over a wide range of salinities, most bacteria live where NaCl concentrations are less than 3%.

Application

This procedure is used to demonstrate bacterial tolerances to NaCl.

In This Exercise...

You are going to grow two human commensal bacteria at a variety of NaCl concentrations to determine the maximum tolerance for each organism. *Staphylococcus aureus* is an inhabitant of human skin and nasal passages.

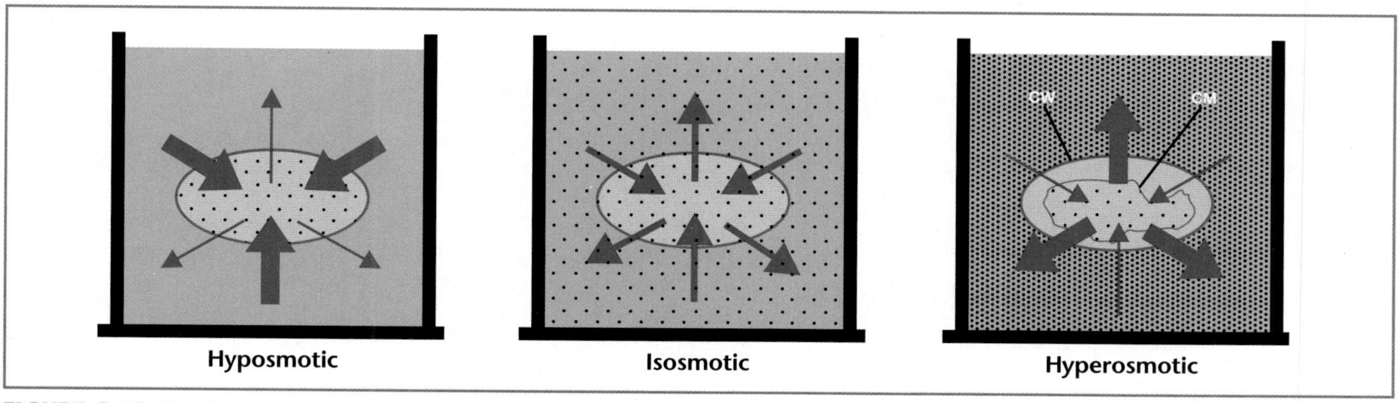

FIGURE 2-42 THE EFFECT OF OSMOTIC PRESSURE ON BACTERIAL CELLS
This osmosis diagram illustrates the movement of water into and out of cells. The labels refer to the osmotic pressure of the solution outside the cell. In a hyposmotic environment, the cell has greater osmotic pressure, so the net movement of water (arrows) will be into the cell. In an isosmotic environment, there is no net movement because the osmotic pressure of the cell and that of the environment are equal. (Actually the water is moving equally in both directions.) In a hyperosmotic environment, the osmotic pressure of the environment is greater, so the net movement is outward and results in plasmolysis. Note the shrinking membrane (CM) and the rigid cell wall (CW) in the hyperosmotic solution.

Escherichia coli is a common bacterial species living in human intestines. See if you can predict a correlation between the osmotolerance of each organism and the areas of the human body where it typically resides.

Materials

Per Student Group

- three tubes each of Nutrient Broths containing 2%, 5%, 8%, and 11% NaCl, respectively
- spectrophotometer (optional)
- cuvettes (optional)
- sterile transfer pipettes
- fresh broth cultures of:
  - *Staphylococcus aureus*
  - *Escherichia coli*

Medium Recipe

NaCl-Adjusted Nutrient Broth

Beef extract	3.0 g
Peptone	5.0 g
Distilled or deionized water	1.0 L
NaCl	20.0, 50.0, 80.0, or 110.0 g (2%, 5%, 8%, and 11% respectively)

Procedure

Lab One

1. Obtain three tubes of each NaCl broth—one of each concentration per organism plus one of each for controls—and label them accordingly.
2. Mix each culture thoroughly before the following transfers. Using a sterile pipette, transfer a *single drop* of each broth culture to its appropriate Nutrient Broth tube. (Because you will be comparing the amount of growth in the Nutrient Broth tubes, you should take care to begin by transferring the same volume of culture to each tibe. Use the same pipette for all transfers with a single organism.)
3. Incubate all tubes at 35 ± 2°C for 48 hours.

Lab Two (without a spectrophotometer)

1. Clean the outside of all tubes with a tissue and place them in a test tube rack organized into groups *by organism*.
2. Gently shake each broth until uniform turbidity is achieved.
3. Compare all tubes in a group to each other and to a control tube. Rate each one as 0, 1, 2, or 3 according to its turbidity (0 is clear and 3 is very turbid). (***Note***: The different NaCl broths may show some color variability; therefore, base your conclusions solely on turbidity and a comparison with the control, not color.)
4. Enter your results in the chart provided on the Data Sheet.
5. On the graph paper provided on the Data Sheet, plot the numeric values versus % NaCl for both organisms.

Lab Two (using a spectrophotometer)

1. Clean the outside of all tubes with a tissue and place them in a test tube rack organized into groups *by NaCl concentration.*
2. Shake each broth gently until uniform turbidity is achieved.
3. Set the spectrophotometer wavelength at 650 nm. If the spectrophotometer is digital, set it to "absorbance" (For instruction on how to operate a spectrophotometer, refer to Appendix E)
4. Blank the spectrophotometer with the uninoculated control for the first group of tubes. Take absorbance readings of all the tubes in the group and enter the results in the chart provided on the Data Sheet. Turbidity (a qualitative measure) increases with higher cell density. Measuring absorbance—which increases proportionately with turbidity—allows us to quantify the differences in growth. (***Note:*** Your instructor may provide you with cuvettes for this portion of the exercise. Use a nonsterile transfer pipette to fill the cuvettes. After use, dispose of the cuvettes and transfer pipettes in an autoclave receptacle.)
5. Using the control for each group to blank the spectrophotometer, continue reading all of the broths and enter the results in the chart provided.
6. On the graph paper provided on the Data Sheet, plot the absorbance versus % NaCl of both organisms.

References

Forbes, Betty A., Daniel F. Sahm, and Alice S. Weissfeld. 2002. Chapter 10 in *Bailey and Scott's Diagnostic Microbiology,* 11th ed. Mosby, Inc., St. Louis.

Holt, John G. (Editor). 1994. *Bergey's Manual of Determinative Bacteriology,* 9th ed. Williams and Wilkins, Baltimore.

Koneman, Elmer W., Stephen D. Allen, William M. Janda, Paul C. Schreckenberger, and Washington C. Winn, Jr. 1997. *Color Atlas and Textbook of Diagnostic Microbiology,* 5th ed. J. B. Lippincott Co., Philadelphia.

Moat, Albert G., John W. Foster, and Michael P. Spector. 2002. Pages 582–587 in *Microbial Physiology,* 4th ed. Wiley-Liss, New York.

Varnam, Alan H., and Malcolm G. Evans. 2000. *Environmental Microbiology.* ASM Press, Washington, DC.

White, David. 2000. Pages 388–394 in *The Physiology and Biochemistry of Prokaryotes,* 2nd ed. Oxford University Press, New York.

Control of Microorganisms

As demonstrated in earlier exercises, knowledge of factors that affect microbial growth are of great importance to microbiologists. Of nearly equal importance is the ability to prevent, stop, or inhibit unwanted growth.

The two most common forms of control are sterilization and disinfection. Sterilization methods kill all vegetative cells and spores, whereas disinfection methods are effective only against pathogens, not spores or even all vegetative cells. The two most common types of sterilization are incineration and moist-heat. Incineration is literally the burning of cells to ash. Moist-heat sterilization, is typically done in an autoclave at 121°C for 15 minutes, and kills cells and spores by denaturing all proteins.

In this unit you will examine both physical and chemical means of microbial control. The following exercises illustrate the lethal effects on bacteria of hand-scrubbing, disinfectants (including antiseptics), and radiation. (For more information on microbial control, refer to Exercises 6-6 Temperature—Lethal Effects, and Exercise 7-3 Antimicrobial Susceptibility Test.)

Exercise 2-13

The Lethal Effect of Ultraviolet Light on Microbial Growth

Theory

Ultraviolet (UV) light is a type of **electromagnetic energy**. Like all electromagnetic energy, UV travels in waves and is distinguishable from all others by its **wavelength**. Wavelength is the distance between adjacent wave crests and is typically measured in nanometers (nm) (Figure 2-43).

Ultraviolet light is divided into three groups categorized by wavelength:

UV-A, the longest wavelengths, ranging from 315 to 400 nm

UV-B, wavelengths between 280 and 315 nm

UV-C, wavelengths ranging from 100 to 280 nm. These wavelengths are most detrimental to bacteria. Bacterial exposure to UV-C for more than a few minutes usually results in irreparable DNA damage and death of the organism.

For a discussion on the mutagenic effects of UV and DNA repair, refer to Exercise 9-2.

Application

Ultraviolet light is commonly used to disinfect laboratory work surfaces.

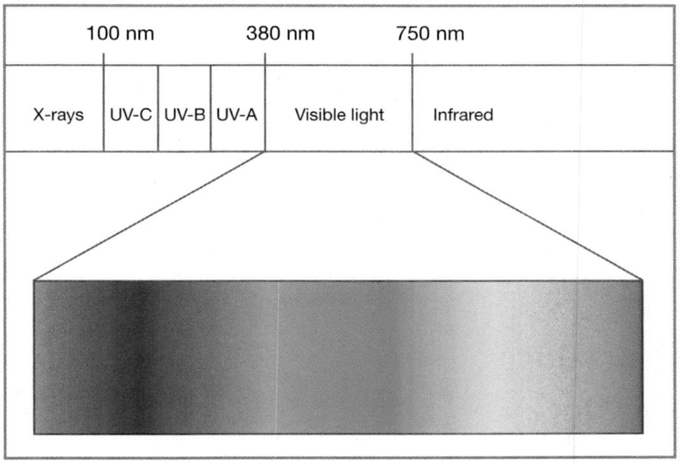

FIGURE 2-43 ELECTROMAGNETIC SPECTRUM
The shortest and highest energy wavelengths are those of gamma rays, starting at about at 10^{-5} nm. Radio waves, at the other end of the spectrum, can be one kilometer or longer. Between about 100 nm and 380 nm (just above visible light) is the sliver known as ultraviolet light.

In This Exercise...

Today you will examine the effect of UV exposure on two *Bacillus subtilis* cultures (24-hour and 7-day) and one 24-hour *Escherichia coli* culture. Because of the large number of plates to be treated, the work will be divided among six groups of students. Refer to Table 2-2 for assignments.

Organism	No UV	5 min.	10 min.	15 min.	20 min.	25 min.	30 min
Bacillus subtilis (24-hour culture)	1	1	2	3	4	5	6
Bacillus subtilis (7-day culture)	1	1	2	3	4	5	6
Escherichia coli (24-hour culture)	1	1	2	3	4	5	6

TABLE 2-2. Group Assignments by Number

Materials

Per Student Group

- ultraviolet lamp with appropriate shielding
- cardboard strips or disks to cover plates
- three Tryptic Soy Agar (TSA) plates (six for group one)
- stopwatch or electronic timer
- sterile Nutrient Broths (three per group; six for group 1)
- sterile cotton swabs (three per group; six for group 1)
- slant cultures of:
  - *Bacillus subtilis* (24-hour culture)
  - *Bacillus subtilis* (7-day culture)
  - *Escherichia coli* (24-hour culture)

Procedure

Lab One

1. Enter your group number and exposure time (from Table 2-2) on the Data Sheet.
2. Obtain three TSA plates and label the bottom of each with the name of the organism to be inoculated and your group number. Draw a line to divide the plates in half, and label the sides "A" and "B."
3. Pour a tube of Nutrient Broth into each slant and mix gently. Be careful to do this aseptically and not to overflow the tube.
4. Dip a sterile cotton swab into the broth of one culture and wipe the excess on the inside of the tube. Inoculate the appropriate plate by spreading the organism over the entire surface of the agar. Do this by streaking the plate surface completely three times, rotating it one-third turn between streaks. When incubated, this will form a bacterial lawn.
5. Repeat step 4 with the other two organisms and plates.
6. Place a paper towel on the table next to the UV lamp and soak it with disinfectant.
7. Place your plates under the UV lamp and set the covers, open side down, on the disinfectant-soaked towel. Cover the B half of the plates with the cardboard as shown in Figure 2-44.
8. Turn on the lamp for the prescribed time. ***Caution: Be sure the protective shield is in place and do not look at the light while it is on!*** Immediately replace the plate covers.
9. Invert and incubate the plate at 35 ± 2°C for 24 to 48 hours.

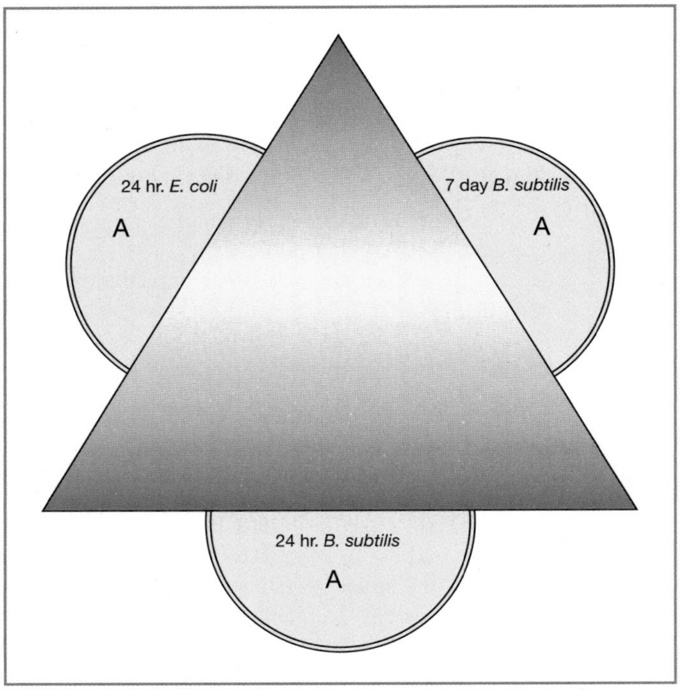

FIGURE 2-44 Plates Shielded for UV Exposure
Place the three plates under the UV lamp with the covers removed and the cardboard shield covering half of each plate as shown. Make sure the Petri dish covers are placed open side down on a disinfectant-soaked towel.

Lab Two

1. Remove your plates from the incubator and observe for growth. (***Note:*** Side B is your control—that is, it gives you an example of good growth for comparison. It should be covered with a bacterial lawn. If this is not the case, see your instructor.)
2. In Table 2-2, under your exposure time, record the growth on side A of each plate. Enter "0" if you observe no growth, "1" for poor growth, "2" for moderate growth, and "3" for abundant growth.
3. Your instructor may provide an overhead transparency or chalkboard space for you to enter your data.
4. Complete the chart on the Data Sheet using the data provided by the rest of the class.
5. On the graph paper provided with the Data Sheet, construct a graph representing growth versus UV exposure time for all three organisms.

References

Lewin, Benjamin. 1990. *Genes IV.* Oxford University Press, Cambridge, MA.

Varnam, Alan H., and Malcolm G. Evans. 2000. *Environmental Microbiology.* ASM Press, Washington, DC.

Exercise 2-14

Use-Dilution: Measuring Disinfectant Effectiveness

Theory

The use-dilution method is a standard procedure used to measure the effectiveness of disinfectants against microorganisms. In this procedure, organisms are exposed to varying concentrations (dilutions) of test disinfectants for 10 minutes, transferred to growth media, and incubated for 48 hours. If a solution is sufficient to prevent microbial growth at least 95% of the time, it meets the required standards and is considered a usable dilution of that disinfectant for a specific application.

Application

This procedure is used to establish appropriate dilutions for disinfectants.

In This Exercise...

Today you will examine the effectiveness of selected concentrations of four common household disinfectants—household bleach, hydrogen peroxide, ethyl alcohol and isopropyl alcohol—on *Staphylococcus aureus* and *Escherichia coli.* (With the exception of bleach, these substances are designed to disinfect skin and, therefore, are referred to more accurately as antiseptics.) The tasks are divided among eight groups of students. Each group will be responsible for one organism and three dilutions of one disinfectant. Refer to Table 2-3 for your assignments.

Materials

Per Student Group

- 100 mL flask of sterile deionized water
- disinfectants.* (Three concentrations of one disinfectant per group. See Table 2-3.)
  - 0.01%, 0.1%, and 1.0% household bleach
  - 0.03 %, 0.3%, and 3% hydrogen peroxide
  - 10%, 30%, and 50% isopropyl alcohol
  - 10%, 30%, and 50% ethyl alcohol
- eight sterile 60 mm Petri dishes
- one 60 mm Petri dish containing five sterile ceramic beads (available from Key Scientific, Round Rock, TX)

*Note to instructor: 15 mL volumes of each disinfectant concentration per group is sufficient.

Disinfectant	*Staphylococcus aureus*	*Escherichia coli*
Bleach	Group 1	Group 2
Hydrogen Peroxide	Group 3	Group 4
Isopropyl Alcohol	Group 5	Group 6
Ethyl Alcohol	Group 7	Group 8

Note: Each group is responsible for one organism and three dilutions of one disinfectant.

TABLE 2-3 GROUP ASSIGNMENTS

- one sterile 100 mm Petri dish containing filter paper
- sterile transfer pipette
- six sterile Nutrient Broth tubes
- needle-nose forceps
- small beaker with alcohol (for flaming forceps)
- fresh broth cultures of (one per group):
  - *Staphylococcus aureus*
  - *Escherichia coli*

Procedure**

Timing is important in this procedure. Read through it and make a plan before you begin so your transfers and soaking times are done uniformly and are consistent with those of other groups.

Lab One

1. Enter the name of your organism here: ______________________.
2. Enter the name of your disinfectant here: ______________________.
3. Obtain a Petri dish of sterile ceramic beads, one broth culture, one plate containing sterile filter paper, eight 60 mm plates, the three dilutions of your assigned disinfectant, and six sterile Nutrient Broths. Prepare a small beaker of alcohol for flaming the forceps.
4. Place the materials properly on your workspace as shown in the procedural diagram in Figure 2-45. Three broths should be labeled according to the organism and the disinfectant used; the other three

**This procedure has been modified for instructional purposes.

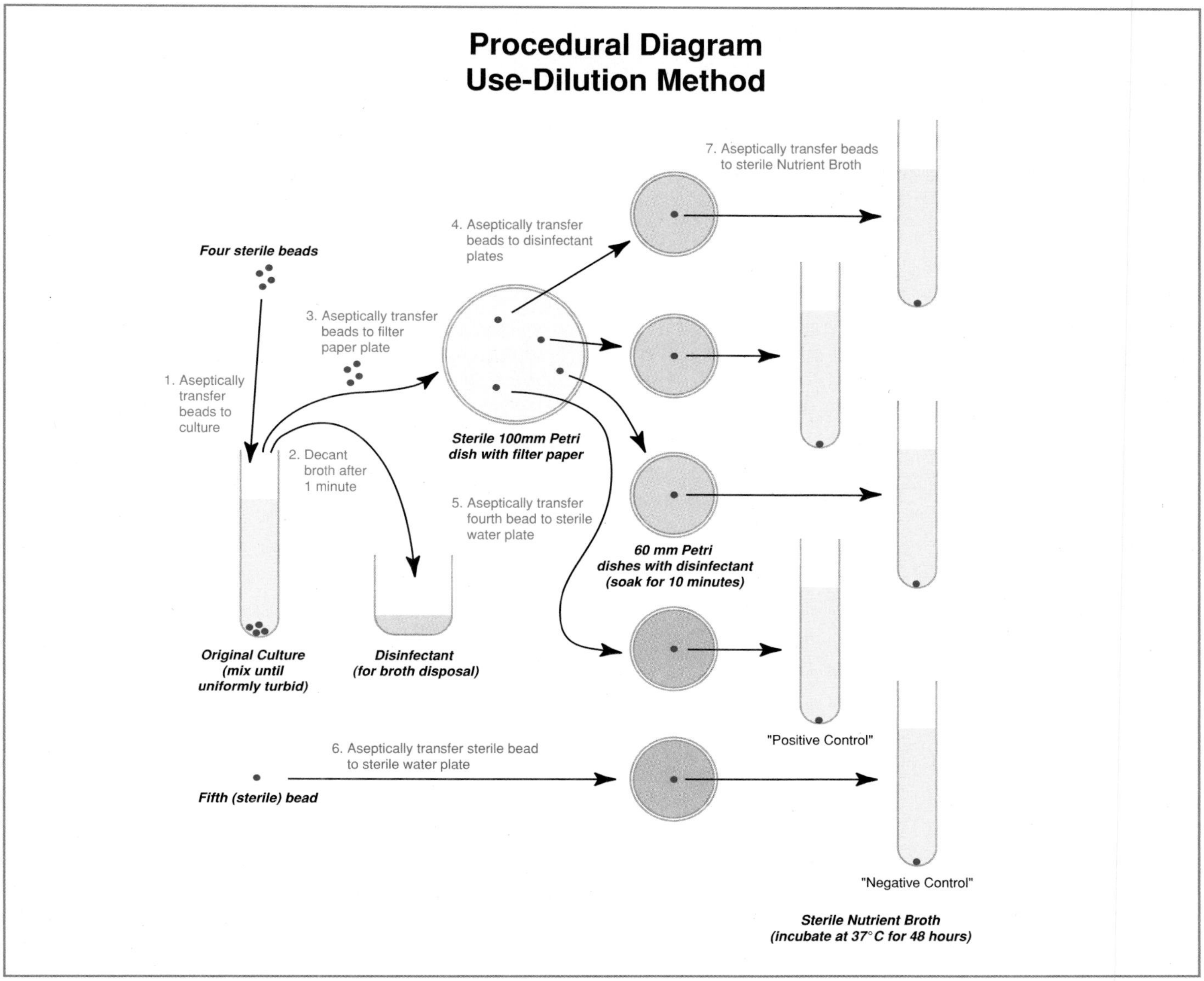
Procedural Diagram
Use-Dilution Method
Four sterile beads
1. Aseptically transfer beads to culture
2. Decant broth after 1 minute
3. Aseptically transfer beads to filter paper plate
4. Aseptically transfer beads to disinfectant plates
5. Aseptically transfer fourth bead to sterile water plate
6. Aseptically transfer sterile bead to sterile water plate
7. Aseptically transfer beads to sterile Nutrient Broth
Sterile 100mm Petri dish with filter paper
60 mm Petri dishes with disinfectant (soak for 10 minutes)
Original Culture (mix until uniformly turbid)
Disinfectant (for broth disposal)
Fifth (sterile) bead
"Positive Control"
"Negative Control"
Sterile Nutrient Broth (incubate at 37°C for 48 hours)

FIGURE 2-45 Procedural Diagram for Use-Dilution

are controls labeled: "Positive Control," "Negative Control," and "Turbidity Control."

5. Mix the broth culture gently until uniform turbidity is achieved.
6. Alcohol-flame your forceps and place four ceramic beads into the broth culture. (*Note:* Always use proper aseptic technique, and *sterilize the forceps before all transfers in this exercise.*)
7. After 1 minute, transfer the broth using a sterile transfer pipette into a flask of disinfectant and dispense the beads onto the sterile filter paper. Remove as much of the broth as you can to avoid wetting the filter paper excessively. You may have to "coax" the beads out of the tube with a sterile inoculating loop. Using sterile forceps, spread the beads apart on the paper and allow them sufficient time to dry. Do not roll them around, as this will remove the organism.
8. When the beads are dry (about 5 minutes), place three beads into the three disinfectant dilutions in 60 mm plates. Mark the time here: _________.
9. Place the fourth culture bead in a sterile water plate.
10. Place the fifth bead (a sterile bead) in the second sterile water plate.
11. After 10 minutes from the time marked above, remove the beads from the disinfectant solutions, in the same order as they were added, and place them in the appropriately labeled Nutrient Broths. Mix the broths immediately to disperse any residual disinfectant on the beads.

12. Transfer the fourth culture bead in the sterile water plate to the Nutrient Broth labeled "Positive Control."
13. Transfer the sterile bead in the sterile water plate to the Nutrient Broth labeled "Negative Control."
14. Incubate all broths (including the uninoculated Turbidity Control) at 35 ± 2°C for 48 hours.

Lab Two

1. Remove all broth tubes from the incubator and closely examine your controls. To be able to proceed, your inoculated bead (Positive Control) must have produced growth and your uninoculated bead (Negative Control) must have produced no growth. If both of these requirements have been met, you may proceed. If not, see your instructor.
2. Comparing to the Turbidity Control, examine the remainder of your beads. Using "G" to indicate growth and "NG" to indicate no growth, enter your results in the charts provided on the Data Sheet.

Reference

Widmer, Andreas F. and Reno Frei. 2003. Chapter 8 in *Manual of Clinical Microbiology*, 8th ed. Edited by Patrick R. Murray, Ellen Jo Baron, James H. Jorgensen, Michael A. Pfaller, and Robert H. Yolken. ASM Press, American Society for Microbiology, Washington, DC.

Exercise 2-15

Effectiveness of Hand Scrubbing

Theory

The origin of the **Germ theory of disease** is not absolutely clear, but its major precept—that microorganisms are the causes of infections—is almost universally accepted. Thanks to this popular belief, the use of antiseptics and disinfectants in medicine and in the laboratory is paramount.

In the laboratory, hand contamination can occur while transferring specimens, handling supplies and equipment, or simply touching work surfaces. In most cases, contamination is not seen, heard, or felt. Therefore, the unwanted transfer of microorganisms always must be considered a possibility, and precautionary measures must be a routine part of laboratory work.

To avoid infection, microbiologists wash their hands thoroughly when entering and prior to leaving the lab, after working in a biological safety cabinet, after removing gloves, and before eating, drinking, or smoking.

Application

This procedure is used to determine the effectiveness of scrubbing the skin to disinfect it.

In This Exercise...

In Exercise 2-14 you studied the relative effectiveness of various chemical agents at disinfecting. Today you will be divided into pairs to examine the effectiveness of normal hand washing as a microbial control mechanism. You will culture your hands before and after washing and then observe the resulting growth.

Materials

Per Student Pair

- two sterile Nutrient Agar plates
- hand soap
- sterile absorbent cotton
- paper towels

Procedure (per student pair)

Lab One: Student #1

1. Divide your agar plates into quadrants with a marking pen by drawing lines across its base. Label the quadrants 1, 2, 3, and 4.
2. Touch quadrants 1 and 2 with the index fingers of your right and left hands, respectively.
3. Go to the sink and wash your hands in your normal way, using any standard hand soap (not antibacterial).
4. With your lab partner holding a paper towel for you, blot your right hand until your index finger is dry. With the help, again, of your lab partner, dry your left index finger with sterile absorbent cotton.
5. Touch quadrants 3 and 4 with your index fingers as in step 2 above.
6. Incubate your plate in an inverted position for 48 hours at 25°C.

Lab One: Student #2

1. Divide your agar plates into quadrants with a marking pen by drawing lines across its base. Label the quadrants 1, 2, 3, and 4.
2. Touch quadrants 1 and 2 with the index fingers of your right and left hands, respectively.
3. Go to the sink and wash your hands in your normal way, but without soap—using only warm water.
4. With your lab partner holding a paper towel for you, blot your right hand until your index finger is dry. With the help, again, of your lab partner, dry your left index finger with sterile absorbent cotton.
5. Touch quadrants 3 and 4 with your index fingers as in step 2 above.
6. Incubate your plate in an inverted position for 48 hours at 25°C.

Lab Two

1. Examine the plates for relative amount and diversity of growth. Record your results on the Data Sheet.

Microscopy and Staining

SECTION 3

Microbiology as a biological discipline would not be what it is today without microscopes and cytological stains. Our ability to visualize, sometimes in great detail, the form and structure of microbes too small or transparent to be seen otherwise is attributable to developments in microscopy and staining techniques. In this section you will learn (or refine) your microscope skills. Then you will learn simple and more sophisticated bacterial staining techniques.

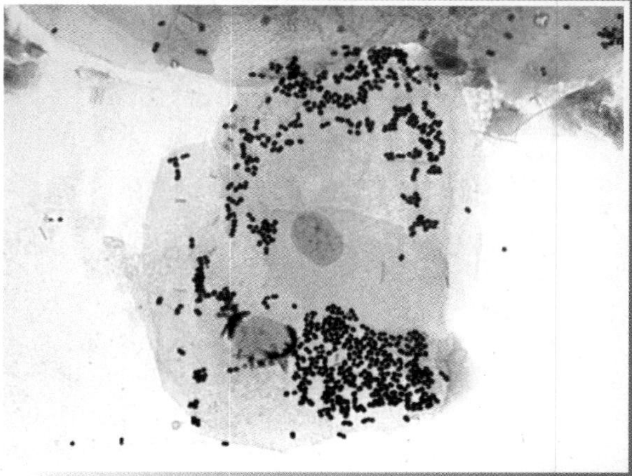

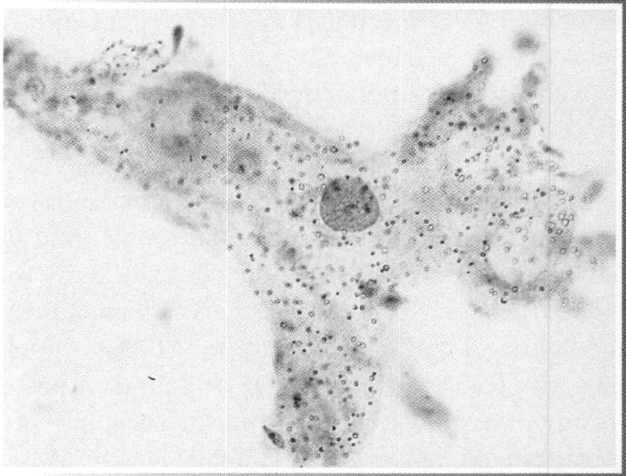

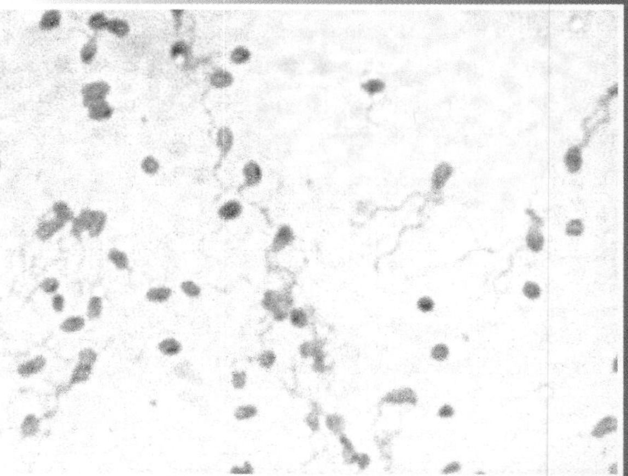

Microscopy

The earliest microscopes used visible light to create images and were little more than magnifying glasses. Today, more sophisticated compound light microscopes (Figure 3-1) are used routinely in microbiology laboratories. The various types of light microscopy include bright-field, dark-field, fluorescence, and phase contrast microscopy (Figure 3-2). Although each method has specific applications and advantages, the one used most commonly in introductory classes and clinical laboratories is bright-field microscopy. Many research applications use electron microscopy because of its ability to produce higher quality images of greater magnification.

Exercise 3-1

Introduction to the Light Microscope

Theory

Bright-field microscopy produces an image made from light that is transmitted through a specimen (Figure 3-2A). The specimen restricts light transmission and appears "shadowy" against a bright background (where light enters the microscope unimpeded). Because most biological specimens are transparent, the contrast between the specimen and the background can be improved with the application of stains to the specimen (Exercises 3-4 through 3-11). The "price" of the improved contrast is that the staining process usually kills cells. This is especially true of bacterial-staining protocols.

Image formation begins with light coming from an internal or an external light source (Figure 3-3). It passes through the **condenser** lens, which concentrates the light and makes illumination of the specimen more uniform. **Refraction** (bending) of light as it passes through the **objective lens** from the specimen produces a magnified **real image.** This image is magnified again as it passes through the **ocular lens** to produce a **virtual image** that appears below or within the microscope. The amount of magnification that each lens produces is marked on the lens (Figure 3-4A and B). Total magnification of the specimen can be calculated by using the following formula:

$$\text{Total Magnification} = \text{Magnification of the Objective Lens} \times \text{Magnification of the Ocular Lens}$$

The practical limit to magnification with a light microscope is around 1300X. Although higher magnifications are possible, image clarity is more difficult to maintain as the magnification increases. Clarity of an image is called **resolution.** The **limit of resolution** (or **resolving power**) is an actual measurement of how far apart two points must be for the microscope to view them as being separate. Notice that resolution improves as resolving power is made smaller.

The best limit of resolution achieved by a light microscope is about 0.2 μm. (That is, at its absolute best, a light microscope cannot distinguish between two points closer together than 0.2 μm.) For a specific

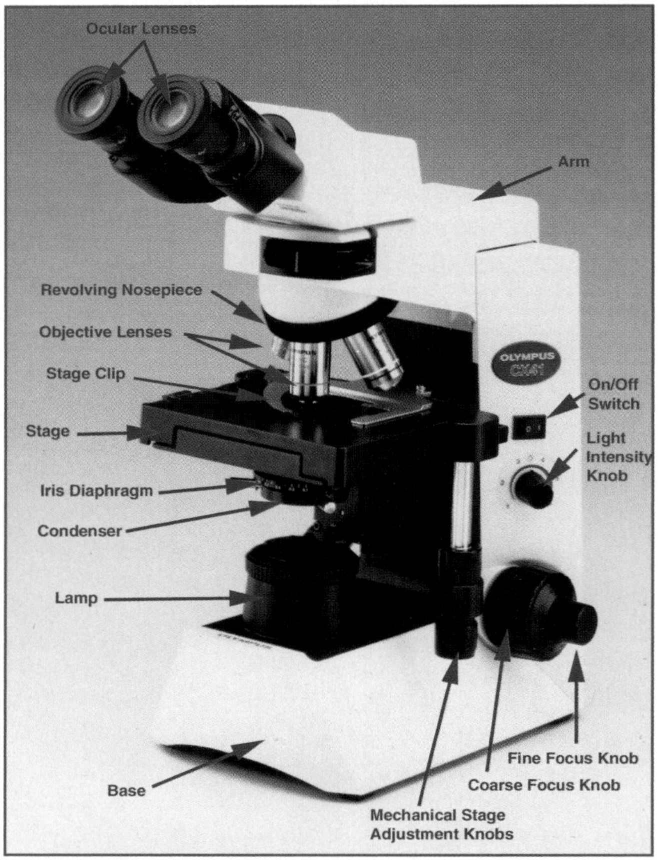

FIGURE 3-1 A BINOCULAR COMPOUND MICROSCOPE
A quality microscope is an essential tool for microbiologists. Most are assembled with exchangeable component parts and can be customized to suit the user's specific needs.
(Photograph courtesy of Olympus America Inc.)

FIGURE 3-2 **Types of Light Microscopy**
(A) Bright-field micrograph of a stained specimen of the fungus *Penicillium*. Notice that in this whole mount, very little of the specimen is in focus because of its thickness. To observe it properly, continually adjust the fine focus while scanning the field. (B) Dark-field micrograph of several *Paramecium*. Notice the contrast between internal components even without stain. (C) Phase contrast micrograph of *Amoeba*. Again, notice the contrast produced without stain. (D) Fluorescence micrograph of *Mycobacterium kansasii*.

(Photograph C courtesy of Gary D. Wisehart, San Diego City College)

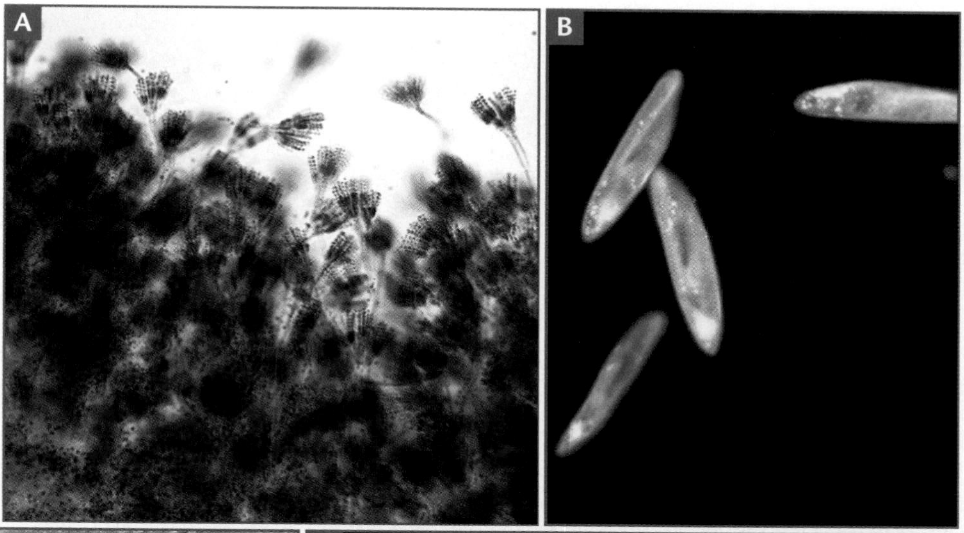

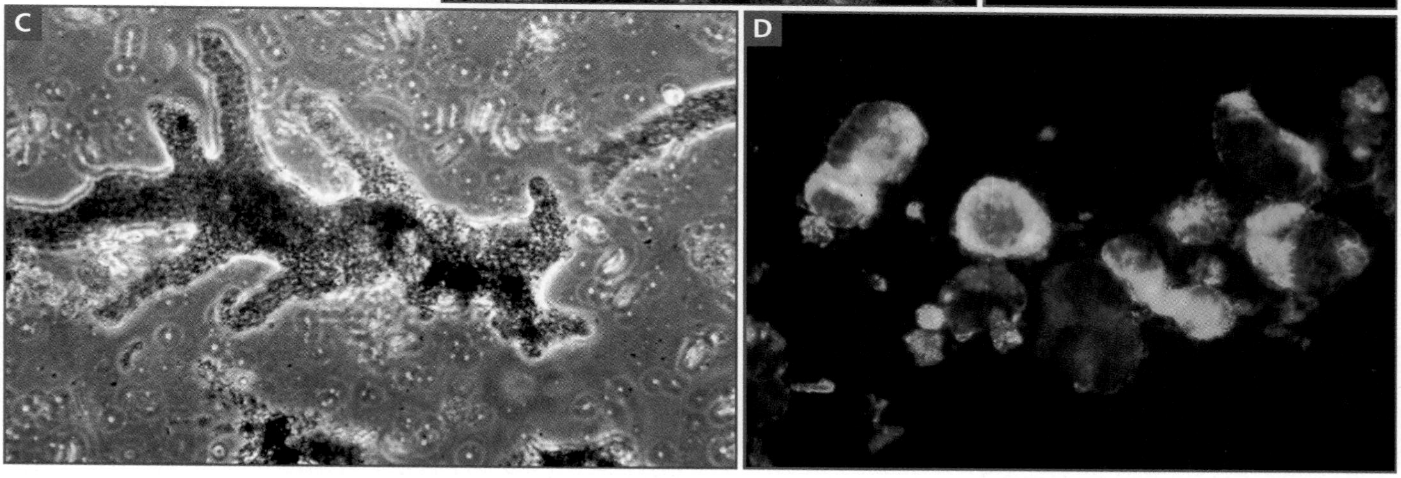

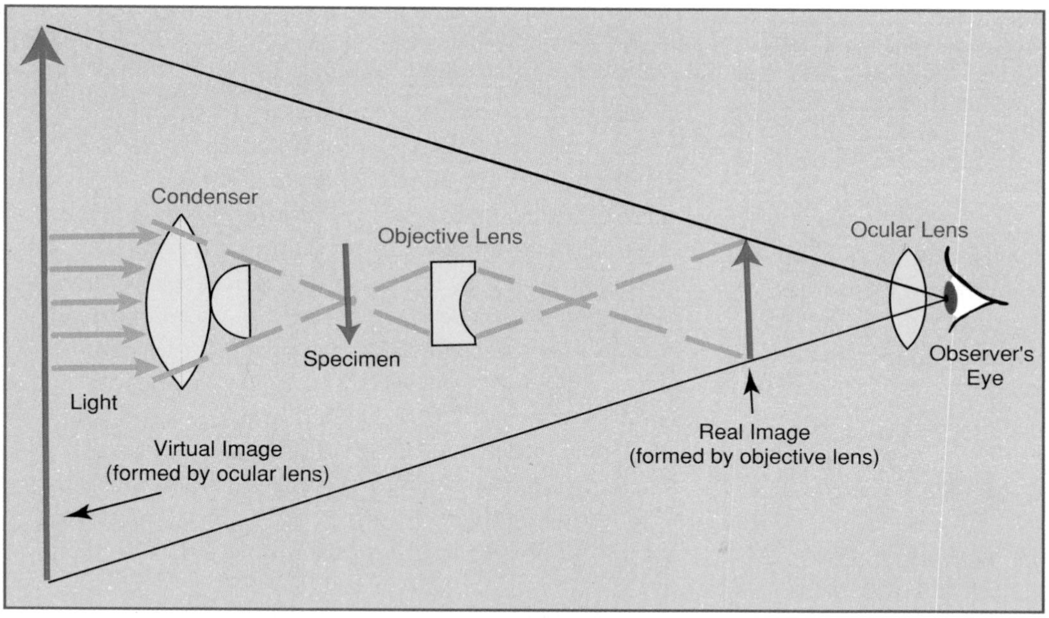

FIGURE 3-3 **Image Production in a Compound Light Microscope**
Light from the source is focused on the specimen by the condenser lens. It then enters the objective lens, where it is magnified to produce a real image. The real image is magnified again by the ocular lens to produce a virtual image that is seen by the eye. (After Chan, *et al.*, 1986)

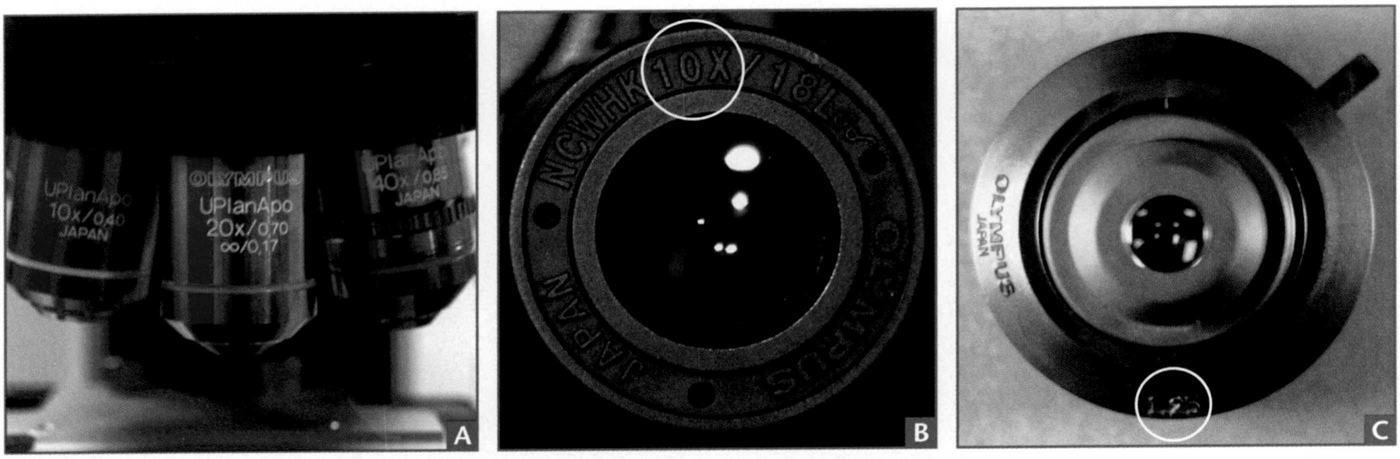

FIGURE 3-4 MARKINGS OF MAGNIFICATION AND NUMERICAL APERTURE ON MICROSCOPE COMPONENTS
(A) Three plan apochromatic objective lenses on the nosepiece of a light microscope. *Plan* means the lens produces a flat field of view. *Apochromatic* lenses are made in such a way that chromatic aberration is reduced to a minimum. From left to right, the lenses magnify 10X, 20X, and 40X, and have numerical apertures of 0.40, 0.70, and 0.85. The 20X lens has other markings on it. The mechanical tube length is the distance from the nosepiece to the ocular and is usually between 160 to 210 mm. However, this 20X lens has been corrected so the light rays are made parallel, effectively creating an infinitely long mechanical tube length (∞). This allows insertion of accessories into the light path without decreasing image quality. The thickness of cover glass to be used is also given (0.17 ± 0.01 mm). Also notice the standard colored rings for each objective: yellow for 10X, green for 20X (or 16X), and light blue for 40X (or 50X). (B) A 10X ocular lens. (C) A condenser (removed from the microscope) with a numerical aperture of 1.25. The lever in the upper right is used to open and close the iris diaphragm and adjust the amount of light entering the specimen.

microscope, the actual limit of resolution can be calculated using the following formula:

$$D = \frac{\lambda}{NA_{condenser} + NA_{objective}}$$

where D is the minimum distance at which two points can be resolved, λ is the wavelength of light used, and $NA_{condenser}$ and $NA_{objective}$ are the numerical apertures of the condenser lens and objective lenses, respectively. Because numerical aperture has no units, the units for D are the same as the units for wavelength, which typically are in nanometers (nm).

Numerical aperture is a measure of a lens's ability to "capture" light coming from the specimen and use it to make the image. As with magnification, it is marked on the lens (Figure 3-4A and C). Using immersion oil between the specimen and the oil objective lens increases its numerical aperture and, in turn, makes its limit of resolution smaller. (If necessary, oil also may be placed between the condenser lens and the slide.) The result is better resolution.

The light microscope may be modified to improve its ability to produce images with contrast without staining, which often distorts or kills the specimen. In **dark-field microscopy** (Figure 3-2B), a special condenser is used so only the light reflected off the specimen enters the objective. The appearance is of a brightly lit specimen against a dark background, and often with better resolution than that of the bright-field microscope.

Phase contrast microscopy (Figure 3-2C) uses special optical components to exploit subtle differences in the refractive indices of water and cytoplasmic components to produce contrast. Light waves that are in phase (that is, their peaks and valleys exactly coincide) reinforce one another, and their total intensity (because of the summed amplitudes) increases. Light waves that are out of phase by exactly one-half wavelength cancel each other and result in no intensity—that is, darkness. Wavelengths that are out of phase by any amount will produce some degree of cancellation and result in brightness that is less than maximum but more than darkness. Thus, contrast is provided by differences in light intensity that result from differences in refractive indices in parts of the specimen that put light waves more or less out of phase. As a result, the specimen appears as various levels of darks against a bright background.

Fluorescence microscopy (Figure 3-2D) uses a fluorescent dye that emits fluorescence when illuminated with ultraviolet light. In some cases, specimens possess naturally fluorescing chemicals and no dye is needed.

Application

Light microscopy (used in conjunction with cytological stains) is used to identify microbes from patient specimens or the environment. It also may be used to visually examine a specimen for the presence of more than one type of bacteria, or for the presence of other cell types that indicate tissue inflammation or contamination by a patient's cells.

In this Exercise…

Today you will become familiar with the operation and limitations of your light microscope. You will examine actual specimens in subsequent lab exercises as assigned by your instructor.

Materials

- compound light microscope
- nonsterile cotton swabs
- lens-cleaning solution or 95% ethanol

Instructions for Using the Microscope

Proper use of the microscope is essential for your success in microbiology. Fortunately, with practice and by following a few simple guidelines, you can achieve satisfactory results quickly. Because student labs may be supplied with a variety of microscopes, your instructor may supplement the following procedures and guidelines with instructions specific to your equipment. Refer to Figure 3-1 as you read the following (if working independently), or follow along on your microscope as your instructor guides you. (*Note:* This is a thorough treatment of microscope use, and not all parts may be immediately relevant to your laboratory. Refer back to this exercise as necessary.)

Transport

1. Carry your microscope to your workstation using both hands—one hand grasping the microscope's arm and the other supporting the microscope beneath its base.
2. Place the microscope *gently* on the table.

Cleaning

1. Use cotton swabs to clean the objective, ocular, and condenser lenses. Lens paper is useful for gently blotting oil from the oil immersion lens, but for the final cleaning you should use cotton swabs dipped in pure ethanol or commercial lens cleaning solution.
2. To clean an ocular, moisten the cotton swab with cleaning solution and gently wipe in a spiral motion starting at the center of the lens and working outward.

Operation

1. Raise the substage condenser to a couple of millimeters below its maximum position nearly even with the stage and open the iris diaphragm.
2. Plug in the microscope and turn on the lamp. Adjust the light intensity slowly to its maximum.
3. Using the nosepiece ring, move the scanning objective (usually 4X) into position. Do not rotate the nosepiece by the objectives as this can damage the objective lenses and cause them to unscrew from the nosepiece.
4. Place a slide on the stage in the mechanical slide holder and center the specimen over the opening in the stage.
5. If using a binocular microscope, adjust the distance between the two oculars to match your own interpupillary distance.
6. Use the coarse-focus adjustment knob to bring the image into focus. (***Note:*** For most microscopes, the distance from the nosepiece opening to the focal plane of each lens has been standardized at 45 mm. This makes the lenses **parfocal** and gives the user an idea of where to begin focusing.) Bring the image into sharpest focus using the fine-focus adjustment knob. Then observe the specimen with your eyes relaxed and slightly above the oculars to allow the images to fuse into one. If you are using a monocular microscope, keep both eyes open anyway to reduce eye fatigue.
7. If you are using a binocular microscope, adjust the oculars' focus to compensate for differences in visual acuity of your two eyes. Close the eye with the adjustable ocular and bring the image into focus with the coarse- and fine-focus knobs. Then, using only the eye with the adjustable ocular, focus the image using the ocular's focus ring.
8. Adjust the iris diaphragm and condenser position to produce optimum illumination, contrast, and image. (As a rule, use the maximum light intensity combined with the smallest aperture in the iris diaphragm that produces optimum illumination.)
9. Scan the specimen to locate a promising region to examine in more detail.
10. If you are observing a nonbacterial specimen, progress through the objectives until you see the degree of structural detail necessary for your purposes. You will have to adjust the fine focus and illumination for each objective. Before advancing to the next objective, be sure to position a desirable portion of the specimen in the center of the field or you will risk "losing" it at the higher magnification.
11. If you are working with a bacterial smear, you will have to use the oil immersion lens.
12. To use the oil immersion lens, work through the low (10X), then high dry (40X) objectives, adjusting

the fine focus and illumination for each. Before advancing to the next objective, be sure to position a desirable portion of the specimen in the center of the field or you risk "losing" it at the higher magnification. When the specimen is in focus under high dry, rotate the nosepiece to a position midway between the high dry and oil immersion lenses. Then place a drop of immersion oil on the specimen. *Be careful not to get any oil on the microscope or its lenses, and be sure to clean it up if you do.* Rotate the oil lens so its tip is submerged in the oil drop. Be careful not to trap any air between the slide and the oil objective. If you do, rotate the oil lens into and out of position a couple of times to pop the bubble. (***Note***: Do not move the stage down to add oil to the slide or the specimen will no longer be in focus. On a properly adjusted microscope, the oil and the high dry lenses have the same focal plane. Therefore, when a specimen is in focus on high dry, the oil lens, although longer, will also be in focus and won't touch the slide when rotated into position.) Focus and adjust the illumination to maximize the image quality.

13. When you are finished, lower the stage (or raise the objective) and remove the slide. Dispose of the freshly prepared slides in a jar of disinfectant or a sharps container; return permanent slides to storage.

Storage

When you are finished for the day:

1. Move the scanning objective into position.
2. Center the mechanical stage.
3. Lower the light intensity to its minimum, then turn off the light.
4. Wrap the electrical cord according to your particular lab rules.
5. Clean any oil off the lenses, stage, *etc.* Be sure to use only cotton swabs or lens paper for cleaning any of the optical surfaces of the microscope (see "Cleaning," above).
6. Return the microscope to its appropriate storage place.

References

Abramowitz, Mortimer. 2003. *Microscope Basics and Beyond.* Olympus America Inc., Scientific Equipment Group, Melville, NY.

Ash, Lawrence R., and Thomas C. Orihel. 1991. Pages 187–190 in *Parasites: A Guide to Laboratory Procedures and Identification.* American Society for Clinical Pathology (ASCP) Press, Chicago.

Bradbury, Savile, and Brian Bracegirdle. 1998. Chapter 1 in *Introduction to Light Microscopy.* BIOS Scientific Publishers Limited, Oxford, United Kingdom.

Forbes, Betty A., Daniel F. Sahm, and Alice S. Weissfield. 2002. Pages 119–121 in *Bailey and Scott's Diagnostic Microbiology,* 11th ed. Mosby, St. Louis.

Exercise 3-2

Calibration of the Ocular Micrometer

Theory

An **ocular micrometer** is a type of ruler installed in the microscope eyepiece, composed of uniform but unspecified graduations (Figure 3-5). As such, it must be calibrated before any viewed specimens can be measured. The device used to calibrate ocular micrometers is called a **stage micrometer**. As illustrated in Figure 3-6, a stage micrometer is a type of microscope slide containing a ruler with 10 µm and 100 µm graduations. (Other measuring instruments may be used in place of a stage micrometer, as shown in Figure 3-7.)

When the stage micrometer is placed on the stage, it is magnified by the objective being used; therefore, the size of the graduations (relative to the ocular micrometer divisions) increases as magnification increases. Consequently, the *value* of ocular micrometer divisions decreases as magnification increases. For this reason, calibration must be done for each magnification.

As shown in Figure 3-8, the stage micrometer is placed on the stage and brought into focus such that it is superimposed by the ocular micrometer. Then the first (left) line of the ocular micrometer is aligned with one of the marks on the stage micrometer. (The line chosen on the stage micrometer depends on the power of the lens being calibrated. Lower powers use the large graduations; higher powers use the smaller graduations on the left. Figure 3-8 illustrates proper alignment with the scanning objective.)

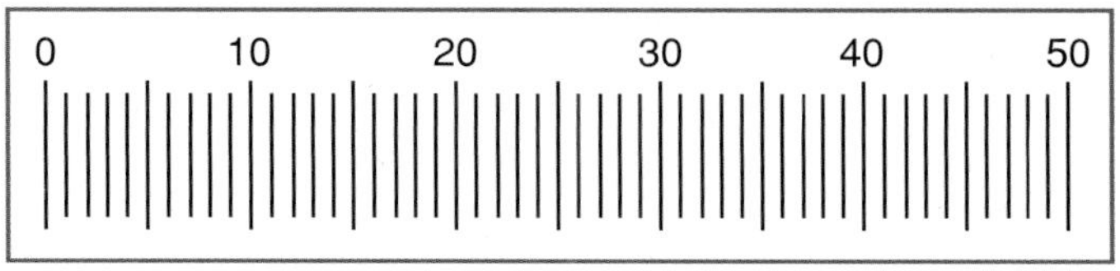

FIGURE 3-5 An Ocular Micrometer
The ocular micrometer is a scale with uniform increments of unknown size. It has to be calibrated for each objective lens on the microscope.

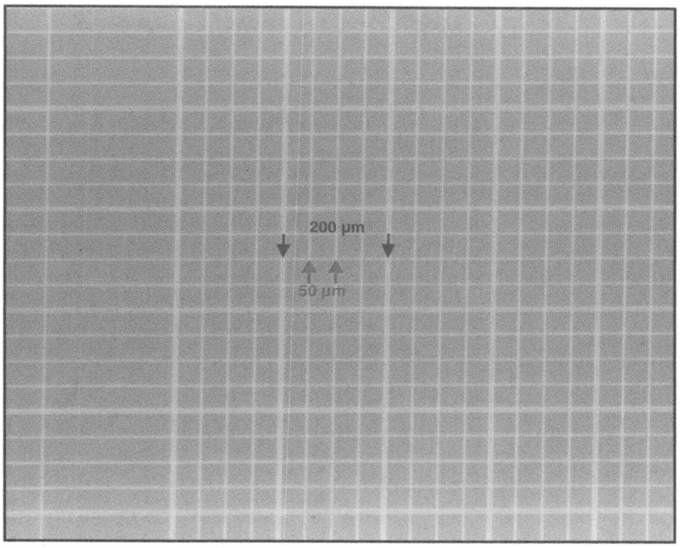

FIGURE 3-7 A Hemacytometer
Any instrument with markings of known distance apart may be used as a stage micrometer. The hemacytometer is a grid with lines 50 µm apart (red arrows). A larger grid is formed by lines 200 µm apart (blue arrows). Use any horizontal line as the micrometer, with the smallest divisions 50 µm apart.

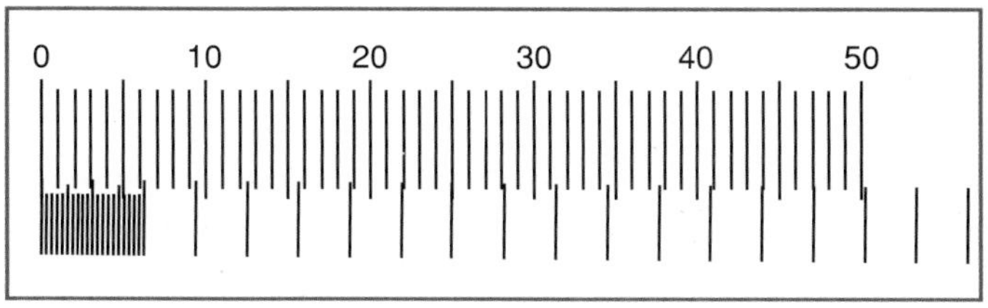

FIGURE 3-8 What They Look Like in Use
When properly aligned, the ocular micrometer scale is superimposed over the stage micrometer scale. Notice that they line up at the left ends.

Notice in Figure 3-8 that line 25 of the ocular micrometer and the eighth major line of the stage micrometer are perfectly aligned. This indicates that 25 ocular micrometer divisions (also called ocular units, or OU) span a distance of 800 µm because the stage micrometer lines are 100 µm apart. Notice also that line 47 of the ocular micrometer is aligned with the fifteenth major stage micrometer line. This means that 47 ocular units span 1500 µm. These values have been entered for you in Table 3-1.

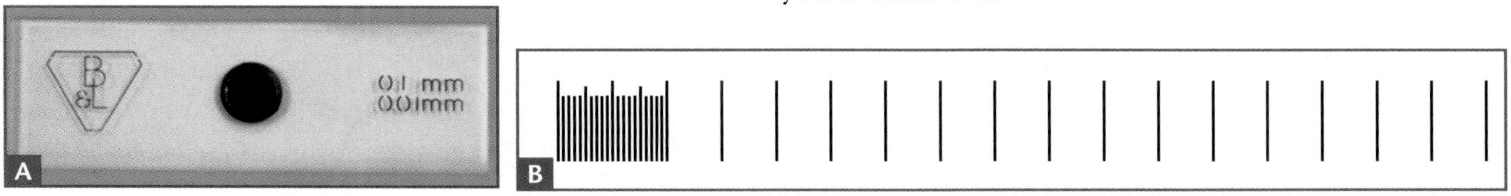

FIGURE 3-6 A Stage Micrometer
(A) The stage micrometer is a microscope slide with a ruler engraved into it. The markings on this micrometer indicate that the major increments are 0.1 mm (100 µm) apart. There is also a section of the scale that is marked off in 0.01 mm (10 µm) increments. (B) This drawing represents what the stage micrometer on the slide in (A) looks like. The micrometer is 2200 µm long. The major divisions are 100 µm apart. The 200 µm at the left are divided into 10 µm increments.

Stage Micrometer	Ocular Micrometer
800 µm	25 OU
1500 µm	47 OU

TABLE 3-1 Sample Data from Figure 3-8

To determine the value of an ocular unit on a given magnification, divide the distance (from the stage micrometer) by the corresponding number of ocular units.

$$\frac{800\ \mu m}{25\ \text{ocular units}} = 32\ \frac{\mu m}{OU}$$

$$\frac{1500\ \mu m}{47\ \text{ocular units}} = 32\ \frac{\mu m}{OU}$$

As shown in this example, it is customary to record more than one measurement. Each measurement is calculated separately. If the calculated ocular unit values differ, use their arithmetic mean as the calibration for that objective lens.

As mentioned previously, each magnification must be calibrated. Because of its short working distance, calibrating the oil immersion lens may be difficult to accomplish using the stage micrometer. If this is the case, its value can be calculated using the calibration value of one of the other lenses. Refer to Table 3-2 for the total magnifications for each objective lens on a typical microscope. Notice that the magnification of the oil immersion lens is 10 times greater than the medium-power lens. This means that objects viewed on the stage (stage micrometer *or* specimens) appear 10 times larger when changing from low power to oil immersion. But, because the magnification of the ocular micrometer does not change, an ocular division now covers only one-tenth the distance. Thus, the size of an ocular unit using the oil immersion lens can be calculated by dividing the calibration for low power by 10.

Ocular micrometer values can be calculated for any lens using values from any other lens, and they provide a good check of measured values. As calculated from Figure 3-8, 32 µm/OU was the calibration for the scanning objective. For practice, calculate the low, high dry and oil immersion calibrations. Write the values in the Table 3-2.

Once you have determined the ocular unit values for each objective lens, use the ocular micrometer as a ruler to measure specimens. For instance, if you determine that under the scanning objective a cell is 5 ocular units long, the cell's actual length would be determined as follows (using the sample values from Table 3-2):

Power	Total Magnification	Calibration (µm/OU)
Scanning	40X	32
Low Power	100X	
High Dry Power	400X	
Oil Immersion	1000X	

TABLE 3-2 Total Magnifications for Different Objective Lenses of a Typical Microscope and the Calibrations of the Ocular Micrometer for Each

$$\text{Cell Dimension} = \text{Ocular Units} \times \text{Calibration}$$

$$\text{Cell Dimension} = 5\ \text{Ocular Units} \times 32\ \mu m/OU$$

$$\text{Cell Dimension} = 160\ \mu m$$

Application

The ability to measure microbes is useful in their identification and characterization.

In This Exercise…

This lab exercise involves calibrating the ocular micrometer on your microscope. Actual measurement of specimens will be done in subsequent lab exercises as assigned by your instructor.

Materials

Per Student

- compound microscope equipped with an ocular micrometer
- stage micrometer

Procedure

Following is the general procedure for calibrating the ocular micrometer on your microscope. Your instructor will notify you of any specific details unique to your laboratory.

1. Check your microscope and determine which ocular has the micrometer in it.
2. Move the scanning objective into position.

3. Place the stage micrometer on the stage and position it so its image is superimposed by the ocular micrometer and the left-hand marks line up.
4. Examine the two micrometers and, as described above, record two or three points where they line up exactly. Record these values on the Data Sheet and calculate the value of each ocular unit.
5. Change to low power and repeat the process.
6. Change to high dry power and repeat the process.
7. Change to the oil immersion lens and repeat the process. If this cannot be done accurately, complete the calibration from the value of another lens.
8. Compute average calibrations for each objective lens and record these on the Data Sheet.
9. As long as you keep this microscope throughout the term, you may use the calibrations you recorded without recalibrating the microscope.

References

Abramoff, Peter, and Robert G. Thompson. 1982. Pages 5 and 6 in *Laboratory Outlines in Biology—III.* W. H. Freeman and Co., San Francisco.

Ash, Lawrence R., and Thomas C. Orihel. 1991. Pages 187–190 in *Parasites: A Guide to Laboratory Procedures and Identification.* American Society for Clinical Pathology (ASCP) Press, Chicago.

Exercise 3-3

Examination of Eukaryotic Microbes

Theory Cells are divided into two major groups—the **prokaryotes** and the **eukaryotes**—based on size and complexity. These and other differences are summarized in Table 3-3 and shown in Figure 3-9. The prokaryotes are further divided into two domains—the Archaea and the Bacteria. The eukaryotes belong to a single domain and are divided into as few as four and as many as eight kingdoms. The four kingdoms are: Protista, Fungi, Animalia, and Plantae. (The eight eukaryotic kingdom system breaks up the protists into five kingdoms.) Figure 3-10 provides a phylogenetic tree of these groups based on RNA comparisons. In this lab, you will observe simple eukaryotic microorganisms of various types: protists (protozoans and algae) and yeasts.

Protist Survey: Protozoans and Algae

Protozoans are unicellular eukaryotic heterotrophic microorganisms. A typical life cycle includes a vegetative **trophozoite** and a resting **cyst** stage. Some have additional stages, making their life cycles more complex.

One protozoan classification scheme recognizes the following groups: Phylum Sarcomastigophora (including Subphylum Mastigophora [the flagellates] and Subphylum Sarcodina [the amoebas]), Phylum Ciliophora (the ciliates), and Phylum Apicomplexa (sporozoans and others). Sarcodines move by forming cytoplasmic extensions called **pseudopods**. Division is by binary fission. Ciliates owe their motility to the numerous **cilia** covering the cell. Reproduction is by transverse fission. Members of Mastigophora are characterized by one or more

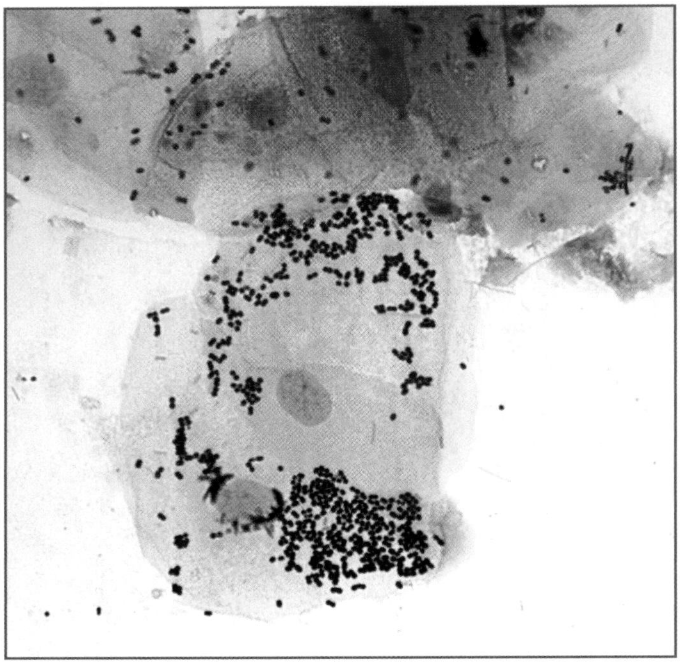

FIGURE 3-9 PROKARYOTIC AND EUKARYOTIC CELLS (GRAM STAIN, X1000)
This is a direct smear specimen taken from around the base of the teeth below the gum line. The large, pink cells are human epithelial cells and are eukaryotic (notice the prominent nuclei). The small, purple cells are prokaryotic bacteria. Typically, prokaryotic cells range in size from 1 to 5 μm, whereas eukaryotic cells are in the 10 to 100 μm range.

Character	Prokaryotes	Eukaryotes
Organismal groups	Bacteria and Archaebacteria	Plants, Animals, Fungi and Protists
Typical size	1–5 μm	10–100 μm
Membrane-bound organelles (including a nucleus)	Absent	Present
Ribosomes	70S (30S and 50S subunits)	80S (40S and 60S subunits)
Microtubules	Absent	Present
Flagellar movement	Rotary	Whip-like
DNA	Single, circular molecule called a chromosome	Two to many linear molecules; each is a chromosome
Introns	Rare	Common
Mitotic division	Absent	Present

TABLE 3-3 Summary of Major Prokaryotic and Eukaryotic Features

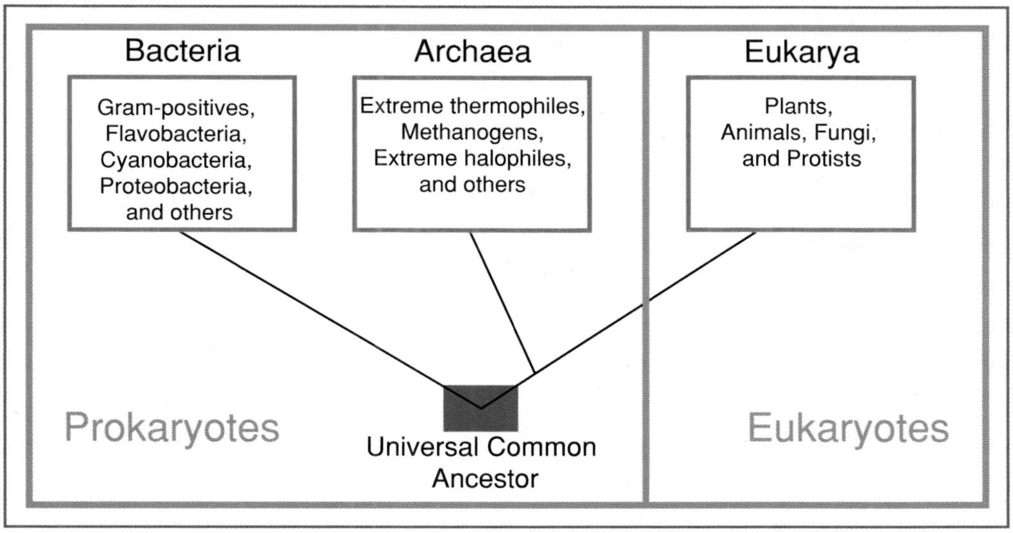

FIGURE 3-10
THE THREE DOMAINS OF LIFE
The domains are based on 16s (in prokaryotes) and 18s (in eukaryotes) rRNA sequencing results. The Archaea and Bacteria are prokaryotic domains, each containing an as yet undetermined number of kingdoms. Kingdom Eukarya includes all the eukaryotic organisms and is divided into the familiar Plant, Animal, and Fungus Kingdoms. The Protists include all the other eukaryotes that don't fit into the first three kingdoms and probably will be broken up into several kingdoms as we learn more.

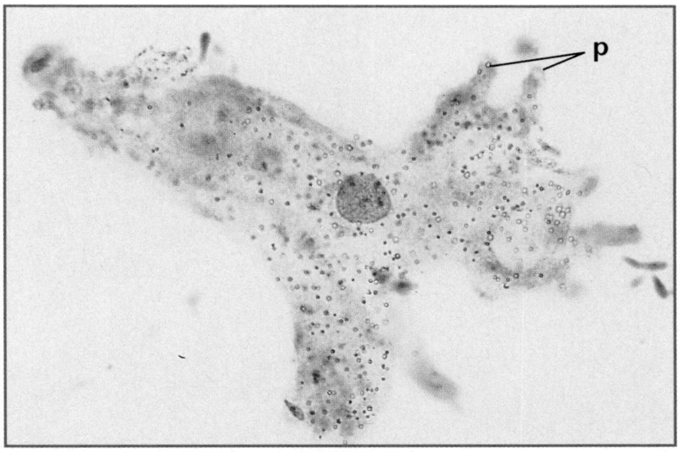

FIGURE 3-11 ***AMOEBA*, A SARCODINE (X53)**
Note the numerous pseudopods (P).

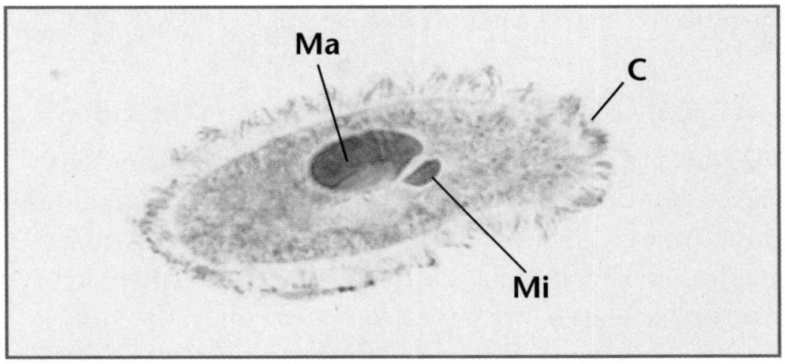

FIGURE 3-12 ***PARAMECIUM BURSARIA*, A CILIATE (X132)**
Note the cilia (C) around the edge of the cell. The macronucleus (Ma) and micronucleus (Mi) are also visible.

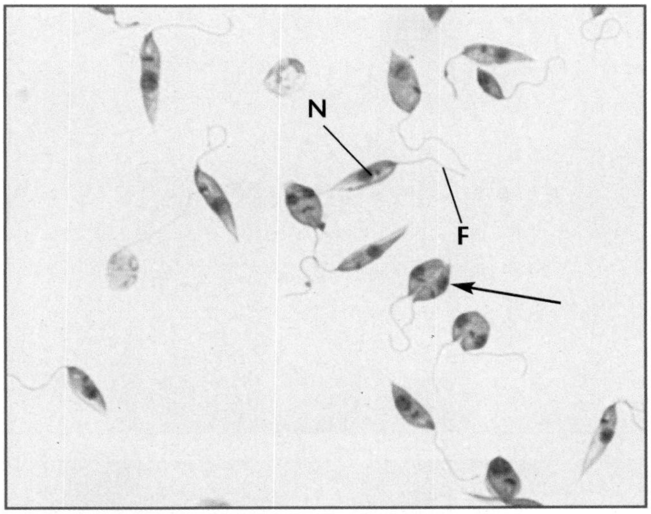

FIGURE 3-13 ***LEISHMANIA DONOVANI* (X1320)**
Notice the anterior flagellum (F) and the nucleus (N). Two of the cells (arrow) are dividing.

flagella and division by longitudinal fission. Sporozoans are typically nonmotile and usually have complex life cycles involving asexual reproduction in one host and sexual reproduction in another. Figures 3-11 through 3-13 show protozoan representatives of Mastigophora, Sarcodina, and Ciliophora.

Algae are a diverse group of simple, photosynthetic eukaryotic organisms with uncertain relatedness. **Green algae** (Division Chlorophyta) are common in freshwater and are usually unicellular or colonial. *Spirogyra* (Figure 3-14) and *Volvox* (Figure 3-15) are examples.

Other algal groups are the red algae, brown algae, golden-brown algae, yellow-green algae and diatoms (Figure 3-16). These groups are differentiated by their photosynthetic pigments, motility, cell wall material and storage carbohydrate.

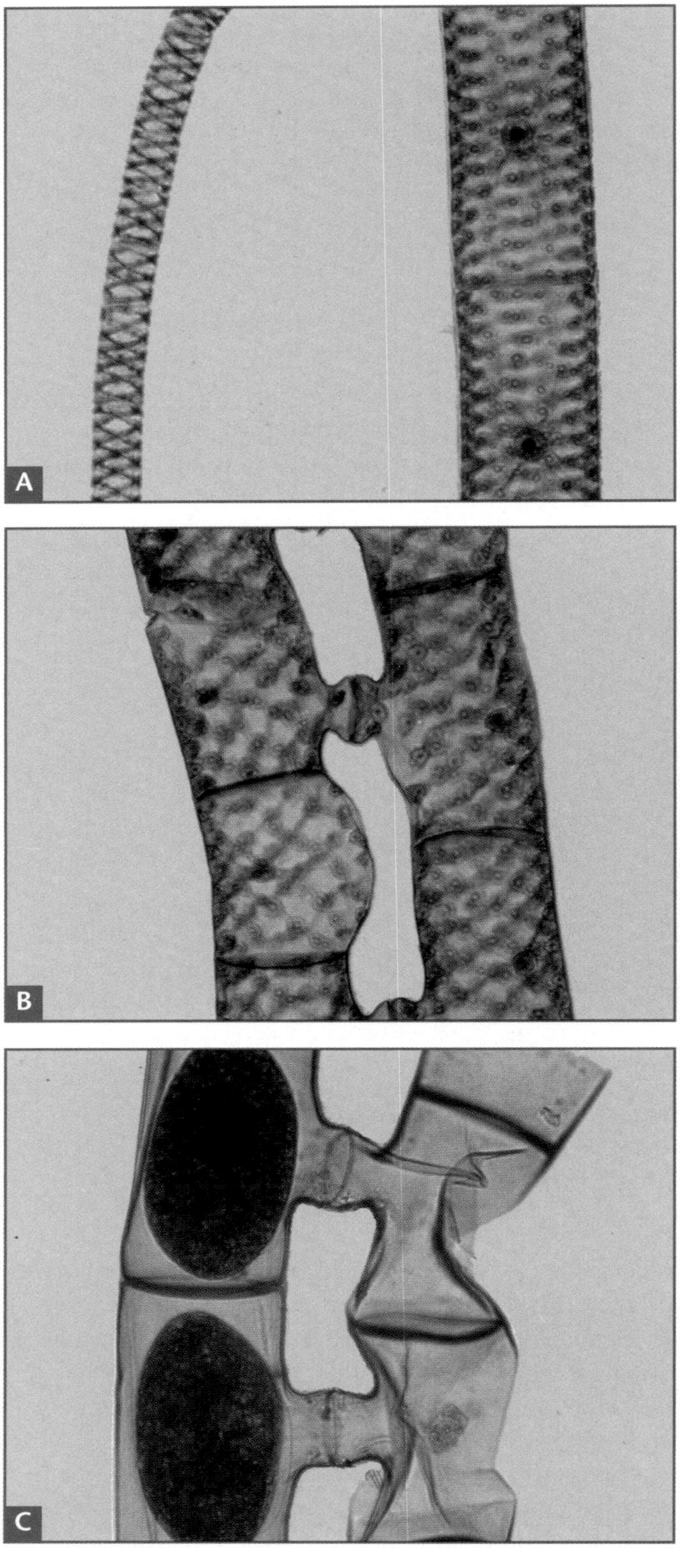

FIGURE 3-14 *SPIROGYRA SPP.*
Notice the spiral chloroplast and the nucleus in each cell. (A) A vegetative filament of cells. (B) Early conjugation between filaments. (C) Conjugation is completed with the formation of zygote.

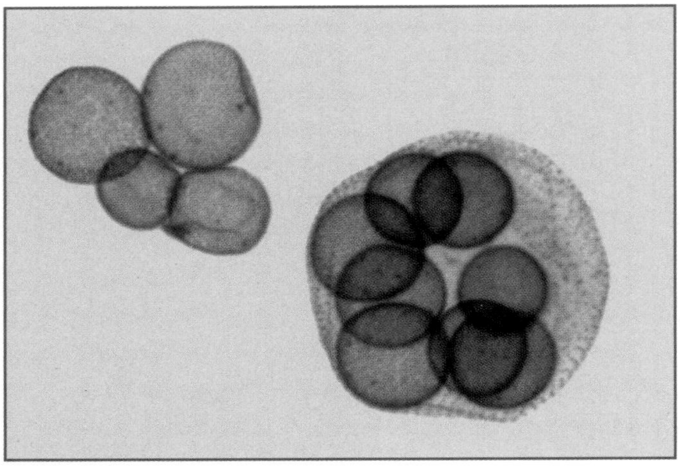

FIGURE 3-15 *VOLVOX* IS A COLONIAL GREEN ALGA
Flagellated cells join together to form the hollow, spherical colony. Daughter colonies are visible within the larger ones.

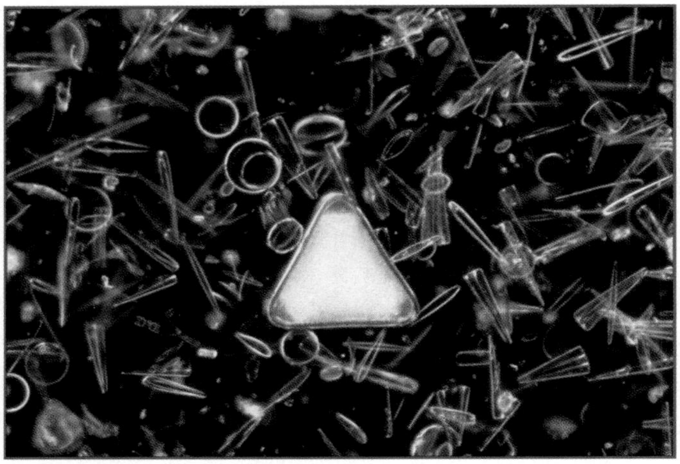

FIGURE 3-16 A VARIETY OF DIATOMS
Diatoms have cell walls composed of silica and organic material. The wall is divided into two halves, one of which fits into the other like the lid and base of a Petri dish.

Fungal Survey: Yeasts and Molds

Members of the Kingdom Fungi are nonmotile eukaryotes. Their cell wall is usually made of the polysaccharide chitin, not cellulose as in plants. Unlike animals (that ingest then digest their food), fungi are **absorptive heterotrophs**. That is, they secrete **exoenzymes** into the environment, then absorb the digested nutrients. Most are **saprophytes** that decompose dead organic matter, but some are **parasites** of plants, animals, or humans. Fungi are informally divided into unicellular **yeasts** and filamentous **molds** based on their overall appearance. Brewer's yeast *(Saccharomyces cerevisiae)* is shown in Figure 3-17.

In this lab you will observe representative protozoans, algae and fungi. Some specimens are on commercially prepared slides, whereas others need to be prepared as **wet mounts** (Figure 3-18) of living cultures.

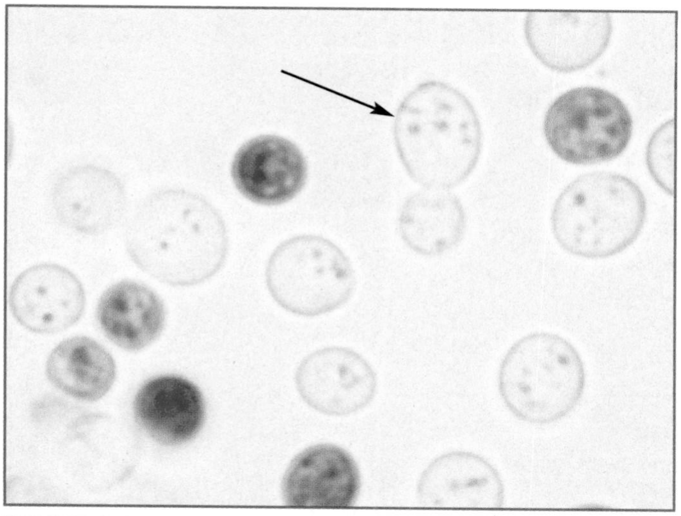

FIGURE 3-17 *SACCHAROMYCES CEREVISIAE*
The brewer's yeast is shown in a wet mount (X660). Note the budding cell in the center of field (arrow).

In a wet mount, a drop of water is placed on the slide and the organisms are introduced into it. Or, if the organism is already in a liquid medium, then a drop of medium is placed on the slide. A cover glass is placed over the preparation to flatten the drop and keep the objective lens from getting wet. A stain may or may not be applied to add contrast.

Application

Familiarity with eukaryotic cells not only rounds out your microbiological experience, it is important to be able to differentiate eukaryotic from prokaryotic cells when examining environmental or clinical specimens.

In This Exercise...

Most of this manual is devoted to prokaryotes, but in this exercise you will be given the opportunity to examine various eukaryotic microorganisms. This will not only serve to familiarize you with simple eukaryotes, but also give you practice using the microscope, measuring specimens, and making wet-mount preparations.

Materials

Per Class

- Prepared slides or living cultures of a variety of protists and fungi (such as, *Amoeba, Paramecium, Leishmania*—prepared only, *Spirogyra, Volvox*, diatoms, and *Saccharomyces*.)

Per Student

- clean glass slides and cover glasses
- compound microscope
- cytological stains (*e.g.*, methylene blue, I_2KI)
- methyl cellulose
- immersion oil
- cotton swabs
- lens paper
- lens cleaning solution

Procedure

1. Obtain a microscope and place it on the table or workspace. Check to be sure the stage is all the way down and the scanning objective is in place.
2. Begin with a prepared slide. Clean it with a tissue if it is dirty, then place it on the microscope stage. Center the specimen under the scanning objective.
3. Follow the instructions given in Exercise 3-1, to bring the specimen into focus at the highest magnification that allows you to see the entire structure you want to view.
4. Practice scanning with the mechanical stage until you are satisfied that you have seen everything interesting to see. Sketch what you see in the table provided on the Data Sheet.
5. Measure cellular dimensions and record these in the table provided on the Data Sheet.
6. Repeat with as many slides as you have time for.
7. Prepare wet mounts of available specimens by following the Procedural Diagram in Figure 3-18. Methyl cellulose may be added to the wet mount if you have fast swimmers. Sketch what you see and record cellular dimensions in the chart provided on the Data Sheet.
8. When you are finished observing specimens, blot the oil from the oil immersion lens (if used) with a lens paper and do a final cleaning with a cotton swab and alcohol or lens cleaning solution. Dry the lens with a clean swab.
9. Return all lenses and adjustments to their storage positions before putting the microscope away.

References

Campbell, Neil A., and Jane B. Reece. 2005. Chapters 25 and 28 in *Biology,* 7th ed. Pearson Education/Benjamin Cummings Publishing Co., San Francisco.

Freeman, Scott. 2005. Chapter 27 in *Biological Science,* 2nd ed. Pearson Education/Prentice Hall. Upper Saddle River, NJ.

Madigan, Michael T. and John M. Martinko. 2006. Chapter 11 in *Brock's Biology of Microorganisms,* 11th ed. Pearson Education/Prentice Hall, Upper Saddle River, NJ.

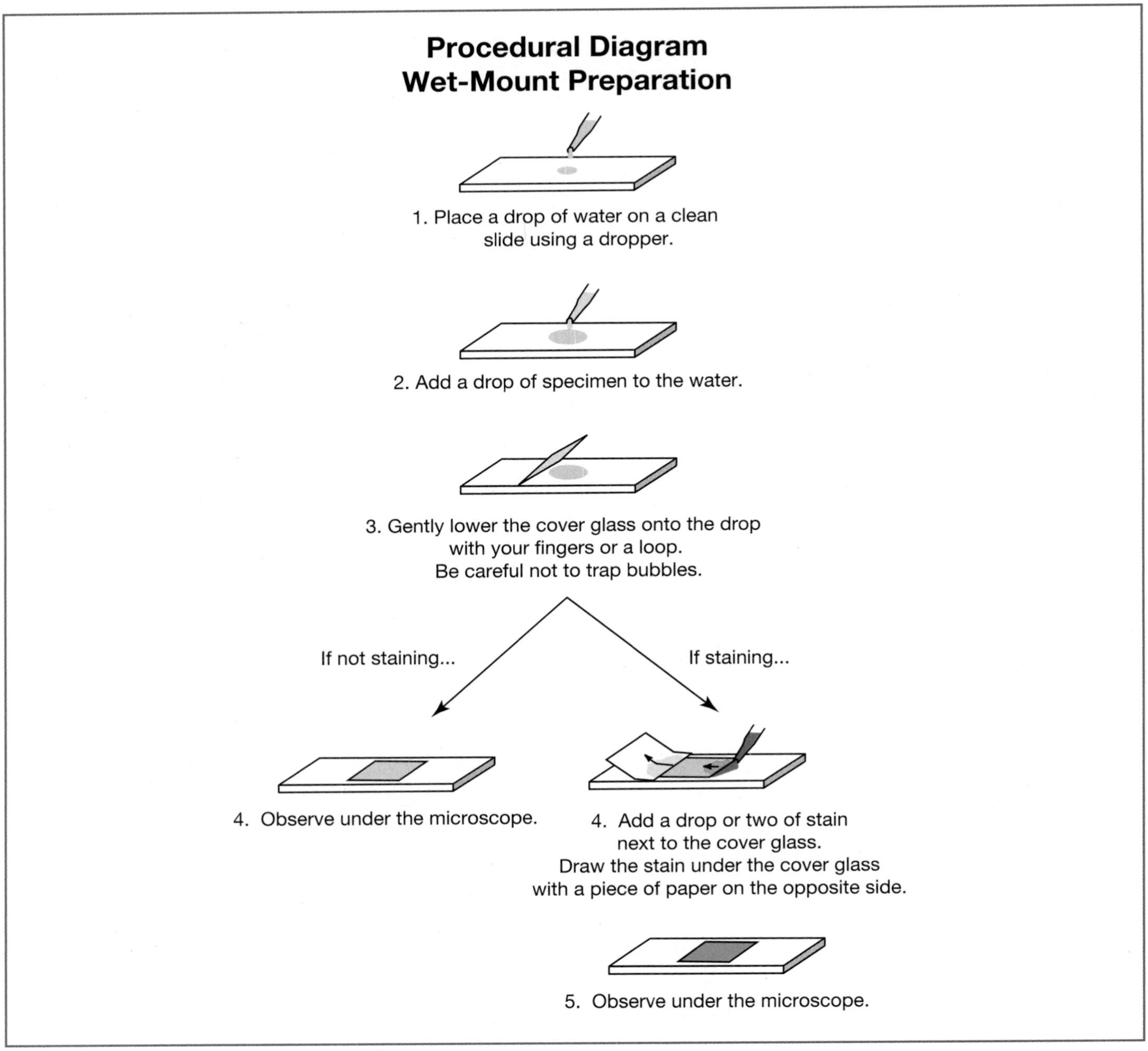
Procedural Diagram
Wet-Mount Preparation
1. Place a drop of water on a clean slide using a dropper.
2. Add a drop of specimen to the water.
3. Gently lower the cover glass onto the drop with your fingers or a loop. Be careful not to trap bubbles.
If not staining...
If staining...
4. Observe under the microscope.
4. Add a drop or two of stain next to the cover glass. Draw the stain under the cover glass with a piece of paper on the opposite side.
5. Observe under the microscope.

FIGURE 3-18 WET MOUNT PROCEDURAL DIAGRAM
Wet mounts are made using living specimens.

Bacterial Structure and Simple Stains

In Exercise 3-1 you were introduced to two of the three important features of a microscope and microscopy: magnification and resolution. A third feature is **contrast**. To be visible, the specimen must contrast with the background of the microscope field. Because cytoplasm is essentially transparent, viewing cells with the light microscope is difficult without stains to provide that contrast. In this set of exercises, you will learn how to correctly prepare a bacterial smear for staining and how to perform simple and negative stains. Cell morphology, size, and arrangement then may be determined. In a medical laboratory these usually are determined with a Gram stain (Exercise 3-6), but you will be using simple stains as an introduction to the staining process, as well as an introduction to these cellular characteristics.

Bacterial cells are much smaller than eukaryotic cells (Figure 3-9) and come in a variety of **morphologies** (shapes) and **arrangements**. Determining cell morphology is an important first step in identifying a bacterial species. Cells may be spheres (**cocci**, singular **coccus**), rods (**bacilli**, singular **bacillus**) or spirals (**spirilla**, singular **spirillum**). Variations of these shapes include slightly curved rods (**vibrios**), short rods (**coccobacilli**) and flexible spirals (**spirochetes**). Examples of cell shapes are shown in Figures 3-19 through 3-26. In Figure 3-26, *Corynebacterium xerosis* illustrates **pleomorphism**. A variety of cell shapes—slender, ellipsoidal, or ovoid rods—may be seen in a given sample.

Cell arrangement, determined by the number of planes in which division occurs and whether the cells separate after division, is also useful in identifying bacteria. Spirilla rarely are seen as anything other than single cells, but cocci and bacilli do form multicellular associations. Because cocci exhibit the most variety in arrangements, they are used for illustration in Figure 3-27. If the two daughter cells remain attached after a coccus divides, a **diplococcus** is formed. The same process happens in bacilli that produce **diplobacilli**. If the cells continue to divide in the same plane and remain attached, they exhibit a **streptococcus** or **streptobacillus** arrangement.

If a second division occurs in a plane perpendicular to the first, a **tetrad** is formed. A third division plane perpendicular to the other two produces a cube-shaped arrangement of eight cells called a **sarcina**. Tetrads and sarcinae are seen only in cocci. If the division planes of a coccus are irregular, a cluster of cells is produced to form a **staphylococcus**. Figures 3-28 through 3-32 illustrate common cell arrangements.

Arrangement and morphology often are easier to see when the organisms are grown in a broth rather than a solid medium, or are observed from a direct smear. If you have difficulty identifying cell morphology or arrangement, consider transferring the organism to a broth culture and trying again.

One last bit of advice: Don't expect nature to conform perfectly to our categories of morphology and cell arrangement. These are convenient descriptive categories that will not be applied easily in all cases. When examining a slide, look for the most common morphology and most complex arrangement. Do not be afraid to report what you see. For instance, it's okay to say, "Cocci in singles, pairs and chains."

Cells are three-dimensional objects with a surface that contacts the environment and a volume made up of cytoplasm. The ability to transport nutrients into the cell and wastes out of the cell is proportional to the amount of surface area doing the transport. The demand for nutrients and production of wastes is proportional to a cell's volume. It's a mathematical fact of life that as an object gets bigger, its volume increases more rapidly than its surface area. Therefore, a cell can achieve a size at which its surface area is not adequate to supply the nutrient needs of its cytoplasm. That is, its **surface-to-volume ratio** is too small. At this point, a cell usually divides its volume to increase its surface area. This phenomenon is a major factor in limiting cell size and determining a cell's habitat.

Bacilli, cocci, and spirilla with the same volume have different amounts of surface area. A sphere has a lower surface-to-volume ratio than a bacillus or spirillum of the same volume. A streptococcus, however, would have approximately the same surface-to-volume ratio as a bacillus of the same volume.

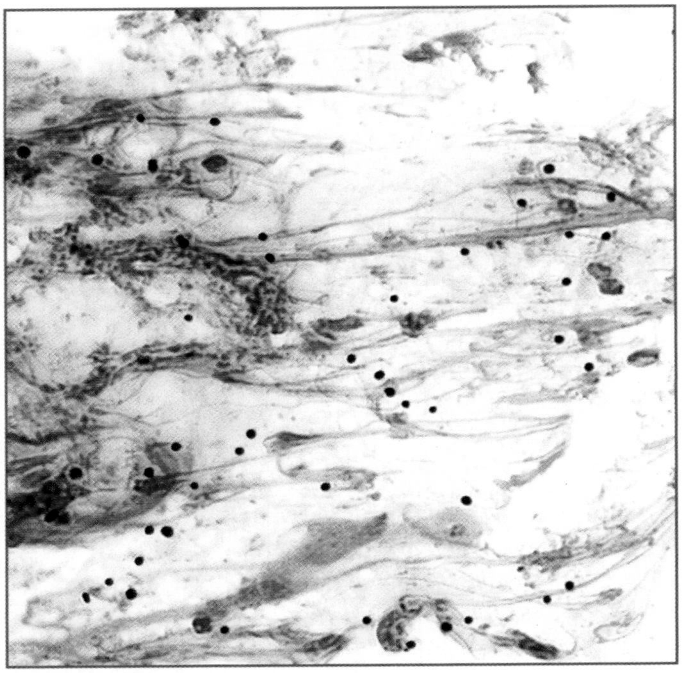

FIGURE 3-19 Single Cocci from a Nasal Swab (Gram Stain, X1000)
This direct smear of a nasal swab illustrates unidentified cocci (dark circles) stained with crystal violet. The red background material is mostly mucus.

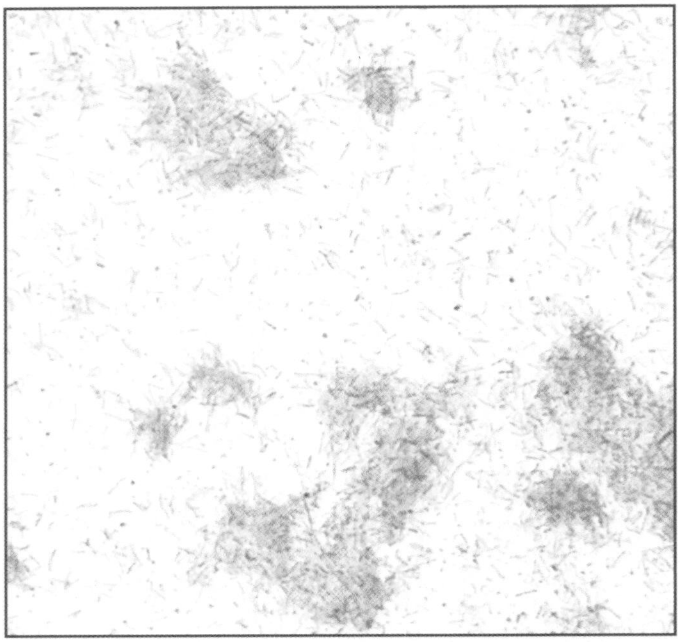

FIGURE 3-21 Long, Thin Bacillus (Gram Stain, X1000)
The cells of *Aeromonas sobria*, a freshwater organism, are considerably longer than they are wide. These cells were grown in culture. Notice that a cell can be a bacillus without being in the genus *Bacillus*.

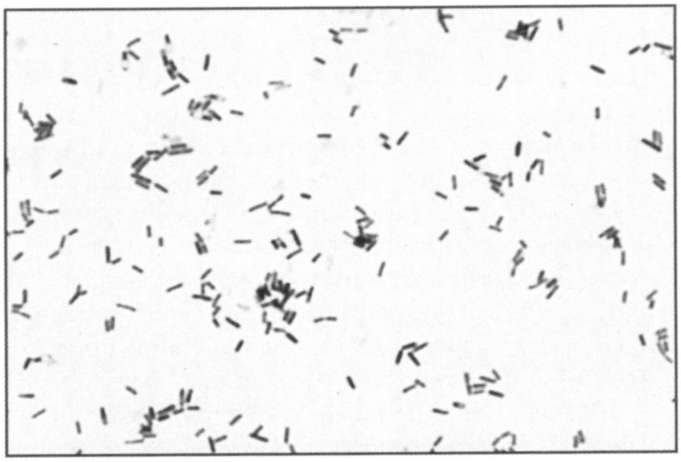

FIGURE 3-20 "Typical" Bacillus (Crystal Violet Stain, X1050)
Notice the variability in rod length (because of different ages of the cells) in this stain of the soil organism *Bacillus subtilis* grown in culture. Also notice the squared ends on the cells, typical of the genus *Bacillus*.

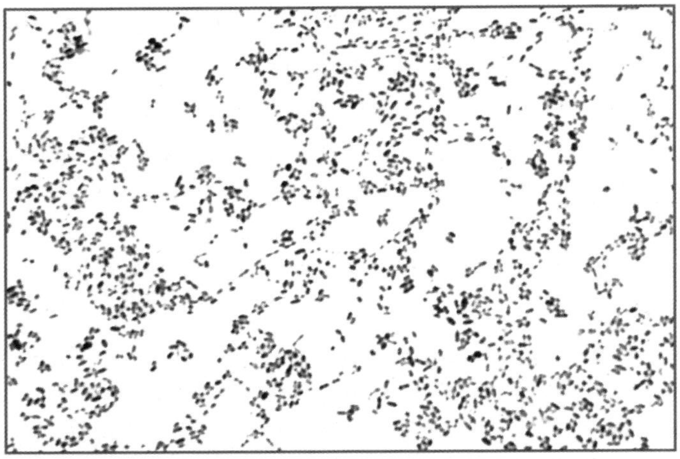

FIGURE 3-22 Coccobacillus (Gram Stain, X1000)
Lactococcus lactis is a short rod that a beginning microbiologist might confuse with a coccus. Notice the slight elongation of the cells, and also that most cells are not more than twice as long as they are wide. *L. lactis* is found naturally in raw milk and milk products, but these cells were grown in culture.

FIGURE 3-23 Spirillum (Crystal Violet, X1000)
Cells of the freshwater bacterium *Spirillum serpens* range from long, straight rods to spirals. This specimen was obtained from culture.

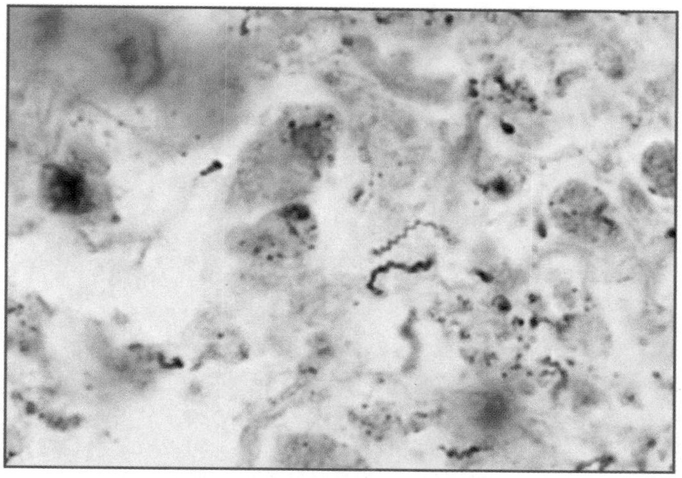

FIGURE 3-24 Spirochete (Silver Stain, X1320)
Spirochaetes are flexible, curved rods. Shown is *Treponema pallidum* in tissue stained with a silver stain that makes the cells appear black. *T. pallidum* is the causative agent of syphilis in humans and cannot be cultured.

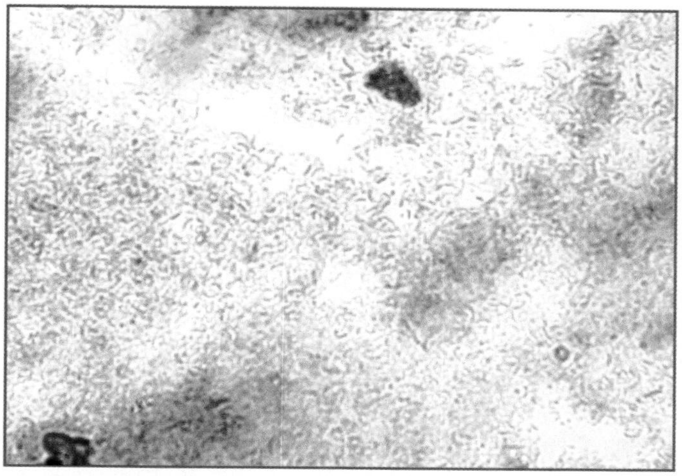

FIGURE 3-25 Vibrio (Gram Stain, X1000)
Vibrio cholerae is the causative agent of cholera in humans. Careful examination of the smear will reveal most rods as curved. These cells are from culture.

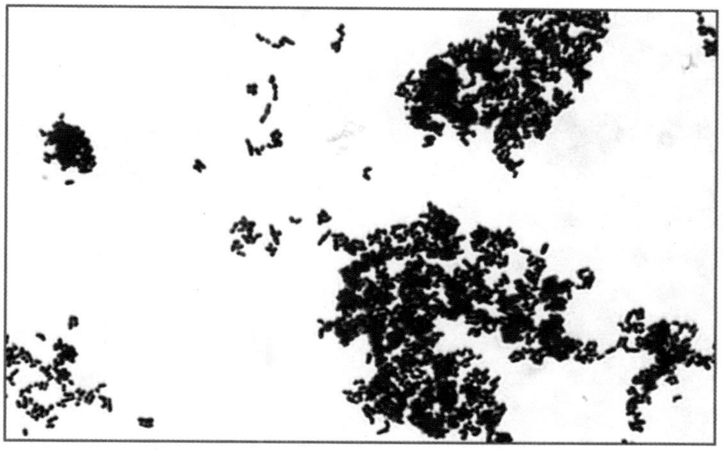

FIGURE 3-26 Bacterial Pleomorphism (Gram stain, X1000)
Some organisms grow in a variety of shapes and are said to be **pleomorphic**. Notice that the rods of *Corynebacterium xerosis* range from almost spherical to many times longer than wide. This organism normally inhabits skin and mucous membranes and may be an opportunistic pathogen in compromised patients.

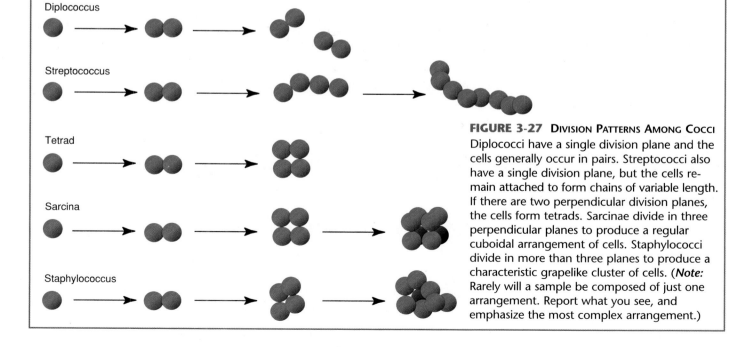

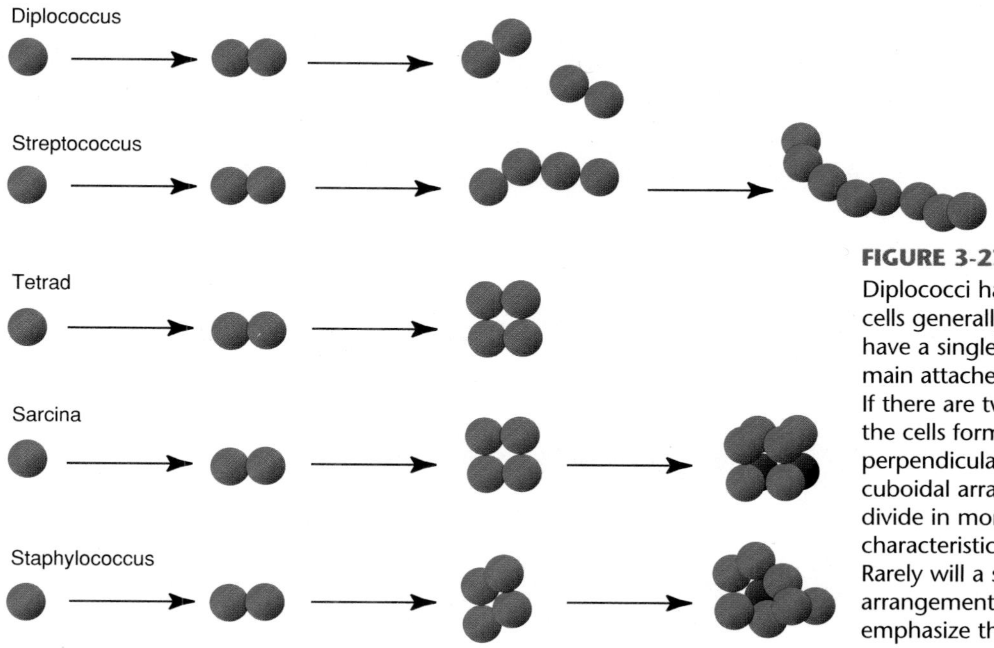

FIGURE 3-27 Division Patterns Among Cocci
Diplococci have a single division plane and the cells generally occur in pairs. Streptococci also have a single division plane, but the cells remain attached to form chains of variable length. If there are two perpendicular division planes, the cells form tetrads. Sarcinae divide in three perpendicular planes to produce a regular cuboidal arrangement of cells. Staphylococci divide in more than three planes to produce a characteristic grapelike cluster of cells. (***Note:*** Rarely will a sample be composed of just one arrangement. Report what you see, and emphasize the most complex arrangement.)

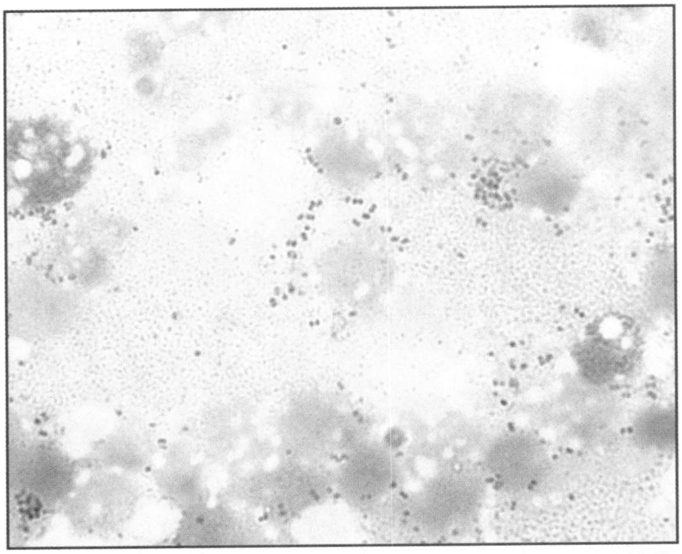

FIGURE 3-28 Diplococcus Arrangement (Gram Stain, X1000)
Neisseria gonorrhoeae, a diplococcus, causes gonorrhea in humans. Members of this genus produce diplococci with flattened adjacent sides.

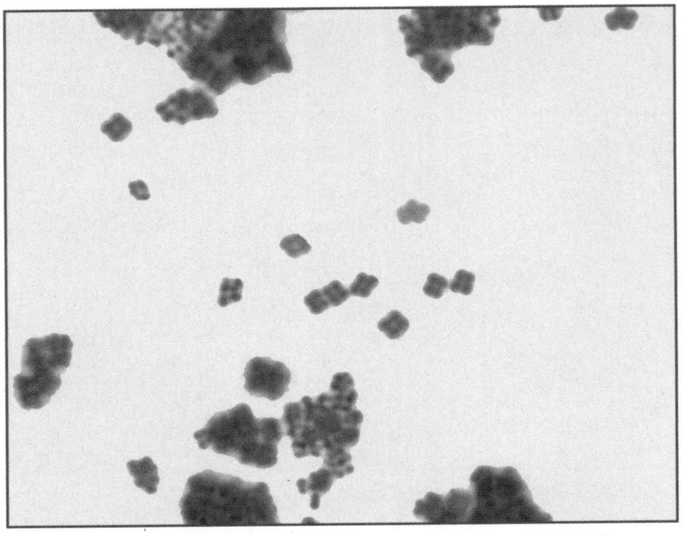

FIGURE 3-29 Tetrad Arrangement (Gram Stain, X1320)
Micrococcus roseus grows in squared packets of cells, even when they are bunched together. The normal habitat for *Micrococcus* species is the skin, but the ones here were obtained from culture.

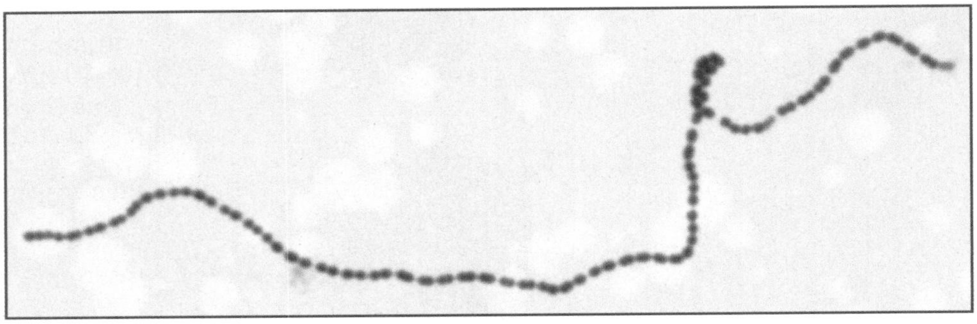

FIGURE 3-30 Streptococcus Arrangement (Gram Stain, X1000)
Enterococcus faecium is a streptococcus that inhabits the digestive tract of mammals. This specimen is from a broth culture (which enables the cells to form long chains) and was stained with crystal violet. Notice the slight elongation of these cells along the axis of the chain.

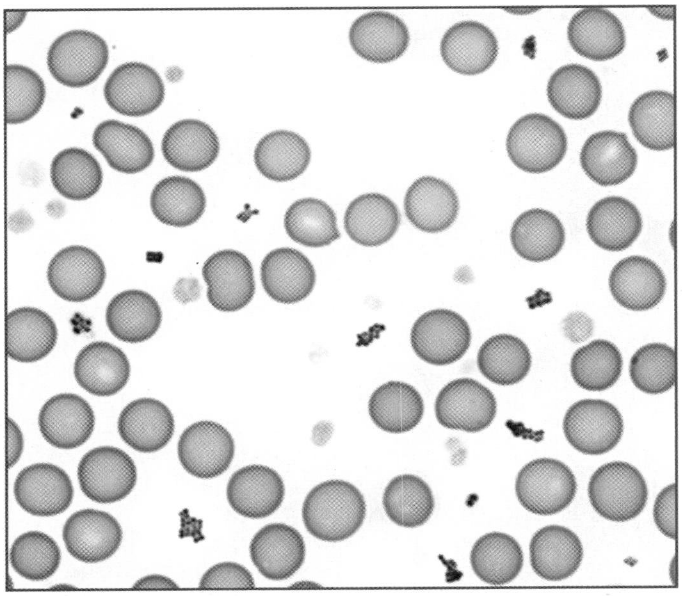

FIGURE 3-31 Staphylococcus Arrangement (X1000)
Staphylococcus aureus is shown in a blood smear. Note the staphylococci interspersed between the erythrocytes. *S. aureus* is a common opportunistic pathogen of humans.

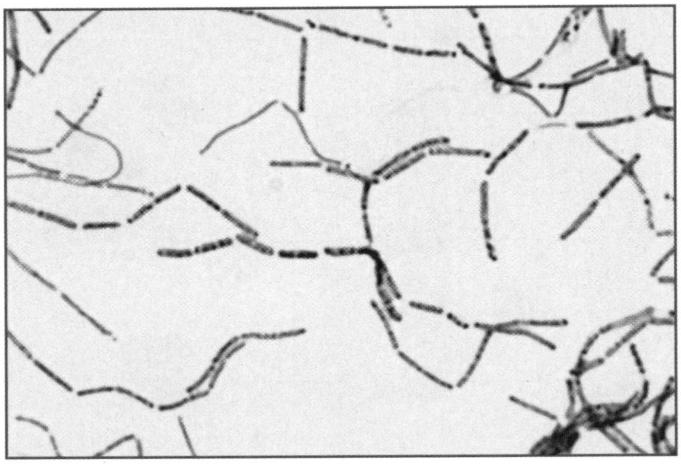

FIGURE 3-32 Streptobacillus Arrangement (Crystal Violet Stain, X1200)
Bacillus megaterium is a streptobacillus. These cells were obtained from culture.

Exercise 3-4

Simple Stains

Theory

Stains are solutions consisting of a solvent (usually water or ethanol) and a colored molecule (often a benzene derivative), the **chromogen.** The portion of the chromogen that gives it its color is the **chromophore.** A chromogen may have multiple chromophores, with each adding intensity to the color. The **auxochrome** is the charged portion of a chromogen and allows it to act as a dye through ionic or covalent bonds between the chromogen and the cell. **Basic stains**[1] (where the auxochrome becomes positively charged as a result of picking up a hydrogen ion or losing a hydroxide ion) are attracted to the negative charges on the surface of most bacterial cells. Thus, the cell becomes colored (Figure 3-33). Common basic stains include methylene blue, crystal violet and safranin. Examples of basic stains may be seen in Figures 3-19 through 3-26, 3-28 through 3-32, and 3-34.

Basic stains are applied to bacterial smears that have been **heat-fixed.** Heat-fixing kills the bacteria, makes them adhere to the slide, and coagulates cytoplasmic proteins to make them more visible. It also distorts the cells to some extent.

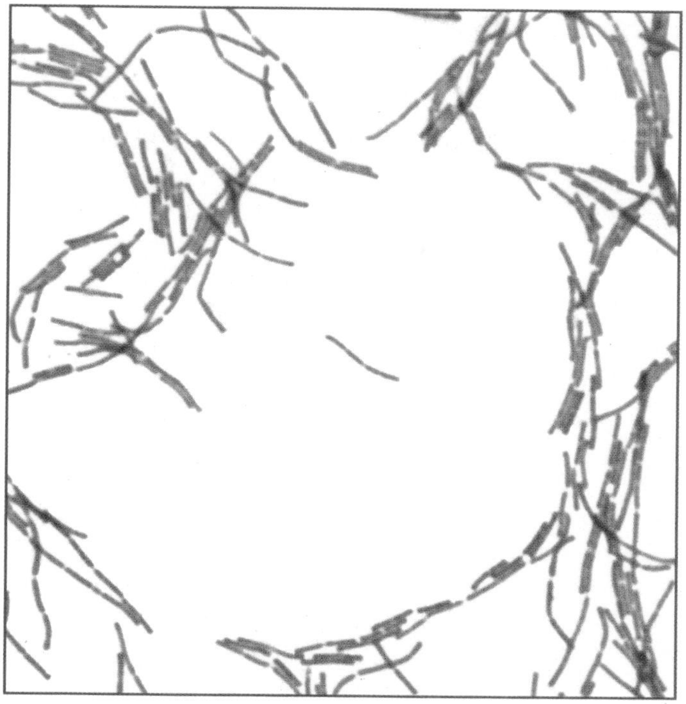

FIGURE 3-34 SAFRANIN DYE IN A SIMPLE STAIN (X1000)
This is a simple stain using safranin, a basic stain. Notice that the stain is associated with the cells and not the background. The organism is *Bacillus subtilis,* the type species for the genus *Bacillus,* grown in culture.

Application

Because cytoplasm is transparent, cells usually are stained with a colored dye to make them more visible under the microscope. Then cell morphology, size, and arrangement can be determined. In a medical laboratory, these are usually determined with a Gram stain (Exercise 3-6), but you will be using simple stains as an introduction to the staining.

In This Exercise...

Today you will learn how to prepare a bacterial smear (emulsion) and perform simple stains. Several different organisms will be supplied so you can begin to see the variety of cell morphologies and arrangements in the bacterial world. We suggest that you perform all the stains on one or two organisms (to get practice) and look at your lab partners' stains to see the variety of cell types. Be sure that you view all the available organisms.

[1] Notice that the term "basic" means "alkaline," not "elementary"; however, coincidentally, *basic* stains can be used for *simple* staining procedures.

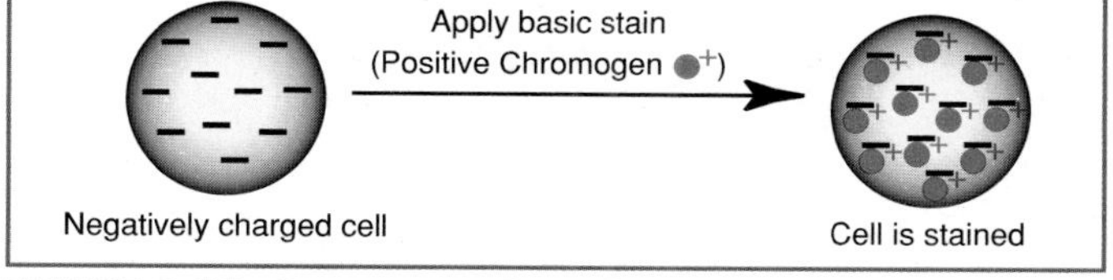

FIGURE 3-33 CHEMISTRY OF BASIC STAINS
Basic stains have a positively charged chromogen (C^+), which forms an ionic bond with the negatively charged bacterial cell, thus colorizing the cell.

Materials

Per Student

- clean glass microscope slides
- methylene blue stain
- safranin stain
- crystal violet stain
- disposable gloves
- staining tray
- staining screen
- bibulous paper tablet (or paper towels)
- slide holder
- recommended organisms:
  - *Bacillus cereus*
  - *Micrococcus luteus*
  - *Moraxella catarrhalis*
  - *Rhodospirillum rubrum*
  - *Staphylococcus epidermidis*
  - *Vibrio harveyi*

Procedure

1. A bacterial smear (emulsion) is made prior to most staining procedures. Follow the Procedural Diagram in Figure 3-35 to prepare bacterial smears of each organism. (*Note:* If working in groups, each student should perform the various stains on one or two organisms, then observe each other's slides to see the variety of cell shapes and arrangements.)

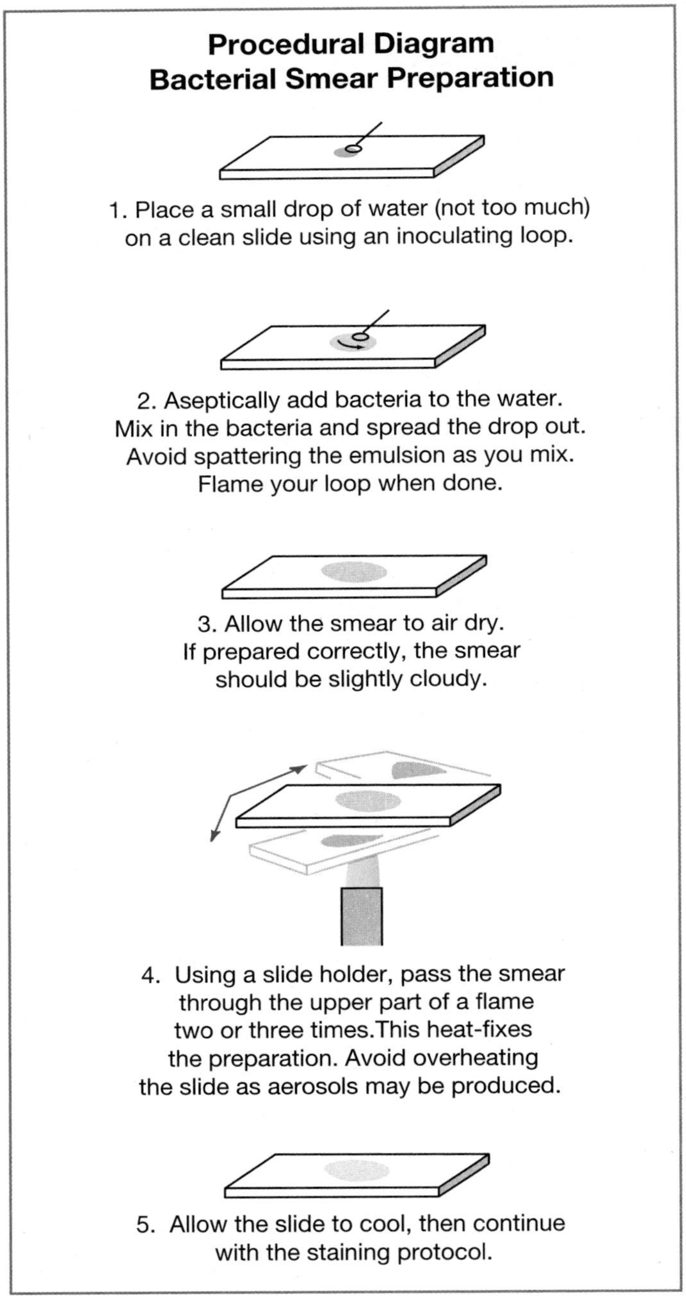

FIGURE 3-35 PROCEDURAL DIAGRAM: MAKING A BACTERIAL SMEAR (EMULSION)
Preparation of uniform bacterial smears will make consistent staining results easier to obtain. Heat-fixing the smear kills the bacteria, makes them adhere to the slide, and coagulates their protein for better staining. ***Caution:*** Avoid producing aerosols. Do not spatter the smear as you mix it, do not blow on or wave the slide to speed-up air-drying, and do not overheat when heat-fixing.

2. Heat-fix each smear as described in Figure 3-35.
3. Following the basic staining procedure illustrated in the Procedural Diagram in Figure 3-36, prepare two slides with each stain using the following times:

crystal violet:	stain for 30 to 60 seconds
safranin:	stain for up to 1 minute
methylene blue:	stain for 30 to 60 seconds

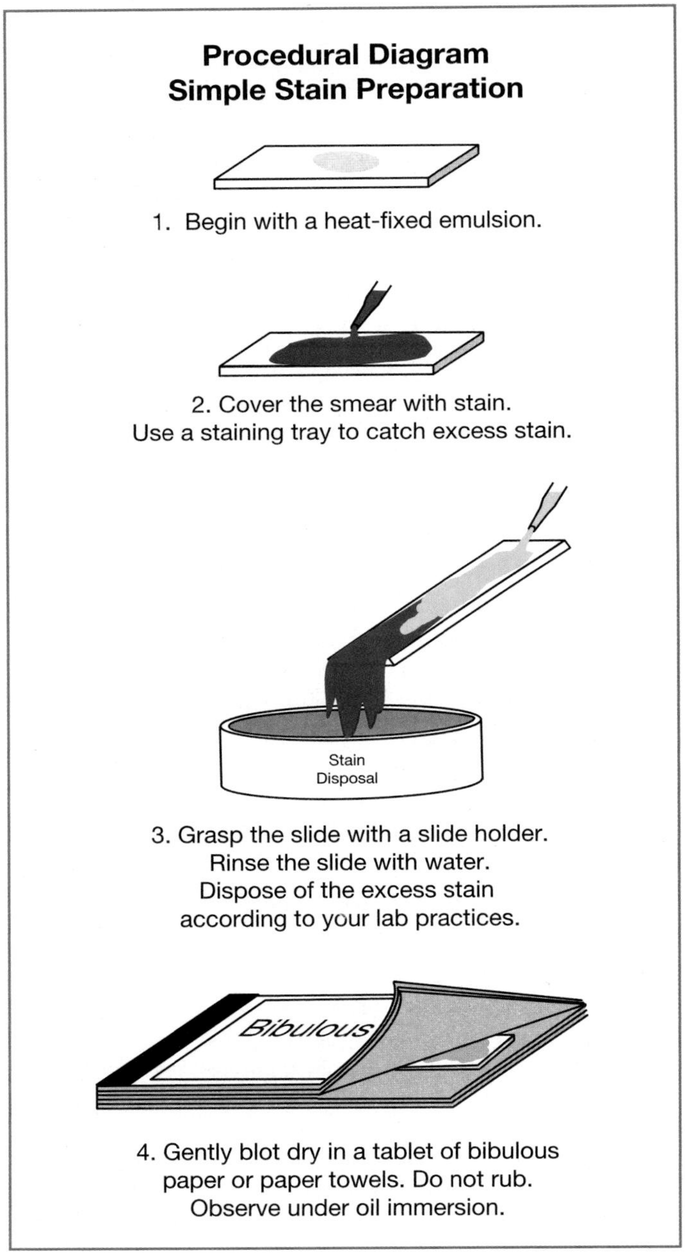

Procedural Diagram
2. Aseptically add bacteria to the water. Mix in the bacteria and spread the drop out. Avoid spattering the emulsion as you mix. Flame your loop when done.
3. Allow the smear to air dry. If prepared correctly, the smear should be slightly cloudy.
4. Using a slide holder, pass the smear through the upper part of a flame two or three times.This heat-fixes the preparation. Avoid overheating the slide as aerosols may be produced.
Simple Stain Preparation
1. Begin with a heat-fixed emulsion.
2. Cover the smear with stain. Use a staining tray to catch excess stain.
Stain Disposal
3. Grasp the slide with a slide holder. Rinse the slide with water. Dispose of the excess stain according to your lab practices.
Bibulous
4. Gently blot dry in a tablet of bibulous paper or paper towels. Do not rub. Observe under oil immersion.

FIGURE 3-36 PROCEDURAL DIAGRAM: SIMPLE STAIN
Staining times differ for each stain, but cell density of your smear also affects staining time. Strive for consistency in making your smears. ***Caution:*** Be sure to flame your loop after cell transfer and properly dispose of the slide when you are finished observing it.

Record your actual staining times in the table provided in the Data Sheet so you can adjust for overstaining or understaining.

4. Using the oil immersion lens, observe each slide. Record your observations of cell morphology, arrangement, and size in the chart provided on the Data Sheet.
5. Dispose of the slides and used stain according to your laboratory's policy.

References

Chapin, Kimberle. 1995. Chapter 4 in *Manual of Clinical Microbiology,* 6th ed., edited by Patrick R. Murray, Ellen Jo Baron, Michael A. Pfaller, Fred C. Tenover, and Robert H. Yolken. American Society for Microbiology, Washington, DC.

Chapin, Kimberle C., and Patrick R. Murray. 2003. Pages 257–259 in *Manual of Clinical Microbiology,* 8th ed., edited by Patrick R. Murray, Ellen Jo Baron, James H. Jorgensen, Michael A. Pfaller, and Robert H. Yolken. American Society for Microbiology, Washington, DC.

Forbes, Betty A., Daniel F. Sahm, and Alice. S. Weissfeld. 2002. Chapter 9 in *Bailey and Scott's Diagnostic Microbiology,* 11th ed. Mosby-Year Book, Inc. St. Louis.

Murray, R. G. E., Raymond N. Doetsch, and C. F. Robinow. 1994. Page 27 in *Methods for General and Molecular Bacteriology,* edited by Philipp Gerhardt, R. G. E. Murray, Willis A. Wood, and Noel R. Krieg. American Society for Microbiology, Washington, DC.

Norris, J. R., and Helen Swain. 1971. Chapter II in *Methods in Microbiology,* Vol. 5A, edited by J. R. Norris and D. W. Ribbons. Academic Press, Ltd., London.

Power, David A., and Peggy J. McCuen. 1988. Page 4 in *Manual of BBL™ Products and Laboratory Procedures,* 6th ed. Becton Dickinson Microbiology Systems, Cockeysville, MD.

Exercise 3-5

Negative Stains

Theory

The negative staining technique uses a dye solution in which the chromogen is acidic and carries a negative charge. (An acidic chromogen gives up a hydrogen ion, which leaves it with a negative charge.) The negative charge on the bacterial surface repels the negatively charged chromogen, so the cell remains unstained against a colored background (Figure 3-37). A specimen stained with the acidic stain nigrosin is shown in Figure 3-38.

Application

The negative staining technique is used to determine morphology and cellular arrangement in bacteria that are too delicate to withstand heat-fixing. A primary example is the spirochete *Treponema*, which is distorted by the heat-fixing of other staining techniques. Also, where determining the accurate size is crucial, a negative stain can be used because it produces minimal cell shrinkage.

In This Exercise…

Today you will perform negative stains on three different organisms. You also will have the opportunity to compare the size of *M. luteus* measured with the negative stain to its size as determined using a simple stain.

Materials

Per Student

- nigrosin stain or eosin stain
- clean glass microscope slides

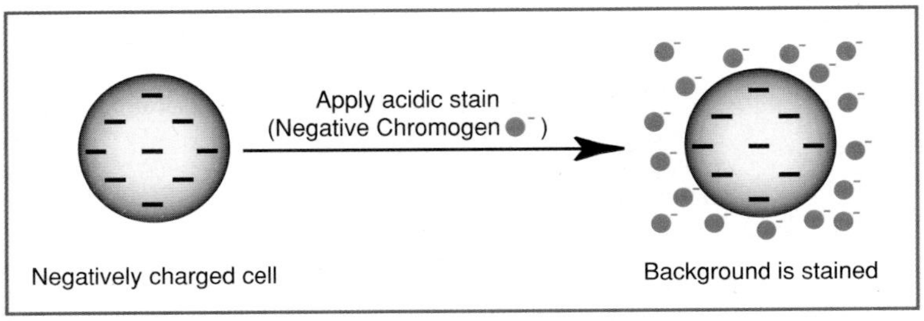

FIGURE 3-37 **Chemistry of Acidic Stains**
Acidic stains have a negatively charged chromophore (C⁻) that is repelled by negatively charged cells. Thus, the background is colored and the cell remains transparent.

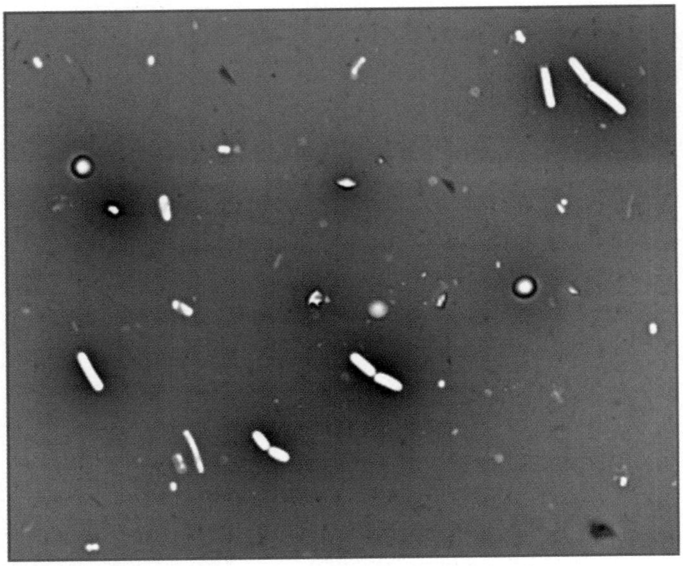

FIGURE 3-38 **A Nigrosin Negative Stain (X1000)**
Notice that the *Bacillus megaterium* cells are unstained against a dark background.

- disposable gloves
- recommended organisms:
  - *Micrococcus luteus*
  - *Bacillus megaterium*
  - *Rhodospirillum rubrum*

Procedure

1. Follow the Procedural Diagram in Figure 3-39 to prepare a negative stain of each organism.
2. Dispose of the spreader slide in a disinfectant jar or sharps container immediately after use.
3. Observe using the oil immersion lens. Record your observations in the chart on the Data Sheet.
4. Dispose of the specimen slide in a disinfectant jar or sharps container after use.

References

Claus, G. William. 1989. Chapter 5 *in Understanding Microbes—A Laboratory Textbook for Microbiology.* W. H. Freeman and Co., New York.

Murray, R. G. E., Raymond N. Doetsch, and C. F. Robinow. 1994. Page 27 in *Methods for General and Molecular Bacteriology,* edited by Philipp Gerhardt, R. G. E. Murray, Willis A. Wood, and Noel R. Krieg. American Society for Microbiology, Washington, DC.

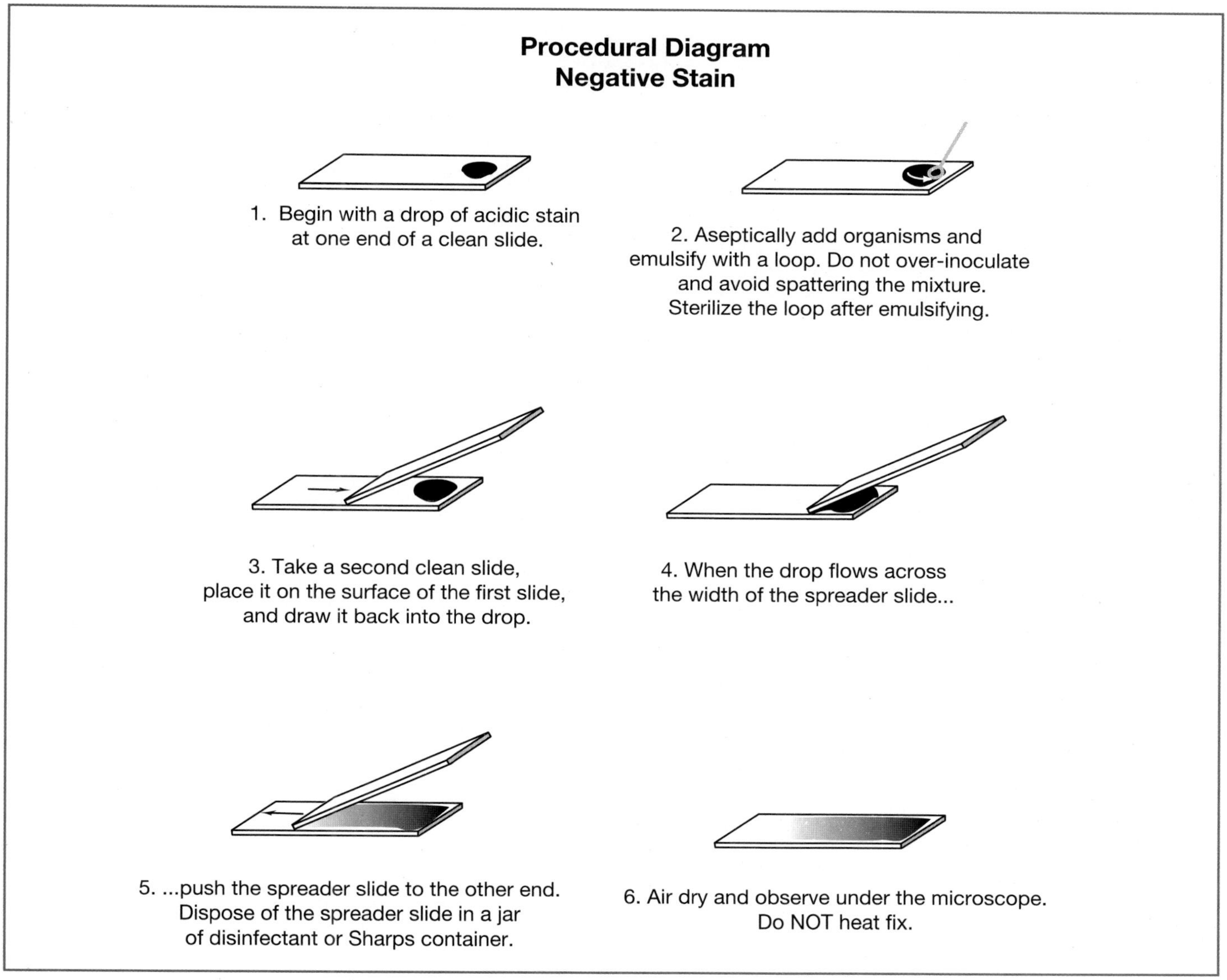
Procedural Diagram
Negative Stain
1. Begin with a drop of acidic stain at one end of a clean slide.
2. Aseptically add organisms and emulsify with a loop. Do not over-inoculate and avoid spattering the mixture. Sterilize the loop after emulsifying.
3. Take a second clean slide, place it on the surface of the first slide, and draw it back into the drop.
4. When the drop flows across the width of the spreader slide...
5. ...push the spreader slide to the other end. Dispose of the spreader slide in a jar of disinfectant or Sharps container.
6. Air dry and observe under the microscope. Do NOT heat fix.

FIGURE 3-39 PROCEDURAL DIAGRAM: NEGATIVE STAIN
Be sure to sterilize your loop after transfer, and to appropriately dispose of the spreader and specimen slides.

Differential Stains

Differential stains allow a microbiologist to detect differences between organisms or differences between parts of the same organism. In practice, these are used much more frequently than simple stains because they not only allow determination of cell size, morphology, and arrangement (as with a simple stain) but information about other features as well.

The Gram stain is the most commonly used differential stain in bacteriology. Other differential stains are used for organisms not distinguishable by the Gram stain and for those that have other important cellular attributes, such as acid-fastness, a capsule, spores, or flagella. With the exception of the acid-fast stain, these other stains sometimes are referred to as **structural stains.**

Exercise 3-6

Gram Stain

Theory The gram stain is a differential stain in which a **decolorization** step occurs between the application of two basic stains. The Gram stain has many variations, but they all work in basically the same way (Figure 3-40). The **primary stain** is crystal violet. Iodine is added as a **mordant** to enhance crystal violet staining by forming a **crystal violet-iodine complex**. Decolorization follows and is the most critical step in the procedure. Gram-negative cells are decolorized by the solution (of variable composition—generally alcohol or acetone) whereas Gram-positive cells are not. Gram-negative cells can thus be colorized by the **counterstain** safranin. Upon successful completion of a Gram stain, Gram-positive cells appear purple and Gram-negative cells appear reddish-pink (Figure 3-41).

Electron microscopy and other evidence indicate that the ability to resist decolorization or not is based on the different wall constructions of Gram-positive and Gram-negative cells. Gram-negative cell walls have a higher lipid content (because of the outer membrane) and a thinner peptidoglycan layer than Gram-positive cell walls. The alcohol/acetone in the decolorizer extracts the lipid, making the Gram-negative wall more porous and incapable of retaining the crystal violet–iodine complex, thereby decolorizing it. The thicker peptidoglycan and greater degree of cross-linking (because of teichoic acids) trap the crystal violet–iodine complex more effectively, making the Gram-positive wall less susceptible to decolorization.

Although some organisms give Gram-variable results, most variable results are a consequence of poor technique. The decolorization step is the most crucial and most likely source of Gram stain inconsistency. It is possible to **over-decolorize** by leaving the alcohol on too long and get reddish Gram-*positive* cells. It also is possible to **under-decolorize** and produce purple Gram-*negative* cells. Neither of these situations changes the actual Gram reaction for the organism being stained. Rather, these are false results because of poor technique.

Until correct results are obtained consistently, it is recommended that control smears of Gram-positive and Gram-negative organisms be stained along with the organism in question (Figure 3-42). As an alternative control, a direct smear made from the gumline may be Gram-stained (Figure 3-43) with the expectation that

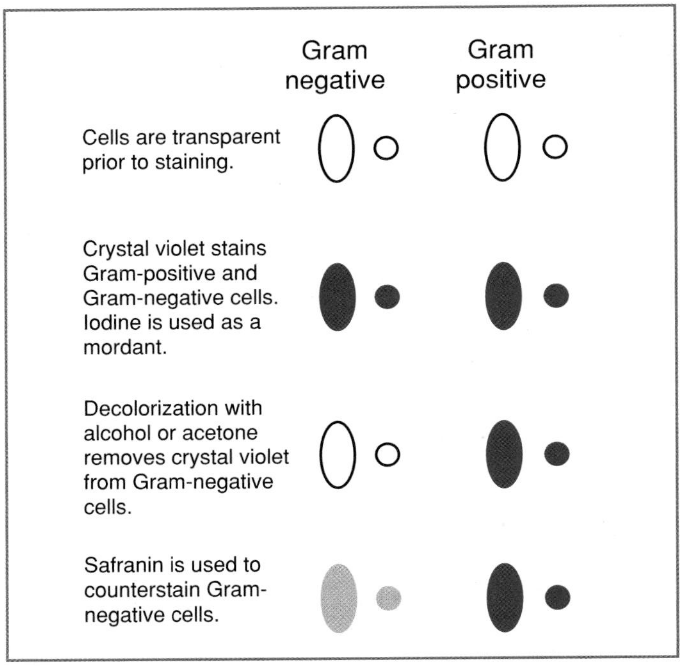

FIGURE 3-40 GRAM STAIN
After application of the primary stain (crystal violet), decolorization, and counterstaining with safranin, Gram-positive cells stain violet and Gram-negative cells stain pink/red. Notice that crystal violet and safranin are both basic stains, and that the decolorization step is what makes the Gram stain differential.

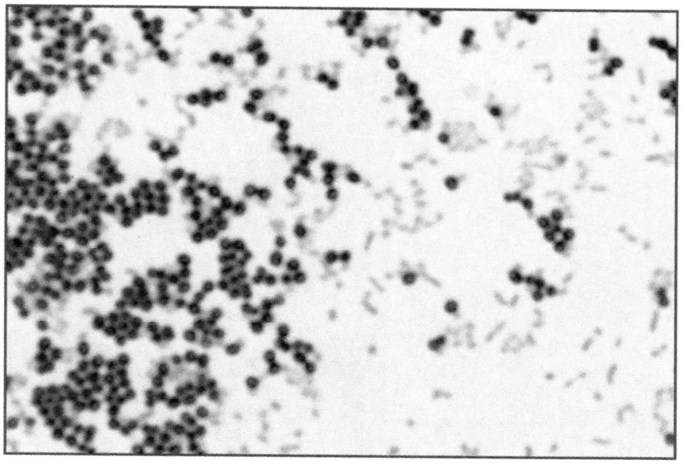

FIGURE 3-41 Gram Stain of *Staphylococcus* (+) and *Citrobacter* (–) (X1320)
Staphylococcus has a staphylococcal arrangement, whereas *Citrobacter* is a bacillus of varying lengths.

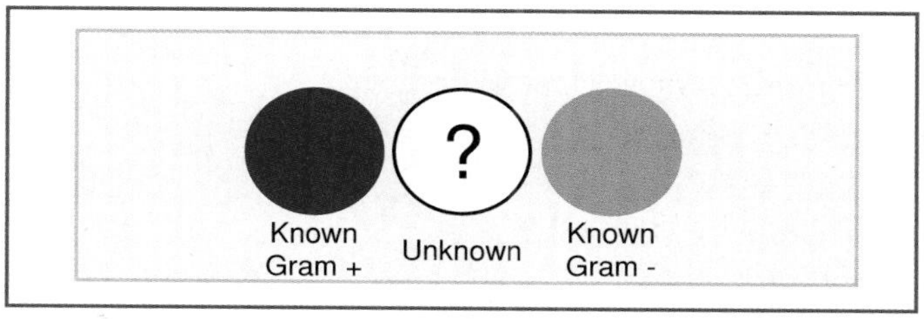

FIGURE 3-42 Positive Controls To Check Your Technique
Staining known Gram-positive and Gram-negative organisms on either side of your unknown organism act as positive controls for your technique. Try to make the emulsions as close to one another as possible. Spreading them out across the slide makes it difficult to stain and decolorize them equally.

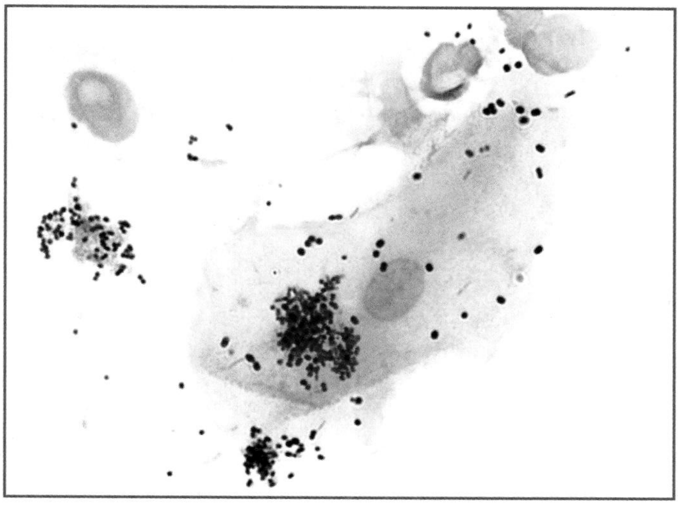

FIGURE 3-43 Direct Smear Positive Control (Gram Stain, X1000)
A direct smear made from the gumline may also be used as a Gram stain control. Expect numerous Gram-positive bacteria (especially cocci) and some Gram-negative cells, including your own epithelial cells. In this slide, Gram-positive cocci predominate, but a few Gram-negative cells are visible, including Gram-negative rods and a Gram-negative diplococcus on the surface of the epithelial cell.

both Gram-positive and Gram-negative organisms will be seen. Over-decolorized and under-decolorized gumline direct smears are shown for comparison (Figures 3-44 and 3-45). Positive controls also should be run when using new reagent batches.

Interpretation of Gram stains can be complicated by nonbacterial elements. For instance, stain crystals from an old or improperly made stain solution can disrupt the field (Figure 3-46) or stain precipitate may be mistakenly identified as bacteria (Figure 3-47).

Age of the culture also affects Gram stain consistency. Older Gram-positive cultures may lose their ability to resist decolorization and give an artifactual Gram-negative result. Cultures 24 hours old or less are best for this procedure.

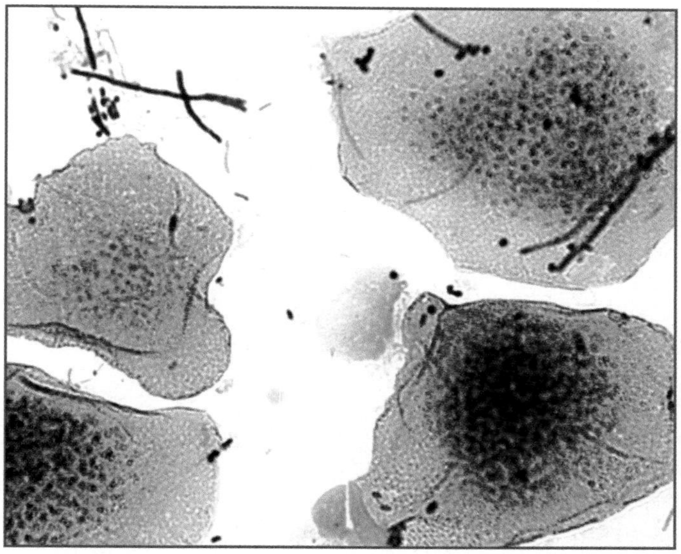

FIGURE 3-44 Under-Decolorized Gram Stain (X1000)
This is a direct smear from the gumline. Notice the purple patches of stain on the epithelial cells. Also notice the variable quality of this stain—the epithelial cell to the left of center is stained better than the others.

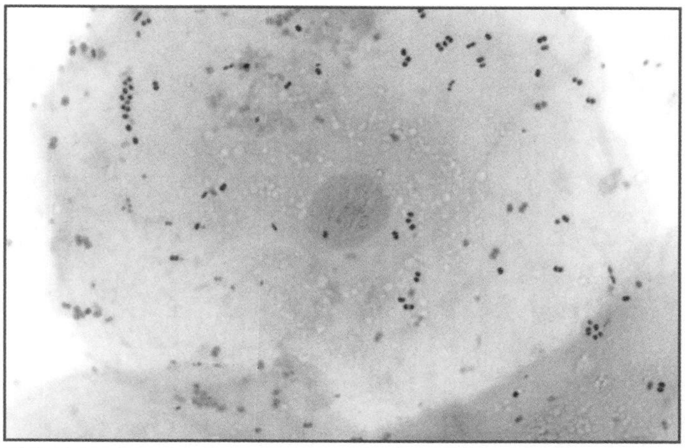

FIGURE 3-45 Over-Decolorized Gram Stain (X1000)
This also is a direct smear from the gumline. Notice the virtual absence of any purple cells.

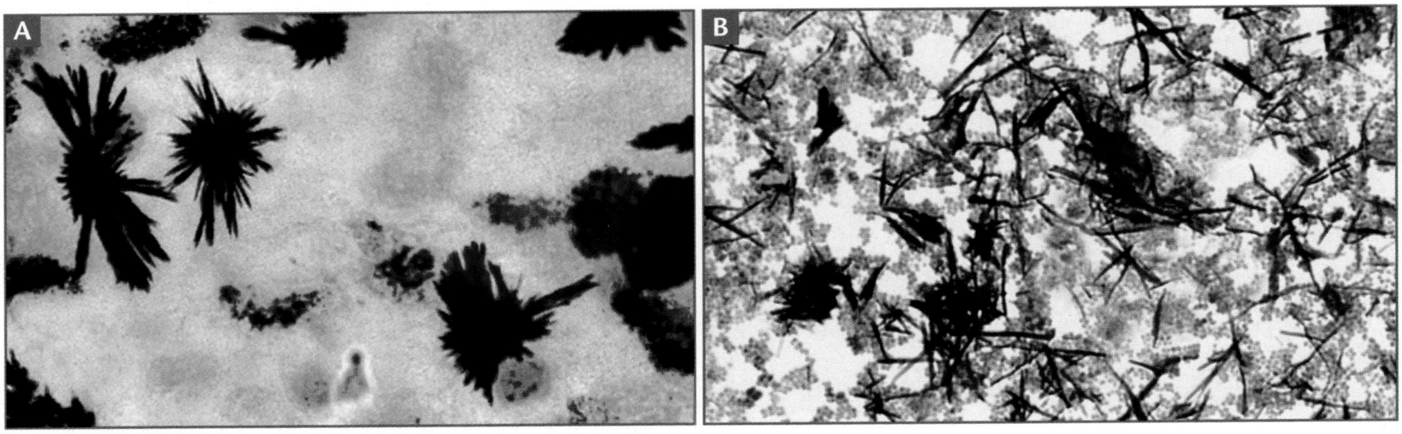

FIGURE 3-46 **Crystal Violet Crystals (Gram Stain, X1000)**
If the staining solution is not adequately filtered or is old, crystal violet crystals may appear. Although they are pleasing to the eye, they obstruct your view of the specimen. Crystals from two different Gram stains are shown here: (A) a gumline direct smear; (B) *Micrococcus roseus* grown in culture.

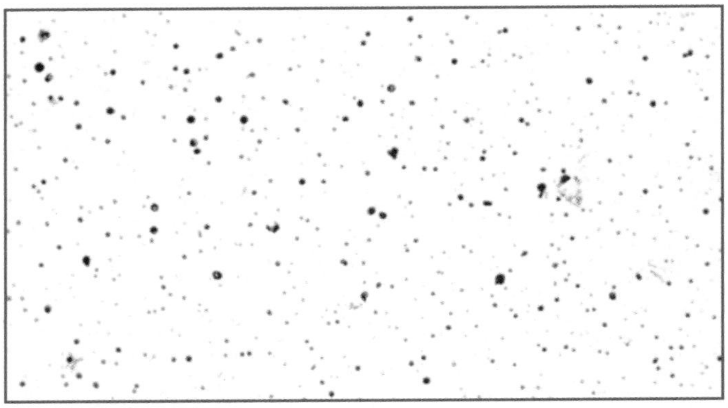

FIGURE 3-47 **Stain Precipitate (Gram Stain, X1000)**
If the slide is not rinsed thoroughly or the stain is allowed to dry on the slide, spots of stain precipitate may form and may be confused with bacterial cells. Their variability in size is a clue that they are not bacteria.

Application

The Gram stain, used to distinguish between Gram-positive and Gram-negative cells, is the most important and widely used microbiological differential stain. In addition to Gram reaction, this stain allows determination of cell morphology, size, and arrangement. It typically is the first differential test run on a specimen brought into the laboratory for identification. In some cases, a rapid, presumptive identification of the organism or elimination of a particular organism is possible.

In This Exercise...

The Gram stain is the single most important differential stain in bacteriology. Therefore, you will have to practice it and practice it some more to become proficient in its execution. The organisms should be used in the combinations given so you will have one Gram-positive and one Gram-negative on each slide.

Materials

Per Student

- clean glass microscope slides
- sterile toothpick
- Gram stain solutions (commercial kits are available)
    - Gram crystal violet
    - Gram iodine
    - 95% ethanol (or ethanol/acetone solution)
    - Gram safranin
- bibulous paper or paper towels
- disposable gloves
- staining tray
- staining screen
- slide holder
- recommended organisms (overnight cultures grown on agar slants):
    - *Staphylococcus epidermidis*
    - *Escherichia coli*
    - *Moraxella catarrhalis*
    - *Bacillus subtilis*

Procedure

1. Follow the procedure illustrated in Figure 3-35 to prepare and heat-fix smears of *Staphylococcus epidermidis* and *Escherichia coli* immediately next to one another on the same clean glass slide. (If you make the emulsions at opposite ends of the slide, you may find it difficult to stain and decolorize each equally.) Strive to prepare smears of uniform thickness, as thick smears risk being under-decolorized.
2. Repeat step 1 for *Moraxella catarrhalis* and *Bacillus subtilis* on a second slide.

3. Because Gram stains require much practice, you may want to prepare several slides of each combination and let them air-dry simultaneously. Then they'll be ready if you need them.
4. Use the sterile toothpick to obtain a sample from your teeth at the gumline. (Do not draw blood! What you want is easily removed from your gingival pockets.) Transfer the sample to a drop of water on a clean glass slide, air-dry, and heat-fix.
5. Follow the basic staining procedure illustrated in Figure 3-48. We recommend staining the pure cultures first. After your technique is consistent, stain the oral sample.

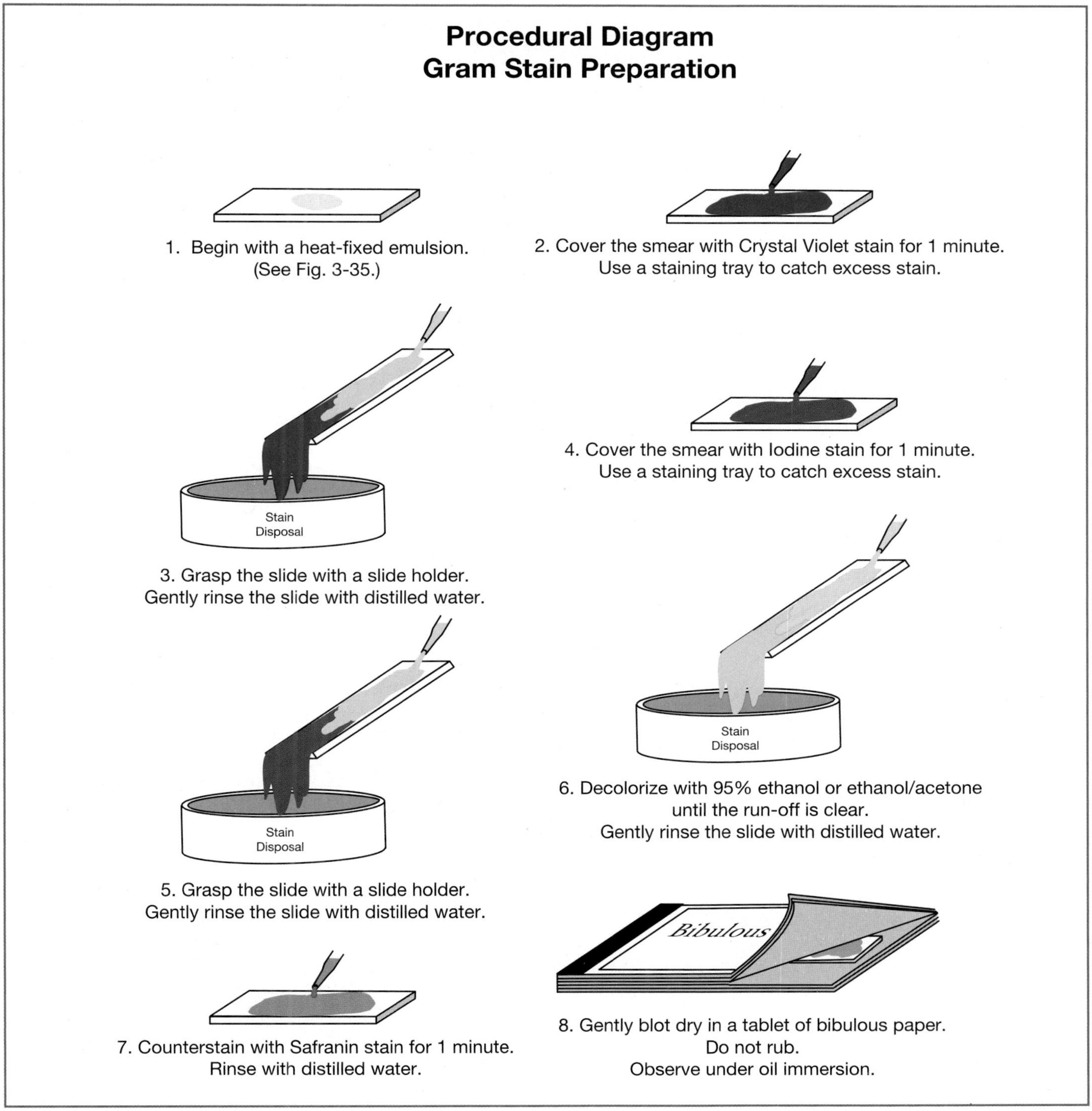
Procedural Diagram
Gram Stain Preparation
1. Begin with a heat-fixed emulsion.
(See Fig. 3-35.)
2. Cover the smear with Crystal Violet stain for 1 minute.
Use a staining tray to catch excess stain.
Stain
Disposal
3. Grasp the slide with a slide holder.
Gently rinse the slide with distilled water.
4. Cover the smear with Iodine stain for 1 minute.
Use a staining tray to catch excess stain.
Stain
Disposal
5. Grasp the slide with a slide holder.
Gently rinse the slide with distilled water.
Stain
Disposal
6. Decolorize with 95% ethanol or ethanol/acetone
until the run-off is clear.
Gently rinse the slide with distilled water.
7. Counterstain with Safranin stain for 1 minute.
Rinse with distilled water.
Bibulous
8. Gently blot dry in a tablet of bibulous paper.
Do not rub.
Observe under oil immersion.

FIGURE 3-48 PROCEDURAL DIAGRAM: GRAM STAIN

Pay careful attention to the staining times. If your preparations do not give "correct" results, the most likely source of error is in the decolorization step. Adjust its timing accordingly on subsequent stains.

6. Observe using the oil immersion lens. Record your observations of cell morphology and arrangement, dimensions, and Gram reactions in the chart provided on the Data Sheet.
7. Dispose of the specimen slides in a jar of disinfectant or a sharps container after use.

References

Chapin, Kimberle C., and Patrick R. Murray. 2003. Pages 258–260 in *Manual of Clinical Microbiology*, 8th ed., edited by Patrick R. Murray, Ellen Jo Baron, James H. Jorgensen, Michael A. Pfaller, and Robert H. Yolken. American Society for Microbiology, Washington, DC.

Forbes, Betty A., Daniel F. Sahm, and Alice. S. Weissfeld. 2002. Chapter 9 in *Bailey and Scott's Diagnostic Microbiology*, 11th ed. Mosby-Year Book, St. Louis.

Koneman, Elmer W., Stephen D. Allen, William M. Janda, Paul C. Schreckenberger, and Washington C. Winn, Jr. 1997. Chapter 14 in *Color Atlas and Textbook of Diagnostic Microbiology*, 5th Ed. J. B. Lippincott Co., Philadelphia.

Murray, R. G. E., Raymond N. Doetsch, and C. F. Robinow. 1994. Pages 31 and 32 in *Methods for General and Molecular Bacteriology*, edited by Philipp Gerhardt, R. G. E. Murray, Willis A. Wood, and Noel R. Krieg. American Society for Microbiology, Washington, DC.

Norris, J. R., and Helen Swain. 1971. Chapter II in *Methods in Microbiology*, Vol 5A, edited by J. R. Norris and D. W. Ribbons. Academic Press, Ltd., London.

Power, David A., and Peggy J. McCuen. 1988. Page 261 in *Manual of BBL™ Products and Laboratory Procedures*, 6th ed. Becton Dickinson Microbiology Systems, Cockeysville, MD.

Exercise 3-7

Acid-Fast Stains

Theory The presence of mycolic acids in the cell walls of acid-fast organisms is the cytological basis for the acid-fast differential stain. Mycolic acid is a waxy substance that gives acid-fast cells a higher affinity for the primary stain and resistance to decolorization by an acid alcohol solution. A variety of acid-fast staining procedures are employed, two of which are the Ziehl-Neelsen (ZN) method and the Kinyoun (K) method. These differ primarily in that the ZN method uses heat as part of the staining process, whereas the K method is a "cold" stain. In both protocols the bacterial smear may be prepared in a drop of serum to help the "slippery" acid-fast cells adhere to the slide. The two methods provide comparable results.

The waxy wall of acid-fast cells repels typical aqueous stains. (As a result, most acid-fast positive organisms are only weakly Gram-positive.) In the ZN method (Figure 3-49), the phenolic compound carbolfuchsin is used as the primary stain because it is lipid-soluble and penetrates the waxy cell wall. Staining by carbolfuchsin is further enhanced by steam-heating the preparation to drive the stain into the cell. Acid alcohol is used to decolorize nonacid-fast cells; acid-fast cells resist this decolorization. A counterstain, such as methylene blue, then is applied. Acid-fast cells are reddish-purple; nonacid-fast cells are blue (Figure 3-50).

The Kinyoun method (Figure 3-51) uses a slightly more lipid-soluble and concentrated carbolfuchsin as the primary stain. These properties allow the stain to penetrate the acid-fast walls without the use of heat but make this method slightly less sensitive than the ZN method. Decolorization with acid alcohol is followed by a contrasting counterstain, such as brilliant green (Figure 3-52) or methylene blue.

Application The acid-fast stain is a differential stain used to detect cells capable of retaining a primary stain when treated with an acid alcohol. It is an important differential stain used to identify bacteria in the genus *Mycobacterium,* some of which are pathogens (*e.g., M. leprae* and *M. tuberculosis*, causative agents of leprosy and tuberculosis, respectively—see below). Members of the actinomycete genus *Nocardia* (*N. brasiliensis* and *N. asteroides* are opportunistic pathogens) are partially acid-fast. Oocysts of coccidian parasites, such as *Cryptosporidium* and *Isospora,* are also acid-fast. Because so few organisms are acid-fast, the acid-fast stain is run only when infection by an acid-fast organism is suspected.

Acid-fast stains are useful in identifying **acid-fast bacilli (AFB)** and rapid, preliminary diagnosis of tuberculosis (with greater than 90% predictive value from sputum samples). It also can be performed on patient samples to track the progress of antibiotic therapy and determine the degree of contagiousness. A prescribed number of microscopic fields is examined and the number of AFB is determined and reported using a standard scoring system.

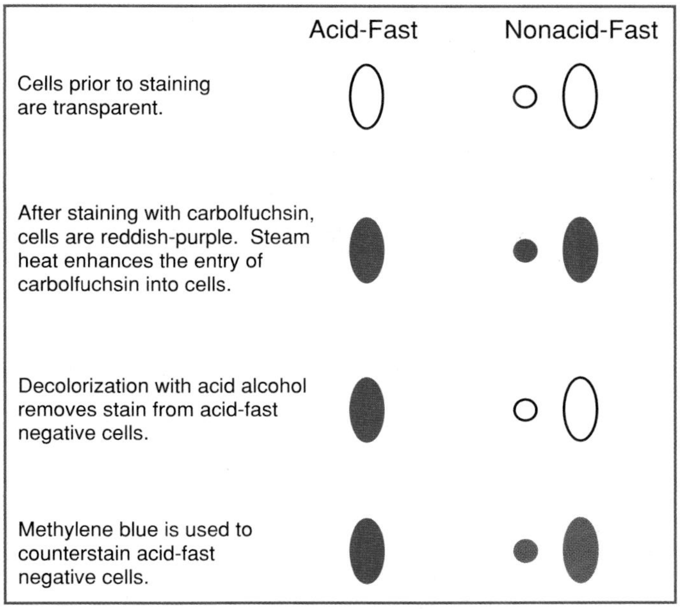
Acid-Fast
Nonacid-Fast
Cells prior to staining are transparent.
After staining with carbolfuchsin, cells are reddish-purple. Steam heat enhances the entry of carbolfuchsin into cells.
Decolorization with acid alcohol removes stain from acid-fast negative cells.
Methylene blue is used to counterstain acid-fast negative cells.

FIGURE 3-49 ZIEHL-NEELSEN ACID-FAST STAIN
Acid-fast cells stain reddish-purple; nonacid-fast cells stain blue or the color of the counterstain if a different one is used.

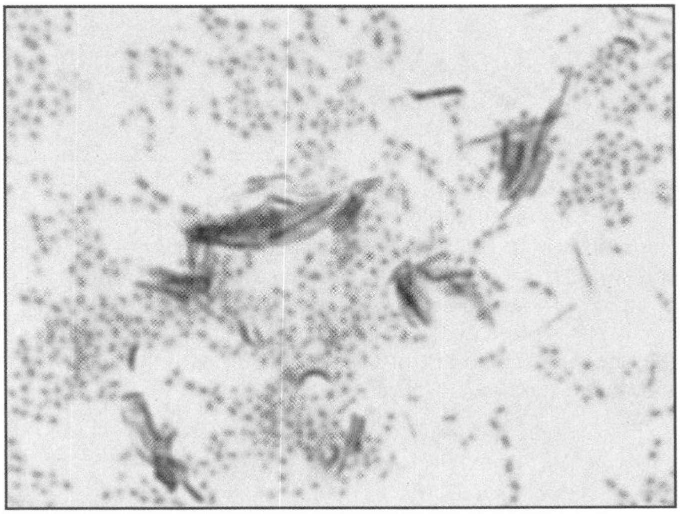

FIGURE 3-50 AN ACID-FAST STAIN USING THE ZN METHOD (X1500)
In this acid-fast stain, notice how the acid-fast positive cells of *Mycobacterium smegmatis* show characteristic clumping. *Enterococcus faecium* provides an example of an acid-fast negative organism. Both organisms were grown in culture. Compare this micrograph with the one in Figure 3-52.

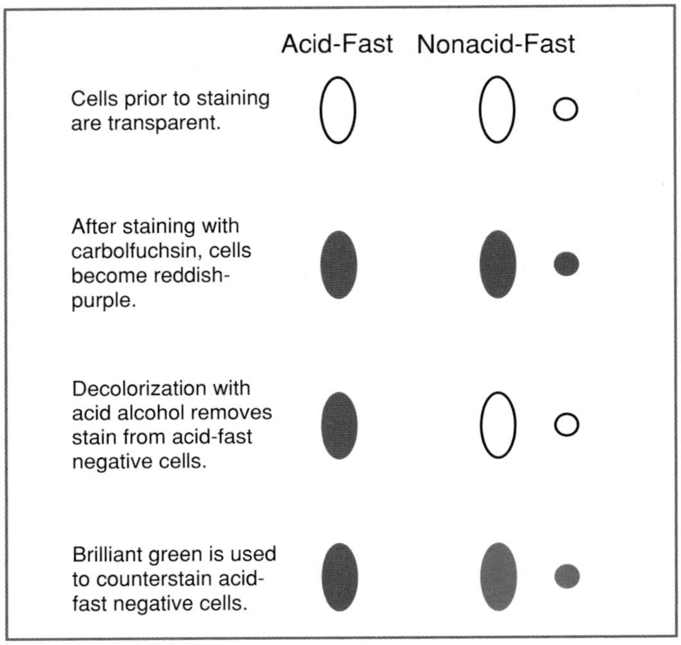
Acid-Fast
Nonacid-Fast
Cells prior to staining are transparent.
After staining with carbolfuchsin, cells become reddish-purple.
Decolorization with acid alcohol removes stain from acid-fast negative cells.
Brilliant green is used to counterstain acid-fast negative cells.

FIGURE 3-51 Kinyoun Acid-Fast Stain
Acid-fast cells stain reddish-purple; nonacid-fast cells stain green or the color of the counterstain if a different one is used.

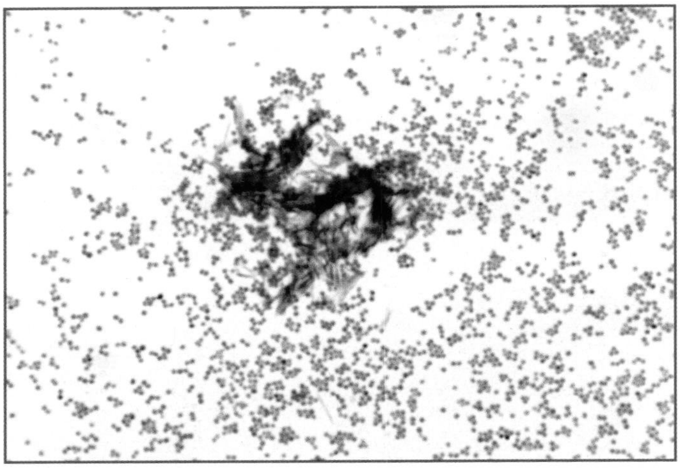

FIGURE 3-52 An Acid-Fast Stain Using the Kinyoun Method (X1000)
This is an acid-fast stain of *Mycobacterium smegmatis* (+) and *Staphylococcus epidermidis* (−) from cultures. Again, notice the clumping of the acid-fast organisms. Brilliant green dye commonly stains more of a gray color, but it still contrasts with the carbolfuchsin of acid-fast positive cells. Compare this micrograph with the one in Figure 3-50.

In This Exercise...

Today you will perform an acid-fast stain designed primarily to identify members of the genus *Mycobacterium*. Because you know ahead of time which organisms should give a positive result and which should give a negative result, it is okay to mix them into a single emulsion.

Materials

Per Student

- clean glass microscope slides
- staining tray
- staining screen
- bibulous paper or paper towel
- slide holder
- Ziehl-Neelsen Stains (complete kits are commercially available)
  - methylene blue stain
  - Ziehl's carbolfuchsin stain
  - acid alcohol (95% ethanol + 3% HCl)
- Kinyoun Stains (Complete kits are commercially available)
  - Kinyoun carbolfuchsin
  - acid alcohol (95% ethanol + 3% HCl)
  - brilliant green stain
- sheep serum
- heating apparatus (steam or hotplate)—for ZN only
- nonsterile Petri dish for transporting slides
- disposable gloves
- lab coat or apron
- eye goggles
- recommended organisms:
  - *Mycobacterium phlei* or *Mycobacterium smegmatis*
  - *Staphylococcus epidermidis*

Procedure: Ziehl-Neelsen (ZN) Method

1. Prepare a smear of each organism on a clean glass slide as illustrated in Figure 3-35, substituting a drop of sheep serum for the drop of water. Air-dry and then heat-fix the smears. ***Note:*** You may make two separate smears right next to one another on the slide or mix the two organisms in one smear.
2. Follow the staining protocol shown in the Procedural Diagram (Figure 3-53). Use a steaming apparatus (such as the one in Figure 3-54) to heat the slide. If the slide must be carried to and from the steaming apparatus, put it in a covered Petri dish.
3. Observe, using the oil immersion lens. Record your observations of cell morphology and arrangement, dimensions, and acid-fast reaction in the chart on the Data Sheet.
4. When finished, dispose of slides in a disinfectant jar or sharps container.

Procedure: Kinyoun Method

1. Prepare a smear of each organism on a clean glass slide as illustrated in Figure 3-35, substituting a drop of sheep serum for the drop of water. Air-dry and then heat-fix the smears. ***Note:*** You may make two separate smears right next to one another on the slide or mix the two organisms in one smear.
2. Follow the staining protocol shown in the Procedural Diagram (Figure 3-55).
3. Observe using the oil immersion lens. Record your observations of cell morphology and arrangement, dimensions, and acid-fast reaction on the Data Sheet.
4. When finished, dispose of slides in a disinfectant jar or a sharps container.

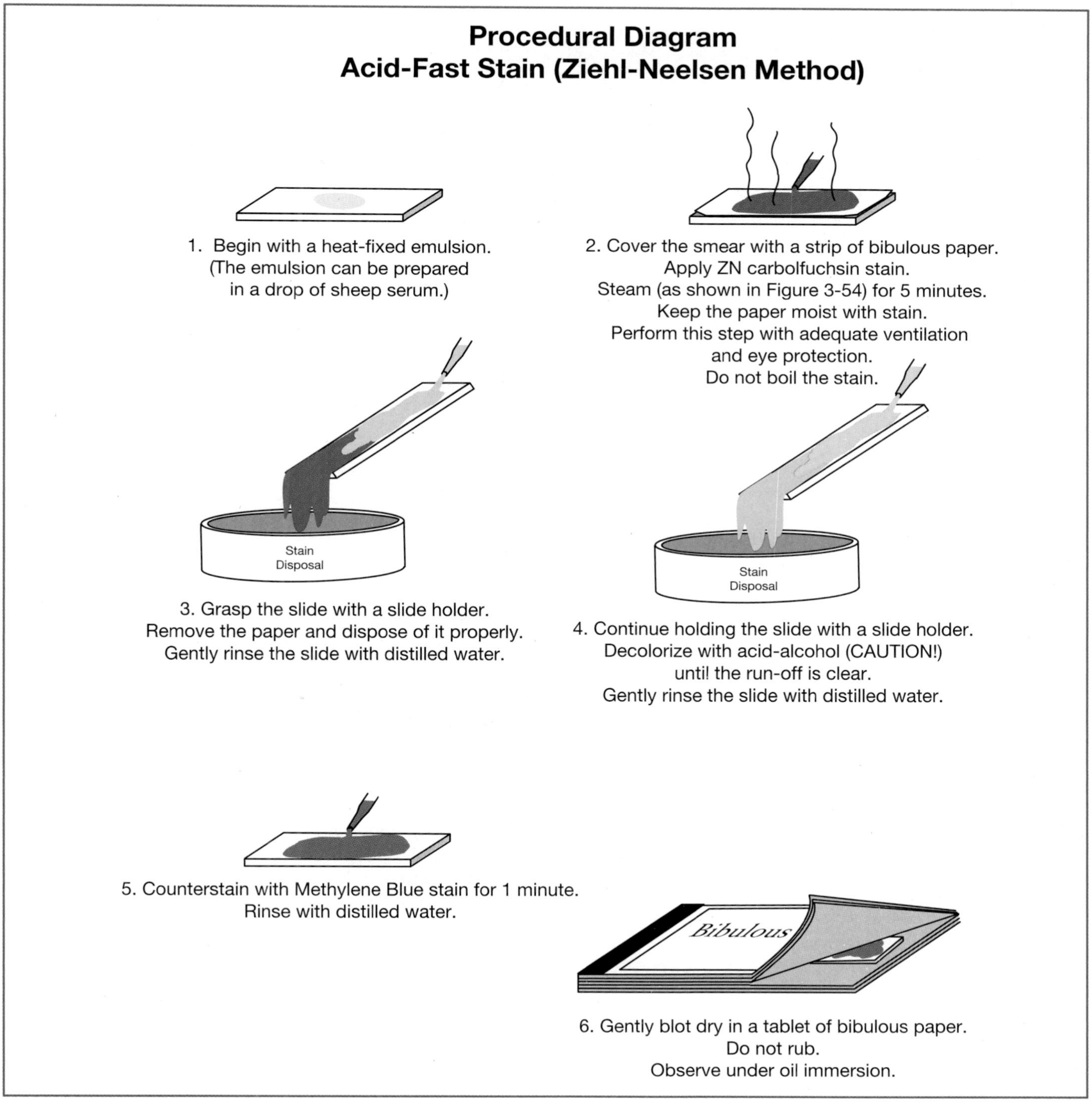

Procedural Diagram
Acid-Fast Stain (Ziehl-Neelsen Method)
1. Begin with a heat-fixed emulsion. (The emulsion can be prepared in a drop of sheep serum.)
2. Cover the smear with a strip of bibulous paper. Apply ZN carbolfuchsin stain. Steam (as shown in Figure 3-54) for 5 minutes. Keep the paper moist with stain. Perform this step with adequate ventilation and eye protection. Do not boil the stain.
Stain Disposal
3. Grasp the slide with a slide holder. Remove the paper and dispose of it properly. Gently rinse the slide with distilled water.
Stain Disposal
4. Continue holding the slide with a slide holder. Decolorize with acid-alcohol (CAUTION!) until the run-off is clear. Gently rinse the slide with distilled water.
5. Counterstain with Methylene Blue stain for 1 minute. Rinse with distilled water.
Bibulous
6. Gently blot dry in a tablet of bibulous paper. Do not rub. Observe under oil immersion.

FIGURE 3-53 Procedural Diagram: ZN Acid-Fast Stain
Be sure to perform this stain in a fume hood or well-ventilated area. Carry the slide to and from the steaming apparatus in a covered Petri dish.

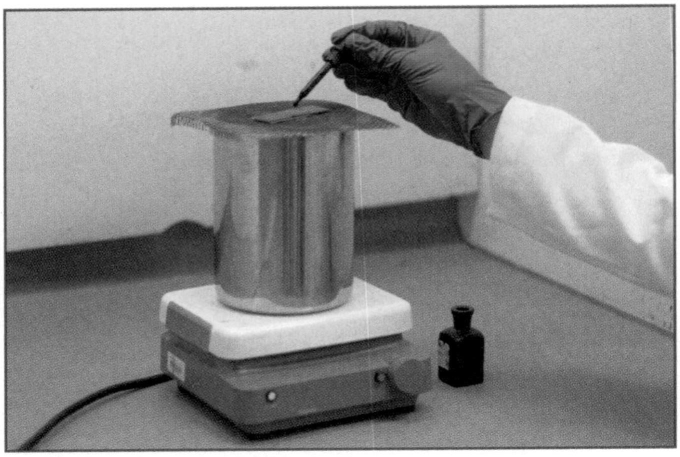

FIGURE 3-54 Steaming the Slide During the Ziehl-Neelsen Procedure
Carefully steam the slide to force the carbolfuchsin into acid-fast cells. Do not boil the slide or let it dry out. Keep it moist with stain for the entire 5 minutes of steaming. ***Caution:*** This should be performed in a well-ventilated area with hand, clothing, and eye protection.

References

Chapin, Kimberle C. and Patrick R. Murray. 2003. Pages 259–261 in *Manual of Clinical Microbiology,* 8th ed., edited by Patrick R. Murray, Ellen Jo Baron, James H. Jorgensen, Michael A. Pfaller, and Robert H. Yolken. American Society for Microbiology, Washington, DC.

Doetsch, Raymond N. and C. F. Robinow. 1994. Page 32 in *Methods for General and Molecular Bacteriology*, edited by Philipp Gerhardt, R. G. E. Murray, Willis A. Wood, and Noel R. Krieg. American Society for Microbiology, Washington, DC.

Forbes, Betty A., Daniel F. Sahm, and Alice. S. Weissfeld. 2002. Chapter 9 in *Bailey and Scott's Diagnostic Microbiology*, 11th ed. Mosby-Year Book, St. Louis.

Norris, J. R., and Helen Swain. 1971. Chapter II in *Methods in Microbiology,* Vol. 5A, edited by J. R. Norris and D. W. Ribbons. Academic Press, Ltd., London.

Power, David A., and Peggy J. McCuen. 1988. Page 5 in *Manual of BBL™ Products and Laboratory Procedures,* 6th ed. Becton Dickinson Microbiology Systems, Cockeysville, MD.

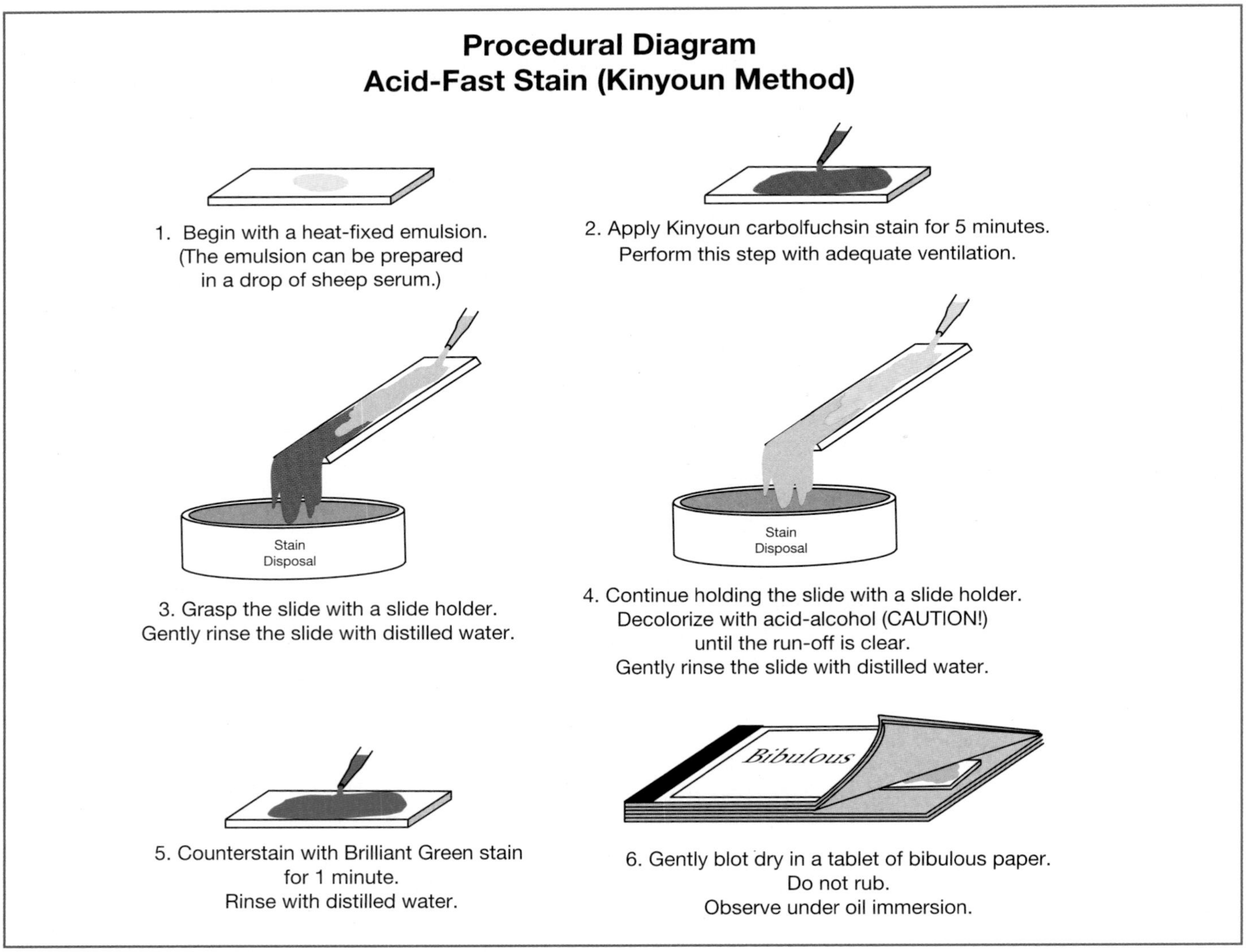

Procedural Diagram
Acid-Fast Stain (Kinyoun Method)
1. Begin with a heat-fixed emulsion. (The emulsion can be prepared in a drop of sheep serum.)
2. Apply Kinyoun carbolfuchsin stain for 5 minutes. Perform this step with adequate ventilation.
Stain Disposal
3. Grasp the slide with a slide holder. Gently rinse the slide with distilled water.
Stain Disposal
4. Continue holding the slide with a slide holder. Decolorize with acid-alcohol (CAUTION!) until the run-off is clear. Gently rinse the slide with distilled water.
5. Counterstain with Brilliant Green stain for 1 minute. Rinse with distilled water.
Bibulous
6. Gently blot dry in a tablet of bibulous paper. Do not rub. Observe under oil immersion.

FIGURE 3-55 Procedural Diagram: Kinyoun Acid-Fast Stain

Exercise 3-9

Endospore Stain

Theory An **endospore** is a dormant form of the bacterium that allows it to survive poor environmental conditions. Spores are resistant to heat and chemicals because of a tough outer covering made of the protein **keratin**. The keratin also resists staining, so extreme measures must be taken to stain the spore. In the Schaeffer-Fulton method (Figure 3-58), a primary stain of malachite green is forced into the spore by steaming the bacterial emulsion. Malachite green is water-soluble and has a low affinity for cellular material, so **vegetative cells** and **spore mother cells** can be decolorized with water and counterstained with safranin (Figure 3-59).

Spores may be located in the middle of the cell (**central**), at the end of the cell (**terminal**), or between the end and middle of the cell (**subterminal**). Spores also may be differentiated based on shape—either **spherical** or **elliptical** (**oval**)—and size relative to the cell (*i.e.*, whether they cause the cell to look swollen or not). These structural features are shown in Figures 3-60 and 3-61.

Application The spore stain is a differential stain used to detect the presence and location of spores in bacterial cells. Only a few genera produce spores. Among them are the genera *Bacillus* and *Clostridium*. Most members of *Bacillus* are soil, freshwater, or marine **saprophytes**, but a few are pathogens, such as *B. anthracis*, the causal agent of anthrax. Most members of *Clostridium* are soil or aquatic saprophytes or inhabitants of human intestines, but four pathogens are fairly well known: *C. tetani*, *C. botulinum*, *C. perfringens*, and *C. difficile,* which produce tetanus, botulism, gas gangrene, and pseudomembranous colitis, respectively.

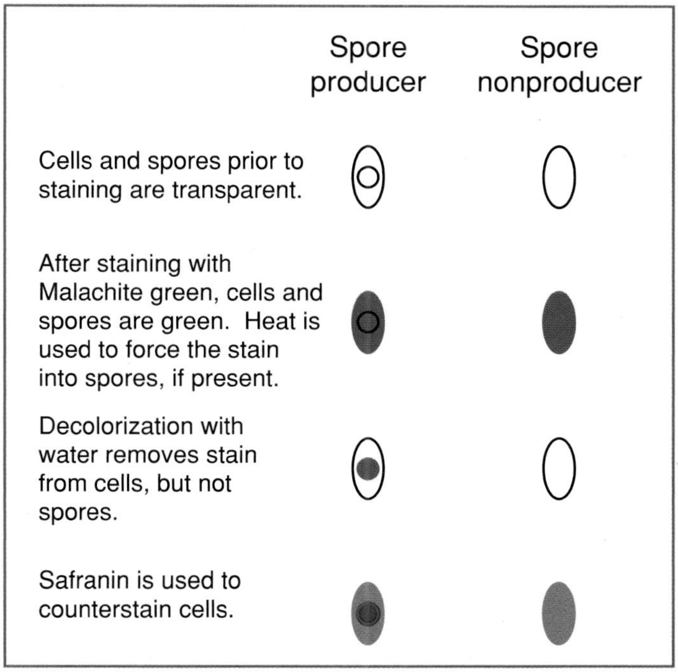

Spore producer
Spore nonproducer
Cells and spores prior to staining are transparent.
After staining with Malachite green, cells and spores are green. Heat is used to force the stain into spores, if present.
Decolorization with water removes stain from cells, but not spores.
Safranin is used to counterstain cells.

FIGURE 3-58 THE SCHAEFFER-FULTON SPORE STAIN
Upon completion, spores are green and vegetative and spore mother cells are red.

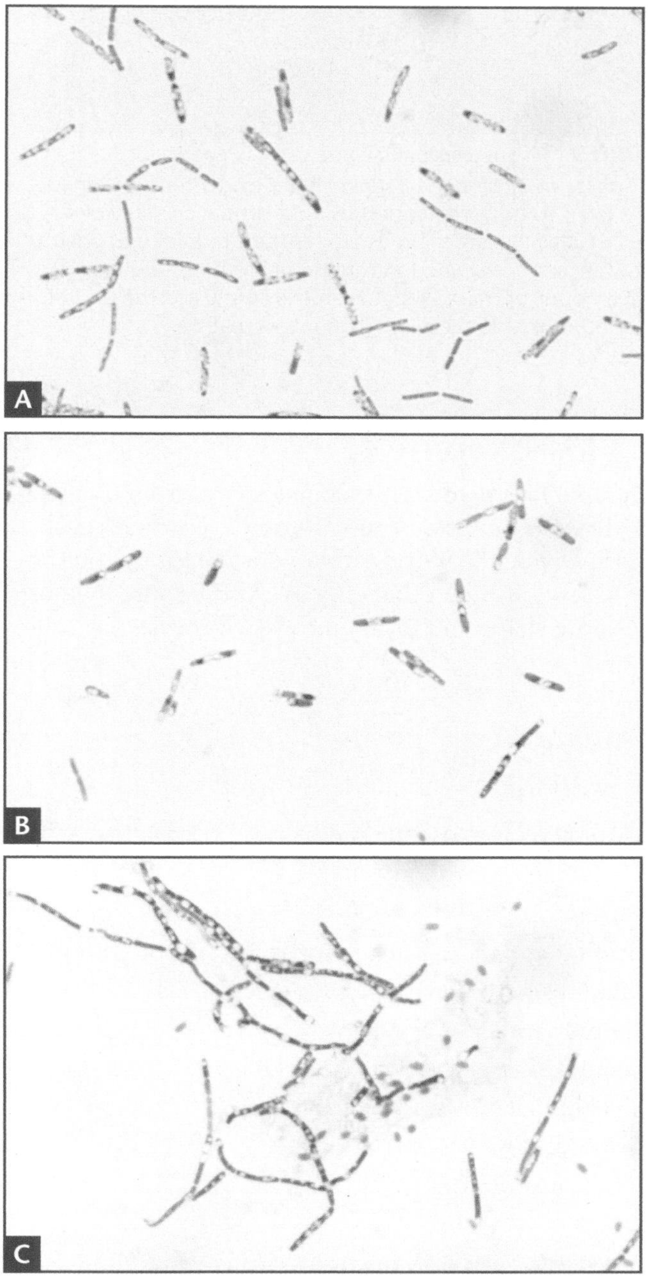

FIGURE 3-59 *BACILLUS CEREUS* CULTURE STAINED AT VARIOUS STAGES OF GROWTH BY THE SCHAEFFER-FULTON TECHNIQUE (X1200)
(A) Vegetative cells of *B. cereus* prior to sporulation. Notice the unstained regions within the cells that could be mistaken for spores if a spore stain had not been done. (B) Central elliptical spores of *B. cereus*. In this specimen, the spores are still within the mother cells. (C) In this specimen of *B. cereus,* the only spores visible have been released.

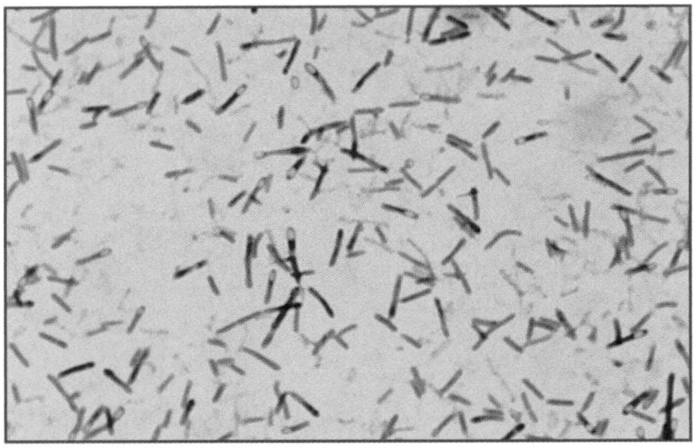

FIGURE 3-60 Subterminal Spores
The spores of *Clostridium botulinum* are evident as unstained, white ovals in this preparation using a simple crystal violet stain (X1000). Because spores are resistant to heat and chemicals, they are not stained in routine procedures. Notice the small amount of material between the spore and the end of the cell that makes these spores subterminal.

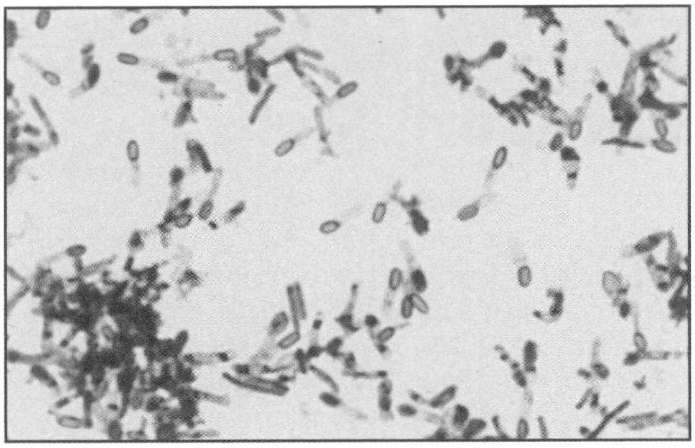

FIGURE 3-61 Elliptical Terminal Spores (X1200)
Clostridium tetani stained by a different spore stain protocol using carbolfuchsin. Notice the swollen ends of the cells because of the spores.

In This Exercise...

In the spore stain you will examine spores from different-aged *Bacillus* cultures. You will not have to stain each species. Divide the work within your lab group, but be sure to look at each others' slides. Also be sure to compare spore shape and location for each species.

Materials

Per Student

- clean glass microscope slides
- malachite green stain
- safranin stain
- heating apparatus (steam apparatus or hot plate)
- bibulous paper or paper towel
- staining tray
- staining screen
- slide holder
- disposable gloves
- lab coat or apron
- goggles
- nonsterile Petri dish for transporting slides
- recommended organisms (per student group):
  - 48 hour and 5 day Nutrient Agar slant pure cultures of *Bacillus cereus*
  - 48 hour and 5 day Nutrient Agar slant pure cultures of *Bacillus coagulans*
  - 48 hour and 5 day Nutrient Agar slant pure cultures of *Bacillus megaterium*
  - 48 hour and 5 day Nutrient Agar slant pure cultures of *Bacillus subtilis*

Procedure

1. Prepare and heat-fix a smear of each culture on the same slide as illustrated in Figure 3-35. Divide the work within your lab group. Minimally, each student should prepare a smear of the 48-hour and 5-day cultures of one species.
2. Follow the instructions in the Procedural Diagram in Figure 3-62. Use a steaming apparatus like the one shown in Figure 3-54. Be sure to have adequate ventilation and eye protection. If you must carry the slide to and from the steaming apparatus, put it in a covered Petri dish.
3. Observe, using the oil immersion lens. Record your observations of cell morphology and arrangement, cell dimensions, and spore presence, position and shape in the chart provided on the Data Sheet.
4. Dispose of specimen slides in a disinfectant jar or a sharps container.

References

Claus, G. William. 1989. Chapter 9 in *Understanding Microbes—A Laboratory Textbook for Microbiology*. W. H. Freeman and Co., New York.

Murray, R. G. E., Raymond N. Doetsch, and C. F. Robinow. 1994. Page 34 in *Methods for General and Molecular Bacteriology*, edited by Philipp Gerhardt, R. G. E. Murray, Willis A. Wood, and Noel R. Krieg. American Society for Microbiology, Washington, DC.

Procedural Diagram
Spore Stain (Schaeffer-Fulton Method)

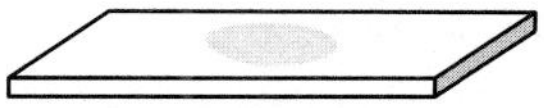

1. Begin with a heat-fixed emulsion.

2. Cover the smear with a strip of bibulous paper.
Apply Malachite Green stain.
Steam (as shown in Figure 3-54) for 5 minutes.
Keep the paper moist with stain.
Perform this step with adequate ventillation
and eye protection.

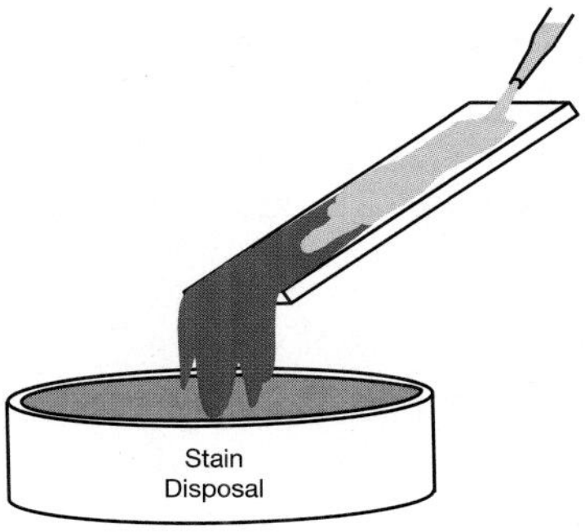

3. Grasp the slide with a slide holder.
Remove the paper and dispose of it properly.
Gently rinse the slide with water.

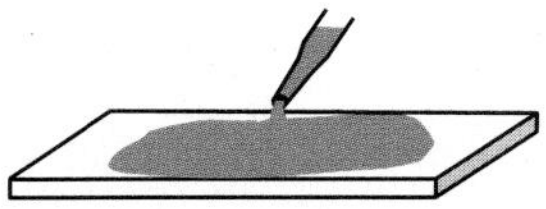

4. Counterstain with Safranin stain for 1 minute.
Rinse with water.

5. Gently blot dry in a tablet of bibulous paper.
Do not rub.
Observe under oil immersion.

FIGURE 3-62 Procedural Diagram: Schaeffer-Fulton Spore Stain
Steam the staining preparation; do not boil it. Be sure to perform this procedure with adequate ventilation (preferably a fume hood) and eye protection.

Exercise 3-10

Wet Mount and Hanging Drop Preparations

Theory

A wet mount preparation is made by placing the specimen in a drop of water on a microscope slide and covering it with a cover glass. Because no stain is used and most cells are transparent, viewing is best done with as little illumination as possible (Figure 3-63). Motility often can be observed at low or high dry magnification, but viewing must be done quickly because of drying of the preparation. As the water recedes, bacteria will appear to be herded across the field. This is not motility. You should look for independent motion of the cells.

A hanging drop preparation allows longer observation of the specimen because it doesn't dry out as quickly. A thin ring of petroleum jelly is applied to the four edges on one side of a cover glass. A drop of water then is placed in the center of the cover glass and living microbes are transferred into it. A depression microscope slide is carefully placed over the cover glass in such a way that the drop is received into the depression and is undisturbed. The petroleum jelly causes the cover glass to stick to the slide.

The preparation then may be picked up, inverted so the cover glass is on top, and placed under the microscope for examination. As with the wet mount, viewing is best done with as little illumination as possible. The petroleum jelly forms an airtight seal that slows drying of the drop, allowing a long period for observation of cell size, shape, binary fission, and motility.

If these techniques are done to determine motility, the observer must be careful to distinguish between true motility and the **Brownian motion** created by collisions with water molecules. In the latter, cells will appear to vibrate in place. With true motility, cells will exhibit independent movement over greater distances.

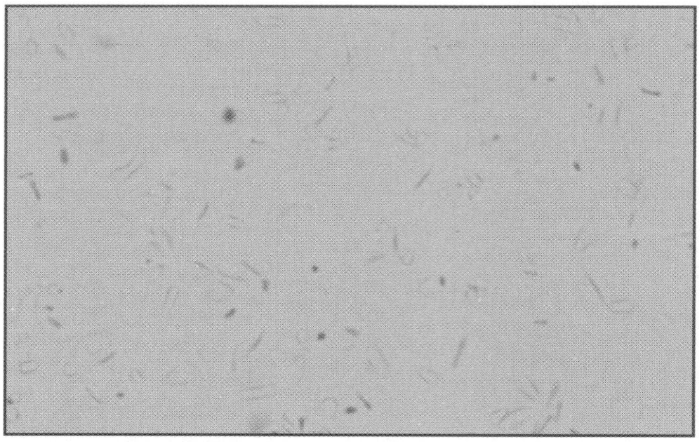

FIGURE 3-63 WET MOUNT (X1000)
Shown is an unstained wet mount preparation of a motile Gram-negative rod. Because of the thickness of the water in the wet mount, cells show up in many different focal planes and are mostly out of focus. To get the best possible image, adjust the condenser height and reduce the light intensity with the iris diaphragm.

Application

Most bacterial microscopic preparations result in death of the microorganisms as a result of heat-fixing and staining. Simple **wet mounts** and the **hanging drop technique** allow observation of living cells to determine motility. They also are used to see natural cell size, arrangement, and shape. All of these characteristics may be useful in identification of a microbe.

In This Exercise...

Today you will have the opportunity to view living bacterial cells swimming. The hanging drop preparation allows longer viewing, whereas the simple wet mount can be used to view motility, and it is the starting point for a flagella stain (Exercise 3-11).

Materials

Per Student

- depression slide and cover glass
- clean microscope slides and cover glasses
- petroleum jelly
- toothpick
- recommended organisms (overnight cultures grown on solid media):
  - *Aeromonas hydrophila* or *Aeromonas sobria*
  - *Pseudomonas aeruginosa*
  - *Staphylococcus epidermidis*

Procedure: Hanging Drop Preparation

1. Follow the procedure illustrated in Figure 3-64 for each specimen.
2. Observe under high dry or oil immersion and record your results in the chart on the Data Sheet.
3. When finished, remove the cover glass from the slide with an inoculating loop and soak both in a disinfectant jar for at least 15 minutes. Flame the loop. After soaking, rinse the slide with 95% ethanol to remove the petroleum jelly, then with water to remove the alcohol. Dry the slide for reuse.

Procedure: Wet Mount Preparation

1. Place a loopful of water on a clean glass slide.
2. Add bacteria to the drop. Don't over-inoculate. Flame the loop after transfer.
3. Gently lower a cover glass with your loop supporting one side over the drop of water. Avoid trapping air bubbles.
4. Observe under high dry or oil immersion and record your results in the chart on the Data Sheet.
5. Dispose of the slide and cover glass in a disinfectant jar or a sharps container when finished.

References

Iino, Tetsuo, and Masatoshi Enomoto. 1969. Chapter IV in *Methods in Microbiology,* Vol 1, edited by J. R. Norris and D. W. Ribbins. Academic Press, Ltd., London.

Murray, R. G. E., Raymond N. Doetsch, and C. F. Robinow. 1994. Page 26 in *Methods for General and Molecular Bacteriology*, edited by Philipp Gerhardt, R. G. E. Murray, Willis A. Wood, and Noel R. Krieg. American Society for Microbiology, Washington, DC.

Quesnel, Louis B. 1969. Chapter X in *Methods in Microbiology,* Vol 1, edited by J. R. Norris and D. W. Ribbins. Academic Press, Ltd., London.

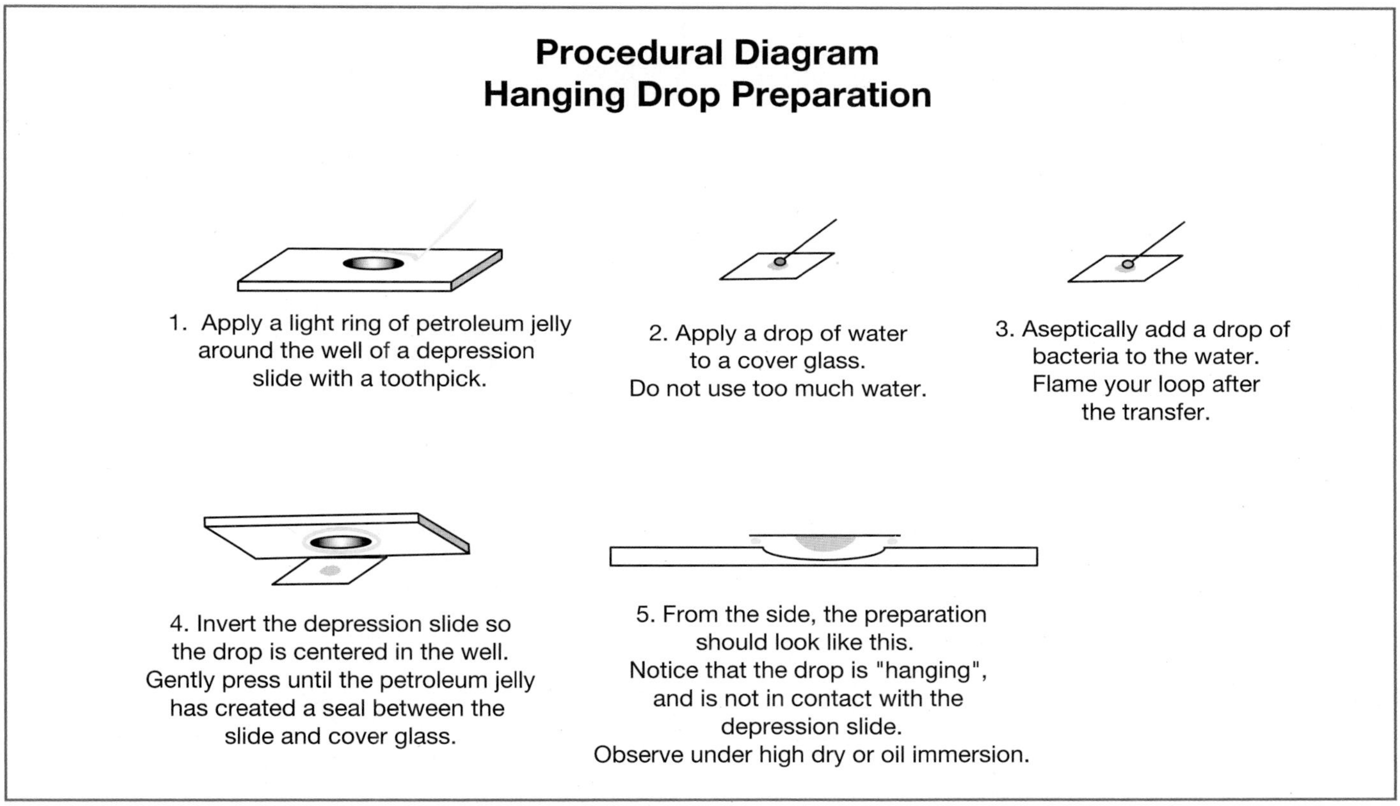

Procedural Diagram
Hanging Drop Preparation
1. Apply a light ring of petroleum jelly around the well of a depression slide with a toothpick.
2. Apply a drop of water to a cover glass. Do not use too much water.
3. Aseptically add a drop of bacteria to the water. Flame your loop after the transfer.
4. Invert the depression slide so the drop is centered in the well. Gently press until the petroleum jelly has created a seal between the slide and cover glass.
5. From the side, the preparation should look like this. Notice that the drop is "hanging", and is not in contact with the depression slide. Observe under high dry or oil immersion.

FIGURE 3-64 **PROCEDURAL DIAGRAM: HANGING DROP PREPARATION**
The hanging drop method is used for long-term observation of a living specimen.

Exercise 3-11

Flagella Stain

Theory

Bacterial flagella typically are too thin to be observed with the light microscope and ordinary stains. Various special flagella stains have been developed that use a **mordant** to assist in encrusting flagella with stain to a visible thickness. Most require experience and advanced techniques, and typically are not performed in beginning microbiology classes. The method provided in this exercise, though comparatively simple, still requires practice and a bit of good luck.

The number and arrangement of flagella may be observed with a flagella stain. A single flagellum is said to be **polar** and the cell has a **monotrichous** arrangement (Figure 3-65). Other arrangements (shown in Figures 3-66 through 3-68) include **amphitrichous**, with flagella at both ends of the cell; **lophotrichous**, with tufts of flagella at the end of the cell; and **peritrichous**, with flagella emerging from the entire cell surface.

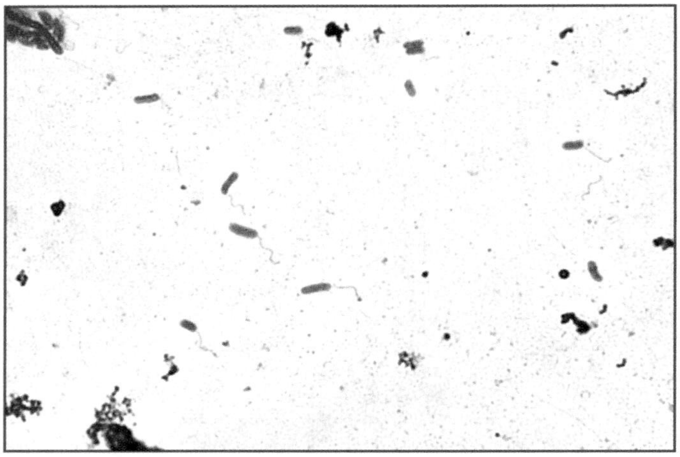

FIGURE 3-65 POLAR FLAGELLA (X1000)
Pseudomonas aeruginosa is often suggested as a positive control for flagella stains. Notice the single flagellum emerging from the ends of many (but not all) cells. This is a result of the fragile nature of flagella, which can be broken from the cells during slide preparation.

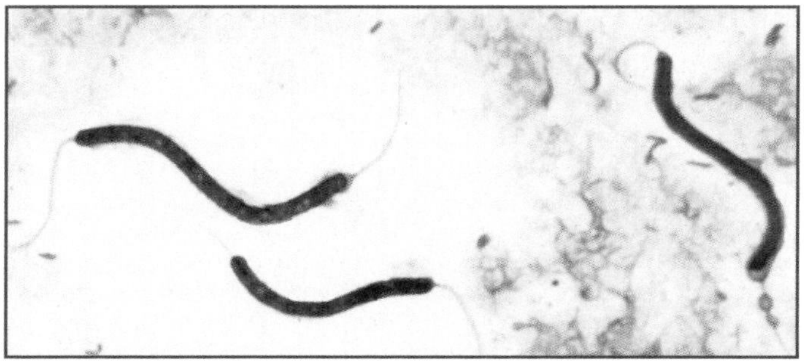

FIGURE 3-66 AMPHITRICHOUS FLAGELLA (X2200)
Spirillum volutans has a flagellum at each end.

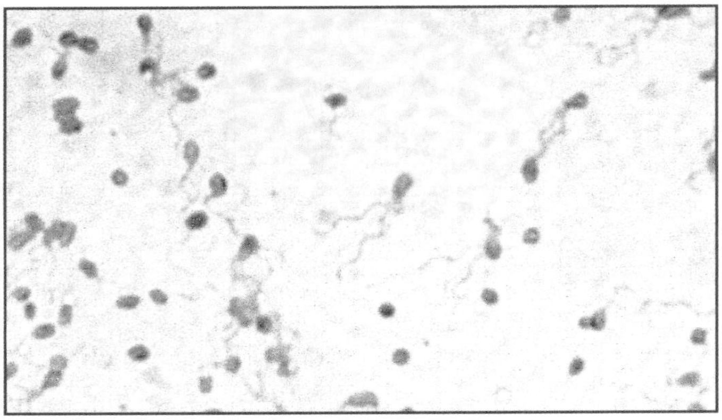

FIGURE 3-67 LOPHOTRICHOUS FLAGELLA (X2000)
Several flagella emerge from one end of this *Pseudomonas* species. Not all cells have flagella because they were too delicate to stay intact during the staining procedure.

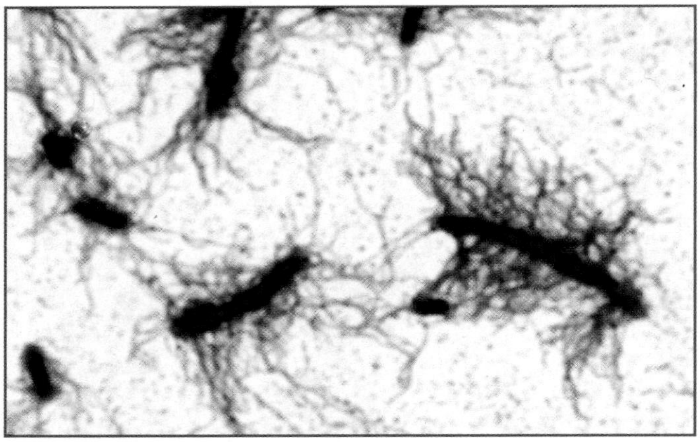

FIGURE 3-68 PERITRICHOUS FLAGELLA (X2640)
This *Proteus* species has flagella emerging from the entire cell surface. It is an intestinal inhabitant of humans and animals.

Application

The flagella stain allows direct observation of flagella. The presence and arrangement of flagella may be useful in identifying bacterial species.

In This Exercise...

You will begin this exercise with a simple wet mount preparation (Exercise 3-10). Once you have determined that the organism is motile, you will stain it to demonstrate its flagella. Be sure to note the flagellar arrangement.

Materials

- inoculating loop
- clean microscope slides and cover glasses
- hypodermic syringe with 0.2 μm filter
- Ryu flagella stain
- commercially prepared slides of motile organisms showing flagella.

- recommended organisms (grown on agar slants):
  - *Aeromonas hydrophila* or *Aeromonas sobria*
  - *Proteus vulgaris*
  - *Staphylococcus epidermidis*

Procedure

1. Use a separate slide for each preparation. Make sure the slide is clean, and handle it by its edges to prevent oil from your hands soiling the surface.
2. Prepare a wet mount of the organism as directed in Exercise 3-10. Because bacterial flagella are extremely fragile, handle the preparations gently to minimize the inevitable damage.
3. Observe the specimen under high dry power and note any motility. If the specimen is motile, continue with the procedure. If the specimen is nonmotile, observe the Brownian motion and then prepare a wet mount of a different organism.
4. If you see motility, allow the slide to dry for about 10–15 minutes. If practical to do so, leave the slide on the microscope stage so you will be able to find the cells again easily. Then apply the Ryu stain to the edge of the coverslip (Figures 3-69 and 3-70) using the syringe. Capillary action will draw the stain under. Be careful not to get stain on the microscope.
5. Allow the preparation to stain for a while. Keep it on the microscope. Check periodically for faint, stained threads emerging from the cells. Record your observations in the chart on the Data Sheet.
6. Dispose of your specimen slide in a disinfectant jar or sharps container after use.
7. Examine prepared slides of motile organisms and record your results on the Data Sheet.

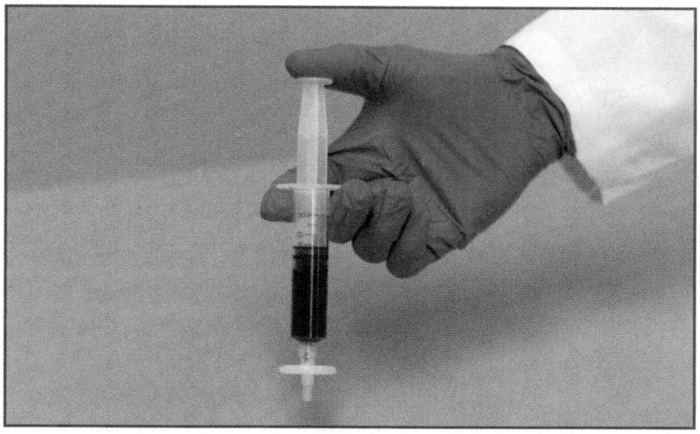

FIGURE 3-69 SYRINGE WITH RYU STAIN
Apply the stain using a syringe fitted with a 0.2 μm filter. For safety, we strongly recommend not using a needle with the syringe, but care must be taken in dispensing the stain not to deliver too much.

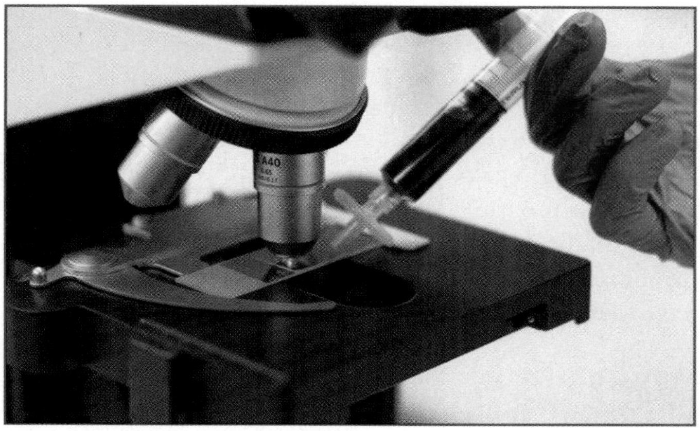

FIGURE 3-70 APPLYING THE STAIN
Carefully place a couple of drops of stain next to the coverslip. Capillary action will draw the stain under. After about 10 minutes of staining, begin examining the slide for flagella. Be careful not to spill stain on the microscope.

References

Heimbrook, Margaret E., Wen Lan L. Wang, and Gail Campbell (1989). *Staining Bacterial Flagella Easily*. J. Clin. Microbiol. 27:2612–2615.

Iino, Tetsuo and Masatoshi Enomoto. 1971. Chapter IV in *Methods in Microbiology*, Vol 5A, edited by J. R. Norris and D. W. Ribbins. Academic Press, Ltd., London.

Murray, R. G. E., Raymond N. Doetsch and C. F. Robinow. 1994. Page 35 in *Methods for General and Molecular Bacteriology*, edited by Philipp Gerhardt, R. G. E. Murray, Willis A. Wood, and Noel R. Krieg. American Society for Microbiology, Washington, DC.

Exercise 3-12

Morphological Unknown

Theory

In this exercise you will be given one pure bacterial culture selected from the organisms listed in Table 3-4. Your job will be to identify it using only the staining techniques covered in Section Three.

Your first task is to convert the information organized in Table 3-4 into flowchart form. A flowchart is simply a visual tool to illustrate the process of elimination that is the foundation of unknown determination. We have started it for you on the Data Sheet by giving you a few of the branches and listing appropriate organisms. You must complete it by adding necessary branches until you have shown a path to identify each of the organisms. The Procedure below contains a detailed explanation of the process.

Once you have designed your flowchart, you will run one stain at a time on your organism. As you match your staining results with those in the flowchart, you will follow a path to identification of your unknown.

A final differential stain will be performed as a **confirmatory test.** It will serve as further evidence that you have identified your unknown correctly.

Application

Schemes employing differential tests are the main strategy for microbial identification.

In This Exercise...

You will practice the stains you have learned to this point, as well as familiarize yourself with the standard process of elimination used in bacterial identification.

Materials

Per Student

- microscope
- clean microscope slides
- clean cover glasses
- Gram stain kit
- acid-fast stain kit
- capsule stain kit
- spore stain kit
- Bunsen burner
- striker
- inoculating loop
- Unknown organisms in numbered tubes (one per student). Fresh slant cultures of[1]
  - *Aeromonas hydrophila*
  - *Bacillus subtilis*
  - *Corynebacterium xerosis*
  - *Klebsiella pneumoniae*
  - *Lactococcus lactis*
  - *Kocuria rosea* (*Micrococcus roseus*)
  - *Mycobacterium smegmatis*
  - *Moraxella catarrhalis*
  - *Rhodospirillum rubrum*
  - *Shigella flexneri*
  - *Staphylococcus aureus*

  (***Note:*** Your instructor will choose organisms appropriate to your specific microbiology course and facilities.)
- Gram-positive and Gram-negative control organisms (one set per group)—18–24 hour Trypticase Soy Agar slant cultures in tubes labeled with Gram reaction.
- A set of organisms (listed above) to be used for positive controls (per class)
- sterile Trypticase Soy Broth tubes (one per student)
- sterile Trypticase Soy Agar slants (one per student)

Procedure

1. Using the information contained in Table 3-4, complete the flowchart on the Data Sheet. Each of the 11 organisms should occupy a solitary position at the end of a branch. Do not include stains in a branch that do not differentiate any organisms.
2. Obtain one unknown slant culture. Record its number in the space labeled "Unknown Number" on the Data Sheet.
3. Perform a Gram stain of the organism. The stain should be run with known Gram-positive and Gram-negative controls to verify your technique (see Figure 3-42). In addition to providing information on Gram reaction, the Gram stain will allow observation of cell size, shape, and arrangement. Enter all these and the date in the appropriate boxes of the chart provided. (***Note:*** Cell size for the unknown candidates is not given and should not be used to differentiate the species. To give you an idea of typical cell sizes, the cocci should be about 1 µm in diameter, and the rods 1 to 5 µm in length and about 1 µm in width.)
4. If you have cocci, or are unclear about cell arrangement, aseptically transfer your unknown to sterile trypticase soy broth and examine it again in 24–48

[1] Cultures should have abundant growth.

TABLE OF RESULTS

Organism	Gram Stain	Cell Morphology	Cell Arrangement (in broth)	Acid-Fast Stain	Motility (Wet Mount)	Capsule	Spore Stain (Run on cultures older than 48 hours)
Aeromonas hydrophila	–	rod	Single cells	–	+	–	–
Bacillus subtilis	+	rod	Usually single cells	–	+	–	+
Corynebacterium xerosis	+	rod	single cells or multiples in angular or palisade arrangement	–	–	–	–
Klebsiella pneumoniae	–	rod	Usually single cells, sometimes pairs or short chains	–	–	+	–
Kocuria rosea (Micrococcus roseus)	+	coccus (1–2.5µm)	pairs or tetrads	–	–	–	–
Lactococcus lactis	+	ovoid cocci (appearing stretched in the direction of the chain or pair)	pairs or chains	–	–	+	–
Mycobacterium smegmatis	weak +	rod	single cells or branched; sometimes in dense clusters	+	–	–	–
Moraxella catarrhalis	–	coccus	pairs with adjacent sides flattened	–	–	+	–
Rhodospirillum rubrum	–	spirillum or bent rod	single	–	+	–	–
Shigella flexneri	–	rod	single	–	–	–	–
Staphylococcus aureus	+	coccus	singles, pairs, tetrads or clusters	–	–	+	–

TABLE 3-4 Table of results for organisms used in the exercise

These results are typical for the species listed. Your specific strains may vary because of their genetics, their age, or the environment in which they are grown. It is important that you record *your* results in case strain variability leads to misidentification.

hours using a crystal violet or carbolfuchsin simple stain. Cell arrangement often is easier to interpret using cells grown in broth.

5. Based on the results of your Gram stain and cell morphology, follow the appropriate branches in the flowchart until you reach the list of possible organisms that matches your unknown.
6. Perform the next stain and determine to which new branch of the flowchart your unknown belongs. Record the result and date of this stain in the chart provided.
7. Repeat Step 6 until you eliminate all but one organism in the flowchart—this is your unknown! (If you are not doing all the stains on the day the unknowns were handed out, be sure to inoculate appropriate media so you will have fresh cultures to work with. If you need to do a spore stain, incubate your original culture for another 24–48 hours, then perform the stain.)
8. After you identify your unknown, perform one more stain to confirm your result. This **confirmatory test** should be one you did not perform previously and should be added to your flowchart. The result of this test should agree with the predicted result (as given in Table 3-4). If your confirmatory test result doesn't match the expected result, repeat any suspect stains and find the source of error. (***Note:*** The source of error may be your technique or bacterial strain variability. The results in Table 3-4 are typical for each organism, but some strains may vary. By rerunning suspect stain(s) *with controls*, you probably will be able to identify the source of error.)
9. Use a colored marker to highlight the path on the flowchart that leads to your unknown. Have your instructor check your work.

Selective Media

SECTION 4

Individual microbial species in a **mixed culture** must be isolated and cultivated as **pure cultures** before they can be further tested and reliably identified. The most common means of separating different organisms in a mixed culture is to streak for isolation, as demonstrated in Exercise 1-3. The **selective media** introduced in this section are designed to enhance that isolation procedure by inhibiting unwanted growth while encouraging others and, in some cases, distinguishing specific characteristics between those that do grow.

Most selective media contain indicators to expose differences between organisms and are also called **differential** selective media. The media illustrated in this section are used specifically to isolate pathogenic Gram-negative bacilli or Gram-positive cocci from human or environmental samples containing a mixture of organisms.

Clinical microbiologists, who are familiar with human pathogens and the types of infections they cause, choose selective media that will screen out normal flora that also are likely to be in the sample. Environmental microbiologists often choose selective and differential media to detect **coliform** bacteria. Coliforms are common inhabitants of the human intestinal tract and, as such, their presence in the environment is strong evidence of fecal contamination.

In the exercises that follow, you will examine some commonly used selective media for the isolation of Gram-positive cocci and Gram-negative rods. In most of the exercises, we have given examples of spot inoculations with pure cultures and streak plates containing individual colonies isolated from mixed samples.

All of the exercises contain a Table of Results identifying the various reactions produced on a specific medium. These include color results, interpretations of results, symbols used to quickly identify the various reactions, and presumptive identification of typical organisms encountered with these media. Presumptive identification is not final identification; it is an "educated guess" based on evidence provided by the selective/differential medium coupled with information about the origin of a sample.

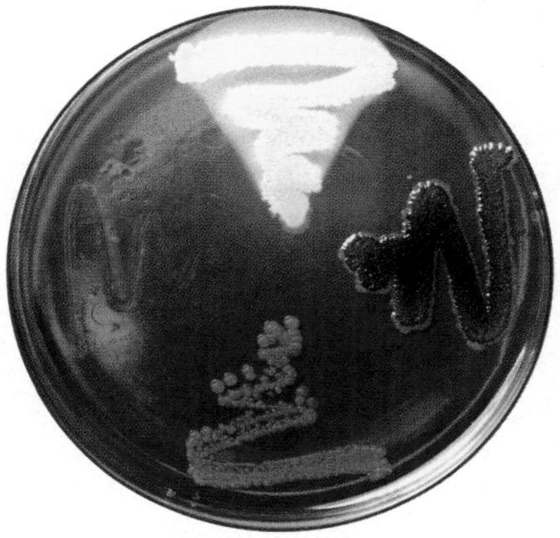

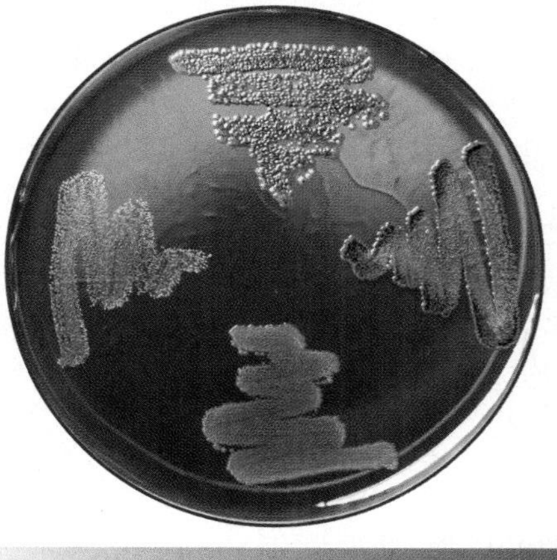

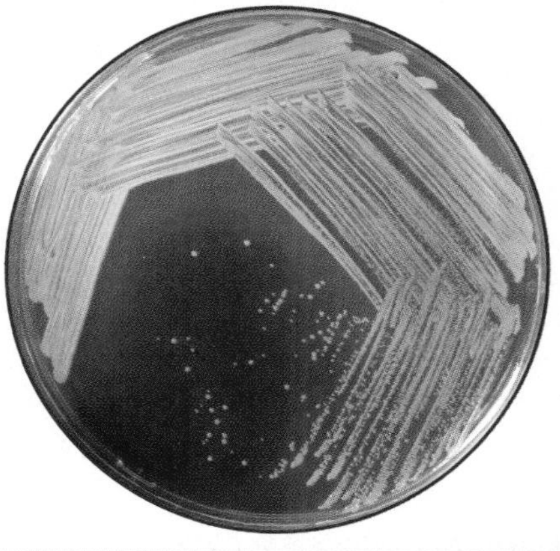

Selective Media for Isolation of Gram-Positive Cocci

Gram-negative organisms frequently inhibit growth of Gram-positives when they are cultivated together. Therefore, when looking for staphylococci or streptococci in a clinical sample, it may be necessary to begin by streaking the unknown mixture onto a selective medium that inhibits Gram-negative growth.

The selective media introduced in this unit are used to isolate (and sometimes presumptively identify) streptococci and staphylococci in human samples. Phenylethyl Alcohol Agar is a simple selective medium designed to inhibit Gram-negative organisms. Mannitol Salt Agar is a selective *and* differential medium developed to favor growth of *Staphylococcus* and to differentiate pathogenic from nonpathogenic members of the genus.

Exercise 4-1

Mannitol Salt Agar

Theory Mannitol Salt Agar (MSA) contains the carbohydrate mannitol, 7.5% sodium chloride (NaCl), and the pH indicator phenol red. Phenol red is yellow below pH 6.8, red at pH 7.4 to 8.4, and pink at pH 8.4 and above. The sodium chloride makes the medium selective for staphylococci because most other bacteria cannot survive in this level of salinity. The pathogenic species of *Staphylococcus* ferment mannitol (Figure 4-1) and produce acid, which turns the pH indicator yellow. Nonpathogenic staphylococcal species grow but produce no color change. For more information on fermentation, refer to Appendix A.

The development of yellow halos around the bacterial growth provides strong evidence that the organism is a pathogenic *Staphylococcus* (usually *S. aureus*). Good growth that produces no color change is evidence for nonpathogenic *Staphylococcus* (Figures 4-2 and 4-3). With few exceptions, organisms that grow poorly on the medium are not staphylococci.

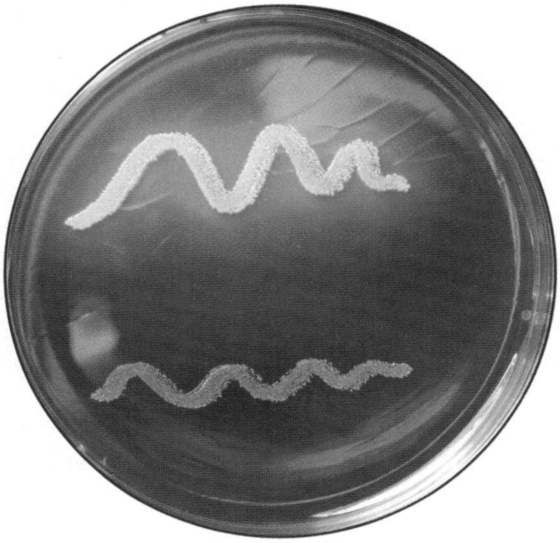

FIGURE 4-2 MANNITOL SALT AGAR
This MSA was inoculated with two of the organisms used in today's lab. Both grew well, but only the top one fermented mannitol and produced acid end products. This is evidenced by the yellow growth and halo surrounding it. Compare this plate with the one in Figure 4-3 and see if you can identify the organisms.

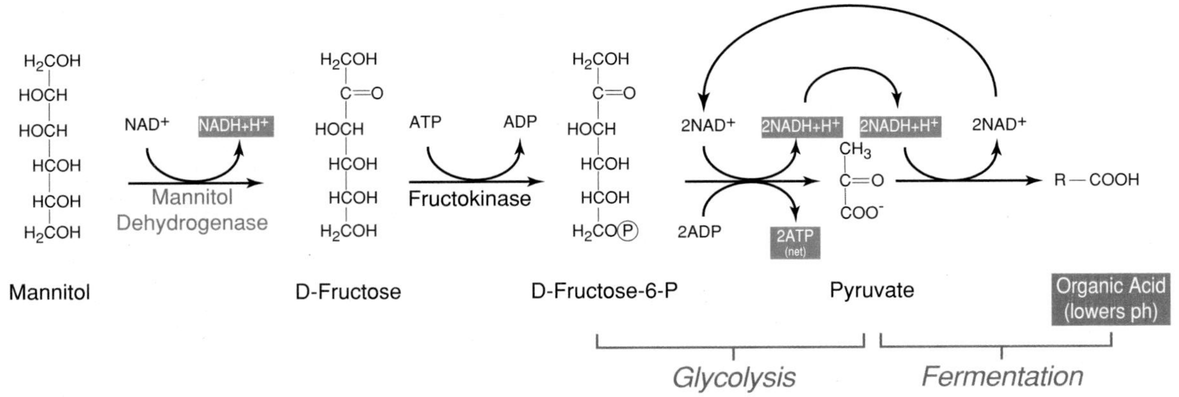

FIGURE 4-1 MANNITOL FERMENTATION WITH ACID END PRODUCTS

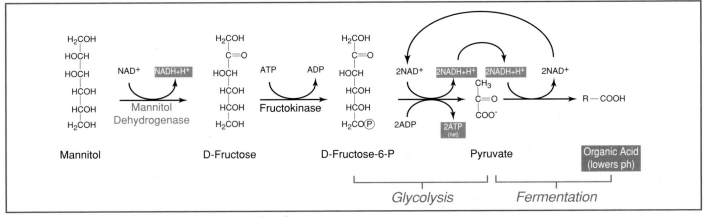

Application

Mannitol Salt Agar is used for isolation and differentiation of pathogenic *Staphylococcus* species, principally *S. aureus*.

In This Exercise...

Today you will spot inoculate one MSA plate and one Nutrient Agar (NA) plate with three test organisms. The NA will serve as a comparison for growth quality on the MSA plate.

Materials

Per Student Group

- one MSA plate
- one NA plate
- fresh broth cultures of:
  - *Staphylococcus aureus*
  - *Staphylococcus epidermidis*
  - *Escherichia coli*

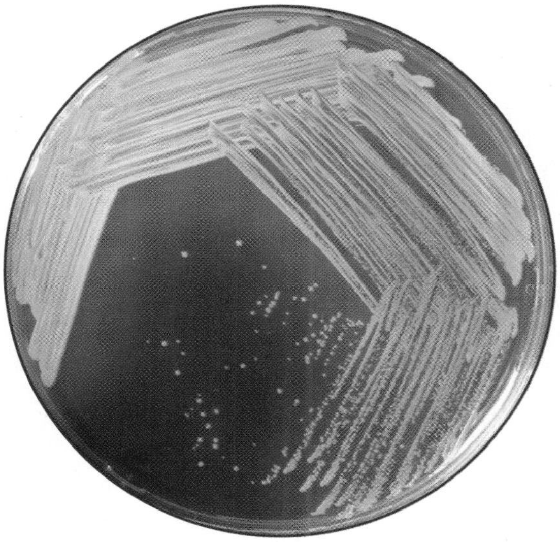

FIGURE 4-3 Mannitol Salt Agar Streaked for Isolation
This MSA was inoculated with the same two organisms as in Figure 4-2. The solid growth in the first three quadrants is still a mixture of the two organisms, so disregard the color in this region. Note the small red colonies and larger yellow colonies in the fourth streak.

Medium Recipes

Mannitol Salt Agar

• Beef extract	1.0 g
• Peptone	10.0 g
• Sodium chloride	75.0 g
• D-Mannitol	10.0 g
• Phenol red	0.025 g
• Agar	15.0 g
• Distilled or deionized water	1.0 L
pH 7.2–7.6 at 25°C	

Nutrient Agar

• Beef extract	3.0 g
• Peptone	5.0 g
• Agar	15.0 g
• Distilled or deionized water	1.0 L
pH 6.6–7.0 at 25°C	

Procedure

Lab One

1. Mix each culture well.
2. Using a permanent marker, divide the bottom of each plate into three sectors.
3. Label the plates with the organisms' names, your name, and the date.
4. Spot-inoculate the sectors on the Mannitol Salt Agar plate with the test organisms. Refer to Appendix B if necessary.
5. Repeat Step 4 with the Nutrient Agar plate.
6. Invert and incubate the plates at 35 ± 2°C for 24 to 48 hours.

TABLE OF RESULTS

Result	Interpretation	Presumptive ID
Poor growth or no growth (P)	Organism is inhibited by NaCl	Not *Staphylococcus*
Good growth (G)	Organism is not inhibited by NaCl	*Staphylococcus*
Yellow growth or halo (Y)	Organism produces acid (A) from mannitol fermentation	Possible pathogenic *Staphylococcus aureus*
Red growth (no halo) (R)	Organism does not ferment mannitol. No reaction (NR)	Nonpathogenic *Staphylococcus*

TABLE 4-1 Mannitol Salt Agar Results and Interpretations

Lab Two

1. Examine and compare the plates for color and quality of growth.
2. Record your results on the Data Sheet.

References

Delost, Maria Danessa. 1997. Page 112 in *Introduction to Diagnostic Microbiology, a Text and Workbook*. Mosby, St. Louis.

Forbes, Betty A., Daniel F. Sahm, and Alice S. Weissfeld. 2002. Chapter 19 in *Bailey and Scott's Diagnostic Microbiology*, 11th ed. Mosby, St. Louis.

Zimbro, Mary Jo, and David A. Power. 2003. Page 349 in *Difco™ & BBL™ Manual, Manual of Microbiological Culture Media*. Becton, Dickinson and Co., Sparks, MD.

Exercise 4-2

Phenylethyl Alcohol Agar

Theory Phenylethyl Alcohol Agar (PEA) is an undefined, selective medium that allows growth of Gram-positive organisms and stops or inhibits growth of most Gram-negative organisms (Figure 4-4). The active ingredient, phenylethyl alcohol, functions by interfering with DNA synthesis in Gram-negative organisms.

Application PEA is used to isolate staphylococci and streptococci (including enterococci and lactococci) from specimens containing mixtures of bacterial flora. It typically is used screen out the common contaminants, *Escherichia coli* and *Proteus spp*. When prepared with 5% sheep blood, it is used for cultivation of Gram-positive anaerobes.

In This Exercise...

You will spot inoculate one PEA plate and one Nutrient Agar (NA) plate with three test organisms. The NA plate will serve as a comparison for growth quality on the PEA plate.

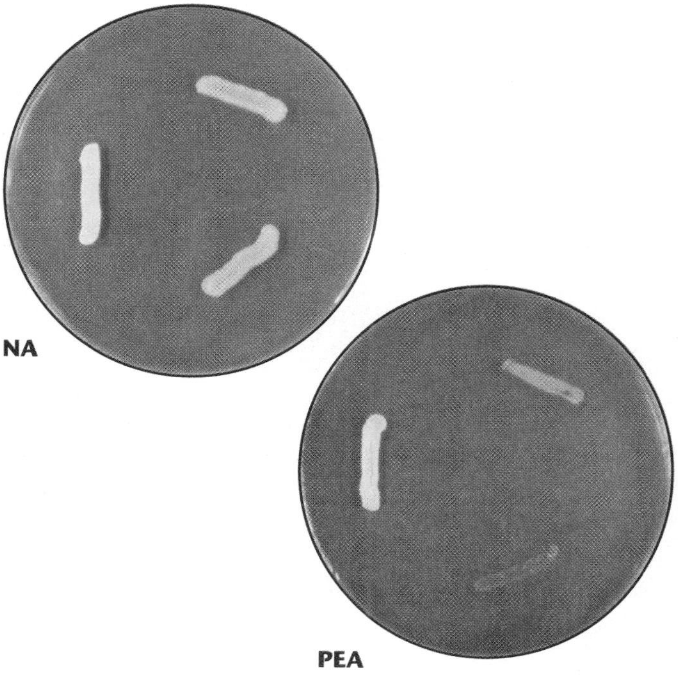

FIGURE 4-4 NUTRIENT AGAR VERSUS PHENYLETHYL ALCOHOL AGAR These plates were inoculated with the same three organisms—one Gram-positive coccus and two Gram-negative rods. All three organisms grow well on the NA, but only the Gram-positive organism (left) grows well on the PEA. Which of the three organisms chosen for today's lab could this one be?

Materials

Per Student Group

- one PEA plate
- one NA plate
- fresh broth cultures of:
  - *Escherichia coli*
  - *Enterococcus faecalis*
  - *Staphylococcus aureus*

Medium Recipes

Phenylethyl Alcohol Agar

• Tryptose	10.0 g
• Beef extract	3.0 g
• Sodium chloride	5.0 g
• Phenylethyl alcohol	2.5 g
• Agar	15.0 g
• Distilled or deionized water	1.0 L
pH 7.1–7.5 at 25°C	

Nutrient Agar

• Beef extract	3.0 g
• Peptone	5.0 g
• Agar	15.0 g
• Distilled or deionized water	1.0 L
pH 6.6 – 7.0 at 25°C	

Procedure

Lab One

1. Mix each culture well.
2. Using a permanent marker, divide the bottom of each plate into three sectors.
3. Label the plates with the organisms' names, your name, and the date.
4. Spot-inoculate the sectors on the PEA plate with the test organisms. Refer to Appendix B if necessary.
5. Repeat Step 4 with the Nutrient Agar plate.
6. Invert and incubate the plates at 35±2°C for 24 to 48 hours.

Lab Two

1. Examine and compare the plates for color and quality of growth.
2. Record your results on the Data Sheet.

TABLE OF RESULTS

Result	Interpretation	Presumptive ID
Poor growth or no growth (P)	Organism is inhibited by phenylethyl alcohol	Probable Gram-negative organism
Good growth (G)	Organism is not inhibited by phenylethyl alcohol	Probable *Staphylococcus, Streptococcus, Enterococcus*, or *Lactococcus*

TABLE 4-2 PEA RESULTS AND INTERPRETATIONS

References

Forbes, Betty A., Daniel F. Sahm, and Alice S. Weissfeld. 2002. Chapter 10 in *Bailey and Scott's Diagnostic Microbiology*, 11th ed. Mosby, St. Louis.

Zimbro, Mary Jo, and David A. Power. 2003. Page 443 in *Difco™ & BBL™ Manual, Manual of Microbiological Culture Media*. Becton, Dickinson and Co., Sparks, MD.

Selective Media for Isolation of Gram-Negative Rods

Members of the family *Enterobacteriaceae*—the enteric "gut" bacteria—are commonly found in clinical samples. The media used to isolate and differentiate these organisms from each other (and other Gram-negative pathogens) frequently combine several selective and differential elements to gather as much information as possible with a single inoculation and incubation.

Two examples are Hektoen Enteric (HE) Agar and Xylose Lysine Desoxycholate (XLD) Agar. HE Agar differentiates *Salmonella* and *Shigella* both from each other and from other enterics based on their ability to overcome the inhibitory effects of bile, reduce sulfur to H_2S, and ferment lactose, sucrose, or salicin. XLD Agar favors growth of *Salmonella, Shigella* or *Providencia* based on their ability to overcome the inhibitory effects of desoxycholate, and differentiates them based on their ability to reduce sulfur to H_2S, decarboxylate the amino acid lysine, and ferment xylose or lactose.

Other media such as Eosin Methylene Blue Agar and Endo Agar test for the presence of coliforms in environmental samples. Loosely defined, coliforms are enteric organisms that produce acid and gas from the fermentation of lactose. Their presence in the environment suggests fecal contamination and the possible presence of more serious pathogens.

Other media included in this unit are Desoxycholate Agar and MacConkey Agar. Each is selective for Gram-negative organisms and differentiates based on lactose fermentation.

Exercise 4-3

Desoxycholate Agar

Theory Desoxycholate (DOC) Agar is a selective and differential medium containing lactose, sodium desoxycholate, and neutral red dye. Lactose is a fermentable carbohydrate, desoxycholate is a Gram-positive inhibitor, and neutral red, which is colorless above pH 6.8 and red below, is added as a pH indicator.

Neutral red will reveal lactose fermentation (Figure 4-5 and Appendix A) by turning the bacterial growth red where acid products have lowered the pH (Figures 4-6 and 4-7). Lactose nonfermenters will remain their natural color or the color of the medium. Thus, the characteristics to look for on DOC are the quality of growth and color production.

Application DOC Agar is used for isolation and differentiation among the *Enterobacteriaceae*. It also is used to detect coliform bacteria in dairy products.

In This Exercise...

You will spot-inoculate one DOC Agar plate and one Nutrient Agar (NA) plate with four test organisms. The NA plate will serve as a comparison for growth quality on the DOC Agar plate.

Materials

Per Student Group

- one DOC plate
- one NA plate
- fresh broth cultures of:
  - *Enterobacter aerogenes*
  - *Enterococcus faecalis*
  - *Escherichia coli*
  - *Shigella flexneri*

Medium Recipes

Desoxycholate Agar (Modified Leifson)

• Peptone	10.0 g
• Lactose	10.0 g
• Sodium desoxycholate	1.0 g
• Sodium chloride	5.0 g
• Dipotassium phosphate	2.0 g
• Ferric citrate	1.0 g
• Sodium citrate	1.0 g
• Agar	16.0 g
• Neutral red	0.033 g
• Distilled or deionized water	1.0 L

pH 7.1–7.5 at 25°C

Nutrient Agar

- Beef extract 3.0 g
- Peptone 5.0 g
- Agar 15.0 g
- Distilled or deionized water 1.0 L

pH 6.6–7.0 at 25°C

Procedure

Lab One

1. Mix each culture well.

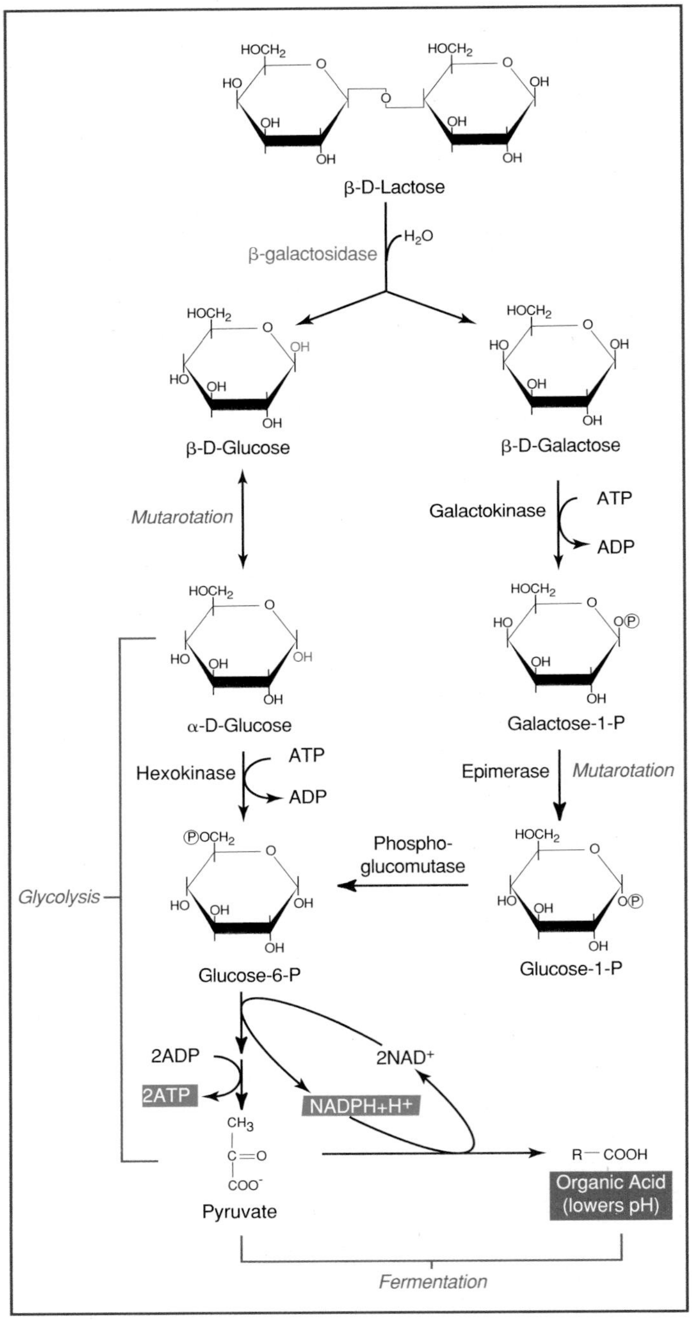

FIGURE 4-5 Lactose Fermentation with Acid End Products

FIGURE 4-6 Desoxycholate Agar
This DOC medium was inoculated with the four organisms selected for this exercise. The top organism and the one on the right are lactose fermenters, as evidenced by the pink color. Note the bile precipitate around the top organism. The bottom organism grew well, but, because of the absence of pink color, does not appear to be a lactose fermenter. The organism on the left did not grow. Compare this figure with your plate and see if you can identify all of the organisms.

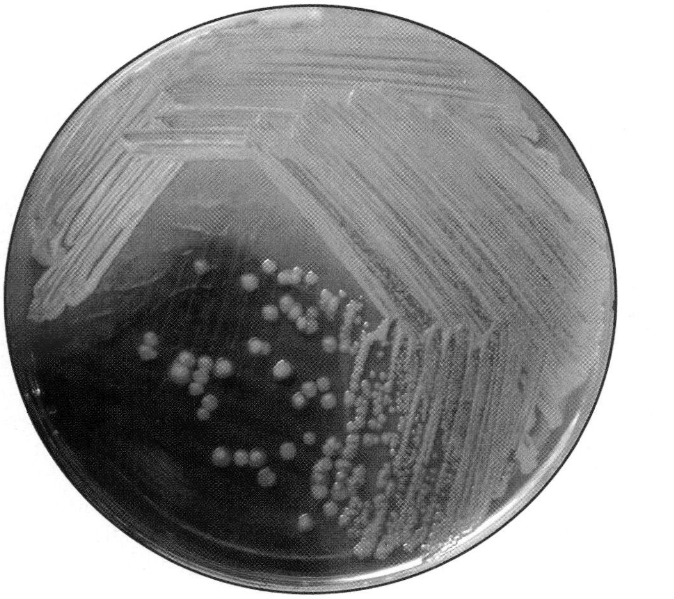

FIGURE 4-7 Desoxycholate Agar Streaked For Isolation
This DOC medium was inoculated with two of today's organisms. Both grew well, but only one fermented the lactose. The solid growth in the first three quadrants is a mixture of the two organisms, so disregard the color produced in this region. Note only the colors of the individual colonies. Can you guess which two organisms are on this plate?

2. Using a permanent marker, divide the bottom of each plate into four sectors.
3. Label the plates with the organisms' names, your name, and the date.
4. Spot-inoculate the sectors on the DOC plate with the test organisms. Refer to Appendix B if necessary.
5. Repeat Step 4 with the Nutrient Agar plate.
6. Invert and incubate the plates at 35 ± 2°C for 24 to 48 hours.

Lab Two

1. Examine and compare the plates for color and quality of growth.
2. Record your results in the space provided on the Data Sheet.

References

Koneman, Elmer W., Stephen D. Allen, William M. Janda, Paul C. Schreckenberger, and Washington C. Winn, Jr. 1997. Chapters 4 and 6 in *Color Atlas and Textbook of Diagnostic Microbiology,* 5th ed. J. B. Lippincott Co., Philadelphia.

Zimbro, Mary Jo, and David A. Power. 2003. Page 181 in *Difco™ & BBL™ Manual, Manual of Microbiological Culture Media.* Becton, Dickinson and Co., Sparks, MD.

TABLE OF RESULTS

Result	Interpretation	Presumptive ID
Poor growth or no growth (P)	Organism is inhibited by desoxycholate	Gram-positive
Good growth (G)	Organism is not inhibited by desoxycholate	Gram-negative
Growth is red or pink (R)	Organism ferments lactose to acid (A) end products	Possible coliform
Growth is colorless (not red or pink) (C)	Organism does not ferment lactose. No reaction (NR)	Noncoliform

TABLE 4-3 DOC RESULTS AND INTERPRETATIONS

Exercise 4-4

Endo Agar

Theory Endo Agar contains the color indicators sodium sulfite and basic fuchsin (which also double as Gram-positive inhibitors). Lactose is included as a fermentable carbohydrate. Lactose fermenters (Figure 4-5) growing on the medium will appear red or pink and will darken the medium slightly because of the reaction of sodium sulfite with the fermentation intermediate acetaldehyde. (Refer to Appendix A for more details on fermentation.) Lactose nonfermenters produce colorless to slightly pink growth (Figure 4-8). Some lactose fermenters, such as *Escherichia coli* and *Klebsiella pneumoniae* (coliforms), produce large amounts of acid, which gives the colonies a gold metallic sheen (Figure 4-9).

Application Endo agar is used to detect fecal contamination in water and dairy products.

In This Exercise...

You will spot-inoculate one Endo Agar plate and one Nutrient Agar (NA) plate with three test organisms. The Nutrient Agar will serve as a comparison for growth quality on the Endo Agar plate.

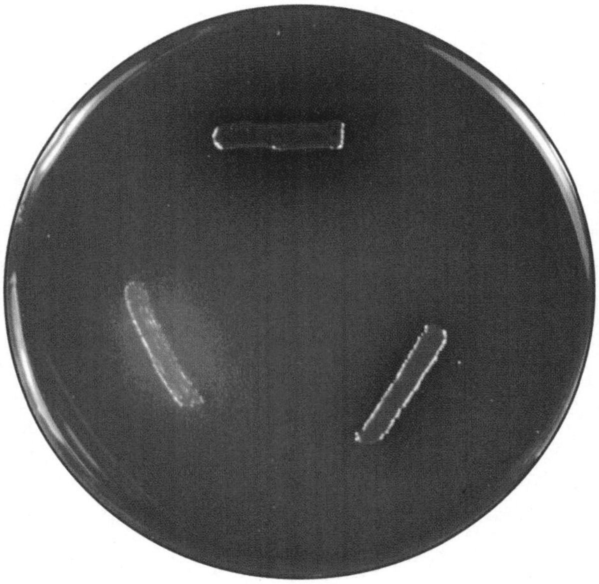

FIGURE 4-8 ENDO AGAR
This Endo Agar was inoculated with three members of *Enterobacteriaceae*—two coliforms (top and lower right) and one noncoliform (lower left). All three organisms exhibit good growth indicative of Gram-negative organisms. Note the gold metallic sheen on the top organism and the darkening of the medium surrounding it.

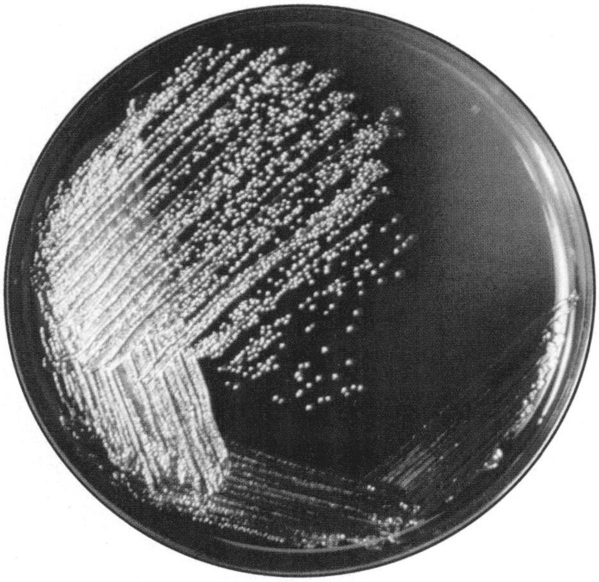

FIGURE 4-9 METALLIC SHEEN
This Endo Agar was streaked with the coliform shown at the top of Figure 4-8 to illustrate the metallic sheen. Note also the darkening of the medium surrounding the growth.

Materials

Per Student Group

- one Endo Agar plate
- one Nutrient Agar plate
- fresh broth cultures of:
  - *Enterococcus faecalis*
  - *Escherichia coli*
  - *Shigella flexneri*

Medium Recipes

Endo Agar

• Peptone	10.0 g
• Lactose	10.0 g
• Dipotassium phosphate	3.5 g
• Agar	15.0 g
• Basic Fuchsin	0.5 g
• Sodium Sulfite	2.5 g
• Distilled or deionized water	1.0 L

pH 7.3–7.7 at 25°C

Nutrient Agar

• Beef extract	3.0 g
• Peptone	5.0 g
• Agar	15.0 g
• Distilled or deionized water	1.0 L

pH 6.6–7.0 at 25°C

Procedure

Lab One

1. Mix each culture well.
2. Using a permanent marker, divide the bottom of each plate into three sectors.
3. Label the plates with the organisms' names, your name, and the date.
4. Spot inoculate the sectors on the Endo Agar plate with the test organisms. Refer to Appendix B if necessary.
5. Repeat step 4 with the Nutrient Agar plate.
6. Invert and incubate the plates at 35 ± 2°C for 24 to 48 hours.

Lab Two

1. Examine and compare the plates for color and quality of growth.
2. Record your results on the Data Sheet.

Reference

Zimbro, Mary Jo and David A. Power. 2003. Page 207 in *Difco™ & BBL™ Manual, Manual of Microbiological Culture Media*. Becton, Dickinson and Co., Sparks, MD.

TABLE OF RESULTS

Result	Interpretation	Presumptive ID
Poor growth or no growth (P)	Organism is inhibited by sodium sulfite and/or basic fuchsin	Gram-positive
Good growth (G)	Organism is not inhibited by sodium sulfite or basic fuchsin	Gram-negative
Red or pink growth (R)	Organism ferments lactose to acid end products	Possible coliform
Red or pink growth with darkened medium and/or metallic sheen (RD)	Organism ferments lactose rapidly and produces large amount of acid (A)	Probable coliform
Growth is "colorless" (not red or pink and does not produce metallic sheen) (C)	Organism does not ferment lactose. No reaction (NR)	Noncoliform

TABLE 4-4 Endo Agar Results and Interpretations

Exercise 4-5

Eosin Methylene Blue Agar

Theory Eosin Methylene Blue (EMB) Agar contains peptone, lactose, sucrose, and the dyes eosin Y and methylene blue. The sugars provide fermentable substrates to encourage growth of fecal coliforms. The dyes inhibit the growth of Gram-positive organisms and, under acidic conditions, also produce a dark purple complex usually accompanied by a green metallic sheen. This sheen serves as an indicator of the vigorous lactose or sucrose fermentation typical of fecal coliforms. Smaller amounts of acid production (typical of *Enterobacter aerogenes* and slow lactose fermenters) result in a pink coloration of the growth. Nonfermenters retain their normal color or take on the coloration of the medium (Figures 4-10 and 4-11).

Application EMB Agar is used for isolation of fecal coliforms. It can be streaked for isolation or used in the Membrane Filter Technique as discussed in Exercise 8-1.

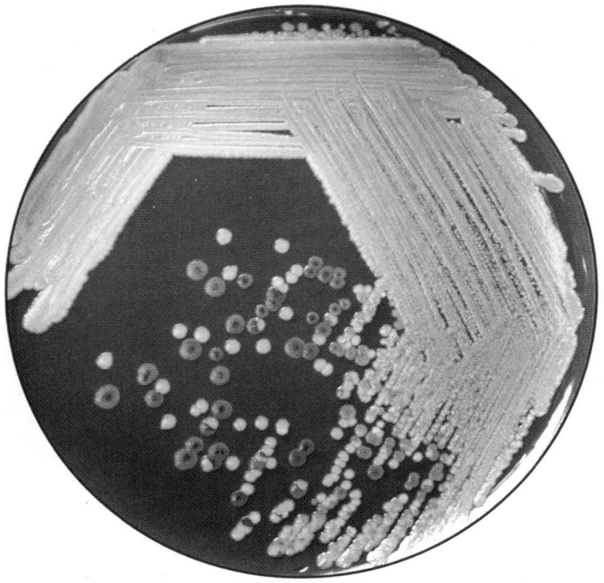

FIGURE 4-11 EOSIN METHYLENE BLUE AGAR STREAKED FOR ISOLATION
This EMB agar was inoculated with two of the Gram-negative rods shown in Figure 4-10. One is a coliform; the other is not. Color produced by mixtures of bacteria is not determinative, so disregard the first three quadrants and pay attention only to the individual colonies. The coliform that produced a green metallic sheen on the plate shown in Figure 4-10 produced pink colonies with dark centers. The noncoliform colonies are beige.

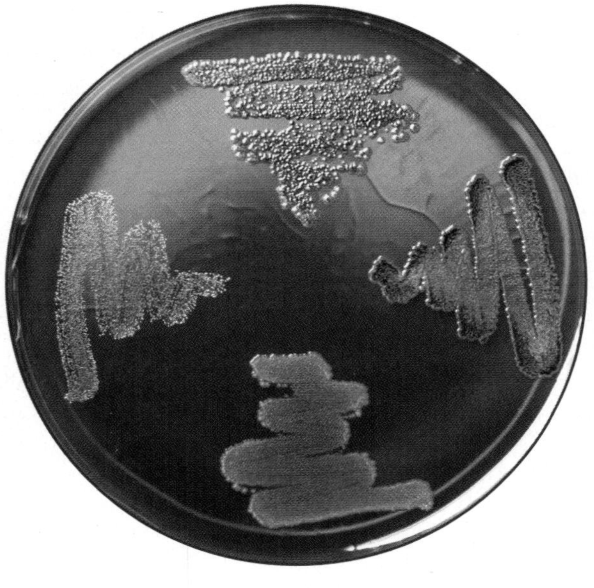

FIGURE 4-10 EOSIN METHYLENE BLUE AGAR
This EMB Agar was inoculated with (clockwise from the top) two coliforms, a Gram-negative noncoliform, and a Gram-positive organism. Note the characteristic green metallic sheen of the coliform at the top and pink coloration of the one at the right. Both organisms are lactose fermenters; the difference in color results from the degree of acid production. The organism on the bottom is a nonfermenter, as indicated by the lack of color. Growth of the Gram-positive organism on the left was inhibited by eosin Y and methylene blue. Compare the effectiveness of Gram-positive inhibition in this medium to those in Figures 4-6 and 4-12.

In This Exercise...

You will spot inoculate one EMB Agar plate and one Nutrient Agar (NA) plate with four test organisms. The NA plate will serve as a comparison for growth quality on the EMB Agar plate.

Materials

Per Student Group

- one EMB plate
- one NA plate
- fresh broth cultures of:
  - *Enterobacter aerogenes*
  - *Enterococcus faecalis*
  - *Escherichia coli*
  - *Salmonella typhimurium*

Medium Recipes

Eosin Methylene Blue Agar

- Peptone 10.0 g
- Lactose 5.0 g

- Sucrose 5.0 g
- Dipotassium phosphate 2.0 g
- Agar 13.5 g
- Eosin Y 0.4 g
- Methylene blue 0.065 g
- Distilled or deionized water 1.0 L

pH 6.9–7.3 at 25°C

Nutrient Agar

- Beef extract 3.0 g
- Peptone 5.0 g
- Agar 15.0 g
- Distilled or deionized water 1.0 L

pH 6.6–7.0 at 25°C

Procedure

Lab One

1. Mix each culture well.
2. Using a permanent marker, divide the bottom of each plate into four sectors.
3. Label the plates with the organisms' names, your name, and the date.
4. Spot inoculate the four sectors on the EMB plate with the test organisms. Refer to Appendix B if necessary.
5. Repeat step 4 with the Nutrient Agar plate.
6. Invert and incubate the plates at 35 ± 2°C for 24 to 48 hours.

Lab Two

1. Examine and compare the plates for color and quality of growth.
2. Record your results on the Data Sheet.

References

Forbes, Betty A., Daniel F. Sahm, and Alice S. Weissfeld. 2002. Chapter 10 in *Bailey and Scott's Diagnostic Microbiology,* 11th ed. Mosby, St. Louis.

Koneman, Elmer W., Stephen D. Allen, William M. Janda, Paul C. Schreckenberger, and Washington C. Winn, Jr. 1997. Chapter 4 in *Color Atlas and Textbook of Diagnostic Microbiology,* 5th ed. J. B. Lippincott Co., Philadelphia.

Zimbro, Mary Jo, and David A. Power. 2003. Page 218 in *Difco™ & BBL™ Manual, Manual of Microbiological Culture Media.* Becton, Dickinson and Co., Sparks, MD.

TABLE OF RESULTS

Result	Interpretation	Presumptive ID
Poor growth or no growth (P)	Organism is inhibited by eosin and methylene blue	Gram-positive
Good growth (G)	Organism is not inhibited by eosin and methylene blue	Gram-negative
Growth is pink and mucoid (Pi)	Organism ferments lactose with little acid production (A)	Possible coliform
Growth is "dark" (purple to black, with or without green metallic sheen) (D)	Organism ferments lactose and/or sucrose with much acid production (A)	Probable coliform
Growth is "colorless" (no pink, purple, or metallic sheen) (C)	Organism does not ferment lactose or sucrose. No reaction (NR)	Noncoliform

TABLE 4-5 EMB Results and Interpretations

Exercise 4-6

Hektoen Enteric Agar

Theory

Hektoen Enteric (HE) Agar is an undefined medium designed to isolate *Salmonella* and *Shigella* species from other enterics based on the ability to ferment lactose, sucrose, or salicin, and to reduce sulfur to hydrogen sulfide gas (H_2S). In addition to the three sugars, sodium thiosulfate is included as a source of sulfur. Ferric ammonium citrate is added to react with H_2S and form a black precipitate. Bile salts are included to inhibit most Gram-positive cocci. Bromthymol blue and acid fuchsin dyes are added as color indicators.

Differentiation is possible as a result of the various colors produced in the colonies and in the agar. Enterics that produce acid from fermentation will produce yellow to salmon-pink colonies. Organisms such as *Salmonella*, *Shigella*, and *Proteus* that do not ferment any of the sugars produce blue-green colonies. *Proteus* and *Salmonella* species that reduce sulfur to H_2S form colonies containing a black precipitate. Refer to Figures 4-12 and 4-13.

Application

HE Agar is used to isolate and differentiate *Salmonella* and *Shigella* species from other Gram-negative enteric organisms.

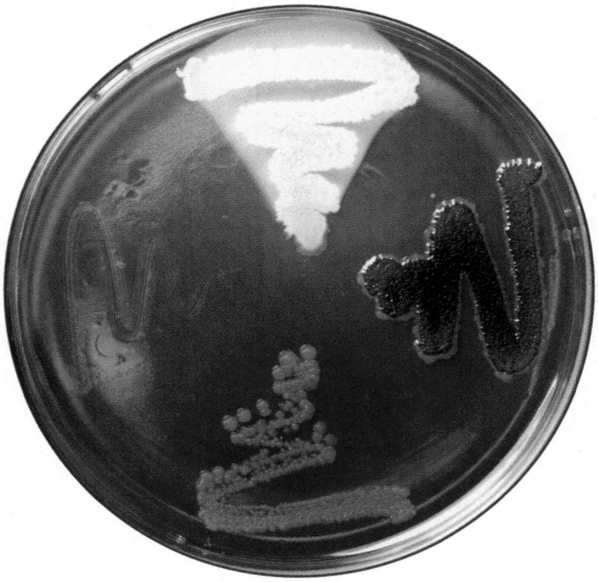

FIGURE 4-12 HEKTOEN ENTERIC AGAR
This HE Agar was inoculated with (clockwise from top), a Gram-negative lactose fermenter (indicated by the yellow growth), two lactose nonfermenters, and a Gram-positive organism. Note the blue-green growth of the lactose nonfermenters. Note also the black precipitate in the growth on the right, indicative of a reaction between ferric ammonium citrate and H_2S produced from sulfur reduction. The Gram-positive organism on the left was severely inhibited by the bile salts.

FIGURE 4-13 HEKTOEN ENTERIC AGAR STREAKED FOR ISOLATION
This HE agar was streaked with two of the organisms chosen for today's exercise. One is a lactose fermenter evidenced by the yellow growth; the other is a sulfur reducer, as indicated by the black precipitate in the colonies. Can you identify these two organisms?

In This Exercise...

You will streak-inoculate one Hektoen Enteric Agar plate and one Nutrient Agar plate with four test organisms. The Nutrient Agar will serve as a comparison for growth quality on the Hektoen Enteric Agar plate.

Materials

Per Student Group

- one HE Agar plate
- one NA plate
- fresh broth cultures of:
  - *Enterococcus faecalis*
  - *Escherichia coli*
  - *Salmonella typhimurium*
  - *Shigella flexneri*

Medium Recipes

Hektoen Enteric Agar

• Yeast extract	3.0 g
• Peptic digest of animal tissue	12.0 g
• Lactose	12.0 g
• Sucrose	12.0 g
• Salicin	2.0 g
• Bile salts	9.0 g
• Sodium chloride	5.0 g
• Sodium thiosulfate	5.0 g

- Ferric ammonium citrate 1.5 g
- Bromthymol blue 0.064 g
- Acid fuchsin 0.1 g
- Agar 13.5 g
- Distilled or deionized water 1.0 L

pH 7.4–7.8 at 25°C

Nutrient Agar

- Beef extract 3.0 g
- Peptone 5.0 g
- Agar 15.0 g
- Distilled or deionized water 1.0 L

pH 6.6–7.0 at 25°C

Procedure

Lab One

1. Mix each culture well.
2. Using a permanent marker, divide the bottom of each plate into four sectors.
3. Label the plates with the organisms' names, your name, and the date.
4. Spot-inoculate the four sectors on the HE plate with the test organisms. Refer to Appendix B if necessary.
5. Repeat Step 4 with the Nutrient Agar plate.
6. Invert and incubate the plates aerobically at 35 ± 2°C for 48 hours.

Lab Two

1. Examine and compare the plates for color and quality of growth.
2. Record your results on the Data Sheet.

References

Forbes, Betty A., Daniel F. Sahm, and Alice S. Weissfeld. 2002. Chapter 10 in *Bailey and Scott's Diagnostic Microbiology,* 11th ed. Mosby-Yearbook, St. Louis.

Koneman, Elmer W., Stephen D. Allen, William M. Janda, Paul C. Schreckenberger, and Washington C. Winn, Jr. 1997. Chapter 4 in *Color Atlas and Textbook of Diagnostic Microbiology,* 5th ed. J. B. Lippincott Co., Philadelphia.

Zimbro, Mary Jo, and David A. Power. 2003. Page 265 in *Difco™ & BBL™ Manual, Manual of Microbiological Culture Media.* Becton, Dickinson and Co., Sparks, MD.

TABLE OF RESULTS

Result	Interpretation	Presumptive ID
Poor growth or no growth (P)	Organism is inhibited by bile and/or one of the dyes included	Gram-positive
Good growth (G)	Organism is not inhibited by bile or any of the dyes included	Gram-negative
Pink to orange growth (Pi)	Organism produces acid from lactose fermentation (A)	Not *Shigella* or *Salmonella*
Blue-green growth with black precipitate (Bppt)	Organism does not ferment lactose, but reduces sulfur to hydrogen sulfide (H_2S)	Possible *Salmonella*
Blue-green growth without black precipitate (B)	Organism does not ferment lactose or reduce sulfur. No reaction (NR)	Possible *Shigella* or *Salmonella*

TABLE 4-6 Hektoen Enteric Agar Results and Interpretations

Exercise 4-7

MacConkey Agar

Theory

MacConkey Agar is a selective and differential medium containing lactose, bile salts, neutral red, and crystal violet. Bile salts and crystal violet inhibit growth of Gram-positive bacteria. Neutral red dye is a pH indicator that is colorless above a pH of 6.8 and red at a pH less than 6.8. Acid accumulating from lactose fermentation turns the dye red. Lactose fermenters turn a shade of red on MacConkey agar, whereas lactose nonfermenters retain their normal color or the color of the medium (Figures 4-14 and 4-15). Formulations without crystal violet allow growth of *Enterococcus* and some species of *Staphylococcus*, which ferment the lactose and appear pink on the medium.

Application

MacConkey Agar is used to isolate and differentiate members of the *Enterobacteriaceae* based on the ability to ferment lactose. Variations on the standard medium include MacConkey Agar w/o CV (without crystal violet) to allow growth of Gram-positive cocci, or MacConkey Agar CS to control swarming bacteria (*Proteus*) that interfere with other results.

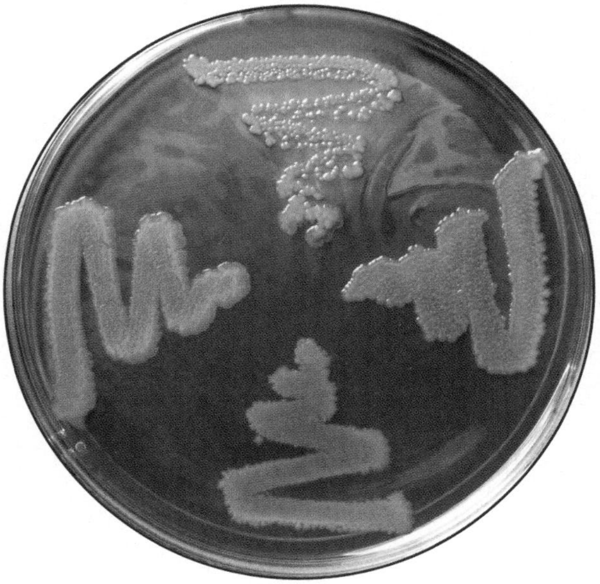

FIGURE 4-14 MacConkey Agar
This MacConkey Agar was inoculated with four members of *Enterobacteriaceae*, all of which show abundant growth. The top organism and the one on the right are lactose fermenters, as evidenced by the pink color. Note the bile precipitate around the top organism. The organisms on the bottom and on the left produced no color, so they do not appear to be lactose fermenters.

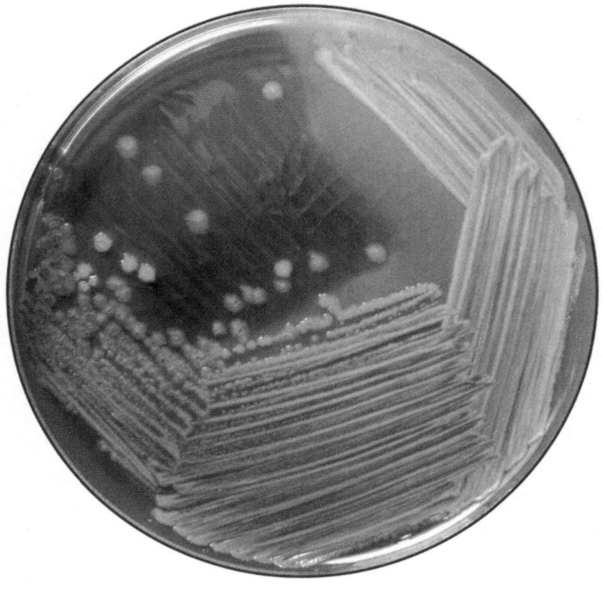

FIGURE 4-15 MacConkey Agar Streaked for Isolation
This MacConkey Agar was inoculated with two of today's organisms. Both grew well, but only one fermented the lactose. The solid growth in the first three quadrants is a mixture of the two organisms, so the color produced in that region is not determinative. Note the colors of the individual colonies.

In This Exercise...

You will spot inoculate one MacConkey Agar plate and one Nutrient Agar (NA) plate with four test organisms. The NA plate will serve as a comparison for growth quality on the MacConkey Agar plate.

Materials

Per Student Group

- one MacConkey Agar plate
- one Nutrient Agar plate
- fresh broth cultures of:
  - *Enterococcus faecalis*
  - *Escherichia coli*
  - *Proteus mirabilis*
  - *Shigella flexneri*

Medium Recipes

MacConkey Agar

• Pancreatic digest of gelatin	17.0 g
• Pancreatic digest of casein	1.5 g
• Peptic digest of animal tissue	1.5 g
• Lactose	10.0 g
• Bile salts	1.5 g
• Sodium chloride	5.0 g
• Neutral red	0.03 g

- Crystal violet 0.001
- Agar 13.5 g
- Distilled or deionized water 1.0 L

pH 6.9–7.3 at 25°C

Nutrient Agar

- Beef extract 3.0 g
- Peptone 5.0 g
- Agar 15.0 g
- Distilled or deionized water 1.0 L

pH 6.6–7.0 at 25°C

Procedure

Lab One

1. Mix each culture well.
2. Using a permanent marker, divide the bottom of each plate into four sectors.
3. Label the plates with the organisms' names, your name, and the date.
4. Spot-inoculate the sectors on the MacConkey Agar plate with the test organisms. Refer to Appendix B if necessary.
5. Repeat Step 4 with the Nutrient Agar plate.
6. Invert and incubate the plates at 35 ± 2°C for 24 to 48 hours.

Lab Two

1. Examine and compare the plates for color and quality of growth.
2. Refer to Table 4-7 when recording your results on the Data Sheet.

References

Forbes, Betty A., Daniel F. Sahm, and Alice S. Weissfeld. 2002. Chapter 10 in *Bailey and Scott's Diagnostic Microbiology,* 11th ed. Mosby-Yearbook, St. Louis.

Koneman, Elmer W., Stephen D. Allen, William M. Janda, Paul C. Schreckenberger, and Washington C. Winn, Jr. 1997. Chapter 4 in *Color Atlas and Textbook of Diagnostic Microbiology,* 5th ed. J. B. Lippincott Co., Philadelphia.

Zimbro, Mary Jo, and David A. Power. 2003. Page 334 in *Difco™ & BBL™ Manual, Manual of Microbiological Culture Media.* Becton, Dickinson and Co., Sparks, MD.

TABLE OF RESULTS

Result	Interpretation	Presumptive ID
Poor growth or no growth (P)	Organism is inhibited by crystal violet and/or bile	Gram-positive
Good growth (G)	Organism is not inhibited by crystal violet or bile	Gram-negative
Pink to red growth with or without bile precipitate (R)	Organism produces acid from lactose fermentation (A)	Probable coliform
Growth is "colorless" (not red or pink) (C)	Organism does not ferment lactose. No reaction (NR)	Noncoliform

TABLE 4-7 MacConkey Agar Results and Interpretations

Exercise 4-8

Xylose Lysine Desoxycholate Agar

FIGURE 4-16 XYLOSE LYSINE DESOXYCHOLATE AGAR
This XLD agar was inoculated with (clockwise from top) three members of *Enterobacteriaceae*—two sulfur-reducing lactose nonfermenters and one lactose fermenter. Note the black precipitate in the growth on top resulting from the reaction of ferric ammonium citrate with the reduced sulfur (H_2S). Note also the yellow color of the growth on the left as a result of acid production from lactose fermentation. The organism on the right is a nonfermenting Gram-negative rod and typically is a sulfur reducer. Lack of black precipitate suggests that it either has either lost this ability or is a contaminant.

Theory

Xylose Lysine Desoxycholate (XLD) Agar is a selective and differential medium containing sodium desoxycholate, xylose, L-lysine, and ferric ammonium citrate. Desoxycholate is a Gram-positive inhibitor, xylose is a fermentable carbohydrate, L-lysine is an amino acid provided for decarboxylation (see Exercise 5-11), and ferric ammonium citrate is an indicator to mark the presence of hydrogen sulfide gas (H_2S) from sulfur reduction. Phenol red, which is yellow when acidic and red or pink when alkaline, is added as a pH indicator.

Organisms that ferment xylose will acidify the medium and produce yellow colonies. Organisms that are able to remove the carboxyl group from (decarboxylate) L-lysine will release alkaline products and produce red colonies. Organisms that are able to reduce sulfur will produce a black precipitate in the growth as a result of the reaction of ferric ammonium citrate with H_2S.

Shigella and *Providencia*, which do not ferment xylose but decarboxylate lysine, appear red on the medium. *Salmonella* species, which ferment xylose but then decarboxylate the lysine, also appear as red colonies but with black centers because of the reduction of sulfur to H_2S. Other enterics that ordinarily would revert to decarboxylation after exhausting the sugar and alkalinize the medium are prevented from doing so by the large amount of sugar included and the short incubation time. These organisms appear yellow on the medium (Figure 4-16).

Application

XLD Agar is a selective and differential medium used to isolate and identify *Shigella* and *Providencia* from stool samples.

In This Exercise...

You will spot inoculate one XLD Agar plate and one Nutrient Agar (NA) plate with five test organisms. The NA plate will serve as a comparison for growth quality on the XLD Agar plate.

Materials

Per Student Group

- one XLD Agar plate
- one NA plate
- fresh broth cultures of:
  - *Enterococcus faecalis*
  - *Escherichia coli*
  - *Providencia stuartii*
  - *Salmonella typhimurium*
  - *Shigella flexneri*

Medium Recipe

Xylose Lysine Desoxycholate Agar

• Xylose	3.5 g
• L-Lysine	5.0 g
• Lactose	7.5 g
• Sucrose	7.5 g
• Sodium chloride	5.0 g
• Yeast extract	3.0 g
• Phenol red	0.08 g
• Sodium desoxycholate	2.5 g
• Sodium thiosulfate	6.8 g
• Ferric ammonium citrate	0.8 g
• Agar	13.5 g
• Distilled or deionized water	1.0 L

pH 7.3–7.7 at 25°C

Nutrient Agar

- Beef extract 3.0 g
- Peptone 5.0 g
- Agar 15.0 g
- Distilled or deionized water 1.0 L

pH 6.6–7.0 at 25°C

Procedure

Lab One

1. Mix each culture well.
2. Using a permanent marker, divide the bottom of each plate into five sectors.
3. Label the plates with the organisms' names, your name, and the date.
4. Spot inoculate the sectors on the XLD Agar plate with the test organisms. Refer to Appendix B if necessary.
5. Repeat Step 4 with the Nutrient Agar plate.
6. Invert and incubate the plates at 35±2°C for 18 to 24 hours. (Be sure to remove the plates from the incubator at 24 hours. Reversions from fermentation to decarboxylation, due to exhaustion of sugar in the medium, with the resulting yellow-to-red color shift may lead to false identification as *Shigella* or *Providencia*.)

Lab Two

1. Examine and compare the plates for color and quality of growth.
2. Record your results on the Data Sheet.

References

Forbes, Betty A., Daniel F. Sahm, and Alice S. Weissfeld. 2002. Chapter 10 in *Bailey and Scott's Diagnostic Microbiology*, 11th ed. Mosby-Yearbook, St. Louis.

Zimbro, Mary Jo, and David A. Power. 2003. Page 625 in *Difco™ & BBL™ Manual, Manual of Microbiological Culture Media*. Becton, Dickinson and Co., Sparks, MD.

TABLE OF RESULTS

Result	Interpretation	Presumptive ID
Poor growth (P)	Organism is inhibited by desoxycholate	Gram-positive
Good growth (G)	Organism is not inhibited by desoxycholate	Gram-negative
Growth is yellow (Y)	Organism produces acid from xylose fermentation (A)	Not *Shigella* or *Providencia*
Red growth with black center (RB)	Organism reduces sulfur to hydrogen sulfide (H_2S)	Not *Shigella* or *Providencia* (probable *Salmonella*)
Red growth without black center (R)	Organism does not ferment xylose or ferments xylose slowly; alkaline products from lyzine decarboxylation (K)	Probable *Shigella* or *Providencia*

TABLE 4-8 XLD Results and Interpretations

Differential Tests

SECTION 5

Over most of the last century, bacteria have been differentiated based on their enormous biochemical diversity. The identification systems microbiologists use for this purpose are called **differential tests** or **biochemical tests** because they differentiate and sometimes identify microorganisms based on specific biochemical characteristics. As you will see repeatedly in this section, microbial biochemical characteristics are brought about by enzymatic reactions that are part of specific and well-understood metabolic pathways. The tests used in this section were selected to illustrate those pathways and the means used to identify them.

The first half of Section Five will introduce you to the tests individually so you can familiarize yourself with the metabolic pathways and the mechanisms by which specific tests expose them. In the second half you will be given a chance to use some multiple test media designed for rapid identification.

The categories of differential tests included in this section are:

- Energy metabolism, including fermentation of carbohydrates and aerobic and anaerobic respiration
- Utilization of a specified medium component
- Decarboxylation and deamination of amino acids
- Hydrolytic reactions requiring intracellular or extracellular enzymes
- Multiple reactions performed in a single combination medium
- Miscellaneous differential tests

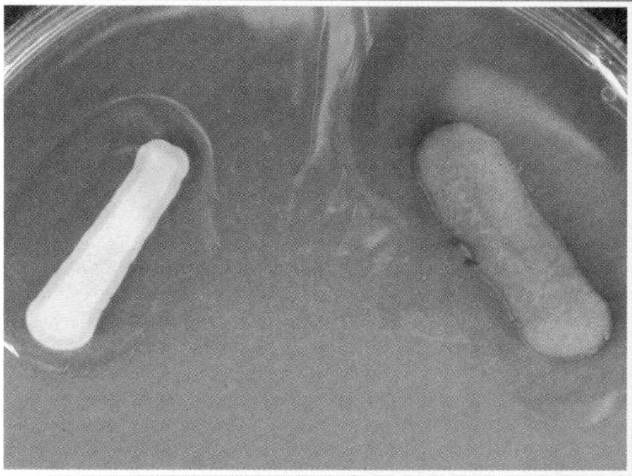

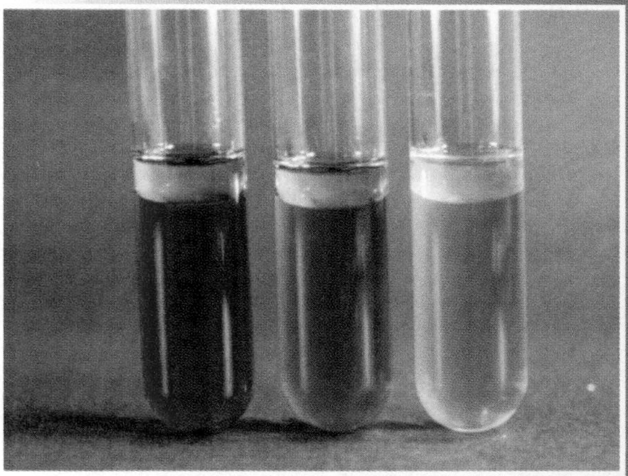

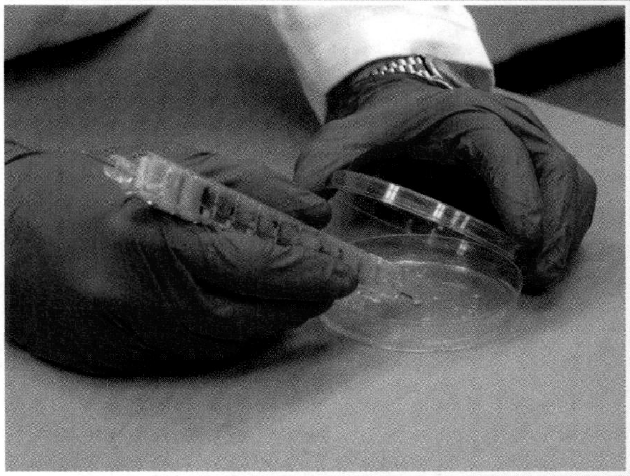

Introduction to Energy Metabolism Tests

Heterotrophic bacteria obtain their energy by means of either **respiration** or **fermentation**. Both catabolic systems convert the chemical energy of organic molecules to high-energy bonds in **adenosine triphosphate (ATP)**. In respiration, glucose is converted to ATP in three distinct phases: (1) **glycolysis**, (2) the **tricarboxylic acid cycle (Krebs cycle)**, and (3) **oxidative phosphorylation** (sometimes called the **electron transport chain**, or ETC).

Glycolysis splits the six-carbon glucose molecule into two three-carbon **pyruvate** molecules with the production of ATP and reduced **coenzymes**. The Krebs cycle is the complex pathway in which **acetyl-CoA** (from the conversion of pyruvate) is oxidized to CO_2 and more coenzymes are reduced. ATP is also a product. The electron transport chain (ETC) is a series of **oxidation–reduction** reactions that receives electrons from the reduced coenzymes produced during glycolysis and the Krebs cycle. At the end of the ETC is an inorganic molecule called the **final electron acceptor (FEA)**. When oxygen is the final electron acceptor, the respiration is **aerobic**. If the FEA is an inorganic molecule other than oxygen (*e.g.*, sulfate or nitrate) the respiration is **anaerobic**.

In contrast to respiration, fermentation is the metabolic process by which glucose acts as an electron donor and one or more of its organic products act as the final electron acceptor. Reduced carbon compounds in the form of acids and organic solvents, as well as CO_2, are the typical end products of fermentation.

In Exercise 5-1, you will perform a simple test to determine the ability of an organism to perform oxidative and/or fermentative metabolism of sugars. In the exercises that follow, you will perform tests that demonstrate microbial fermentative characteristics and tests designed to detect specific respiratory constituents or pathways.

Exercise 5-1

Oxidation–Fermentation Test

Theory

The Oxidation–Fermentation (O–F) Test is designed to differentiate bacteria on the basis of fermentative or oxidative metabolism of carbohydrates. In oxidation pathways, a carbohydrate is directly oxidized to pyruvate and further converted to CO_2 and energy by way of the Krebs cycle and the electron transport chain. As mentioned above, an inorganic molecule such as oxygen is required to act as the final electron acceptor. Fermentation also converts carbohydrates to pyruvate but uses it to produce one or more acids (as well as other compounds). Consequently, fermenters identified by this test acidify O–F medium to a greater extent than do oxidizers.

Hugh and Leifson's O–F medium includes a high sugar-to-peptone ratio to reduce the possibility that alkaline products from peptone utilization will neutralize weak acids produced by oxidation of the carbohydrate. Bromthymol blue dye, which is yellow at pH 6.0 and green at pH 7.1, is added as a pH indicator. A low agar concentration produces a semi-solid medium that allows determination of motility.

The medium is prepared with glucose, lactose, sucrose, maltose, mannitol or xylose and is not slanted. Two tubes of the specific sugar medium are stab-inoculated several times with the test organism. After inoculation, one tube is sealed with a layer of sterile mineral oil to promote anaerobic growth and fermentation (Figure 5-1). The other tube is left unsealed to allow aerobic growth and oxidation.

Organisms that are able to ferment the carbohydrate or ferment *and* oxidize the carbohydrate will turn the sealed and unsealed media yellow throughout. Organisms that are able to oxidize only will turn the unsealed medium yellow (or partially yellow) and leave the sealed medium green or blue. Slow or weak fermenters will turn both tubes slightly yellow at the top. Organisms that are not able to metabolize the sugar will either produce no color change or turn the medium blue because of alkaline products from amino acid degradation. The results are summarized in Table 5-1 and shown in Figure 5-2.

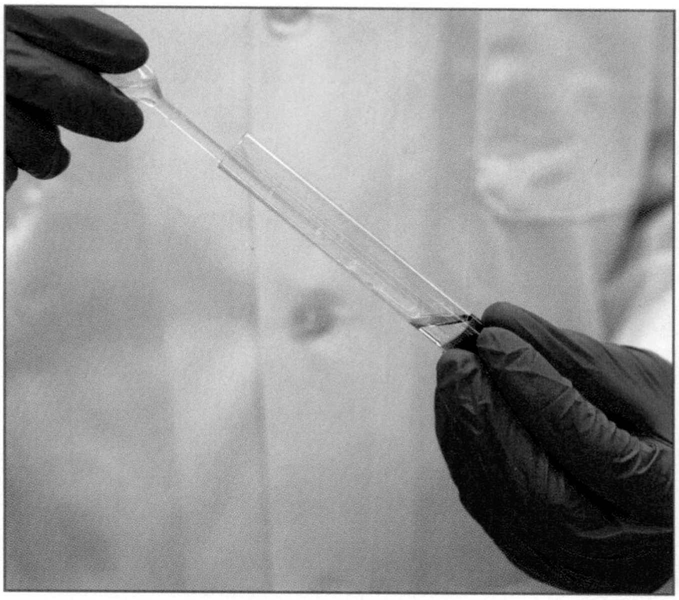

FIGURE 5-1 Adding the Mineral Oil Layer
Tip the tube slightly to one side and gently add 3–4 mm of sterile mineral oil. Be sure to use a sterile pipette for each tube.

FIGURE 5-2 Oxidation–Fermentation Test
The two tubes on the left were inoculated with an oxidative (O) organism, evidenced by the slight yellowing of the unsealed tube and lack of color change in the sealed tube. The two tubes on the right were inoculated with an organism capable of both oxidative and fermentative metabolism (O/F). Mineral oil in the first and third tubes creates an anaerobic environment. This reveals the organism on the right as a glucose fermenter because without oxygen diffusing into the medium, there can be no oxidative use of the carbohydrate. When both tubes are yellow, the organism is said to have fermented in the sealed tube and either oxidized or oxidized and fermented the carbohydrate in the unsealed tube.

Application

The O–F Test is used to differentiate bacteria based on their ability to oxidize or ferment specific sugars. It allows presumptive separation of the fermentative *Enterobacteriaceae* from the oxidative *Pseudomonas* and *Bordetella*, and the nonreactive *Alcaligenes* and *Moraxella*.

In This Exercise...

Six O–F media tubes will be used for this exercise—four for inoculation (two for each organism) and two for controls. One of each pair will receive a mineral oil overlay to create an anaerobic environment and force fermentation if the organism is capable of doing so. You will not do a motility determination as motility will be covered in Exercise 5-28. For best results, use a heavy inoculum and several stabs for each tube. Use only sterile pipettes to add the mineral oil.

Materials

Per Student Group

- six O–F glucose tubes
- sterile mineral oil
- sterile transfer pipettes
- fresh agar slants of:
  - *Escherichia coli*
  - *Pseudomonas aeruginosa*

Medium Recipe

Hugh and Leifson's O-F Medium with Glucose

• Pancreatic digest of casein	2.0 g
• Sodium chloride	5.0 g
• Dipotassium phosphate	0.3 g
• Agar	2.5 g
• Bromthymol blue	0.08 g
• Glucose	10.0 g
• Distilled or deionized water	1.0 L

pH = 6.6–7.0 at 25°C

Procedure

Lab One

1. Obtain six O–F tubes. Label six (in three pairs) with the names of the organisms, your name, and the date. Label the last pair "control."

TABLE OF RESULTS			
Sealed	**Unsealed**	**Interpretation**	**Symbol**
Green or blue	Any amount of yellow	Oxidation	O
Yellow throughout	Yellow throughout	Oxidation and fermentation or fermentation only	O/F or F
Slightly yellow at the top	Slightly yellow at the top	Oxidation and slow fermentation or slow fermentation only	O/F or F
Green or blue	Green or blue	No sugar metabolism; organism is nonsaccharolytic	N

TABLE 5-1 O-F Medium results and interpretations.

2. Stab-inoculate two O–F tubes with each test organism. Stab several times to a depth of about one cm from the bottom of the agar. Do not inoculate the controls.
3. Overlay one of each pair of tubes (including the controls) with 3–4 mm of sterile mineral oil (Figure 5-1).
4. Incubate all tubes at 35 ± 2°C for 48 hours.

Lab Two

1. Examine the tubes for color changes. Re-incubate any O–F tubes that have not changed color for an additional 48 hours.
2. Record your results in the chart provided on the Data Sheet.

References

Collins, C. H., Patricia M. Lyne, J. M. Grange. 1995. Page 112 in *Collins and Lyne's Microbiological Methods*, 7th ed. Butterworth-Heinemann, United Kingdom.

Delost, Maria Dannessa. 1997. Pages 218–219 in *Introduction to Diagnostic Microbiology.* Mosby, St. Louis.

Forbes, Betty A., Daniel F. Sahm, and Alice S. Weissfeld. 2002. Pages 154–155 in *Bailey & Scott's Diagnostic Microbiology,* 11th ed. Mosby, St. Louis.

MacFaddin, Jean F. 2000. Page 379 in *Biochemical Tests for Identification of Medical Bacteria*, 3rd ed. Lippincott Williams & Wilkins, Philadelphia.

Smibert, Robert M., and Noel R. Krieg. 1994. Page 625 in *Methods for General and Molecular Bacteriology*, edited by Philipp Gerhardt, R. G. E. Murray, Willis A. Wood, and Noel R. Krieg, American Society for Microbiology, Washington, DC.

Zimbro, Mary Jo and David A. Power, editors. 2003. Page 410 in *Difco™ and BBL™ Manual—Manual of Microbiological Culture Media.* Becton Dickinson and Co., Sparks, MD.

Fermentation Tests

Carbohydrate fermentation, as defined at the beginning of this section, is the metabolic process by which an organic molecule acts as an electron donor (becoming oxidized in the process) and one or more of its organic products act as the final electron acceptor (FEA). In actuality, the term "carbohydrate fermentation" is used rather broadly to include hydrolysis of disaccharides prior to the fermentation reaction. Thus, a "lactose fermenter" is an organism that splits the disaccharide lactose into the monosaccharides glucose and galactose and then ferments the monosaccharides. In this section you will frequently see the term "fermenter." Unless it is expressly used otherwise, the term should be assumed to include the initial hydrolysis and/or conversion reactions.

Fermentation of glucose begins with the production of pyruvate. Although some organisms use alternate pathways, most bacteria accomplish this by glycolysis. The end products of pyruvate fermentation include a variety of organic acids, alcohols, and hydrogen or carbon dioxide gas. The specific end products depend on the specific organism and the substrate fermented. Refer to Figure 5-3 and Appendix A.

In this unit you will perform tests using three differential media formulated to detect fermentation. General-purpose fermentation media typically include one or more carbohydrates and a pH indicator to detect acid formation. Purple Broth and Phenol Red Broth are examples of general-purpose fermentation media. MR-VP broth is a more specialized medium in that it allows detection of two specific fermentation pathways. The Methyl Red (MR) Test detects bacteria capable of performing a **mixed acid fermentation.** The Voges-Proskauer (VP) Test identifies bacteria that are able to produce acetoin as part of a **2,3-butanediol fermentation.**

Exercise 5-2

Phenol Red Broth

Theory

Phenol Red (PR) Broth is a differential test medium prepared as a base to which a carbohydrate is added. Also included in the broth are peptone and the pH indicator phenol red. Phenol red is yellow below pH 6.8, pink to magenta above pH 7.4, and red in between. During preparation the pH is adjusted to approximately 7.3 so it appears red. Finally, an inverted Durham tube is added to each tube as an indicator of gas production.

Acid production from fermentation of the carbohydrate lowers the pH below the neutral range of the indicator and turns the medium yellow (Figure 5-3). Deamination of peptone amino acids produces ammonia (NH_3), which raises the pH and turns the broth pink. Gas production, also from fermentation, is indicated by a bubble or pocket in the Durham tube where the broth has been displaced (Figure 5-4).

Application

PR broth is used to differentiate members of *Enterobacteriaceae* and to distinguish them from other Gram-negative rods.

In This Exercise...

Today you will inoculate PR broths with four organisms. Use Table 5-2 as a guide to interpretation of and correct symbol use for your results.

Materials

Per Student Group

- five PR Glucose Broths with Durham tubes
- five PR Lactose Broths with Durham tubes
- five PR Sucrose Broths with Durham tubes
- five PR Base Broths with Durham tubes
- fresh cultures of:
  - *Escherichia coli*
  - *Shigella flexneri*
  - *Proteus vulgaris*
  - *Alcaligenes faecalis*

Medium Recipe

PR (Carbohydrate) Broth*

- Pancreatic digest of casein 10.0 g
- Sodium chloride 5.0 g
- Carbohydrate (glucose, lactose, sucrose) 10.0 g
- Phenol red 0.018 g
- Distilled or deionized water 1.0 L

pH = 7.1–7.5 at 25°C

*This formulation without carbohydrate is PR Base Broth.

Procedure

Lab One

1. Obtain five of each PR broth. Label each with the name of the organism, your name, the medium name, and the date. The base broths without carbohydrate are negative controls for color comparison.
2. Inoculate one of each broth with each test organism.
3. Incubate all the tubes at 35 ± 2°C for 48 hours.

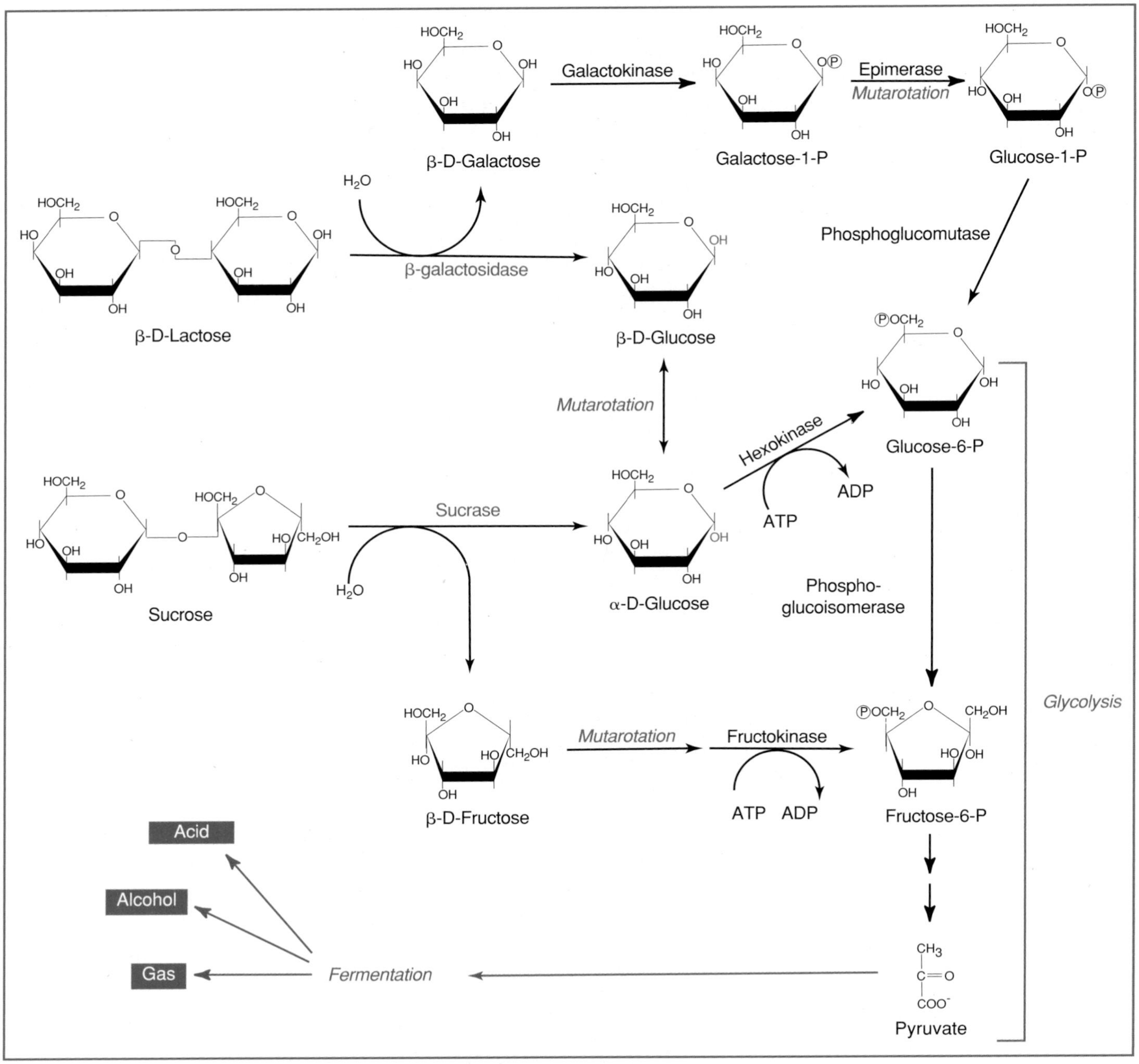

FIGURE 5-3 FERMENTATION OF THE DISACCHARIDES LACTOSE AND SUCROSE
Notice that fermentation of the disaccharides lactose and sucrose relies on enzymes to hydrolyze them into two monosaccharides. Once the monosaccharides are formed, they are fermented via glycolysis, just as glucose is.

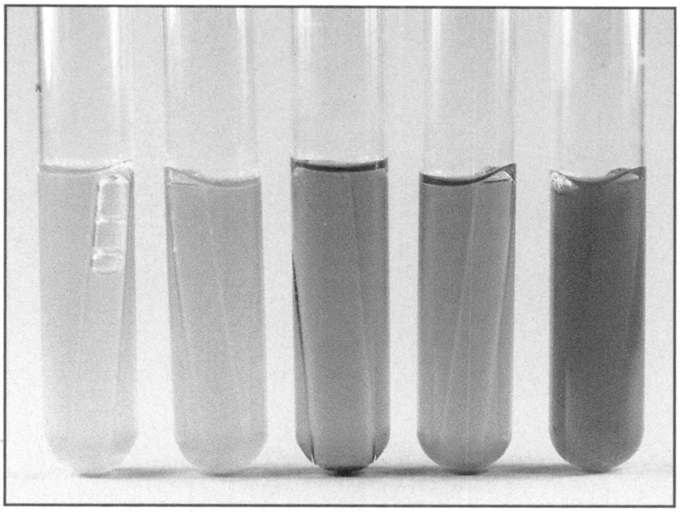

FIGURE 5-4 PR GLUCOSE BROTH RESULTS
From left to right, these five PR Glucose Broths demonstrate acid and gas production (A/G), acid production without gas (A/ −), uninoculated control tube for comparison, no reaction (− / −), and alkaline condition as a result of peptone degradation (K).

Lab Two

1. Using the uninoculated broths for color comparison and Table 5-2 as a guide, examine all the tubes and enter your results in the chart provided on the Data Sheet. When indicating the various reactions, use the standard symbols as shown, with the acid reading first, followed by a slash and then the gas reading. K indicates alkalinity and –/– symbolizes no reaction. ***Note:*** *Do not try to read these results without control tubes for comparison.*

References

Lányi, B. 1987. Page 44 in *Methods in Microbiology*, Vol. 19, edited by R. R. Colwell and R. Grigorova. Academic Press, New York.

MacFaddin, Jean F. 2000. Page 57 in *Biochemical Tests for Identification of Medical Bacteria*, 3rd ed. Lippincott Williams & Wilkins, Philadelphia.

Zimbro, Mary Jo and David A. Power, editors. 2003. Page 440 in *Difco™ and BBL™ Manual—Manual of Microbiological Culture Media*. Becton Dickinson and Co., Sparks, MD.

TABLE OF RESULTS

Result	Interpretation	Symbol
Yellow broth, bubble in tube	Fermentation with acid and gas end products	A/G
Yellow broth, no bubble in tube	Fermentation with acid end products; no gas produced	A/ −
Red broth, no bubble in tube	No fermentation	− / −
Pink broth, no bubble in tube	Degradation of peptone; alkaline end products	K

TABLE 5-2 PR Broth results and interpretations.

Exercise 5-3

Purple Broth

Theory Purple Broth, like Phenol Red Broth (Exercise 5-2), is a differential medium that consists of a basal recipe to which carbohydrate is added, allowing determination of fermentation characteristics for a variety of carbohydrates (Figure 5-3). The base formulation contains peptone and the pH indicator bromcresol purple, which is yellow below pH 6.8 and purple above (Figure 5-5). The 6.8 final pH of the medium is slightly more acidic than Phenol Red Broth, making it ideal for testing enteric bacteria. The addition of an inverted Durham tube to the medium allows determination of gas production.

Application Purple Broth is used to differentiate members of *Enterobacteriaceae* and to distinguish them from other Gram negative rods.

FIGURE 5-5 PURPLE LACTOSE BROTH RESULTS
These four Purple Lactose Broths illustrate, from left to right, no reaction (–/–) [or alkaline reaction (K) from degradation of peptone], an uninoculated control for comparison, acid and gas production (A/G), and acid production without gas (A/–).

In This Exercise...

You will inoculate Purple Broths with four organisms. Use Table 5-3 as a guide to interpret your results. Be sure to use the control tubes for comparison when reading the results.

Materials

Per Student Group

- five Purple Glucose Broths with Durham tubes
- five Purple Lactose Broths with Durham tubes
- five Purple Sucrose Broths with Durham tubes
- five Purple Base Broths with Durham tubes
- fresh cultures of:
  - *Escherichia coli*
  - *Shigella flexneri*
  - *Proteus vulgaris*
  - *Alcaligenes faecalis*

Medium Recipe

Purple Broth*

• Peptone	10.0 g
• Beef extract	1.0 g

*This formulation without carbohydrate is Purple Base Broth.

TABLE OF RESULTS		
Result	**Interpretation**	**Symbol**
Yellow broth; bubble in Durham tube	Fermentation with acid and gas end products	A / G
Yellow broth; no bubble in Durham tube	Fermentation with acid end products; no gas produced	A / –
Purple broth; no bubble in Durham tube; turbid	No fermentation; degradation of peptone with alkaline end products	K
Purple broth; no bubble in Durham tube; not turbid	No fermentation; slow or no degradation of peptone with alkaline end products	– / –

TABLE 5-3 Purple Broth results and interpretations.

- Sodium chloride 5.0 g
- Bromcresol purple 0.02 g
- Carbohydrate (glucose, lactose or sucrose) 10.0 g
- Distilled or deionized water 1.0 L

pH 6.6–7.0 at 25°C

Procedure

Lab One

1. Obtain five of each Purple Broth. Label them with the names of the organisms, your name, the name of the medium, and the date. The base broths without carbohydrate are negative controls for color comparison.
2. Inoculate one of each broth with each test organism.
3. Incubate all tubes aerobically at 35 ± 2°C for 48 hours.

Lab Two

Using the controls for color comparison, examine all tubes and enter your results in the chart provided on the Data Sheet. When indicating the various reactions, use the standard symbols as shown, with the acid reading first, followed by a slash and then the gas reading. K indicates alkalinity, and –/– symbolizes no reaction. ***Note:*** *Do not try to read these results without control tubes for comparison.*

References

Forbes, Betty A., Daniel F. Sahm, and Alice S. Weissfeld. 2002. Page 269 in *Bailey & Scott's Diagnostic Microbiology*, 11th ed. Mosby, Inc., St. Louis.

Lányi, B. 1987. Page 44 in *Methods in Microbiology*, Vol. 19, edited by R. R. Colwell and R. Grigorova, Academic Press, New York.

MacFaddin, Jean F. 2000. Page 57 in *Biochemical Tests for Identification of Medical Bacteria*, 3rd ed. Lippincott Williams & Wilkins, Philadelphia.

Zimbro, Mary Jo, and David A. Power, editors. 2003. Page 465 in *Difco™ and BBL™ Manual—Manual of Microbiological Culture Media*. Becton Dickinson and Co., Sparks, MD.

Exercise 5-4

Methyl Red and Voges-Proskauer Tests

Theory

Methyl Red and Voges-Proskauer (MR-VP) Broth is a combination medium used for both Methyl Red (MR) *and* Voges-Proskauer (VP) tests. It is a simple solution containing only peptone, glucose, and a phosphate buffer. The peptone and glucose provide protein and fermentable carbohydrate, respectively, and the potassium phosphate resists pH changes in the medium.

The MR test is designed to detect organisms capable of performing a **mixed acid fermentation**, which overcomes the phosphate buffer in the medium and lowers the pH (Figures 5-6 and 5-7). The acids produced by these organisms tend to be stable, whereas acids produced by other organisms tend to be unstable and subsequently are converted to more neutral products.

Mixed acid fermentation is verified by the addition of methyl red indicator dye following incubation. Methyl red is red at pH 4.4 and yellow at pH 6.2. Between these two pH values it is various shades of orange. Red color is the only true indication of a positive result; orange is negative or inconclusive. Yellow is negative (Figure 5-8).

The Voges-Proskauer test was designed for organisms that are able to ferment glucose, but quickly convert their acid products to acetoin and 2, 3-butanediol (Figures 5-7 and 5-9). Addition of VP reagents to the medium oxidizes the **acetoin** to **diacetyl**, which in turn reacts with **guanidine nuclei** from peptone to produce a red color (Figure 5-10). A positive VP result, therefore, is red. No color change (or development of copper color) after the addition of reagents is negative. Copper color is a result of interactions between the reagents and should not be confused with the true red color of

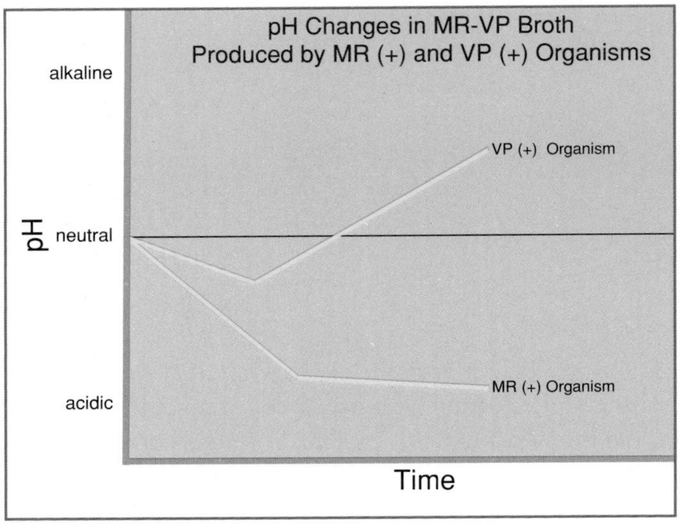

FIGURE 5-7 pH CHANGES IN MR-VP BROTH
The MR test identifies organisms that perform a mixed acid fermentation and produce stable acid end products. MR (+) organisms permanently lower the broth's pH. The VP test is used to identify organisms that perform a 2,3-butanediol fermentation. VP (+) organisms initially may produce acid and temporarily lower the pH, but because the 2,3-butanediol fermentation end products are neutral, the pH at completion of the test is near neutral.

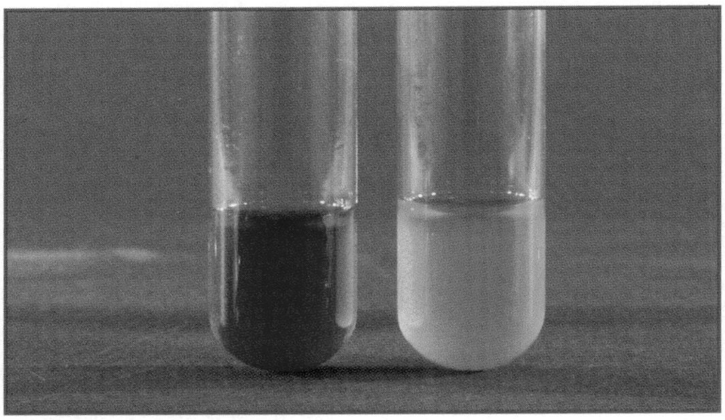

FIGURE 5-8 METHYL RED TEST
The tube on the left is MR-positive; the tube on the right is MR-negative.

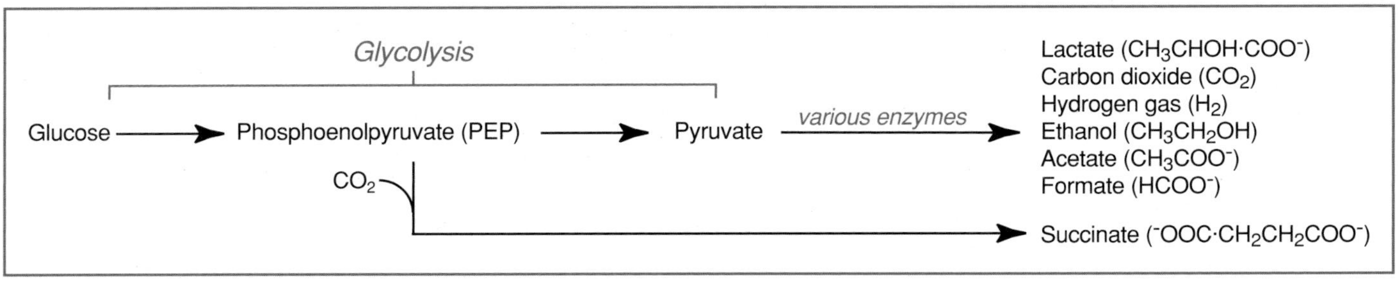

FIGURE 5-6 MIXED ACID FERMENTATION OF *E. COLI*
E. coli is a representative Methyl Red positive organism and is recommended as a positive control for the test. Its mixed acid fermentation produces the end products listed in order of abundance. Most of the formate is converted to H_2 and CO_2 gases. ***Note:*** The amount of succinate falls between acetate and formate, but is derived from PEP, not pyruvate. *Salmonella* and *Shigella* are also Methyl Red positive.

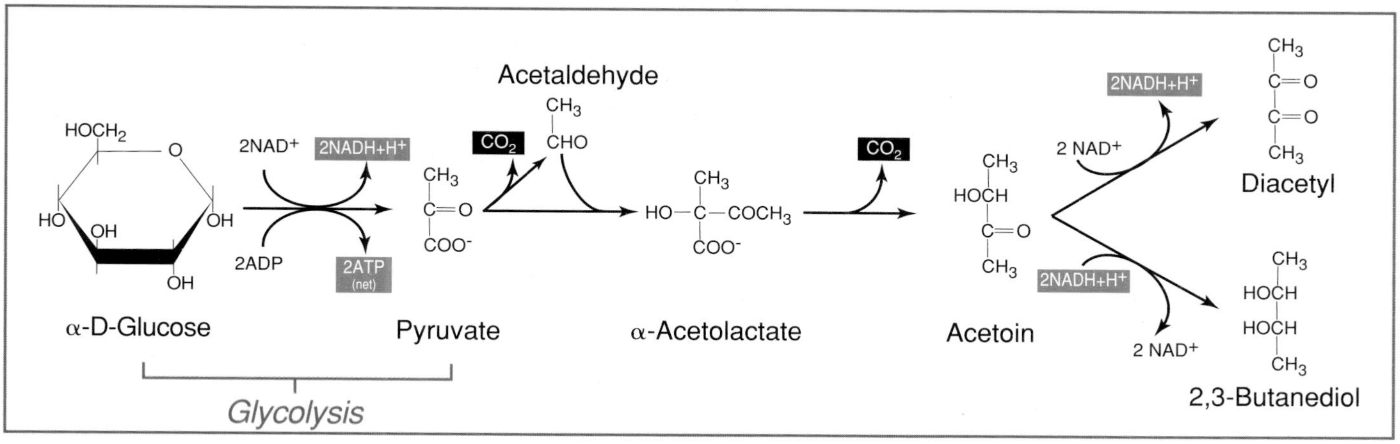

FIGURE 5-9 2, 3-Butanediol Fermentation
Acetoin is an intermediate in this fermentation. Reduction of acetoin by NADH leads to the end product 2,3-butanediol. Oxidation of acetoin produces diacetyl, as in the indicator reaction for the VP test (see Figure 5-10).

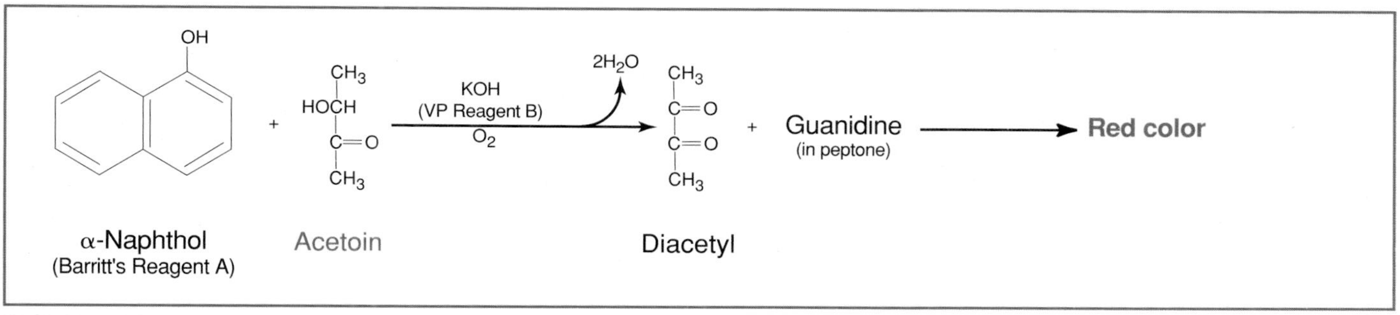

FIGURE 5-10 Indicator Reaction of Voges-Proskauer Test
Reagents A and B are added to VP broth after 48 hours' incubation. These reagents react with acetoin and oxidize it to diacetyl, which in turn reacts with guanidine (from the peptone in the medium) to produce a red color.

a positive result (Figure 5-11). Use of positive and negative controls for comparison is usually recommended.

After incubation, two 1 mL volumes are transferred from the MR-VP broth to separate test tubes. Methyl red indicator reagent is added to one tube, and VP reagents are added to the other. Color changes then are observed and documented. Refer to the procedural diagram in Figure 5-11.

Application

The Methyl Red and Voges-Proskauer tests are components of the *IMViC* battery of tests (Indole, Methyl red, Voges-Proskauer, and Citrate) used to distinguish between members of the family *Enterobacteriaceae* and differentiate them from other Gram-negative rods. For more information, refer to Appendix A.

In This Exercise…

Today you will perform Methyl Red and Voges-Proskauer tests on *Escherichia coli* and *Enterobacter aerogenes*. After a 48 hour incubation period, you will observe the results of a mixed-acid fermentation and a 2, 3-butanediol

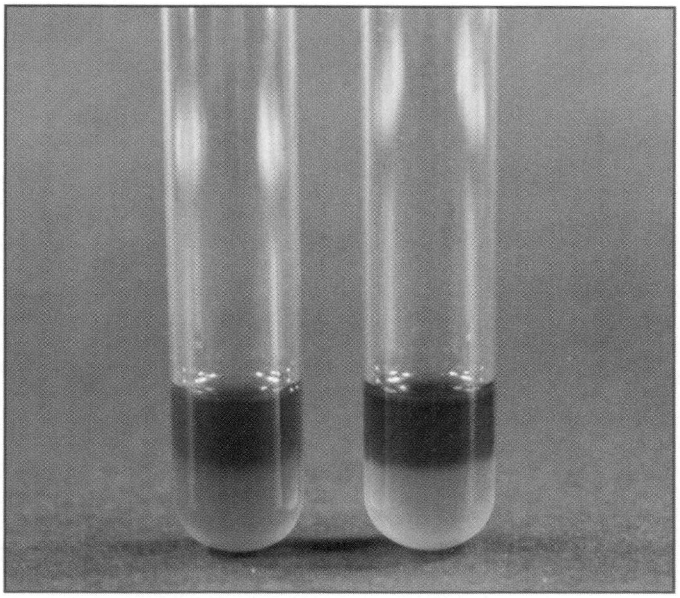

FIGURE 5-11 The Voges-Proskauer Test
The tube on the left is VP-negative; the tube on the right is VP-positive. The copper color at the top of the VP-negative tube is the result of the reaction of KOH and α-naphthol and should not be confused with a positive result.

fermentation. The exercise involves transfer of a portion of broth containing living organisms and the measured addition of reagents to complete the observable reactions. Please use care in your transfers and in your measurements. For help with the procedure, refer to Figure 5-12. For help interpreting your results, refer to Figures 5-8 and 5-11, and Tables 5-4 and 5-5.

Materials

Per Student Group

- three MR-VP broths
- Methyl Red reagent
- VP reagents A and B
- six nonsterile test tubes

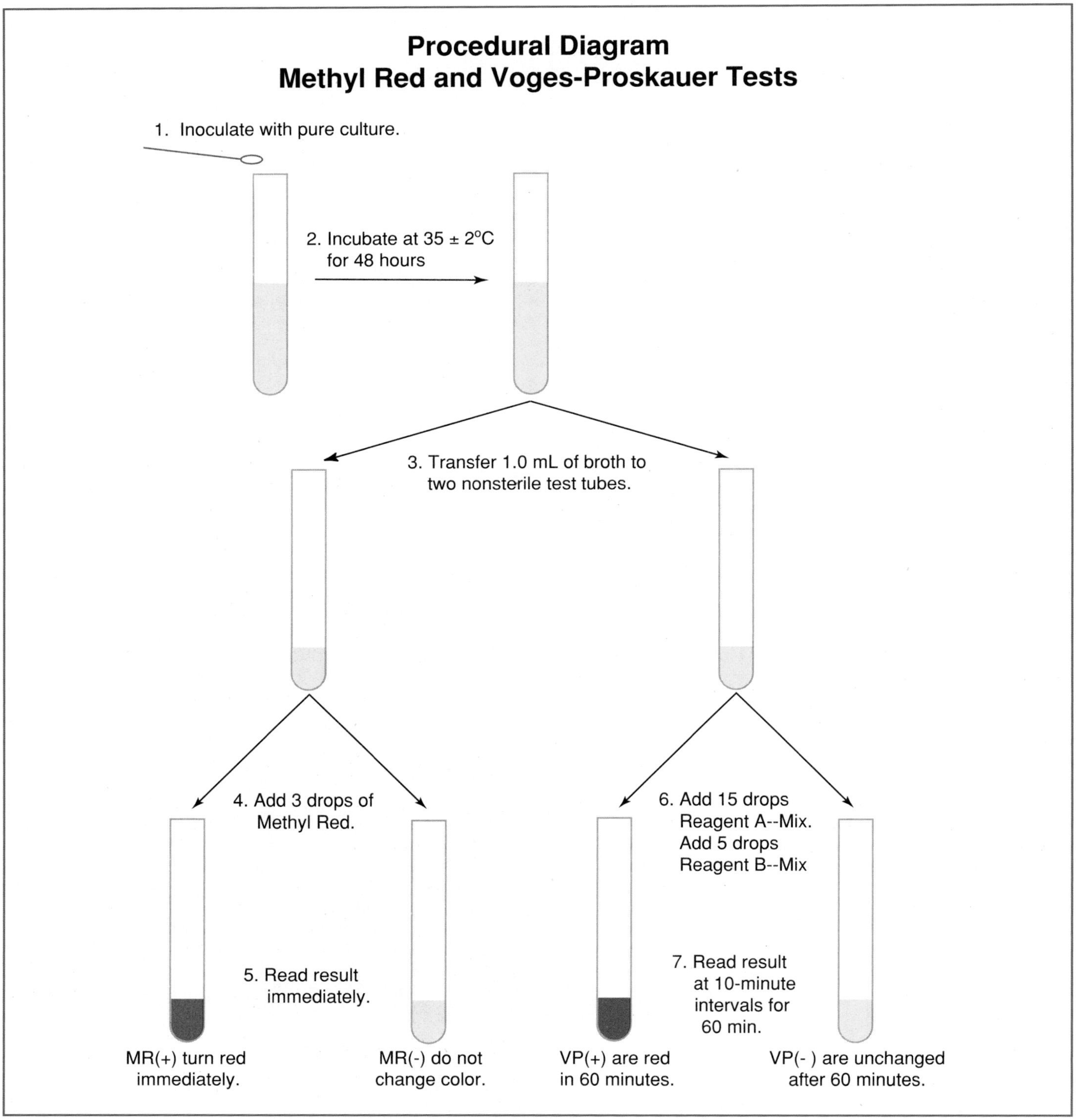
Procedural Diagram
Methyl Red and Voges-Proskauer Tests
1. Inoculate with pure culture.
2. Incubate at 35 ± 2°C for 48 hours
3. Transfer 1.0 mL of broth to two nonsterile test tubes.
4. Add 3 drops of Methyl Red.
6. Add 15 drops Reagent A--Mix. Add 5 drops Reagent B--Mix
5. Read result immediately.
7. Read result at 10-minute intervals for 60 min.
MR(+) turn red immediately.
MR(-) do not change color.
VP(+) are red in 60 minutes.
VP(-) are unchanged after 60 minutes.

FIGURE 5-12 Procedural Diagram for the Methyl Red and Voges-Proskauer Tests

TABLE OF RESULTS		
Result	**Interpretation**	**Symbol**
Red	Mixed acid fermentation	+
No color change	No mixed acid fermentation	–

TABLE 5-4 Methyl Red Test results and interpretations.

TABLE OF RESULTS		
Result	**Interpretation**	**Symbol**
Red	2,3-butanediol fermentation (acetoin produced)	+
No color change	No 2,3-butanediol fermentation (acetoin is not produced)	–

TABLE 5-5 Voges-Proskauer Test results and interpretations.

- three nonsterile 1 mL pipettes
- fresh cultures of:
  - *Escherichia coli*
  - *Enterobacter aerogenes*

Medium and Reagent Recipes

MR-VP Broth

• Buffered peptone	7.0 g
• Dipotassium phosphate	5.0 g
• Dextrose (glucose)	5.0 g
• Distilled or deionized water	1.0 L

pH 6.7–7.1 at 25°C

Methyl Red Reagent

• Methyl Red dye	0.1 g
• Ethanol	300.0 mL
• Distilled water to bring volume to	500.0 mL

VP Reagent A

• α-naphthol	5.0 g
• Absolute alcohol to bring volume to	100.0 mL

VP Reagent B*

• Potassium hydroxide	40.0 g
• Distilled water to bring volume to	100.0 mL

Procedure

Lab One

1. Obtain three MR-VP broths. Label two of them with the names of the organisms, your name, and the date. Label the third tube "control."
2. Inoculate two broths with the test cultures. Do not inoculate the control.
3. Incubate all tubes at 35 ± 2°C for 48 hours.

Lab Two

1. Mix each broth well. Transfer 1.0 mL from each broth to two non-sterile tubes. The pipettes should be sterile, but the tubes need not be. Refer to the procedural diagram in Figure 5-12.
2. For the three pairs of tubes, add the reagents as follows:

Tube #1: Methyl Red Test

a. Add three drops of Methyl Red reagent. Observe for red color formation immediately.

b. Negative results should be reincubated and tested again in 2–3 days as lab time permits.

c. Using Table 5-4 as a guide, record your results in the chart provided on the Data Sheet.

Tube #2: VP Test

a. Add 15 drops (0.6 mL) of VP reagent A. Mix well to oxygenate the medium.

b. Add 5 drops (0.2 mL) of VP reagent B. Mix well again to oxygenate the medium.

c. Place the tubes in a test tube rack, and do not move them for 1 hour. Observe for red color formation at regular intervals starting at 10 minutes for up to 1 hour.

d. Using Table 5-5 as a guide, record your results in the chart provided on the Data Sheet. It is sometimes difficult to differentiate weak positive reactions from negative reactions. VP (–) reactions often produce a copper color; weak VP (+) reactions produce a pink color. The strongest color change should be at the surface.)

*This reagent must be stored in a darkened glass container.

References

Chapin, Kimberle C., and Tsai-Ling Lauderdale. 2003. Page 361 in *Manual of Clinical Microbiology,* 8th ed. Edited by Patrick R. Murray, Ellen Jo Baron, James H. Jorgensen, Michael A. Pfaller, and Robert H. Yolken. ASM Press, American Society for Microbiology, Washington, DC.

Delost, Maria Dannessa. 1997. Pages 187–188 in *Introduction to Diagnostic Microbiology.* Mosby, St. Louis.

Forbes, Betty A., Daniel F. Sahm, and Alice S. Weissfeld. 2002. Page 275 in *Bailey & Scott's Diagnostic Microbiology,* 11th ed. Mosby, Inc., St. Louis.

MacFaddin, Jean F. 2000. Pages 321 and 439 in *Biochemical Tests for Identification of Medical Bacteria,* 3rd ed. Williams & Wilkins, Baltimore.

Zimbro, Mary Jo, and David A. Power, editors. 2003. Page 330 in *Difco™ and BBL™ Manual—Manual of Microbiological Culture Media.* Becton Dickinson and Co. Sparks, MD.

Tests Identifying Microbial Ability to Respire

In this unit we will examine several techniques designed to differentiate bacteria based on their ability to respire. As mentioned earlier in this section, respiration is the conversion of glucose to energy in the form of ATP by way of glycolysis, the Krebs cycle, and oxidative phosphorylation in the electron transport chain (ETC). In respiration, the final electron acceptor at the end of the ETC is always an inorganic molecule.

Tests that identify an organism as an aerobic or an anaerobic respirer generally are designed to detect specific products of (or constituent enzymes used in) the reduction of the final electron acceptor (FEA). Aerobic respirers reduce oxygen to water or other compounds. Anaerobic respirers reduce other inorganic molecules such as nitrate or sulfate. Nitrate is reduced to nitrogen gas (N_2) or other nitrogenous compounds such as nitrite (NO_2), and sulfate is reduced to hydrogen sulfide gas (H_2S). Metabolic pathways are described in detail in Appendix A.

The exercises and organisms chosen for this section directly or indirectly demonstrate the presence of an ETC. The catalase test detects organismal ability to produce catalase—an enzyme that detoxifies the cell by converting hydrogen peroxide produced in the ETC to water and molecular oxygen. The oxidase test identifies the presence of cytochrome c oxidase in the ETC. The nitrate reduction test examines bacterial ability to transfer electrons to nitrate at the end of the ETC and thus respire anaerobically. For more information on biochemical tests that demonstrate anaerobic respiration, refer to Exercise 5-20 SIM and 5-21 Triple Sugar Iron Agar.

Exercise 5-5

Catalase Test

Theory The electron transport chains of aerobic and facultatively anaerobic bacteria are composed of molecules capable of accepting and donating electrons as conditions dictate. As such, these molecules alternate between the oxidized and reduced forms, passing electrons down the chain to the final electron acceptor (FEA). Energy lost by electrons in this sequential transfer is used to perform oxidative phosphorylation (*i.e.*, produce ATP from ADP).

One carrier molecule in the ETC called **flavoprotein** can bypass the next carrier in the chain and transfer electrons directly to oxygen (Figure 5-13). This alternate pathway produces two highly potent cellular toxins — hydrogen peroxide (H_2O_2) and superoxide radical (O_2^-).

Organisms that produce these toxins also produce enzymes capable of breaking them down. **Superoxide dismutase** catalyzes conversion of superoxide radicals (the more lethal of the two compounds) to hydrogen peroxide (Figure 5-13). **Catalase** converts hydrogen peroxide into water and gaseous oxygen (Figure 5-14).

Bacteria that produce catalase can be easily detected using typical store-grade hydrogen peroxide. When hydrogen peroxide is added to a catalase-positive culture, oxygen gas bubbles form immediately (Figures 5-15 and 5-16). If no bubbles appear, the organism is catalase-negative. This test can be performed on a microscope slide or by adding hydrogen peroxide directly to the bacterial growth.

$$\underset{\text{Reduced Flavoprotein}}{FPH_2} + O_2 \longrightarrow \underset{\text{Oxidized Flavoprotein}}{FP} + \underset{\text{Hydrogen Peroxide}}{H_2O_2}$$

$$2H^+ + \underset{\text{Superoxide Radical}}{2O_2^-} \xrightarrow{\text{Superoxide dismutase}} \underset{\text{Hydrogen Peroxide}}{H_2O_2} + O_2$$

FIGURE 5-13 MICROBIAL PRODUCTION OF H_2O_2
Hydrogen peroxide may be formed through the transfer of electrons from reduced flavoprotein to oxygen or from the action of superoxide dismutase.

$$\underset{\text{Hydrogen Peroxide}}{2H_2O_2} \xrightarrow{\text{Catalase}} 2H_2O + O_2\uparrow$$

FIGURE 5-14 CATALASE MEDIATED CONVERSION OF H_2O_2
Catalase is an enzyme of aerobes, microaerophiles, and facultative anaerobes that converts hydrogen peroxide to water and oxygen gas.

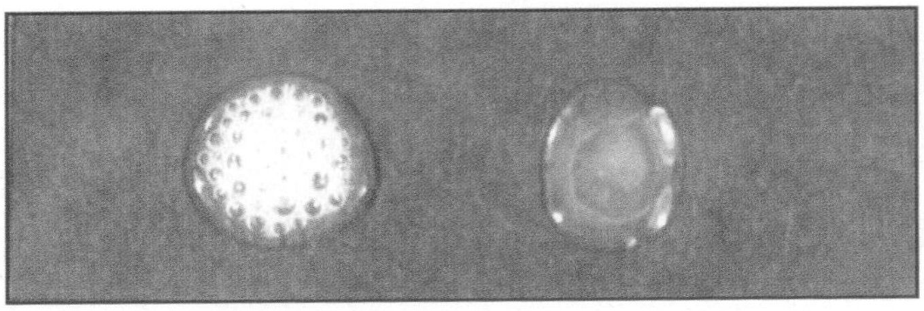

FIGURE 5-15 CATALASE SLIDE TEST
Visible bubble production indicates a positive result in the catalase slide test. A catalase-positive organism is on the left; a catalase-negative organism is on the right.

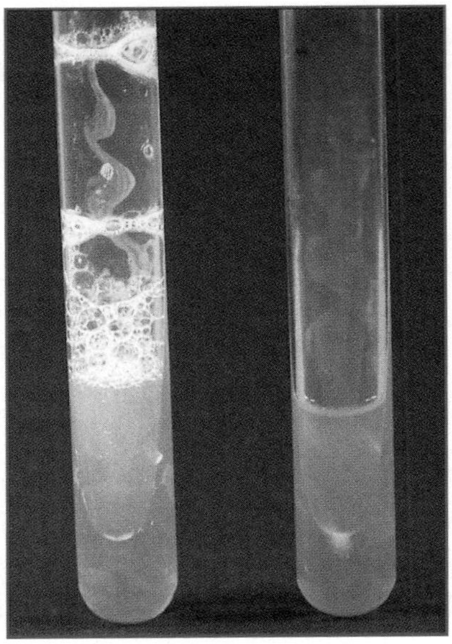

FIGURE 5-16
CATALASE TUBE TEST
The catalase test also may be performed on an agar slant. A catalase-positive organism is on the left; a catalase-negative organism is on the right.

Application

This test is used to identify organisms that produce the enzyme catalase. It is used most commonly to differentiate members of the catalase-positive *Micrococcaceae* from the catalase-negative *Streptococcaceae*. Variations on this test also may be used in identification of *Mycobacterium* species.

In This Exercise...

You will perform the Catalase Test. Materials will be provided so you can perform both the slide test and the slant test. It is important to do the slide test first because the tube test requires addition of hydrogen peroxide directly to the slant, after which the culture will be unusable.

Materials

Per Student Group

- three Nutrient Agar slants
- hydrogen peroxide (3% solution)
- transfer pipettes
- microscope slides
- fresh cultures of:
  - *Staphylococcus epidermidis*
  - *Enterococcus faecalis*

Procedure

Slide Test

1. Transfer a large amount of growth to a microscope slide. (Be sure to perform this test in the proper order. Placing the metal loop into H_2O_2 could catalyze a false positive reaction.)
2. Aseptically place one or two drops of hydrogen peroxide directly onto the bacteria and observe for the formation of bubbles immediately. (When running this test on an unknown, a positive control should be run simultaneously to verify the quality of the peroxide.)
3. If you get a negative result, observe the slide, using the scanning objective on your microscope.
4. Record your results in the chart provided on the Data Sheet.

Slant test

1. Add approximately 1 mL of hydrogen peroxide to each of the three slants. Do this *one slant at a time*, observing and recording the results as you go.
2. Record your results in the chart provided on the Data Sheet.

References

Collins, C. H., Patricia M. Lyne, J. M. Grange. 1995. Page 110 in *Collins and Lyne's Microbiological Methods,* 7th ed., Butterworth-Heinemann, Oxford, United Kingdom.

Forbes, Betty A., Daniel F. Sahm, and Alice S. Weissfeld. 2002. Pages 151 and 166 in *Bailey & Scott's Diagnostic Microbiology,* 11th ed. Mosby, St. Louis.

Lányi, B. 1987. Page 20 in *Methods in Microbiology,* Vol. 19, edited by R. R. Colwell and R. Grigorova. Academic Press, New York.

MacFaddin, Jean F. 2000. Page 78 in *Biochemical Tests for Identification of Medical Bacteria,* 3rd ed., Williams & Wilkins, Baltimore.

Smibert, Robert M., and Noel R. Krieg. 1994. Page 614 in *Methods for General and Molecular Bacteriology*, edited by Philipp Gerhardt, R. G. E. Murray, Willis A. Wood, and Noel R. Krieg, American Society for Microbiology, Washington, DC.

TABLE OF RESULTS

Result	Interpretation	Symbol
Bubbles	Catalase is present	+
No bubbles	Catalase is absent	–

TABLE 5-6 Catalase Test results and interpretations.

Exercise 5-6
Oxidase Test

Theory

Bacterial respiratory electron transport chains capture energy in the form of high-energy bonds in ATP via the oxidation of NADH and $FADH_2$ produced in the Krebs cycle and glycolysis (see Appendix A). The electron transport chains of many aerobes, facultative anaerobes, and microaerophiles contain a carrier molecule called **cytochrome c oxidase.** Cytochrome c oxidase is responsible for the transfer of electrons to oxygen, thus reducing it to water (Figure 5-17).

The oxidase test is performed using a **chromogenic reducing agent** called tetramethyl-*p*-phenylenediamine to detect bacteria that produce cytochrome c oxidase. Chromogenic reducing agents are chemicals that change or produce color as they become oxidized (Figure 5-18).

In the oxidase test, the chromogenic reducing reagent is added directly to bacterial growth on solid media (Figure 5-19), or (more conveniently) a colony is transferred to paper saturated with reagent (Figure 5-20). Within seconds the reagent, which acts as an artificial electron donor, will change color if oxidized cytochrome c is present. No color change within the allotted time means no cytochrome oxidase is present and is a negative result (Table 5-7).

Application

This test is used to identify bacteria containing the respiratory enzyme cytochrome c oxidase. Among its many uses is the presumptive identification of the oxidase-positive *Neisseria*. It also can be useful in differentiating the oxidase-negative *Enterobacteriaceae* from the oxidase-positive *Pseudomonadaceae*.

In This Exercise...

You will be performing the oxidase test (likely) using one of several commercially available systems. Your instructor will have chosen one that is best for your laboratory. Generally, test systems produce a color change

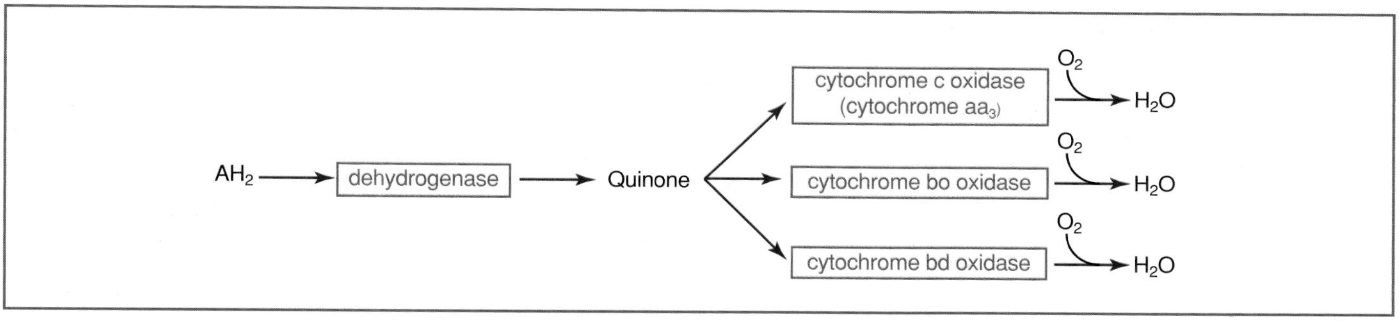

FIGURE 5-17 Aerobic Electron Transport Chains (ETCs)
There are a variety of different bacterial aerobic electron transport chains. All begin with a dehydrogenase that picks up electrons from electron carriers, such as NADH or $FADH_2$ (represented by AH_2). Chains that resemble the mitochondrial ETC have cytochrome c oxidase at the end and give a positive result for the oxidase test. Other bacteria, such as members of the *Enterobacteriaceae*, are capable of aerobic respiration but have a different terminal oxidase system and give a negative result for the cytochrome c oxidase test. The two lower paths in the diagram are both found in *E. coli*. The amount of available O_2 determines which one is most active. (After White, 2000.)

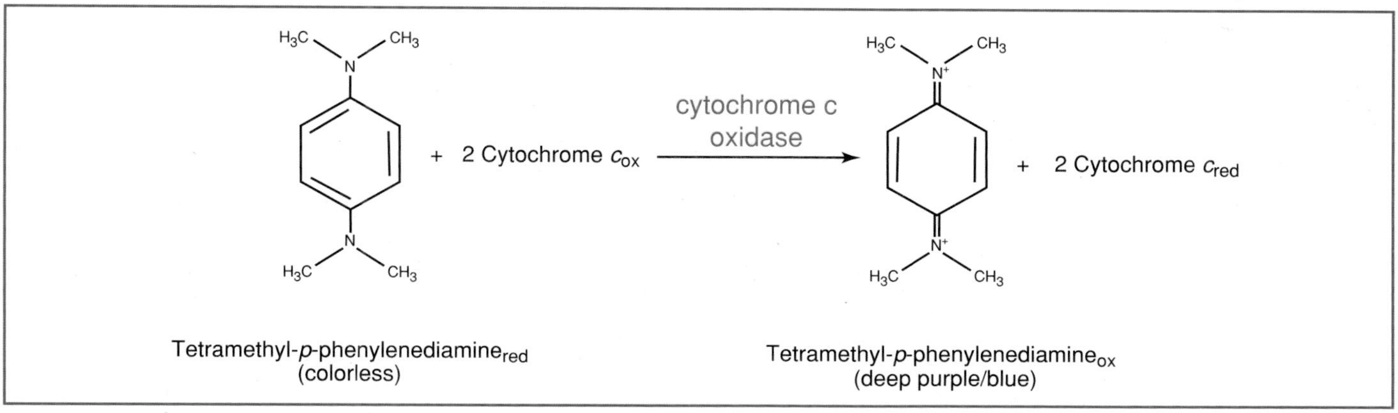

FIGURE 5-18 Chemistry of the Oxidase Reaction
The oxidase enzyme shown is not directly involved in the indicator reaction as shown. Rather, it removes electrons from cytochrome c, making it available to react with the phenylenediamine reagent.

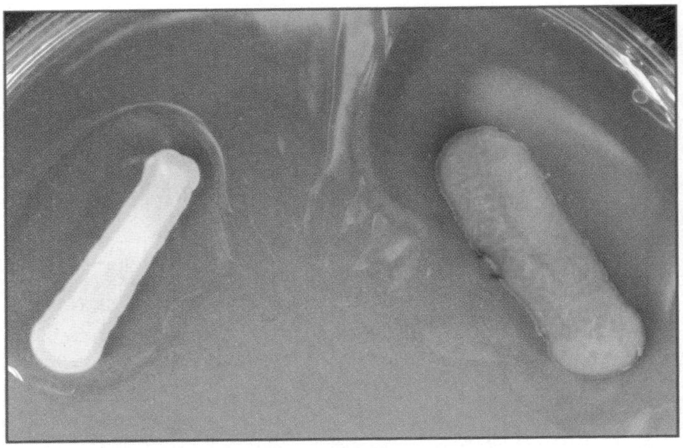

FIGURE 5-19 **Oxidase Test on Bacterial Growth**
A few drops of reagent on oxidase-positive bacteria will produce a purple-blue color immediately.

from colorless to blue/purple. The system shown in Figure 5-20 is BBL™ *Dry*Slide™, which calls for a transfer of inoculum to the slide and a reading within 20 seconds. Reagents used for this test are unstable and will independently oxidize shortly after they become wet. Therefore, it is important not to take your readings beyond the time recommended by the manufacturer.

Materials

Per Student Group

- sterile cotton tipped applicator or plastic loop
- sterile water
- fresh slant cultures of:
  - *Escherichia coli*
  - *Moraxella catarrhalis*

Procedure

1. Place a drop of sterile water on the reagent paper. Allow 20 seconds for a reaction.
2. Following your instructor's guidelines, transfer some of the culture to the reagent slide. Get a fair amount of growth on the end of a sterile wooden applicator, and roll it on the reagent slide.
3. Observe for color change within 20 seconds.
4. Repeat steps one and two for the second organism.
5. Enter your results in the chart provided on the Data Sheet.
6. When performing this test on an unknown, run a positive control simultaneously to verify the quality of the test system.

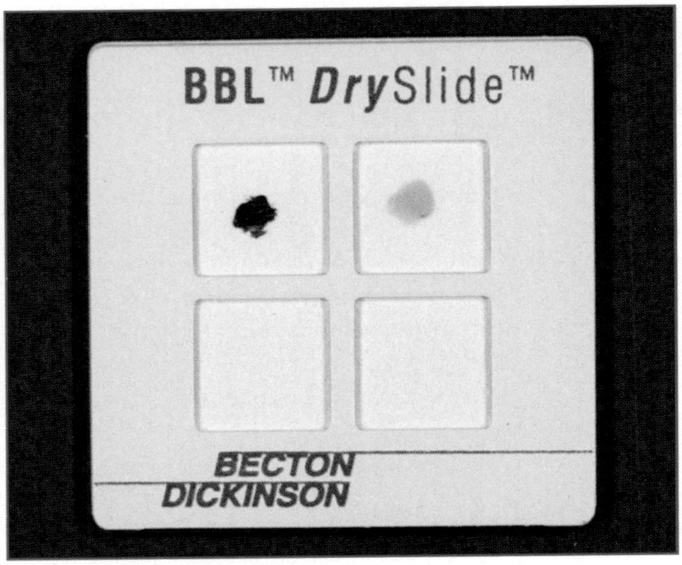

FIGURE 5-20 **Oxidase Slide Test**
Positive results with this test should appear within 20 seconds. The dark blue is a positive result; no color change is a negative result. The organism in the top right panel demonstrates a negative result (not blue), but appears yellow because it is pigmented. (BBL™ *Dry*Slide™ systems available from Becton Dickinson, Sparks, MD.)

References

BBL™ *Dry*Slide™ package insert.

Collins, C. H., Patricia M. Lyne, J. M. Grange. 1995. Page 116 in Collins and Lyne's *Microbiological Methods,* 7th ed. Butterworth-Heinemann, Oxford, United Kingdom.

Forbes, Betty A., Daniel F. Sahm, Alice S. Weissfeld. 2002. Pages 151–152 in *Bailey & Scott's Diagnostic Microbiology,* 11th ed. Mosby, St. Louis.

Lányi, B. 1987. Page 18 in *Methods in Microbiology,* Vol. 19, edited by R. R. Colwell and R. Grigorova. Academic Press, New York.

MacFaddin, Jean F. 2000. Page 368 in *Biochemical Tests for Identification of Medical Bacteria,* 3rd ed. Lippincott Williams & Wilkins, Philadelphia.

Smibert, Robert M., and Noel R. Krieg. 1994. Page 625 in *Methods for General and Molecular Bacteriology*, edited by Philipp Gerhardt, R. G. E. Murray, Willis A. Wood, and Noel R. Krieg. American Society for Microbiology, Washington, DC.

White, David, 2000. Chapter 4 in *The Physiology and Biochemistry of Prokaryotes,* 2nd ed. Oxford University Press, New York.

TABLE OF RESULTS

Result	Interpretation	Symbol
Dark blue within 20 seconds	Cytochrome c oxidase is present	+
No color change to blue or blue after 20 seconds	Cytochrome c oxidase is absent	–

TABLE 5-7 Catalase Test results and interpretations.

Exercise 5-7
Nitrate Reduction Test

Theory As mentioned in the introduction to this unit, anaerobic respiration involves the reduction of (*i.e.*, transfer of electrons to) an inorganic molecule other than oxygen. As the name implies, nitrate reduction is one such example. Many Gram negative bacteria (including most *Enterobacteriaceae*) contain the enzyme **nitrate reductase** and perform a single-step reduction of nitrate to nitrite (($NO_3 \rightarrow NO_2$). Other bacteria, in a multi-step process known as **denitrification**, are capable of enzymatically converting nitrate to molecular nitrogen (N_2). Some products of nitrate reduction are shown in Figure 5-21.

Nitrate broth is an undefined medium of beef extract, peptone, and potassium nitrate (KNO_3). An inverted Durham tube is placed in each broth to trap a portion of any gas produced. In contrast to many differential media, no color indicators are included. The color reactions obtained in nitrate broth take place as a result of reactions between metabolic products and reagents added after incubation (Figure 5-22).

Before a broth can be tested for nitrate reductase activity (nitrate reduction to nitrite), it must be examined for evidence of denitrification. This is simply a visual inspection for the presence of gas in the Durham tube (Figure 5-23). If there is gas in the Durham tube, and the organism is known not to be a fermenter (as evidenced by a fermentation test), the test is complete. Denitrification has taken place. Gas produced in a nitrate reduction test by an organism capable of fermenting is not determinative because the source of the gas is unknown. For more information on fermentation, see Exercises 5-2 through 5-4 in this section.

If there is no visual evidence of denitrification, sulfanilic acid (nitrate reagent A) and naphthylamine (nitrate reagent B) are added to the medium to test for nitrate reduction to nitrite. If present, nitrite will form

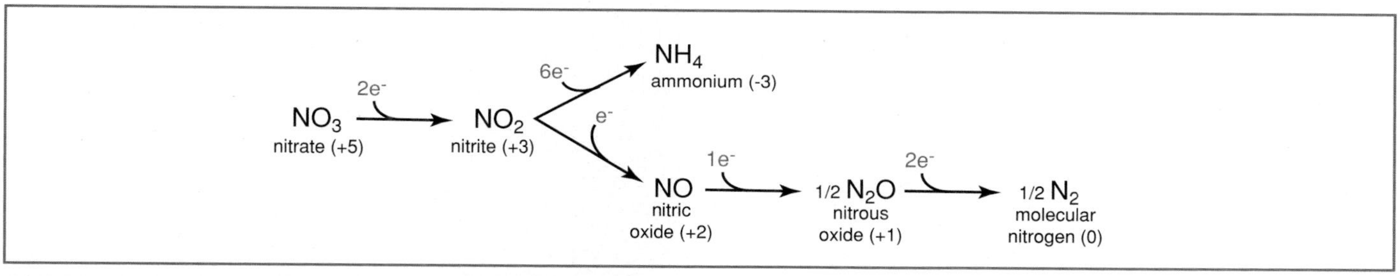

FIGURE 5-21 POSSIBLE END PRODUCTS OF NITRATE REDUCTION
Nitrate reduction is complex. Many different organisms under many different circumstances perform nitrate reduction with many different outcomes. Members of the *Enterobacteriaceae* simply reduce NO_3 to NO_2. Other bacteria, functionally known as "denitrifiers," reduce NO_3 all the way to N_2 via the intermediates shown, and are important ecologically in the nitrogen cycle. Both of these are anaerobic respiration pathways (also known as "nitrate respiration"). Other organisms are capable of assimilatory nitrate reduction, in which NO_3 is reduced to NH_4, which can be used in amino acid synthesis. The oxidation state of nitrogen in each compound is shown in parentheses.

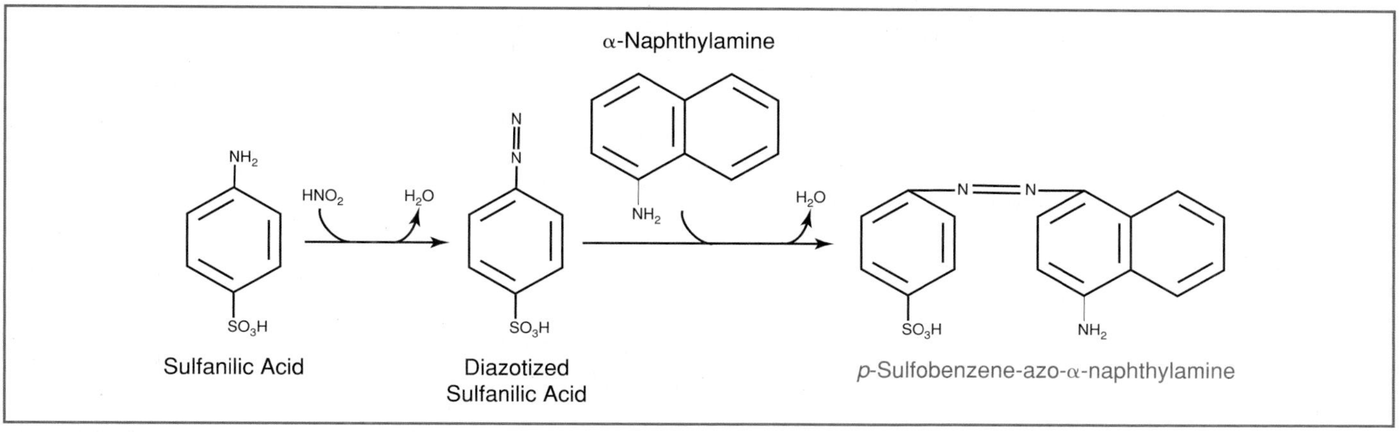

FIGURE 5-22 INDICATOR REACTION
If nitrate is reduced to nitrite, nitrous acid (HNO_2) will form in the medium. Nitrous acid then reacts with sulfanilic acid to form diazotized sulfanilic acid, which reacts with the α-naphthylamine to form *p*-sulfobenzene-azo-α-naphthylamine, which is red. Thus, a red color indicates the presence of nitrite and is considered a positive result for nitrate reduction to nitrite.

FIGURE 5-23 Incubated Nitrate Broth Before the Addition of Reagents
Numbered 1 through 5 from left to right, these are Nitrate Broths immediately after incubation before the addition of reagents. Tube 2 is an uninoculated control used for color comparison. Note the gas produced by tube 5. It is a known nonfermenter and, therefore, will receive no reagents. The gas produced is an indication of denitrification and a positive result. Tubes 1 through 4 will receive reagents. See Figure 5-24.

FIGURE 5-24 Incubated Nitrate Broth After Addition of Reagents
After the addition of reagents, tube 1 shows a positive result. Tube 3 and tube 4 are inconclusive because they show no color change. Zinc dust must be added to tubes 2 (control), 3, and 4 to verify the presence or absence of nitrate. See Figure 5-25.

FIGURE 5-25 Incubated Nitrate Broth After Addition of Reagents and Zinc
Finally, a pinch of zinc is added to tubes 2, 3, and 4 because they have been colorless up to this point. Tube 2 (the control tube) and tube 3 turned red. This is a negative result because it indicates that nitrate is still present in the tube. Tube 4 did not change color, which indicates that the nitrate was reduced by the organism beyond nitrite to some other nitrogenous compound. This is a positive result.

nitrous acid (HNO_2) in the aqueous medium. Nitrous acid reacts with the added reagents to produce a red, water-soluble compound (Figure 5-24). Therefore, formation of red color after the addition of reagents indicates that the organism reduced nitrate to nitrite. If no color change takes place with the addition of reagents, the nitrate either was not reduced or was reduced to one of the other nitrogenous compounds shown in Figure 5-21. Because it is visually impossible to tell the difference between these two occurrences, another test must be performed.

In this stage of the test, a small amount of powdered zinc is added to the broth to catalyze the reduction of any nitrate (which may still be present as KNO_3) to nitrite. If nitrate is present at the time zinc is added, it will be converted immediately to nitrite, and the above-described reaction between nitrous acid and reagents will follow and turn the medium red. In this instance, the red color indicates that nitrate was *not* reduced by the organism (Figure 5-25). No color change after the addition of zinc indicates that the organism reduced the nitrate to NH_3, NO, N_2O, or some other nongaseous nitrogenous compound.

In This Exercise...

After inoculation and incubation of the Nitrate Broths, you will first inspect the Durham tubes included in the media for bubbles—an indication of gas production (Figure 5-23). One or more of the organisms selected for this exercise are fermenters, and possible producers of gas other than molecular nitrogen; therefore, you will proceed with adding reagents to all tubes. Record any gas observed in the Durham tubes and see if there is a correlation between that and the color results after the reagents and zinc dust have been added. Follow the procedure carefully, and use Table 5-8 as a guide when interpreting the results.

Application

This test is used to distinguish members of *Enterobacteriaceae*, all of which are nitrate reducers, from other Gram negative rods.

Materials

Per Student Group

- four Nitrate Broths
- nitrate test reagents A and B
- empty nonsterile test tubes
- zinc powder
- fresh cultures of:
  - *Erwinia amylovora*
  - *Escherichia coli*
  - *Pseudomonas aeruginosa*

Medium and Reagent Recipes

Nitrate Broth

• Beef extract	3.0 g
• Peptone	5.0 g
• Potassium nitrate	1.0 g
• Distilled or deionized water	1.0 L

pH 6.8–7.2 at 25°C

Reagent A

• *N,N*-Dimethyl-α-naphthylamine	0.6 g
• 5N Acetic acid (30%)	100.0 mL

Reagent B

• Sulfanilic acid	0.8 g
• 5N Acetic acid (30%)	100.0

Procedure

Lab One

1. Obtain four Nitrate Broths. Label three of them with the names of the organisms, your name, and the date. Label the fourth tube "control."
2. Inoculate three broths with the test organisms. Do not inoculate the control.
3. Incubate all tubes at 35 ± 2°C for 24 to 48 hours.

Lab Two

1. Examine each tube for evidence of gas production. Record your results in the chart provided on the Data Sheet. Refer to Table 5-8 when making your interpretations. (Be methodical when recording your results. Red color can have opposite interpretations depending on where you are in the procedure.)
2. Using Figure 5-26 as a guide, proceed to the addition of reagents to *all* tubes as follows:
   a. Add eight drops each (approximately 0.5 mL) of reagent A and reagent B to each tube. Mix well and let the tubes stand undisturbed for 10 minutes. Using the control as a comparison, record your test results in the chart provided in the Data Sheet. Set aside any test broths that are positive.
   b. Where appropriate, add a pinch of zinc dust. (A very small amount is all that is needed. Dip a wooden applicator into the zinc dust and transfer only the amount that clings to the wood.) Let the tubes stand for 10 minutes. Record your results in the chart provided on the Data Sheet.

TABLE OF RESULTS

Result	Interpretation	Symbol
Gas (Nonfermenter)	Denitrification—production of nitrogen gas ($NO_3 \rightarrow NO_2 \rightarrow N_2$)	+
Gas (Fermenter, or status is unknown)	Source of gas is unknown; requires addition of reagents	
Red color (after addition of reagents A and B)	Nitrate reduction to nitrite ($NO_3 \rightarrow NO_2$)	+
No color (after the addition of reagents)	Incomplete test; requires the addition of zinc dust	
No color change (after addition of zinc)	Nitrate reduction to non-gaseous nitrogenous compounds ($NO_3 \rightarrow NO_2 \rightarrow$ non gaseous nitrogenous products)	+
Red color (after addition of zinc dust)	No nitrate reduction	–

TABLE 5-8 Nitrate Test results and interpretations.

References

Atlas, Ronald M., and Richard Bartha. 1998. Pages 423–425 in *Microbial Ecology—Fundamentals and Applications,* 4th ed. Benjamin/Cummings Science Publishing, Menlo Park, CA.

Forbes, Betty A., Daniel F. Sahm, and Alice S. Weissfeld. 2002. Page 277 in *Bailey & Scott's Diagnostic Microbiology,* 11th ed. Mosby, St. Louis.

Lányi, B. 1987. Page 21 in *Methods in Microbiology,* Vol. 19, edited by R. R. Colwell and R. Grigorova. Academic Press, New York.

MacFaddin, Jean F. 2000. Page 348 in *Biochemical Tests for Identification of Medical Bacteria,* 3rd ed. Williams & Wilkins, Philadelphia.

Moat, Albert G., John W. Foster, and Michael P. Spector. Chapter 14 in *Microbial Physiology,* 4th ed. Wiley-Liss, New York.

Zimbro, Mary Jo, and David A. Power, editors. 2003. Page 400 in *Difco™ and BBL™ Manual—Manual of Microbiological Culture Media.* Becton Dickinson and Co., Sparks, MD.

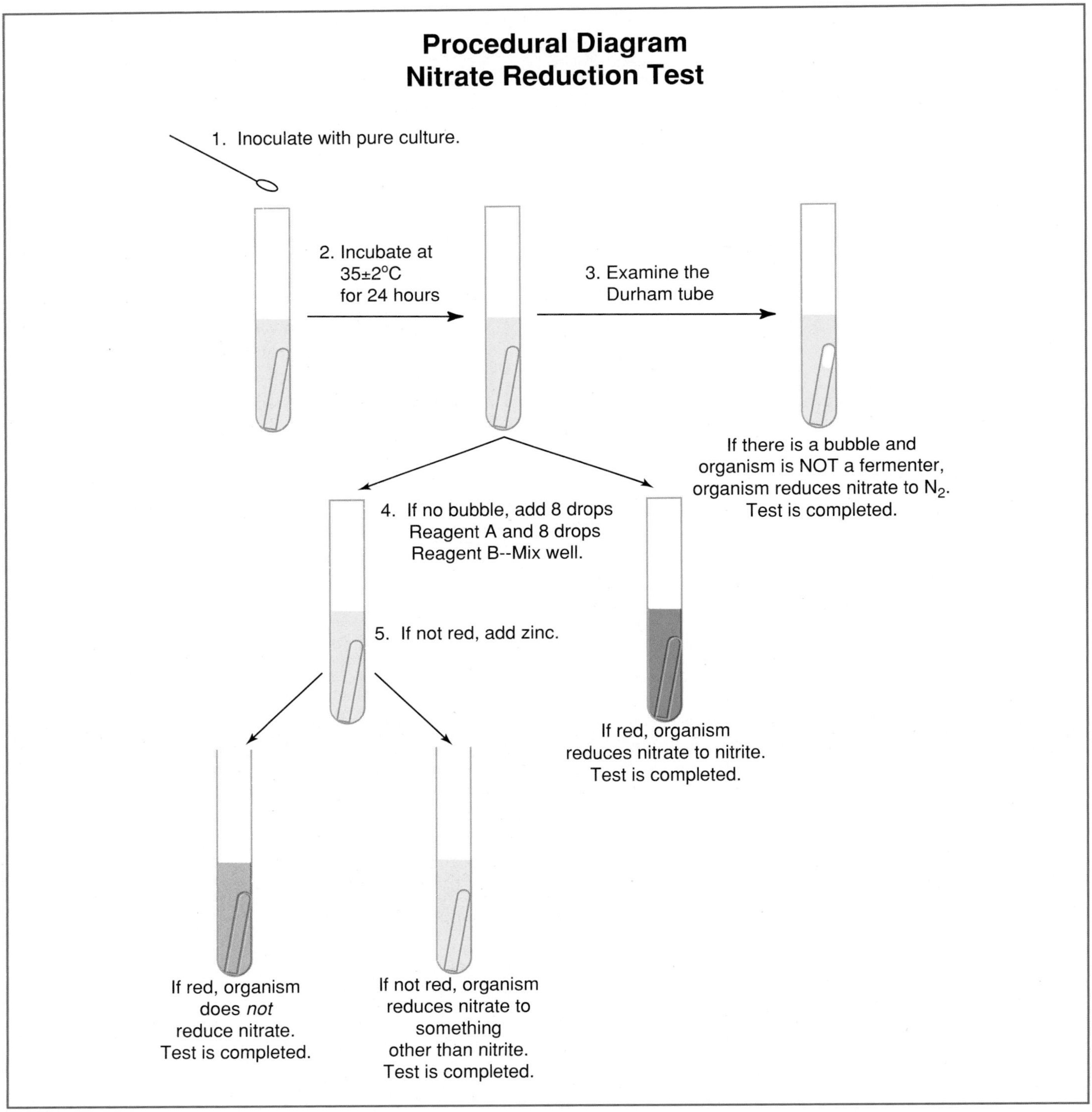

FIGURE 5-26 PROCEDURAL DIAGRAM FOR THE NITRATE REDUCTION TEST

Utilization Media

In this unit you will perform tests using two examples of utilization media—Simmons Citrate Medium and Malonate Broth. Utilization media are highly defined formulations designed to differentiate organisms based on their ability to grow when an essential nutrient (*e.g.*, carbon or nitrogen) is strictly limited. For example, citrate medium contains sodium citrate as the only carbon-containing compound and ammonium ion as the only nitrogen source. Malonate broth contains three sources of carbon but prevents utilization of all but one by competitive inhibition of a specific enzyme.

Exercise 5-8

Citrate Test

Theory In many bacteria, citrate (citric acid) produced as acetyl coenzyme A (from the oxidation of pyruvate or the β-oxidation of fatty acids) reacts with oxaloacetate at the entry to the Krebs cycle. Citrate then is converted back to oxaloacetate through a complex series of reactions, which begins the cycle anew. Refer to Appendix A and Figure 5-58 for more information on the Krebs cycle and fatty acid metabolism.

In a medium in which citrate is the only available carbon source, bacteria that possess **citrate-permease** can transport the molecules into the cell and produce pyruvate (pyruvic acid) by a reversal of the above-described reaction. Pyruvate then can be converted to a variety of products depending on the pH of the environment (Figure 5-27).

Simmons Citrate Agar is a defined medium that contains sodium citrate as the sole carbon source and ammonium phosphate as the sole nitrogen source. Bromthymol blue dye, which is green at pH 6.9 and blue at pH 7.6, is added as an indicator. Bacteria that survive in the medium and utilize the citrate also convert the ammonium phosphate to ammonia (NH_3) and ammonium hydroxide (NH_4OH), both of which tend to alkalinize the agar. As the pH goes up, the medium changes from green to blue (Figure 5-28). Thus, conversion of the medium to blue is a positive citrate test result (Table 5-9).

Occasionally a citrate-positive organism will grow on a Simmons Citrate slant without producing a change in color. In most cases, this is because of incomplete incubation. In the absence of color change, growth on the slant indicates that citrate is being utilized and is evidence of a positive reaction. To avoid confusion between actual growth and heavy inoculum, which may appear as growth, citrate slants typically are

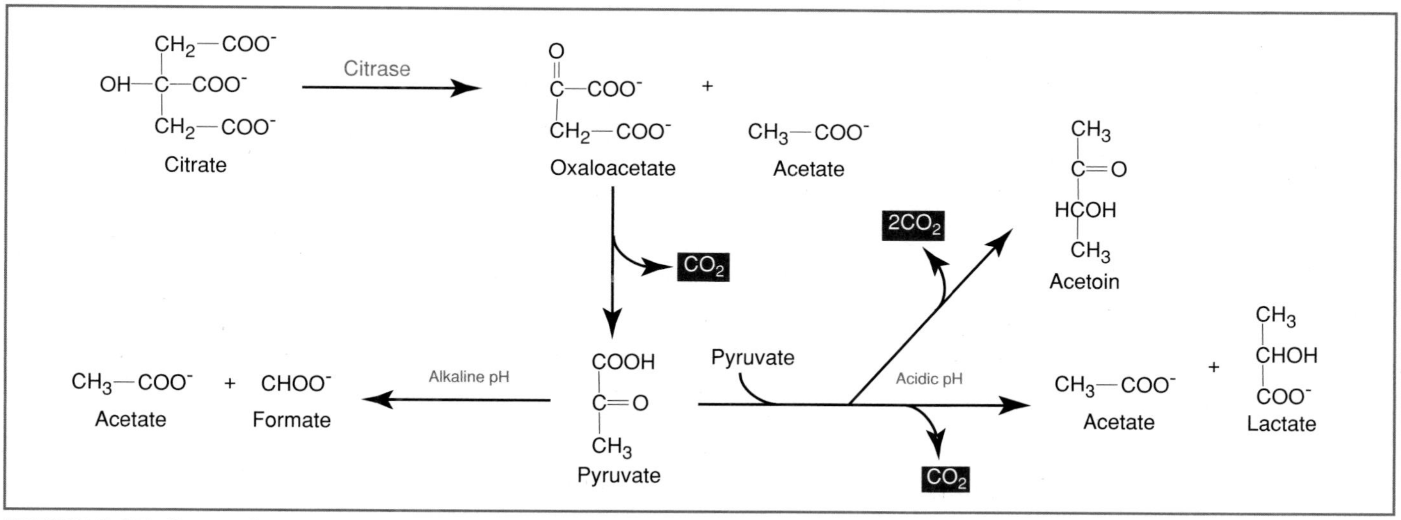

FIGURE 5-27 CITRATE CHEMISTRY
In the presence of citrate-permease enzyme, citrate enters the cell and is converted to pyruvate. The pyruvate then is converted to a variety of products depending on the pH of the environment.

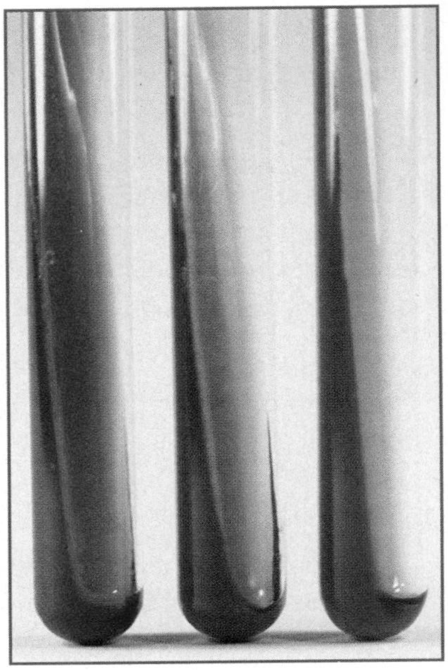

FIGURE 5-28
Citrate Test Results
These Simmons Citrate slants were inoculated with a citrate-positive (+) organism on the left and a citrate-negative (−) organism in the center. The slant on the right is uninoculated.

TABLE OF RESULTS

Result	Interpretation	Symbol
Blue (even a small amount)	Citrate is utilized	+
No color change; growth	Citrate is utilized	+
No color change; no growth	Citrate is not utilized	−

TABLE 5-9 Citrate Test results and interpretations.

inoculated lightly with an inoculating needle rather than a loop.

Application

The citrate utilization test is used to determine the ability of an organism to use citrate as its sole source of carbon. Citrate utilization is one part of a test series referred to as the IMViC (*I*ndole, *M*ethyl Red, *V*oges-Proskauer and Citrate tests) that is used to distinguish between members of the family *Enterobacteriaceae* and differentiate them from other Gram negative rods.

In This Exercise...

You will be inoculating Simmons Citrate slants with a citrate-positive and a citrate-negative organism. To avoid confusing growth on the slant with a heavy inoculum, be sure to inoculate using a needle instead of a loop.

Materials

Per Student Group

- three Simmons Citrate slants
- fresh slant cultures of:
  - *Enterobacter aerogenes*
  - *Escherichia coli*

Medium Recipe

Simmons Citrate Agar

- Ammonium dihydrogen phosphate 1.0 g
- Dipotassium phosphate 1.0 g
- Sodium chloride 5.0 g
- Sodium citrate 2.0 g
- Magnesium sulfate 0.2 g
- Agar 15.0 g
- Bromthymol blue 0.08 g
- Distilled or deionized water 1.0 L

pH 6.7–7.1 at 25°C

Procedure

Lab One

1. Obtain three Simmons Citrate tubes. Label two with the names of the organisms, your name, and the date. Label the third tube "control."
2. Using an inoculating needle and *light inoculum*, streak the slants with the test organisms. Do not inoculate the control.
3. Incubate all tubes at 35 ± 2°C for 48 hours.

Lab Two

1. Observe the tubes for color changes and/or growth.
2. Record your results in the chart provided on the Data Sheet.

References

Collins, C. H., Patricia M. Lyne, and J. M. Grange. 1995. Page 111 in *Collins and Lyne's Microbiological Methods,* 7th ed. Butterworth-Heinemann, Oxford, United Kingdom.

Forbes, Betty A., Daniel F. Sahm, and Alice S. Weissfeld. 2002. Page 266 in *Bailey & Scott's Diagnostic Microbiology,* 11th ed. Mosby, St. Louis.

MacFaddin, Jean F. 2000. Page 98 in *Biochemical Tests for Identification of Medical Bacteria,* 3rd ed. Lippincott Williams & Wilkins, Philadelphia.

Smibert, Robert M., and Noel R. Krieg. 1994. Page 614 in *Methods for General and Molecular Bacteriology*, edited by Philipp Gerhardt, R. G. E. Murray, Willis A. Wood and Noel R. Krieg, American Society for Microbiology, Washington, DC.

Zimbro, Mary Jo, and David A. Power, editors. 2003. Page 514 in *Difco™ and BBL™ Manual—Manual of Microbiological Culture Media.* Becton Dickinson and Co., Sparks, MD.

Exercise 5-9

Malonate Test

Theory

One of the many enzymatic reactions of the Krebs cycle, as illustrated in Appendix A, is the oxidation of succinate to fumarate. In the reaction, which requires the enzyme **succinate dehydrogenase**, the coenzyme FAD is reduced to $FADH_2$. Refer to the upper reaction in Figure 5-29.

Malonate (malonic acid), which can be added to growth media, is similar enough to succinate to replace it as the substrate in the reaction (Figure 5-29, lower reaction). This **competitive inhibition** of succinate dehydrogenase, in combination with the subsequent buildup of succinate in the cell, shuts down the Krebs cycle and will kill the organism unless it can ferment or utilize malonate as its sole remaining carbon source.

Malonate Broth is the medium used to make this determination. It contains a high concentration of sodium malonate, yeast extract, and a very small amount of glucose to promote growth of organisms that otherwise are slow to respond. Buffers are added to stabilize the medium at pH 6.7. Bromthymol blue dye, which is green in uninoculated media, is added to indicate any shift in pH. If an organism cannot utilize malonate but manages to ferment a small amount of glucose, it may turn the medium slightly yellow or produce no color change at all. These are negative results. If the organism utilizes malonate, it will alkalinize the medium and change the indicator from green to deep blue (Figure 5-30). Deep blue is positive (Table 5-10).

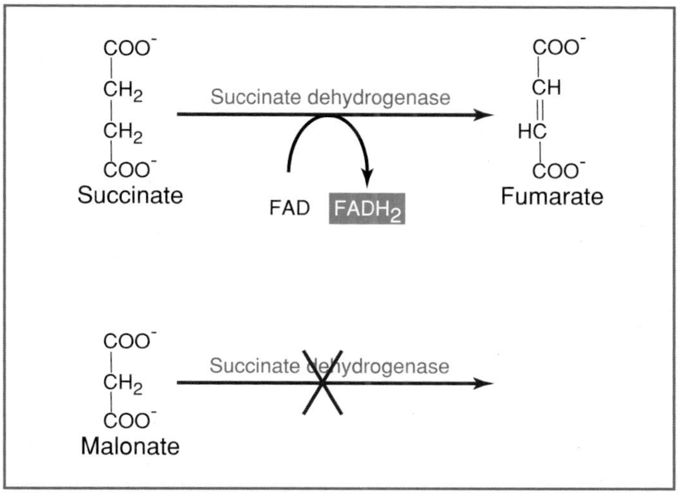

FIGURE 5-29 COMPETITIVE INHIBITION OF SUCCINATE DEHYDROGENASE BY MALONATE
The attachment of malonate to the enzyme succinate dehydrogenase prevents the attachment of succinate and, thus, the conversion of succinate to fumarate.

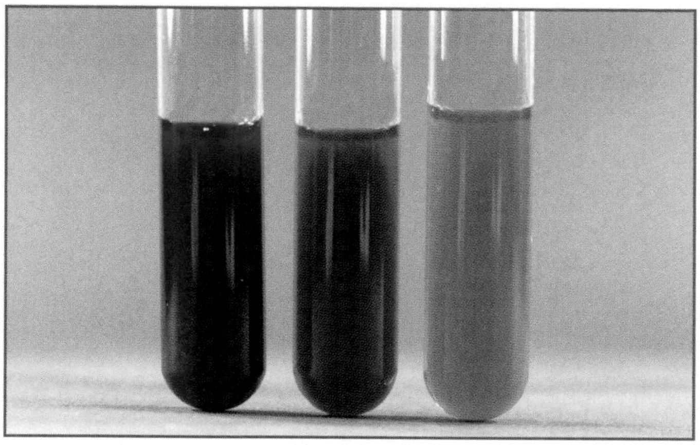

FIGURE 5-30 MALONATE TEST RESULTS
Malonate broths illustrating malonate-positive (+) on the left, an uninoculated control in the center, and malonate-negative (−) on the right.

TABLE OF RESULTS

Result	Interpretation	Symbol
Dark blue	Malonate is utilized	+
No color change or slightly yellow	Malonate is not utilized	−

TABLE 5-10 Malonate Test results and interpretations.

Application

The Malonate Test was designed originally to differentiate between *Escherichia*, which will not grow in the medium, and *Enterobacter.* Its use as a differential medium has now broadened to include other members of *Enterobacteriaceae.*

In This Exercise…

Today, you will inoculate two Malonate Broths and observe for color changes. Blue color is the only positive for this test; all others are negative.

Materials

Per Student Group

- three Malonate Broths
- fresh cultures of:
  - *Enterobacter aerogenes*
  - *Escherichia coli*

Medium Recipe

Malonate Broth

- Yeast extract 1.0 g
- Ammonium sulfate 2.0 g
- Dipotassium phosphate 0.6 g
- Monopotassium phosphate 0.4 g
- Sodium chloride 2.0 g
- Sodium malonate 3.0 g
- Glucose 0.25 g
- Bromthymol blue 0.025 g
- Distilled or deionized water 1.0 L

pH 6.5–6.9 at 25°C

Procedure

Lab One

1. Obtain three Malonate Broths. Label two of them with the names of the organisms, your name, and the date. Label the third tube "control."
2. Inoculate two broths with the test organisms. Do not inoculate the control.
3. Incubate all tubes at 35 ± 2°C for 48 hours.

Lab Two

1. Observe the tubes for color changes.
2. Record your results in the chart provided on the Data Sheet.

References

MacFaddin, Jean F. 2000. Page 310 in *Biochemical Tests for Identification of Medical Bacteria,* 3rd ed. Lippincott Williams & Wilkins, Philadelphia.

Zimbro, Mary Jo, and David A. Power, editors. 2003. Page 343 in *Difco™ and BBL™ Manual—Manual of Microbiological Culture Media.* Becton Dickinson and Co., Sparks, MD.

Decarboxylation and Deamination Tests

Decarboxylation and deamination tests were designed to differentiate members of *Enterobacteriaceae* and to distinguish them from other Gram negative rods. Most members of *Enterobacteriaceae* produce one or more enzymes that are necessary to break down amino acids. Enzymes that catalyze the removal of an amino acid's carboxyl group (COOH) are called **decarboxylases.** Enzymes that catalyze the removal of an amino acid's amine group (NH_2) are called **deaminases.**

Each decarboxylase and deaminase is specific to a particular substrate. Decarboxylases catalyze reactions that produce alkaline products. Thus, we are able to identify an organism's ability to produce a specific decarboxylase by preparing base medium, adding a known amino acid, and including a pH indicator to mark the shift to alkalinity. Differentiation of an organism in deamination media employs the same principle of substrate exclusivity by including a single known amino acid but requires the addition of a chemical reagent to produce a readable result.

In the next two exercises you will be introduced to the most common of both types of tests. In Exercise 5-10 you will test bacterial ability to decarboxylate the amino acids arginine, lysine, and ornithine. In Exercise 5-11 you will test bacterial ability to deaminate the amino acid phenylalanine.

Exercise 5-10

Decarboxylation Test

Theory Møller's Decarboxylase Base Medium contains peptone, glucose, the pH indicator bromcresol purple, and the **coenzyme** pyridoxal phosphate. Bromcresol purple is purple at pH 6.8 and above, and yellow below pH 5.2. Base medium can be used with one of a number of specific amino acid substrates, depending on the decarboxylase to be identified.

After inoculation, an overlay of mineral oil is used to seal the medium from external oxygen and promote fermentation (Figure 5-31). Glucose fermentation in the anaerobic medium initially turns it yellow because of the accumulation of acid end products. The low pH and presence of the specific amino acid induces **decarboxylase-positive** organisms to produce the enzyme.

Decarboxylation of the amino acid results in accumulation of alkaline end products that turn the medium purple (refer to Figures 5-32 through 5-35). If the organism is a glucose fermenter (as are all of the *Enterobacteriaceae*) but does not produce the appropriate decarboxylase, the medium will turn yellow and remain so. If the organism does not ferment glucose, the medium will exhibit no color change. Purple color is the only positive result; all others are negative (Figure 5-36).

Application Decarboxylase tests are used to differentiate organisms in the family *Enterobacteriaceae* and to distinguish them from other Gram negative rods.

In This Exercise...

You will perform decarboxylation tests on three organisms in Møller's arginine, lysine, ornithine, and base decarboxylase media. This procedure requires the aseptic addition of mineral oil. Be careful not to introduce contaminants while adding the oil. Refer to Figure 5-36 and Table 5-11 when interpreting your results.

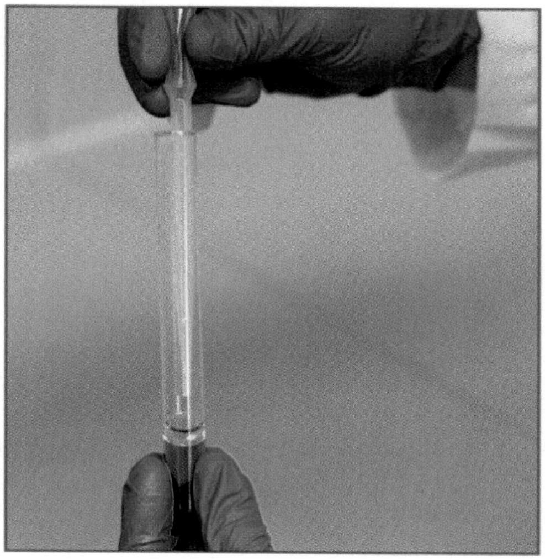

FIGURE 5-31 Adding Mineral Oil to the Tubes

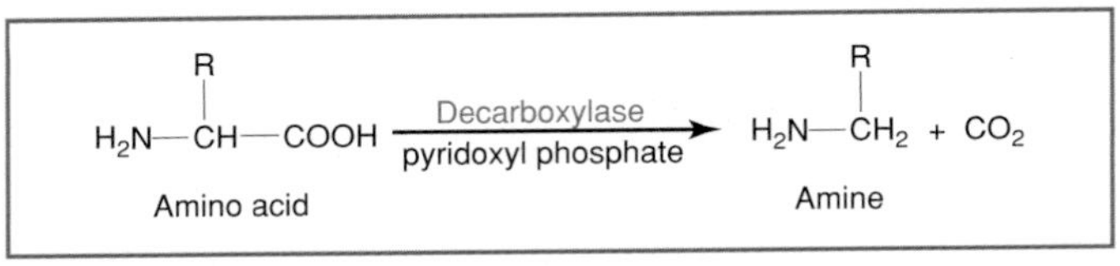

FIGURE 5-32 AMINO ACID DECARBOXYLATION
Removal of an amino acid's carboxyl group results in the formation of an amine and carbon dioxide.

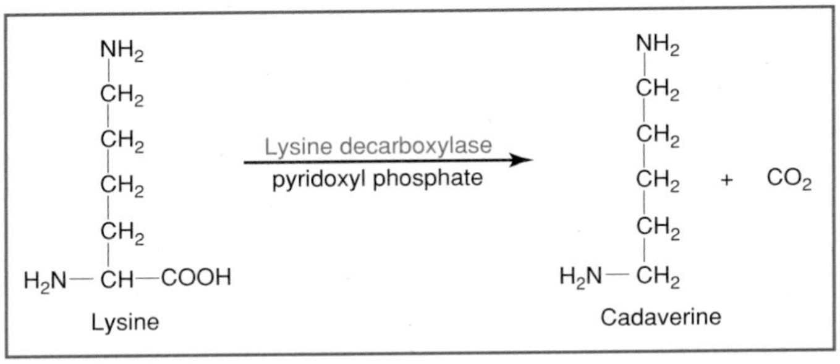

FIGURE 5-33 LYSINE DECARBOXYLATION
Decarboxylation of the amino acid lysine produces cadaverine and CO_2.

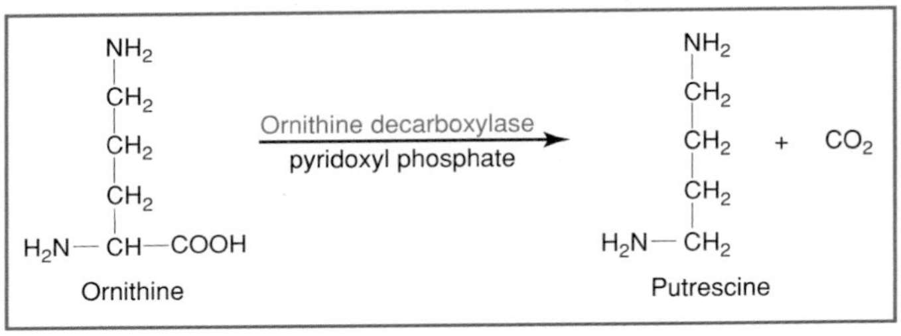

FIGURE 5-34 ORNITHINE DECARBOXYLATION
Decarboxylation of the amino acid ornithine produces putrescine and CO_2.

Materials

Per Student Group

- four Lysine Decarboxylase Broths
- four Ornithine Decarboxylase Broths
- four Arginine Decarboxylase Broths
- four Decarboxylase Base Media
- sterile mineral oil
- sterile transfer pipettes
- recommended organisms:
  - *Klebsiella pneumoniae*
  - *Enterobacter aerogenes*
  - *Proteus vulgaris*

FIGURE 5-36 DECARBOXYLATION TEST RESULTS
A lysine decarboxylase-positive (+) organism is on the left, an uninoculated control is in the center, and a lysine decarboxylase negative (−) is on the right. The color results shown here are representative of arginine and ornithine decarboxylase media as well.

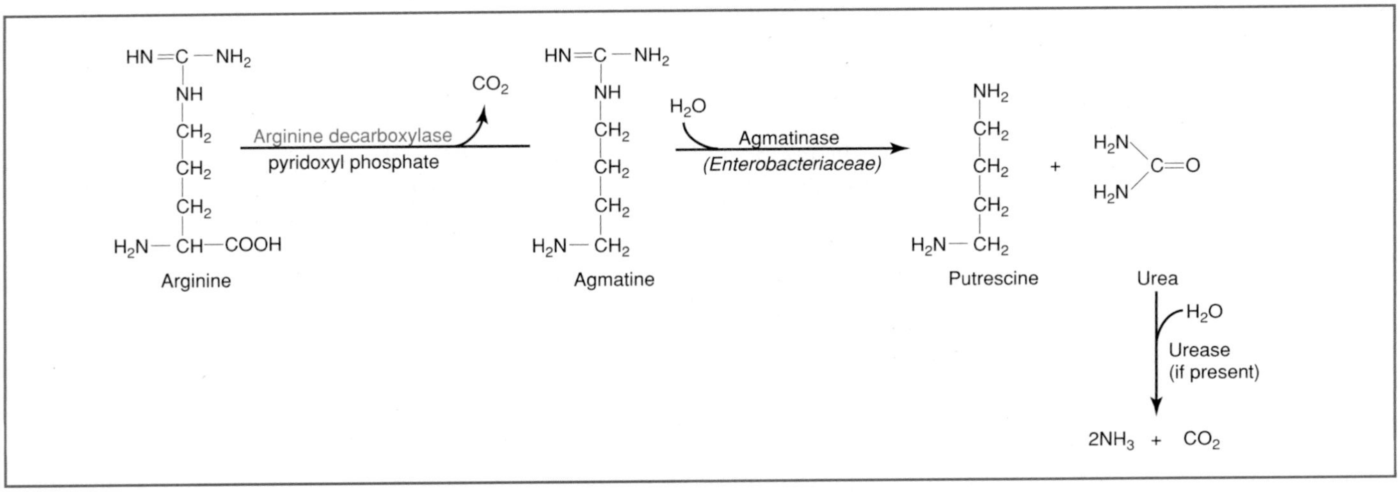

FIGURE 5-35 ARGININE DECARBOXYLATION
Decarboxylation of the amino acid arginine produces the amine agmatine. Members of *Enterobacteriaceae* are capable of degrading agmatine into putrescine and urea. Those strains with urease can break down the urea further into ammonia and carbon dioxide. Thus, the end products of arginine catabolism are carbon dioxide, putrescine and urea, *or* (in the presence of urease) carbon dioxide, putrescine and ammonia.

TABLE OF RESULTS

Result	Interpretation	Symbol
No color change	No decarboxylation	–
Yellow	Fermentation; no decarboxylation	–
Purple (may be slight)	Decarboxylation; oganism produces the specific decarboxylase enzyme	+

TABLE 5-11 Decarboxylase Test results and interpretations.

Medium Recipe

Decarboxylase Medium (Møller)

• Peptone	5.0 g
• Beef extract	5.0 g
• Glucose (dextrose)	0.5 g
• Bromcresol purple	0.01 g
• Cresol red	0.005 g
• Pyridoxal	0.005 g
• L-Lysine, L-Ornithine, or L-Arginine	10.0 g
• Distilled or deionized water	1.0 L

pH 5.8–6.2 at 25°C

Procedure

Lab One

1. Obtain four of each decarboxylase broth. Label three of each with the name of the organism, your name, and the date. Label the fourth of each broth "control."
2. Inoculate the media with the test organisms. Do not inoculate the control.
3. Overlay all tubes with 3 to 4 mm sterile mineral oil (Figure 5-31).
4. Incubate all tubes aerobically at 35 ± 2°C for one week.

Lab Two

1. Remove the tubes from the incubator and examine them for color changes.
2. Record your results in the chart provided on the Data Sheet.

References

Collins, C. H., Patricia M. Lyne, and J. M. Grange. 1995. Page 111 in *Collins and Lyne's Microbiological Methods,* 7th ed. Butterworth-Heinemann, Oxford, United Kingdom.

DIFCO Laboratories. 1984. Page 268 in *DIFCO Manual,* 10th ed. DIFCO Laboratories, Detroit.

Forbes, Betty A., Daniel F. Sahm, and Alice S. Weissfeld. 2002. Page 267 in Bailey & Scott's *Diagnostic Microbiology,* 11th ed. Mosby, St. Louis.

Lányi, B. 1987. Page 29 in *Methods in Microbiology*, Vol. 19, edited by R. R. Colwell and R. Grigorova. Academic Press, New York.

MacFaddin, Jean F. 2000. Page 120 in *Biochemical Tests for Identification of Medical Bacteria,* 3rd ed. Lippincott Williams & Wilkins, Philadelphia.

Exercise 5-11

Phenylalanine Deaminase Test

Theory Organisms that produce phenylalanine deaminase can be identified by their ability to remove the amine group (NH_2) from the amino acid phenylalanine. The reaction, as shown in Figure 5-37, splits a water molecule and produces ammonia (NH_3) and phenylpyruvic acid. Deaminase activity is, therefore, evidenced by the presence of phenylpyruvic acid.

Phenylalanine agar provides a rich source of phenylalanine. A reagent containing ferric chloride ($FeCl_3$) is added to the medium after incubation. The normally colorless phenylpyruvic acid reacts with the ferric chloride and turns a dark green color almost immediately (Figure 5-38). Formation of green color indicates the presence of phenylpyruvic acid and, hence, the presence of phenylalanine deaminase. Yellow is negative (Figure 5-39).

Application This medium is used to differentiate the genera *Morganella*, *Proteus*, and *Providencia* from other members of the *Enterobacteriaceae*.

In This Exercise...

You will inoculate two phenylalanine agar slants with test organisms. After incubation you will add 12% $FeCl_3$ (Phenylalanine Deaminase Test Reagent) and watch for a color change. Refer to Figure 5-39 and Table 5-12 when making your interpretations and recording your results.

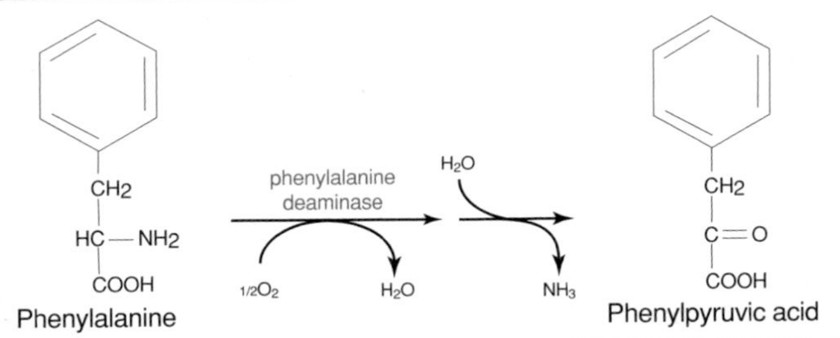

FIGURE 5-37 Deamination of Phenylalanine
Deamination is the removal of an amine (NH_2) from an amino acid.

$$\text{Phenylpyruvic Acid} + FeCl_3 \longrightarrow \text{Green Color}$$

FIGURE 5-38 Indicator Reaction
Phenylpyruvic acid produced by phenylalanine deamination reacts with $FeCl_3$ to produce a green color and indicates a positive phenylalanine test. Be sure to read the result immediately, as the color may fade.

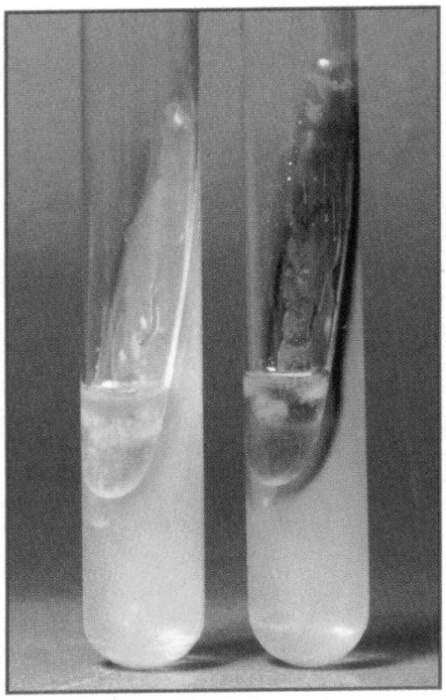

FIGURE 5-39
Phenylalanine Agar Tubes
A phenylalanine deaminase-negative (−) organism is on the left, and a phenylalanine deaminase-positive (+) is on the right.

TABLE OF RESULTS

Result	Interpretation	Symbol
Green color	Phenylalanine deaminase is present	+
No color change	Phenylalanine deaminase is absent	−

TABLE 5-12 Phenylalanine Test results and interpretations.

Materials

Per Student Group

- three Phenylalanine Agar slants
- 12% ferric chloride solution
- fresh slant cultures of:
    - *Escherichia coli*
    - *Proteus vulgaris*

Medium and Reagent Recipes

Phenylalanine Agar

• DL-Phenylalanine	2.0 g
• Yeast extract	3.0 g
• Sodium chloride	5.0 g
• Sodium phosphate	1.0 g
• Agar	12.0 g
• Distilled or deionized water	1.0 L

pH 7.1–7.5 at 25°C

Phenylalanine Deaminase Test Reagent

• Ferric chloride	12.0 g
• Concentrated hydrochloric acid	2.5 mL
• Deionized water (to bring the total to)	100.0 mL

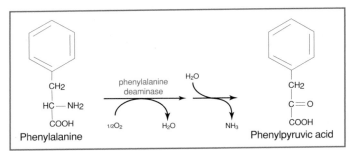

Procedure

Lab One

1. Obtain three Phenylalanine Agar slants. Label two slants with the name of the organism, your name, and the date. Label the third slant "control."
2. Streak two slants with heavy inocula of the test organisms. Do not inoculate the control.
3. Incubate all slants aerobically at 35 ± 2°C for 18 to 24 hours.

Lab Two

1. Add a few drops of 12% ferric chloride solution to each tube and observe for color change. (***Note:*** This color may fade quickly, so read and record your results immediately.)
2. Record your results in the chart provided on the Data Sheet.

References

Forbes, Betty A., Daniel F. Sahm, and Alice S. Weissfeld. 2002. Page 280 in *Bailey & Scott's Diagnostic Microbiology,* 11th ed. Mosby, St. Louis.

Lányi, B. 1987. Page 28 in *Methods in Microbiology,* Vol. 19, edited by R. R. Colwell and R. Grigorova. Academic Press, New York.

MacFaddin, Jean F. 2000. Page 388 in *Biochemical Tests for Identification of Medical Bacteria,* 3rd ed. Lippincott Williams & Wilkins, Philadelphia.

Zimbro, Mary Jo, and David A. Power, editors. 2003. Page 442 in *Difco™ and BBL™ Manual—Manual of Microbiological Culture Media.* Becton Dickinson and Co., Sparks, MD.

Tests Detecting the Presence of Hydrolytic Enzymes

Reactions that use water to split complex molecules are called **hydrolysis** (or hydrolytic) reactions. The enzymes required for these reactions are called hydrolytic enzymes. When the enzyme catalyzes its reaction inside the cell, it is referred to as **intracellular.** Enzymes secreted from the cell to catalyze reactions in the external environment are called **extracellular** enzymes, or **exoenzymes.** In this unit you will be performing tests that identify the actions of both intracellular and extracellular types of enzymes.

The hydrolytic enzymes detected by the Urease and Bile Esculin Tests are intracellular and produce identifiable color changes in the medium. Nutrient Gelatin detects gelatinase—an exoenzyme that liquefies the solid medium. DNase Agar, Milk Agar, Tributyrin Agar, and Starch Agar are plated media that identify extracellular enzymes capable of diffusing into the medium and producing a distinguishable halo of clearing around the bacterial growth.

Exercise 5-12

Bile Esculin Test

Theory Bile Esculin Azide Agar is both a selective and a differential medium. It contains esculin, peptone, bile, sodium azide, and ferric citrate. Esculin is a **glycoside** composed of glucose and esculetin. Peptone and esculin provide nutrients, and bile inhibits growth of Gram positive organisms (except Group D streptococci and enterococci), and sodium azide inhibits growth of Gram negative organisms. Ferric citrate is included as an indicator.

Esculin can be easily hydrolyzed under acidic conditions (Figure 5-40), but few bacteria outside of Group D streptococci and enterococci are able to do so in the presence of bile. When the esculin molecules are split, esculetin reacts with ferric citrate and forms a dark brown phenolic iron complex (Figure 5-41).

Group D streptococci and enterococci grow well on the medium and usually darken it in 18 to 24 hours. Therefore, an organism that has not darkened more than half the medium after 48 hours is considered negative. A slant that is more than half darkened at any time within 48 hours can be recorded as a completed positive (Figure 5-42).

Application This test is most commonly used for presumptive identification of enterococci and members of the *Streptococcus bovis* group, all of which are positive.

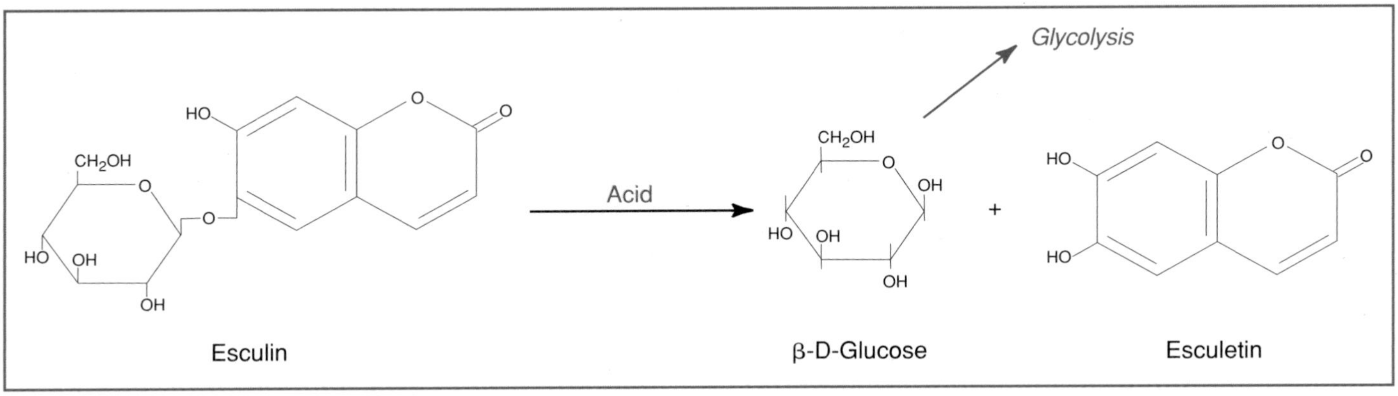

FIGURE 5-40 ACID HYDROLYSIS OF ESCULIN
Many organisms can hydrolyze esculin, but the group D streptococci and enterococci are unique in their ability to do this in the presence of bile salts.

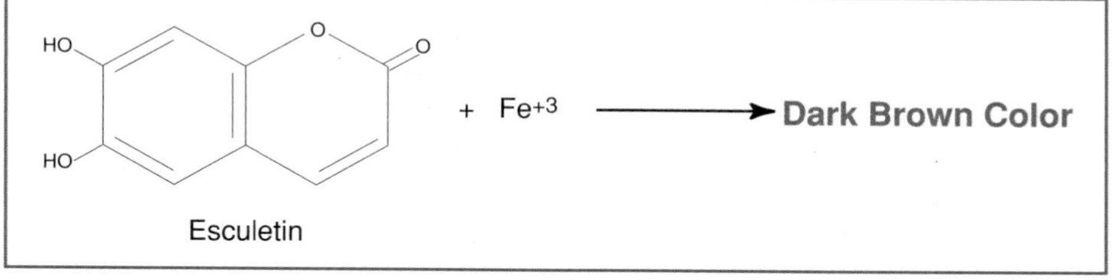

FIGURE 5-41 Bile Esculin Test Indicator Reaction
Esculetin, produced during the hydrolysis of esculin, reacts with Fe^{+3} in the medium to produce a dark brown to black color in the medium.

FIGURE 5-42 Bile Esculin Slant Test Results
From left to right are two Bile Esculin-positive results, one negative result, and an uninoculated control. A positive result is indicated if more than half the tube turns black within 48 hours.

In This Exercise...

You will streak inoculate two Bile Esculin Agar slants with test organisms. We recommend examining them at 24 hours if time permits. Any tubes more than half blackened, even before completion of 48 hours, are recorded as positive. Use Table 5-13 as a guide.

Materials

Per Student Group

- three Bile Esculin Agar slants
- fresh cultures of:
  - *Lactococcus lactis*
  - *Enterococcus faecalis*
  - *Proteus mirabilis*

Medium Recipe

Bile Esculin Azide Agar

- Yeast extract 3.0 g
- Peptone 3.0 g
- Tryptone 17.0 g
- Oxgall 10.0 g
- Esculin 1.0 g
- Ferric ammonium citrate 0.5 g
- Sodium chloride 5.0 g
- Sodium azide 0.15 g
- Agar 15.0 g
- Distilled or deionized water 1.0 L

pH 6.4–6.8 at 25°C

Procedure

Lab One

1. Obtain four Bile Esculin Agar slants. Label three with the name of the organism, your name, and the date. Label the fourth "control."
2. Streak-inoculate two slants with the test organisms. Do not inoculate the control.
3. Incubate all slants at 35 ± 2°C for 48 hours.

Lab Two

1. Observe the slants and record any that are more than half darkened as positive. Any slants that are less than half darkened are negative. Use Table 5-13 as a guide in interpreting the results.
2. Record your results in the chart provided on the Data Sheet.

TABLE OF RESULTS

Result	Interpretation	Symbol
More than half of the medium is darkened within 48 hours	Presumptive identification as a member of Group D *Streptococcus* or *Enterococcus*	+
No color change or less than half of the medium is darkened after 48 hours	Presumptive determination as not a member of Group D *Streptococcus* or *Enterococcus*	–

TABLE 5-13 Bile Esculin Test results and interpretations.

References

Delost, Maria Dannessa. 1997. Page 132 in *Introduction to Diagnostic Microbiology*. Mosby, St. Louis.

Forbes, Betty A., Daniel F. Sahm, and Alice S. Weissfeld. 2002. Page 264 in *Bailey & Scott's Diagnostic Microbiology,* 11th ed. Mosby, St. Louis.

Lányi, B. 1987. Page 56 in *Methods in Microbiology,* Vol. 19, edited by R. R. Colwell and R. Grigorova. Academic Press, New York.

MacFaddin, Jean F. 2000. Page 8 in *Biochemical Tests for Identification of Medical Bacteria,* 3rd ed. Lippincott Williams & Wilkins, Philadelphia.

Zimbro, Mary Jo, and David A. Power, editors. 2003. Page 76 in *Difco™ and BBL™ Manual—Manual of Microbiological Culture Media.* Becton Dickinson and Co., Sparks, MD.

Exercise 5-13

Starch Hydrolysis

Theory

Starch is a polysaccharide made up of α-D-glucose subunits. It exists in two forms—linear (amylose) and branched (amylopectin)—usually as a mixture with the branched configuration being predominant. The α-D-glucose molecules in both amylose and amylopectin are bonded by 1,4-α-glycosidic (acetal) linkages (Figure 5-43). The two forms differ in that amylopectin contains polysaccharide side chains connected to approximately every 30th glucose in the main chain. These side chains are identical to the main chain except that the number 1 carbon of the first glucose in the side chain is bonded to carbon number 6 of the main chain glucose. The bond is, therefore, a 1,6-α-glycosidic linkage.

Starch is too large to pass through the bacterial cell membrane. Therefore, to be of metabolic value to the bacteria, it first must be split into smaller fragments or individual glucose molecules. Organisms that produce and secrete the extracellular enzymes **α-amylase** and **oligo-1,6-glucosidase** are able to hydrolyze starch by breaking the glycosidic linkages between sugar subunits. As shown in Figure 5-43, the end result of these reactions is the complete hydrolysis of the polysaccharide to its individual α-glucose subunits.

Starch agar is a simple plated medium of beef extract, soluble starch, and agar. When organisms that produce α-amylase and oligo-1,6-glucosidase are cultivated on starch agar, they hydrolyze the starch in the area surrounding their growth. Because both starch and its sugar subunits are virtually invisible in the medium, the reagent iodine is used to detect the presence or absence of starch in the vicinity around the bacterial growth. Iodine reacts with starch and produces a blue or dark brown color; therefore, any microbial starch hydrolysis will be revealed as a clear zone surrounding the growth (Figure 5-44 and Table 5-14).

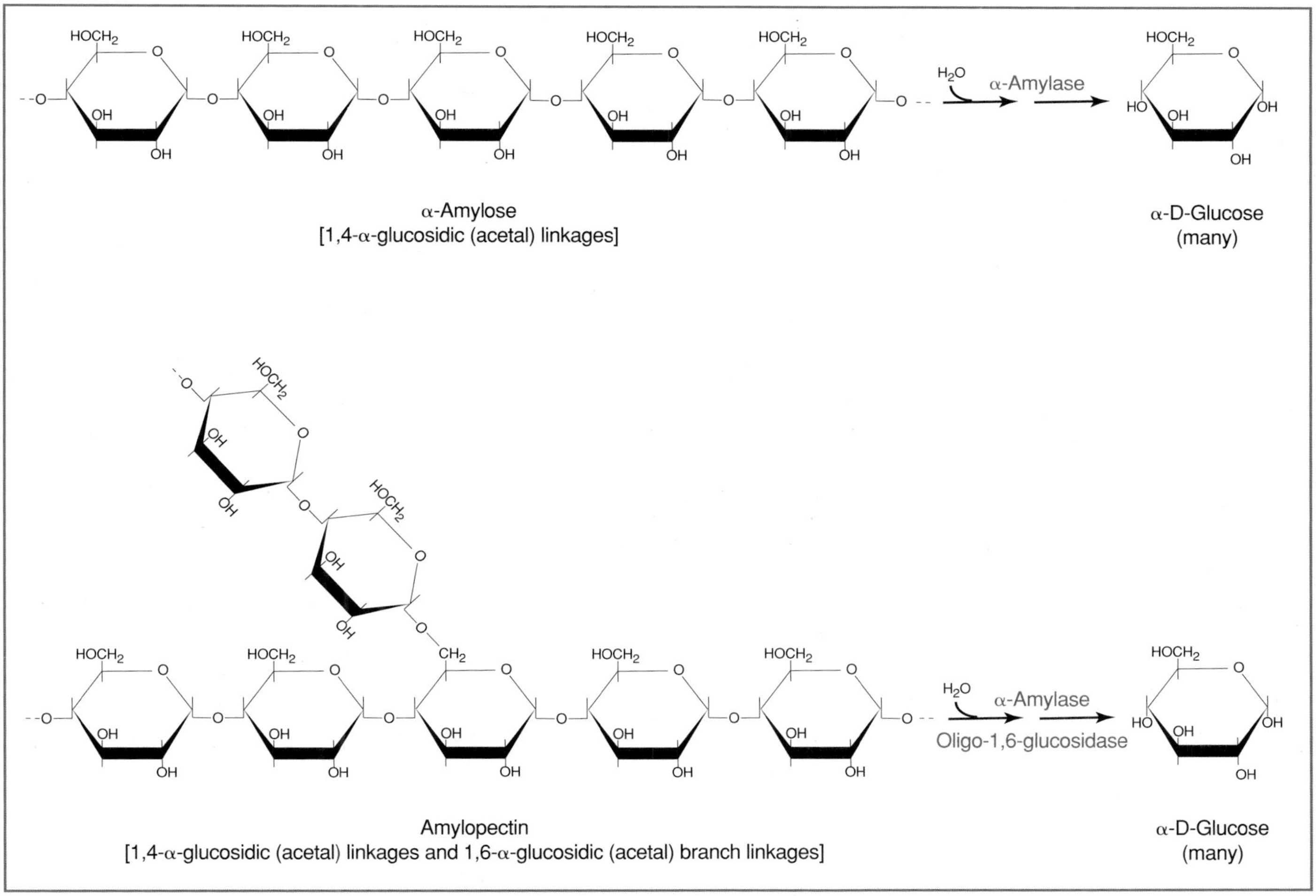

FIGURE 5-43 STARCH HYDROLYSIS BY α-AMYLASE AND OLIGO-1,6-GLUCOSIDASE
The enzymes α-amylase and oligo-1,6-glucosidase hydrolyze starch by breaking glycosidic linkages, resulting in the release of the component glucose subunits.

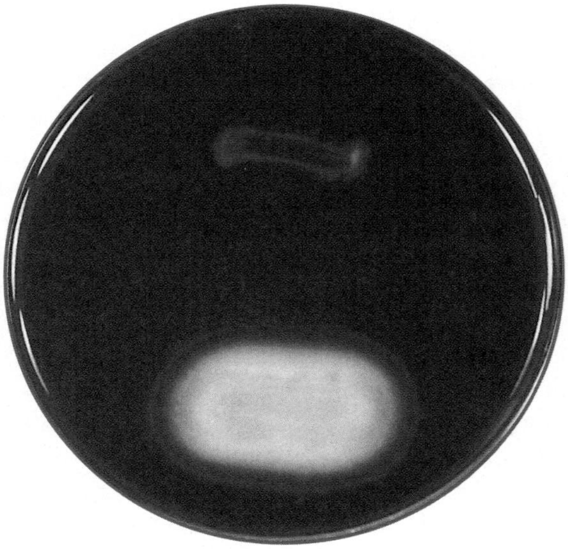

FIGURE 5-44 STARCH HYDROLYSIS TEST
This is a starch agar plate with iodine added to detect amylase activity. The organism above shows no clearing and is negative (−). The organism below is surrounded by a halo of clearing and is positive (+).

TABLE OF RESULTS

Result	Interpretation	Symbol
Clearing around growth	Amylase is present	+
No clearing around growth	No amylase is present	−

TABLE 5-14 Amylase Test results and interpretations.

Application

Starch agar originally was designed for cultivating *Neisseria*. It no longer is used for this, but with pH indicators it is used to isolate and presumptively identify *Gardnerella vaginalis*. It aids in differentiating species from the genera *Corynebacterium, Clostridium, Bacillus, Bacteroides, Fusobacterium*, and members of *Enterococcus*.

In This Exercise...

You will inoculate a Starch Agar plate with two organisms. After incubation you will add a few drops of iodine to reveal clear areas surrounding the growth. Iodine will not color the agar where the growth is, only the area surrounding the growth. Before you add the iodine, make a note of the growth patterns on the plate so you don't confuse thinning growth at the edges with the halo produced by starch hydrolysis.

Materials

Per Student Group

- one Starch Agar plate
- Gram iodine (from your Gram stain kit)
- recommended organisms:
  - *Bacillus subtilis*
  - *Staphylococcus aureus*

Medium Recipe

Starch Agar

• Beef extract	3.0 g
• Soluble starch	10.0 g
• Agar	12.0 g
• Distilled or deionized water	1.0 L

pH 7.3–7.7 at 25°C

Procedure

Lab One

1. Using a marking pen, divide the Starch Agar plate into three equal sectors. Be sure to mark on the bottom of the plate.
2. Label the plate with the names of the organisms, your name, and the date.
3. Spot-inoculate two sectors with the test organisms.
4. Invert the plate and incubate it aerobically at 35 ± 2°C for 48 hours.

Lab Two

1. Remove the plate from the incubator, and note the location and appearance of the growth before adding the iodine. (Growth that is thinning at the edge may give the appearance of clearing in the agar after iodine is added.)
2. Cover the growth and surrounding areas with Gram iodine. Immediately examine the areas surrounding the growth for clearing. (Usually the growth on the agar prevents contact between the starch and the iodine so no color reaction takes place at that point. Beginning students sometimes look at this lack of color change and incorrectly judge it as a positive result. Therefore, when examining the agar for clearing, look for a halo *around* the growth, not at the growth itself.)
3. Record your results in the chart provided on the Data Sheet.

References

Collins, C. H., Patricia M. Lyne, and J. M. Grange. 1995. Page 117 in *Collins and Lyne's Microbiological Methods,* 7th ed. Butterworth-Heinemann, Oxford, United Kingdom.

Lányi, B. 1987. Page 55 in Methods in Microbiology, Vol. 19, edited by R. R. Colwell and R. Grigorova, Academic Press Inc., New York.

MacFaddin, Jean F. 2000. Page 412 in *Biochemical Tests for Identification of Medical Bacteria*, 2nd ed. Lippincott Williams & Wilkins, Philadelphia.

Smibert, Robert M., and Noel R. Krieg. 1994. Page 630 in *Methods for General and Molecular Bacteriology*, edited by Philipp Gerhardt, R. G. E. Murray, Willis A. Wood, and Noel R. Krieg. American Society for Microbiology, Washington, DC.

Zimbro, Mary Jo, and David A. Power, editors. 2003. *Difco™ and BBL™ Manual—Manual of Microbiological Culture Media.* Becton Dickinson and Co., Sparks, MD.

Exercise 5-14

ONPG Test

Theory

For bacteria to ferment lactose, they must possess two enzymes: **β-galactoside permease**, a membrane-bound transport protein; and **β-galactosidase**, an intracellular enzyme that hydrolyzes the disaccharide into β-glucose and β-galactose (Figure 5-45).

Bacteria possessing both enzymes are active β-lactose fermenters. Bacteria that cannot produce β-galactosidase, even if their membranes contain β-galactoside permease, cannot ferment β-lactose. Bacteria that possess β-galactosidase but no β-galactoside permease may mutate and, over a period of days or weeks, begin to produce the permease. Distinguishing these **late lactose fermenters** from nonfermenters is made possible by *o*-nitrophenyl-β-D-galactopyranoside (ONPG).

When ONPG is made available to bacteria, it freely enters the cells without the aid of a permease. Because of its similarity to β-lactose, it then can become the substrate for any β-galactosidase present. In the reaction that occurs, ONPG is hydrolyzed to β-galactose and *o*-nitrophenol (ONP), which is yellow (Figures 5-46 and 5-47, and Table 5-15).

It should be noted that β-galactosidase is an inducible enzyme; it is produced in response to the presence of an appropriate substrate. Therefore, organisms to be tested with ONPG are typically prepared by growing them overnight in a lactose-rich medium, such as Kligler's Iron Agar or Triple Sugar Iron Agar, to ensure that they will be actively producing β-galactosidase, given that capability.

Application

The ONPG test is used to differentiate late lactose fermenters from lactose nonfermenters in the family *Enterobacteriaceae*.

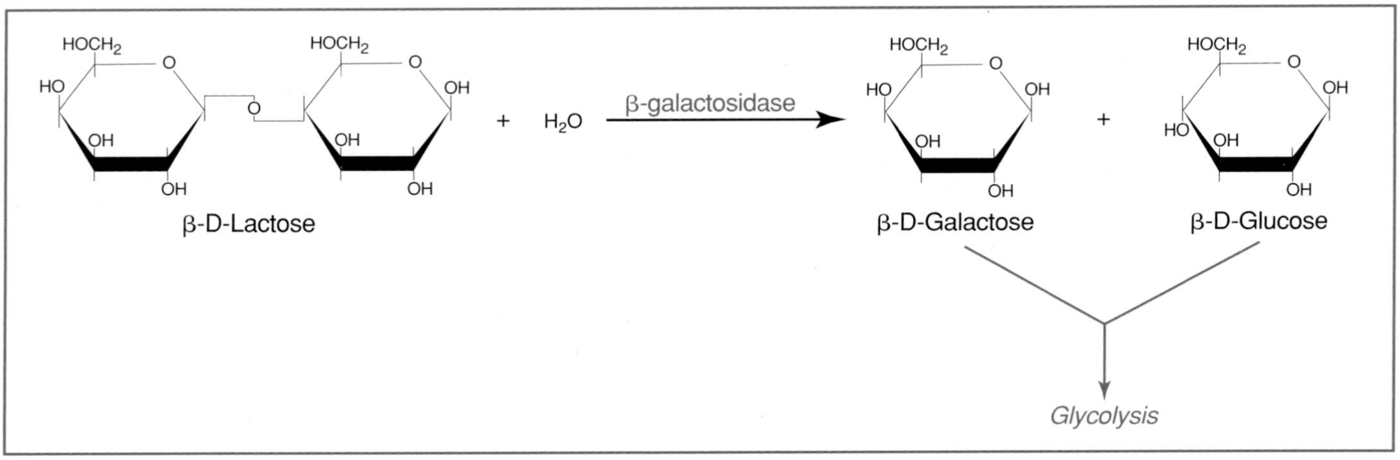

FIGURE 5-45 Hydrolysis of Lactose By β-Galactosidase
β-galactosidase is an inducible enzyme in many bacteria; that is, it is produced only in the presence of the inducer lactose. Its activity is to hydrolyze the disaccharide lactose into its component monosaccharides.

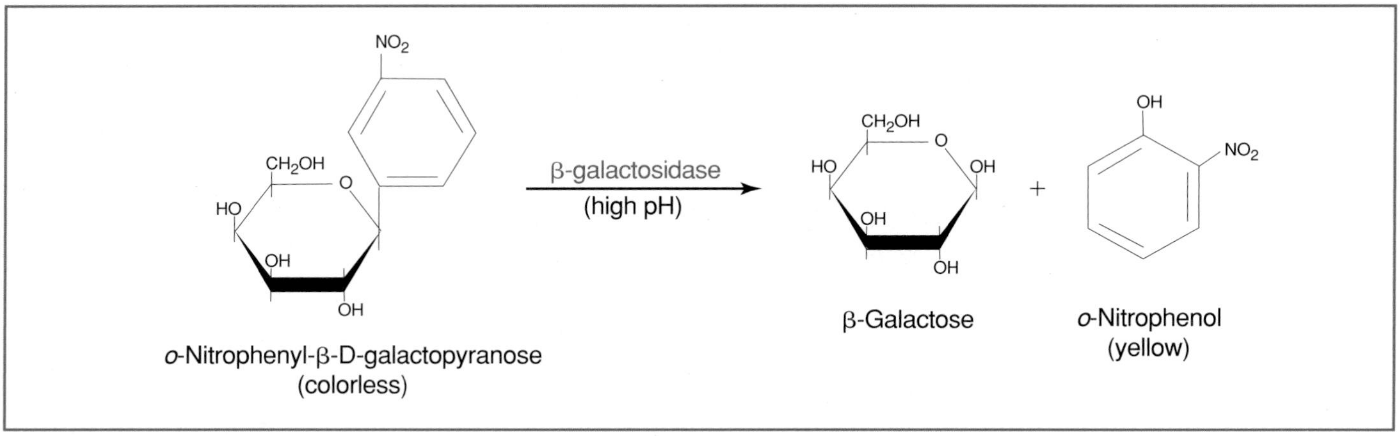

FIGURE 5-46 Conversion of ONPG to β-Galactose and *o*-Nitrophenol by β-Galactosidase
ONPG is similar enough to lactose that β-galactosidase will catalyze its hydrolysis into β-galactose and *o*-nitrophenol. The latter compound is yellow and is indicative of a positive ONPG test.

FIGURE 5-47 THE ONPG TEST
ONPG-positive (+) is on the left, and ONPG-negative (−) is on the right.

In This Exercise...

You will be performing the ONPG test using Key Scientific Products' K1490 ONPG WEETABS. If your instructor uses a different version of the test, he/she will give you instructions to follow based on the manufacturer's recommendations. The organisms chosen for this exercise are two of the enterics that this test was designed to differentiate.

Materials

Per Student Group

- Key Scientific Products' K1490 ONPG WEETABS
- sterile distilled or deionized water
- sterile transfer pipettes
- fresh Kligler's Iron Agar or Triple Sugar Iron Agar cultures of:
  - *Citrobacter diversus*
  - *Salmonella typhimurium*

TABLE OF RESULTS

Result	Interpretation	Symbol
Yellow color formation	Organism produces β-galactosidase	+
No color change	Organism does not produce β-galactosidase	−

TABLE 5-15 ONPG Test results and interpretations.

Procedure

1. Dispense 0.5 mL volumes of sterile water into three WEETAB tubes, and allow the tablets to dissolve.
2. Transfer a heavy inoculum of the test organisms to two tubes. Do not inoculate the third tube.
3. Incubate all tubes at 35 ± 2°C for 2 hours.
4. Remove all tubes from the incubator and examine them for color changes. Record your results on the Data Sheet.

References

DIFCO Laboratories. 1984. Page 294 in *DIFCO Manual,* 10th ed., DIFCO Laboratories, Detroit.

Forbes, Betty A., Sahm, Daniel F., and Alice S. Weissfeld. 1998. Page 442 in *Bailey & Scott's Diagnostic Microbiology,* 10th ed. Mosby, St. Louis.

Key Scientific Products' K1490 ONPG WEETABS Package Insert. Key Scientific Products. Round Rock, TX.

MacFaddin, Jean F. 2000. Page 160 in *Biochemical Tests for Identification of Medical Bacteria,* 2nd ed. Lippincott Williams & Wilkins, Philadelphia.

Exercise 5-15

Urease Tests

Theory

Urea is a product of decarboxylation of certain amino acids. It can be hydrolyzed to ammonia and carbon dioxide by bacteria containing the enzyme **urease**. Many enteric bacteria (and a few others) possess the ability to metabolize urea, but only members of *Proteus, Morganella*, and *Providencia* are considered rapid urease-positive organisms.

Urea Agar was formulated to differentiate rapid urease-positive bacteria from slower urease-positive and urease-negative bacteria. It contains urea, peptone, potassium phosphate, glucose, and phenol red. Peptone and glucose provide essential nutrients for a broad range of bacteria. Potassium phosphate is a mild buffer used to resist alkalinization of the medium from peptone metabolism. Phenol red, which is yellow or orange below pH 8.4 and red or pink above, is included as an indicator.

Urea hydrolysis (Figure 5-48) to ammonia by urease-positive organisms will overcome the buffer in the medium and change it from orange to pink. The agar must be examined daily during incubation. Rapid urease-positive organisms will turn the entire slant pink within 24 hours. Weak positives may take several days (Table 5-16). Urease-negative organisms either produce no color change in the medium or turn it yellow from acid products (Figure 5-49).

Urea broth differs from urea agar in two important ways. First, its only nutrient source is a trace (0.0001%) of yeast extract. Second, it contains buffers strong enough to inhibit alkalinization of the medium by all but the rapid urease-positive organisms mentioned above. Other organisms, even some that ordinarily would be able to metabolize urea, cannot survive the severe nutrient limitations or overcome the buffers in urea broth. Pink color in the medium in less than 24 hours indicates a rapid urease-positive organism. Orange or yellow is negative (Figure 5-50 and Table 5-17).

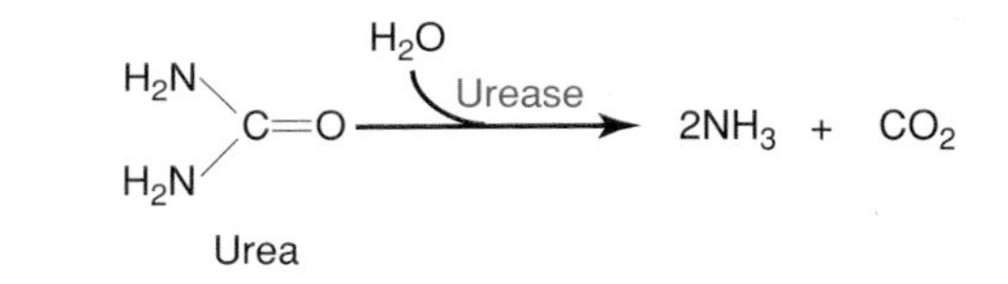

FIGURE 5-48 UREA HYDROLYSIS
Urea hydrolysis produces ammonia, which raises the pH in the medium and turns the pH indicator pink.

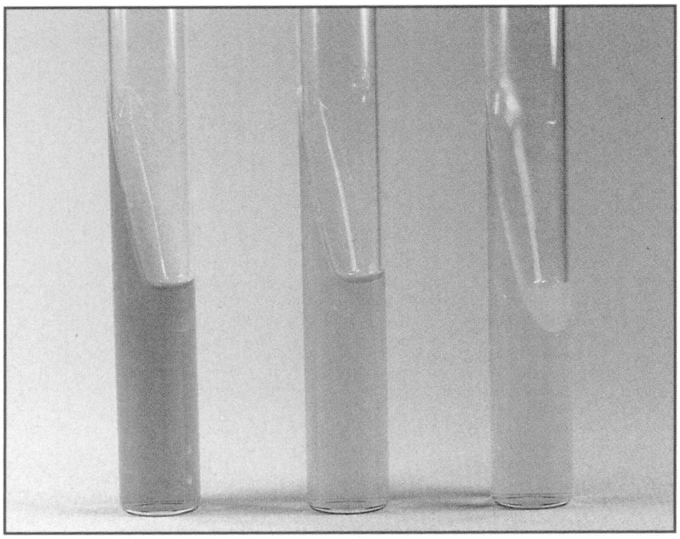

FIGURE 5-49 UREASE AGAR TEST RESULTS
Urease agar tubes after a 24-hour incubation illustrate a rapid urea splitter (+) on the left and a urease-negative organism on the right. An uninoculated control is in the center.

Application

This test is used to differentiate organisms based on their ability to hydrolyze urea with the enzyme urease. Urinary tract pathogens from the genus *Proteus* may be distinguished from other enteric bacteria by their rapid urease activity.

In This Exercise...

You will perform both forms of the test to determine urease activity of three organisms. Urease slants may

TABLE OF RESULTS

Result		Interpretation	Symbol
24 hours	24 hours to 6 days		
All pink		Rapid urea hydrolysis; strong urease production	+
Partially pink	All pink or partially pink	Slow urea hydrolysis; weak urease production	w+
Orange or yellow	All pink or partially pink	Slow urea hydrolysis; weak urease production	w+
Orange or yellow	Orange or yellow	No urea hydrolysis; urease is absent	–

TABLE 5-16 Urease Agar Test results and interpretations.

FIGURE 5-50 Urease Broth Test Results
Urease broth tubes after a 24-hour incubation illustrate a urease-positive (+) organism on the left, an uninoculated control in the middle, and urease-negative (−) organism on the right.

require incubation up to 6 days, but broths should be read at 24 hours. If your labs are scheduled more than 24 hours apart, arrange to have someone place your broths in a refrigerator until you can examine them.

Materials

Per Student Group

- four Urea Broths
- four Urea Agar slants
- fresh slant cultures of:
  - *Escherichia coli*
  - *Proteus vulgaris*
  - *Klebsiella pneumoniae*

Medium Recipes

Christensen's Urea Agar

• Peptone	1.0 g
• Dextrose (glucose)	1.0 g
• Sodium chloride	5.0 g
• Potassium phosphate, monobasic	2.0 g
• Urea	20.0 g
• Agar	15.0 g
• Phenol red	0.012 g
• Distilled or deionized water	1.0 L

pH 6.8 at 25°C

Rustigian and Stuart's Urea Broth

• Yeast extract	0.1 g
• Potassium phosphate, monobasic	9.1 g
• Potassium phosphate, dibasic	9.5 g
• Urea	20.0 g
• Phenol red	0.01 g
• Distilled or deionized water	1.0 L

pH 6.8 at 25°C

Procedure

Lab One

1. Obtain four tubes of each medium. Label three with the names of the organisms, your name, and the date. Label the fourth tube of each medium "control."
2. Inoculate three broths with heavy inocula from the test organisms. Do not inoculate the control.
3. Streak-inoculate three slants with the test organisms, covering the entire agar surface with a heavy inoculum. Do not stab the agar butt. Do not inoculate the control.
4. Incubate all tubes aerobically at 35 ± 2°C for 24 hours. (If your labs are scheduled more than 1 day apart, try to read these media at 24 hours or arrange to have someone else read them for you. Positive results can be removed and placed in the refrigerator for later examination.)

Lab Two

1. Remove all tubes from the incubator and examine them for color changes. Record any rapid positive and slow positive test results in the chart provided on the Data Sheet. Discard all broth tubes and positive agar tubes in an appropriate autoclave container.
2. Return any negative agar tubes to the incubator. Inspect them daily for pink color formation for up to 6 days. (Negative tubes must not be recorded as negative until they have incubated the full 6 days.)
3. Enter your agar results daily in the chart provided. Refer to Tables 5-16 and 5-17 when interpreting your results.

TABLE OF RESULTS

Result	Interpretation	Symbol
Pink	Rapid urea hydrolysis; strong urease production	+
Orange or yellow	No urea hydrolysis; organism does not produce urease or cannot live in broth	−

TABLE 5-17 Urease Broth Test results and interpretations.

References

Collins, C. H., Patricia M. Lyne, and J. M. Grange. 1995. Page 117 in *Collins and Lyne's Microbiological Methods,* 7th ed. Butterworth-Heinemann, Oxford, United Kingdom.

Delost, Maria Dannessa. 1997. Page 196 in *Introduction to Diagnostic Microbiology*. Mosby, St. Louis.

DIFCO Laboratories. 1984. Page 1040 in *DIFCO Manual,* 10th ed. DIFCO Laboratories, Detroit.

Forbes, Betty A., Daniel F. Sahm, and Alice S. Weissfeld. 2002. Page 283 in *Bailey & Scott's Diagnostic Microbiology,* 11th ed. Mosby, St. Louis.

Lányi, B. 1987. Page 24 in *Methods in Microbiology,* Vol. 19, edited by R. R. Colwell and R. Grigorova. Academic Press, New York.

MacFaddin, Jean F. 2000. Page 298 in *Biochemical Tests for Identification of Medical Bacteria,* 3rd ed. Lippincott Williams & Wilkins, Philadelphia.

Smibert, Robert M., and Noel R. Krieg. 1994. Page 630 in *Methods for General and Molecular Bacteriology*, edited by Philipp Gerhardt, R. G. E. Murray, Willis A. Wood, and Noel R. Krieg. American Society for Microbiology, Washington, DC.

Zimbro, Mary Jo and David A. Power, editors. 2003. Page 606 in *Difco™ and BBL™ Manual—Manual of Microbiological Culture Media.* Becton Dickinson and Co., Sparks, MD.

Exercise 5-16

Casease Test

Theory Many bacteria require proteins as a source of amino acids and other components for synthetic processes. Some bacteria have the ability to produce and secrete enzymes into the environment that catalyze the break-down of large proteins to smaller peptides or individual amino acids, thereby enabling their uptake across the membrane (Figure 5-51).

Casease is an enzyme some bacteria produce to hydrolyze the milk protein **casein**, the molecule that gives milk its white color. When broken down to smaller fragments, the ordinarily white casein loses its opacity and becomes clear.

The presence of casease can easily be detected with the test medium Milk Agar (Figure 5-52). Milk Agar is an undefined medium containing tryptone, yeast extract, dextrose, and powdered milk. When plated Milk Agar is inoculated with a casease-positive organism, secreted casease will diffuse into the medium around the colonies and create a zone of clearing where the casein has been hydrolyzed. Casease-negative organisms do not secrete casease and, thus, do not produce clear zones around the growth.

Application The casease test is used for the enumeration of bacteria in milk, ice cream, and whey.

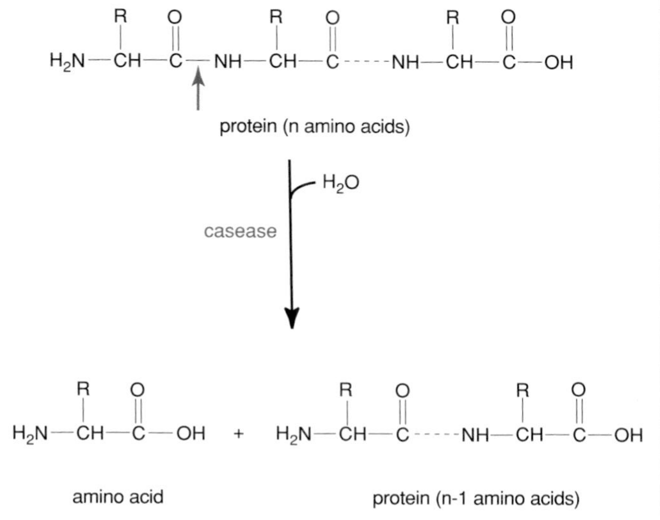

FIGURE 5-51 CASEIN HYDROLYSIS
Hydrolysis of any protein occurs by breaking peptide bonds (red arrow) between adjacent amino acids to produce short peptides or individual amino acids.

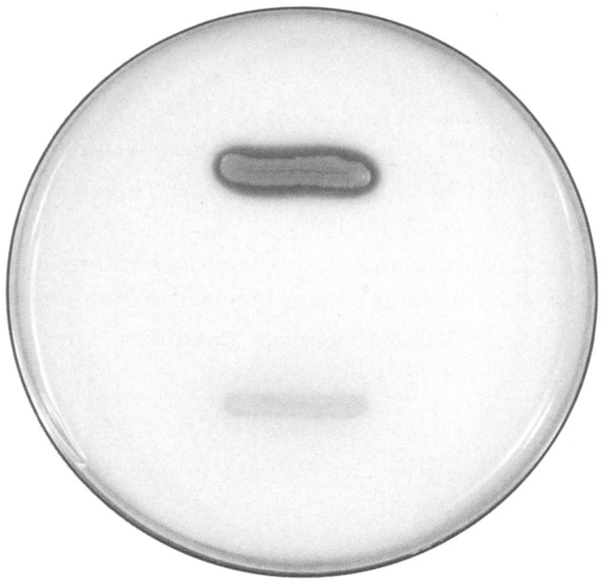

FIGURE 5-52 MILK AGAR PLATE
This plate was inoculated with a casease-positive organism above and a casease-negative organism below.

TABLE OF RESULTS

Result	Interpretation	Symbol
Clearing in agar	Casease is present	+
No clearing in agar	Casease is absent	−

TABLE 5-18 Casease Test results and interpretations.

In This Exercise...

You will inoculate a single milk agar plate with a casease-positive and a casease-negative organism. Use Table 5-18 as a guide when making your interpretations.

Materials

Per Student Group

- one Milk Agar plate
- fresh cultures of:
  - *Lactococcus lactis*
  - *Staphylococcus aureus*
  - *Escherichia coli*

Medium Recipe

Milk Agar

• Beef extract	3.0 g
• Peptone	5.0 g
• Agar	15.0 g
• Powdered nonfat milk	100.0 g
• Distilled or deionized water	1.0 L

pH 7.0–7.4 at 25°C

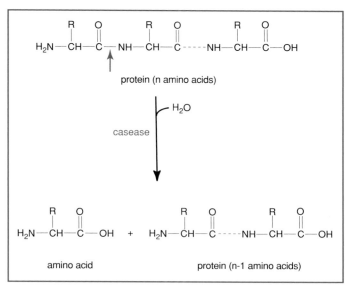

Procedure

Lab One

1. Using a marking pen, divide the plate into four equal sectors. Be sure to mark on the bottom of the plate.
2. Label the plate with the names of the organisms, your name, and the date.
3. Spot-inoculate three sectors with the test organisms and leave the fourth sector uninoculated as a control.
4. Invert the plate and incubate it aerobically at 35 ± 2°C for 24 hours.

Lab Two

1. Examine the plates for clearing around the bacterial growth.
2. Record your results in the chart provided on the Data Sheet.

References

Chan, E. C. S., Michael J. Pelczar, Jr. and Noel R. Krieg. 1986. Page 137 in *Laboratory Exercises In Microbiology,* 5th ed. McGraw-Hill, New York.

Holt, John G., editor. 1994. *Bergey's Manual of Determinative Bacteriology,* 9th ed. Williams and Wilkins, Baltimore.

Smibert, Robert M., and Noel R. Krieg. 1994. Page 613 in *Methods for General and Molecular Bacteriology*, edited by Philipp Gerhardt, R. G. E. Murray, Willis A. Wood, and Noel R. Krieg. American Society for Microbiology, Washington, DC.

Zimbro, Mary Jo, and David A. Power, editors. 2003. Page 364 in *Difco™ and BBL™ Manual—Manual of Microbiological Culture Media.* Becton Dickinson and Co., Sparks, MD.

Exercise 5-17
Gelatinase Test

Theory Gelatin is a protein derived from collagen—a component of vertebrate connective tissue. **Gelatinases** comprise a family of extracellular enzymes produced and secreted by some microorganisms to hydrolyze gelatin. Subsequently, the individual amino acids can be taken up by the cell and used for metabolic purposes. Bacterial hydrolysis of gelatin occurs in two sequential reactions as shown in Figure 5-53.

The presence of gelatinases can be detected using Nutrient Gelatin, a simple test medium composed of gelatin, peptone, and beef extract. Nutrient Gelatin differs from most other solid media in that the solidifying agent (gelatin) is also the substrate for enzymatic activity. Consequently, when a tube of Nutrient Gelatin is stab-inoculated with a gelatinase-positive organism, secreted gelatinase (or gelatinases) will liquefy the medium. Gelatinase-negative organisms do not secrete the enzyme and do not liquefy the medium (Figures 5-54 and 5-55). It should be noted that gelatinase activity is sometimes very slow, producing only partial liquefaction after a 7-day incubation period.

A small disadvantage of Nutrient Gelatin is that it melts at 28°C (82°F). Therefore, inoculated stabs are typically incubated at 25°C along with an uninoculated control to verify that any liquefaction is not temperature related.

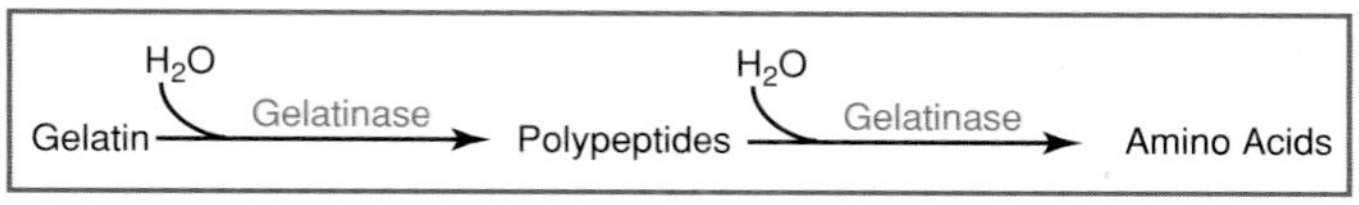

FIGURE 5-53 Gelatin Hydrolysis
Gelatin is hydrolyzed by the gelatinase family of enzymes.

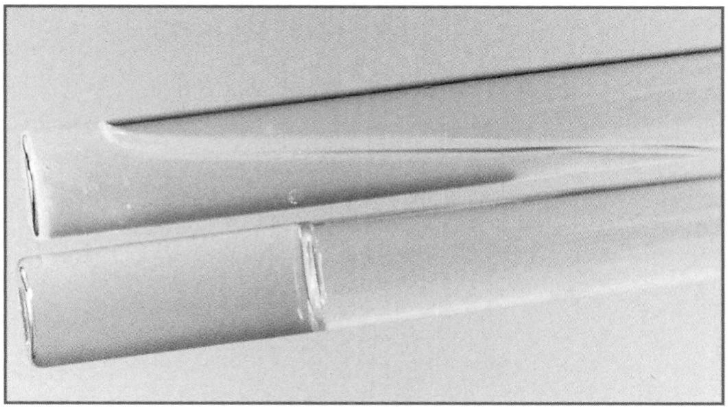

FIGURE 5-54 Nutrient Gelatin Stabs
A gelatinase-positive (+) organism is above, and a gelatinase–negative (−) organism is below.

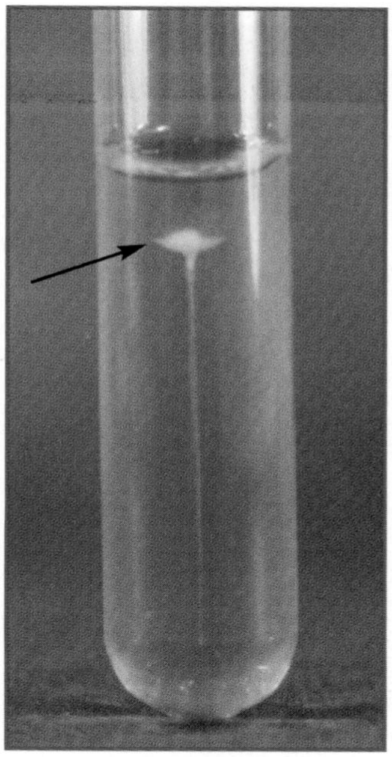

FIGURE 5-55
Crateriform Liquefaction
This form of liquefaction also may be of diagnostic use because not all gelatinase-positive microbes liquefy the gelatin completely. Shown here is an organism liquefying the gelatin in the shape of a crater (arrow).

Application This test is used to determine the ability of a microbe to produce gelatinases. *Staphylococcus aureus,* which is gelatinase-positive, can be differentiated from *S. epidermidis*. *Serratia* and *Proteus* species are positive members of *Enterobacteriaceae* whereas most others in the family are negative. *Bacillus anthracis, B. cereus*, and several other members of the genus are gelatinase-positive, as are *Clostridium tetani* and *C. perfringens*.

In This Exercise...

You will inoculate three Nutrient Gelatin tubes and incubate them for up to a week. Gelatin liquefaction is a positive result but is difficult to differentiate from gelatin that has melted as a result of temperature. Therefore, be sure to incubate a control tube along with the others. If the control melts as a result of temperature, refrigerate all tubes until it resolidifies. Use Table 5-19 as a guide to interpretation.

TABLE OF RESULTS

Result	Interpretation	Symbol
Gelatin is liquid (control is solid)	Gelatinase is present	+
Gelatin is solid	No gelatinase is present	−

TABLE 5-19 Gelatinase Test results and interpretations.

Materials

Per Student Group

- three Nutrient Gelatin stabs
- fresh cultures of:
  - *Staphylococcus aureus*
  - *Escherichia coli*

Medium Recipe

Nutrient Gelatin

• Beef Extract	3.0 g
• Peptone	5.0 g
• Gelatin	120.0 g
• Distilled or deionized water	1.0 L

pH 6.6–7.0 at 25°C

Procedure

Lab One

1. Obtain three Nutrient Gelatin stabs. Label each of two tubes with the name of the organism, your name, and the date. Label the third tube "control."
2. Stab-inoculate two tubes with heavy inocula of the test organisms. Do not inoculate the control.
3. Incubate all tubes at 25°C for up to 1 week.

Lab Two

1. Examine the control tube. If the gelatin is solid, the test can be read. If it is liquefied, it likely is because of the temperature. All tubes must be refrigerated until the control has solidified.
2. When the control has solidified, examine the inoculated media for gelatin liquefaction.
3. Record your results in the chart provided on the Data Sheet.

References

Collins, C. H., Patricia M. Lyne, and J. M. Grange. 1995. Page 112 in *Collins and Lyne's Microbiological Methods,* 7th ed. Butterworth-Heinemann, Oxford, United Kingdom.

Forbes, Betty A., Daniel F. Sahm, and Alice S. Weissfeld. 2002. Page 271 in *Bailey & Scott's Diagnostic Microbiology,* 11th ed. Mosby, St. Louis.

Lányi, B. 1987. Page 44 in *Methods in Microbiology,* Vol. 19, edited by R. R. Colwell and R. Grigorova. Academic Press, New York.

MacFaddin, Jean F. 2000. Page 128 in *Biochemical Tests for Identification of Medical Bacteria,* 3rd ed. Lippincott Williams & Wilkins, Philadelphia.

Smibert, Robert M., and Noel R. Krieg. 1994. Page 617 in *Methods for General and Molecular Bacteriology,* edited by Philipp Gerhardt, R. G. E. Murray, Willis A. Wood, and Noel R. Krieg. American Society for Microbiology, Washington, DC.

Zimbro, Mary Jo, and David A. Power, editors. 2003. Page 408 in *Difco™ and BBL™ Manual—Manual of Microbiological Culture Media.* Becton Dickinson and Co., Sparks, MD.

Exercise 5-18

DNase Test

Theory

An enzyme that catalyzes the depolymerization of DNA into small fragments is called a **deoxyribonuclease** or **DNase** (Figure 5-56). Ability to produce this enzyme can be determined by culturing and observing an organism on a DNase Test Agar Plate.

DNase Test Agar contains an emulsion of DNA, peptides as a nutrient source, and methyl green dye. The dye and polymerized DNA form a complex that gives the agar a blue-green color. Bacterial colonies that secrete DNase will hydrolyze DNA in the medium into smaller fragments unbound from the methyl green dye. This results in clearing around the growth (Figure 5-57).

Application

DNase Test Agar is used to distinguish *Serratia* species from *Enterobacter* species, *Moraxella catarrhalis* from *Neisseria* species, and *Staphylococcus aureus* from other *Staphylococcus* species.

In This Exercise...

You will inoculate one DNase plate with two of the organisms the test was designed to differentiate. For best results, limit incubation time to 24 hours. Depending on the age of the plates and the length of incubation, the clearing around a DNase-positive organism may be subtle and difficult to see. When reading the plate, it might be helpful to look through it while holding it several inches above a white piece of paper. Use Table 5-20 to assist in interpretation.

Materials

Per Student Group

- one DNase Test Agar plate
- fresh cultures of:
  - *Staphylococcus aureus*
  - *Staphylococcus epidermidis*

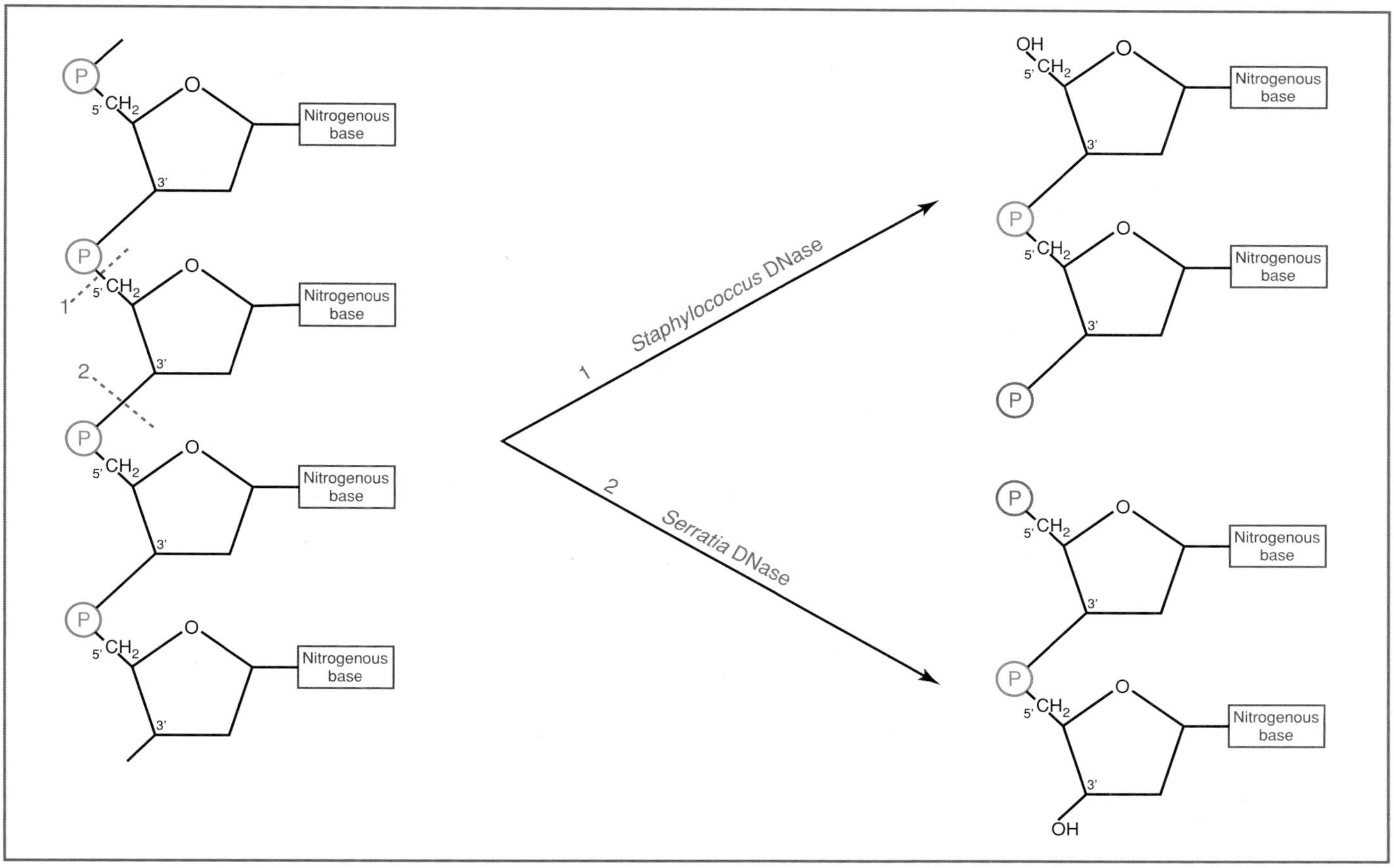

FIGURE 5-56 Two Patterns of DNA Hydrolysis
DNase from *Staphylococcus* hydrolyzes DNA at the bond between the 5'-carbon and the phosphate (illustrated by line 1), thereby producing fragments with a free 3'-phosphate (shown in red on the upper fragment). Most fragments are one or two nucleotides long. A dinucleotide is shown here. *Serratia* DNase cleaves the bond between the phosphate and the 3'-carbon (illustrated by line 2) and produces fragments with free 5'-phosphates (shown in red on the lower fragment). Most fragments are two to four nucleotides in length. A dinucleotide is shown here.

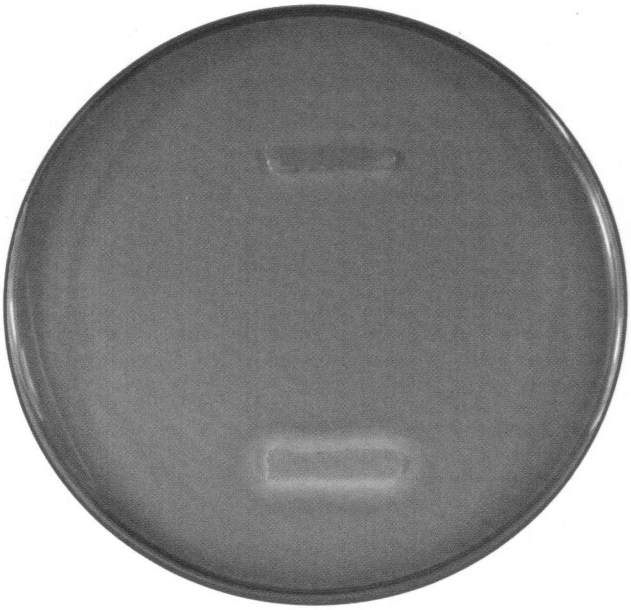

FIGURE 5-57 DNase Test Results
DNase agar plate is inoculated with a DNase-positive (+) organism below and a DNase-negative (−) organism above.

TABLE OF RESULTS

Result	Interpretation	Symbol
Clearing in agar around growth	DNase is present	+
No clearing in agar around growth	DNase is absent	−

TABLE 5-20 DNase Test results and interpretations.

Medium Recipe

DNase Test Agar with Methyl Green

- Tryptose 20.0 g
- Deoxyribonucleic acid 2.0 g
- Sodium chloride 5.0 g
- Agar 15.0 g
- Methyl green 0.05 g
- Distilled or deionized water 1.0 L

pH 7.1–7.4 at 25°C

Procedure

Lab One

1. Using a marking pen, divide the DNase Test Agar plate into three equal sectors. Be sure to mark the bottom of the plate.
2. Label the plate with the names of the organisms, your name, and the date.
3. Spot-inoculate two sectors with the test organisms and leave the third sector as a control.
4. Invert the plate and incubate it aerobically at 35 ± 2°C for 24 hours. (If your labs are scheduled more than 24 hours apart, arrange to have someone place your plates in a refrigerator until you can examine them.)

Lab Two

1. Examine the plates for clearing around the bacterial growth.
2. Record your results in the table provided on the Data Sheet.

References

Collins, C. H., Patricia M. Lyne, and J. M. Grange. 1995. Page 114 in *Collins and Lyne's Microbiological Methods*, 7th ed. Butterworth-Heinemann, Oxford, United Kingdom.

Delost, Maria Dannessa. 1997. Page 111 in *Introduction to Diagnostic Microbiology*. Mosby, St. Louis.

Forbes, Betty A., Daniel F. Sahm, and Alice S. Weissfeld. 2002. Page 268 in *Bailey & Scott's Diagnostic Microbiology*, 11th ed. Mosby, St. Louis.

Lányi, B. 1987. Page 33 in *Methods in Microbiology*, Vol. 19, edited by R. R. Colwell and R. Grigorova. Academic Press, New York.

MacFaddin, Jean F. 2000. Page 136 in *Biochemical Tests for Identification of Medical Bacteria*, 3rd ed. Lippincott Williams & Wilkins, Philadelphia.

Zimbro, Mary Jo, and David A. Power, editors. 2003. Page 170 in *Difco™ and BBL™ Manual—Manual of Microbiological Culture Media*. Becton Dickinson and Co., Sparks, MD.

Exercise 5-19

Lipase Test

Theory

Lipid is the word generally used to describe all types of fats. The enzymes that hydrolyze fats are called **lipases.** Bacteria can be differentiated based on their ability to produce and secrete lipases. Although a variety of simple fats can be used for this determination, tributyrin oil is the most common constituent of lipase testing media because it is the simplest triglyceride found in natural fats and oils.

Simple fats are known as **triglycerides,** or **triacylglycerols** (Figure 5-58). Triglycerides are composed of glycerol and three long chain fatty acids. As is true of many biochemicals, tributyrin is too large to enter the cell, so a lipase is released to break it down prior to cellular uptake. After lipolysis (hydrolysis), the glycerol can be converted to dihydroxyacetone phosphate, an intermediate of glycolysis (see Appendix A). The fatty acids are catabolized by a process called **β-oxidation.** Two carbon fragments from the fatty acid are combined with Coenzyme A to produce Acetyl-CoA, which then may be used in the Krebs cycle to produce energy. Each Acetyl-CoA produced by this process also yields one NADH and one $FADH_2$ (important coenzymes in the electron transport chain). Glycerol and fatty acids may be used alternatively in anabolic pathways.

Tributyrin Agar is prepared as an emulsion which makes the agar appear opaque. When the plate is inoculated with a lipase-positive organism, clear zones will appear around the growth as evidence of lipolytic activity. If no clear zones appear, the organism is lipase-negative (Figure 5-59). Table 5-21 summarizes results and interpretations.

Application

The lipase test is used to detect and enumerate lipolytic bacteria especially in high-fat dairy products. A variety of other lipid substrates, including

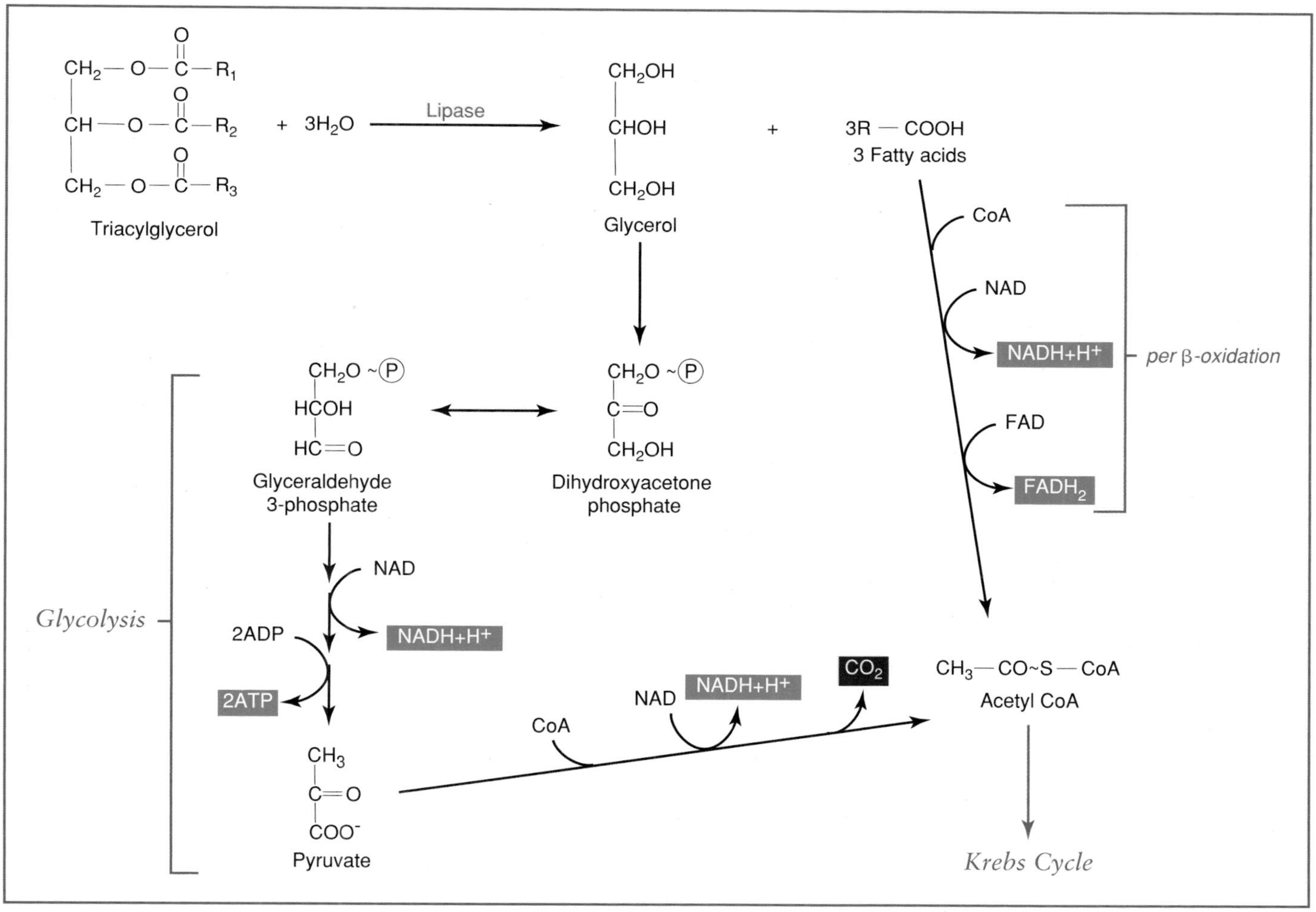

FIGURE 5-58 LIPID METABOLISM

A triacylglycerol is a simple fat molecule, composed of a three-carbon alcohol (glycerol) bonded to three fatty acid chains (represented by R_1, R_2, and R_3 in this diagram. The products of lipid catabolism can be used in glycolysis (glycerol) and the Krebs cycle (acetyl CoA).

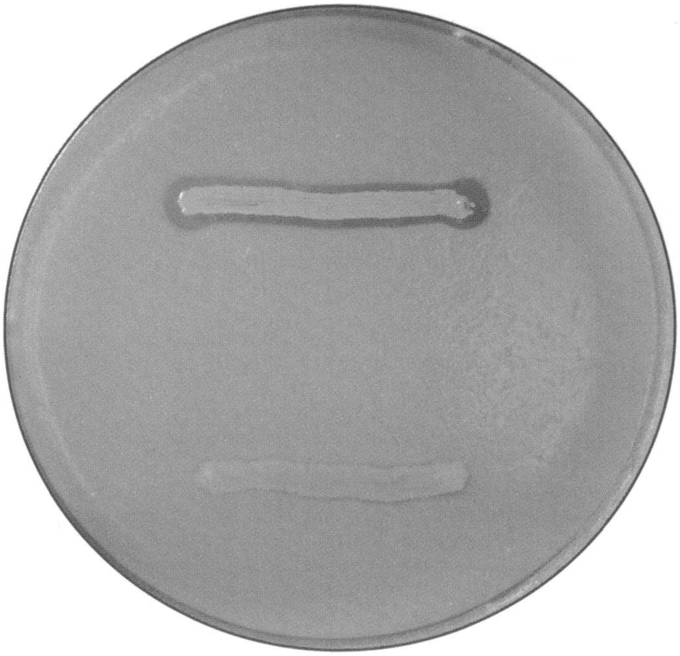

FIGURE 5-59 TRIBUTYRIN AGAR
This Tributyrin Agar plate was inoculated with a lipase-positive (+) organism above and a lipase-negative (−) organism below.

corn oil, olive oil, and soybean oil, are used to detect differential characteristics among members of *Enterobacteriaceae*, *Clostridium*, *Staphylococcus*, and *Neisseria*. Several fungal species also demonstrate lipolytic ability.

In This Exercise...

You will inoculate a Tributyrin Agar plate with *Staphylococcus aureus* and *Proteus mirabilis*, one of which will produce a clearing around the growth. If no clearing appears within 48 hours, return the plate to the incubator for an additional 24–48 hours.

Materials

Per Student Group

- one Tributyrin Agar plate
- fresh cultures of:
  - *Staphylococcus aureus*
  - *Proteus mirabilis*

Media Recipe

Tributyrin Agar

• Beef extract	3.0 g
• Peptone	5.0 g
• Agar	15.0 g
• Tributyrin oil	10.0 mL
• Distilled or deionized water	1.0 L

pH 5.8–6.2 at 25°C

TABLE OF RESULTS

Result	Interpretation	Symbol
Clearing in agar around growth	Lipase is present	+
No clearing in agar around growth	Lipase is absent	−

TABLE 5-21 Lipase Test results and interpretations.

Procedure

Lab One

1. Using a marking pen, divide the bottom of the plate into three equal sectors. ***Note:*** Make sure the medium is opaque prior to inoculation. If it is not, obtain a plate that is.
2. Label the plate with the names of the organisms, your name, and the date.
3. Spot-inoculate two sectors with the test organisms, leaving the third sector uninoculated as a control.
4. Invert the plate and incubate it aerobically at 35 ± 2°C for 48 hours.

Lab Two

Examine the plates for clearing around the bacterial growth and record your results in the chart provided on the Data Sheet. If no clearing has developed, reincubate for 24 to 48 hours.

References

Collins, C. H., Patricia M. Lyne, and J. M. Grange. 1995. Page 114 in *Collins and Lyne's Microbiological Methods,* 7th ed. Butterworth-Heinemann, Oxford, United Kingdom.

Haas, Michael J. 2001. Chapter 15 in *Compendium of Methods for the Microbiological Examination of Foods,* 4th ed. Edited by Frances Pouch Downes and Keith Ito. American Public Health Association, Washington, DC.

Knapp, Joan S., and Roselyn J. Rice. 1995. Page 335 in *Manual of Clinical Microbiology,* 6th ed., edited by Patrick R. Murray, Ellen Jo Baron, Michael A. Pfaller, Fred C. Tenover, and Robert H. Yolken. ASM Press, Washington, DC.

MacFaddin, Jean F. 2000. Page 286 in *Biochemical Tests for Identification of Medical Bacteria,* 3rd ed. Lippincott Williams & Wilkins, Philadelphia.

Zimbro, Mary Jo, and David A. Power, editors. 2003. Page 521 in *Difco™ and BBL™ Manual—Manual of Microbiological Culture Media.* Becton Dickinson and Co., Sparks, MD.

Combination Differential Media

Combination differential media are, as the name implies, combinations of ingredients put together in such a way as to allow multiple biochemical determinations with a single inoculation. They typically include core tests (some of which can be simply observed, others that require addition of reagents) to differentiate members of specific bacterial groups. For example, SIM medium tests for sulfur reduction, indole production and motility—tests recommended to differentiate between *Salmonella* and *Shigella*.

As with most biochemical tests, combination test media are frequently used in tandem with a selective differential medium (Section 4). Selective media promote isolation of desired organisms and provide valuable evidence used to determine the sequence of tests to follow. MacConkey Agar (Exercise 4-7), for example, may be streaked initially to isolate any Gram-negative bacilli followed by the combination medium Kligler's Iron Agar (KIA) or Triple Sugar Iron Agar (TSI). KIA and TSI, which detect fermentation and hydrogen sulfide production, are recommended for differentiation among Gram-negative bacilli.

Other examples of combination media considered in this unit are Lysine Iron Agar (LIA) and Litmus Milk. LIA is used to differentiate enteric bacilli based on their ability to decarboxylate or deaminate lysine, and to respire anaerobically by reducing sulfur. Litmus milk detects up to seven different bacterial biochemical functions, including, lactose fermentation, casein hydrolysis, and deamination.

Exercise 5-20

SIM Medium

Theory SIM medium is used for determination of three bacterial activities: sulfur reduction, indole production from tryptophan, and motility. The semisolid medium includes casein and animal tissue as sources of amino acids, an iron-containing compound, and sulfur in the form of sodium thiosulfate.

Sulfur reduction to H_2S can be accomplished by bacteria in two different ways, depending on the enzymes present.

1. The enzyme **cysteine desulfurase** catalyzes the putrefaction of the amino acid cysteine to pyruvate (Figure 5-60).
2. The enzyme **thiosulfate reductase** catalyzes the reduction of sulfur (in the form of sulfate) at the end of the anaerobic respiratory electron transport chain (Figure 5-61).

Both systems produce hydrogen sulfide (H_2S) gas. When either reaction occurs in SIM medium, the H_2S produced combines with iron, in the form of ferrous ammonium sulfate, to form ferric sulfide (FeS), a black precipitate (Figure 5-62). Any blackening of the medium is an indication of sulfur reduction and a positive test. No blackening of the medium indicates no sulfur reduction and a negative reaction (Figure 5-63).

Indole production in the medium is made possible by the presence of tryptophan (contained in casein and animal protein). Bacteria possessing the enzyme **tryptophanase** can hydrolyze tryptophan to pyruvate, ammonia (by deamination) and indole (Figure 5-64).

The hydrolysis of tryptophan in SIM medium can be detected by the addition of Kovacs' reagent after a period of incubation. Kovacs' reagent contains **dimethylaminobenzaldehyde** (DMABA) and HCl dissolved in amyl alcohol. When a few drops of Kovacs' reagent are added to the tube, DMABA reacts with any indole present and produces a quinoidal compound that turns the reagent layer red (Figures 5-65 and 5-66). The formation of red color in the reagent layer indicates a positive reaction and the presence of tryptophanase. No red color is indole-negative.

Motility determination in SIM medium is made possible by the reduced agar concentration and the method of inoculation. The medium is inoculated with a single stab from an inoculating needle. Motile organisms are

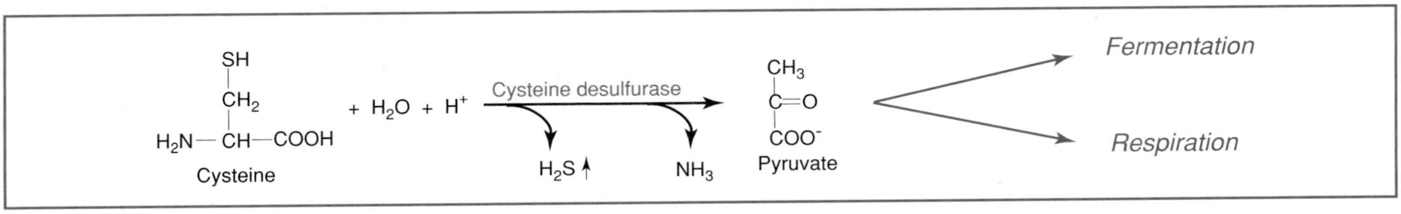

FIGURE 5-60 Putrefaction of Cysteine
Putrefaction involving cysteine desulfurase produces H_2S. This reaction is a mechanism for getting energy out of the amino acid cysteine.

$$3S_2O_3^{=} + 4H^+ + 4e^- \xrightarrow{\text{Thiosulfate reductase}} 2SO_3^{=} + 2H_2S\uparrow$$

FIGURE 5-61 Reduction of Thiosulfate
Anaerobic respiration with thiosulfate as the final electron acceptor also produces H_2S.

$$H_2S + FeSO_4 \longrightarrow H_2SO_4 + FeS\downarrow$$

FIGURE 5-62 Indicator Reaction
Hydrogen sulfide, a colorless gas, can be detected when it reacts with ferrous sulfate in the medium to produce the black precipitate ferric sulfide.

FIGURE 5-63 Sulfur Reduction in SIM Medium
The organism on the left is H_2S-negative; the organism on the right is H_2S-positive.

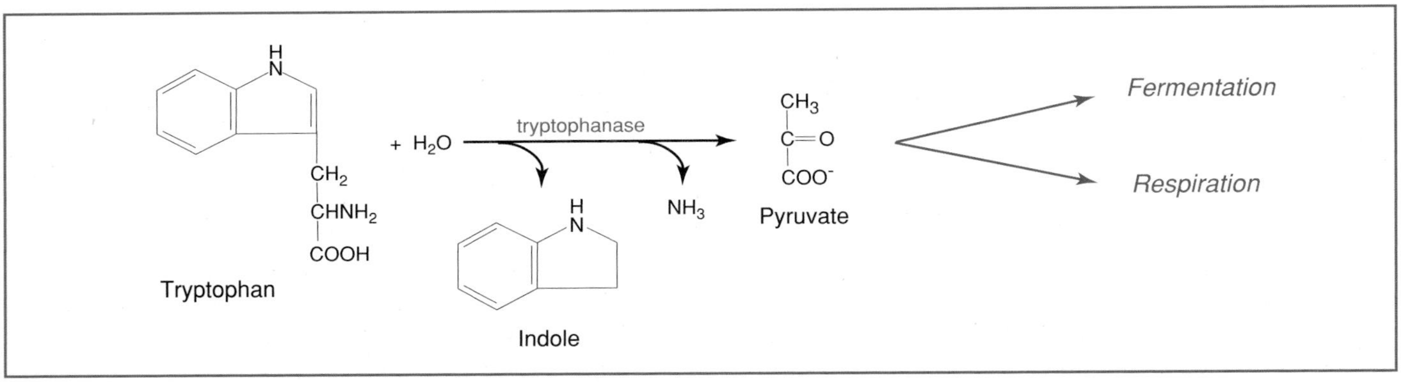

FIGURE 5-64 Tryptophan Catabolism in Indole-Positive Organisms
This reaction is a mechanism for getting energy out of the amino acid tryptophan.

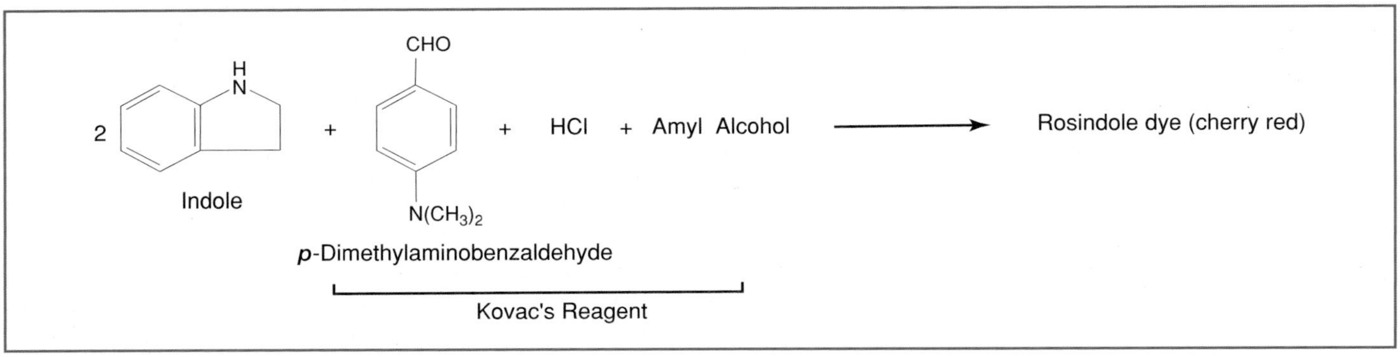

FIGURE 5-65 Indole Reaction with Kovac's Reagent
If present, indole reacts with the DMABA in Kovac's reagent to produce a red color.

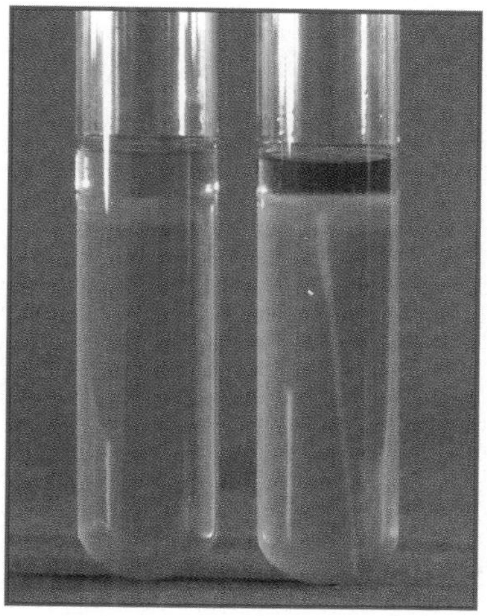

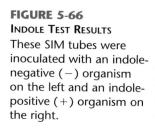

FIGURE 5-66
INDOLE TEST RESULTS
These SIM tubes were inoculated with an indole-negative (−) organism on the left and an indole-positive (+) organism on the right.

FIGURE 5-67
MOTILITY IN SIM
These SIM tubes were inoculated with a motile organism on the left and a nonmotile organism on the right.

able to move about in the semi-solid medium and can be detected by the radiating growth pattern extending outward in all directions from the central stab line. Growth that radiates in *all* directions and appears slightly fuzzy is an indication of motility (Figure 5-67). This should not be confused with the (seemingly) spreading growth produced by lateral movement of the inoculating needle when stabbing.

Application

SIM medium is used to identify bacteria that are capable of producing indole, using the enzyme tryptophanase. The Indole Test is one component of the IMViC battery of tests (*I*ndole, *M*ethyl red, *V*oges-Proskauer, and *C*itrate) used to differentiate the *Enterobacteriaceae*. SIM medium also is used to differentiate sulfur-reducing members of *Enterobacteriaceae*, especially members of the genera *Salmonella, Francisella,* and *Proteus* from the negative *Morganella morganii* and *Providencia rettgeri*. In addition to the first two functions of SIM, motility is an important differential characteristic of *Enterobacteriaceae*.

In This Exercise...

You will stab inoculate SIM medium with three organisms demonstrating a variety of results. When reading your test results, make the motility and H_2S determinations before adding the indole reagent. Use Tables 5-22, 5-23 and 5-24, and Figures 5-63, 5-66, 5-67 as guides when making your interpretations.

Materials

Per Student Group

- three SIM tubes
- Kovac's reagent
- fresh cultures of:
  - *Escherichia coli*
  - *Salmonella typhimurium*
  - *Shigella flexneri*

TABLE OF RESULTS

Result	Interpretation	Symbol
Black in the medium	Sulfur reduction (H_2S production)	+
No black in the medium	Sulfur is not reduced	−

TABLE 5-22 Sulfur Reduction results and interpretations.

TABLE OF RESULTS

Result	Interpretation	Symbol
Red in the alcohol layer of Kovac's reagent	Tryptophan is broken down into indole and pyruvate	+
Reagent color is unchanged	Tryptophan is not broken down into indole and pyruvate	−

TABLE 5-23 Indole Production results and interpretations.

TABLE OF RESULTS

Result	Interpretation	Symbol
Growth radiating outward from the stab line	Motility	+
No radiating growth	Nonmotile	−

TABLE 5-24 Motility results and interpretations.

Medium and Reagent Recipes

SIM (Sulfur-Indole-Motility) Medium

- Pancreatic digest of casein 20.0 g
- Peptic digest of animal tissue 6.1 g
- Ferrous ammonium sulfate 0.2 g
- Sodium thiosulfate 0.2 g
- Agar 3.5 g
- Distilled or deionized water 1.0 L

pH 7.1–7.4 at 25°C

Kovacs' Reagent

- Amyl alcohol 75.0 mL
- Hydrochloric acid, concentrated 25.0 mL
- *p*-dimethylaminobenzaldehyde 5.0 g

Procedure

Lab One

1. Obtain four SIM tubes. Label two with the names of the organisms, your name, and the date. Label one tube "control."
2. Stab-inoculate three tubes with the test organisms. Insert the needle only about two-thirds the depth of the agar. Be careful to remove the needle along the original stab line. Do not inoculate the control.
3. Incubate all tubes aerobically at 35 ± 2°C for 24 to 48 hours.

Lab Two

1. Examine the tubes for spreading from the stab line *and* formation of black precipitate in the medium. Record any H_2S production and/or motility in the chart provided on the Data Sheet.
2. Add Kovacs' reagent to each tube (to a depth of 2–3 mm). After several minutes, observe for the formation of red color in the reagent layer.
3. Record your results in the chart on the Data Sheet.

References

Delost, Maria Dannessa. 1997. Page 186 in *Introduction to Diagnostic Microbiology*. Mosby, St. Louis.

MacFaddin, Jean F. 1980. Page 162 in *Biochemical Tests for Identification of Medical Bacteria*, 2nd ed. Williams & Wilkins, Baltimore.

Zimbro, Mary Jo and David A. Power, editors. 2003. Page 490 in *Difco™ and BBL™ Manual—Manual of Microbiological Culture Media*. Becton Dickinson and Co., Sparks, MD.

Exercise 5-21

Triple Sugar Iron Agar

Theory

Triple Sugar Iron Agar (TSI) is a rich medium designed to differentiate bacteria on the basis of glucose fermentation, lactose fermentation, sucrose fermentation, and sulfur reduction. In addition to the three carbohydrates, it includes beef extract, yeast extract, and peptone as sources of carbon and nitrogen, and sodium thiosulfate as a source of reducible sulfur. Phenol red is the pH indicator, and ferrous sulfate is the hydrogen sulfide indicator.

The medium is prepared as an agar slant with a deep butt, thereby providing both aerobic and anaerobic growth environments. It is inoculated by a stab in the agar butt followed by a fishtail streak of the slant. The incubation period is 18 to 24 hours for carbohydrate fermentation and up to 48 hours for hydrogen sulfide reactions. Many reactions in various combinations are possible (Figure 5-68 and Table 5-25).

When TSI is inoculated with a glucose-only fermenter, acid products lower the pH and turn the entire medium yellow within a few hours. Because glucose is in short supply (0.1%), it will be exhausted within about 12 hours. The organism then will begin to break down amino acids in the medium, producing NH_3 and raising the pH. This process is called a **reversion**. After 18 to 24 hours the alkaline products will be sufficient to turn the slant red but will not overcome the acid conditions in the butt, which remains yellow. Thus, a TSI with a red slant and yellow butt indicates that glucose was the only carbohydrate fermented.

Organisms that are able to ferment lactose and/or sucrose also turn the medium yellow throughout. However, because the concentrations of lactose and sucrose in the medium are much higher (1.0%) than that of glucose, resulting in greater acid production, both slant

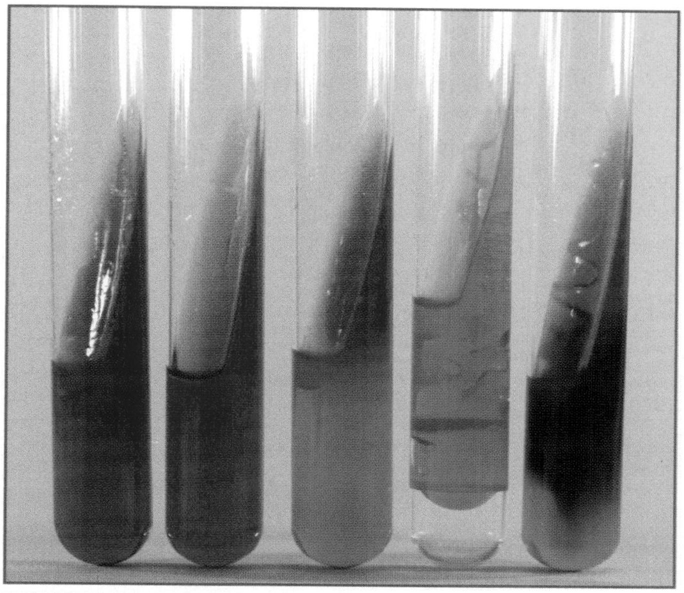

FIGURE 5-68 TSI Agar Slants
These TSI Agar slants from left to right illustrate alkaline slant/no change in the butt (K/NC); uninoculated control; alkaline slant/acid butt (K/A); acid slant/acid butt, gas (A/A, G); and alkaline slant/acid butt, hydrogen sulfide present (K/A, H_2S). Refer to Table 5-25 for interpretations.

TABLE OF RESULTS

Result	Interpretation	Symbol
Yellow slant/yellow butt	Glucose and lactose and/or sucrose fermentation with acid accumulation in slant and butt.	A/A
Red slant/yellow butt	Glucose fermentation with acid production. Peptone catabolized aerobically (in the slant) with alkaline products (reversion).	K/A
Red slant/red butt	No fermentation. Peptone catabolized aerobically and anaerobically with alkaline products. Not from *Enterobacteriaceae*.	K/K
Red slant/no change in the butt	No fermentation. Peptone catabolized aerobically with alkaline products. Not from *Enterobacteriaceae*.	K/NC
No change in slant/no change in butt	Organism is growing slowly or not at all. Not from *Enterobacteriaceae*.	NC/NC
Black precipitate in the agar	Sulfur reduction. (An acid condition, from fermentation of one or more of the sugars, exists in the butt even if the yellow color is obscured by the black precipitate.)	H_2S
Cracks in or lifting of agar	Gas production.	G

TABLE 5-25 TSI results and interpretations.

and butt will remain yellow after a 24-hour incubation period. Therefore, a tube after 24 hours that has a yellow slant and butt indicates that glucose and *either or both* of the other two carbohydrates were fermented. (**Note:** Remember that lactose and sucrose are disaccharides and that all organisms that ferment them also ferment glucose.) Gas produced by fermentation of any of the three carbohydrates will appear as fissures in the medium or will lift the agar off the bottom of the tube.

Hydrogen sulfide (H_2S) may be produced by the reduction of thiosulfate in the medium or by the breakdown of cysteine in the peptone. Ferrous sulfate reacts with the H_2S to form a black precipitate, usually seen in the butt. Acid conditions must exist for thiosulfate reduction; therefore, black precipitate in the medium is an indication of sulfur reduction *and* fermentation. If the black precipitate obscures the color of the butt, the color of the slant determines which carbohydrates have been fermented (*i.e.*, red slant = glucose fermentation, yellow slant = glucose and lactose and/or sucrose fermentation).

An organism that does not ferment any of the three carbohydrates but utilizes peptone and amino acids will alkalinize the medium and turn it red. If the organism can use the peptone aerobically and anaerobically, both the slant and butt will appear red. An obligate aerobe will turn only the slant red. (**Note:** The change may be subtle. Be sure to compare to the uninoculated control.)

Timing is critical in reading TSI results. An early reading could reveal yellow throughout the medium, leading you to believe that the organism is a lactose and/or sucrose fermenter when it simply has not yet exhausted the glucose. A reading after the lactose and sucrose have been depleted could reveal a yellow butt and red slant leading you to falsely believe that the organism is a glucose-only fermenter. Tubes that have been interpreted for carbohydrate fermentation can be re-incubated for 24 hours before H_2S determination. Refer to Table 5-25 for information on the correct symbols and method of reporting the various reactions.

Application

TSI is primarily used to differentiate members of *Enterobacteriaceae* and to distinguish them from other Gram-negative bacilli such as *Pseudomonas* or *Alcaligenes*.

In This Exercise…

Today, you will inoculate four TSI agar slants. Use a large inoculum and try not to introduce excessive air when stabbing the agar butt. Also be sure to remove all tubes from the incubator after no more than 24 hours to take fermentation readings.

Materials

Per Student Group

- five TSI slants
- recommended organisms (grown on solid media):
  - *Pseudomonas aeruginosa*
  - *Escherichia coli*
  - *Morganella morganii*
  - *Proteus vulgaris*

Medium Recipe

Triple Sugar Iron Agar

• Beef extract	3.0 g
• Yeast extract	3.0 g
• Peptone	15.0 g
• Proteose peptone	5.0 g
• Lactose	10.0 g
• Sucrose	10.0 g
• Dextrose (glucose)	1.0 g
• Ferrous sulfate	0.2 g
• Sodium chloride	5.0 g
• Sodium thiosulfate	0.3 g
• Agar	12.0 g
• Phenol red	0.024 g
• Distilled or deionized water	1.0 L

pH 7.2–7.6 at 25°C

Procedure

Lab One

1. Obtain five TSI slants. Label four of the slants with the names of the organisms, your name, and the date. Label the fifth slant "control."
2. Inoculate four TSI slants with the test organisms. Using a heavy inoculum, stab the agar butt and then streak the slant. Do not inoculate the control.
3. Incubate all slants aerobically at 35 ± 2°C for 18 to 24 hours.

Lab Two

Examine the tubes for characteristic color changes and gas production. Use Table 5-25 as a guide while recording your results on the Data Sheet. The proper format for recording results is: slant reaction/butt reaction, gas production, hydrogen sulfide production. For example, an acid slant and acid butt with gas and black precipitate would be recorded as: A/A, G, H_2S.

References

Delost, Maria Dannessa. 1997. Pages 184–185 in *Introduction to Diagnostic Microbiology.* Mosby, St. Louis.

Forbes, Betty A., Daniel F. Sahm, and Alice S. Weissfeld. 2002. Page 282 in *Bailey & Scott's Diagnostic Microbiology,* 11th ed. Mosby, St. Louis.

MacFaddin, Jean F. 2000. Page 239 in *Biochemical Tests for Identification of Medical Bacteria,* 3rd ed. Lippincott Williams & Wilkins, Philadelphia.

Zimbro, Mary Jo, and David A. Power, editors. 2003. Page 574 in *Difco™ and BBL™ Manual—Manual of Microbiological Culture Media.* Becton Dickinson and Co., Sparks, MD.

Exercise 5-22

Lysine Iron Agar

Theory Lysine Iron Agar (LIA) is a combination medium that detects bacterial ability to decarboxylate or deaminate lysine and to reduce sulfur. It contains peptone and yeast extract to support growth, the amino acid lysine for deamination and decarboxylation reactions, and sodium thiosulfate—a source of reducible sulfur. A small amount of glucose (0.1%) is included as a fermentable carbohydrate. Ferric ammonium citrate is included as a sulfur reduction indicator and bromcresol purple is the pH indicator. Bromcresol purple is purple at pH 6.8 and yellow at or below pH 5.2.

LIA is prepared as a slant with a deep butt. This results in an aerobic zone in the slant and an anaerobic zone in the butt. After it is inoculated with two stabs of the butt and a fishtail streak of the slant, the tube is tightly capped and incubated for 18 to 24 hours.

If the medium has been inoculated with a lysine decarboxylase-positive organism, acid production from glucose fermentation will induce production of decarboxylase enzymes. The acidic pH will turn the medium yellow, but subsequent decarboxylation of the lysine will alkalinize the agar and return it to purple. Purple color throughout indicates lysine decarboxylation. Purple color in the slant with a yellow (acidic) butt indicates glucose fermentation, but no lysine decarboxylation took place; degradation of peptone alkalinized the slant.

If the organism produces lysine deaminase, the resulting deamination reaction will produce compounds that react with the ferric ammonium citrate and produce a red color. Deamination reactions require the presence of oxygen. Therefore, any evidence of deamination will be seen only in the slant. A red slant with yellow (acidic) butt indicates lysine deamination.

Hydrogen sulfide (H_2S) is produced in Lysine Iron Agar by the anaerobic reduction of thiosulfate. Ferric ions in the medium react with the H_2S to form a black precipitate in the butt. Refer to Figure 5-69 and Table 5-26 for the various reactions and symbols used to record them.

Application LIA is used to differentiate enterics based on their ability to decarboxylate or deaminate lysine and produce hydrogen sulfide (H_2S). LIA also is used in combination with Triple Sugar Iron Agar to identify members of *Salmonella* and *Shigella*.

In This Exercise...

You will be inoculating 4 LIA slants. Use heavy inocula and tighten the caps before incubating, to maintain anaerobic conditions in the butt. Because the decarboxylase/H_2S–positive organisms being differentiated produce results rapidly, incubation time will be 18 to 24 hours. If your labs are scheduled more than 24 hours apart, have someone remove the tubes from the incubator at the appropriate time and refrigerate them until time to record the results. When entering your data, be sure to write the slant information first, followed by the butt and hydrogen sulfide results (example: K/K, H_2S).

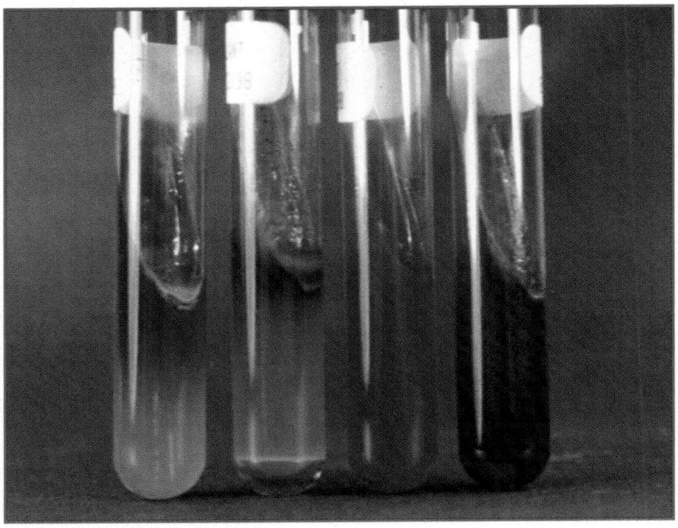

FIGURE 5-69 LIA RESULTS
Lysine Iron Agar tubes illustrating from left to right: (R/A); (K/A, H_2S, [note the small amount of black precipitate near the middle and the gas production from glucose fermentation at the base]); uninoculated control; and (K/K [obscured by the black precipitate], H_2S).

TABLE OF RESULTS

Result	Interpretation	Symbol
Purple slant/purple butt	Lysine deaminase negative; Lysine decarboxylase positive	K/K
Purple slant/yellow butt	Lysine deaminase negative; Lysine decarboxylase negative; Glucose fermentation	K/A
Red slant/yellow butt	Lysine deaminase positive; Lysine decarboxylase negative; Glucose fermentation	R/A
Black precipitate	Sulfur reduction	H_2S

TABLE 5-26 LIA results and interpretations.

Refer to Table 5-26 and Figure 5-69 when making your interpretations.

Materials

Per Student Group

- five LIA slants
- fresh cultures on solid media:
  - *Citrobacter freundii*
  - *Proteus mirabilis*
  - *Escherichia coli*
  - *Salmonella typhimurium*

Medium Recipe

Lysine Iron Agar

• Peptone	5.0 g
• Yeast Extract	3.0 g
• Dextrose (glucose)	1.0 g
• L-Lysine hydrochloride	10.0 g
• Ferric ammonium citrate	0.5 g
• Sodium thiosulfate	0.04 g
• Bromcresol purple	0.02 g
• Agar	15.0 g

pH 6.5–6.9 at 25°C

Procedure

Lab One

1. Obtain five LIA slants. Label four slants with the names of the organisms, your name, and the date. Label the fifth slant "control."
2. Inoculate four slants with the test organisms. Using heavy inocula, stab the agar butt twice and then streak the slant. Do not inoculate the control.
3. Incubate all slants with the caps tightened at 35 ± 2°C for 18 to 24 hours.

Lab Two

1. Examine the tubes for characteristic color changes.
2. Record your results in the chart provided on the Data Sheet.

References

Delost, Maria Dannessa. 1997. Page 194 in *Introduction to Diagnostic Microbiology.* Mosby, St. Louis.

Forbes, Betty A., Daniel F. Sahm, and Alice S. Weissfeld. 2002. Page 274 in *Bailey & Scott's Diagnostic Microbiology,* 11th ed. Mosby, St. Louis.

Zimbro, Mary Jo, and David A. Power, editors. 2003. *Difco™ and BBL™ Manual—Manual of Microbiological Culture Media.* Becton Dickinson and Co., Sparks, MD.

Exercise 5-23

Litmus Milk Medium

Theory Litmus Milk is an undefined medium consisting of skim milk and the pH indicator azolitmin. Skim milk provides nutrients for growth, lactose for fermentation, and protein in the form of casein. Azolitmin (litmus) is pink at pH 4.5 and blue at pH 8.3. Between these extremes it is light purple.

Four basic reactions occur in litmus milk: lactose fermentation, reduction of litmus, casein coagulation, and casein hydrolysis. In combination these reactions yield a variety of results, each of which can be used to differentiate bacteria. Several possible combinations are described in Table 5-27.

Lactose fermentation acidifies the medium and turns the litmus pink (Figure 5-70, second tube from the right). This acid reaction typically begins with the splitting of the disaccharide into the monosaccharides glucose and galactose by the enzyme β-galactosidase. Refer to Figure 5-71 and Appendix A as necessary. Accumulating acid may cause the casein to precipitate and form an acid clot (Figures 5-72 and 5-73). Acid clots solidify the medium and can appear pink or white with a pink band at the top (Figure 5-70, second tube from the left) depending on the oxidation-reduction status of litmus. Reduced litmus is white; oxidized litmus is purple. Acid clots can be dissolved in alkaline conditions. Fissures or cracks in the clot are evidence of gas production (Figure 5-74). Heavy gas production that breaks up the clot is called stormy fermentation.

In addition to being a pH indicator, litmus is an E_h (oxidation-reduction) indicator. As mentioned above,

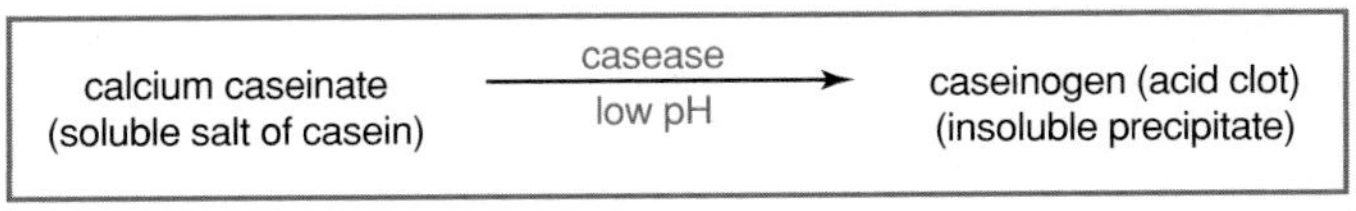

FIGURE 5-72 Acid Clot Formation
An acid clot is the result of casease catalyzing the formation of caseinogen, an insoluble precipitate, under acidic conditions.

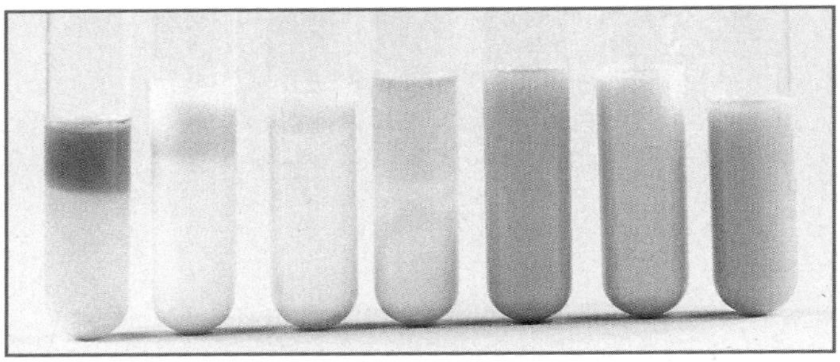

FIGURE 5-70 Reactions in Litmus Milk
These Litmus Milk tubes illustrate, from left to right: digestion, alkaline reaction (DK); acid clot formation, reduction of litmus (ACR); acid clot formation, reduction of litmus, gas production from fermentation (ACRG—note small gas fissure in clot); curd formation, reduction of litmus (CR); uninoculated control; acid reaction (A); and alkaline reaction (K).

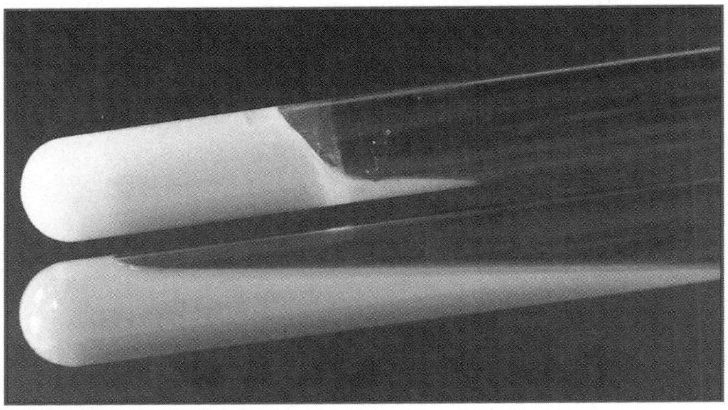

FIGURE 5-73 Acid Clot
An acid clot appears in the top tube; an uninoculated control is below. Note the reduced litmus (white) at the bottom of the clot.

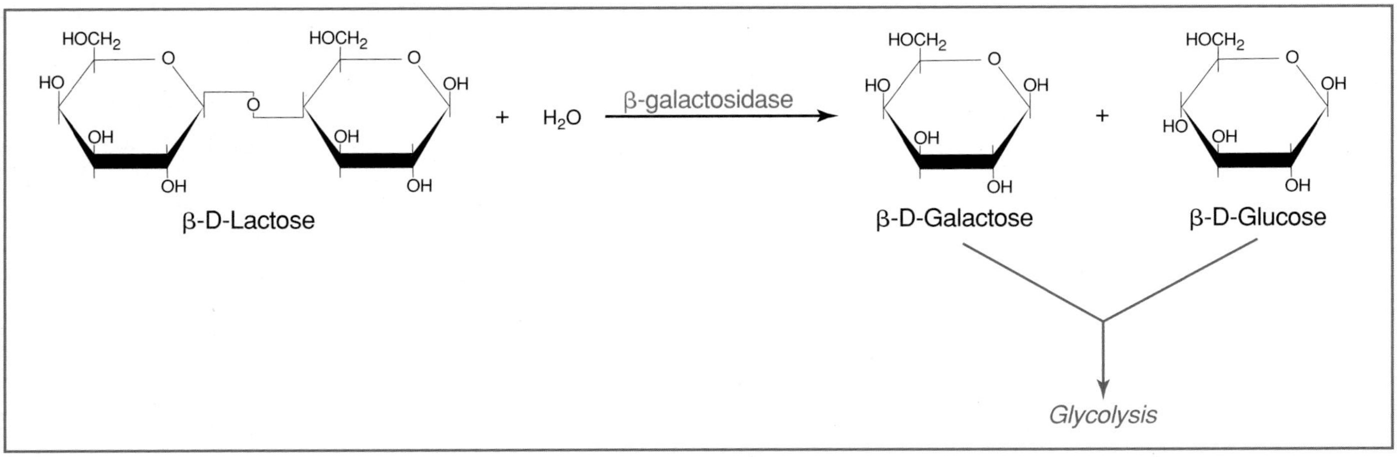

FIGURE 5-71 Lactose Hydrolysis
Lactose hydrolysis requires the enzyme β-galactosidase and produces glucose and galactose—two fermentable sugars.

reduced litmus is white. If litmus is reduced during lactose fermentation, it will turn the medium white in the lower portion of the tube where the reduction rate is greatest.

Some bacteria produce proteolytic enzymes such as rennin, pepsin, or chymotrypsin that coagulate casein and produce a curd (Figure 5-70, center tube, and Figure 5-75). A curd differs from an acid clot in that it will not dissolve in alkaline conditions and tends to retract from the sides of the tube, revealing a straw colored fluid called whey.

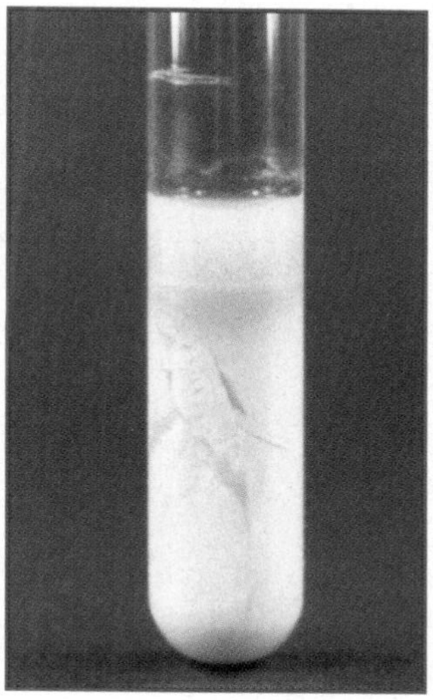

FIGURE 5-74 Gas Production from Fermentation This organism has produced gas fissures in the clot.

$$\underset{\text{(soluble)}}{\text{casein}} \xrightarrow[\text{Ca}^{++}]{\text{rennin}} \underset{\text{(insoluble precipitate)}}{\text{paracasein (soft curd)}}$$

FIGURE 5-75 Curd Formation Rennin converts casein to paracasein to form a curd.

Certain enzymes can digest both acid clots and curds. A digestion reaction leaves only a clear to brownish fluid behind (Figure 5-70, tube on the left). Bacteria that are able only to partially digest the casein typically produce ammonia (NH_3), which raises the pH of the medium and turns the litmus blue. Formation of a blue or purple ring at the top of the clear fluid or bluing of the entire medium indicates an alkaline reaction.

Application

Litmus milk is used primarily to differentiate members within the genus *Clostridium*. It differentiates *Enterobacteriaceae* from other Gram negative bacilli based on the ability of enterics to reduce litmus. Litmus milk also is used to cultivate and maintain cultures of lactic acid bacteria.

In This Exercise...

You will observe several different reactions in Litmus Milk. Most tubes will illustrate more than one reaction, so refer to Figure 5-70 and Table 5-27 when making your interpretations.

Materials

Per Student Group

- six Litmus Milk tubes
- fresh cultures of:
  - *Alcaligenes faecalis*
  - *Pseudomonas aeruginosa*
  - *Klebsiella pneumoniae*
  - *Lactococcus lactis*
  - *Enterococcus faecium*

TABLE OF RESULTS

Result	Interpretation	Symbol
Pink color	Acid reaction	A
Pink and solid (white in the lower portion if the litmus is reduced); clot not movable	Acid clot	AC
Fissures in clot	Gas	G
Clot broken apart	Stormy fermentation	S
White color (lower portion of medium)	Reduction of litmus	R
Semisolid and not pink; clear to gray fluid at top	Curd	C
Clarification of medium; loss of "body"	Digestion	D
Blue medium or blue band at top	Alkaline reaction	K
No change	None of the above reactions	NC

TABLE 5-27 Litmus Milk results and interpretations. (These results may appear together in a variety of combinations.)

Medium Recipe

- Litmus Milk Medium
- Skim milk 100.0 g
- Azolitmin 0.5 g
- Sodium sulfite 0.5 g
- Distilled or deionized water 1.0 L

pH 6.3–6.7 at 25°C

Procedure

Lab One

1. Obtain six Litmus Milk tubes. Label five tubes with the names of the organisms, your name, and the date. Label the sixth tube "control."
2. Inoculate five tubes with the test cultures. Do not inoculate the control.
3. Incubate all tubes aerobically at 35 ± 2°C for 7 to 14 days.

Lab Two

1. Examine the tubes for color changes, gas production, and clot formation. Be sure to compare all tubes to the control, and refer to Table 5-27 when making your interpretations.
2. Record your results in the chart provided on the Data Sheet.

References

Forbes, Betty A., Daniel F. Sahm, Alice S. Weissfeld. 2002. Pages 272–273 in *Bailey & Scott's Diagnostic Microbiology,* 11th ed. Mosby, St. Louis.

MacFaddin, Jean F. 2000. Page 294 in *Biochemical Tests for Identification of Medical Bacteria,* 3rd ed. Lippincott Williams & Wilkins, Philadelphia.

Zimbro, Mary Jo, and David A. Power, editors. 2003. Page 309 in *Difco™ and BBL™ Manual—Manual of Microbiological Culture Media.* Becton Dickinson and Co., Sparks, MD.

Antimicrobial Susceptibility and Resistance

Any microorganism that can be cultivated outside of its host in the laboratory can be tested for susceptibility to antimicrobial agents. This is important to know for therapeutic *and* diagnostic reasons. As you will see in the Kirby-Bauer Test (Exercise 7-3), establishing antibiotic susceptibility is an important step in planning a therapeutic course of action. In this unit you will perform two diagnostic tests—Bacitracin Test (for susceptibility) and β-Lactamase Test (for penicillin resistance)—used to differentiate members of *Staphylococcus*, *Streptococcus*, and *Enterococcus*.

Exercise 5-24

Bacitracin Susceptibility Test

Theory Antibiotics are antimicrobial substances produced by microorganisms. Bacitracin, produced by *Bacillus licheniformis*, is a powerful peptide antibiotic that inhibits bacterial cell wall synthesis (Figure 5-76). Thus, it is effective only on bacteria that have cell walls and are in the process of growing.

The Bacitracin Test is a simple test performed by placing a bacitracin-impregnated disk on an agar plate inoculated to produce a bacterial lawn. The bacitracin diffuses into the agar and, where its concentration is sufficient, inhibits growth of susceptible bacteria. Inhibition of bacterial growth will appear as a clearing on the agar plate. Any zone of clearing 10 mm or greater around the disk is interpreted as bacitracin susceptibility (Figure 5-77). For more information on antimicrobial susceptibility, refer to Exercise 7-3, Antimicrobial Susceptibility Test.

Application This test is used to differentiate and presumptively identify β-hemolytic group A streptococci (*Streptococcus pyogenes*) from other β-hemolytic streptococci. It also differentiates the genus *Staphylococcus* (resistant) from the susceptible *Micrococcus* and *Stomatococcus*.

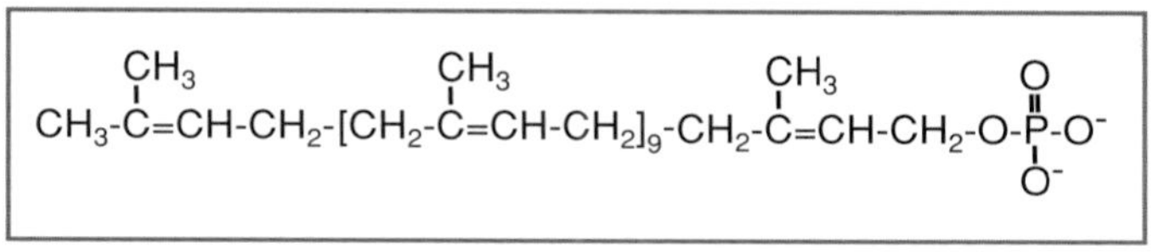

FIGURE 5-76 UNDECAPRENYL PHOSPHATE
Undecaprenyl phosphate is involved in transporting peptidoglycan subunits across the cell membrane during cell wall synthesis. Bacitracin interferes with its release from the peptidoglycan subunit. It is a C_{55} molecule derived from 11 isoprene subunits plus a phosphate.

In This Exercise...

You will be inoculating a Blood Agar plate with two of the organisms this test was designed to detect. You will be inoculating from broth with a cotton swab. Be careful not to overdo it, as it is important to keep the two organisms separate on the plate. Wipe the excess broth off the swab on the inside of the culture tube before you make the transfer.

FIGURE 5-77 BACITRACIN SUSCEPTIBILITY ON A SHEEP BLOOD AGAR PLATE
The organism above has no clear zone and is resistant (R); the organism below has a clear zone larger than 10 mm and is susceptible (S).

TABLE OF RESULTS

Result	Interpretation	Symbol
Zone of clearing 10 mm or greater	Organism is sensitive to bacitracin	+
Zone of clearing less than 10 mm	Organism is resistant to bacitracin	–

TABLE 5-28 Bacitracin Test results and interpretations.

Materials

Per Student Group

- one Blood Agar plate (commercial preparation of TSA containing 5% sheep blood)
- sterile cotton applicators
- 0.04 unit bacitracin disks
- beaker of alcohol with forceps
- fresh broth cultures of:
  - *Staphylococcus aureus*
  - *Micrococcus luteus*

Procedure

Lab One

1. Obtain one Blood Agar plate. Using a sterile cotton applicator, inoculate half of the plate with *S. aureus*. (Make the inoculum as light as possible by wiping and twisting the wet cotton swab on the inside of the culture tube before removing it.) Inoculate the plate by making a single streak nearly halfway across its diameter. Turn the plate 90° and spread the organism evenly to produce a bacterial lawn covering nearly half the agar surface. Refer to the photo in Figure 5-77.
2. Being careful not to mix the organisms, repeat the process on the other half of the plate using *M. luteus*. Allow the broth to be absorbed by the agar for 5 minutes before proceeding to step 3.
3. Sterilize the forceps by placing them in the Bunsen burner flame long enough to ignite the alcohol. (***Note:*** Do not hold the forceps in the flame; you are simply burning off the excess alcohol.) Once the alcohol has burned off, use the forceps to place a bacitracin disk in the center of the half of the plate containing *S. aureus*. Gently tap the disk into place to ensure that it makes full contact with the agar surface. Return the forceps to the alcohol.
4. Repeat step 3, placing a bacitracin disk on the half of the plate containing *M. luteus*. Tap the disk into place and return the forceps to the alcohol.
5. Invert the plate, label it appropriately, and incubate it for 24 to 48 hours at room temperature.

Lab Two

1. Remove the plate from the incubator and examine it for clearing around the disks.
2. Record your results in the chart on the Data Sheet.

References

Baron, Ellen Jo, Lance R. Peterson, and Sydney M. Finegold. 1994. Page 329 in *Bailey & Scott's Diagnostic Microbiology*, 9th ed. Mosby–Year Book, St. Louis.

Delost, Maria Dannessa. 1997. Page 107 in *Introduction to Diagnostic Microbiology*. Mosby, St. Louis.

DIFCO Laboratories. 1984. Page 292 in *DIFCO Manual*, 10th ed. DIFCO Laboratories, Detroit.

Forbes, Betty A., Daniel F. Sahm, and Alice S. Weissfeld. 2002. Page 290 in *Bailey & Scott's Diagnostic Microbiology*, 11th ed. Mosby, St. Louis.

Koneman, Elmer W., *et al.* 1997. Pages 551 and 1299 in *Color Atlas and Textbook of Diagnostic Microbiology*, 5th ed. Lippincott-Raven Publishers, Philadelphia.

MacFaddin, Jean F. 2000. Page 3 in *Biochemical Tests for Identification of Medical Bacteria*, 3rd ed. Lippincott Williams & Wilkins, Philadelphia.

Exercise 5-25

β-Lactamase Test

Theory Penicllins and cephalosporins—called **β-lactam antibiotics** because of the β-lactam ring in their chemical structure—kill bacteria by interfering with cell wall synthesis. The bacterial enzyme **transpeptidase** catalyzes cross-linking between peptidoglycan subunits, thereby adding rigidity to the cell wall. By competing for sites on transpeptidase, β-lactam antibiotics prevent essential cross-linking of the peptidoglycan. Many bacteria have developed resistance to these antibiotics by producing enzymes called **β-lactamases,** which hydrolyze the β-lactam ring, thus destroying the structure of the antibiotic. Partial chemical structures of β-lactam antibiotics and the site of action for β-lactamases are shown in Figure 5-78.

The β-Lactamase Test is one of many tests used to identify β-lactamase production by measuring resistance to β-lactam antibiotics. In this test, a paper disc containing **nitrocefin** is smeared with the test organism. Nitrocefin is a cephalosporin, susceptible to most β-lactamases, that turns pink when it is hydrolyzed. Therefore, if the test organism produces β-lactamase, it will hydrolyze the nitrocefin and produce a pink spot on the disc (Figure 5-79).

Application The β-Lactamase Test is used to quickly identify if patient isolates are resistant to penicillins and cephalosporins. It is especially useful in identifying resistant strains of *Neisseria gonorrhoeae*, *Staphylococcus spp*., and members of genus *Enterococcus*.

In This Exercise...

You will perform a simple transfer to a chemically impregnated disc and watch for red color to develop. Use a heavy inoculum and Table 5-29 to assist in interpretation.

Materials

Per Student Group

- Cefinase® discs (available from Becton Dickinson Microbiology Systems, Sparks, MD 21152)
- fresh slant cultures of:
  - *Staphylococcus epidermidis*
  - *Enterococcus faecalis*

Procedure

1. Place two Cefinase® discs in a sterile Petri dish or on a microscope slide.

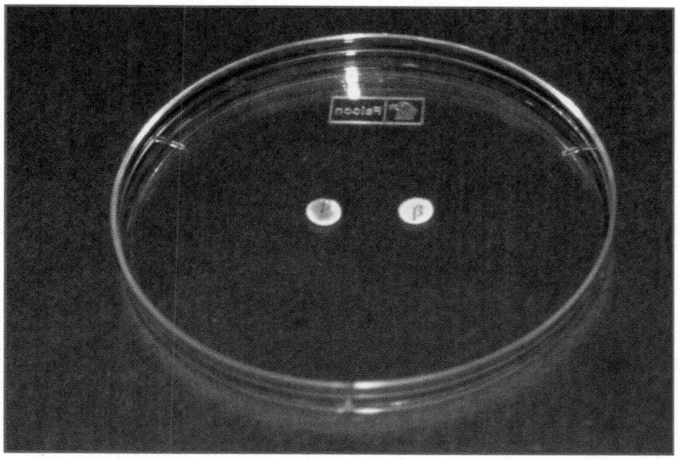

FIGURE 5-79 CEFINASE® DISCS
A β-lactam resistant strain is on the left, and a susceptible strain is on the right. Cefinase® Discs are available from Becton Dickinson Microbiology Systems, Sparks, MD.

TABLE OF RESULTS

Result	Interpretation	Symbol
Red or pink	β-lactamase production	+
Yellow or no color change	No β-lactamase production	–

TABLE 5-29 β-Lactamase Test results and interpretations.

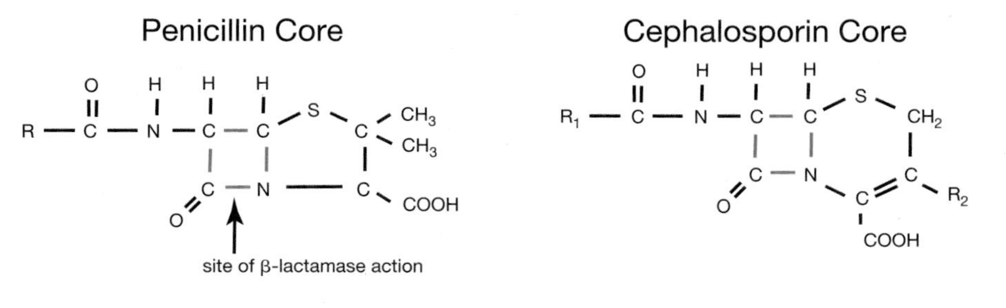

FIGURE 5-78 β-LACTAM ANTIBIOTIC STRUCTURE
Although penicillins and cephalosporins have unique structures, their bactericidal effect is similar, in that they both contain a β-lactam ring and interfere with peptidoglycan cross-linking. The arrow indicates the site of β-lactamase activity.

2. Inoculate each with a test organism and observe for color changes. Use a heavy inoculum.
3. Record your results in the chart provided on the Data Sheet.

References

Delost, Maria Dannessa. 1997. Page 79 in *Introduction to Diagnostic Microbiology.* Mosby, St. Louis.

Forbes, Betty A., Daniel F. Sahm, and Alice S. Weissfeld. 2002. Pages 246–247 in *Bailey & Scott's Diagnostic Microbiology,* 11th ed. Mosby, St. Louis.

Koneman, Elmer W., *et al.* 1997. Chapter 15 in *Color Atlas and Textbook of Diagnostic Microbiology,* 5th ed. Lippincott, Philadelphia.

MacFaddin, Jean F. 2000. Page 254 in *Biochemical Tests for Identification of Medical Bacteria*, 3rd ed. Lippincott Williams & Wilkins, Philadelphia.

Other Differential Tests

This unit includes tests that do not fit elsewhere but are important to consider. Blood Agar is especially useful for detecting hemolytic ability of Gram positive cocci—typically *Streptococcus* species. It also is used as a general-purpose growth medium appropriate for fastidious and nonfastidious microorganisms alike. The coagulase tests are commonly used to presumptively identify pathogenic *Staphylococcus* species. Motility agar is used to detect bacterial motility, especially in differentiation of *Enterobacteriaceae* and other Gram negative rods.

Exercise 5-26

Blood Agar

Theory

Several species of Gram-positive cocci produce exotoxins called **hemolysins**, which are able to destroy red blood cells (RBCs) and hemoglobin. Blood Agar, which is a mixture of Tryptic Soy Agar and sheep blood, allows differentiation of bacteria based on their ability to hemolyze RBCs.

The three major types of hemolysis are β-hemolysis, α-hemolysis, and γ-hemolysis. β-hemolysis is the complete destruction of RBCs and hemoglobin, and results in a clearing of the medium around the colonies (Figure 5-80). α-hemolysis is the partial destruction of RBCs and produces a greenish discoloration of the agar around the colonies (Figure 5-81). γ-hemolysis is actually nonhemolysis and appears as simple growth with no change to the medium (Figure 5-82).

Hemolysins produced by streptococci are called **streptolysins.** They come in two forms—type O and type S. **Streptolysin O** is oxygen-labile and expresses maximal activity under anaerobic conditions. **Streptolysin S** is oxygen-stable but expresses itself optimally under anaerobic conditions as well. The easiest method of providing an environment favorable for streptolysins on Blood Agar is what is called the **streak–stab technique.** In this procedure the Blood Agar plate is streaked for isolation and then stabbed with a loop. The stabs encourage streptolysin activity because of the reduced oxygen concentration of the subsurface environment (Figure 5-83).

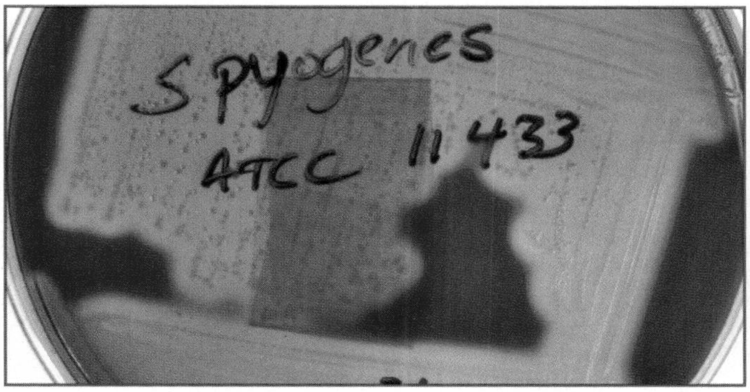

FIGURE 5-80 β-Hemolysis
Streptococcus pyogenes demonstrates β-hemolysis. The clearing around the growth is a result of complete lysis of red blood cells. This photograph was taken with transmitted light.

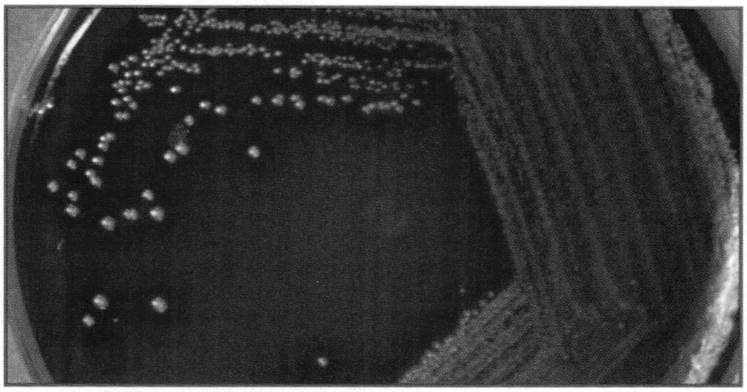

FIGURE 5-81 α-Hemolysis
An unidentified throat culture isolate demonstrating α-hemolysis. The greenish zone around the colonies results from incomplete lysis of red blood cells.

FIGURE 5-82 γ-Hemolysis
This streak plate of *Staphylococcus epidermidis* on a Sheep Blood Agar illustrates no hemolysis.

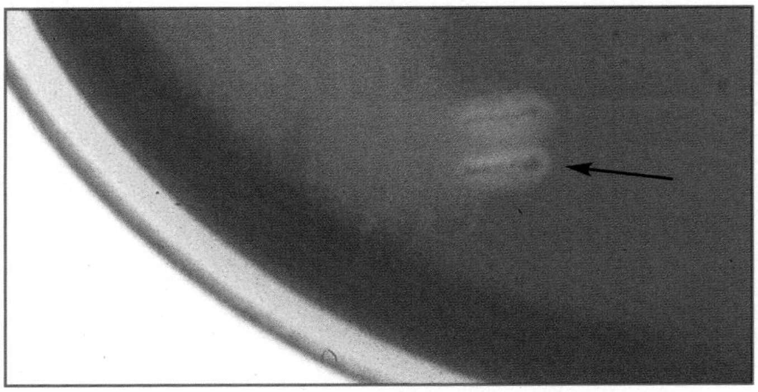

FIGURE 5-83 Aerobic Versus Anaerobic Hemolysis
An unidentified throat culture isolate demonstrates α-hemolysis when growing on the surface, but β-hemolysis beneath the surface surrounding the stabs (arrow). This results from production of an oxygen-labile hemolysin.

Application

Blood Agar is used for isolation and cultivation of many types of fastidious bacteria. It also is used to differentiate bacteria based on their hemolytic characteristics, especially within the genera *Streptococcus*, *Enterococcus*, and *Aerococcus*.

In This Exercise...

You will not be provided with organisms to inoculate your plate. Your partner instead will take a swab of your throat. Then you'll inoculate the blood agar plate in one quadrant to begin a streak for isolation. Isolation of the different colonies is the only way to properly observe the different forms of hemolysis. (***Note:*** When you streak your plate, do only the first streak with the swab, and do streaks two, three, and four with your inoculating loop. If you are not sure how to do this, refer to Exercise 1-3. Because of the potential for cultivating a pathogen, tape the plate lid down immediately after streaking. Upon removing your plate from the incubator, do not open it until your instructor says it is safe to do so.)

Materials

Per Student Group

- one Blood Agar plate (commercially available—TSA containing 5% sheep blood)
- Sterile cotton swabs
- Sterile tongue depressors

Medium Recipe

5% Sheep Blood Agar (TSA + 5% Sheep Blood)

- infusion from beef heart (solids) 2.0 g
- pancreatic digest of casein 13.0 g
- sodium chloride 5.0 g
- yeast extract 5.0 g
- agar 15.0 g
- defibrinated sheep blood 50.0 mL
- distilled or deionized water 1.0 L

pH 7.1–7.5 at 25°C

Procedure

Lab One

1. Have your lab partner obtain a culture from your throat. Follow the procedure in Appendix B.
2. Immediately transfer the specimen to a Blood Agar Plate. Use the swab to begin a streak for isolation. Refer to Exercise 1-3 if necessary.
3. Dispose of the swab in a container designated for autoclaving.
4. Complete the streaking with your loop as described in Exercise 1-3.
5. After completing the streak, use your loop to stab the agar in two or three places in the first streak pattern, and then in two or three places not previously inoculated.
6. Label the plate with your name, the specimen source ("throat culture"), and the date.
7. Tape the lid down to prevent it from opening accidentally. Invert and incubate the plate aerobically at 35 ± 2°C for 24 hours.

Lab Two

1. After incubation, do not open your plate until your instructor has seen it and given permission to do so.

TABLE OF RESULTS

Result	Interpretation	Symbol
Clearing around growth	Organism hemolyzes RBCs completely	β-hemolysis
Greening around growth	Organism hemolyzes RBCs partially	α-hemolysis
No change in the medium	Organism does not hemolyze RBCs	no (γ) hemolysis

TABLE 5-30 Blood Agar results and interpretations.

2. Observe for color changes and clearing around the isolated growth, using transmitted light. This can be done using a colony counter or by holding the plate up to a light. Record your results in the chart on the Data Sheet.

References

Delost, Maria Dannessa. 1997. Page 103 in *Introduction to Diagnostic Microbiology.* Mosby, Inc. St. Louis.

Forbes, Betty A., Daniel F. Sahm, and Alice S. Weissfeld. 2002. Page 16 in *Bailey & Scott's Diagnostic Microbiology,* 11th ed. Mosby, St. Louis.

Koneman, Elmer W., Stephen D. Allen, William M. Janda, Paul C. Schreckenberger, and Washington C. Winn, Jr. 1997. Chapter 12 in *Color Atlas and Textbook of Diagnostic Microbiology,* 5th ed. J. B. Lippincott, Philadelphia.

Krieg, Noel R. 1994. Page 619 in *Methods for General and Molecular Bacteriology*, edited by Philipp Gerhardt, R. G. E. Murray, Willis A. Wood, and Noel R. Krieg, American Society for Microbiology, Washington, DC.

Power, David A., and Peggy J. McCuen. 1988. Page 115 in *Manual of BBL™ Products and Laboratory Procedures,* 6th ed. Becton Dickinson Microbiology Systems, Cockeysville, MD.

Zimbro, Mary Jo, and David A. Power, editors. 2003. *Difco™ and BBL™ Manual—Manual of Microbiological Culture Media.* Becton Dickinson and Co., Sparks, MD.

Exercise 5-27

Coagulase Tests

Theory

Staphylococcus aureus is an opportunistic pathogen that can be highly resistant to both the normal immune response and antimicrobial agents. Its resistance results, in part, from the production of a coagulase enzyme. Coagulase works in conjunction with normal plasma components to form protective fibrin barriers around individual bacterial cells or groups of cells, shielding them from phagocytosis and other types of attack.

Coagulase enzymes occur in two forms—**bound coagulase** and **free coagulase**. Bound coagulase, also called "**clumping factor**," is attached to the bacterial cell wall and reacts directly with fibrinogen in plasma. The fibrinogen then precipitates, causing the cells to clump together in a visible mass. Free coagulase is an extracellular enzyme (released from the cell) that reacts with a plasma component called coagulase-reacting factor (CRF). The resulting reaction is similar to the conversion of prothrombin and fibrinogen in the normal clotting mechanism.

Two forms of the Coagulase Test have been devised to detect the enzymes: the Tube Test and the Slide Test. The Tube Test detects the presence of either bound or free coagulase, and the Slide Test detects only bound coagulase. Both tests utilize rabbit plasma treated with anticoagulant to interrupt normal clotting mechanisms.

The Tube Test is performed by adding the test organism to rabbit plasma in a test tube. Coagulation of the plasma (including any thickening or formation of fibrin threads) within 24 hours indicates a positive reaction (Figure 5-84). The plasma typically is examined for clotting (without shaking) periodically for about 4 hours. After 4 hours, coagulase-negative tubes can be incubated overnight but no more than a total of 24 hours, because coagulation can take place early and revert to liquid within 24 hours.

In the Slide Test, bacteria are transferred to a slide containing a small amount of plasma. Agglutination of the cells on the slide within 1 to 2 minutes indicates the presence of bound coagulase (Figure 5-85). Equivocal or negative slide test results typically are given the tube test for confirmation.

Application

The Coagulase Test typically is used to differentiate *Staphylococcus aureus* from other Gram-positive cocci.

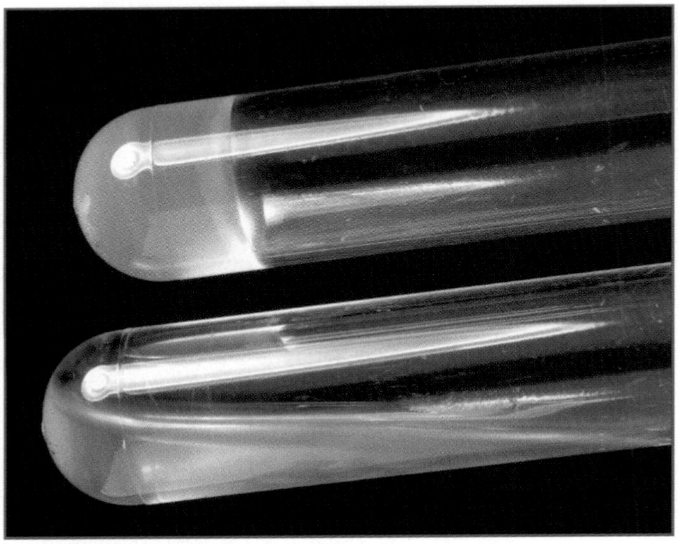

FIGURE 5-84 Coagulase Tube Test
These coagulase tubes illustrate a coagulase-negative (−) organism, below, and a coagulase-positive (+) organism, above. The Tube Test identifies both bound and free coagulase enzymes. Coagulase increases bacterial resistance to phagocytosis and antibodies by surrounding infecting organisms with a clot.

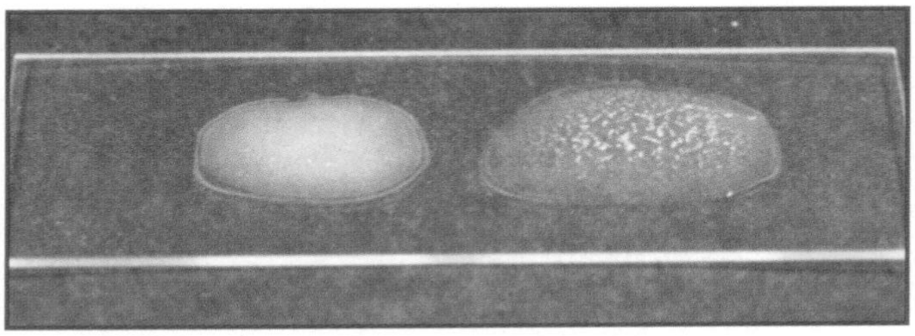

FIGURE 5-85 Coagulase Slide Test (Clumping Factor)
This slide illustrates a coagulase-positive (+) organism on the right and a coagulase-negative (−) organism on the left. Agglutination of the coagulase plasma is indicative of a positive result for bound coagulase.

In This Exercise...

You will perform both Tube and Slide Coagulase Tests. Enter your Slide Test results on the Data Sheet immediately. Read the Tube Test the following day.

Materials

Per Student Group

- three sterile rabbit plasma tubes (0.5 mL in 12 mm × 75 mm test tubes)
- sterile 1 mL pipettes
- sterile saline (0.9% NaCl)
- microscope slides
- fresh slant cultures of:
  - *Staphylococcus aureus*
  - *Staphylococcus epidermidis*

Procedure

Lab One: Tube Test

1. Obtain three coagulase tubes. Label two tubes with the names of the organisms, your name, and the date. Label the third tube "control."
2. Inoculate two tubes with the test organisms. Mix the contents by gently rolling the tube between your hands. Do not inoculate the control.
3. Incubate all tubes at 35 ± 2°C for up to 24 hours, checking for coagulation periodically for the first 4 hours.

Lab One: Slide Test (Clumping Factor)

1. Obtain two microscope slides and divide them in half with a marking pen. Label the sides A and B.
2. Place a drop of sterile saline on side A and a drop of coagulase plasma on side B of each slide.
3. Transfer a loopful of *S. aureus* to each half of one slide, making sure to completely emulsify the bacteria in the solutions. Observe for agglutination within 2 minutes. Clumping after 2 minutes is not a positive result.
4. Repeat step 3 using the other slide and *S. epidermidis*.
5. Record your results in the chart on the Data Sheet. Refer to Figure 5-85 and Table 5-31 when making your interpretations. Confirm any negative results by comparing with the completed Tube Test in 24 hours.

Lab Two

1. Remove all tubes from the incubator no later than 24 hours after inoculation. Examine for clotting of the plasma.
2. Record your results on the Data Sheet. Refer to Figure 5-84 and Table 5-32 when making your interpretations.

TABLE OF RESULTS

Result	Interpretation	Symbol
Clumping of cells	Plasma has been coagulated	+
No clumping of cells	Plasma has not been coagulated	–

TABLE 5-31 Coagulase Slide Test results and interpretations.

TABLE OF RESULTS

Result	Interpretation	Symbol
Medium is solid	Plasma has been coagulated	+
Medium is liquid	Plasma has not been coagulated	–

TABLE 5-32 Coagulase Tube Test results and interpretations.

References

Collins, C. H., Patricia M. Lyne, and J. M. Grange. 1995. Page 111 in *Collins and Lyne's Microbiological Methods,* 7th ed. Butterworth-Heinemann, Oxford, United Kingdom.

Delost, Maria Dannessa. 1997. Pages 98–99 in *Introduction to Diagnostic Microbiology*. Mosby, St. Louis.

DIFCO Laboratories. 1984. Page 232 in *DIFCO Manual,* 10th ed. DIFCO Laboratories, Detroit.

Forbes, Betty A., Daniel F. Sahm, and Alice S. Weissfeld. 2002. Pages 266–267 in *Bailey & Scott's Diagnostic Microbiology,* 11th ed. Mosby, St. Louis.

Holt, John G. (Editor). 1994. *Bergey's Manual of Determinative Bacteriology,* 9th ed. Williams and Wilkins, Baltimore.

Lányi, B. 1987. Page 62 in *Methods in Microbiology,* Vol. 19, edited by R. R. Colwell and R. Grigorova. Academic Press, New York.

MacFaddin, Jean F. 2000. Page 105 in *Biochemical Tests for Identification of Medical Bacteria,* 3rd ed. Lippincott Williams & Wilkins, Philadelphia.

Exercise 5-28

Motility Test

Theory Motility Test Medium is a semisolid medium designed to detect bacterial motility. Its agar concentration is reduced from the typical 1.5% to 0.4%—just enough to maintain its form while allowing movement of motile bacteria. It is inoculated by stabbing with a straight transfer needle. Motility is detectable as diffuse growth radiating from the central stab line.

A tetrazolium salt (TTC) is sometimes added to the medium to make interpretation easier. TTC is used by the bacteria as an electron acceptor. In its oxidized form, TTC is colorless and soluble; when reduced it is red and insoluble (Figure 5-86). A positive result for motility is indicated when the red (reduced) TTC is seen radiating outward from the central stab. A negative result shows red only along the stab line (Figure 5-87).

Application This test is used to detect bacterial motility. Motility is an important differential characteristic of *Enterobacteriaceae*.

In This Exercise...

You will inoculate two Motility Stabs with an inoculating needle. Straighten the needle before you stab the medium. It is important, also, to stab straight into the medium and remove the needle along the same line. Lateral movement of the needle will make interpretation more difficult.

Materials

Per Student Group

- three Motility Test Media stabs
- fresh cultures of:
  - *Enterobacter aerogenes*
  - *Klebsiella pneumoniae*

Medium Recipe

Motility Test Medium

• Beef extract	3.0 g
• Pancreatic digest of gelatin	10.0 g
• Sodium chloride	5.0 g
• Agar	4.0 g
• Triphenyltetrazolium chloride (TTC)	0.05 g
• Distilled or deionized water	1.0 L

pH 7.1–7.4 at 25°C

Procedure

Lab One

1. Obtain three motility stabs. Label two tubes, each with the name of the organism, your name, and the date. Label the third tube "control."
2. Stab-inoculate two tubes with the test organisms. (Motility can be obscured by careless stabbing

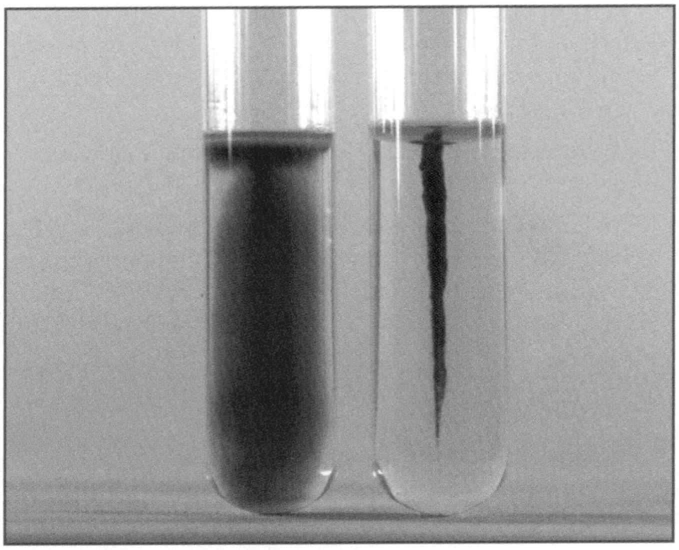

FIGURE 5-87 MOTILITY TEST RESULTS
These Motility Test Media were inoculated with a motile (+) organism on the left and a nonmotile (−) organism on the right.

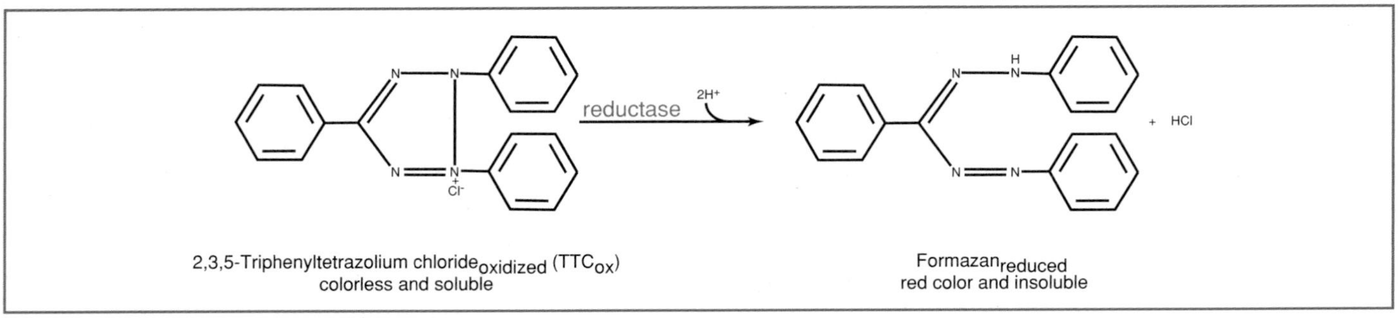

FIGURE 5-86 REDUCTION OF TTC
Reduction of 2,3,5-Triphenyltetrazolium chloride by metabolizing bacteria results in its conversion from colorless and soluble to the red and insoluble compound formazan. The location of growing bacteria can be determined easily by the location of the formazan in the medium.

technique. Try to avoid lateral movement when performing this stab.) Do not inoculate the control.

3. Incubate the tubes aerobically at 35 ± 2°C for 24 to 48 hours.

TABLE OF RESULTS		
Result	**Interpretation**	**Symbol**
Red diffuse growth radiating outward from the stab line	The organism is motile	+
Red growth only along the stab line	The organism is nonmotile	–

TABLE 5-33 Motility Test results and interpretations.

Lab Two

1. Examine the growth pattern for characteristic spreading from the stab line. Growth will appear red because of the additive in the medium.
2. Record your results in the chart provided on the Data Sheet.

References

Forbes, Betty A., Daniel F. Sahm, and Alice S. Weissfeld. 2002. Page 276 in *Bailey & Scott's Diagnostic Microbiology,* 11th ed. Mosby, St. Louis.

MacFaddin, Jean F. 2000. Page 327 in *Biochemical Tests for Identification of Medical Bacteria,* 3rd ed. Lippincott Williams & Wilkins, Philadelphia.

Zimbro, Mary Jo, and David A. Power, editors. 2003. Page 374 in *Difco™ and BBL™ Manual—Manual of Microbiological Culture Media.* Becton Dickinson and Co., Sparks, MD.

Multiple Test Systems

Multiple test systems are systems designed to run an entire battery of tests simultaneously. They employ the same biochemical principles discussed earlier in this section and are read in basically the same manner. All conditions possible with standard tubed media can also be achieved with multiple test media, including aerobic or anaerobic growth conditions and the addition of reagents for confirmatory testing.

The savings in time and money alone make these systems enormously valuable. Even more importantly, what might take days or weeks to do with media preparation and individual tests, multiple test systems can do in as few as 18 hours. Additionally, the media chosen for this unit come with booklets for fast and easy identification. The API 20 E system even offers software and a phone number for help with difficult identifications!

These exercises will give you an opportunity to have a little fun while using some of the skills you have learned. The organisms chosen for these tests will be provided as unknowns for you to identify. In addition, your instructor may allow you to isolate an environmental sample or use one of these systems as a confirmatory test for your unknown project (Exercise 5-31).

Exercise 5-29

API 20 E Identification System for *Enterobacteriaceae* and Other Gram-Negative Rods

Theory The API 20 E system is a plastic strip of 20 microtubes and cupules, partially filled with different dehydrated substrates. Bacterial suspension is added to the microtubes, rehydrating the media and inoculating them at the same time. As with the other biochemical tests in this section, color changes take place in the tubes either during incubation or after addition of reagents. These color changes reveal the presence or absence of chemical action and, thus, a positive or negative result (Figure 5-88).

After incubation, spontaneous reactions—those that do not require addition of reagents—are evaluated first. Tests that require addition of reagents then are performed in a specific order (to avoid complications from gas production) and the results are entered on the Result Sheet (Figure 5-89). An oxidase test is performed separately and constitutes the 21st test.

As shown in Figure 5-89, the Result Sheet divides the tests into groups of three, with the members of a group having numerical values of 1, 2, or 4 respectively. These numbers are assigned for positive (+) results only. Negative (−) results are not counted. The values for positive results in each group are added together to produce a number from 0 to 7, which is entered in the oval below the three tests. The totals from each group are combined sequentially to produce a seven-digit code, which can then be interpreted on the Analytical Profile Index* (Figure 5-90).

In rare instances, the information from the 21 tests (and the 7-digit code) is not discriminatory enough to identify an organism. When this occurs, the organism is grown and examined on MacConkey Agar and supplemental tests are performed for nitrate reduction, oxidation/reduction of glucose, and motility. The results are entered separately in the supplemental spaces on the Results Sheet and used for final identification.

Application The API 20 E multitest system (available from BioMérieux, Inc.) is used clinically for the rapid identification of *Enterobacteriaceae* (over 5500 strains) and other Gram-negative rods (more than 2300 strains).

*The index is now available only as a subscription through BioMérieux-USA.com.

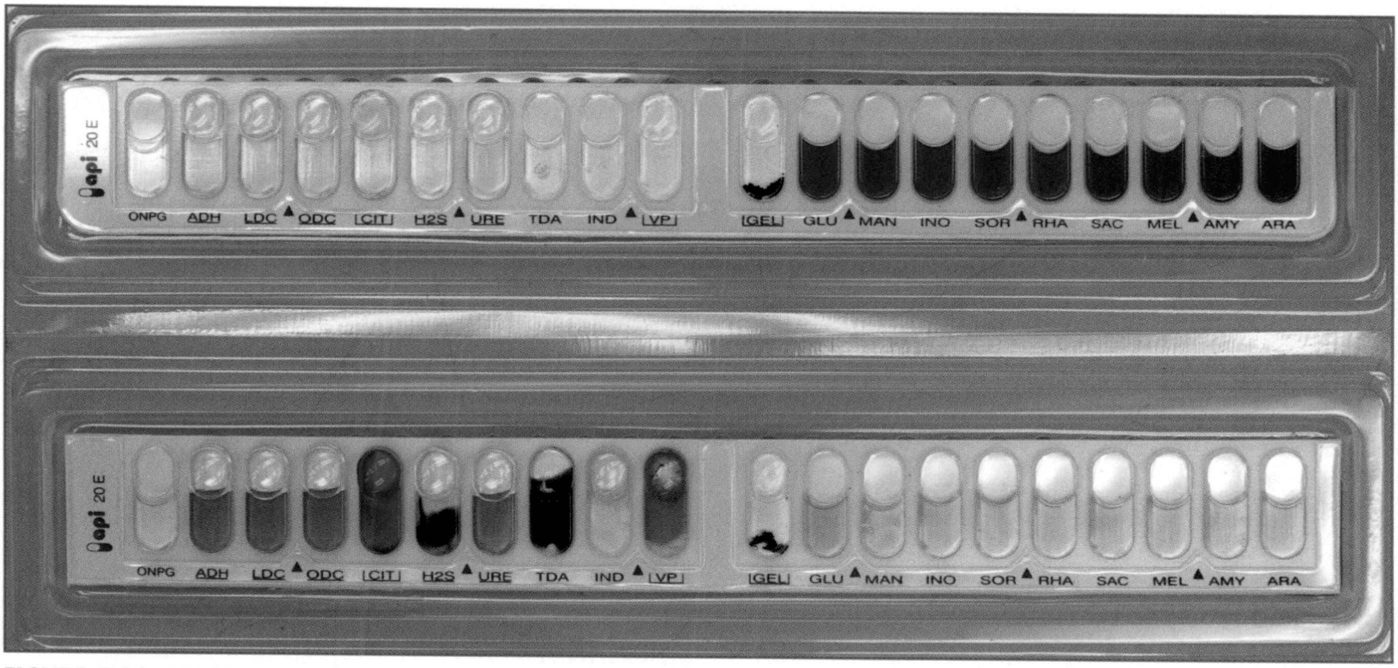

FIGURE 5-88 API 20 E Test Strips
The top strip is uninoculated; the bottom strip (with the exception of GEL) illustrates all positive results.

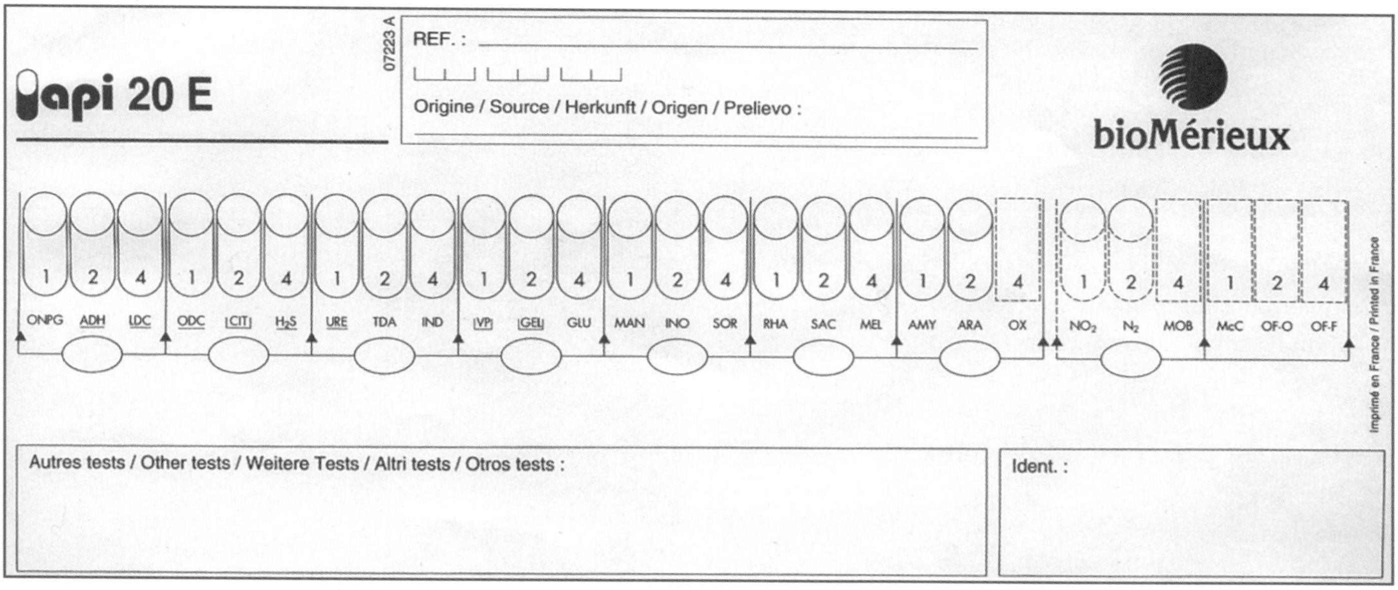

FIGURE 5-89 API 20 E Result Sheet
The API 20 E Result Sheet divides the tests into groups of three, assigning each member of a group a numerical value of 1, 2, or 4. The values for positive results in each group are added together to produce a single-digit number. This number, when combined with the numbers from other groups, produces a seven-digit code that then is interpreted using the Analytical Profile Index supplied by the company (visit http//:www.BioMérieux-USA.com).

In This Exercise...

Today you will use a multitest system to run an entire battery of tests designed to identify a member of the *Enterobacteriaceae*.

Materials

- API 20 E identification system for *Enterobacteriaceae* and other Gram-negative rods.
- API Analytical Profile Index or online subscription

```
7 046 124     TRES BONNE IDENTIFICATION       VERY GOOD IDENTIFICATION       SEHR GUTE IDENTIFIZIERUNG       7 046 124
              A L'ESPECE                      TO THE SPECIES LEVEL           AUF SPEZIES EBENE
                                                     |                       GLUCOSEg  ESC (HYD.)  O/129 R   RM/MR
 Aer.hydrophila gr. 1   %id=56.1 T=0.77 (LDC  25%)(AMY  75%) |Aeromonas caviae        -       +         +        +
                                        (ARA  75%)           |Aeromonas hydrophila    +       +         +       86%
 Aer.hydrophila gr. 2   %id=43.1 T=0.73 (CIT  80%)(VP   80%) |Vibrio fluvialis       0%      NT         -       NT
                                        (AMY  75%)           |Aeromonas sobria        +       -         +        -
 -POSSIBILITE DE Vibrio fluvialis      -POSSIBILITY OF Vibrio fluvialis        -MOEGLICHKEIT VON Vibrio fluvialis
-------------------------------------------------------------------------------------------------------------------
7 046 125     EXCELLENTE IDENTIFICATION       EXCELLENT IDENTIFICATION       AUSGEZEICHNETE IDENTIFIZIERUNG  7 046 125
              A L'ESPECE                      TO THE SPECIES LEVEL           AUF SPEZIES EBENE
                                                     |                       GLUCOSEg  ESC (HYD.)  O/129 R   RM/MR
 Aer.hydrophila gr. 1   %id=56.5 T=0.84 (LDC  25%)(ARA  75%) |Aeromonas caviae        -       +         +        +
 Aer.hydrophila gr. 2   %id=43.4 T=0.80 (CIT  80%)(VP   80%) |Aeromonas hydrophila    +       +         +       86%
                                                             |Vibrio fluvialis       0%      NT         -       NT
                                                             |Aeromonas sobria        +       -         +        -
 -POSSIBILITE DE Vibrio fluvialis      -POSSIBILITY OF Vibrio fluvialis        -MOEGLICHKEIT VON Vibrio fluvialis
-------------------------------------------------------------------------------------------------------------------
7 046 126     BONNE IDENTIFICATION            GOOD IDENTIFICATION            GUTE IDENTIFIZIERUNG            7 046 126
                                                     |                       GLUCOSEg  ESC (HYD.)  O/129 R   RM/MR
 Aer.hydrophila gr. 1   %id=97.9 T=0.84 (LDC  25%)(AMY  75%) |Aeromonas caviae        -       +         +        +
                                                             |Aeromonas hydrophila    +       +         +       86%
                                                             |Vibrio fluvialis       0%      NT         -       NT
                                                             |Aeromonas sobria        +       -         +        -
 -POSSIBILITE DE Vibrio fluvialis      -POSSIBILITY OF Vibrio fluvialis        -MOEGLICHKEIT VON Vibrio fluvialis
```

FIGURE 5-90 ANALYTICAL PROFILE INDEX
Identification is made by locating the seven-digit number in the Analytical Profile Index.

- sterile suspension medium
- ferric chloride reagent (see Exercise 5-11)
- Kovac's reagent (see Exercise 5-20)
- potassium hydroxide reagent (see Exercise 5-4)
- alpha-naphthol reagent (see Exercise 5-4)
- sulfanilic acid reagent (see Exercise 5-7)
- N,N-Dimethyl-1-naphthylamine reagent (see Exercise 5-7)
- oxidase test slides (see Exercise 5-6)
- hydrogen peroxide
- zinc dust
- sterile mineral oil
- sterile Pasteur pipettes
- distilled water
- agar plates containing unidentified bacterial colonies

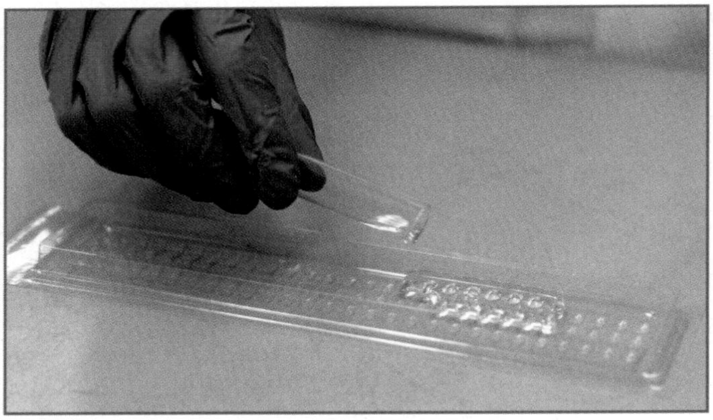

FIGURE 5-91 ADD WATER TO HONEYCOMBED WELLS
The purpose of the water is to humidify the strip during incubation. Spread the water as uniformly as possible in the honeycomb, but it is not necessary to add identical amounts to each well.

Procedure

Lab One

1. Perform a Gram stain on your unknown organism to confirm it is a Gram-negative rod.
2. Open an incubation box and distribute about 5 mL of distilled water into the honeycombed wells of the tray to create a humid atmosphere (Figure 5-91).
3. Record the unknown number on the elongated flap of the tray.
4. Remove the strip from its packaging and place it in the tray.
5. Perform the oxidase test on a colony identical to the colony that will be tested. Refer to Exercise 5-6 if you need help with this. Record the result on the score sheet as the 21st test.

6. Using a sterile Pasteur pipette, remove a single, well-isolated colony from the plate (Figure 5-92) and fully emulsify it in a tube of Suspension Medium or sterile 0.85% saline (Figure 5-93). Try not to pick up any agar.
7. Using the same pipette, fill both tube and cupule of test CIT, VP, and GEL with bacterial suspension (Figure 5-94). Fill only the tubes (not the cupules) of all other tests.

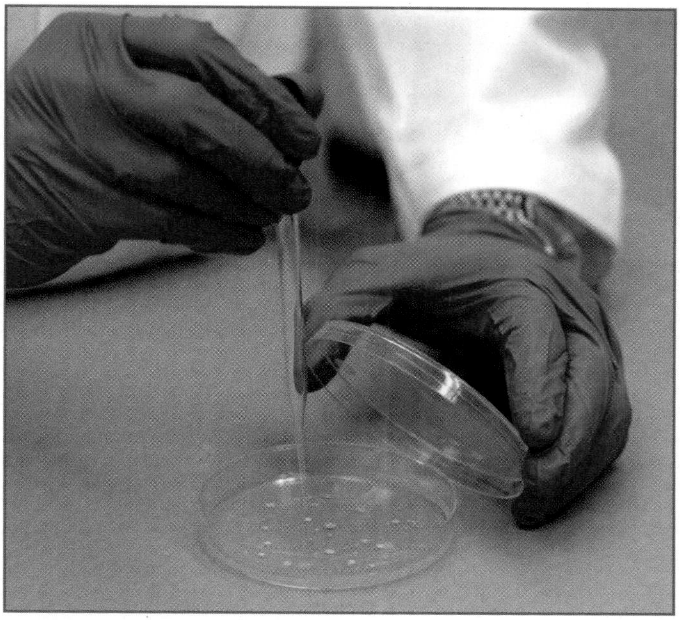

FIGURE 5-92 PICK A COLONY
Being careful not to disturb the agar surface, pick up one isolated colony with a Pasteur pipette.

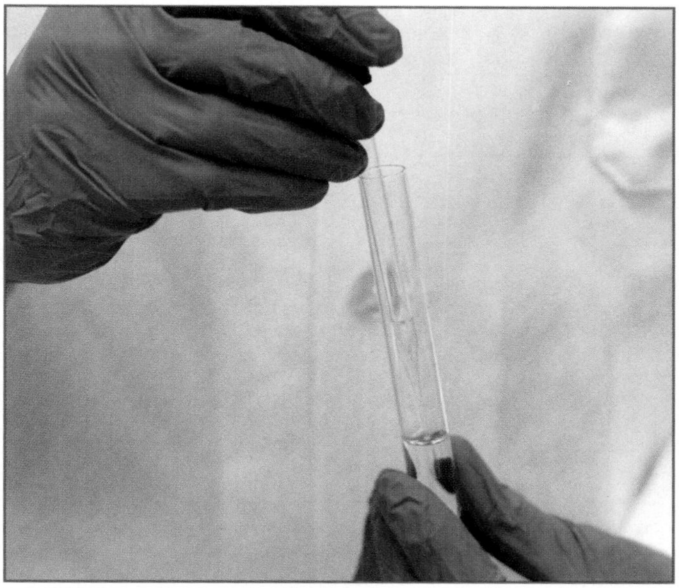

FIGURE 5-93 EMULSIFY THE COLONY
Emulsify the colony in 5 mL Suspension Medium or sterile 0.85% saline. Do this by filling and emptying the pipette several times until uniform turbidity is achieved in the medium.

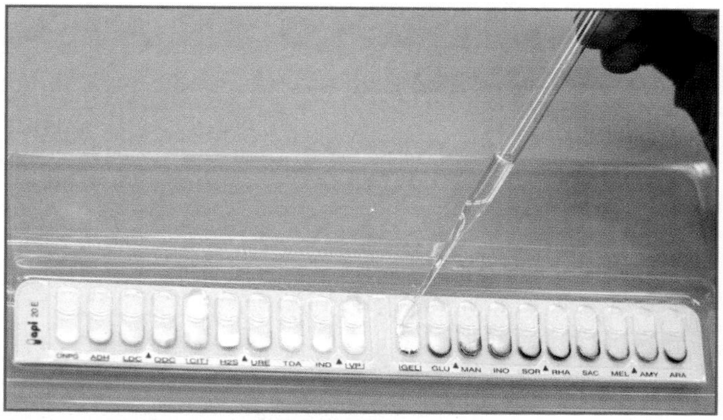

FIGURE 5-94 INOCULATE THE TUBES
Inoculate the tubes by placing the pipette at the side of the tube and *gently* filling with the suspension. If necessary to avoid creating bubbles, tilt the strip slightly. Tap gently with your index finger if necessary to remove bubbles.

8. Overlay the ADH, LDC, ODC, H_2S, and URE microtubes by filling the cupules with sterile mineral oil.
9. Close the incubation box and incubate at 35 ± 2°C for 24 hours. (***Note:*** If the tests cannot be read at 24 hours, have someone place the incubation box in a refrigerator until they can be read.)

Lab Two

1. Refer to Table 5-34 (Table 2 on the package insert) as you examine the test results on the strip.
2. Record all spontaneous reactions on the result sheet. (Spontaneous reactions are reactions completed without the addition of reagents.)
3. Add 1 drop of ferric chloride to the TDA microtube (Figure 5-95). A dark brown color indicates a positive reaction to be recorded on the result sheet.
4. Add 1 drop of potassium hydroxide reagent and 1 drop of alpha-naphthol reagent to the VP microtube. A pink or red color indicates a positive reaction to be recorded on the result sheet. If a slightly pink color appears in 10 to 12 minutes, the reaction should be considered negative.
5. Add 2 drops of sulfanilic acid reagent and 2 drops of N,N dimethyl-1-naphthylamine reagent to the GLU microtube. Wait 2 to 3 minutes. A red color indicates a positive reaction ($NO_3 \rightarrow N0_2$). Enter this on the result sheet as positive. Yellow color at this point is inconclusive because the nitrate may have been reduced to nitrogen gas ($NO_3 \rightarrow N_2$, sometimes evidenced by gas bubbles). If the substrate in the tube remains yellow after adding reagents, add 2 to 3 mg of zinc dust to the tube. A yellow tube after 5 minutes is positive for N_2 and is recorded on the

TABLE OF RESULTS

#	Test	Substrate/Activity	Result	Interpretation	Symbol
1	ONPG	O-nitophenyl-β-D-galactopyranoside	Yellow[1]	Organism produces beta-galactosidase	+
			Colorless	Organism does not produce beta-galactosidase	−
2	ADH	Arginine	Red/Orange	Organism produces arginine dehydrolase	+
			Yellow	Organism does not produce arginine dehydrolase	−
3	LDC	Lysine	Red/Orange[2]	Organism produces lysine decarboxylase	+
			Yellow	Organism does not produce lysine decarboxylase	−
4	ODC	Ornithine	Red/Orange[2]	Organism produces ornithine decarboxylase	+
			Yellow	Organism does not produce ornithine decarboxylase	−
5	CIT	Sodium citrate	Blue-green/Blue[3]	Organism utilizes citrate as sole carbon source	+
			Pale green/Yellow	Organism does not utilize citrate	−
6	H_2S	Sodium thiosulfate	Black	Organism reduces sulfur	+
			Colorless/grayish	Organism does not reduce sulfur	−
7	URE	Urea	Red/Orange[2]	Organism produces urease	+
			Yellow	Organism does not produce urease	−
8	TDA	Tryptophan	Brown-red	Organism produces tryptophan deaminase	+
			Yellow	Organism does not produce tryptophan deaminase	−
9	IND	Tryptophane	Red ring	Organism produces indole	+
			Yellow	Organism does not produce indole	−
10	VP	Creatine Sodium pyruvate	Pink/red[4]	Organism produces acetoin	+
			Colorless	Organism does not produce acetoin	−
11	GEL	Kohn's charcoal gelatin	Diffusion of black pigment	Organism produces gelatinase	+
			No black pigment diffusion	Organism does not produce gelatinase	−
12	GLU	Glucose[5]	Yellow	Organism ferments glucose	+
			Blue/blue-green	Organism does not ferment glucose	−
13	MAN	Mannitol[5]	Yellow	Organism ferments mannitol	+
			Blue/blue-green	Organism does not ferment mannitol	−
14	INO	Inositol[5]	Yellow	Organism ferments inositol	+
			Blue/blue-green	Organism does not ferment inositol	−
15	SOR	Sorbitol[5]	Yellow	Organism ferments sorbitol	+
			Blue/blue-green	Organism does not ferment sorbitol	−

[1] A very pale yellow is also positive.
[2] Orange after 36 hours is negative.
[3] Reading made in the cupule.
[4] A slightly pink color after 5 minutes is negative.
[5] Fermentation begins in the lower portion of the tube; oxidation begins in the cupule.

TABLE 5-34 API 20 E results and interpretations.

TABLE OF RESULTS					
#	Test	Substrate/Activity	Result	Interpretation	Symbol
16	RHA	Rhamnose[5]	Yellow	Organism ferments rhamnose	+
			Blue/blue-green	Organism does not ferment rhamnose	–
17	SAC	Sucrose (saccharose)[5]	Yellow	Organism ferments sucrose	+
			Blue/blue-green	Organism does not ferment sucrose	–
18	MEL	Melibiose[5]	Yellow	Organism ferments melibiose	+
			Blue/blue-green	Organism does not ferment melibiose	–
19	AMY	Amygdalin[5]	Yellow	Organism ferments amygdalin	+
			Blue/blue-green	Organism does not ferment amygdalin	–
20	ARA	Arabinose[5]	Yellow	Organism ferments arabinose	+
			Blue/blue-green	Organism does not ferment arabinose	–
21	OX	Separate test done on paper test strip	Violet	Organism possesses cytochrome-oxidase	+
			Colorless	Organism does not possess cytochrome-oxidase	–
22	GLU (nitrate reduction)	Potassium nitrate	Red after addition of reagents	Organism reduces nitrate to nitrite	+
			Yellow after addition of reagents	Organism does not reduce nitrate to nitrite	–
			Yellow after addition of zinc	Organism reduces nitrate to N_2 gas	+
			Orange-red after addition of zinc	Organism does not reduce nitrate to N_2 gas	–
23	MOB	Motility Medium or wet mount slide	Motility	Organism is motile	+
			Nonmotility	Organism is not motile	–
24	McC	MacConkey Medium	Growth	Organism is probably Enterobacteriaceae	+
			No growth	Organism is not Enterobacteriaceae	–
25	OF	Glucose	Yellow under mineral oil	Organism ferments glucose	+
			Green under mineral oil	Organism does not ferment glucose (not *Enterobacteriaceae*)	–
			Yellow without mineral oil	Organism either ferments glucose or utilizes it oxidatively	+
			Green without mineral oil	Organism does not utilize glucose (not *Enterobacteriaceae*)	–

TABLE 5-34 *(Continued).*

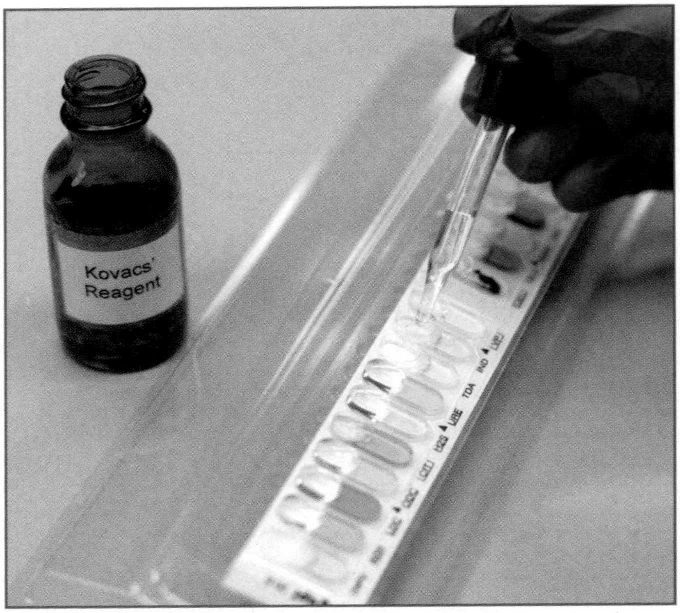

FIGURE 5-95 Add Reagents
Follow the instructions in the text.

result sheet as positive. If the test turns pink-red after the addition of zinc, this is a negative reaction—the nitrates *still present* in the tube have been reduced by the zinc.

6. Add 1 drop of Kovac's reagent to the IND microtube. Wait 2 minutes. A red ring indicates a positive reaction to be recorded on the result sheet.
7. Add the positive results within each group on the result sheet, using the values given, and enter the number in the circle (as shown in Figure 5-89).
8. Locate the seven-digit code in the Analytical Profile Index, and identify your organism. Enter the name on the Result Sheet and tape it on the Data Sheet. Answer the questions.

References

API 20 E Identification System for *Enterobacteriaceae* and Other Gram-Negative Rods package insert.

API 20 E Analytical Profile Index.

Exercise 5-30

Enterotube® II

Theory The Enterotube® II is a multiple test system designed to identify enteric bacteria based on: glucose, adonitol, lactose, arabinose, sorbitol, and dulcitol fermentation, lysine and ornithine decarboxylation, sulfur reduction, indole production, acetoin production from glucose fermentation, phenylalanine deamination, urea hydrolysis, and citrate utilization.

The Enterotube® II, as diagramed in Figure 5-96, is a tube containing 12 individual chambers. Inside the tube, running lengthwise through its center, is a removable wire. After the end-caps are removed (aseptically), one end of the wire is touched to an isolated colony (on a streak plate) and drawn back through the tube to inoculate the media in each chamber (Figures 5-97 through 5-101).

After 18 to 24 hours' incubation, the results are interpreted, the indole test is performed (Figure 5-102), and the tube is scored on an Enterotube® II Results Sheet (Figure 5-103). As shown in the figure, the combination of positives entered on the score sheet results in a five-digit numeric code. This code is used for identification in the Enterotube® II Computer Coding and Identification System (CCIS) (Figure 5-104).

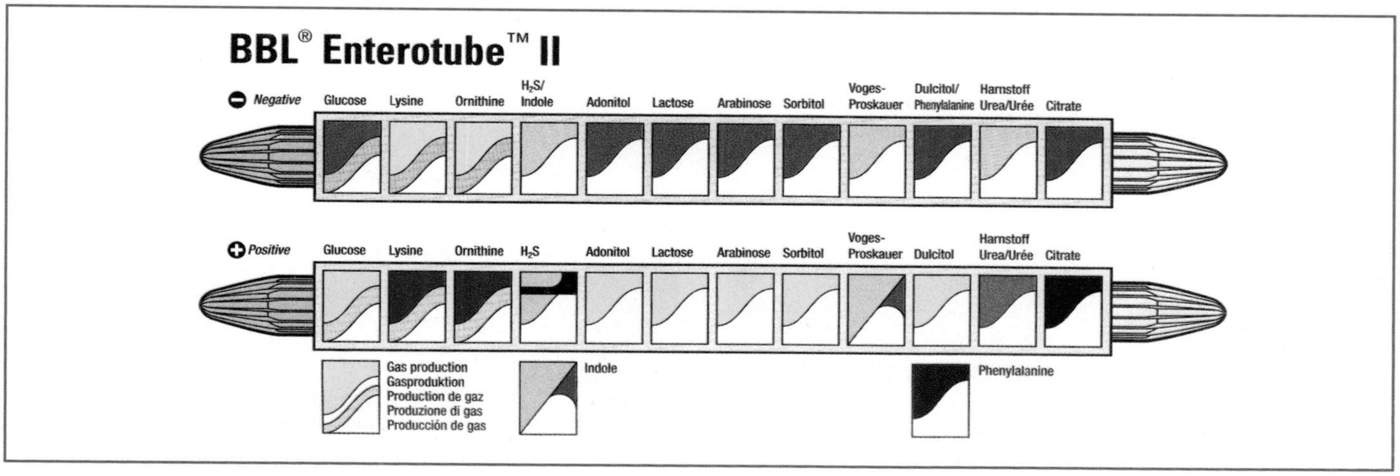

FIGURE 5-96 ENTEROTUBE® II TEST RESULT DIAGRAM

Illustration courtesy Becton Dickinson and Co.

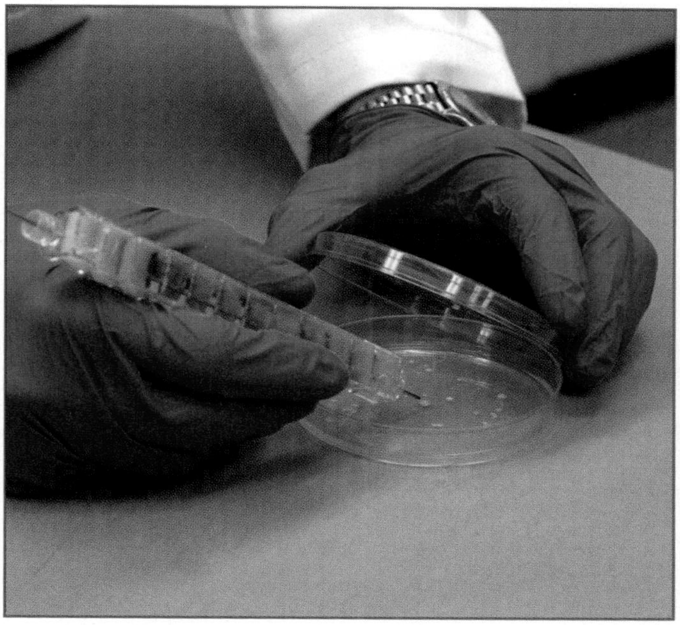

FIGURE 5-97 HARVEST GROWTH
Using the sterile tip of the wire, aseptically remove growth from a colony on the agar surface. Do not dig into the agar. The inoculum should be large enough to be visible.

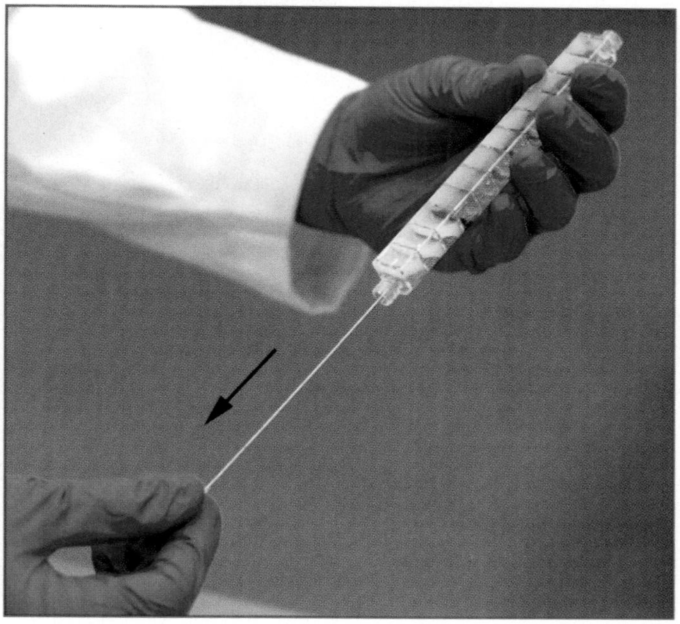

FIGURE 5-98 PULL WIRE THROUGH
Loosen the wire by turning it slightly. While continuing to rotate it withdraw the wire until its tip is inside the last compartment (glucose).

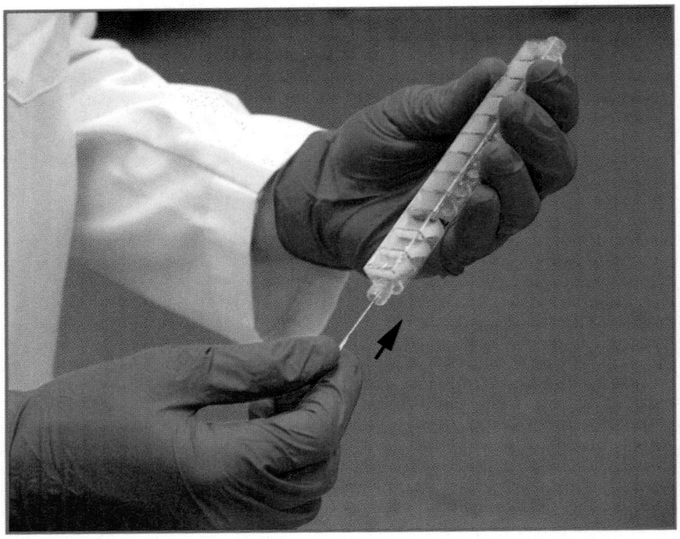

FIGURE 5-99 REINSERT WIRE
Slide the wire back into the Enterotube® II until you can see the tip inside the citrate compartment. The notch in the wire should be lined up with the end of the tube nearest the glucose compartment.

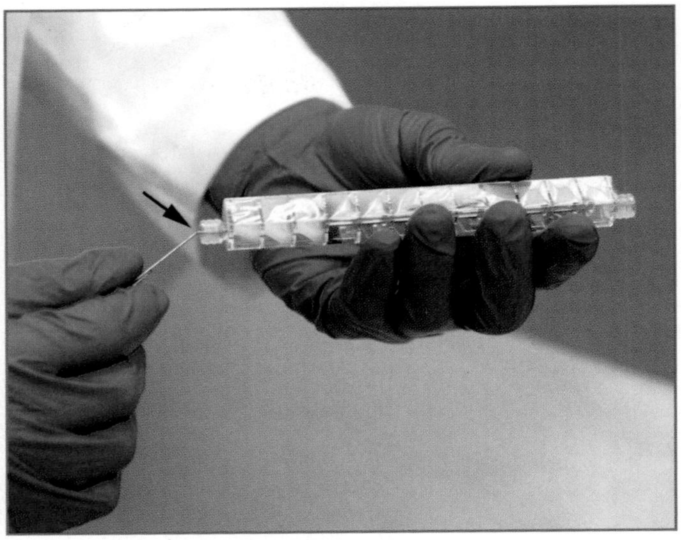

FIGURE 5-100 BREAK WIRE
Bend the wire until it breaks off at the notch.

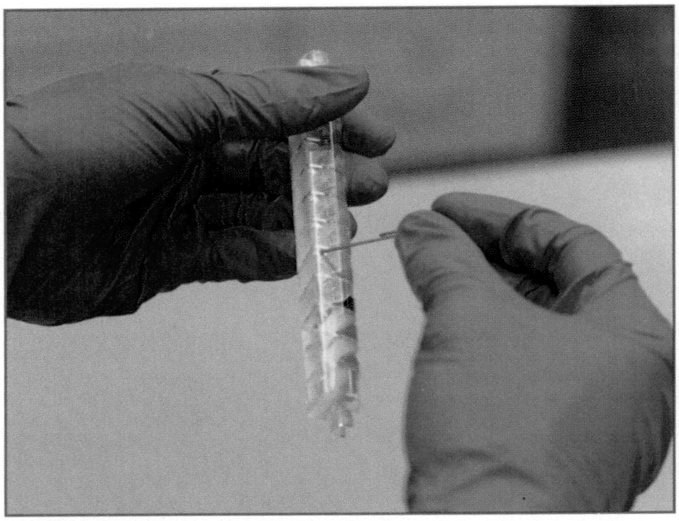

FIGURE 5-101 PUNCTURE AIR INLETS
Using the broken wire, puncture the plastic covering the eight air inlets on the back side of the tube.

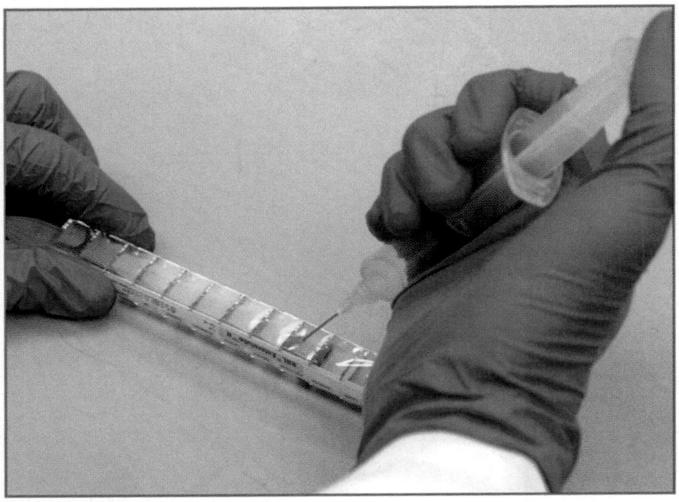

FIGURE 5-102 ADD REAGENTS
Use a needle and syringe to add the reagents through the plastic film on the flat side of the tube. Do this with the tube in a test tube rack and the glucose end pointing down. The indole test should be done only after other tests have been read. Add the Voges-Proskauer reagents only if directed to do so by the CCIS.

The CCIS is a master list of all enterics and their assigned numeric codes. In most cases, the five-digit number applies to a single organism, but when two or more species share the same code, a confirmatory VP test is performed to further differentiate the organisms. An Enterotube® II before and after inoculation is shown in Figure 5-105.

Application

The Enterotube® II is a multiple test system used for rapid identification of bacteria from the family *Enterobacteriaceae*.

In This Exercise...

You will use a multitest system to run an entire battery of tests designed to identify a member of the *Enterobacteriaceae*.

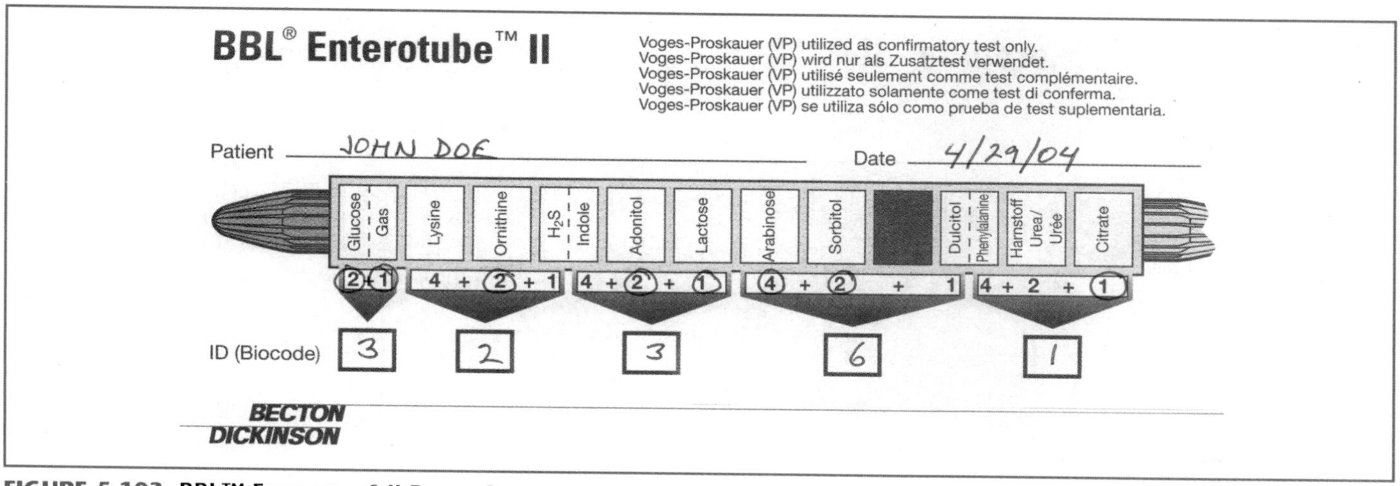

FIGURE 5-103 BBL™ ENTEROTUBE® II RESULT SHEET
This is the result sheet for the tube in Figure 5-105B. The ID value obtained was 32361, which identifies the organism as *Enterobacter aerogenes*.

ID VALUE	ORGANISM IDENTIFICATION	ATYPICAL TESTS	CONFIRMATORY TESTS				
			MAL	TRE			
20406	PROTEUS VULGARIS	H2S-	+	V(30%)			
	MORGANELLA MORGANII	ORN-	–	–			
	PROVIDENCIA STUARTII	PAD-	–	+			
20407	PROVIDENCIA STUARTII	NONE					
			SHA				
20410	*SHIGELLA SEROGROUPS A, B OR C	DUL+	+				
			SHA	ADA			
20420	*SHIGELLA SEROGROUPS A, B OR C	SOR+	+	–			
	ESCHERICHIA COLI (AD)	ARA-	–	+			
20425	PROVIDENCIA STUARTII	SOR+					
20427	PROVIDENCIA STUARTII	SOR+					
20430	ESCHERICHIA COLI (AD)	ARA-					
			SHA	ADA	MOT	VP	YEL
20440	*SHIGELLA SEROGROUPS A, B OR C	NONE	+	–	–	–	–
	ESCHERICHIA COLI (AD)	SOR-	–	+	–	–	–
	ENTEROBACTER AGGLOMERANS	IND+	–	–	+	V(70%)	+(75%)
20441	ENTEROBACTER AGGLOMERANS	IND+					

FIGURE 5-104 ENTEROTUBE® II COMPUTER CODING AND IDENTIFICATION SYSTEM
The five-digit numbers on the Enterotube® II Results Pad are matched against the Data Base in the CCIS to get identification.

Materials

- Becton Dickinson Microbiology Systems' Enterotube® II Identification System for *Enterobacteriaceae*
- Enterotube® II Computer Coding and Identification System (CCIS) identification booklet from Becton Dickinson Microbiology Systems
- Kovac's reagent
- VP reagents (KOH and α-naphthol)
- Needles and syringes or disposable pipettes for addition of reagents
- Streak plate containing unidentified colonies

Reagent Recipes

Kovac's Reagent

- Amyl alcohol 75.0 mL
- Hydrochloric acid, concentrated 25.0 mL
- p-dimethylaminobenzaldehyde 5.0 g

KOH Reagent

- Potassium hydroxide 20.0 g
- Distilled water to bring volume to 100.0 mL

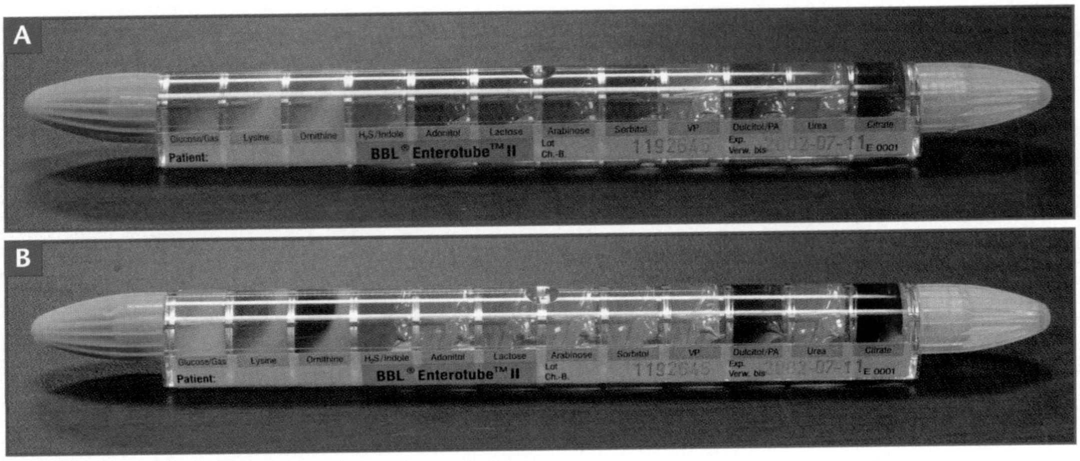

FIGURE 5-105
ENTEROTUBE® II TEST RESULTS
(A) Uninoculated tube.
(B) Tube inoculated with *Enterobacter aerogenes* after 24 hours incubation. This tube shows atypical negative results for lysine decarboxylase; after an additional 24 hours incubation the lysine medium turned purple (+).

α-Naphthol Reagent

- α-naphthol 5.0 g
- Absolute Ethanol to bring volume to 100.0 mL

Procedure

Lab One

1. Perform a Gram stain on your isolate to verify that it is a Gram-negative rod.
2. Place a paper towel on the tabletop and soak it with disinfectant.
3. Remove the blue cap and then the white cap from the Enterotube® II, being careful not to contaminate the sterile wire tip. Place the caps open end down on the paper towel.
4. Aseptically remove a large amount of growth from one of the plated colonies (Figure 5-97). Try not to remove any of the agar with it.
5. Grasp the looped end of the wire and, while turning, gently pull it back through all of the Enterotube® II compartments (Figure 5-98). You do not have to completely remove the wire, but be careful to pull it back far enough to inoculate the last compartment.
6. Using the same turning motion as described above, slide the wire back into the tube (Figure 5-99). Push it in until the tip of the wire is inside the citrate compartment and the notch in the wire lines up with the opposite end of the tube.
7. Bend the wire at the notch until it breaks off (Figure 5-100).
8. Locate the air inlets on the side of the tube opposite the Enterotube® II label. Using the removed piece of inoculating wire, puncture the plastic membrane in the adonitol, lactose, arabinose, sorbitol, Voges-Proskauer, dulcitol/PA, urea, and citrate compartments (Figure 5-101). These openings will create the necessary aerobic conditions for growth. Be careful not to perforate the plastic film covering the flat side of the tube.
9. Discard the wire in disinfectant or a sharps container and replace the Enterotube® II caps.
10. Incubate the tube lying flat at 35 ± 2°C for 18 to 24 hours. (***Note:*** If the tube cannot be read at 24 hours, have someone place it in a refrigerator until it can be read.)

Lab Two

1. Examine the tube and circle the appropriate positive results on the Enterotube® II Results Pad.
2. *After* circling all positive test results on the Enterotube® II Results Pad, place the tube in a rack with the glucose compartment on the bottom. Puncture the plastic membrane of the H_2S/Indole compartment and (with a needle and syringe or disposable pipette) add one or two drops Kovac's reagent to the compartment (Figure 5-102).
3. Observe the compartment for the formation of a red color within 10 seconds. Record a positive result on the Enterotube® II Results Pad.
4. Add the numbers in each bracketed section on the Enterotube® II Results Pad to obtain a five-digit number. Find the number in the CCIS. If not directed to perform a VP test, record the name of your organism on the score sheet and skip to #7 below. If directed to run a confirmatory Voges-Proskauer test, proceed to #5.
5. Puncture the plastic membrane of the VP compartment and (with a needle and syringe or disposable pipette) add two drops of the KOH reagent and three drops of the α-Naphthol reagent. Observe for the formation of red color within 20 minutes. Interpret the result using the CCIS.

6. Discard the Enterotube® in an appropriate autoclave container.
7. Enter all results on the Data Sheet and answer the questions.

References

Enterotube® II Identification System for Enterobacteriaceae package insert.

Enterotube® II Computer Coding and Identification System.

TABLE OF RESULTS

Compartment	Test	Result	Interpretation	Symbol
1	Glucose	Red	Organism does not ferment glucose	–
		Yellow, wax not lifted	Organism ferments glucose to acid	+
		Yellow, wax lifted	Organism ferments glucose to acid and gas	+
2	Lysine	Purple	Organism decarboxylates lysine	+
		Yellow	Organism does not decarboxylate lysine	–
3	Ornithine	Purple	Organism decarboxylates ornithine	+
		Yellow	Organism does not decarboxylate ornithine	–
4	H_2S	Beige	Organism does not reduce sulfur	–
		Black	Organism reduces sulfur	+
	Indole	Red after Kovacs' reagent	Organism produces indole	+
		No red after Kofacs' reagent	Organism does not produce indole	–
5	Adonitol	Red	Organism does not ferment adonitol	–
		Yellow	Organism ferments adonitol	+
6	Lactose	Red	Organism does not ferment lactose	–
		Yellow	Organism ferments lactose	+
7	Arabinose	Red	Organism does not ferment arabinose	–
		Yellow	Organism ferments arabinose	+
8	Sorbitol	Red	Organism does not ferment sorbitol	–
		Yellow	Organism ferments sorbitol	+
9	VP	Colorless	Organism does not produce acetoin	–
		Red	Organism produces acetoin	+
10	Dulcitol	Green	Organism does not ferment dulcitol	–
		Yellow	Organism ferments dulcitol	+
	PA	Green	Organism does not deaminate phenylalanine	–
		Black-smoky-gray	Organism deaminates phenylalanine	+
11	Urea	Beige	Organism does not hydrolyze urea	–
		Red-purple	Organism hydrolyzes urea	+
12	Citrate	Green	Organism does not utilize citrate	–
		Blue	Organism utilizes citrate	+

TABLE 5-35 Enterotube® II results and interpretations.

Bacterial Unknowns Project

The exercise in this unit is actually a term project to be done when you have completed Sections One through Five. These sections prepare you with the skills needed to successfully do what professional laboratory microbiologists do daily—that is, *isolate* from mixed culture, *grow* in pure culture, and *identify* unknown species of bacteria. Scared? Don't worry. You will not be asked to perform any task not introduced in previous sections. If you have any uncertainty regarding performance of these activities, refer to the appropriate exercises.

Although this project may seem intimidating, it is manageable if you don't try to do too much too fast. Take your time and think about what you want to do next, based on earlier results. Use the information from the lessons learned in Exercise 3-12 and closely follow the procedures listed.

This exercise gives you a unique opportunity (and responsibility) to set your own course in a learning project. Take advantage of it, as there is great fun and satisfaction to be had. Our experience tells us that students traditionally do one of three things. They

1. treat the project as a fun puzzle to solve,
2. become completely stressed out and hate every minute of it,
3. get lost early on and, for whatever reason, don't ask for help until it is too late to receive meaningful assistance.

Many of you will be nervous, perhaps even confused at first. Your instructor understands this. It is normal and actually can help you to focus if you use it in a positive way. For your success and overall satisfaction with this project, *ask your instructor for help if you need it*. Happy hunting!

Exercise 5-31

Bacterial Unknowns Project

In This Exercise...

You will be given a mixture of two bacteria in broth culture. Your first job is to get the bacteria isolated and growing in pure culture, using the techniques of Sections Two and Four. Next, you will employ Section Three techniques as you perform Gram stains on each to determine Gram reaction, as well as cell morphology and size. Then, results from the differential biochemical tests of Section Five will lead you to identification of your unknowns. And, of course, all work must be done safely and aseptically using the methods covered in Lab Safety and Section One.

As in Exercise 3-12 (the Morphological Unknown), you will be expected to keep accurate records of all activities, including mistakes and unexpected or equivocal results. You also will be expected to construct a flowchart for each to show the tests you ran and the organisms eliminated by each until you have eliminated all but your unknown. The flowchart is a visual presentation of your thought processes in solving this problem.

Materials

Recommended organisms (to be mixed in pairs in a broth by the lab technician immediately prior to use—generally, a few drops of the Gram-negative added to an overnight culture of the Gram-positive provides an appropriate ratio for isolation and growth).

- **Gram-positives**
  - *Bacillus cereus*
  - *Bacillus coagulans*
  - *Corynebacterium xerosis*
  - *Enterococcus faecalis*
  - *Kocuria rosea*
  - *Lactobacillus plantarum*
  - *Lactococcus lactis*
  - *Micrococcus luteus*
  - *Mycobacterium smegmatis*
  - *Staphylococcus aureus*
  - *Staphylococcus epidermidis*

- **Gram-negatives**
  - *Aeromonas hydrophila*
  - *Alcaligenes faecalis*
  - *Chromobacterium violaceum*
  - *Citrobacter amalonaticus*
  - *Enterobacter aerogenes*
  - *Erwinia amylovora*
  - *Escherichia coli*
  - *Hafnia alvei*
  - *Moraxella catarrhalis*
  - *Morganella morganii*
  - *Neisseria sicca*
  - *Proteus mirabilis*
  - *Pseudomonas aeruginosa*
- Appropriate stains and biochemical media

Procedure

1. **Preliminary duties**
   a. Your instructor will tell you which organisms will actually be used based on your lab's inventory. You also will be advised as to the optimum temperature for the specific strain your lab has of each organism.
   b. Your instructor will tell you which media and stains will be available for testing.
   c. Working as a class, you will determine the results of each organism for each test available. This information will provide a database of results that you can use to compare against your unknown's results. We recommend running and using these "class controls" in lieu of referring to results in *Bergey's Manual of Systematic Bacteriology* or some other standard reference, for the following reasons:
      - As many as 10% of the strains of a species listed as positive or negative on a test give the opposite result.
      - Many species have even higher strain variability.
      - Not all test results are listed for all organisms.
   d. Class control tests should be run for the standard times, unless the timing of class sessions makes this impractical. It is imperative that incubation times for tests on unknowns be exactly the same as was used for the controls.
   e. You are to incubate your class controls at the optimum temperature for each species as given to you by your instructor.
   f. Class control results will be collected, tabulated, and distributed to each student. Your instructor will provide details on this process.
2. **Isolation of the Unknown**
   a. You will be given a broth containing a *fresh* mixture of two unknown bacteria selected from the list of possible organisms. Enter the number of your unknown on your Data Sheets.
   b. Mix the broth, then streak for isolation on two agar plates. Your instructor will tell you what media are available for use. You may be supplied with an undefined medium, such as Trypticase Soy Agar, or a selective medium, such as Phenylethyl Alcohol Agar or Desoxycholate Agar. Enter all relevant information concerning your isolation procedures on the Data Sheets, including date, medium, source of bacteria, type of inoculation, and incubation temperature.
   c. Incubate one agar plate at 25°C and the other at 35 ± 2°C for at least 24 hours.
   d. After incubation, check for isolated colonies that have different morphologies. If you have isolation of both, go on to step 3a. If you do *not* have isolation of *either*, continue with step 2e. If you have isolation of only one, go to step 3a for the isolated species and step 2e for the one not isolated. Be sure to enter relevant information on the Data Sheets.
   e. If you do not have isolation, follow the advice that best matches your situation. (Be sure to record everything you do in the isolation. Include the date, source of inoculum, medium to which it is being transferred, type of inoculation and incubation temperature.)
      - Look for growth with different colony morphologies, even if they're not separate. If you see different growth, use a portion of each and streak more plates. Then incubate them for either a shorter time or at a suboptimal temperature, because they grew *too* well the first time.
      - If you see only one type of growth, ask for a selective medium that favors growth of the one you're missing. (This will require a Gram stain of the one you do have.) Streak the mixture and incubate again. Also, reincubate your original plate. Some species are slow growers, so their absence may be a result of a slow growth rate.

   After incubation, observe the plate(s) and repeat or go to step 3a, whichever is appropriate. You also may consult with your instructor for guidance in particularly difficult situations.

3. **Growing the Unknown in Pure Culture**
   a. Once you have isolation of an unknown, transfer a portion of its colony to an agar slant to produce a pure culture. Use the rest of the colony for a Gram stain. (If you don't have enough of the colony left for a Gram stain, do the Gram stain on your pure culture after incubation.)
   b. After Gram staining, label your pure culture accordingly.
   c. Enter the following information on your Data Sheet: optimum growth temperature, colony morphology, cell morphology and arrangement, and cell size. Also, complete the description of your isolation procedure by noting the source of the isolate, the medium to which you are transferring it, and the incubation temperature.

4. **Identification of the Unknown**
   a. Follow this procedure for each unknown. Be sure to enter on the Data Sheet the inoculation and reading dates for each test. Also include the test result and any comments about the test that explain any deviation from standard procedure (*e.g.*, reading tests before or after the accepted incubation time, running a test and not using it in the flowchart).
   b. Construct a flowchart that divides all the organisms first by Gram reaction, then by cellular morphology. Use the flowchart in Figure 5-106 as a style guide even though the actual organisms you use will be different.
   c. You will have two options for proceeding at this point. Your instructor will let you know which to use.
      (1) Find the group of organisms on the flowchart that matches the results of your unknown. Then, look at your class controls results for *just those organisms*, and choose a test that will divide these organisms into at least two groups. (Also consider your ability to return and read the test after the appropriate incubation time. That is, don't inoculate a 48-hour test on Thursday if you can't get into the lab on Saturday!) Continue the flowchart from the branch with the remaining candidates for your unknown.
      (2) Rather than allowing you to choose just any test that works, you may be required to follow some form of standard approach to identification. Your instructor will provide you with relevant information based on the organism inventory you are working with.
   d. Inoculate the test medium you have chosen, and incubate it for the appropriate time. Use the optimum temperature for growth as given by your instructor. *It is important to run your tests at that optimum temperature because this is the temperature at which the class controls were run.*
   e. While the test is incubating, you should begin planning what your next test will be. A final decision about the next test cannot be made until you have results from the first, but you can decide which test to run if the result is positive, and which one to run if it is negative. This applies to all stages of the flowchart: Because you won't know if a subsequent test is relevant or not until you have results from the current one, *you should not inoculate a medium until you have those results.* It's really easy: Run one test at a time for each unknown. (***Note:*** In a clinical situation, where rapid identification of a pathogen may save a patient's life, tests routinely are run concurrently. But remember that correct identification is only one objective of this project. More important is for you to demonstrate an understanding of the logic behind the process and execute it in the most efficient manner.)
   f. Repeat the process of inoculating a medium, getting the results, and then choosing a subsequent test until you eliminate all but one organism. This *should* be your unknown. Then continue with step 4g.
   g. When you have eliminated all but one organism, you will run one more test—the confirmatory test. This must be a test that has not been run previously on your organism. It's also nice (but not necessary) if the test you choose gives a positive result. (In general, we have more confidence in positive results than in negative results, because false positives are usually harder to get than false negatives.) The confirmatory test provides you with the unique opportunity to predict the result before you run the test. If it matches, you are more certain that you have correctly identified your unknown. Continue with step 4j. If it doesn't match, continue with step 4h.
   h. If your confirmatory test doesn't match the result you expected for your unknown, check with your instructor to see where your organism was eliminated incorrectly. In most cases, it will be difficult at this point to determine what was responsible—you, the class controls, or the organism itself.

Misidentification may be a result of one or any combination of factors:

- The test procedure could have been done incorrectly by you or the person responsible for running class controls on your organism.
- The test may have been interpreted incorrectly by you or the person responsible for running class controls on your organism.
- The inoculum in your test or the class controls might have been too small to give a positive result in the limited incubation time.
- The wrong organism could have been inoculated at the time of the test or the class controls (many look alike in a tube, and once the label goes on, as far as the microbiologist is concerned, that culture becomes the labeled organism regardless of what's really in there!)
- For whatever reason, the organism maybe didn't react "correctly" during your test or during the class controls.

i. Based on your instructor's advice, you will do one of the following:
   - Rerun the test where you incorrectly eliminated your unknown (and perhaps rerun the test on the remaining organisms to check the class control results), or
   - Rerun the test where you incorrectly eliminated your unknown without running the controls again, or
   - Eliminate the problematic test from your flowchart, but use the other tests you've already done. If these don't allow you to identify your unknown, more tests will need to be run. Continue with step 4f.

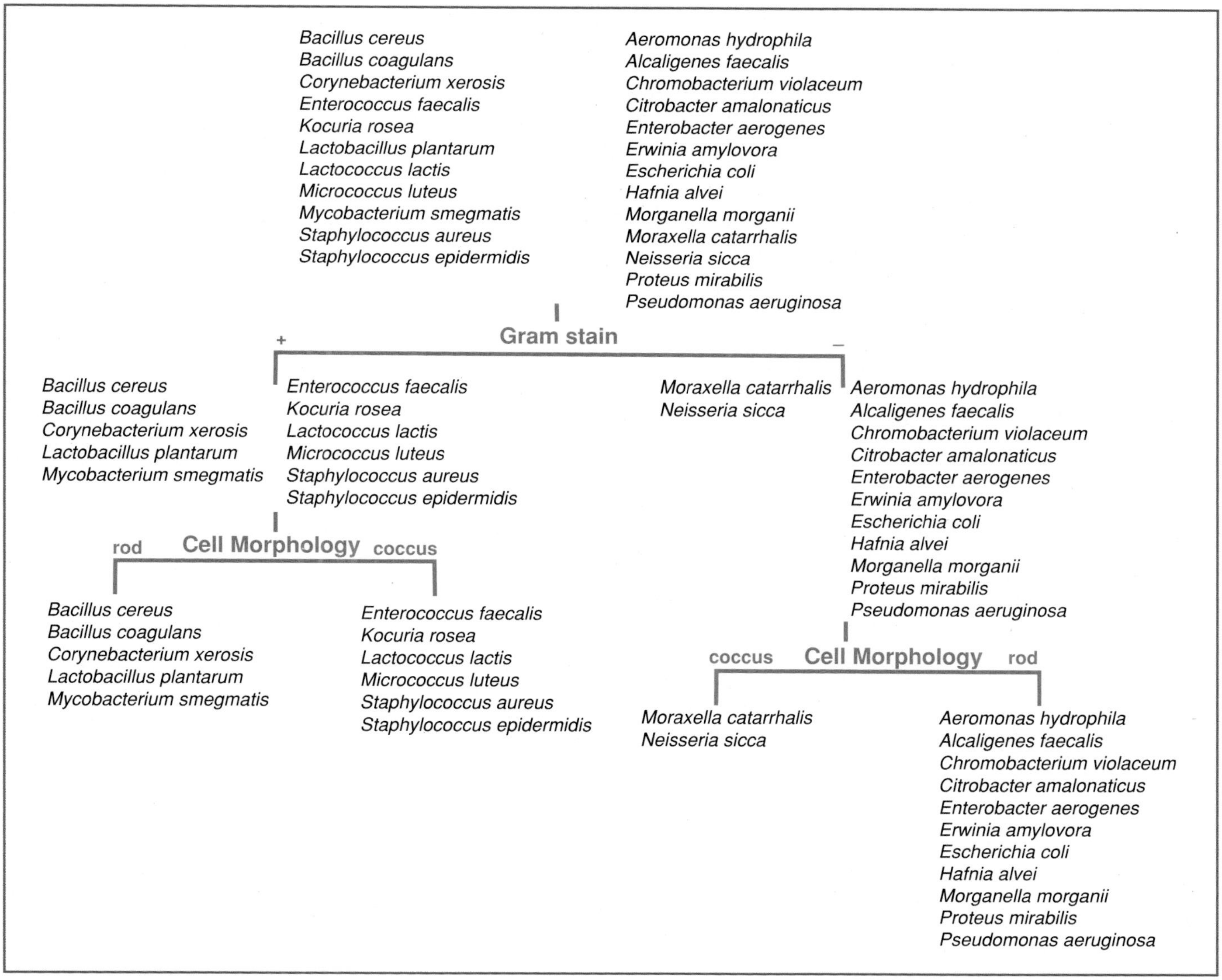

FIGURE 5-106 SAMPLE FLOW CHART
Use this flow chart as a guide in constructing your own containing the organisms available in your laboratory.

(*Note:* There will undoubtedly be some "mistakes" uncovered in the class controls as they get repeated. These have to be reported to the class so classmates can incorporate the correct information into their flowcharts.)

j. When you have correctly identified your unknown, complete the Data Sheet and turn it in. Your instructor will advise you as to the point value of each section, the grading scale, and any other items that are required.

Quantitative Techniques

SECTION 6

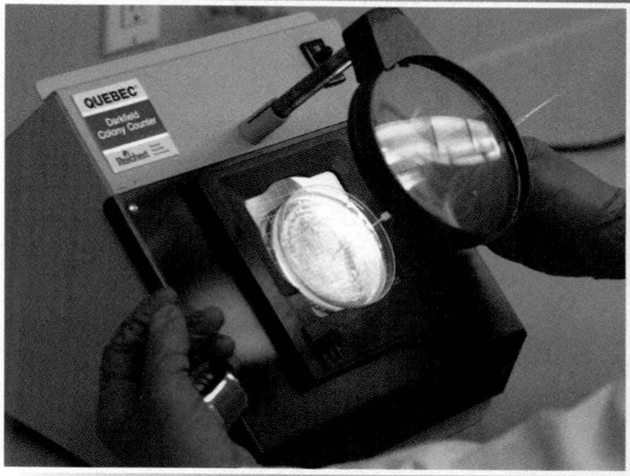

With the exception of Exercise 6-6, Thermal Death Time Versus Decimal Reduction Value, all of the microbiological quantitative techniques included in this section are used to estimate bacterial or viral population density in a given specimen. Each of these techniques, with minor variations, does this by extracting and measuring a very small portion of a sample (*i.e.,* counts the number of cells in it or colonies formed by it), and then extrapolating those data to obtain the overall population density. In conformance with recognized standards, population density is recorded in units per milliliter.

Due in part to the need to work with such small volumes and the need to standardize data, microbiologists make extensive use of a procedure known as **serial dilution.** Serial dilutions, which first appear in Exercise 6-1, are important because they systematize the conversion of raw data (*i.e.,* colonies counted on a spread plate) to usable information (cells/mL, or **original cell density**).

A necessary component of the serial dilution (and a term you will see many times in the following exercises) is **dilution factor,** usually abbreviated to **DF.** As you will see, dilution factors are critically important in keeping track of the multiple interdependent transfers characteristic of serial dilutions.

Another term you will see in the following exercises is **colony forming units,** or **CFU.** In plate counts, CFU describes the origin of colonies more accurately than "cells." For example, when determining the cell density of a sample, we dilute it (usually in a serial dilution), spread it onto a plate, and incubate it. After incubation, what we actually see on the plate is not individual cells but, rather, colonies of cells. Mindful of the fact that bacteria do not always exist as single cells but may arrange themselves in pairs, chains or clusters, we must conclude that the colonies formed by them began as pairs, chains, or clusters as well—hence the term CFU.

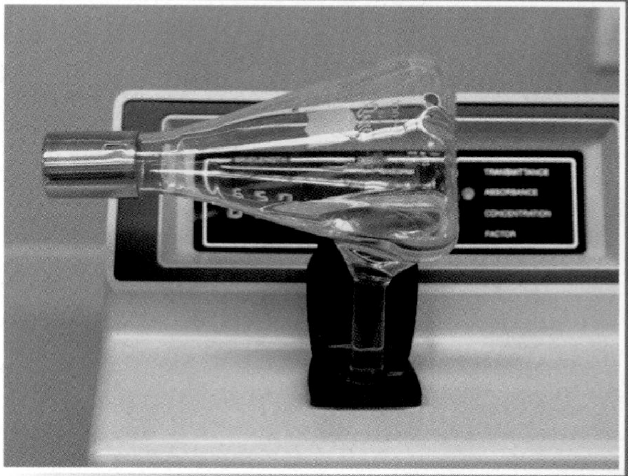

As you proceed through the exercises in this section, you will gain understanding and proficiency with serial dilutions. Because these exercises involve math, some of you will be a little (or a lot) intimidated. But take heart; these calculations are much simpler than they appear! If you are nervous about the math, we encourage you to see your instructor and to work as many practice problems as you can get. Before you know it, you will see these equations as a welcome solution to the very real problem of how to manage otherwise unmanageable numbers!

Most of the quantitative techniques in this section were originally designed for measurements and calculations in milliliters. Many school laboratories now are equipped with digital micropipettes that have the ability to deliver volumes as small as 1.0 microliter (1.0 µL = 0.001 mL). To accommodate schools with modest budgets and for ease of instruction, we have written the following exercises for measurements and calculations in milliliters. Alternative Standard Plate Count, Plaque Assay, and Thermal Death Time procedures for micropipette measurements are included in Appendix F.

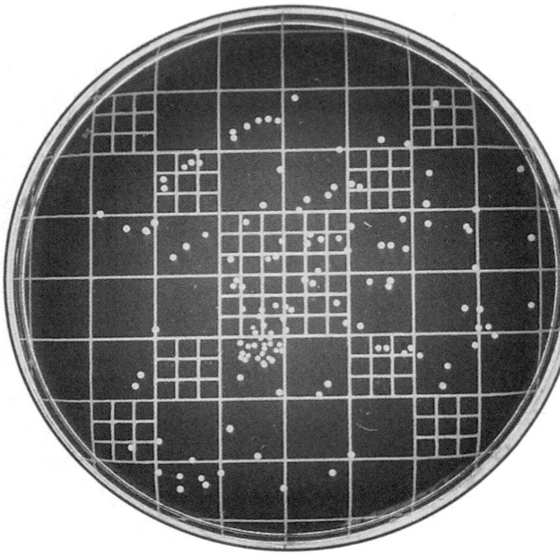

FIGURE 6-1 COUNTABLE PLATE
A countable plate has between 30 and 300 colonies. Therefore, this plate with approximately 130 colonies is countable and can be used to calculate cell density in the original sample. Plates with fewer than 30 colonies are TFTC ("too few to count"). Plates with more than 300 colonies are TMTC ("too many to count").

Exercise 6-1

Standard Plate Count (Viable Count)

Theory A standard plate count is an indirect means of estimating microbial cell density (in cells per milliliter) of a liquid sample using the number of colonies a portion of the sample produces when spread onto an agar plate. Because undiluted microbial samples tend to produce confluent growth when plated, a **serial dilution** is required to sufficiently reduce the cell density to achieve **countable** plates. A countable plate is one that contains between 30 and 300 individual colonies (Figure 6-1).

As shown in Figure 6-2, a serial dilution is simply a series of controlled transfers—several small dilutions, each one compounding the previous one, until the sample is reduced to one millionth or less of its original density. Diluting a sample in this way does two important things.

1. It systematically reduces cell density, thus producing at least one dilution that will yield countable plates.
2. It provides a mathematical framework by which to link the unknown (original cell density) with the known (number of colonies on the plate).

Examining Figure 6-2, you can see that each dilution is given a **dilution factor**, or **DF**. The dilution factor is simply a means of keeping track of how much original sample is still present after the dilution. In other words, a dilution containing 1 mL of sample and 99 mL of **diluent** (water or saline) is given the dilution factor 10^{-2}, because it is now only 1/100 its original concentration.

The term 10^{-2} is used because scientific notation is the preferred (and mathematically simplest) method of documenting and calculating dilution factors. Simple dilution factors are calculated using the following formula.

$$D = \frac{V_1}{V_2}$$

V_1 is the volume of broth being diluted and V_2 is the total combined volume of broth and diluent. If you add 1.0 mL broth to 9.0 mL diluent, the dilution factor is 10^{-1} as shown below.

$$D = \frac{1.0 \text{ mL}}{10.0 \text{ mL}} = \frac{1}{10} = 10^{-1}$$

The numerator represents the volume of broth added to the diluent. The denominator represents the total volume after the broth was added.

In a serial dilution, each transfer further reduces the proportion of original broth in the solution and the dilution factor of each new solution is compounded by the dilution factor of the previous tube. This can be demonstrated with the following formula:

$$V_1 D_1 = V_2 D_2$$

In this expression, V_1 and D_1 are the volume and dilution factor of a sample before the dilution takes place (undiluted samples have a dilution factor of 1). V_2

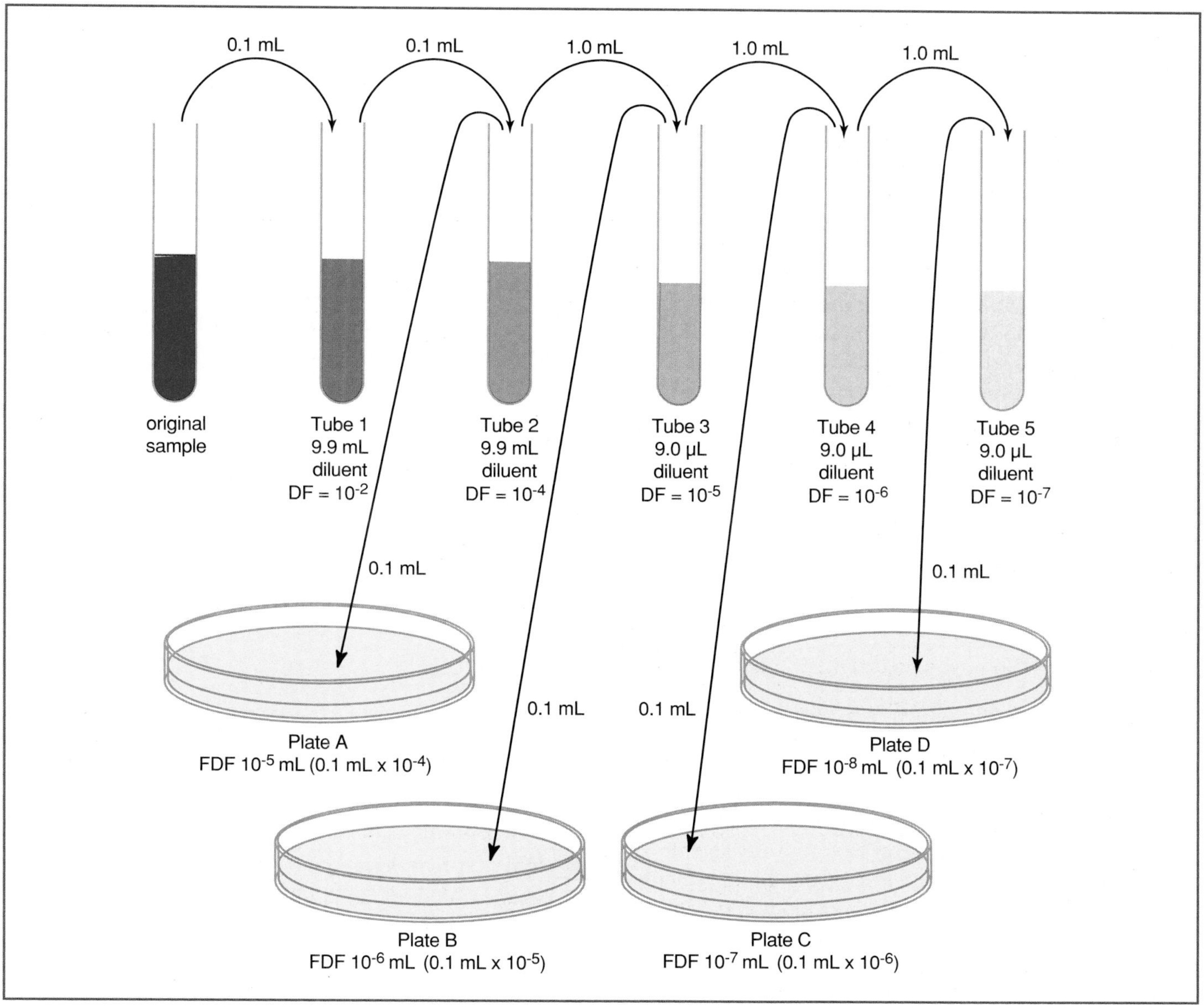

FIGURE 6-2 Serial Dilution Procedural Diagram
This is an illustration of the dilution scheme outlined in the Procedure. The dilution factors assigned to the tubes and their contents indicate the proportion of original sample present in the dilution tube. Note that the final dilution factor of the plates is 10 times greater than that in the source tube. This is because cell density is calculated in CFU/mL and the plates were inoculated with 0.1 mL.

is the combined volume of sample and diluent after the dilution takes place. D_2 is the dilution factor of the new solution. Solving for D_2, the formula is:

$$D_2 = \frac{V_1 D_1}{V_2}$$

or

$$D_2 = \frac{V_1}{V_2} \times D_1$$

Expressing the formula in this latter version illustrates that calculation of dilution factors in a serial dilution is nothing more than calculating each single dilution factor (V_1/V_2) and multiplying it by the previous one (D_1). Suppose, for example, that 0.1 mL of the 10^{-1} dilution above was added to 9.9 mL of diluent. The dilution factor of the second solution would be:

$$D_2 = \frac{V_1}{V_2} \times D_1$$

$$D_2 = \frac{0.1 \text{ mL}}{10.0 \text{ mL}} \times 10^{-1}$$

$$D_2 = 10^{-2} \times 10^{-1} = 10^{-3}$$

As mentioned above, the serial dilution provides the link between colonies produced on the plates and density of the original broth. It does so in the form of the following equation:

$$\text{Original cell density (OCD)} = \frac{\text{Colonies counted}}{(\text{Volume plated})(\text{Dilution factor})}$$

This equation includes two critical factors from the dilution series: the dilution factor and the volume of dilution transferred to the plate. The dilution factor describes what proportion of the dilution is original sample. The volume plated is a necessary correction addressing the fact that the number of cells transferred to the plate is dependent on the volume of dilution transferred.

You will recall from the introduction to this section that colonies don't necessarily develop from single cells. Bacteria frequently are arranged in pairs, chains, or clusters. Therefore, each colony is actually the development of a colony forming unit, or CFU—a term that includes all cell configurations. To simplify the formula and assure that the calculated result will include the correct units (CFU/mL), "colonies counted" can be replaced with CFU as follows:

$$\text{OCD} = \frac{\text{CFU}}{(\text{Volume plated})(\text{Dilution factor})}$$

The convention among microbiologists is to make one final simplification to the formula by combining volume plated and dilution factor to form a single term known as the Final Dilution Factor or FDF. Virtually all plates in the standard plate count are inoculated with 0.1 mL, because this volume can be uniformly spread and is absorbed quickly. Therefore, if the dilution factor of the source tube is 10^{-5} and the inoculum is 0.1 mL, the FDF is 10^{-6} mL ($10^{-5} \times 0.1$ mL $= 10^{-6}$ mL). On rare occasions, a plate is inoculated with 1.0 mL and the FDF is the same as the source tube ($10^{-5} \times 1.0$ mL $= 10^{-5}$ mL). Note that the FDF includes "mL" units where DF does not. This change to FDF is typically done automatically by simply increasing the dilution factor by a factor of 10 when writing it on the plate. Then, after one or two days' incubation, the plates are removed, colonies are counted, and original density is calculated with a simple division problem. In its most condensed form, the formula is written as follows:

$$\text{OCD} = \frac{\text{CFU}}{\text{FDF}}$$

Suppose, for example, after performing the dilution series in Figure 6-2 and incubating the plates, you counted 47 colonies on plate B. Combining the terms 10^{-5} (from the tube) and 0.1 mL (transferred to the plate) to obtain FDF 10^{-6}, the calculation would be:

$$\text{OCD} = \frac{\text{CFU}}{\text{FDF}}$$

$$\text{OCD} = \frac{47\ \text{CFU}}{10^{-6}\ \text{mL}} = 47 \times 10^{6}\ \text{CFU/mL}$$

$$\text{OCD} = 4.7 \times 10^{7}\ \text{CFU/mL}$$

Application The viable count is one method of determining the density of a microbial population. It provides an estimate of actual *living* cells in the sample.

In This Exercise...

You will perform a dilution series and determine the population density of a broth culture of *Escherichia coli*. You will inoculate the plates using the **spread plate technique** as illustrated in Figures 1-27 to 1-30. As described in the figure, the inocula from the dilution tubes will be evenly dispersed over the agar surface with a bent glass rod. You will be sterilizing the glass rod between inoculations by immersing it in alcohol and igniting it. Be careful to organize your work area properly and *at all times keep the flame away from the alcohol beaker.*

Materials

Per Student Group

- sterile 0.1 mL, 1.0 mL, and 10.0 mL pipettes
- sterile dilution tubes
- flask of sterile water
- eight Nutrient Agar plates
- beaker containing ethanol and a bent glass rod
- hand tally counter
- colony counter
- 24-hour broth culture of *Escherichia coli* (This culture will have between 10^7 and 10^{10} CFU/mL.)

Procedure

Refer to the procedural diagram in Figure 6-2 and Exercise 1-4 as needed. Appendix F includes an alternate procedure for digital micropipettes using µL volumes.

Lab One

1. Obtain eight plates, organize them into four pairs, and label them A_1, A_2, B_1, B_2, *etc.*
2. Obtain five dilution tubes and label them 1–5 respectively. Make sure they remain covered until needed.
3. Aseptically add 9.9 mL sterile water to dilution tubes 1 and 2. Cover when finished. Aseptically add 9.0 mL sterile water to dilution tubes 3, 4, and 5. Cover when finished.

4. Mix the broth culture and aseptically transfer 0.1 mL to dilution tube 1; mix well. This is dilution factor 10^{-2} (DF 10^{-2})
5. Aseptically transfer 0.1 mL from dilution tube 1 to dilution tube 2; mix well. This is DF 10^{-4}.
6. Aseptically transfer 1.0 mL from dilution tube 2 to dilution tube 3; mix well. This is DF 10^{-5}.
7. Aseptically transfer 1.0 mL from dilution tube 3 to dilution tube 4; mix well. This is DF 10^{-6}.
8. Aseptically transfer 1.0 mL from dilution tube 4 to dilution tube 5; mix well. This is DF 10^{-7}.
9. Aseptically transfer 0.1 mL from dilution tube 2 to plate A_1. Using the spread plate technique, disperse the sample evenly over the entire surface of the agar. Repeat the procedure with plate A_2 and label both plates "FDF 10^{-5}".
10. Following the same procedure, transfer 0.1 mL volumes from dilution tubes 3, 4, and 5 to plates B, C, and D respectively. Label the plates with their appropriate FDF.
11. Invert the plates and incubate at 35°C for 24 to 48 hours.

Lab Two

1. After incubation, examine the plates and determine the countable pair—plates with 30 to 300 colonies. Only one pair of plates *should* be countable.
2. Count the colonies on both plates, and calculate the average (Figure 6-3). Record it in the chart provided on the Data Sheet. (*Note*: In error, you may have more than one pair that is countable. Count *all* plates that have between 30 and 300 colonies for the practice, and try to identify which plate(s) you have the most confidence in. If *no* plates are in the 30–300 colony range, count the pair that is closest just for the practice and for purposes of the calculations.)

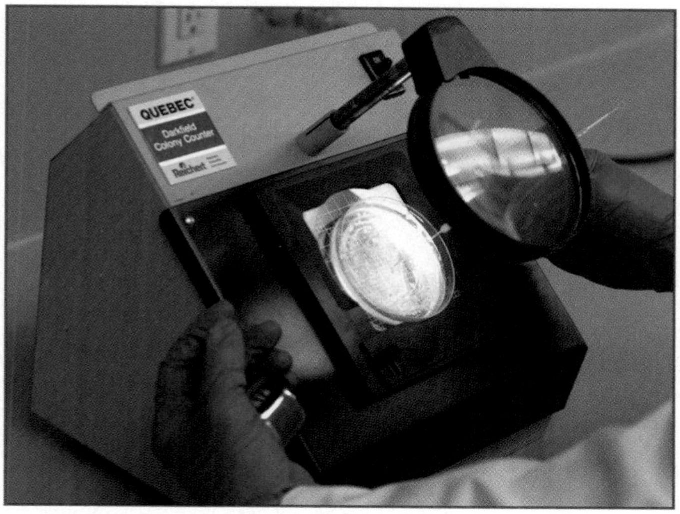

FIGURE 6-3 Counting Bacterial Colonies
Place the plate upside down on the colony counter. Turn on the light and adjust the magnifying glass until all the colonies are visible. Using the grid in the background as a guide, count colonies one section at a time. Mark each colony with a felt-tip marker as you record with a hand tally counter.

3. Using the formula provided in Data and Calculations on the Data Sheet, calculate the density of the original sample and record it in the space provided.

References

Collins, C. H., Patricia M. Lyne, and J. M. Grange. 1995. Page 149 in *Collins and Lyne's Microbiological Methods*, 7th ed. Butterworth-Heinemann, United Kingdom.

Koch, Arthur L. 1994. Page 254 in *Methods for General and Molecular Bacteriology*, edited by Philipp Gerhardt, R.G. E. Murray, Willis A. Wood, and Noel R. Krieg, American Society for Microbiology, Washington, DC.

Postgate, J. R. 1969. Page 611 in *Methods in Microbiology*, Vol. 1., edited by J. R. Norris and D. W. Ribbons. Academic Press, Inc., New York.

Exercise 6-2

Urine Culture

Theory

Urine culture is a semiquantitative method that uses a volumetric loop (not a serial dilution) to reduce the number of cells to a countable level. A volumetric loop is an inoculating loop calibrated to hold a specific volume of liquid. Available in 0.001 mL and 0.01 mL sizes, volumetric loops are useful in situations where population density is not likely to exceed 10^5 CFU/mL. They are not appropriate where cell density is expected to be higher.

In this standard procedure, a loopful of urine is carefully transferred to a Blood Agar plate. The initial inoculation is a single streak across the diameter of the agar plate. The plate then is turned 90° and (without flaming the loop) streaked again, this time across the original line in a zigzag pattern to evenly disperse the bacteria over the entire plate (Figures 6-4 and 6-5). Following a period of incubation, the resulting colonies are counted and population density, usually referred to as "original cell density," or OCD, is calculated.

OCD typically is recorded in "colony forming units," or CFU per milliliter (CFU/mL), as described in the introduction to this section. CFU/mL is determined by dividing the number of colonies on the plate by the volume of the loop. For example, if 150 colonies are counted on a plate inoculated with a 0.001 mL loop, the calculation would be as follows:

$$\text{OCD} = \frac{\text{CFU}}{\text{loop volume}}$$

$$\text{OCD} = \frac{150 \text{ CFU}}{0.001 \text{ mL}}$$

$$\text{OCD} = 1.5 \times 10^5 \text{ CFU/mL}$$

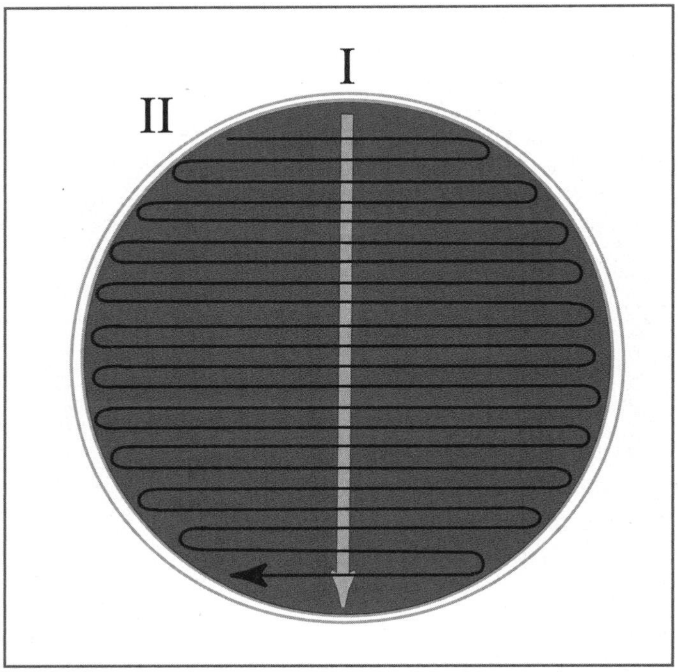

FIGURE 6-4 SEMIQUANTITATIVE STREAK METHOD
Streak 1 is a simple streak line across the diameter of the plate. Streak 2 is a multiple streak at right angles to the first streak.

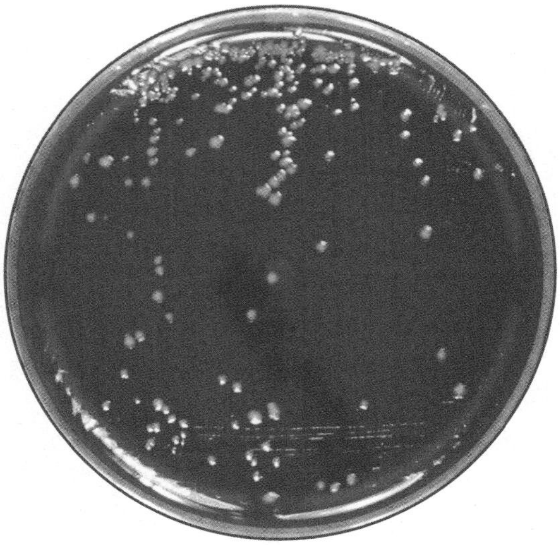

FIGURE 6-5 URINE STREAK ON SHEEP BLOOD AGAR
This plate was inoculated with a 0.01 mL volumetric loop. The cell density can be determined by multiplying the number of colonies by 0.01.

Application

Urine culture is a common method of detecting and quantifying urinary tract infections. It frequently is combined with selective media for specific identification of members of *Enterobacteriaceae* or *Streptococcus*.

In This Exercise...

You will estimate cell density in a urine sample using a volumetric loop and the above formula. Be sure to hold the loop vertically and transfer slowly.

Materials

Per Student Group

- one blood agar plate (TSA with 5% sheep blood)
- one sterile volumetric inoculating loop (either 0.01 mL or 0.001 mL)
- a fresh urine sample

Procedure

Lab One

1. Holding the loop vertically, immerse it in the urine sample. Then carefully withdraw it to obtain the

correct volume of urine. This loop is designed to fill to capacity in the vertical position. Do not tilt it until you get it in position over the plate.

2. Inoculate the blood agar by making a single streak across the diameter of the plate.
3. Turn the plate 90° and, without flaming the loop, streak the urine across the entire surface of the agar, as shown in Figure 6-4.
4. Invert, label and incubate the plate for 24 hours at 35 ± 2°C.

Lab Two

1. Remove the plate from the incubator and count the colonies. Also note any differing colony morphologies, which would indicate possible colonization by more than one species. Enter the data in the chart on the Data Sheet.
2. Calculate the original cell density of the sample using the formula on the Data Sheet. If two species are present, calculate each.
3. Enter the cell density(-ies) in the chart.

References

Forbes, Betty A., Daniel F. Sahm, and Alice S. Weissfeld. 2002. Pages 933–934 in Bailey and Scott's *Diagnostic Microbiology*, 11th ed. Mosby-Yearbook, St. Louis.

Koneman, Elmer W., Stephen D. Allen, William M. Janda, Paul C. Schreckenberger, and Washington C. Winn, Jr. 1997. Page 94 in *Color Atlas and Textbook of Diagnostic Microbiology*, 5th ed. J. B. Lippincott Co., Philadelphia.

Exercise 6-3

Direct Count (Petroff-Hausser)

Theory

Microbial direct counts, like plate counts, take a small portion of a sample and use the data gathered from it to calculate the overall population cell density. This is made possible with a device called a **Petroff-Hausser counting chamber.** The Petroff-Hausser counting chamber is very much like a microscope slide with a 0.02 mm deep chamber or "well" in the center containing an etched grid (Figure 6-6). The grid is one square millimeter and consists of 25 large squares, each of which contains 16 small squares, making a total of 400 small squares. Figures 6-7 and 6-8 illustrate the counting grid.

When the well is covered with a cover glass and filled with a suspension of cells, the volume of liquid above each small square is 5×10^{-8} mL. This may seem like an extremely small volume, but the space above each small square is large enough to hold about 50,000 average-size cocci! Fortunately, dilution procedures prevent this scenario from occurring and cell counting is easily done using the microscope.

As mentioned in the introduction to this section, cell density usually is referred to as "original cell density," or OCD, because most samples must be diluted before attempting a count. OCD is determined by counting the cells found in a predetermined group of small squares (Cell count) and dividing by the number of squares counted (Squares) multiplied by the dilution factor[1] (DF) and the volume of sample above one small square (Volume).

The following is a standard formula for calculating original cell density in a direct count.

$$\text{OCD} = \frac{\text{Cell count}}{(\text{Squares})(\text{DF})(\text{Volume})}$$

To maintain accuracy, some experts recommend a minimum overall count of 600 cells in one or more samples taken from a single population. Optimum density for counting is between 5 and 15 cells per small square.

If, for example, 200 cells from a sample with a dilution factor of 10^{-2} were counted in 16 squares (remembering that the volume above a single small square is 5×10^{-8} mL), the cell density in the original sample would be calculated as follows:

$$\text{OCD} = \frac{\text{cell count}}{(\text{squares})\ (\text{DF})\ (\text{volume})}$$

$$\text{OCD} = \frac{200 \text{ cells}}{(16)(10^{-2})(5 \times 10^{-8} \text{ mL})}$$

$$\text{OCD} = 2.5 \times 10^{10} \text{ cells/mL}$$

FIGURE 6-6 PETROFF-HAUSSER COUNTING CHAMBER
The Petroff-Hausser counting chamber is a device used for the direct counting of bacterial cells. Bacterial suspension is drawn by capillary action from a pipette into the chamber enclosed by a coverslip. The cells are then counted against the grid of small squares in the center of the chamber.

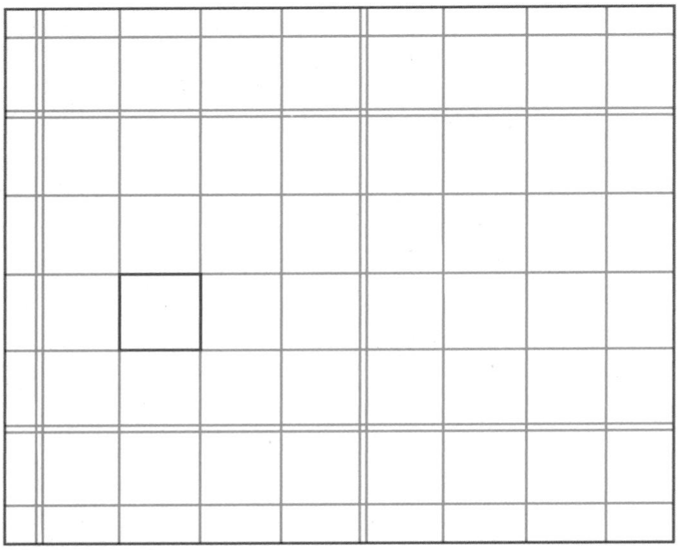

FIGURE 6-7 PETROFF-HAUSSER COUNTING CHAMBER GRID
Shown is a portion of the Petroff-Hausser counting chamber grid. The smallest squares are the ones referred to in the formula. The volume above a small square is 5×10^{-8} mL. When cells land on a line, count them with the square below or to the right.

[1] Dilution factor is calculated using the following formula.

$$D_2 = \frac{V_1 D_1}{V_2}$$

D_2 is the new dilution factor to be determined. V_1 is the volume of sample being diluted. D_1 is the dilution factor of the sample before the dilution (undiluted samples have a dilution factor of 1). V_2 is the new combined volume of sample and diluent after the dilution is completed.

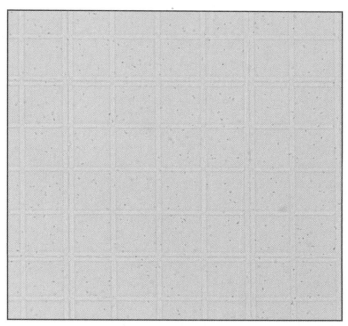

FIGURE 6-8 ORGANISM IN A PETROFF-HAUSSER COUNTING CHAMBER
This is a 10^{-4} dilution of *Vibrio natriegens* on the grid. Can you determine the original cell density?

Application

The direct count method is used to determine bacterial cell density in a sample.

In This Exercise...

You will employ the direct count method to estimate cell density in a bacterial sample.

Materials

Per Student Group

- Petroff-Hausser counting chamber with coverslip
- 1 mL pipettes with pipettor
- Pasteur pipettes with bulbs or disposable transfer pipettes
- hand counter
- staining agents A and B (Appendix G)
- test tubes
- overnight broth culture of:
  - *Proteus vulgaris*

Procedure

1. Transfer 0.1 mL from the original 24-hour culture tube to a nonsterile test tube.
2. Add 0.4 mL Agent A and 0.5 mL Agent B; mix well. (This dilution may not be suitable in all situations. There should be 5 to 15 cells per small square for optimal results. Adjust the proportions of the broth culture and agents A and B if necessary to obtain a countable dilution, but try to keep the total solution volume at 1.0 mL for easier calculation of the dilution factor and cell density.)[2]
3. Place a coverslip on the Petroff-Hausser counting chamber and add a drop of the mixture at the edge of the coverslip. Capillary action will fill the chamber.
4. Place the counting chamber on the microscope and locate the grid in the center, using the low-power objective lens. Do not increase the magnification until you have found the grid on low power. (*Hint*: Close the iris diaphragm to bring the grid lines into view. Too much light will make them impossible to see.) Increase the magnification, focusing one objective at a time, until you have the cells and the grid in focus with the oil immersion lens.
5. Count the number of cells in at least 5 but no more than 16 small squares. Be consistent in your counting of cells that are on a line between squares. By convention, cells on lines belong to the square below or the square to the right.
7. Enter your data in the space provided on the Data Sheet and answer the questions.

References

Koch, Arthur L. 1994. Page 251 in *Methods for General and Molecular Bacteriology*, edited by Philipp Gerhardt, R. G. E. Murray, Willis A. Wood, and Noel R. Krieg. American Society for Microbiology, Washington, DC.

Postgate, J. R. 1969. Page 611 in *Methods in Microbiology*, Vol. 1, edited by J. R. Norris and D. W. Ribbons. Academic Press, Inc., New York.

[2] If the total solution volume results in an odd volume that is less than or exceeds 1.0 mL, adjust your dilution factor as follows. For example, if you added 1.0 mL of sample to 0.4 mL Agent A and 1.0 mL Agent B, your dilution would be 1/15. In scientific notation, this DF is expressed as 1.5×10^{-1} calculated as follows:

$$\frac{1.0 \times 10^{0}}{1.5 \times 10^{1}} = 1.0 \times 10^{0} \times 1.5 \times 10^{-1} = 1.5 \times 10^{-1}$$

Exercise 6-4

Plaque Assay

Theory

Viruses that attack bacteria are called **bacteriophages,** or simply **phages**. Some viruses attach to the bacterial cell wall and inject viral DNA into the bacterial cytoplasm. The viral **genome** then commands the cell to produce more viral DNA and viral proteins, which are used for the assembly of more phages. Once assembly is complete, the cell lyses and releases the phages, which then attack other bacterial cells and begin the replicative cycle all over again. This process, called the **lytic cycle,** is shown in Figure 6-9.

Lysis of bacterial cells growing in a lawn on an agar plate produces a clearing that can be viewed with the naked eye. These clearings are called **plaques.** Plaque assay uses this phenomenon as a means of calculating the phage concentration in a given sample. When a sample of bacteriophage (generally diluted by means of a serial dilution) is added to a plate inoculated with enough bacterial **host** to produce a lawn of growth, the number of plaques formed can be used to calculate the original phage **titer** or density. Refer to the procedural diagram in Figure 6-10.

Plaque assay technique is similar to the standard plate count, in that it employs a serial dilution to produce countable plates needed for later calculations. (***Note:*** Refer to Exercise 6-1 Standard Plate Count as needed for a description of serial dilutions, dilution factors, and calculations.) One key difference is that standard plate count cell dilutions are spread onto agar plates, whereas plaque assay typically employs the **pour-plate technique,** in which the cells and viruses are first incorporated into molten agar and then poured into the plate.

In this procedure, diluted phage is added directly to a small amount of broth culture and allowed a 15-minute **preadsorption period** during which the viral particles attach to the bacterial cells. Then this phage–host mixture is added to a tube of **soft agar,** mixed, and poured onto prepared nutrient agar plates as an **agar overlay.** The consistency of the solidified soft agar is sufficient to immobilize the bacteria while allowing the smaller bacteriophages to diffuse short distances and infect surrounding cells. During incubation, the phage host produces a lawn of growth on the plate in which plaques appear where contiguous cells have been lysed by the virus (Figure 6-11).

The procedure for counting plaques is the same as that of the standard plate count. To be statistically reliable, countable plates must have between 30 and 300 plaques. Calculating phage titer (original phage density) uses the same formula as other plate counts

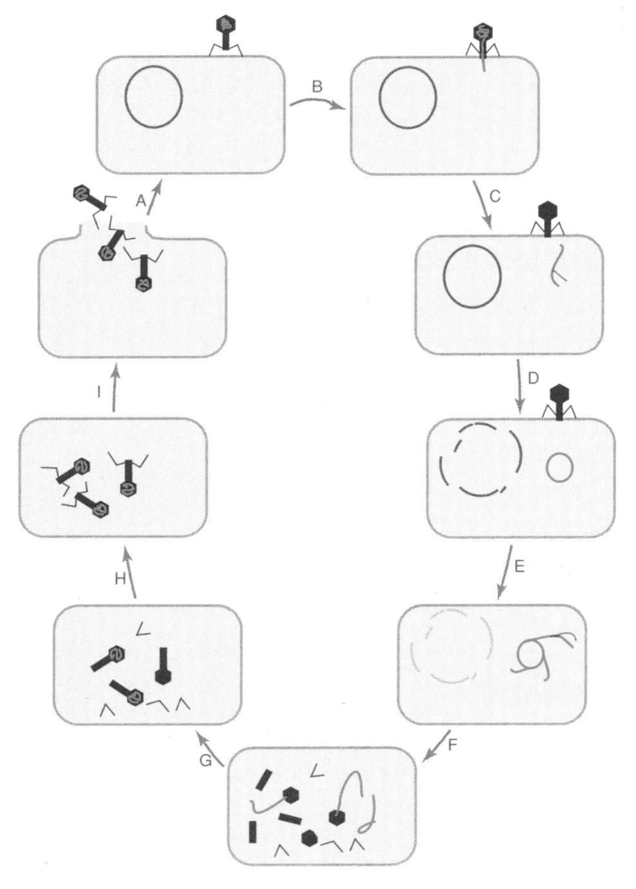

FIGURE 6-9 A Schematic Diagram Of The T4 Replicative Cycle
(A) The virus particle (shown in blue) attaches to the host cell. Each virus will infect only a specific host and this specificity is based on the ability of the virus to make attachment with viral receptors on the host. (B) The virus particle acts like a syringe and injects its DNA (shown in red) into the host cell. (C) Viral DNA is not transcribed all at once. Rather, the genes necessary for the early events in replication are transcribed first, with other genes being transcribed as the appropriate time arises. In this diagram, **early** and **middle genes** are being transcribed into mRNA (shown in green). (D) Viral DNA circularizes and host DNA is broken apart. (E) Viral DNA is replicated by a **rolling circle** mechanism. Host DNA is further degraded into nucleotides, which are used for viral DNA replication. **Late genes** are also transcribed. (F) Capsid, tail and tail fiber proteins are made and assemble into their respective components in separate assembly lines. DNA enters the capsids. (G & H) Assembly continues as first tails and then tail fibers attach to form a complete virus particle (**virion**). (I) The host cell is lysed and releases the completed virus particles, each capable of infecting another host cell. The typical **burst size** for T4 is about 300 viruses. The whole process from attachment to host lysis takes less than 25 minutes!

except that PFU (plaque forming unit) instead of CFU (colony forming unit) becomes the numerator in the equation. Phage titer, therefore, is, expressed in PFU/mL and the formula is written as follows:

$$\text{Original phage density} = \frac{\text{\# PFU}}{(\text{Volume plated})(\text{Dilution factor})}$$

The convention is to make one final simplification to the formula by combining Volume Plated and Dilution Factor to form a single term known as the Final Dilution Factor, or FDF. For convenience, plates are usually inoculated with 0.1 mL. If the dilution factor of the source tube is 10^{-5} and the inoculum is 0.1 mL, the FDF is 10^{-6} mL (10^{-5} x 0.1 mL = 10^{-6} mL). Note that the FDF

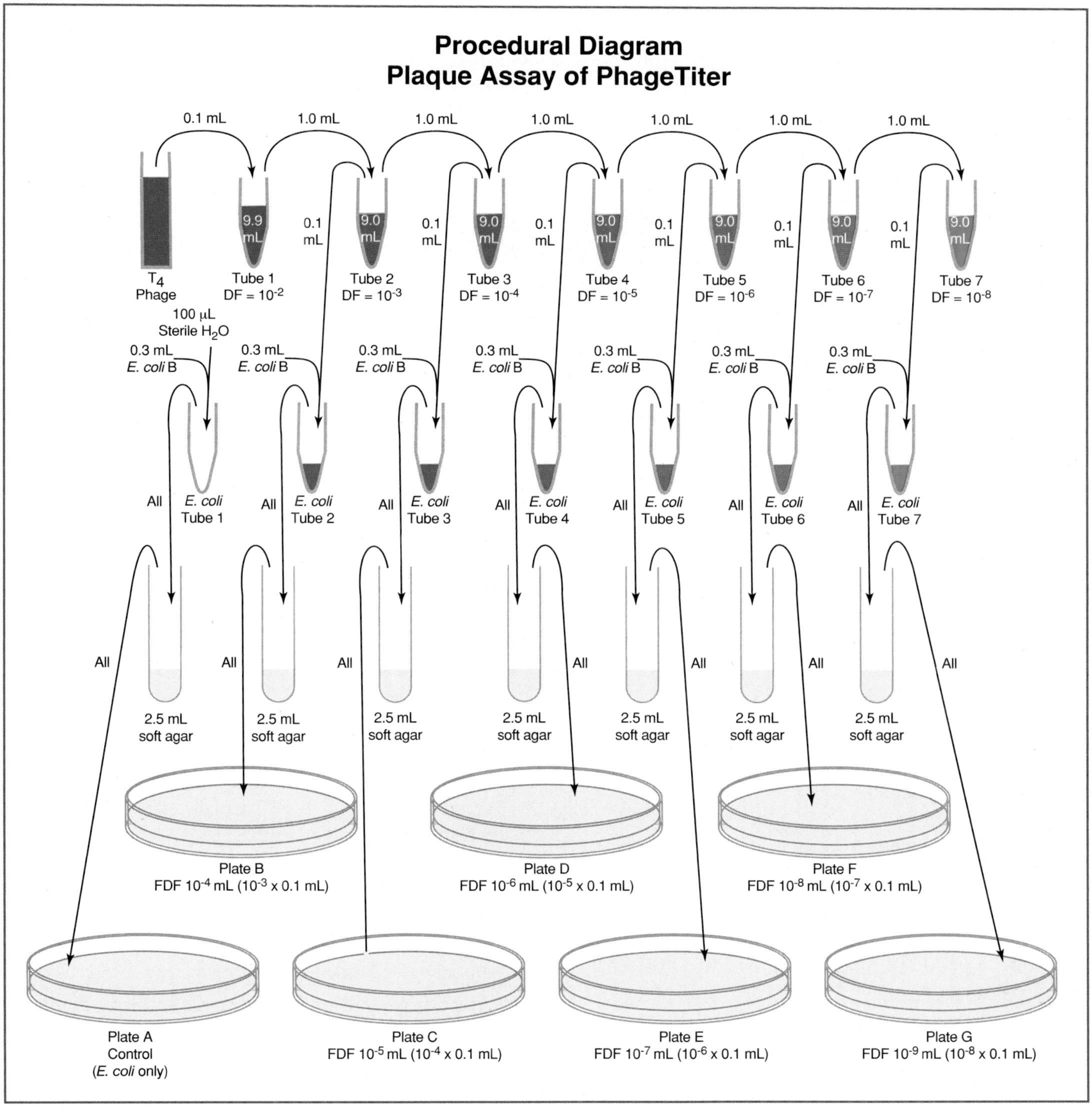

FIGURE 6-10 PROCEDURAL DIAGRAM FOR PLAQUE ASSAY
Use this diagram as a guide while performing the dilutions and transfers outlined in the procedure.

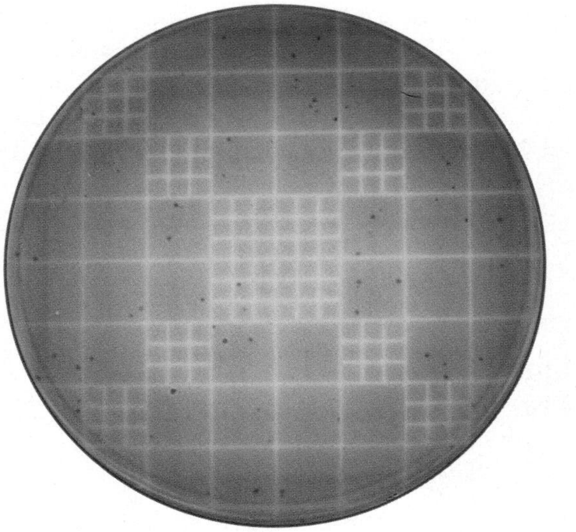

FIGURE 6-11 Countable Plate
This plaque assay plate has between 30 and 300 plaques; therefore, it is countable.

includes "mL" units where DF does not. This change to FDF typically is done automatically by simply increasing the dilution factor by a factor of 10 when writing it on the plate. Then, after one or two days' incubation, the plates are removed, plaques are counted, and the original density is calculated with a simple division problem. In its most condensed form, the formula is written as follows:

$$\text{Original phage density} = \frac{\text{PFU}}{\text{FDF}}$$

If 150 plaques were counted on a plate that received 0.1 mL of a 10^{-4} phage dilution, the calculation would be,

$$\text{Original phage density} = \frac{\text{PFU}}{\text{FDF}}$$

$$\text{Original phage density} = \frac{150\ \text{PFU}}{10^{-5}\ \text{mL}}$$

$$\text{Original phage density} = 1.5 \times 10^{7}\ \text{PFU/mL}$$

Application

This technique is used to determine the concentration of viral particles in a sample. Samples taken over a period of time can be used to construct a viral growth curve.

In This Exercise...

You will be estimating the density (titer) of a T4 coliphage sample using a strain of *Escherichia coli* (*E. coli* B) as the host organism.

Materials

Per Student Group

- sterile 0.1 mL, 1.0 mL, and 10.0 mL pipettes
- 14 sterile dilution tubes
- 7 Nutrient Agar plates
- 7 tubes containing 2.5 mL Soft Agar
- flask of sterile water
- hot water bath set at 50°C
- T4 coliphage
- 24-hour broth culture of *Escherichia coli B* (T-series phage host)

Medium Recipe

Soft Agar

• Beef extract	3.0 g
• Peptone	5.0 g
• Sodium chloride	5.0 g
• Tryptone	2.5 g
• Yeast extract	2.5 g
• Agar	7.0 g
• Distilled or deionized water	1.0 L

Procedure

Refer to the procedural diagram in Figure 6-10 as needed. Appendix F includes an alternative procedure for digital micropipettes and μL volumes.

Lab One

1. Obtain all materials except for the Soft Agar tubes. To keep the agar tubes liquefied, leave them in the water bath and take them out one at a time as needed.
2. Label seven tubes 1 through 7. Label the other seven tubes *E. coli* 1–7. Place all tubes in a rack, pairing like-numbered tubes.
3. Label the Nutrient Agar plates A through G. Place them in the 35°C incubator to warm them. Take them out one at a time as needed. This will keep the Soft Agar (added at step 14) from solidifying too quickly and result in a smoother agar surface.
4. Aseptically transfer 9.9 mL sterile water to dilution tube 1.
5. Aseptically transfer 9.0 mL sterile water to dilution tubes 2–7.
6. Mix the *E. coli* culture and aseptically transfer 0.3 mL into each of the *E. coli* tubes.

7. Mix the T4 suspension and aseptically transfer 0.1 mL to dilution tube 1. Mix well. This is DF (dilution factor) 10^{-2}.
8. Aseptically transfer 1.0 mL from dilution tube 1 to dilution tube 2. Mix well. This is DF 10^{-3}.
9. Aseptically transfer 1.0 mL from dilution tube 2 to dilution tube 3. Mix well. This is DF 10^{-4}.
10. Continue in this manner through dilution tube 10. The dilution factor of tube 7 should be 10^{-8}.
11. Aseptically transfer 0.1 mL sterile water to *E. coli* tube 1. This will be mixed with 2.5 mL Soft Agar and used to inoculate a control plate.
12. Aseptically transfer 0.1 mL from dilution tube 2 to its companion *E. coli* tube. Repeat this procedure with the remaining five tubes.
13. This is the beginning of the preadsorption period. Let all seven tubes stand undisturbed for 15 minutes.
14. Remove one Soft Agar tube from the hot water bath and add the contents of *E. coli* tube 1. Mix well and immediately pour onto plate A. Gently tilt the plate back and forth until the soft agar mixture is spread evenly across the solid medium. Label the plate "Control."
15. Remove a second Soft Agar tube from the water bath and add the contents of *E. coli* tube 2. Mix well and immediately pour onto plate B. Tilt back and forth to cover the agar and label it FDF 10^{-4}.
16. Repeat this procedure with dilutions 10^{-4} thru 10^{-8} and plates C thru G. Label the plates with the appropriate FDF.
17. Allow the agar to solidify completely.
18. Invert the plates and incubate aerobically at 35 ± 2°C for 24 to 48 hours.

Lab Two

1. After incubation, examine the control plate for growth and the absence of plaques.
2. Examine the remainder of your plates and determine which one is countable (30 to 300 plaques). Count the plaques and record the number in the chart provided on the Data Sheet. Record all others as either TMTC (too many to count) or TFTC (too few to count).
3. Using the FDF on the countable plate and the formula provided on the Data Sheet, calculate the original phage density. Record your results in the space provided on the Data Sheet.

References

Collins, C. H., Patricia M. Lyne, J. M. Grange. 1995. Page 149 in *Collins and Lyne's Microbiological Methods*, 7th ed. Butterworth-Heinemann, United Kingdom.

DIFCO Laboratories. 1984. Page 619 in *DIFCO Manual*, 10th ed. DIFCO Laboratories, Detroit.

Province, David L., and Roy Curtiss III. 1994. Page 328 in *Methods for General and Molecular Bacteriology*, edited by Philipp Gerhardt, R. G. E. Murray, Willis A. Wood, and Noel R. Krieg. American Society for Microbiology, Washington, DC.

Exercise 6-5

Closed-System Growth

Theory A closed system—in this exercise, a broth culture in a flask—is one in which no nutrients are added beyond those present in the original medium and no wastes are removed. Bacteria grown in a closed system demonstrate four distinct growth phases: lag phase, exponential phase, stationary phase, and death phase.

The four phases together form a characteristic shape known as a microbial growth curve (Figure 6-12). **Lag phase**, the first phase, constitutes what might be called an adjustment period. Initially, there is no cell division. It is believed that microorganisms use this time to repair damaged cellular components and synthesize enzymes to begin using the resources of the new environment. The duration of lag phase can be quite variable and depends on many factors. Actively reproducing cells transferred to a medium identical to the one from which they are removed, will undergo a short lag period. Cells that are near death, have been refrigerated, or must adjust to a completely different medium will demonstrate a long lag phase.

Eventually, as cells begin to use resources of the new environment and start reproducing, lag phase gradually gives way to the **exponential** (or log) **phase**. This phase is a time of maximum growth that is limited almost exclusively by the organism's reproductive potential. In a closed system, conditions usually are adjusted to be optimal for a specific organism; however, even under the best of conditions, the medium and other physical factors influence the growth rate slightly. Exponential growth is characterized by cellular division and doubling of the population size at regular intervals depending on the organism's generation time. Because not all cells are dividing at exactly the same time, exponential growth is marked by a smooth and dramatic upward sweep of growth, as shown in Figure 6-12.

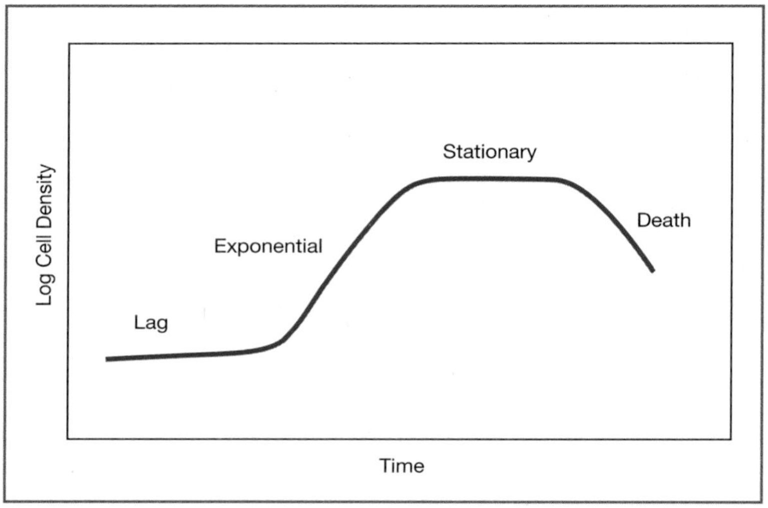

FIGURE 6-12 Microbial Growth Curve
Note the different phases of growth.

As nutrients in the medium decrease and toxic waste products increase, the growth rate gradually declines to where the population's death rate is more or less equal to its reproductive rate. This leveling of growth is called the **stationary phase**. Stationary phase will last until the nutrients are depleted or the medium becomes toxic to the organism. **Death phase**, the final phase, is marked by the decline of the organism. As discussed in Exercise 6-6, microbial death is the reverse of microbial growth and is exponential. That is, a fixed proportion of the population will die in a given time (the time is specific to the organism), regardless of population size.

Several measurements are possible in a closed-system growth experiment. These measurements include the duration of each phase, the **mean growth rate constant, generation time**, and the organism's **mininum, maximum and optimum (cardinal) temperatures.**

Mean growth rate constant (k) is calculated using the increase in population size between two points in time during exponential growth. The following is the standard formula:

$$k = \frac{\log N_2 - N_1}{(0.301)\, t}$$

where N_1 is the population size early in exponential phase, N_2 is the population size later in exponential phase, (t) is the elapsed time between the N_1 and N_2 readings, and 0.301 is the log of 2. Log 2 is used because bacterial reproduction typically occurs by binary fission; therefore, each reproductive cycle results in a doubling of the population size. Calculations of organisms using a different reproductive pattern require the log of a value appropriate to their typical number of offspring.

If population density calculation is not required, absorbance readings taken with the spectrophotometer early and late in the exponential phase (A_1 and A_2) can be substituted in the numerator as follows:

$$k = \frac{A_2 - A_1}{(0.301)\, t}$$

For example, the mean growth rate constant of an organism whose absorbance readings increase from 0.195 to 0.815 in 2 hours would be:

$$k = \frac{0.815 - 0.195}{(0.301)\ 2\ \text{hr}}$$

$$k = \frac{0.62}{0.602 \text{ hr}}$$

$$k = 1.03 \text{ generations/hr}$$

Whereas (k) is the number of generations per unit time, (g) is mean generation time—the amount of time required for a population to double. Mean generation time is simply the inverse of (k). It can be easily calculated using the following formula:

$$g = \frac{1}{k}$$

Taking the above example, the generation time of the organism would be:

$$g = \frac{1}{1.03 \text{ generations/hr}}$$

$$g = 0.97 \text{ hr/generation or } 58 \text{ minutes/generation}$$

Application

This exercise is designed to demonstrate the pattern of microbial growth in a closed system, expose you to the methods involved in determining growth, and demonstrate the calculation of mean growth rate and mean generation time.

In This Exercise...

You will use a spectrophotometer to measure the growth of *Vibrio natriegens* incubated at five different temperatures. The class will be split into five groups—each group incubating at a different temperature. Absorbance readings will be taken every 15 minutes. (As cell density increases, turbidity increases, and so does absorbance of the solution.) Using the data collected for your group, you will plot the growth curve either by computer or on the graph paper provided, with absorbance substituting for population size. You then will calculate the mean growth rate constant and generation time. Finally, using the class data, you will plot the mean growth rate constants over the range of temperatures to estimate the minimum, maximum, and optimum growth (cardinal) temperatures for the organism.

Materials

One of Each Per Student Group

- five water baths set at 20°C, 25°C, 30°C, 35°C, and 40°C
- thermometers (to monitor actual temperatures in the ice and water baths)
- five sterile side-arm flasks containing 47 mL of sterile Brain Heart Infusion (BHI) Broth + 2% NaCl
- five sterile side-arm flasks containing 50 mL BHI + 2% NaCl to be used as controls
- spectrophotometers
- lab tissues
- sterile 5 mL pipettes and pipettor or micropipettes (10–100 µL and 100–1000 µL) with sterile tips
- Bunsen burner
- 20 mL Overnight broth culture of *Vibrio natriegens* in BHI + 2% NaCl (one per class)

Medium Recipe

Brain Heart Infusion Agar

• Infusion from calf brains	200 g
• Infusion from beef hearts	250 g
• Peptone	10 g
• Dextrose	2 g
• Sodium chloride	5 g
• Disodium Phosphate	2.5 g
• Agar	15.0 g
• Distilled or deionized water	1.0 L

pH 7.2–7.6 at 25°C

Procedure

1. Collect all necessary materials: a spectrophotometer, a sterile 5 mL pipette and mechanical pipettor (or micropipette and tips), the side-arm flasks containing 47 mL and 50 mL of broth, and lab tissues. Label the 50 mL flask "control."
2. Immediately label your 47 mL growth flask; place it into the appropriate water bath and allow 15 minutes for the temperature to equilibrate. Turn on the spectrophotometer and let it warm up for a few minutes.
3. If the spectrophotometer is digital, set it to "absorbance." Set the wavelength to 650 nm. (For instruction on spectrophotometer operation, refer to Appendix E.)
4. When all groups are ready to start (all groups *must* start within minutes of each other), mix the *V. natriegens* culture and aseptically add 3.0 mL to your growth flask. Mix it, wipe it dry making sure the side arm is clean, and immediately take a turbidity reading (Figure 6-13). This time is T_0. (***Caution:*** Be careful not to do this too quickly, as the side-arm flask is *very easy to spill* when it is tipped on its side for reading. Further, if it is pulled out on an angle, you risk breaking the side-arm and spilling the culture.)
5. Record the absorbance next to your temperature under T_0 in the chart provided on the Data Sheet.

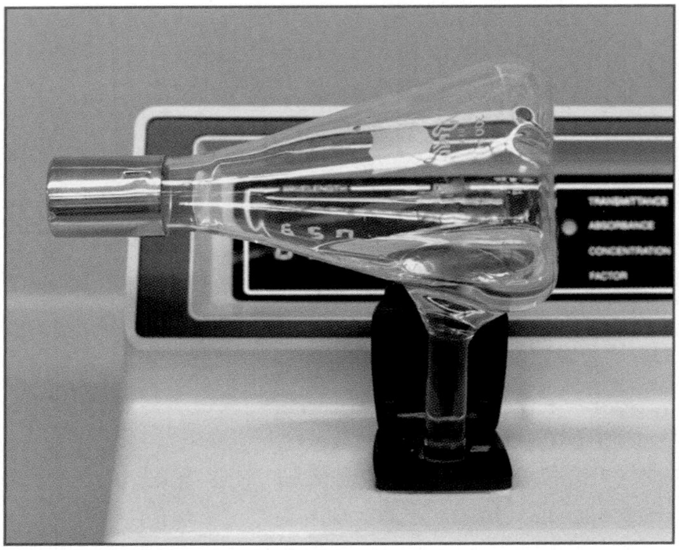

FIGURE 6-13 SIDE-ARM FLASK IN THE SPECTROPHOTOMETER
Insert and remove the flask carefully to avoid spilling the broth. Place the flask in the spectrophotometer the same direction each time to avoid inconsistent readings caused by aberrations in the glass. Cover the port with your fingers to reduce light entering the machine.

Remove the growth flask carefully and return it to the water bath.

6. Monitor the temperature in your water/ice bath frequently. It is more important to have a constant temperature than to have exactly the assigned temperature. Record your *actual* temperature in the chart on the Data Sheet.
7. Repeat the above procedure every 15 minutes (T_{15}, T_{30}, T_{45}, etc), taking turbidity readings and recording them in the chart until you have completed and recorded the T_{240} reading. Make sure to place the flask in the spectrophotometer the same direction each time. Consistency in your readings will help compensate for irregularities in the glass.)

References

Collins, C. H., Patricia M. Lyne, J. M. Grange. 1995. Page 149 in Collins and Lyne's *Microbiological Methods*, 7th ed. Butterworth-Heinemann, Oxford, United Kingdom.

Gerhardt, Philipp, et. al. 1994. Chapter 11 in *Methods for General and Molecular Bacteriology*. American Society for Microbiology, Washington DC.

Koch, Arthur L. 1994. Page 251 in *Methods for General and Molecular Bacteriology*, edited by Philipp Gerhardt, R. G. E. Murray, Willis A. Wood, and Noel R. Krieg. American Society for Microbiology, Washington, DC.

Postgate, J. R. 1969. Page 611 in *Methods in Microbiology*, Vol. 1, edited by J. R. Norris and D. W. Ribbons. Academic Press, New York.

Prescott, Lansing M., John P. Harley, and Donald A. Klein. 1999. Page 114 in *Microbiology*, 4th ed. WCB/McGraw-Hill, Boston.

Exercise 6-6

Thermal Death Time Versus Decimal Reduction Value

Theory The effects of excessive heat on bacteria range from disruption of cell membrane and protein function to complete combustion of the cell and all its components. Temperatures above 60°C are sufficient to kill most vegetative cells, whereas microbial spores may survive temperatures greater than 100°C.

The time needed to kill a specific number of microorganisms at a specific temperature is called **thermal death time (TDT)**. Because microbial death occurs exponentially—that is, the number of *dead* cells increases exponentially—TDT can be expressed in logarithmic terms. Figure 6-14 illustrates a typical death curve where the log of population size (in cells per milliliter) is plotted against time at a specific temperature. This hypothetical organism, in a suspension of 10^5 cells/mL, survives 25 minutes at 65°C.

One variation of TDT, commonly used by food processors, is the **decimal reduction value (D value)**. The D value is the time (in minutes at a specific temperature) needed to reduce the number of viable organisms in a given population by 90%—one logarithmic cycle. (In other words, a population that, upon heating, is reduced from a density of 1.2×10^7 to 1.2×10^6 cells/mL, converting to base 10 logarithmic terms, is reduced from $10^{7.08}$ to $10^{6.08}$ cells/mL—1 logarithmic cycle.) The temperature for the D value is customarily added as a subscript, (*e.g.*, D_{60}, D_{121}, etc).

D value is useful because it mathematically represents death rate as a constant for a species under particular circumstances and is independent of population size. An organism with a D_{60} value of 4 takes 4 minutes at 60°C to reduce a population by 90%, regardless of the starting number of cells. Mathematically, D value is defined as:

$$D_T = \frac{t}{\log_{10} x - \log_{10} y}$$

where

T = temperature
t = time in minutes
x = cell density (cells per milliliter) before exposure to heat
y = cell density after exposure to heat.

For example, if a population of 1.7 x 10^5 cells heated at 60°C for 10 minutes was reduced to 2.6×10^3 cells, the D_{60} value would be 5.49 minutes.

$$D_{60} = \frac{10 \text{ min.}}{\log_{10} 1.7\times10^5 - \log_{10} 2.6\times10^3} = \frac{10 \text{ min.}}{5.23\times3.41} = 5.49 \text{ min.}$$

D value also can be plotted on a TDT standard curve, as shown in Figure 6-15. Either value (D value or TDT) can be determined from the other. A calculated D

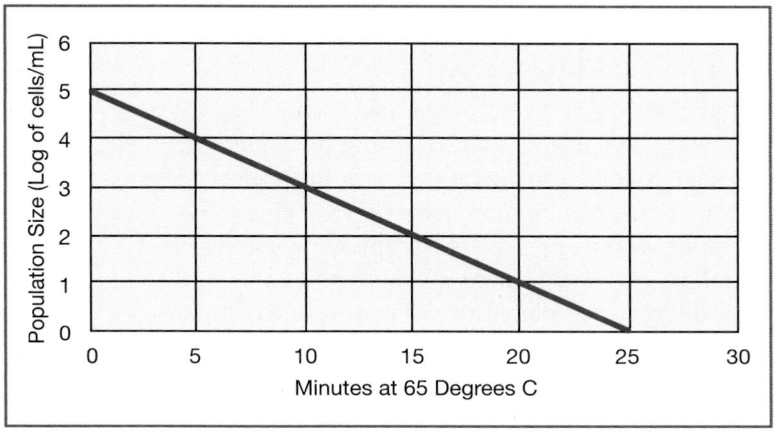

FIGURE 6-14 Thermal Death Time (TDT) Curve
With an original cell density of 105 cells/mL, this hypothetical population survives 25 minutes at 65°C. Notice that it takes 5 minutes to reduce the population size by a factor of 10, *regardless of the magnitude of the population.* What would be the TDT if the population had started at a density of 10^6 cells/mL?

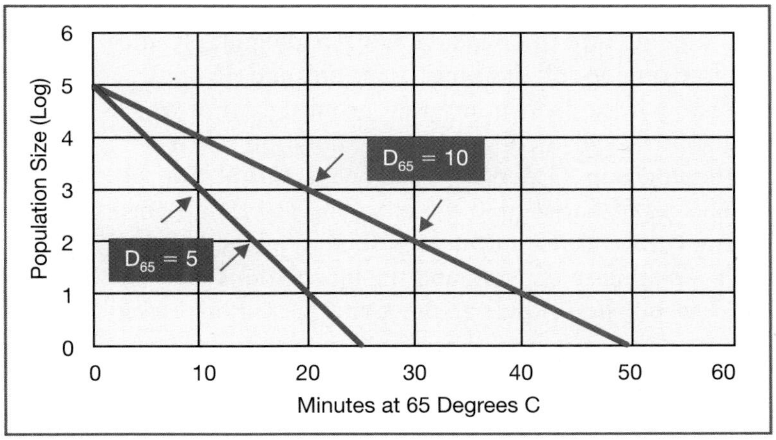

FIGURE 6-15 TDT Curves of Two Populations
Species A (purple) takes 5 minutes to reduce its population by 90% (10^3 to 10^2 cells) whereas species B (blue) takes 10 minutes. The graph can be constructed either by calculating the D value and extrapolating from two data points (shown by the arrows) or by plotting the TDT curve and determining D value from the points of intersection with log values. Both methods produce a straight line. As you can see, D value is independent of population size.

value requires two population density measurements—one before heating and one after heating. This produces two data points on the graph that can be used to construct a straight line TDT curve. Conversely, a TDT curve (determined by measuring the total time needed to kill the entire population) reveals the D value at the points (shown by arrows) where the curve intersects log values from the *y*-axis. The log values on the *y*-axis represent exponents for a $base_{10}$ logarithmic scale. In other words, 1 is interpreted as 10^1, 2 is 10^2, 3 is 10^3 cells per milliliter, *etc.*

Application

This exercise is used to determine microorganismal thermal death time (TDT) and decimal reduction value at 60°C.

In This Exercise...

You will determine the D_{60} values of *Escherichia coli* and *Staphylococcus aureus* cultures using both methods described above. Your instructor will begin by adding bacteria to a broth maintained at 60°C. This is T_0—the beginning of the experiment. At this time and again at 10 minutes (T_{10}), a sample will be removed from the flask, diluted, and plated to perform viable counts to establish the density of the population.

In addition, beginning at T_1, samples will be removed from the flask at 1-minute intervals for 30 minutes (T_{30}). The samples will be used to inoculate Nutrient Broths. After a 48-hour incubation period, the Nutrient Broths will be examined for growth. The first broth in the sequence to produce no bacterial growth will establish the thermal death time of the population. The accumulated data then will be used to plot and calculate the D_{60} values of each organism.

In an attempt to divide the workload equitably and to allow time for all dilutions, transfers, and platings, the class will be divided into four groups, each of which will be further divided into a Plate Subgroup and a Broth Subgroup. Group One will be responsible for the T_0 platings of *E. coli,* and the inoculations of Nutrient Broths 1 through 15. Group 2 will be responsible for the T_{10} platings of *E. coli*, and the inoculations of Nutrient Broths 16 through 30. Group 3 will be responsible for the T_0 platings of *S. aureus,* and inoculations of Nutrient Broths 1 through 15. Group 4 will be responsible for the T_{10} platings of *S. aureus*, and the inoculations of Nutrient Broths 16 through 30.

Materials

Per Class

- stopwatch
- thermometers
- two flasks each containing 49.0 mL Nutrient Broth, maintained at 60°C in a water bath
- 24-hour broth cultures of:
  - *Escherichia coli*
  - *Staphylococcus aureus*

Per Group

- 7 Nutrient Agar plates
- 16 Nutrient Broth tubes
- sterile 0.1 mL, 1.0 mL, 5.0 mL, and 10.0 mL pipettes sterile dilution tubes (large enough to hold 10.0 mL)
- sterile deionized water in a small flask or beaker
- beaker of alcohol and bent glass rod for spreading organisms

Procedure

[*Note:* This procedure uses the Spread-Plate Technique described in Exercise 1-4. Refer to it as needed.] The Plate Subgroups will need at least two or three students for labeling tubes and plates, performing the serial dilution, and spreading plates. The Broth Subgroups will need one or two students to label and inoculate the 15 Nutrient Broths, keep track of time, and mix the broth prior to all transfers. Appendix F includes an alternate procedure for digital micropipettes.

Lab One: Group 1 (T_0)

Plate Subgroup

1. Obtain all plating materials. Place seven sterile test tubes in a rack. Label them 1 through 7. Add 9.0 mL of sterile water to tubes 2 through 7 and cover them until needed.
2. Label seven Nutrient Agar plates 1 through 7 respectively.
3. When your instructor adds 1.0 mL of the *E. coli* culture to the heated flask, he/she will start the stopwatch. This is T_0. Immediately swirl the flask to disperse the sample and remove 1.5 mL with your pipette. If possible, do this step without removing the flask from the water bath. Begin the following serial dilution:
   - Add the 1.5 mL of broth from the flask to tube 1.
   - Transfer 1.0 mL from tube 1 to tube 2; mix. Discard the pipette. Use a clean pipette for each transfer from here forward.
   - Transfer 1.0 mL from tube 2 to tube 3; mix.
   - Transfer 1.0 mL from tube 3 to tube 4; mix.
   - Transfer 1.0 mL from tube 4 to tube 5; mix.
   - Transfer 1.0 mL from tube 5 to tube 6; mix.
   - Transfer 1.0 mL from tube 6 to tube 7; mix.

- Transfer 0.1 mL from tube 1 to plate 1 and spread according to the spread-plate procedure.
- Transfer 0.1 mL from tube 2 to plate 2 and spread.
- Transfer 0.1 mL from tube 3 to plate 3 and spread.
- Transfer 0.1 mL from tube 4 to plate 4 and spread.
- Transfer 0.1 mL from tube 5 to plate 5 and spread.
- Transfer 0.1 mL from tube 6 to plate 6 and spread.
- Transfer 0.1 mL from tube 7 to plate 7 and spread.

4. Calculate the final dilution factor* for each plate and enter it in Chart 1 on the Data Sheet.
5. Tape the plates together, invert, and incubate them at 35 ± 2°C for 48 hours.

Broth Subgroup

1. Obtain all broth transfer materials. Label 15 Nutrient Broth tubes T_1 through T_{15}. Label the 16th tube "Control."
2. When your instructor adds 1.0 mL of the *E. coli* culture to the heated flask, he/she will start the stopwatch. This is T_0. Slightly before T_1, gently swirl the flask to mix the broth. At exactly T_1, transfer 0.1 mL from the flask to the appropriately labeled Nutrient Broth tube. Discard the pipette. Use a clean pipette for each transfer.
3. Repeat this procedure at T_2 through T_{15}. Be sure to mix the broth before each transfer, and use a clean pipette.
4. Incubate all broths at 35 ± 2°C for 48 hours.

Lab One: Group 2 (T_{10})

Plate Subgroup

1. Obtain all plating materials. Place six sterile test tubes in a rack and label them 1 through 6. Add 9.0 mL of sterile water to tubes 2 through 6 and cover them until needed.
2. Label seven Nutrient Agar plates 1 through 7 respectively.
3. When your instructor adds 1.0 mL of the *E. coli* culture to the heated flask, he/she will start the stopwatch. This is T_0. At T_{10} remove 2.5 mL with your pipette. It may work best to wait for students in the broth group to withdraw their sample; then make your transfers as quickly as possible.
4. Begin the following serial dilution:
   - Add the 2.5 mL of broth from the flask to tube 1.
   - Transfer 1.0 mL from tube 1 to tube 2; mix. Discard the pipette. Use a clean pipette for each transfer from here forward.
   - Transfer 1.0 mL from tube 2 to tube 3; mix.
   - Transfer 1.0 mL from tube 3 to tube 4; mix.
   - Transfer 1.0 mL from tube 4 to tube 5; mix.
   - Transfer 1.0 mL from tube 5 to tube 6; mix.
   - Transfer 1.0 mL from tube 1 to plate 1 and spread according to the spread-plate procedure.
   - Transfer 0.1 mL from tube 1 to plate 2 and spread.
   - Transfer 0.1 mL from tube 2 to plate 3 and spread.
   - Transfer 0.1 mL from tube 3 to plate 4 and spread.
   - Transfer 0.1 mL from tube 4 to plate 5 and spread.
   - Transfer 0.1 mL from tube 5 to plate 6 and spread.
   - Transfer 0.1 mL from tube 6 to plate 7 and spread.
5. Calculate the final dilution factor for each plate and enter it in Chart 1 on the Data Sheet.
6. Tape the plates together, invert, and incubate them at 35 ± 2°C for 48 hours.

Broth Subgroup

1. Obtain all broth transfer materials. Label 15 Nutrient Broth tubes T_{16} through T_{30} respectively. Label the 16th tube "Control." Enter these numbers in the appropriate boxes in Chart 2 on the Data Sheet.
2. When your instructor adds 1.0 mL of the *E. coli* culture to the heated flask, he/she will start the stopwatch. This is T_0. Slightly before T_{16}, gently swirl the broth flask and at exactly T_{16}, transfer 0.1 mL from the flask to the appropriately labeled Nutrient Broth tube. Discard the pipette tip.
3. Repeat this procedure at T_{17} through T_{30}. Be sure to mix the broth before each transfer, and use a clean pipette tip.
4. Incubate all broths at 35 ± 2°C for 48 hours.

Lab One: Group 3 (T_0)

Plate Subgroup

Using the *S. aureus* culture, follow the Group 1 Plate Subgroup instructions.

Broth Subgroup

Using the *S. aureus* culture, follow the Group 1 Broth Subgroup instructions.

Lab One—Group 4 (T_{10})

Plate Subgroup

Using the *S. aureus* culture, follow the Group 2 Plate Subgroup instructions.

* Use of final dilution factors and the formula given in the Lab Two Procedure require knowledge of dilutions and calculations, explained fully in Exercise 6-1, Standard Plate Count. If you haven't completed a standard plate count, or you have any doubt about your ability to complete this exercise, we suggest that you read Exercise 6-1 carefully and do some or all of the practice problems.

Broth Subgroup

Using the *S. aureus* culture, follow the Group 2 Broth Subgroup instructions.

Lab Two: All Groups

1. Remove all broths and plates from the incubator.
2. Pick the plate containing between 30 and 300 colonies. There should be only one. It is the countable plate. All others are either TFTC (too few to count) or TMTC (too many to count). Label them as such, and enter this information in Chart 1 on the Data Sheet.
3. If you haven't already done so, count the colonies on the countable plate and enter the number in Chart 1 on the Data Sheet.
4. Examine the broths for growth, comparing each one to the uninoculated control (tube #16). Any turbidity is read as positive for growth. Enter your results in Chart 2 on the Data Sheet. Circle the earliest time that no turbidity appears (*i.e.*, growth did not occur).
5. Calculate the population density of your group's organism for your specified time (T_0 or T_{10}), using the following formula:

$$\text{Cell density (cells/mL)} = \frac{\text{\# Colonies counted}}{\text{Final dilution factor}}$$

6. Enter your results in Chart 1 and again in Chart 3 on the Data Sheet. Also, be sure to convert your results to logarithmic form, and enter that in Chart 3 as well.
7. Collect the data from other groups and enter it in Chart 3 on the Data Sheet. Your instructor will likely provide a transparency or chalkboard space for class data.
8. Follow the directions on the Data Sheet to plot *and* calculate the D values of both organisms.
9. Answer the questions on the Data Sheet.

References

Ray, Bibek. 2001. Chapter 31 in *Fundamental Food Microbiology*, 2nd ed. CRC Press, Boca Raton.

National Canners Association Research Laboratories. 1968. *Laboratory Manual for Food Canners and Processors*, Vol. 1: *Microbiology and Processing*. AVI Publishing Co., Westport, CT.

Stumbo, Charles Raymond. 1973. Chapter 7 in *Thermobacteriology in Food Processing*, 2nd ed. Academic Press, New York.

Medical Microbiology

SECTION 7

The study and application of microbiological principles and those of medicine are inseparable. So, although this is not a medical microbiology manual, it nonetheless is devoted largely to the study of microorganisms and their relationship to human health. Our superficial treatment of medical microbiology in this section is not meant to represent a balanced body of information but, rather, is composed of exercises to augment those you have done already related to medical microbiology.

In this section you will examine lysozyme, a natural antimicrobial agent produced by the body, perform a test to detect susceptibility to dental decay, and check the effectiveness of various antibiotics on sample bacteria. These exercises are followed by two epidemiological exercises. In the first, you will use a standard resource—*Morbidity and Mortality Weekly Report* available from the Centers for Disease Control and Prevention (CDC) Web site—to follow the incidence of a disease of your choice over the course of two years. In the second exercise, you will simulate an epidemic outbreak in your class, determine the source of the outbreak, and perform epidemiological analyses of your class population. The final two exercises utilize flowcharts using standard biochemical tests to identify common members of the *Enterobacteriaceae* and Gram-positive cocci.

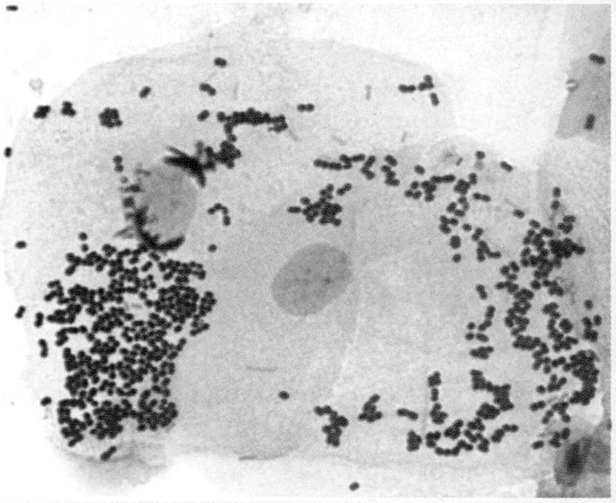

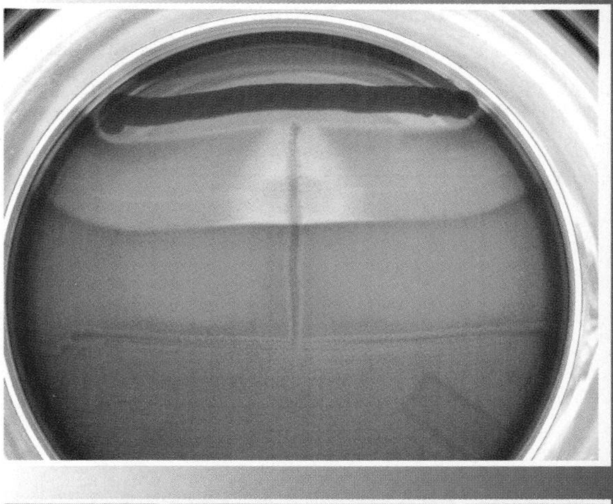

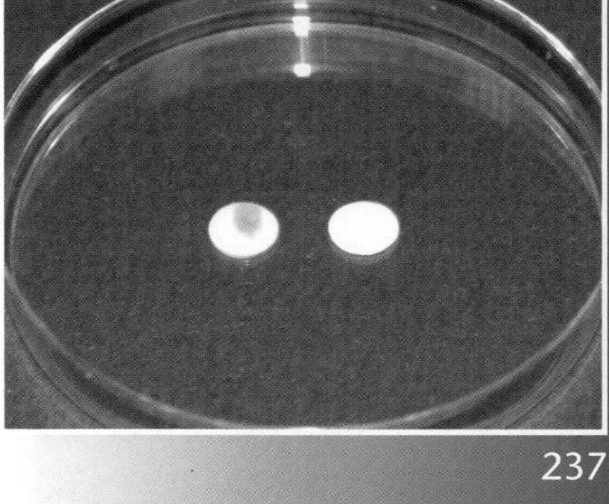

Exercise 7-1

Snyder Test

Theory Snyder test medium is formulated to favor the growth of oral bacteria (Figure 7-1) and discourage the growth of other bacteria. This is accomplished by lowering the pH of the medium to 4.8. Glucose is added as a fermentable carbohydrate, and bromcresol green is the pH indicator. Lactobacilli and oral streptococci survive these harsh conditions and also ferment the glucose and lower the pH even further. The pH indicator, which is green at or above pH 4.8 and yellow below, turns yellow in the process. Development of yellow color in this medium, therefore, is evidence of fermentation and, further, is highly suggestive of the presence of dental decay-causing bacteria (Figure 7-2).

The medium is autoclaved for sterilization, cooled to just over 45°C, and maintained in a warm water bath until needed. The molten agar then is inoculated with a small amount of saliva, mixed well, and incubated for up to 72 hours. The agar tubes are checked at 24-hour intervals for any change in color. High susceptibility to dental caries is indicated if the medium turns yellow within 24 hours. Moderate and slight susceptibility are indicated by a change within 48 and 72 hours, respectively. No change within 72 hours is considered a negative result. These results are summarized in Table 7-1.

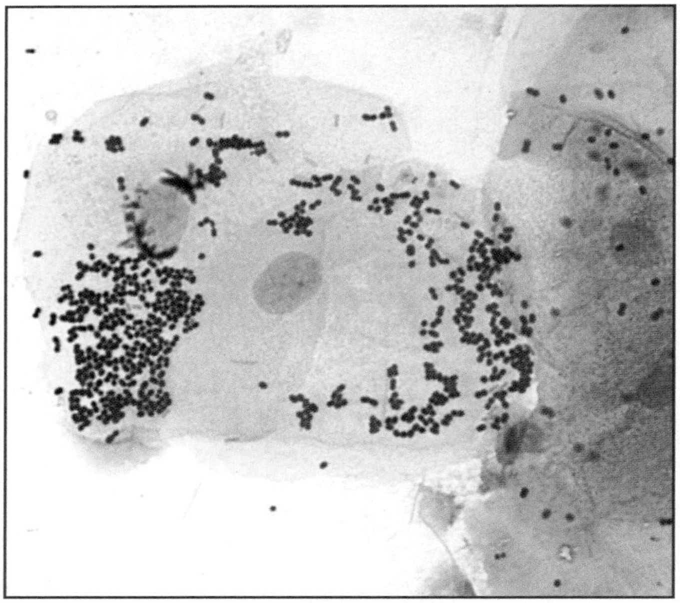

FIGURE 7-1 Gram Stain of Oral Bacteria
Note the predominance of Gram-positive cocci. Also present are a few Gram-negative rods and very few Gram-positive rods (*Lactobacillus*). Note also the size difference between the large epithelial cells in the center and the bacteria.

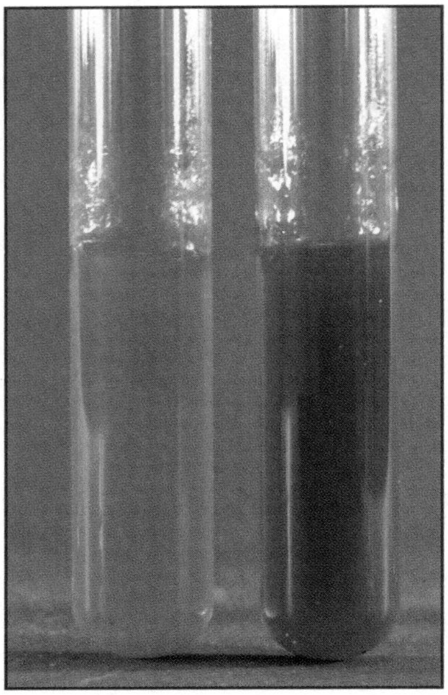

FIGURE 7-2
Snyder Test Results
A positive result is on the left, and a negative result is on the right.

TABLE OF RESULTS

Result	Interpretation
Yellow at 24 hours	High susceptibility to dental caries
Yellow at 48 hours	Moderate susceptibility to dental caries
Yellow at 72 hours	Slight susceptibility to dental caries
Yellow at >72 hours	Negative

TABLE 7-1 Snyder Test Results and Interpretations.

Application The Snyder test is designed to measure susceptibility to dental caries (tooth decay), caused primarily by lactobacilli and oral streptococci.

In This Exercise...

You will check your own oral flora for its ability to produce dental caries.

Materials

Per Student Group

- hot water bath set at 45°C
- small sterile beakers
- sterile 1 mL pipettes with bulbs
- two Snyder agar tubes

Medium Recipe

Snyder Test Medium

• Pancreatic digest of casein	13.5 g
• Yeast extract	6.5 g
• Dextrose	20.0 g
• Sodium chloride	5.0 g
• Agar	16.0 g
• Bromcresol green	0.02 g
• Distilled or deionized water	1.0 L

pH 4.6–5.0 at 25°C

Procedure

Lab One

1. Collect a small sample of saliva (about 0.5 mL) in the sterile beaker.
2. Aseptically add 0.2 mL of the sample to a molten Snyder Agar tube (from the water bath), and roll it between your hands until the saliva is distributed uniformly throughout the agar.
3. Allow the agar to cool to room temperature. Do not slant.
4. Incubate with an uninoculated control at 35 ± 2°C for up to 72 hours.

Lab Two

1. Examine the tubes at 24-hour intervals for color changes.
2. Record your results in the chart provided on the Data Sheet.

Reference

Zimbro, Mary Jo, and David A. Power, editors. 2003. Page 517 in *Difco™ and BBL™ Manual—Manual of Microbiological Culture Media.* Becton Dickinson and Co., Sparks, MD.

Exercise 7-2

Lysozyme Assay

Theory Lysozyme is an enzyme that occurs naturally in egg albumin, and normal body secretions such as tears, saliva, and urine. The enzyme provides limited protection from bacterial infection by breaking bonds in the cell wall's peptidoglycan. **Peptidoglycan** is made up of the alternating repeating glycan subunits *N*-acetylglucosamine (NAG) and *N*-acetylmuramic acid (NAM), cross-linked by peptides. Lysozyme functions by breaking the β-1,4 glycosidic linkages between the NAG and NAM subunits of the glycan polymers.

A lysozyme assay measures the ability of a sample to lyse cells of the substrate organism *Micrococcus lysodeikticus*. Cell lysis occurs as a result of damage caused by the lysozyme and the hypotonic diluent used to dilute the samples.

In this exercise you will measure the lysozyme concentrations in a variety of body fluids. To do this you will first construct a standard curve (Figure 7-3) for comparison. First, you will mix diluted lysozyme samples of known concentration with an equal part of the bacterial substrate solution, then take an absorbance reading (with a spectrophotometer) of each after 20 minutes. You will then plot absorbance vs. lysozyme concentration to produce the standard curve. Dilutions of natural fluid samples can then be prepared, mixed with bacterial substrate solution and read for turbidity after 20 minutes. The light absorbance values of the samples can then be used to interpolate the lysozyme concentrations from the standard curve.

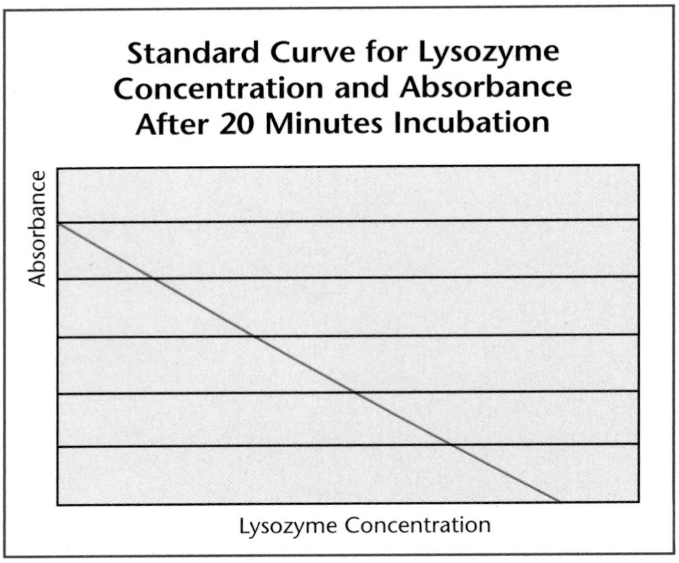

FIGURE 7-3 Lysozyme Standard Curve
This is a typical standard curve illustrating an inverse correlation between light absorbance and lysozyme concentration. As the concentration of lysozyme increases and lysis of substrate cells increases, the solution in the tube absorbs less light. Lysozyme concentration in a sample can be estimated by finding the lysozyme concentration that corresponds to the absorbance value of the sample.

Application This exercise is used to estimate the relative concentrations of lysozyme in body fluids.

In This Exercise...

The work for this exercise will be divided among several groups. One group will perform the dilutions of known lysozyme concentrations for construction of the standard curve. Other groups will dilute and test samples of tears, saliva, urine and/or other body fluids as determined by your instructor. The data then will be shared among lab groups to complete the Data Sheet.

Materials*

Per Student Group

- spectrophotometer with cuvettes
- timer
- micropipettes (100-1000 µL) with sterile tips
- 1 mL and 5 mL pipettes
- propipettes
- parafilm
- lysozyme buffer (20 mL for Group 1, 10 mL for other groups)
- lysozyme substrate—*Micrococcus lysodeikticus*—in solution measuring 10.0% transmittance (15 mL for Group 1, 5 mL for other groups)
- lysozyme dilutions shown in Table 7-2 (2.5 mL of each for Group 1)

*__Note to instructor:__ One liter of buffer is sufficient to make the lysozyme dilutions and substrate solution and for a class of 30 to 35 students.

TABLE OF RESULTS

Tube	Contains	IN
A	0.15625 mg Lysozyme	100 mL Buffer
B	0.3125 mg Lysozyme	100 mL Buffer
C	0.625 mg Lysozyme	100 mL Buffer
D	1.25 mg Lysozyme	100 mL Buffer
E	2.5 mg Lysozyme	100 mL Buffer
F	5.0 mg Lysozyme	100 mL Buffer

TABLE 7-2 Lysome Dilutions

Procedure

Group 1: Standard Curve

1. Turn on the spectrophotometer and allow it to warm up for a few minutes. Set the wavelength to 540 nm and, if it is digital, set it to absorbance. Be sure the filter is set properly for the wavelength used. (For directions on Spectrophotometer use, see Appendix E.)
2. Obtain seven cuvettes. Transfer 2.0 mL of lysozyme substrate (*M. lysodeikticus*) solution to each of six cuvettes; transfer 4.0 mL straight lysozyme buffer to the seventh.
3. Blank the spectrophotometer with the cuvette containing the straight buffer. Continue to check the setting throughout the procedure and blank the machine as needed.
4. Transfer 2.0 mL of the 0.15625 mg/100 mL lysozyme dilution to a substrate-containing cuvette. Mark the time in the Standard Curve table on the Data Sheet. This is T_0 for this concentration. (***Note:*** Time is critical for this part of the exercise. The absorbance of each lysozyme/substrate mixture must be read at ***exactly 20 minutes.*** Therefore, begin timing each transfer at the moment it is done.)
5. Continue transferring 2.0 mL each of the lysozyme dilutions to substrate-containing cuvettes, respectively. Mark each time on the Data Sheet.
6. Take the absorbance readings at T_{20} for each mixture. Record your results in the chart on the Data Sheet.
7. Share these data with the class.

Other Groups: Determination of Lysozyme in Body Fluid Samples

1. Turn on the spectrophotometer and allow it to warm up for a few minutes. Set the wavelength to 540 nm and, if it is digital, set it to absorbance. Be sure the filter is set properly for the wavelength used. (For directions on spectrophotometer use, see Appendix E.)
2. Obtain four cuvettes and all other necessary material for your body fluid sample.
3. Collect your assigned body fluid according to your teacher's instructions.
4. Add 0.2 mL (200 µL) undiluted sample to a cuvette. Add 3.8 mL lysozyme buffer. Mix well. Use this mixture to blank the spectrophotometer. Check this setting frequently throughout the exercise, and adjust as necessary.
5. Add 0.2 mL (200 µL) undiluted sample to a cuvette. Add 1.8 mL lysozyme buffer. Mix. This is a 10^{-1} dilution.
6. Transfer 0.2 mL from the 10^{-1} dilution to the third cuvette. Add 1.8 mL lysozyme buffer. Mix. This is 10^{-2}.
7. Add 1.8 mL lysozyme substrate to the 10^{-1} cuvette and mix well. Mark the time on the Data Sheet. This is T_0 for this mixture. (***Note:*** Time is critical for this part of the exercise. The absorbance of each body fluid/substrate mixture must be read at *exactly 20 minutes*. Therefore, begin timing each transfer at the moment it is done.)
7. Add 2.0 mL lysozyme substrate to the 10^{-1} cuvette. Mix well. Mark the time on the Data Sheet. This is T_0 for this mixture.
8. Take the absorbance readings at T_{20} for each mixture. Record your results in the chart on the Data Sheet.
10. Share these data with the class.

References

DIFCO Laboratories. 1984. Page 515 in *DIFCO Manual,* 10th ed. DIFCO Laboratories, Detroit.

Sprott, G. Dennis, Susan F. Koval, and Carl A. Schnaitman. 1994. Page 78 in *Methods for General and Molecular Bacteriology*, edited by Philipp Gerhardt, R. G. E. Murray, Willis A. Wood, and Noel R. Krieg, American Society for Microbiology, Washington, DC.

Exercise 7-3

Antimicrobial Susceptibility Test (Kirby-Bauer Method)

Theory

Antibiotics are natural antimicrobial agents produced by microorganisms. One type of penicillin, for example, is produced by the mold *Penicillium notatum*. Today, because many agents used to treat infections are synthetic, the terms **antimicrobials** or **antimicrobics** are applied to all substances used for this purpose.

The Kirby-Bauer test, also called the disk diffusion test, is a valuable standard tool for measuring the effectiveness of antimicrobics against pathogenic microorganisms. In the test, antimicrobic-impregnated paper disks are placed on a plate inoculated to form a bacterial lawn. The plates are incubated to allow growth of the bacteria and time for the agent to diffuse into the agar. As the substance moves through the agar, it establishes a concentration gradient. If the organism is susceptible to the agent, a clear zone will appear around the disk where growth has been inhibited (Figure 7-4). The size of this **zone of inhibition** depends upon the sensitivity of the bacteria to the specific antimicrobial agent and the point at which the chemical's **minimum inhibitory concentration** (MIC) is reached.

All aspects of the Kirby-Bauer procedure are standardized to ensure reliable results. Therefore, care must be taken to adhere to these standards. Mueller-Hinton agar, which has a pH between 7.2 and 7.4, is poured to a depth of 4 mm in either 150 mm or 100 mm Petri dishes. The depth is important because of the effect it has upon the diffusion. Thick agar slows lateral diffusion and thus produces smaller zones than plates held to the 4 mm standard. Inoculation is made with a broth culture diluted to match a 0.5 McFarland turbidity standard (Figure 7-5).

The disks, which contain a specified amount of the antimicrobial agent (printed on the disk) are dispensed onto the inoculated plate and incubated at $35 \pm 2°C$ (Figure 7-6). After 16 to 18 hours incubation, the plates are removed and the clear zones measured (Figure 7-7).

Application

Antimicrobial susceptibility testing is a standardized testing method used to measure the effectiveness of antibiotics and other chemotherapeutic agents on pathogenic microorganisms. In many cases, it is an essential tool in prescribing appropriate treatment.

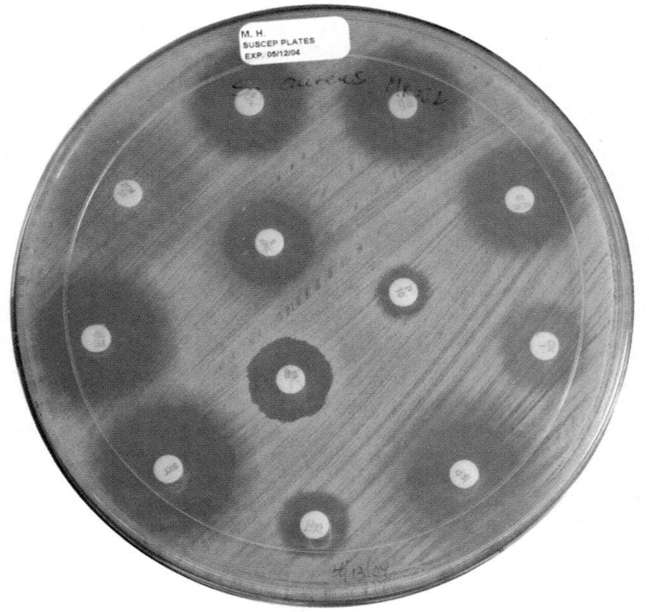

FIGURE 7-4 DISK DIFFUSION TEST OF METHICILLIN-RESISTANT *STAPHYLOCOCCUS AUREUS* (MRSA)
This plate illustrates the effect of (clockwise from top outer right) Nitrofurantoin (F/M300), Norfloxacin (NOR 10), Oxacillin (OX 1), Sulfisoxazole (G .25), Ticarcillin (TIC 75), Trimethoprim-Sulfamethoxazole (SXT), Tetracycline (TE 30), Ceftizoxime (ZOX 30), Ciprofloxacin (CIP 5), and (inner circle from right) Penicillin (P 10), Vancomycin (VA 30), and Trimethoprim (TMP 5) on Methicillin-resistant *Staphylococcus aureus*.

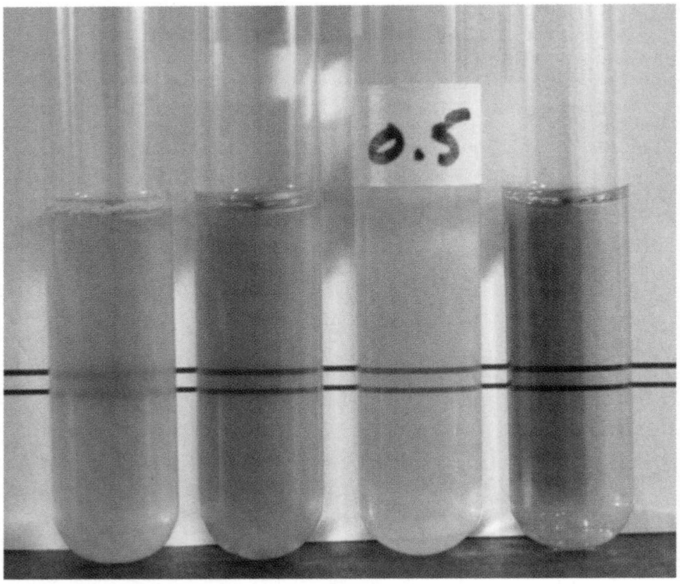

FIGURE 7-5 MCFARLAND STANDARDS
This is a comparison of a McFarland turbidity standard to three broths with varying degrees of turbidity. There are 11 McFarland standards (0.5 to 10), each of which contains a specific percentage of precipitated barium sulfate to produce turbidity. In the Kirby-Bauer procedure, the test culture is diluted to match the 0.5 McFarland standard (roughly equivalent to 1.5×10^8 cells per mL) before inoculating the plate. Comparison is made visually by placing a card containing sharp black lines behind the tubes.

FIGURE 7-6 Disk Dispenser
This antibiotic disk dispenser is used to uniformly deposit disks on a Mueller-Hinton agar plate.

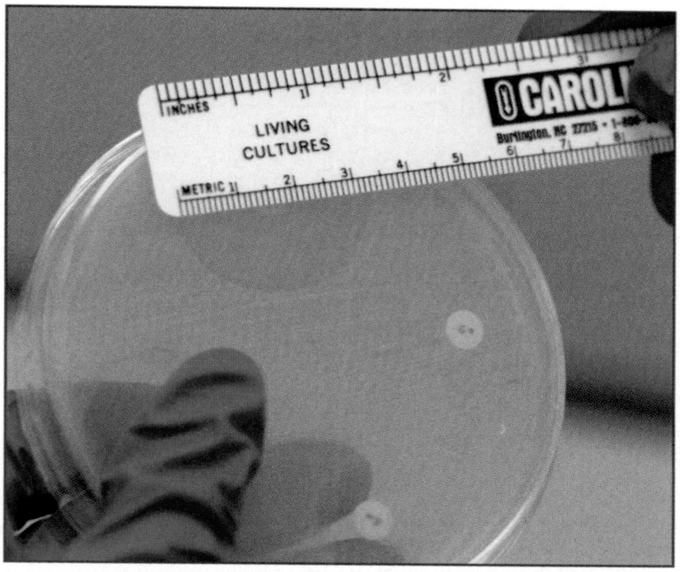

FIGURE 7-7 Measuring the Antimicrobial Susceptibility Zones
Using a metric ruler, measure the diameter of each clearing.

In This Exercise...

Today, you will test the susceptibility of *Escherichia coli* and *Staphylococcus aureus* strains to penicillin, streptomycin, tetracycline, and chloramphenicol.

Materials

Per Student Pair

- two Mueller-Hinton agar plates
- streptomycin, tetracycline, penicillin, and chloramphenicol antibiotic disks
- antibiotic disk dispenser or forceps for placement of disks
- small beaker of alcohol (for sterilizing forceps)
- sterile cotton swabs
- one metric ruler
- sterile saline (0.85%)
- sterile transfer pipettes
- one McFarland 0.5 standard with card
- fresh broth cultures of:
  - *Escherichia coli*
  - *Staphylococcus aureus*

Medium Recipe

Mueller-Hinton II Agar

• Beef extract	2.0 g
• Acid hydrolysate of casein	17.5 g
• Starch	1.5 g
• Agar	17.0 g
• Distilled or deionized water	1.0 L

pH = 7.2–7.4 at 25°C

Procedure

Lab One

1. Gently mix a broth culture and the McFarland standard until they reach their maximum turbidity.
2. Holding the culture and McFarland standard upright in front of you, place the card behind them so you can see the black line through the liquid in the tubes. As you can see in Figure 7-5, the line becomes distorted by the turbidity in the tubes. Use the black line to compare the turbidity level of the two tubes. Dilute the broth with the sterile saline until it appears to have the same level of turbidity as the standard.
3. Repeat this process with the other broth.
4. Dip a sterile swab into the *E. coli* broth and wipe off the excess on the inside of the tube.
5. Inoculate a Mueller-Hinton plate with *E. coli* by streaking the entire surface of the agar three times with the swab. (***Note:*** Rotate the plate 1/3 turn between streaks.)
6. Using a fresh sterile swab, inoculate the other plate with *S. aureus*.
7. Label the plates with the names of the organisms, your name, and the date.
8. Apply the streptomycin, tetracycline, penicillin, and chloramphenicol disks to the agar surface of each plate. You can apply the disks either singly

using sterile forceps, or with a dispenser (Figure 7-6). Be sure to space the disks sufficiently (4 to 5 cm) to prevent overlapping zones of inhibition.

9. Press each disk gently with sterile forceps so it makes full contact with the agar surface.
10. Invert the plates and incubate them aerobically at 35 ± 2°C for 16 to 18 hours. Have a volunteer in the group remove and refrigerate the plates at the appropriate time.

Lab Two

1. Remove the plates from the incubator and measure the diameter of the inhibition zones (Figure 7-7). It may be helpful to use a colony counter when taking measurements.
2. Using Table 7-3 and those provided with your antibiotic disks, record your results in the chart provided on the Data Sheet.

References

Collins, C. H., Patricia M. Lyne, and J. M. Grange, 1995. Page 128 in *Collins and Lyne's Microbiological Methods*, 7th ed. Butterworth-Heinemann, Oxford: United Kingdom.

Ferraro, Mary Jane, and James H. Jorgensen. 2003. Chapter 15 in *Manual of Clinical Microbiology*, 8th ed., edited by Patrick R. Murray, Ellen Jo Baron, James H. Jorgensen, Michael A. Pfaller, and Robert H. Yolken, ASM Press, Washington, DC.

Forbes, Betty A., Daniel F. Sahm, and Alice S. Weissfeld. 2002. Pages 236–240 in *Bailey and Scott's Diagnostic Microbiology*, 11th ed. Mosby-Yearbook, St. Louis.

Koneman, Elmer W., Stephen D. Allen, William M. Janda, Paul C. Schreckenberger, and Washington C. Winn, Jr. 1997. Page 818–822 in *Color Atlas and Textbook of Diagnostic Microbiology*, 5th ed. J. B. Lippincott Company, Philadelphia.

Zimbro, Mary Jo, and David A. Power, editors. 2003. Page 376 in *Difco™ and BBL™ Manual—Manual of Microbiological Culture Media*. Becton Dickinson and Co., Sparks, MD.

Antimicrobial Agent	Code	Disk Potency	Zone Diameter Interpretive Standards (mm)		
			Resistant	Intermediate	Susceptible
Chloramphenicol (for non-*Haemophilus* species	C–30	30µg	≤12	13–17	≥18
Penicillin (for staphylococci)	P–10	10 U	≤28		≥29
Streptomycin	S–10	10 µg	≤11	12–14	≥15
Tetracycline (for most organisms)	TE–30	30 µg	≤14	15–18	≥19

Note: This zone diameter chart contains selected antibiotics on data provided by the National Committee for Clinical Laboratory Standards (NCCLS). Permission to use portions of M100-S5 (Performance Standards for Antimicrobial Susceptibility Testing; Fifth Informational Supplement) has been granted by NCCLS. The interpretive data are valid only if the methodology in M2-A5 (Performance Standards for Antimicrobial Disk Susceptibility Tests, Fifth Edition; Approved Standard) is followed. NCCLS frequently updates the interpretive tables through new editions of the standard and supplements. Users should refer to the most recent editions. The current standard may be obtained from NCCLS, 940 West Valley Road, Suite 1400, Wayne, PA 19087.

TABLE 7-3 ZONE DIAMETER INTERPRETIVE CHART

Exercise 7-4

MMWR Report

Theory

Epidemiology is the study of the causes, occurrence, transmission, distribution, and prevention of diseases in a population. The Centers for Disease Control and Prevention (CDC) in Atlanta, GA, is the national clearinghouse for epidemiological data. The CDC receives reports related to the occurrence of 58 notifiable diseases (Table 7-4) from the United States and its territories, and compiles the data into tabular form, available in the publication *Morbidity and Mortality Weekly Report* (MMWR).

Two important disease measures that **epidemiologists** collect are **morbidity** (sickness) and **mortality** (death). Morbidity relative to a specific disease is the number of susceptible people who have the disease within a defined population during a specific time period. It usually is expressed as a rate. Because population size fluctuates constantly, it is conventional to use the population size at the midpoint of the study period. Also, the usually small decimal fraction obtained for the rate is multiplied by some power of 10 ("K") to achieve a value that is a whole number. Thus, a morbidity rate of 0.00002 is multiplied by 100,000 (10^5) so it can be reported as 2 cases per 100,000 people rather than 0.00002 cases per person. Morbidity rate is calculated using the following equation:

$$\text{Morbidity Rate} = \frac{\text{number of existing cases in a time period}}{\text{size of at-risk population at midpoint of time period}} \times K$$

Mortality, also expressed as a rate, is the number of people who die from a specific disease out of the total population afflicted with that disease in a specified time period. It, too, is multiplied by a factor "K" so the rate can be reported as a whole number of cases. The equation is:

$$\text{Mortality Rate} = \frac{\text{number of deaths due to a disease in a time period}}{\text{number of people with that disease in the time period}} \times K$$

Minimally, an epidemiological study evaluates morbidity or mortality data in terms of **person** (age, sex, race, *etc.*), **place**, and **time**. Sophisticated analyses require training in biostatistics, but the simple epidemiological calculation you will be doing can be performed with little mathematical background. You will calculate **incidence rate,** which is the occurrence of new cases of a disease within a defined population during a specific period of time. As before, "K" is some power of 10 so the rate can be reported as a whole number of cases.

$$\text{Incidence Rate} = \frac{\text{number of new cases in a time period}}{\text{size of at-risk populations at midpoint of time period}} \times K$$

Because our focus is microbiology, we will deal only with **infectious diseases**—those caused by biological agents such as bacteria and viruses. **Noninfectious diseases**, such as stroke, heart disease, and emphysema, also are studied by epidemiologists but are not within the scope of microbiology.

Application

An understanding of the causes and distribution of diseases in a population is useful to health-care providers in a couple of ways. First, awareness of what diseases are prevalent during a certain period of time aids in diagnosis. Second, an understanding of the disease and its cause(s) can be useful in implementing strategies for preventing it.

In This Exercise…

First you will choose a disease from the list of notifiable diseases in Table 7-4. Then you will collect data for the United States over the past two years. Once you have collected data, you will construct graphs illustrating the cumulative totals and incidence values for each year.

Materials

- a computer with Internet access or access to copies of *Morbidity and Mortality Weekly Report.*

Procedure

1. Go to the CDC Web site http://www.cdc.gov. Then follow these links. (***Note:*** Web sites often are revised, so if the site doesn't match this description exactly, it still is probably close. Improvise, and you'll find what you need. These links are current as of December 2005):
   - Click on MMWR under "Publications and Products" in the menu bar at the left of the window.
   - Click on "State Health Statistics" in the menu bar at the left. This will drop down a new menu.
   - Click on "Morbidity Tables."
   - Examine the morbidity tables by choosing the appropriate year and week. Table II is divided into nine parts.
   - Choose a notifiable disease in Table II, and record, on the Data Sheet, the cumulative number of cases

each week for the last two complete years. Each Table displays two consecutive years. It is okay to take both years from the same table. (***Note:*** The numbers in Table II at the CDC site are subject to change because of revisions and late reporting by the states. For more information, follow the link "About These Tables" at the bottom of the page.)

2. Answer the questions and complete the activities on the Data Sheet.

Acquired Immunodeficiency Syndrome (AIDS)	Meningococcal Disease, All Serogroups
Chlamydia	Meningococcal Disease, Serogroup A, C, Y, and W-135
Coccidioidomycosis	Meningococcal Disease, Serogroup B
Cryptosporidiosis	Meningococcal Disease, Other Serogroup
Escherichia coli, Enterohemorrhagic O157:H7	Meningococcal Disease, Serogroup Unknown
Escherichia coli, Enterohemorrhagic, Shiga Toxin Positive, Non-O157	Pertussis
Escherichia coli, Enterohemorrhagic, Shiga Toxin Positive, Not Serogrouped	Rabies, Animal
Giardiasis	Rocky Mountain Spotted Fever
Gonorrhea	Salmonellosis
Haemophilus influenzae, Invasive Disease, All Ages	Shigellosis
Haemophilus influenzae, Invasive Disease, Less Than 5 Years, Serotype B	Streptococcal Disease, Invasive, Group A
Haemophilus influenzae, Invasive Disease, Less Than 5 Years, Non-Serotype B	*Streptococcus pneumoniae*, Invasive, Drug-resistant, All Ages
Haemophilus influenzae, Invasive Disease, Less Than 5 Years, Unknown Serotype	*Streptococcus pneumoniae*, Invasive, <5 years
Hepatitis A, Viral, Acute	Syphilis, Primary and Secondary
Hepatitis B, Viral, Acute	Syphilis, Congenital
Hepatitis C, Viral, Acute	Tuberculosis
Legionellosis	Typhoid fever
Listeriosis	Varicella (Chickenpox)
Lyme Disease	West Nile Virus, Neuroinvasive
Malaria	West Nile Virus, Non-neuroinvasive

TABLE 7-4 40 Notifiable Diseases in the United States and Its Territories Posted on the CDC Web Site (as of December 2005)

Exercise 7-5

Epidemic Simulation

Theory

As mentioned in Exercise 7-4, epidemiology is the study of the causes, occurrence, transmission, distribution, and prevention of diseases in a population. Infectious diseases are transmitted by ingestion, inhalation, direct skin contact, open wounds or lesions in the skin, animal bites, direct blood-to-blood contact as in blood transfusions, and sexual contact. Infectious diseases can be transmitted by way of sick people or animals, healthy people or animals carrying the infectious organism or virus, water contaminated with human or animal feces, contaminated objects (**fomites**), aerosols, or biting insects (**vectors**).

When a disease is transmitted from an area such as the heating or cooling system of a building or from contaminated water that infects many people at once, it is called a **common source epidemic. Propagated transmission** is a disease transmitted from person to person. The first case of such a disease is called the **index case.** Determining the index case is the object of today's lab exercise.

A second objective of today's lab is to gather data to be used in performing simple epidemiological calculations. You will be calculating **incidence rate** and **prevalence** rate for the "disease" spreading through your class.

Incidence rate is the number of new cases of a disease reported in a defined population during a specific period of time. Incidence rate is calculated by the following equation:

$$\text{Incidence rate} = \frac{\text{number of new cases in a time period}}{\text{size of at-risk population at midpoint of time period}} \times K$$

Because population size fluctuates constantly, it is conventional to use the population size at the midpoint of the study period. Also, the usually small decimal fraction obtained for the rate is multiplied by some power of 10 (K in the equation) to get its value up to a number bigger than one. That is, an incidence rate of 0.00002 is multiplied by 100,000 (10^5) so it can be reported as two cases per 100,000 people, rather than 0.00002 cases per person.

You also will be calculating one form of prevalence rate called **point prevalence**, the number of cases of a disease at a specific point in time in a defined population. It is calculated by the following equation:

$$\text{Point prevalence} = \frac{\text{number of existing cases at a point in time}}{\text{total population}} \times K$$

As with incidence rate, the calculated prevalence is multiplied by some power of 10 (K in the equation) to bring the number up to one or more.

Application

Epidemiologists have the task of identifying the source of a disease and establishing the mode of its transmission. Further, epidemiologists characterize diseases quantitatively, using measures such as incidence and prevalence.

In This Exercise...

One of you will become the "index case" in a simulation of an epidemic, and many of you who contact this person directly and indirectly will become "infected." Your job, besides having a little fun, will be to collect the data and determine which one of you is the index case.

Materials

Per Student

- examination glove (latex or synthetic)
- numbered Petri dish containing a piece of candy (one will be contaminated with *Serratia marcescens* or other microbe as chosen by your instructor; the others will be moistened with water)[1]
- Nutrient Agar plate
- two sterile cotton applicators and sterile saline
- marking pen

Procedure

Lab One

1. Obtain a Petri dish containing candy, sterile cotton applicators, and a Nutrient Agar plate. Record the number of your candy dish here. ______________
2. Mark a line on the bottom of the agar plate to divide it into two halves. Label the sides "A" and "B." Also record the number of your candy dish on your plate.
3. Open a package of sterile, cotton-tipped applicators so they can be easily removed.
4. Put a glove on your nondominant hand.
5. Using your ungloved hand, remove one cotton applicator, moisten it in the sterile saline, and sample the palm of the gloved hand. Be sure not

[1] Petri dishes should be numbered according to the number of students in the class. There should be no gaps in the sequence.

to touch anything else with the palm of the glove. When finished, inoculate side A of the plate with the cotton applicator. Discard the applicator in an appropriate autoclave container.

6. Using the gloved hand, pick up the candy and smear it around until a good amount of it is on the glove. Drop the candy back into the Petri dish and close it.
7. Using gloved hands only, student #1 rubs hands with student #2, making sure to transfer anything that may be on the palm of the glove to the palm of the neighbor's glove. Then #2 rubs gloved hands with #3, and #3 and #4, and so on until all students have contacted gloved hands. Be careful not to snap the gloves when you separate, as this will produce aerosols. *Gently slide your hands apart.*
8. When finished, #1 rub hands with the last person in line (highest number).
9. With a second sterile applicator, sample the palm of your gloved hand as before and transfer it to side B of the plate. Be sure to sample the area of the glove that contacted other students' gloves.
10. Discard all materials in an appropriate container(s).
11. Invert the plate and incubate it at 25°C for 24 to 48 hours.
12. Wash your hands.

Lab Two

1. Remove the plates from the incubator, and examine them for characteristic reddish growth of *Serratia marcescens*.
2. Enter your results on the Data Sheet and follow the instructions for establishing the index case, incidence rate, and point prevalence.

Identification Flowcharts

The following two exercises use identification flowcharts to be used with two medically important groups of bacteria: the *Enterobacteriaceae* and the Gram-positive cocci. Although multitest systems (such as the API 20E in Exercise 5-29) are used in clinical settings, it still is useful to see how the results from a few crucial tests can lead to identification of an isolate.

No identification scheme is perfect. Strain variability, limitations of sensitivity and/or specificity of the tests used, the media used for isolation and culture maintenance, and human fallibility combine to make identification more of a probability statement than an absolute.

Standard references list test results as positive or negative when 90% or more of the strains tested give that result. The possibility always exists for an isolate to be in the less than 10% minority that gives an atypical result. Because of this, identification often involves comparing a battery of test results to published results and seeing which organism matches the isolate most closely. Where possible, the flowcharts in this section include multiple differential test results to improve your chances of successful identification. Further, if you misidentify your unknown, your instructor will assist you in finding the source of error.

The charts are in standard flowchart form with dichotomous branching. Organisms with variable results for a specific test are shown on both branches. For easier interpretation, an icon accompanies the results for each test, to remind you of what the results look like. They are not intended to be used as a basis for color comparison. That is the job of your control! Keys to these icons is given in Tables 7-5 and 7-6.

Note: These tables contain organisms that are commonly isolated from medical specimens and can be used only for identification of these organisms. To eliminate the possibility of your testing and falsely identifying an organism that is not included in the table, your instructor will provide "unknown" organisms for you to identify.

Exercise 7-6

Identification of *Enterobacteriaceae*

Theory The *Enterobacteriaceae* is a family of Gram-negative rods that mostly inhabit the intestinal tract. Major characteristics defining the group are as follows:

- Growth on MacConkey agar
- Gram-negative rods
- Oxidase-negative
- Acid production from glucose (with or without gas)
- Most are catalase-positive
- Most reduce NO_3 to NO_2
- Peritrichous flagella, if motile

Many **enterics** are pathogens, but just as many are harmless gut **commensals,** or opportunists. The organisms for which the flowcharts work are listed in Figure 7-8, and are commonly associated with human infections. (***Note:*** Your instructor will choose enteric unknowns appropriate to your specific microbiology course and facilities.)

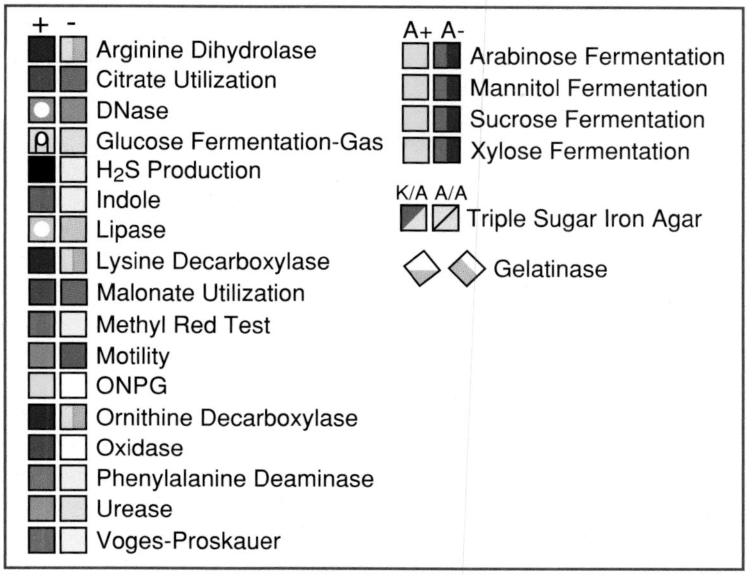

TABLE 7-5 Key to Icons Used in This Exercise
Colors are to remind you what results look like. They should not be used for comparison to your results.

Application Identification of enterics from human specimens requires coordinated and integrated use of biochemical tests and stains. Although several multitest systems are available that utilize computer databases, flowcharts are still a useful way to visualize the process of identification by elimination. It often is striking how few test results, when taken in combination, identify an organism.

In This Exercise...

You will be given an unknown enteric and run biochemical tests, as indicated by the flowcharts, to identify it over a span of several lab periods.

Materials

Per Class

- media (as listed in the charts) for biochemical testing
- pure cultures of organisms to be used for positive controls (appropriate to the course level)

Per Student

- one unknown organism[1]
- Gram Stain kit

Procedure

1. Obtain an unknown.
2. Streak the sample for isolation on two MacConkey agar plates. Incubate one at 25°C and the other at 35 ± 2°C for 24 hours. Record your isolation procedure on the Data Sheet.
3. After incubation, record colony morphology and optimum growth temperature on the Data Sheet. (If you fail to get isolation, ask your instructor if you should restreak for isolation or let the original plates continue to incubate.)
4. Perform a Gram stain and an oxidase test on an isolated colony. Record the results on the Data Sheet. If the isolate is an oxidase-negative, Gram-negative rod, it probably is a member of the *Enterobacteriaceae*. At the discretion of your instructor, further tests chosen from the characteristics above may be run to assure it is an organism from *Enterobacteriaceae*.
5. The next step is to perform the IMViC series of tests: Indole Production, Methyl Red, Voges-Proskauer, and Citrate Utilization. Record the results on the Data Sheet. Match your isolate's IMViC results with one shown in Figure 7-8 and proceed to the appropriate identification chart in Figures 7-9 through 7-17.
6. Follow the tests in the appropriate flowchart to identify your unknown. Do not run more than one test at a time unless your instructor tells you to do so. Where multiple tests are listed at a branch point, run only one. Record the results on the Data Sheet as you go. Keep accurate records of what you have

List of Commonly Isolated Enterobacteriaceae

Citrobacter amalonaticus
*Citrobacter freundii**
Citrobacter koseri
Edwardsiella tarda
Enterobacter aerogenes
Enterobacter cloacae
Enterobacter agglomerans group*
Escherichia coli
*Hafnia alvei**
Klebsiella oxytoca
*Klebsiella ozaenae**
Klebsiella pneumoniae
Morganella morganii
*Proteus mirabilis**
Proteus penneri
*Proteus vulgaris**
Providencia rettgeri
Providencia stuartii
Salmonella (most strains)
Serratia liquefaciens group
*Serratia marcescens**
Shigella dysenteriae (Group A)*
Shigella flexneri (Group B)*
Shigella boydii (Group C)*
Shigella sonnei (Group D)
*Yersinia enterocolitica**
*Yersinia frederiksenii**
Yersinia intermedia
Yersinia pseudotuberculosis

*Organism shows up in more than one figure due to variable results of one or more IMViC tests

Indole	MR	VP	Citrate	
-	+	+	+	Go to Figure 7-9
-	+	+	-	Go to Figure 7-10
-	+	-	-	Go to Figure 7-11
-	+	-	+	Go to Figure 7-12
-	-	+	+	Go to Figure 7-13
-	-	-	+/-	Go to Figure 7-14
-	-	+	-	Go to Figure 7-14
+	+	-	-	Go to Figure 7-15
+	+	-	+	Go to Figure 7-16
+	-	+/-	+/-	Go to Figure 7-17
+	+	+	+/-	Go to Figure 7-17

FIGURE 7-8 IMViC RESULTS OF *ENTEROBACTERIACEAE*
Match the IMViC results of your isolate with one of the eight shown, then proceed with the flowchart in the figure indicated. The organisms listed are commonly isolated *Enterobacteriaceae*. These are the only ones for which the identification flowcharts will work. Some organisms are listed in more than one flowchart because of variable results for a given test. A box divided diagonally means either result for that test is a match. (Results after Farmer *et al.*, 2003.)

[1] *Note:* This exercise can be combined with Exercise 7-7 by passing out a mixture of an enteric and a Gram-positive coccus as unknowns. The student must isolate the two organisms from a mixed culture, and then proceed to identify the two independently.

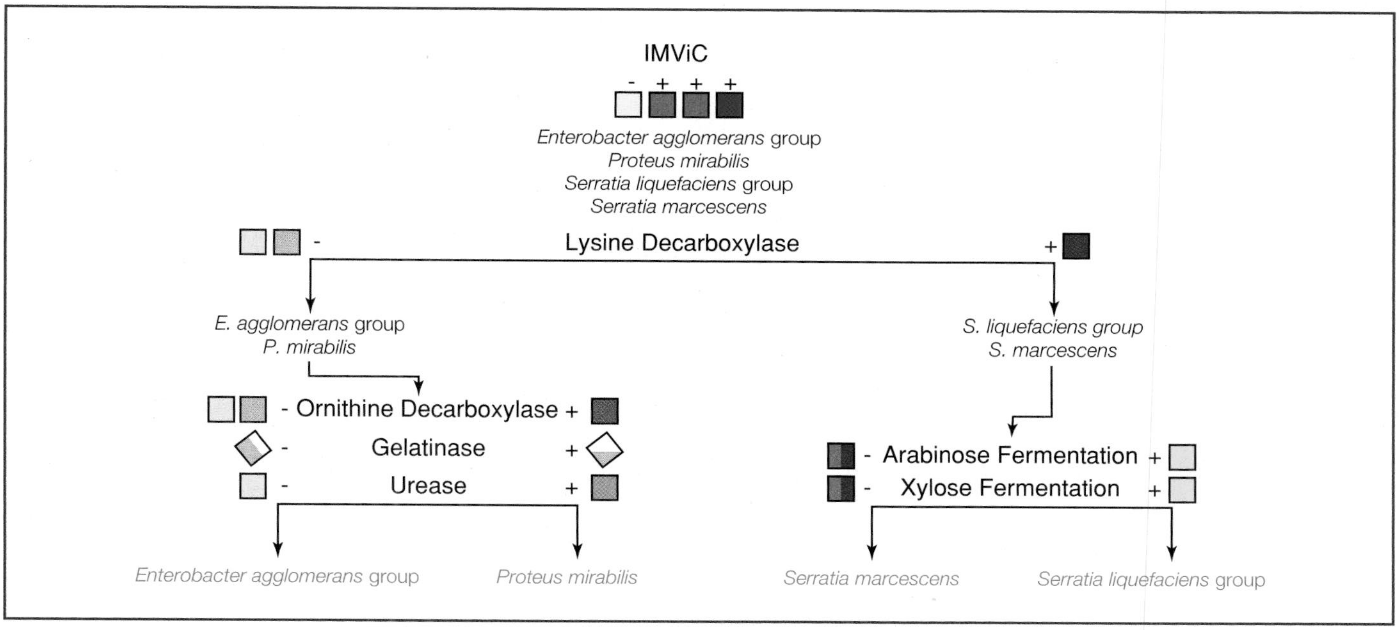

FIGURE 7-9 FLOWCHART FOR IDENTIFYING INDOLE-NEGATIVE, MR-POSITIVE, VP-POSITIVE, CITRATE-POSITIVE *ENTEROBACTERIACEAE*
Please see the appropriate exercises for instructions on how to run each test. Colors in the icons are not intended to match media exactly. Use controls for this. (Results after Farmer *et al.,* 2003.)

done. Do not enter all your data at the end of the project.

7. When you have identified your organism, write its name on the back of the Data Sheet and check with your instructor to see if you are correct. If you are, congratulations! If you aren't, your instructor will advise you as to which tests gave "incorrect" results by writing those test names in the "rerun" space next to your identification. These should be rerun with appropriate positive controls from your school's inventory of organisms. Record the test(s) and positive control(s) on your Data Sheet. Checking the results with positive controls and your unknown will indicate if the "incorrect" test result was truly incorrect or if the strain of organism your school is using doesn't match the majority of strains for that test result.

References

Farmer III, J. J. 2003. Chapter 41 in *Manual of Clinical Microbiology,* 8th ed., edited by Patrick R. Murray, Ellen Jo Baron, James H. Jorgensen, Michael A. Pfaller, and Robert H. Yolken. American Society for Microbiology Press, Washington, DC.

Forbes, Betty A., Daniel F. Sahm, and Alice S. Weissfield. 2002. Chapter 25 in *Bailey and Scott's Diagnostic Microbiology,* 11th ed. Mosby, St. Louis.

Koneman, Elmer W., Stephen D. Allen, William M. Janda, Paul C. Schreckberger, and Washington C. Winn, Jr. 1997. Chapter 4 in *Color Atlas and Textbook of Diagnostic Microbiology,* 5th ed., Lippincott-Raven Publishers, Philadelphia.

MacFaddin, Jean F. 2000. *Biochemical Tests for Identification of Medical Bacteria,* 3rd ed. Lippincott Williams & Wilkins, Philadelphia.

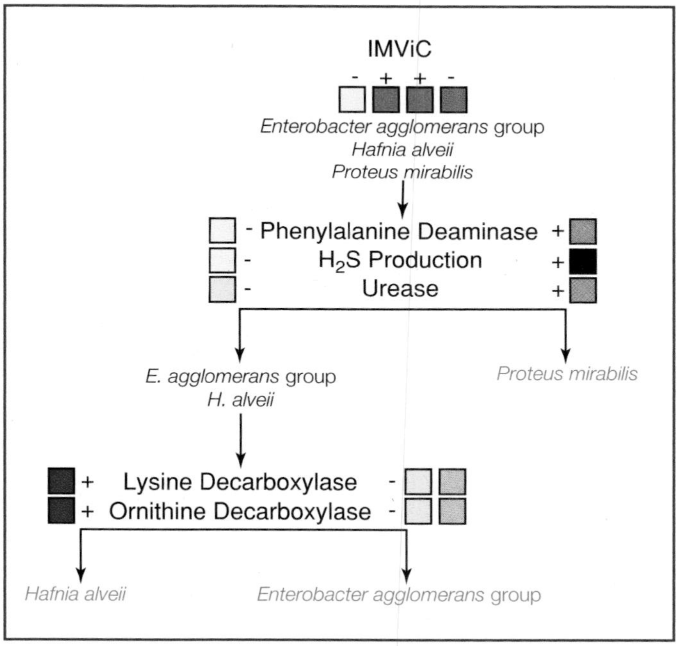

FIGURE 7-10 FLOWCHART FOR IDENTIFYING INDOLE-NEGATIVE, MR-POSITIVE, VP-POSITIVE, CITRATE-NEGATIVE *ENTEROBACTERIACEAE*
Please see the appropriate exercises for instructions on how to run each test. Colors in the icons are not intended to match media exactly. Use controls for this. (Results after Farmer *et al.,* 2003.)

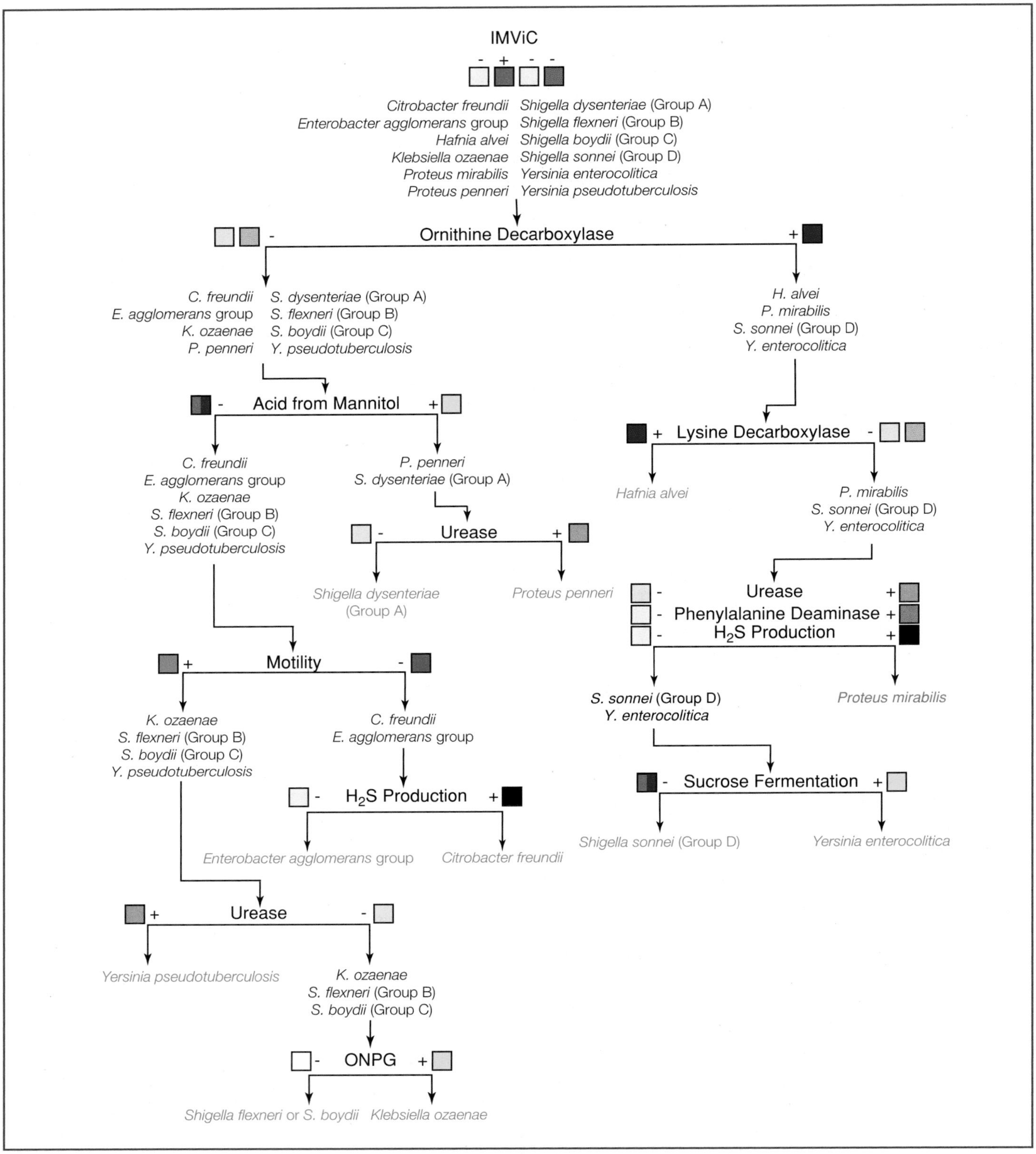

FIGURE 7-11 FLOWCHART FOR IDENTIFYING INDOLE-NEGATIVE, MR-POSITIVE, VP-NEGATIVE, CITRATE-NEGATIVE *ENTEROBACTERIACEAE*
Please see the appropriate exercises for instructions on how to run each test. Colors in the icons are not intended to match media exactly. Use controls for this. (Results after Farmer *et al.,* 2003.)

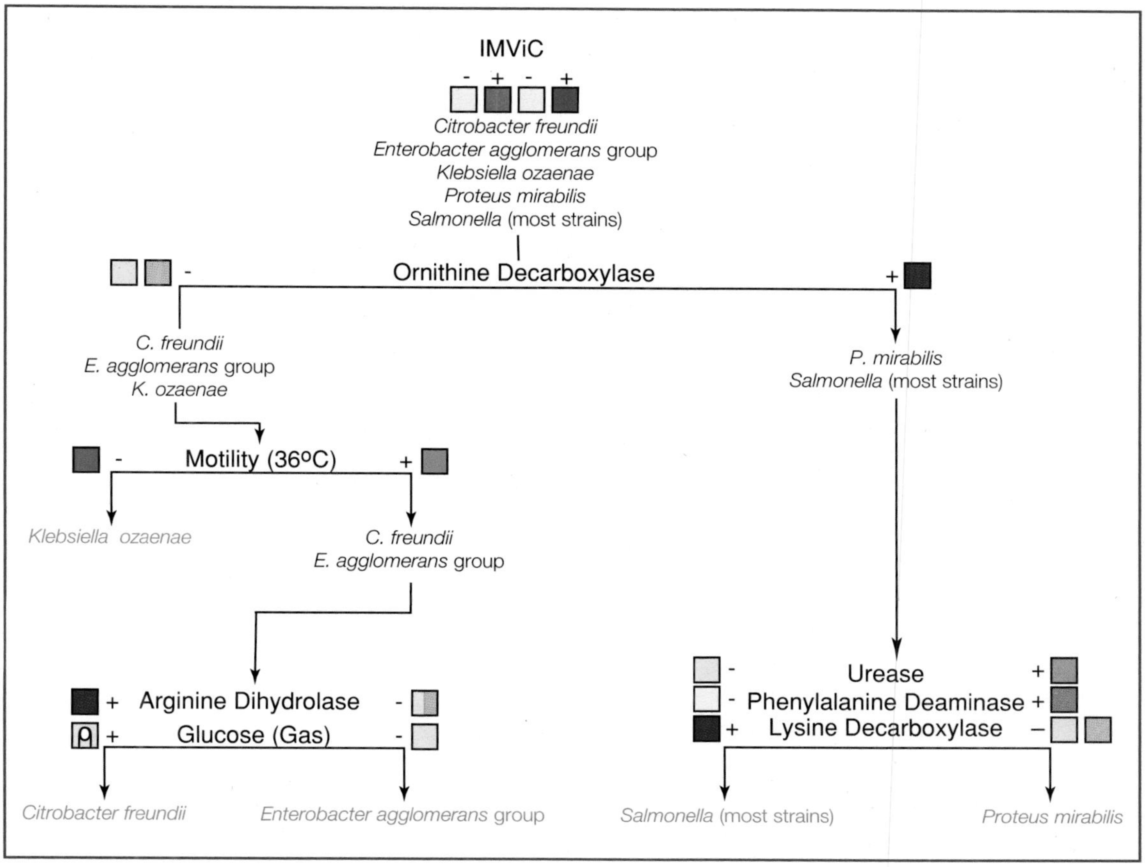

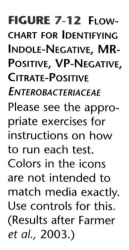

FIGURE 7-12 Flowchart for Identifying Indole-Negative, MR-Positive, VP-Negative, Citrate-Positive *Enterobacteriaceae* Please see the appropriate exercises for instructions on how to run each test. Colors in the icons are not intended to match media exactly. Use controls for this. (Results after Farmer *et al.*, 2003.)

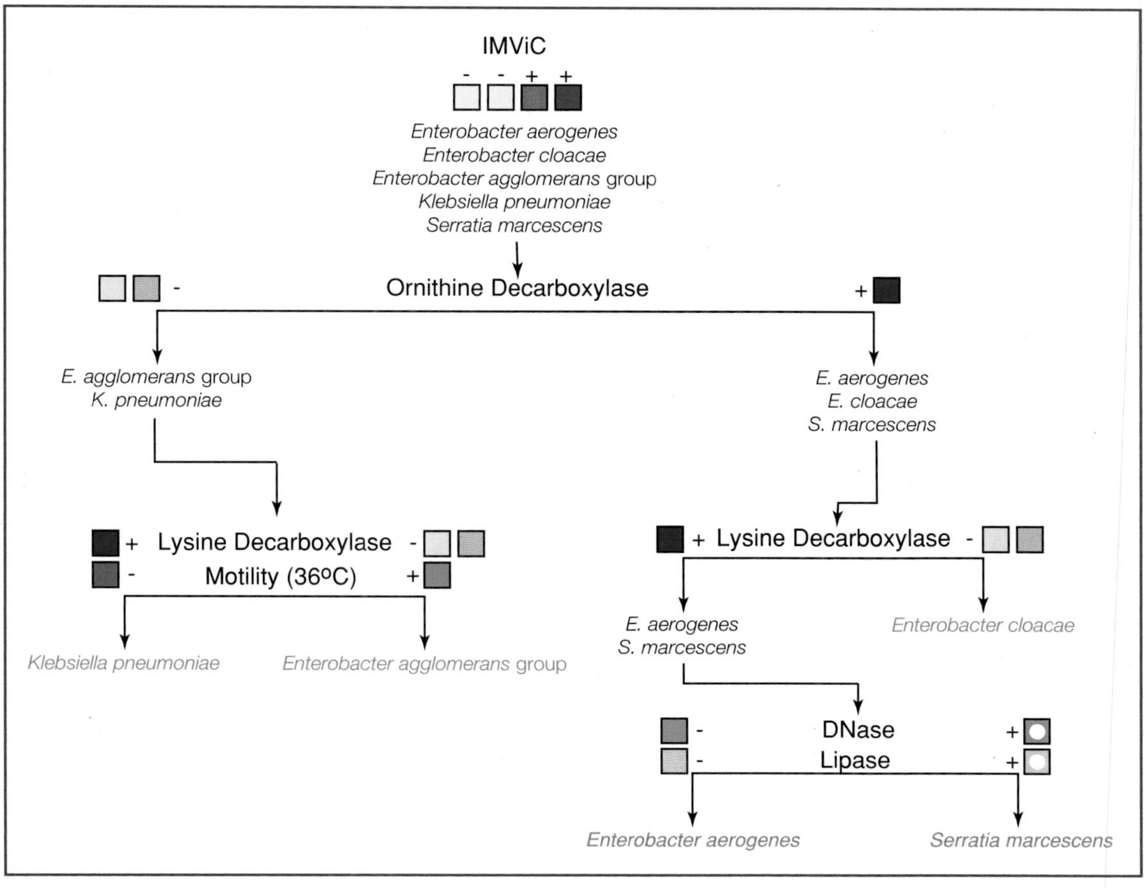

FIGURE 7-13 Flowchart for Identifying Indole-Negative, MR-Negative, VP-Positive, Citrate-Positive *Enterobacteriaceae* Please see the appropriate exercises for instructions on how to run each test. Colors in the icons are not intended to match media exactly. Use controls for this. (Results after Farmer *et al.*, 2003.)

FIGURE 7-14 Flowchart for Identifying Indole-Negative, MR-Negative, VP-Negative, Citrate-Positive or Citrate-Negative, and Indole-Negative, MR-Negative, VP-Positive, Citrate-Negative Enterobacteriaceae
Please see the appropriate exercises for instructions on how to run each test. Colors in the icons are not intended to match media exactly. Use controls for this. (Results after Farmer *et al.*, 2003.)

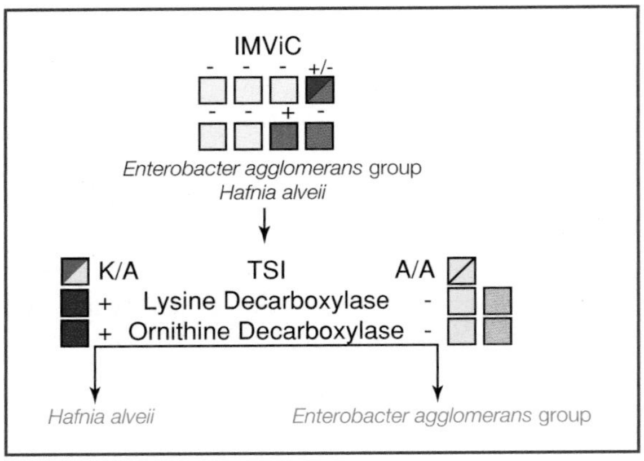

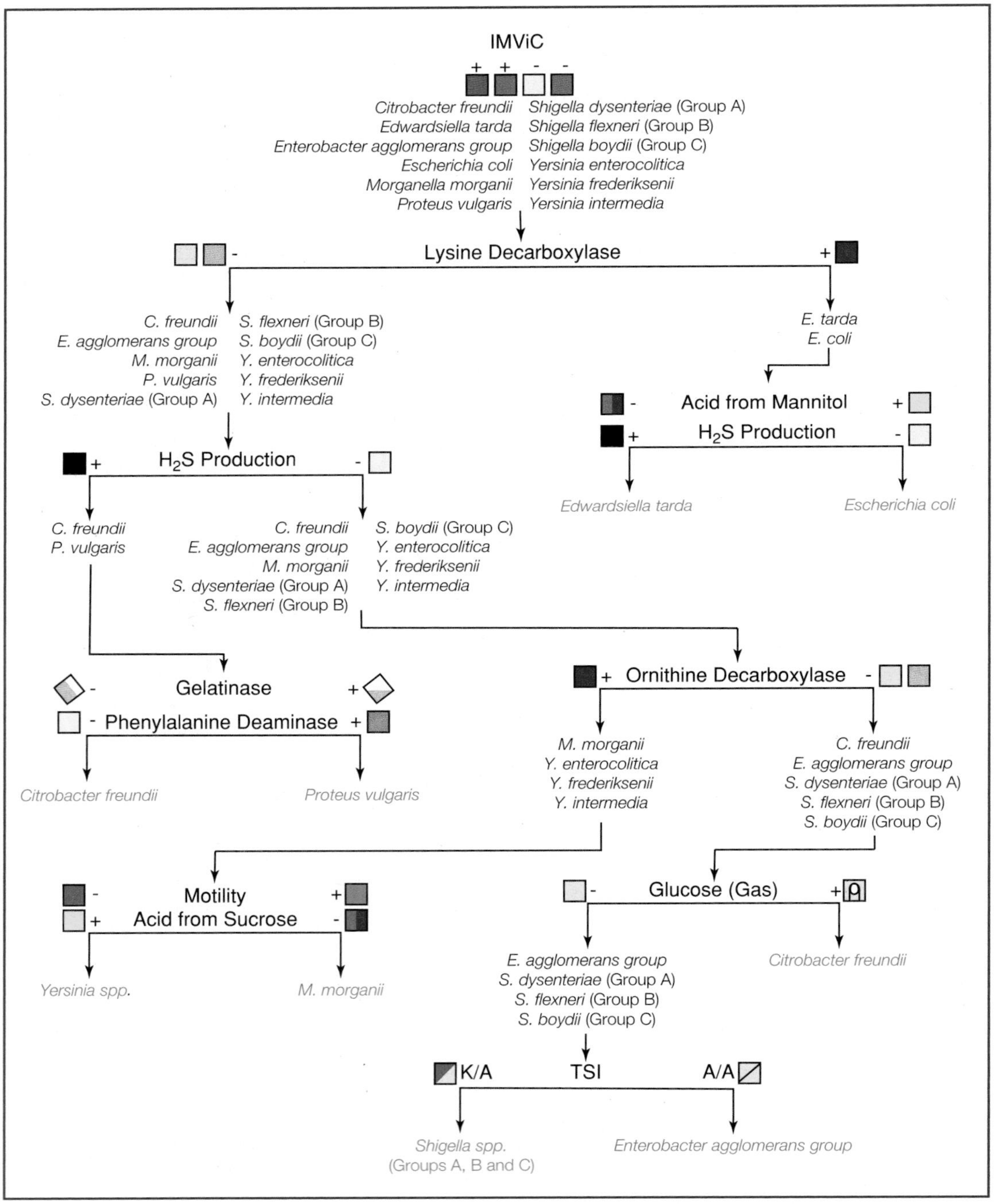

FIGURE 7-15 Flowchart for Identifying Indole-Positive, MR-Positive, VP-Negative, Citrate-Negative Enterobacteriaceae
Please see the appropriate exercises for instructions on how to run each test. Colors in the icons are not intended to match media exactly. Use controls for this. (Results after Farmer *et al.*, 2003.)

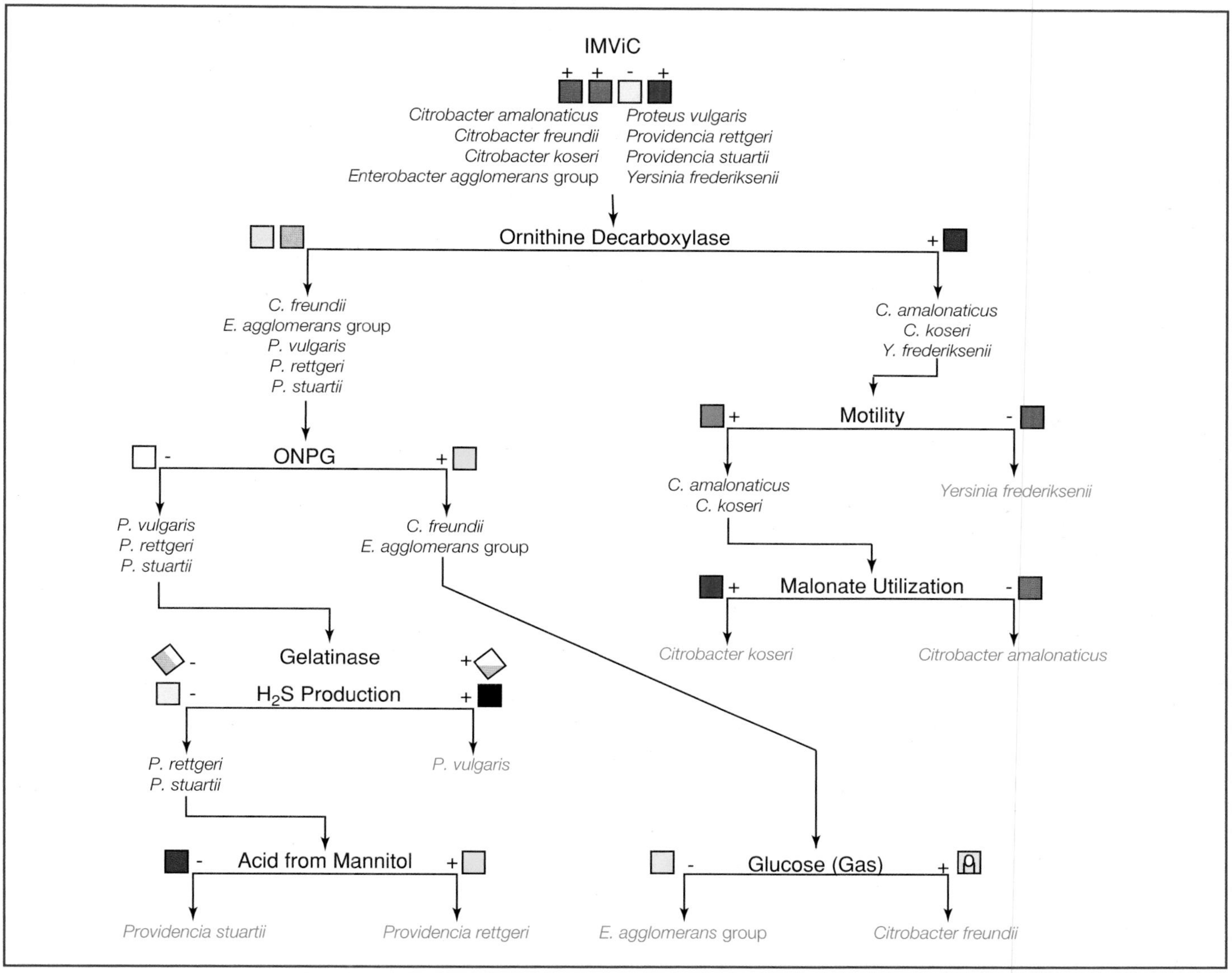

FIGURE 7-16 Flowchart for Identifying Indole-Positive, MR-Positive, VP-Negative, Citrate-Positive *Enterobacteriaceae*
Please see the appropriate exercises for instructions on how to run each test. Colors in the icons are not intended to match media exactly. Use controls for this. (Results after Farmer *et al.*, 2003.)

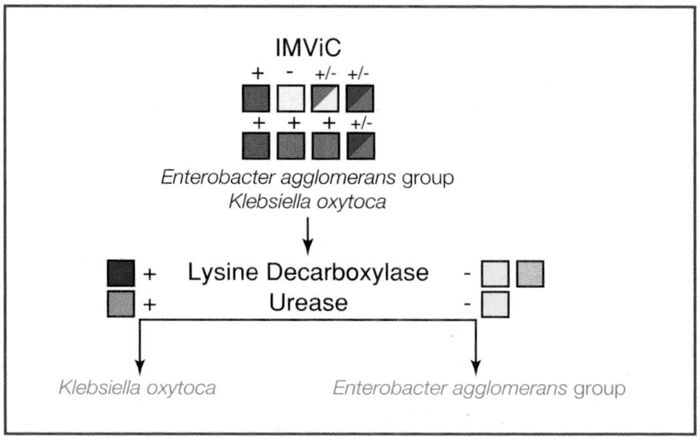

FIGURE 7-17 Flowchart for Identifying Indole-Positive, MR-Negative, VP-Positive or VP-Negative, Citrate-Positive or Citrate-Negative, and Indole-Positive, MR-Positive, VP-Positive, Citrate-Positive or Citrate-Negative *Enterobacteriaceae*
Please see the appropriate exercises for instructions on how to run each test. (Results after Farmer *et al.*, 2003.)

Exercise 7-7

Identification of Gram-positive Cocci

Theory

Gram-positive cocci are frequent isolates in a clinical setting. They are common inhabitants of skin and mucous membranes. Four main genera, briefly described below, will be used in this lab exercise.

1. *Enterococcus*
   - Gram-positive cocci to coccobacilli in singles or short chains
   - Catalase-negative
   - Facultatively anaerobic
   - Grow in 6.5% NaCl
   - Grow in Bile Esculin
   - Most are PYR-positive
2. *Micrococcus*
   - Gram-positive cocci in pairs, tetrads, or clusters
   - Catalase-positive
   - Strictly aerobic
   - Oxidase-positive
   - Do not produce acid from glucose
   - Bacitracin-sensitive
3. *Staphylococcus*
   - Gram-positive cocci in singles, pairs, tetrads, or clusters
   - Catalase-positive
   - Facultatively anaerobic
   - Oxidase-negative
   - Grow in 6.5% NaCl
   - Most produce acid from glucose
   - Most are resistant to bacitracin
4. *Streptococcus*
   - Gram-positive cocci to ovoid cocci in singles or short chains
   - Catalase-negative
   - Facultatively anaerobic
   - Nutritionally fastidious
   - Ferment glucose to acid, but not gas

One species of *Kocuria* is also included in the flowcharts. *Kocuria* species were once placed in *Micrococcus* but have been removed as a result of DNA and rRNA dissimilarities. The specific organisms are listed in Figures 7-18 and 7-19. Table 7-6 lists the tests to be used in their identification. (***Note:*** Your instructor will choose Gram-positive unknowns appropriate to your microbiology course and facilities.)

Application

Identification of Gram-positive cocci from human specimens requires coordinated and integrated use of biochemical tests and stains. Although several serological tests allow rapid identification, flowcharts are still a useful way to visualize the process by elimination.

In This Exercise…

You will be given an unknown Gram-positive coccus from the organisms listed in Figures 7-18 and 7-19. Then you will run biochemical tests as indicated by the flowcharts to identify it, spanning several lab periods.

Materials

Per Class

- media (as listed in Table 7-6) for biochemical testing
- pure cultures of organisms listed in Figures 7-18 and 7-19 to be used for positive controls (appropriate to the course level)
- Todd-Hewitt Broth or Brain-Heart Infusion Broth
- candle jar

Per Student

- Gram stain kit
- 3% H_2O_2
- One unknown organism[1]

[1] ***Note:*** This exercise can be combined with Exercise 7-6 by passing out a mixture of a Gram-negative enteric and a Gram-positive coccus as unknowns. The student must isolate the two organisms from a mixed culture, and then proceed to identify the two independently.

+ -		A+ A-	
	Arginine Dihydrolase		GlucoseFermentation
	Bile Esculin Test	S R	
	CAMP Test	>10 mm / <9 mm	Bacitracin Susceptibility
	NO_3 reduced to NO_2	>16 mm / ≤16 mm	Novobiocin Susceptibility
	Ornithine Decarboxylase	>14 mm / <14 mm	Optochin Susceptibility
	Oxidase	> 0 mm / = 0 mm	SXT Susceptibility
	PYR Test		
	Voges-Proskauer		
β α γ	Hemolysis	+ -	Coagulase

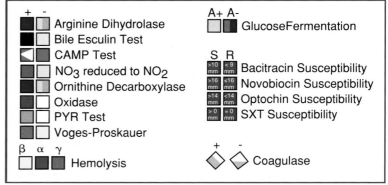

TABLE 7-6 Key to Icons Used in This Exercise
Colors are to remind you what results look like. They should not be used for comparison to your results.

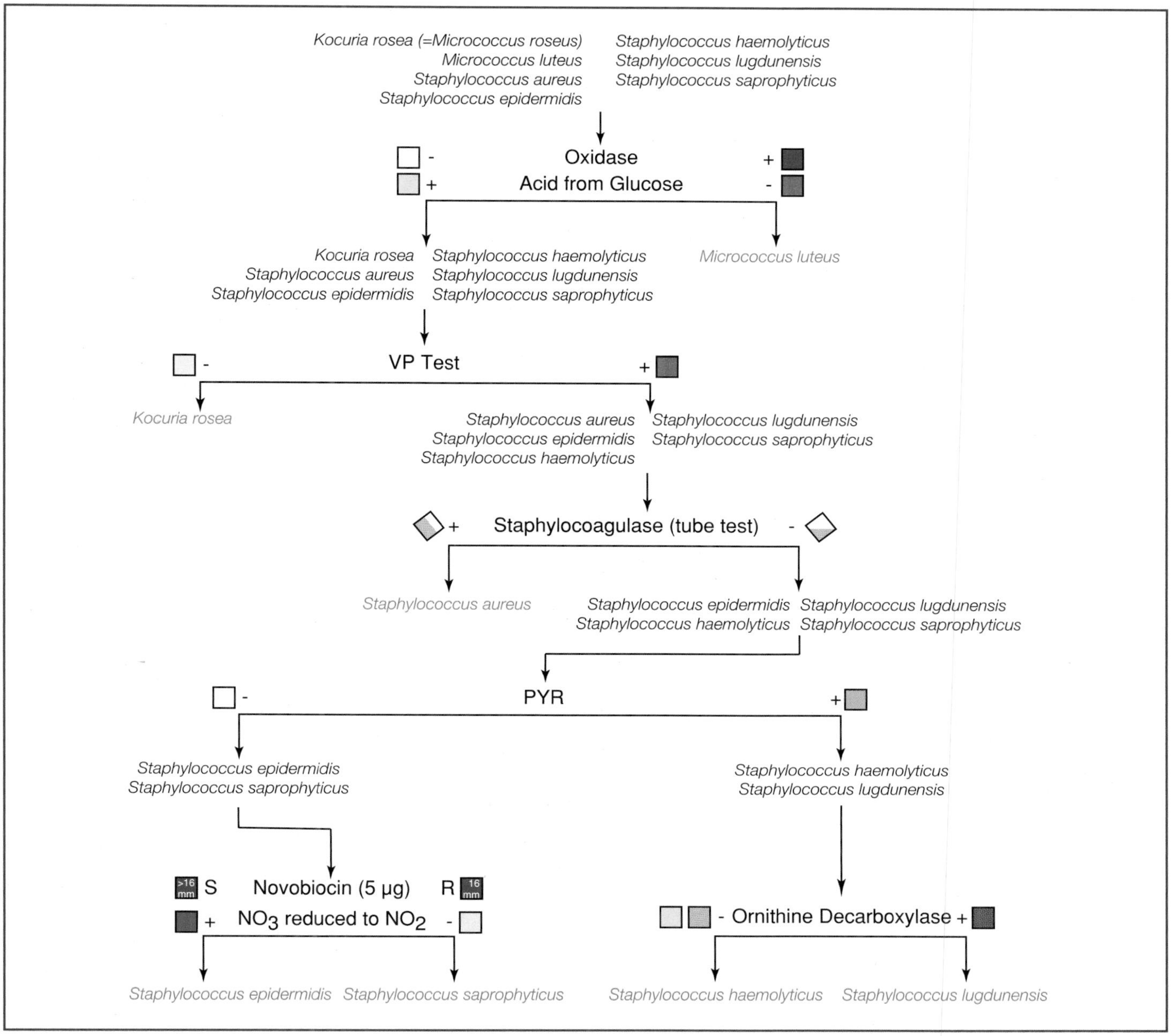

FIGURE 7-18 IDENTIFICATION FLOWCHART FOR GRAM-POSITIVE, CATALASE-POSITIVE COCCI IN TETRADS OR CLUSTERS
Please see the appropriate exercises for instructions on how to run each test.

Procedure

1. Obtain an unknown.
2. Streak for isolation on two Sheep Blood Agar or PEA plates. Incubate one at 25°C and the other at 35 ± 2°C. Record your isolation procedure on the Data Sheet.
3. After incubation, pick an isolated colony and perform a Gram stain. If the isolate is a Gram-positive coccus, run a catalase test. (***Note:*** Be sure to use growth from the top of the colony if isolation was performed on a sheep blood agar plate. This minimizes the possibility of a false positive from the catalase-positive erythrocytes in the medium.) Continue testing colonies until you find one that is a Gram-positive coccus.
4. Use the chart in Figure 7-18 for identification of catalase-positive cocci in tetrads or clusters. Use the chart in Figure 7-19 if the isolate is catalase-negative, with cocci in pairs or chains. If your isolate is catalase-negative, ask your instructor if you would get better growth of your pure culture using Todd-Hewitt Broth or Brain-Heart Infusion Broth and incubating in a candle jar or CO_2 incubator.
5. Follow the tests in the appropriate flowchart to identify your unknown. Do not run more than one

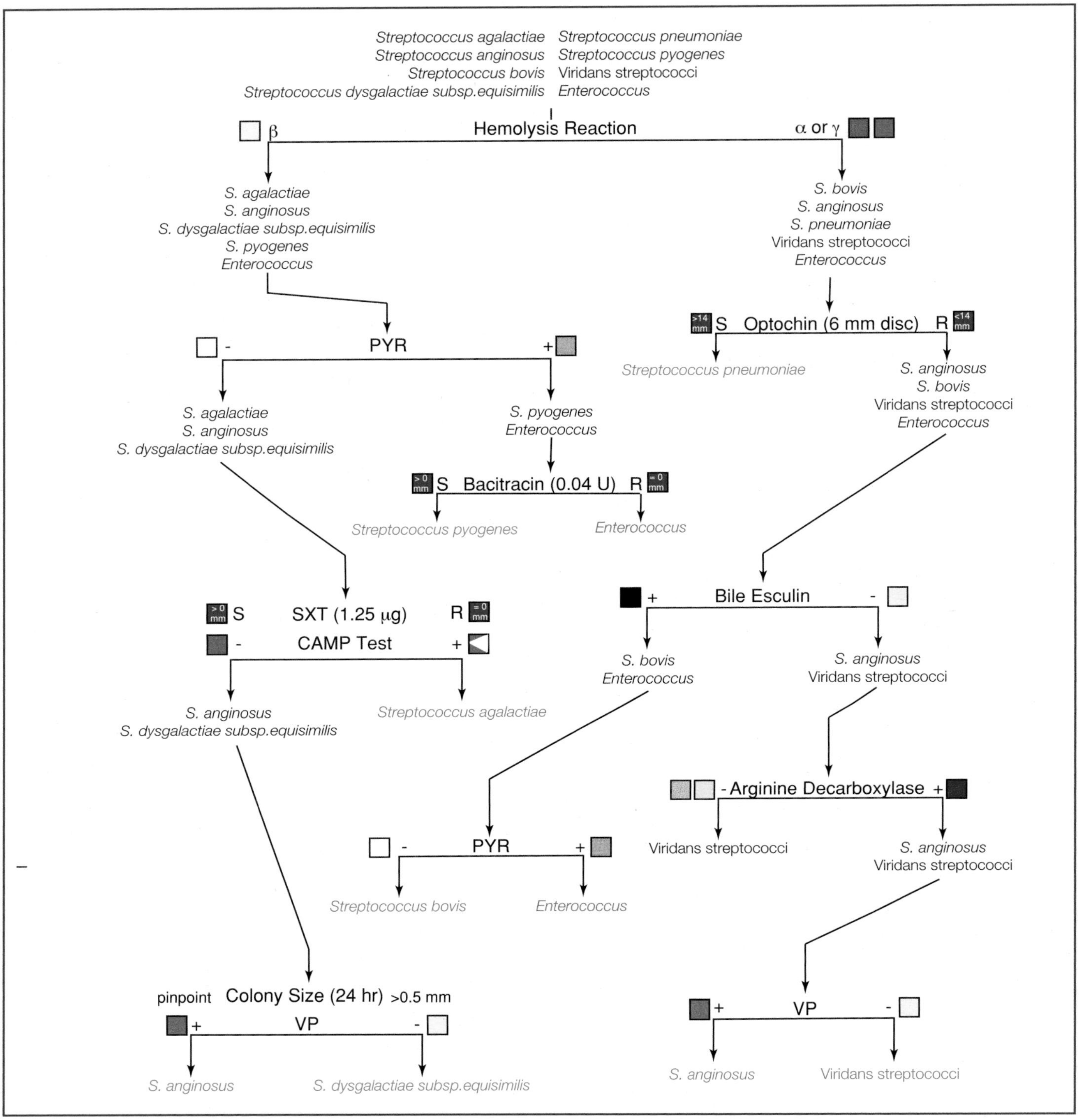

FIGURE 7-19 IDENTIFICATION FLOWCHART FOR GRAM-POSITIVE, CATALASE-NEGATIVE COCCI IN PAIRS OR CHAINS
Please see the appropriate exercises for instructions on how to run each test. The "*S. milleri*" group (or anginosus group) includes streptococci of varying hemolysis reactions but are distinctive in their pinpoint-sized colonies and relatively large zones of hemolysis (when present). The viridans streptococci are a large group of α-hemolytic organisms that are difficult to identify to species.

test at a time unless your instructor tells you to do so. Where multiple tests are listed at a branch point, run only one. Record the results on the Data Sheet as you go. Keep accurate records of what you have done. Do not enter all your data at the end of the project.

6. When you have identified your organism, write its name on the back of the Data Sheet and check with your instructor to see if you are correct. If you are, congratulations! If you aren't, your instructor will advise you as to which tests gave "incorrect" results by writing those test names in the "rerun" space next to your identification. These should be rerun with appropriate positive controls from your school's inventory of organisms. Record the test(s) and positive control(s) on your Data Sheet. Checking the results with positive controls and your unknown will indicate if the "incorrect" test result was truly incorrect or if the strain of organism your school is using doesn't match the majority of strains for that test result.

Some simple differential tests that were not covered in Section Five must be employed to identify these organisms. Following is a brief background and protocol for each.

CAMP Test

Theory Group B *Streptococcus agalactiae* produces the CAMP factor, a hemolytic protein that acts synergistically with the β-hemolysin of *Staphylococcus aureus* subsp. *aureus*. When streaked perpendicularly to an *S. aureus* subsp. *aureus* streak on blood agar (Figure 7-20), an arrowhead zone of hemolysis forms and is a positive result (Figure 7-21).

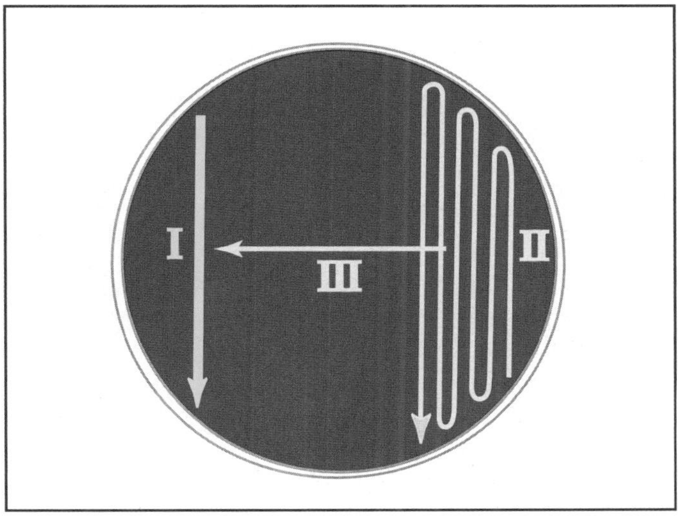

FIGURE 7-20 CAMP Test Inoculation
Two inoculations are made. First, *Staphylococcus aureus* subsp. *aureus* is streaked along one edge of a fresh Blood Agar plate (I). Then the isolate is inoculated densely in the other half of the plate opposite *S. aureus* (II). A single streak then is made out of the inoculated isolate toward the *S. aureus* but not touching it (III).

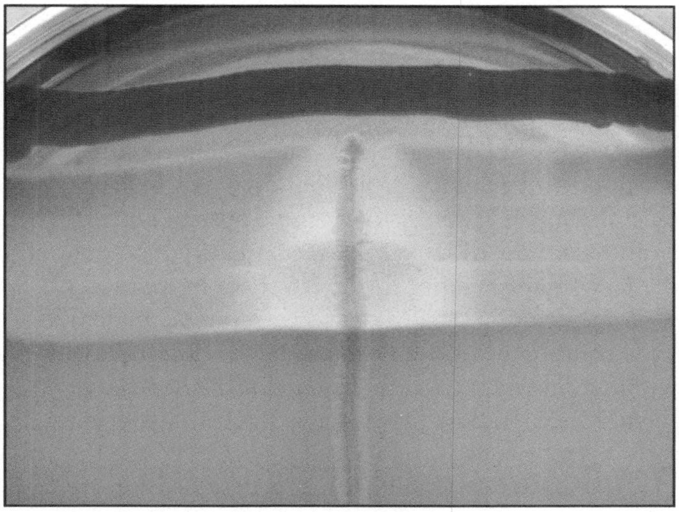

FIGURE 7-21 Positive CAMP Test Result
Note the arrowhead zone of clearing in the region where the CAMP factor of *Streptococcus agalactiae* acts synergistically with the β-hemolysin of *Staphylococcus aureus* subsp. *aureus*.

Application The CAMP test (an acronym of the developers of the test: Christie, Atkins, and Munch-Peterson) is used to differentiate Group B *Streptococcus agalactiae* (+) from other *Streptococcus* species (–).

Procedure

1. Make a single streak of *Staphylococcus aureus* subsp. *aureus* along edge of a fresh Blood Agar plate (Figure 7-20).
2. Inoculate your isolate densely in the other half of the plate opposite the *S. aureus*.
3. Make a single streak out of where you inoculated the isolate towards the *S. aureus* but not touching it.
4. Incubate inverted at 35 ± 2°C for 24 hours.
5. After incubation, look for an arrowhead zone of clearing at the junction of the two inocula. This is a positive CAMP reaction.

PYR Test

Theory Group A streptococci (*S. pyogenes*) and enterococci produce the enzyme **L-pyrrolidonyl arylamidase.** This enzyme hydrolyzes the amide pyroglutamyl-β-naphthylamide to produce L-pyrrolidone and β-naphthylamine.The presence β-naphthylamine can be determined by the formation of a deep red color when the indicator reagent *p*-dimethylaminocinnamaldehyde is added.

PYR may be performed as an 18-hour agar test, a 4-hour broth test, or as a rapid disk test. In each case, the medium (or disk) contains pyroglutamyl-β-naphthylamide, to which is added a heavy inoculum of the test organism.After the appropriate incubation or waiting period, a 0.01% *p*-dimethylaminocinnamaldehyde solution is added. Formation of a deep red color within a few minutes is interpreted as PYR-positive (Figure 7-22).Yellow or orange is PYR-negative.

Application

The PYR test is designed for presumptive identification of group A streptococci (*S. pyogenes*) and enterococci by determining the presence of the enzyme L-pyrrolidonyl arylamidase.

Protocol

1. Place a PYR disk (available commercially) in an empty Petri dish or on a microscope slide.
2. Inoculate the PYR disk with a 24-hour culture of the organism to be tested.
3. Add a drop of the *p*-dimethylaminocinnamaldehyde solution to the disk.
4. A red color within 5 minutes is a positive result. Orange or no color change are considered negative.

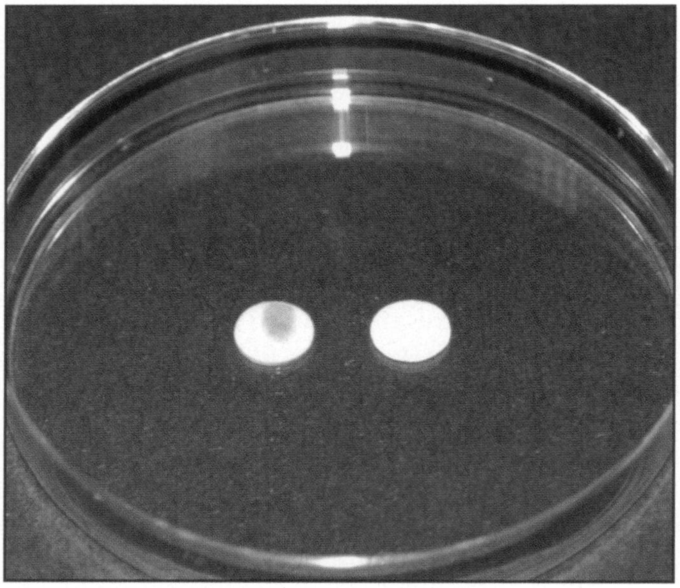

FIGURE 7-22 PYR DISK TEST
The disk on the left was inoculated with *Streptococcus pyogenes* (PYR-positive); the disk on the right contains *Streptococcus agalactiae* (PYR–negative). A red color is positive; yellow or orange are negative.

Optochin, Novobiocin, and SXT Antibiotic Disk Sensitivity Tests

Theory

Paper disks impregnated with known concentrations of antibiotics can be used to determine an organism's susceptibility or resistance to that antibiotic when placed on an inoculated plate (as in Exercises 5-24 and 7-3). A zone of inhibition of certain size indicates susceptibility.

Application

The optochin test is used to presumptively differentiate *Streptococcus pneumoniae* from other α-hemolytic streptococci. The novobiocin test is used to differentiate coagulase-negative staphylococci (CoNS). Most frequently, it is used to presumptively identify the novobiocin-resistant *Staphylococcus saprophyticus*. SXT is used in conjunction with bacitracin to presumptively identify group A and group β-hemolytic streptococci.

Protocol

1. Inoculate a Blood Agar or Trypticase Soy Agar plate with the organism to produce confluent growth (Figure 7-23). (Because you need to use only a small amount of the agar surface, you might share the plate with other students who have to perform antibiotic sensitivity tests.)
2. Using sterile forceps, place the antibiotic disk in the center of your inoculation. More than one disk may be placed on a plate if they are sufficiently separated.

FIGURE 7-23 NOVOBIOCIN DISK TEST
Staphylococcus saprophyticus (R) is on the left; *Staphylococcus epidermidis* (S) is on the right. Note the zone of clearing around *S. epidermidis*.

3. Incubate for 24 hours.
4. Check the plate for a zone of inhibition. Susceptibility (S) is indicated by the following results:
   SXT disk (1.25 µg/23.75 µg)—any zone of inhibition
   Novobiocin disk (5 µg)—clearing of 16 mm or more
   Optochin disk (6 mm)—clearing of 14 mm or more
   Optochin disk (10 mm)—clearing of 16 mm or more

References

Bannerman, Tammy L. 2003. Chapter 28 in *Manual of Clinical Microbiology,* 8th ed., edited by Patrick R. Murray, Ellen Jo Baron, James H. Jorgensen, Michael A. Pfaller, and Robert H. Yolken. American Society for Microbiology Press, Washington, DC.

Koneman, Elmer W., Stephen D. Allen, William M. Janda, Paul C. Schreckberger, and Washington C. Winn, Jr. 1997. Chapters 11 and 12 in *Color Atlas and Textbook of Diagnostic Microbiology,* 5th ed., Lippincott-Raven Publishers. Philadelphia.

MacFaddin, Jean F. 2000. *Biochemical Tests for Identification of Medical Bacteria,* 3rd ed. Lippincott Williams & Wilkins, Philadelphia.

Ruoff, Kathryn L., R.A. Whiley, and D. Beighton. 2003. Chapter 29 in *Manual of Clinical Microbiology,* 8th ed., edited by Patrick R. Murray, Ellen Jo Baron, James H. Jorgensen, Michael A. Pfaller, and Robert H. Yolken. American Society for Microbiology Press, Washington, DC.

Teixeira, Lucia Martins, and Richard R. Facklam. 2003. Chapter 30 in *Manual of Clinical Microbiology,* 8th ed., edited by Patrick R. Murray, Ellen Jo Baron, James H. Jorgensen, Michael A. Pfaller, and Robert H. Yolken. American Society for Microbiology Press, Washington, DC.

Food and Environmental Microbiology

SECTION 8

The microbiological disciplines examined in this section have distinctly different scopes of practice while overlapping to a significant extent with the public health domain (see Section Seven). Environmental microbiology is the study, utilization and control of microorganisms in the environment. Many environmental organisms are used beneficially by industry, such as in the production of vinegar, leaching of low-grade ores, and sewage purification. Other environmental organisms are of human or animal origin (usually fecal) and threaten us with the potential for contaminating food or water.

Food microbiology is devoted to the study and utilization of beneficial microbes, as well as the control of many common, yet potentially deadly, contaminants. Many molds, yeasts, and bacteria are responsible for the production, preservation, and flavoring of foods we enjoy, and others produce toxins so powerful that as little as one nanogram per kilogram of body weight can be lethal.

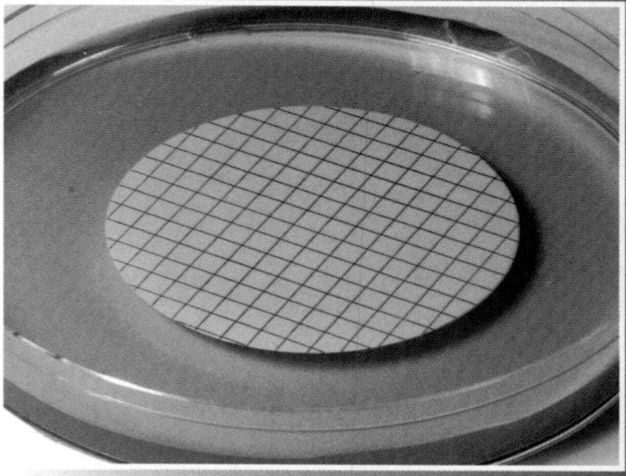

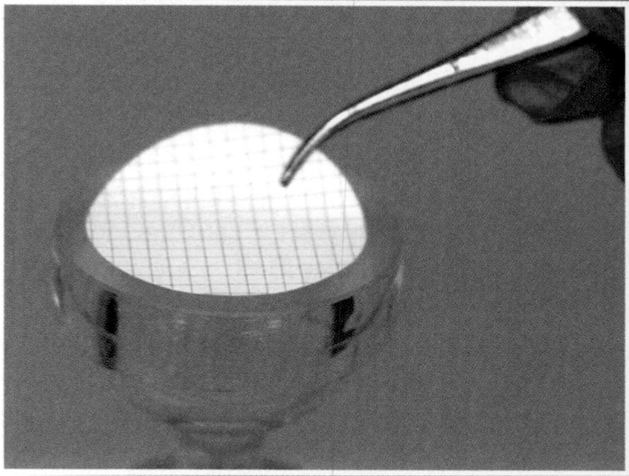

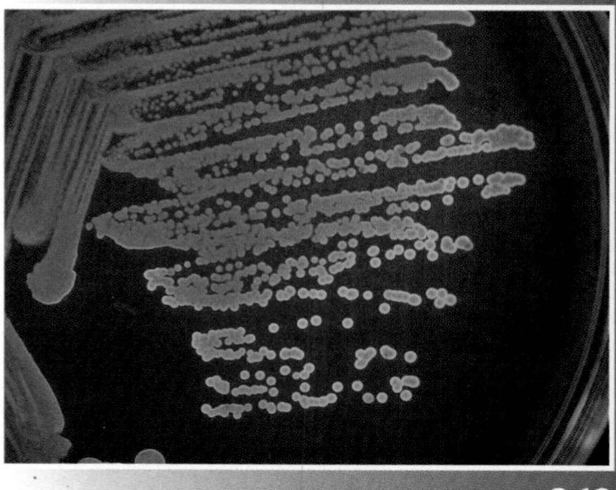

Environmental Microbiology

Environmental microbiology is the study, utilization, and control of microorganisms living in marine, freshwater, or terrestrial habitats. In efforts to keep our environment safe from disease-causing contamination, governmental regulatory agencies conduct tests daily that are similar to those introduced here. The first two tests—the membrane filter technique and the multiple tube fermentation method—are used to measure fecal contamination of water. The microbial soil count is used to isolate and demonstrate the variety of populations residing in soil. Bioluminescence, the final exercise in this unit, is the light-emitting phenomenon expressed by certain marine bacteria.

Exercise 8-1

Membrane Filter Technique

Theory Fecal contamination is a common pollutant in open water and a potential source of serious disease-causing organisms. Because fecal pathogens tend to be fairly short-lived in water and relatively difficult to detect, the membrane filter technique tests for the presence of the much more abundant **coliform** bacteria—*Escherichia coli*, *Klebsiella pneumoniae*, and *Enterobacter aerogenes* (Figure 8-1).

In the membrane filter technique, a test (water) sample is drawn through a special micropore membrane filter designed to trap microorganisms larger than 0.45 µm. The filter then is placed onto an Endo Agar plate and incubated (Figure 8-2).

Endo Agar (Exercise 4-4) is a selective medium that encourages growth of Gram-negative organisms and inhibits Gram-positives. Among other things, it contains lactose and a dye for detecting pH changes. As a result of acid from lactose fermentation, coliforms produce red or mucoid colonies, typically with a gold or green metallic sheen resulting from the additional production of aldehydes. Noncoliform bacteria (including several dangerous pathogens) tend to be colorless, light pink, or the color of the medium.

After incubation, all red or metallic colonies are counted and are used to calculate "coliform colonies/100mL" using the following formula:

$$\frac{\text{coliform colonies}}{100\text{ mL}} = \frac{\text{coliform colonies} \times 100}{\text{volume of original sample in mL}}$$

A "countable" plate contains between 20 and 80 coliform colonies with a total colony count no larger

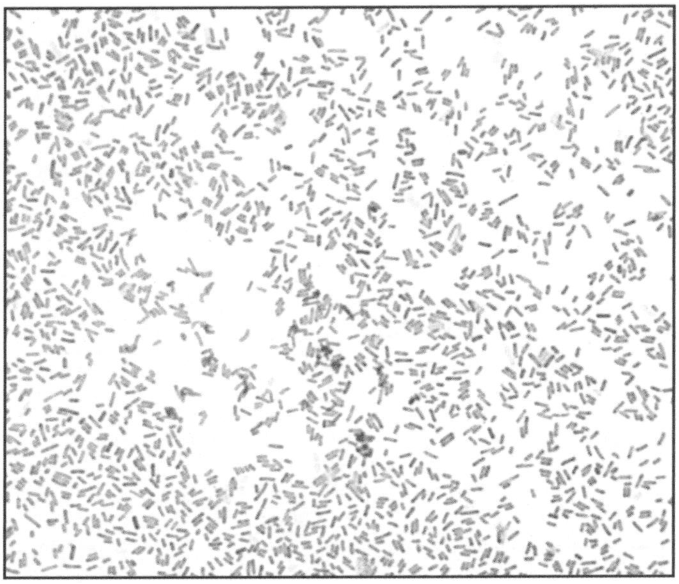

FIGURE 8-1 Gram stain of *Escherichia coli*
E. coli is a principal coliform detected by the membrane filter technique.

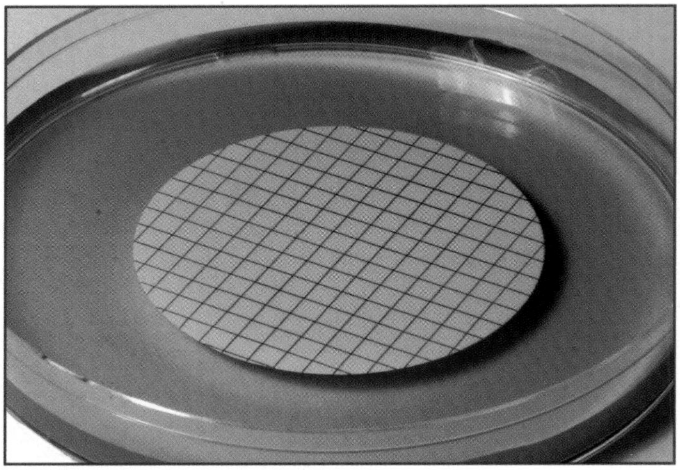

FIGURE 8-2 Membrane Filter
This porous membrane will allow water to pass through but will trap bacteria and particles larger than 0.45 µm.

than 200 (Figure 8-3). To fall within this range, it is customary to dilute heavily polluted samples, thereby reducing the number of cells collecting on the membrane. When dilution is necessary, it is important to record the volume of ***original sample*** only, not any added water. Potable water contains less than one coliform per 100 milliliters of sample.

Application

The membrane filter technique is commonly used in combination with other tests to identify the presence of fecal coliforms in water.

In This Exercise...

You will be using collected water samples and performing a membrane filter technique to determine total coliform population density. Your results will be recorded in coliforms per 100 mL of sample.

Materials

Per Student Group

- one Endo Agar plate
- one sterile membrane filter (pore size 0.45 µm)
- sterile membrane filter suction apparatus (Figure 8-4)
- 100 mL water dilution bottle (to be distributed in the preceding lab)
- 100 mL water sample (obtained by student)

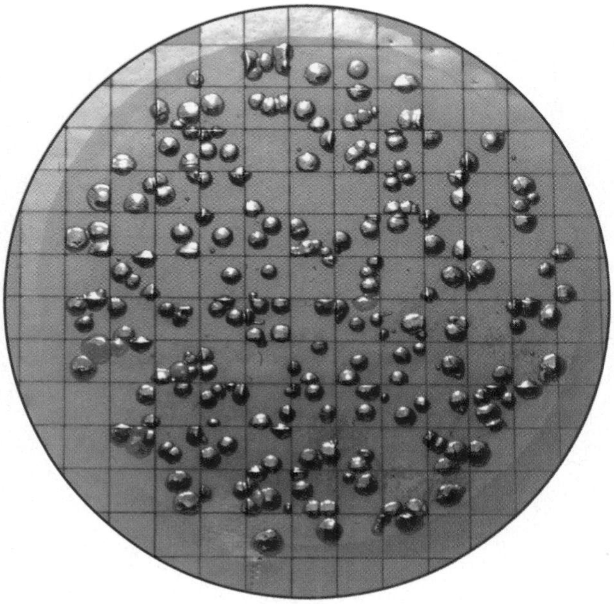

FIGURE 8-3 COLIFORM COLONIES ON A MEMBRANE FILTER
Note the characteristic dark colonies with gold metallic sheen indicating that this water sample is contaminated with fecal coliforms. Potable water has less than one coliform per 100 mL of sample tested.

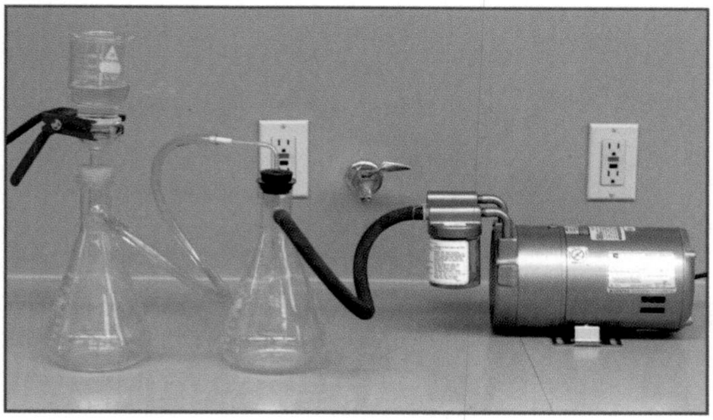

FIGURE 8-4 MEMBRANE FILTER APPARATUS
Assemble the membrane filter apparatus as shown in this photograph. It is important to use two suction flasks (as shown) to avoid getting water into the vacuum source. Secure the flasks on the table, as the tubing may make them top-heavy.

- gloves
- household disinfectant and paper towels
- small beaker containing alcohol and forceps
- vacuum source (pump or aspirator)

Medium Recipe

Endo Agar

• Peptone	10.0 g
• Lactose	10.0 g
• Dipotassium phosphate	3.5 g
• Agar	15.0 g
• Basic Fuchsin	0.5 g
• Sodium Sulfite	2.5 g
• Distilled or deionized water	1.0 L

pH 7.3–7.7 at 25°C

Procedure

Prelab

1. Obtain a 100 mL water dilution bottle from your instructor.
2. Choose an environmental source to sample. (Your instructor may decide to approve your choice to avoid duplication among lab groups.)
3. Visit the environmental site as close to your lab period as possible. Bring your water dilution bottle, a pair of gloves, some household disinfectant, and paper towels. While wearing gloves, fill the bottle to the white line (100 mL) and replace the cap.
4. Wipe the outside of the bottle with disinfectant. Dispose of the towels and gloves in the trash.
5. Store the sample in the refrigerator until your lab period. If the sample must sit for a while before your

lab, leave the cap loose to allow some aeration. Be careful not to spill!

Lab One

1. Alcohol-flame the forceps (Figure 8-5) and place the membrane filter (grid facing up) between the two halves of the filter housing (Figure 8-6).
2. Clamp the two halves of the filter housing together and insert the filter housing into the suction flask as shown in Figure 8-7. (This assembly can be a little top heavy; so have someone hold it or otherwise secure it to prevent tipping.)
3. Pour the appropriate volume of water sample into the funnel. (Refer to Table 8-1 for suggested sample volumes. If the sample size is smaller than 10.0 mL, add 10 to 20 mL of sterile water before filtering to help distribute the cells evenly on the surface of the membrane filter.)
4. Turn on the suction pump (or aspirator) and filter the sample into the flask.
5. Sterilize the forceps again and carefully transfer the filter to the Endo Agar plate, being careful not to fold it or create air pockets (Figure 8-8).
6. Wait a few minutes to allow the filter to adhere to the agar, then invert the plate and incubate it aerobically at 35 ± 2°C for 24 to 48 hours.

FIGURE 8-5 ALCOHOL-FLAMING THE FORCEPS
Dip the forceps into the beaker of alcohol. Remove the forceps and quickly pass them through the Bunsen burner flame—just long enough to ignite the alcohol. *Be sure to keep the flame away from the beaker of alcohol.*

Lab Two

1. Remove the plate and count the colonies on the membrane filter that are dark purple, have a black center, or produce a green metallic sheen.
2. Record your data in the chart on the Data Sheet.
3. Calculate the coliform CFU per 100 milliliters using the following formula:

$$\text{Coliforms/100 mL} = \frac{\text{(coliform colonies counted)(100)}}{\text{(mL of original sample filtered)}}$$

4. Record your results in the chart on the Data Sheet.

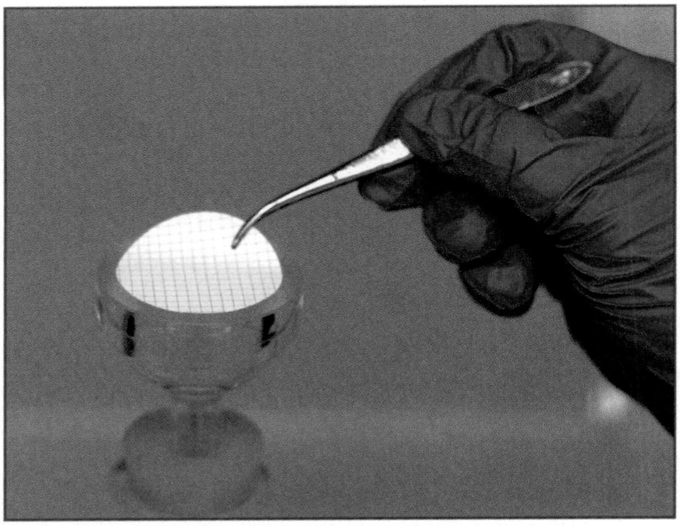

FIGURE 8-6 PLACING THE FILTER ON THE FILTER HOUSING
Carefully place the filter on the bottom half of the filter housing. Clamp the filter funnel over the filter.

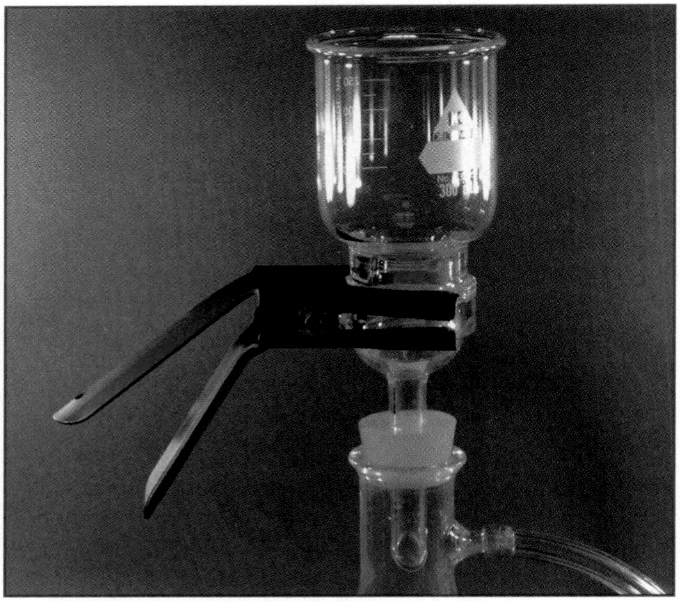

FIGURE 8-7 MEMBRANE FILTER ASSEMBLY
The membrane filter assembly is made up of a two-piece funnel and clamp. The membrane filter is inserted between the two funnel halves and the whole assembly is clamped together.

Volume To Be Filtered (mL)								
Source								
	100	50	10	1.0	0.1	0.01	0.001	0.0001
Drinking water	●							
Swimming pool	●							
Lake	●	●	●					
Well or spring	●	●	●					
Public beach			●	●	●			
River				●	●	●	●	
Raw sewage					●	●	●	●

TABLE 8-1 SUGGESTED SAMPLE VOLUMES FOR MEMBRANE FILTER TEST (Table courtesy of American Public Health Association)

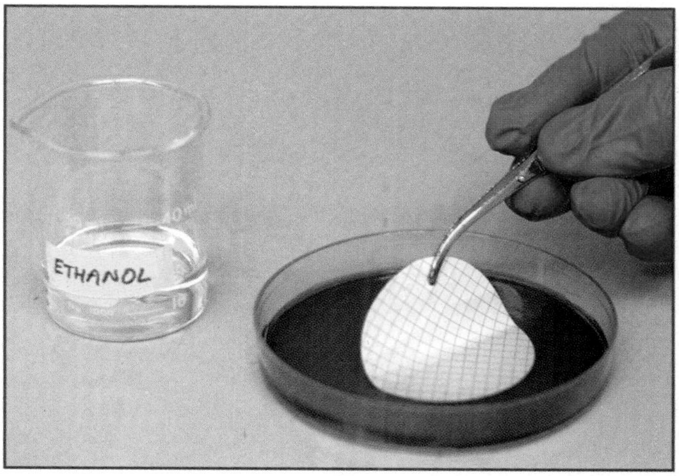

FIGURE 8-8 PLACING THE FILTER ON THE AGAR PLATE
Using sterile forceps, carefully place the filter onto the agar surface with the grid facing up. Try not to allow any air pockets under the filter, because contact with the agar surface is essential for growth of bacteria. Allow a few minutes for the filter to adhere to the agar before inverting the plate.

References

Chan, E. C. S., Pelczar, Jr., and Krieg, Noel R. 1986. Page 291 in *Laboratory Exercises in Microbiology*. McGraw-Hill, New York.

Clesceri, W.E. F, Arnold E. Greenberg, Andrew D. Eaton, and Mary Ann H. Franson. 1998. Chapter 9 in *Standard Methods for the Examination of Water and Wastewater,* 20th ed. American Public Health Association, American Water Works Association, and Water Environment Federation. APHA Publication Office, Washington, DC.

Collins, C. H., Patricia M. Lyne, and J. M. Grange. 1995. Page 270 in Collins and Lyne's *Microbiological Methods,* 7th ed. Butterworth-Heinemann, Oxford: United Kingdom.

Mulvany, J. G. 1969. Page 205 in *Methods in Microbiology,* Vol. 1, edited by J. R. Norris and D. W. Ribbons, Academic Press, New York.

Zimbro, Mary Jo, and David A. Power, eds. 2003. Page 207 in *Difco™ and BBL™ Manual—Manual of Microbiological Culture Media.* Becton Dickinson and Co., Sparks, MD.

Exercise 8-2

Multiple Tube Fermentation Method for Total Coliform Determination

Theory

The **multiple tube fermentation method**, also called **most probable number** or **MPN**, is a common means of calculating the number of coliforms present in 100 mL of a sample. The procedure determines both **total coliform** counts and *E. coli* counts.

The three media used in the procedure are Lauryl Tryptose Broth (LTB), Brilliant Green Lactose Bile (BGLB) Broth, and EC (*E. coli*) Broth. LTB, which includes lactose and lauryl sulfate, is selective for the coliform group. Because it does not screen out all noncoliforms, it is used to *presumptively* determine the presence or absence of coliforms. BGLB broth, which includes lactose and 2% bile, inhibits noncoliforms and is used to *confirm* the presence of coliforms. EC broth, which includes lactose and bile salts, is selective for *E. coli* when incubated at 45.5°C.

All broths are prepared in 10 mL volumes and contain an inverted Durham tube to trap any gas produced by fermentation. The LTB tubes are arranged in six groups of five tubes, as shown in Figure 8-9. Each tube in the first set of five receives 1.0 mL of the original sample. Each tube in the second group receives 1.0 mL of 10^{-1} dilution. Each tube in the third, fourth, fifth, and sixth group receives 1.0 mL of 10^{-2}, 10^{-3}, 10^{-4}, and 10^{-5} respectively. (***Note:*** The appropriate dilutions of the water sample are made beforehand using a technique similar to the one diagrammed in Figure 6-4. This series differs in that all five dilution tubes contain 9.0 mL sterile water. One mL of sample is added to tube 1 to produce a 10^{-1} dilution; one mL from tube 1 is added to tube 2 to produce a 10^{-2} dilution, *etc.*)

After inoculation, the LTB tubes are incubated at 35 ± 2°C for up to 48 hours, then examined for gas production (Figure 8-10). Any positive LTB tubes then are subcultured to BGLB tubes. Again, the cultures are incubated 48 hours at 35 ± 2°C and examined for gas production (Figure 8-11). Positive BGLB cultures then are transferred to EC broth and incubated at 45.5°C for

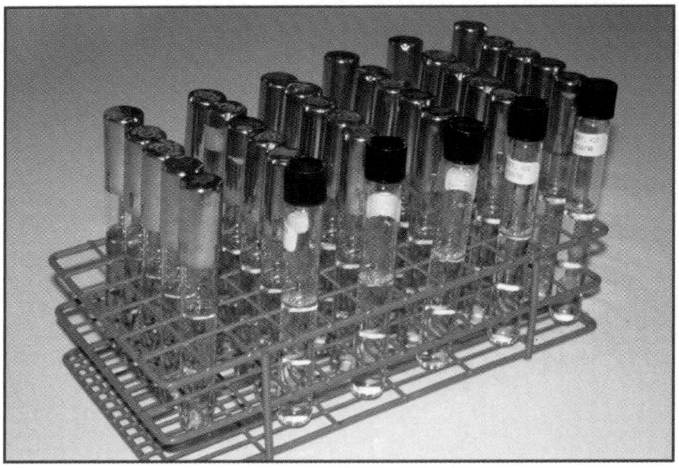

FIGURE 8-9 Multiple Tube Fermentation
This is a multiple tube fermentation of a seawater sample contaminated by sewage. The test photographed contains six groups of five tubes (the standard for heavily contaminated samples) rather than three, as described in Table 8-1. The tubes contain Lauryl Tryptose Broth and a measured volume of water sample as described in the text. Following incubation, each positive broth (based on gas production) will be used to inoculate a BGLB broth and an EC broth.

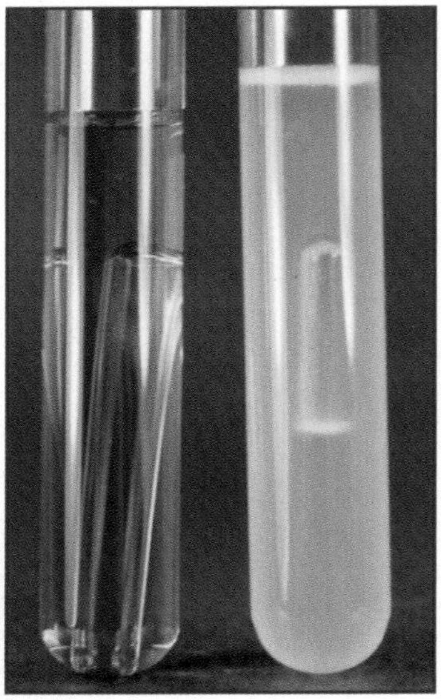

FIGURE 8-10 LTB Tubes
The bubble in the Durham tube on the right is *presumptive* evidence of coliform contamination. The tube on the left is negative.

FIGURE 8-11 BGLB Broth
The bubble in the Durham tube on the right is seen as *confirmation* of coliform contamination. The tube on the left is negative.

48 hours (Figure 8-12). After incubation the EC tubes with gas are counted. Calculation of MPN is based on the combinations of positive results in the BGLB and EC broths using the following formula:

$$\text{MPN/100 mL} = \frac{100P}{\sqrt{V_n V_a}}$$

where

P = total number of positive results,
V_n = combined volume of sample in negative tubes, and
V_a = combined volume of sample in all tubes.

It is customary to calculate and report *both* total coliform *and E. coli* densities. Total coliform MPN is calculated using BGLB broth results, and *E. coli* MPN is based on EC broth results.

Using the data from Table 8-2 and the formula, the calculation would be as follows:

$$\text{MPN/100 mL} = \frac{100 \times 9}{\sqrt{(0.24 \times 5.55)}} = \frac{900}{\sqrt{1.332}} = \frac{900}{1.154} = 780$$

MPN/100 mL = 780 coliforms/100 mL

FIGURE 8-12 EC BROTH
The bubble in the Durham tube on the right is seen as *confirmation* of *E. coli* contamination. The tube on the left is negative.

	Group	A	B	C	Totals (A+B+C)
1	Dilution Factor (DF)	10^0	10^{-1}	10^{-2}	NA
2	Portion of dilution added to LTB tubes that is original sample (1.0 mL × DF)	1.0 mL	0.1 mL	0.01 mL	NA
3	# tubes in group	5	5	5	NA
4	# positive results (gas)	5	3	1	9
5	# negative results (no gas)	0	2	4	NA
6	Volume of original sample in negative sample in negative LTB tubes (DF × 1.0 mL × # negative tubes)	0 mL	0.2 mL	0.04 mL	0.24 mL
7	Volume of original sample in all LTB tubes (DF × 1.0 mL × # tubes)	5.0 mL	0.5 mL	0.05 mL	5.55 mL

TABLE 8-2 BGLB TEST RESULTS
The results shown here are of a hypothetical BGLB test. For simplicity, it was shortened to include three groups of 5 tubes (A, B and C) rather than six groups. The original water sample was diluted to 10^0, 10^{-1} and 10^{-2}. The three dilutions then were used to inoculate the broths in groups A, B, and C, respectively. The first row contains the dilution factor of inoculum used per group. The second row shows the actual amount of original sample that went into each broth. The third row contains the number of tubes in each group (five in this example). The fourth row shows the number of tubes from each group of five that showed evidence of gas production. This total (in red) inserts into the equation. The fifth row shows the number of tubes from each group that did *not* show evidence of gas production. The sixth row is used for calculating the "combined volume of sample in negative tubes" and refers to the inoculum that went into the LTB tubes. This total (in red) inserts into the equation. The seventh row is used for calculating the "combined volume of sample in all tubes" and refers to the inoculum that went into the LTB tubes. This total (in red) inserts into the equation. As you can see, the undiluted inoculation (Group A) produced five positive results and 0 negative results; the 10^{-1} dilution (Group B) produced three positive results and two negative; and the 10^{-2} dilution (Group C) produced one positive and four negative results. The total volume of original sample that went into LTB tubes was 5.55 mL, 0.24 mL of which produced no gas (shown in red in rows 7 and 6, respectively).

Application

This standardized test is used to measure coliform density (cells/100 mL) in water. It may be used to calculate the density of all coliforms present (total coliforms) or to calculate the density of *Escherichia coli* specifically.

In This Exercise...

You will be using collected water samples to perform a multiple tube fermentation. The data collected then will be used to calculate a total coliform count and an *E. coli* count. Both counts will be recorded in coliforms per 100 mL of sample.

Materials

Per Student Group

- 15 lauryl tryptose broth (LTB) tubes (containing 10 mL broth and an inverted Durham tube)
- up to 15 brilliant green lactose bile (BGLB) broth tubes (The number of tubes required will be determined by the results of the LTB test.)
- up to 15 EC broth tubes (The number of tubes required will be determined by the results of the LTB test.)
- two 9.0 mL dilution tubes
- sterile 1.0 mL pipettes and pipettor
- water bath set at 45.5°C
- test tube rack
- labeling tape
- 100 mL water dilution bottle (to be distributed in the preceding lab)
- 100 mL water sample (obtained by student)
- gloves
- household disinfectant and paper towels

Medium Recipes

Brilliant Green Lactose Bile Broth

- Peptone 10.0 g
- Lactose 10.0 g
- Oxgall 20.0 g
- Brilliant green dye 0.0133 g
- Distilled or deionized water 1.0 L

pH 7.0–7.4 at 25°C

Lauryl Tryptose Broth

- Tryptose 20.0 g
- Lactose 5.0 g
- Dipotassium phosphate 2.75 g
- Monopotassium phosphate 2.75 g
- Sodium chloride 5.0 g
- Sodium lauryl sulfate 0.1 g
- Distilled or deionized water 1.0 L

pH 6.6–7.0 at 25°C

EC Broth

- Tryptose 20.0 g
- Lactose 5.0 g
- Dipotassium phosphate 4.0 g
- Monopotassium phosphate 1.5 g
- Sodium chloride 5.0 g
- Distilled or deionized water 1.0 L

pH 6.7–7.1 at 25°C

Procedure [Refer to the procedural diagram in Figure 8-13 as needed.]

Lab One

1. Make a dilution by adding 1.0 mL of the water sample to one of the 9.0 mL dilution tubes. This dilution is only 1/10 original sample; therefore, the DF is 1/10 or 0.1 or 10^{-1} (preferred). For help with dilutions and dilution factors, refer to Exercise 6-1.
2. Make a second dilution by adding 1.0 mL of the 10^{-1} dilution to one of the 9.0 mL dilution tubes. Mix well. This dilution contains 1/100 original sample and has a DF of 10^{-2}.
3. Arrange the 15 LTB tubes into three groups of five in a test tube rack. Label the groups A, B, and C, respectively.
4. Aseptically transfer 1.0 mL of the (undiluted) water sample to each LTB tube in the group labeled A. Mix well. This is undiluted sample, so its dilution factor (calculated before adding it to the broth) is 10^0 or 1. (***Note:*** The dilution factors are based on the dilution performed before making any additions to the broth. The LTB is not part of the dilution.)
5. Add 1.0 mL of the 10^{-1} dilution to each of the LTB tubes in group B. Mix well.
6. Add 1.0 mL of the 10^{-2} dilution to each of the LTB tubes in the C group.
7. Incubate the LTB tubes at 35 to 37°C for 48 hours.

Lab Two

1. Remove the broths from the incubator and, one group at a time, examine the Durham tubes for accumulation of gas. Gas production is a positive result; absence of gas is negative. Record positive and negative results in the chart provided on the Data Sheet. If necessary, refer to the example in Table 8-2.

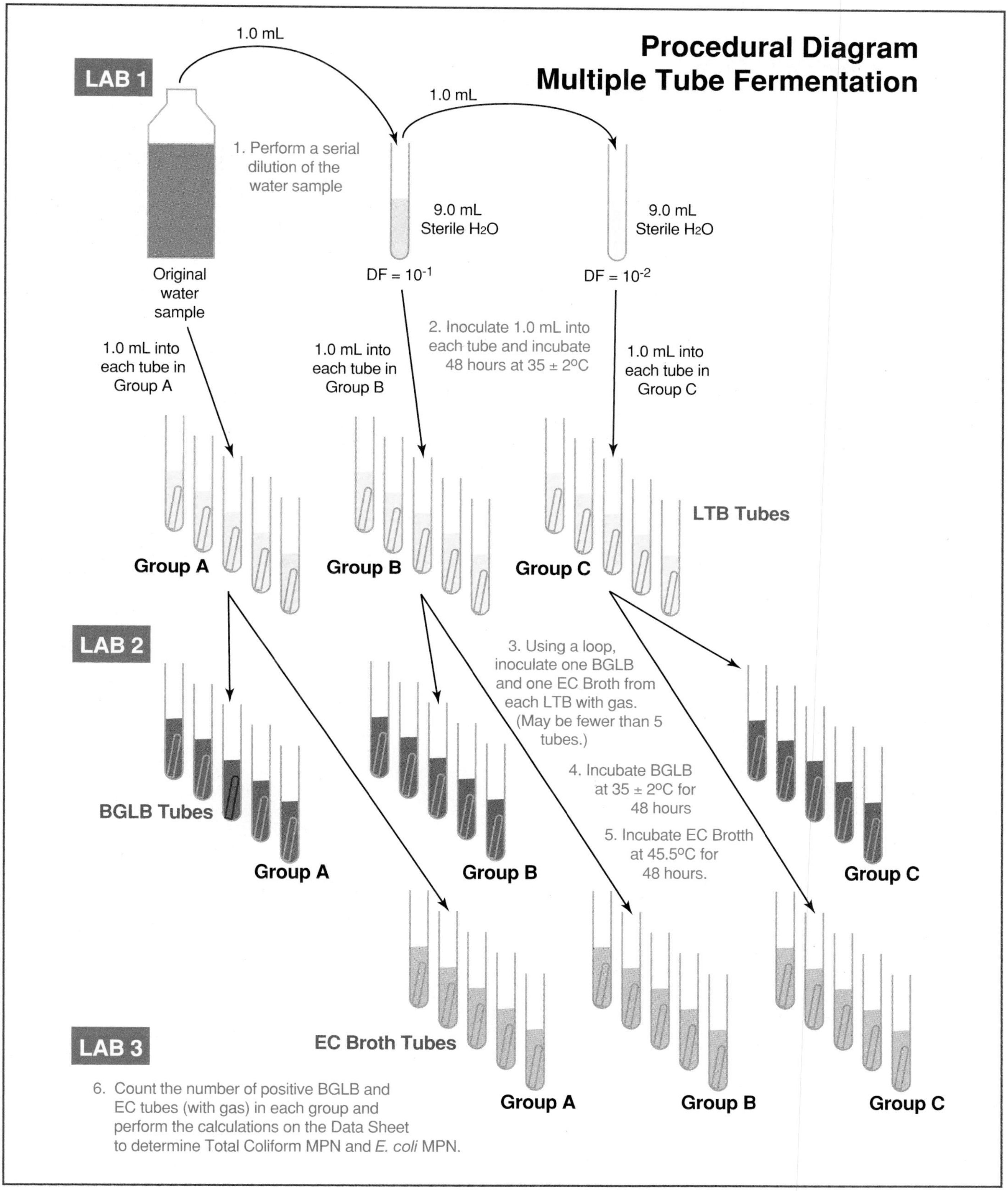
Procedural Diagram
Multiple Tube Fermentation
LAB 1
1.0 mL
1.0 mL
1. Perform a serial dilution of the water sample
9.0 mL Sterile H2O
9.0 mL Sterile H2O
Original water sample
DF = 10-1
DF = 10-2
1.0 mL into each tube in Group A
1.0 mL into each tube in Group B
2. Inoculate 1.0 mL into each tube and incubate 48 hours at 35 ± 2ºC
1.0 mL into each tube in Group C
LTB Tubes
Group A
Group B
Group C
LAB 2
3. Using a loop, inoculate one BGLB and one EC Broth from each LTB with gas. (May be fewer than 5 tubes.)
4. Incubate BGLB at 35 ± 2ºC for 48 hours
5. Incubate EC Brotth at 45.5ºC for 48 hours.
BGLB Tubes
Group A
Group B
Group C
LAB 3
EC Broth Tubes
Group A
Group B
Group C
6. Count the number of positive BGLB and EC tubes (with gas) in each group and perform the calculations on the Data Sheet to determine Total Coliform MPN and E. coli MPN.

FIGURE 8-13 PROCEDURAL DIAGRAM OF THE MULTIPLE TUBE FERMENTATION METHOD FOR TOTAL COLIFORM DETERMINATION

2. Using an inoculating loop, inoculate one BGLB broth with each positive LTB tube showing evidence of gas production. (Make sure that each BGLB tube is *clearly labeled* A, B, or C according to the LTB tube from which it is inoculated.)
3. Inoculate EC broths with the positive LTB tubes in the same manner as the BGLB above. Again, be sure to clearly label all EC tubes appropriately.
4. Incubate the BGLB at 35 to 37°C for 48 hours. Incubate the EC tubes in the 45.5°C water bath for 48 hours.

Lab Three

1. Remove all tubes from the incubator and water bath and examine the Durham tubes for gas accumulation. Count the positive BGLB tubes and enter your results in the chart on the Data Sheet.
2. Using Table 8-2 as a guide, complete the BGLB Data chart on the Data Sheet.
3. Using the data in the BGLB Data chart, calculate the total coliform MPN using the following formula. (This formula is used to calculate both Total Coliform MPN and *E. coli* MPN.)

$$\text{MPN/100 mL} = \frac{(\text{Total number of positive results})\ (100)}{\sqrt{\begin{pmatrix}\text{Combined volume of}\\ \text{sample in negative tubes}\end{pmatrix}\begin{pmatrix}\text{Combined volume of}\\ \text{sample in all tubes}\end{pmatrix}}}$$

Where P = total number of positive results,

V_n = combined volume of sample in negative tubes, and

V_a = combined volume of sample in all tubes.

4. Count the positive EC broth results in the same manner as the BGLB test. Record the results in the chart on the Data Sheet.
5. Determine the *E. coli* MPN in the same manner as described for total coliform count, and record your results in the chart on the Data Sheet.

Reference

Clesceri, Lenore S., Arnold E. Greenberg, and Andrew D. Eaton. 1998. *Standard Methods for the Examination of Water and Wastewater*, 20th ed. Prepared and published jointly by American Public Health Association, American Water Works Association, and Water Environment Federation. APHA Publication Office, Washington, DC.

Exercise 8-3

Bioluminescence

Theory A few marine bacteria from genera *Vibrio* and *Photobacterium* are able to emit light by a process known as **bioluminescence.** Many of these organisms maintain mutualistic relationships with other marine life. For example, *Photobacterium* species living in the Flashlight Fish receive nutrients from the fish and in return provide a unique device for frightening would-be predators.

Bioluminescent bacteria are able to emit light because of an enzyme called **luciferase** (Figure 8-14). In the presence of oxygen and a long-chain aldehyde, luciferase catalyzes the oxidation of reduced flavin mononucleotide ($FMNH_2$). In the process, outer electrons surrounding FMN become excited. Light is emitted when the electronically excited FMN returns to its ground state (Figure 8-15).

It is estimated that a single *Vibrio* cell burns between 6000 and 60000 molecules of ATP per second emitting light. (ATP hydrolysis occurs in conjunction with synthesis of the aldehyde.) It also is known that their luminescence occurs only when a certain threshold population size is reached in a phenomenon called **quorum sensing.** This system is controlled by a genetically produced **autoinducer** that must be in sufficient concentration to trigger the reaction.

In This Exercise...

You will inoculate Seawater Complete Medium (SWC) with *Vibrio fischeri*, incubate it for a day or two, then observe for bioluminescence.

Materials

Per Student Group

- overnight culture of *Vibrio fischeri*

Per Student

- one SWC plate

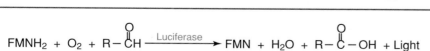

$$FMNH_2 + O_2 + R-\overset{O}{\overset{\|}{C}}H \xrightarrow{\text{Luciferase}} FMN + H_2O + R-\overset{O}{\overset{\|}{C}}-OH + \text{Light}$$

FIGURE 8-14 CHEMISTRY OF BIOLUMINESCENT BACTERIA
The enzyme luciferase catalyzes the oxidation of reduced flavin mononucleotide in the presence of an aldehyde. During the reaction, an electron becomes excited. When it returns to its ground state, light is emitted.

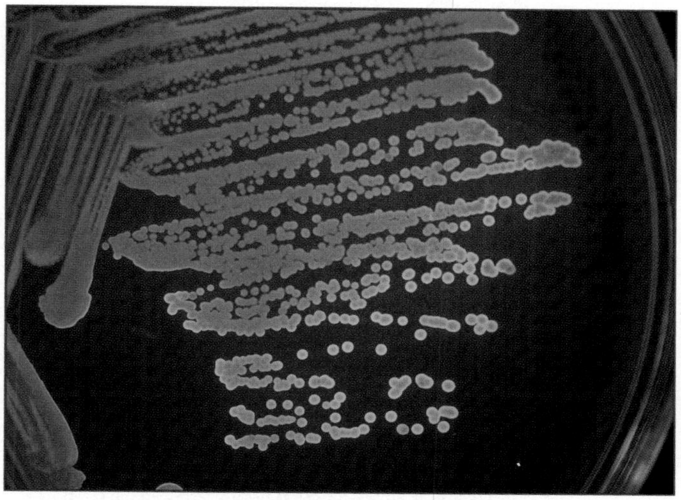

FIGURE 8-15 BIOLUMINESCENCE ON AN AGAR PLATE
This is an unknown bioluminescent bacterium growing on SWC Agar.

Medium Recipe

Seawater Complete Medium

• Pancreatic Digest of Casein	5.0 g
• Yeast Extract	3.0 g
• Glycerol	3.0 mL
• Seawater	750.0 mL

Procedure

Lab One

1. Obtain a culture of *Vibrio fischeri*, and aseptically inoculate it with *V. fischeri*. Any pattern will do.
2. Invert and incubate the plate at 25°C for 24 to 48 hours.

Lab Two

1. Remove the plate from the incubator and examine it in a dark room for light emission. It may take awhile for your eyes to adjust to the dark and see the bioluminescence.

References

DIFCO Laboratories. 1984. *DIFCO Manual*, 10th ed. DIFCO Laboratories, Detroit.

Krieg, Noel R., and John G. Holt (Editor-in-Chief). 1984. Page 518 in *Bergey's Manual of Systematic Bacteriology*, Vol. 1. Lippincott Williams and Wilkins, Baltimore.

Power, David A., and Peggy J. McCuen. 1988. *Manual of BBL™ Products and Laboratory Procedures*, 6th ed. Becton Dickinson Microbiology Systems, Cockeysville, MD.

White, David. 2000. Pages 504–509 in *The Physiology and Biochemistry of Prokaryotes*. Oxford University Press. New York.

Exercise 8-4

Soil Microbial Count

Theory Ordinary soil is a storehouse of microorganisms, both beneficial and pathogenic. The major groups of antibiotic producing genera—*Bacillus*, *Cephalosporium*, *Penicillium*, and *Streptomyces*—can be found in soil. Nitrogen cycling microorganisms, protozoans, worms, and all manner of bacterial and fungal spores also can be isolated from soil.

The media used for this exercise are designed specifically for the organisms to be isolated. Glycerol Yeast Extract Agar is designed for actinomycetes, and doesn't offer enough nutritive value for typical bacteria or fungi. Nutrient Agar is designed for a broad range of bacteria, but not fungi. Sabouraud Dextrose Agar is designed for fungi but also will support bacteria; therefore, penicillin and streptomycin have been added to discourage bacterial growth.

Application This technique is used to determine the relative concentrations of bacteria, actinomycetes, and fungi in soil. It also is a good exercise to demonstrate diversity in soil samples.

In This Exercise...

You will perform a serial dilution of a soil sample and calculate the densities of three very prominent soil residents: bacteria, actinomycetes, and fungi. Colony counts will be performed on plates produced using the pour plate technique described in Exercise 6-4, Plaque Assay. To save time and material, the class will be divided into six groups. Refer to Table 8-3 for group assignments.

Group Number	Organism	Media (See materials)
1	Actinomycetes	GYEA
2	Actinomycetes	GYEA
3	Bacteria	NA
4	Bacteria	NEA
5	Fungi	SDA
6	Fungi	SDA

TABLE 8-3 GROUP ASSIGNMENTS

Materials

Per Student Group

- soil sample
- water bath set at 45°C containing sterile molten:
  - Sabouraud Dextrose Agar
  - Nutrient Agar
  - Glycerol Yeast Extract Agar
- capped bottle containing 100 mL sterile water
- micropipettes (10–100 μL and 100–1000 μL) with sterile tips
- sterile microtubes
- flask of sterile water
- five Sterile Petri dishes
- hand tally counter
- colony counter

Medium Recipes

Sabouraud Dextrose Agar (with antibiotics added to inhibit bacteria)

- Peptone 10.0 g
- Dextrose 40.0 g
- Agar 15.0 g
- Penicillin 20000.0 units
- Streptomycin 0.00004 g
- Distilled or deionized water 1.0 L

pH 5.4–5.8 at 25°C

Nutrient Agar

- Beef extract 3.0 g
- Peptone 5.0 g
- Agar 15.0 g
- Distilled or deionized water 1.0 L

pH 6.6 ± 7.0 at 25°C

Glycerol Yeast Extract Agar

- Glycerol 5.0 mL
- Yeast extract 2.0 g
- Dipotassium phosphate 1.0 g
- Agar 15.0 g
- Distilled or deionized water 1.0 L

Procedure

Note: This is a quantitative exercise employing serial dilutions, dilution factors, and calculations that are explained fully in Exercise 6-1, Standard Plate Count. If you haven't completed a standard plate count, or you have any doubt about your ability to complete this exercise, we suggest that you read Exercise 6-1 carefully and do some or all of the practice problems.

Lab One: Groups 1 and 2

1. Obtain all materials including five Petri dishes. Label the plates GYEA—A through E, respectively. Plate E will be the control.
2. Obtain five microtubes and label them 1 through 5. These are your dilution tubes; make sure they remain covered until needed.
3. Aseptically add 900 µL (0.9 mL) sterile water to dilution tubes 2, 3, 4, and 5. Cover when finished.
4. Add 10 g of the soil sample to the 100 mL water bottle. Shake vigorously for several minutes. This is DF (dilution factor) 10^{-1}. If you don't remember how to calculate DF, refer to Exercise 6-1.
5. Aseptically transfer 1000 µL of the suspended solution from the bottle to dilution tube 1. Remember—this is DF 10^{-1}.
6. Aseptically transfer 100 µL from dilution tube 1 to dilution tube 2; mix well. This is DF 10^{-2}.
7. Aseptically transfer 100 µL from dilution tube 2 to dilution tube 3; mix well. This is DF 10^{-3}.
8. Aseptically transfer 100 µL from dilution tube 3 to dilution tube 4; mix well. This is DF 10^{-4}.
9. Tube 5 will remain sterile and used as a control.
10. Remove one molten GYEA tube from the water bath. Aseptically add 100 µL from dilution tube 1, mix well, and pour into plate A. This is FDF 10^{-2}. Repeat the process using 100 µL from tubes 2 through 5 to plates B through E, respectively. Label the plates with the FDF (final dilution factors). Label plate E "control."
11. Allow the plates time to cool and solidify. Invert and incubate them at 25°C for 2 to 7 days.

Lab One: Groups 3 and 4

1. Obtain all materials including five Petri dishes. Label the plates NA—A through E respectively. Plate E will be the control.
2. Obtain 5 microtubes and label them 1 through 5. These are your dilution tubes; make sure they remain covered until needed.
3. Aseptically add 900 µL (0.9 mL) sterile water to each dilution tube. Cover when finished.
4. Group 1 will have added 10 g of the soil sample to the 100 mL water bottle. (This is DF 10^{-1}. If you don't remember how to calculate dilution factors, refer to Exercise 6-1.)
5. Obtain the bottle from Group 1, mix the sample well, and aseptically transfer 100 µL of the suspended solution to dilution tube 1. Mix the contents of tube 1. This is DF 10^{-2}.
6. Aseptically transfer 100 µL from dilution tube 1 to dilution tube 2; mix well. This is DF 10^{-3}.
7. Aseptically transfer 100 µL from dilution tube 2 to dilution tube 3; mix well. This is DF 10^{-4}.
8. Aseptically transfer 100 µL from dilution tube 3 to dilution tube 4; mix well. This is DF 10^{-5}.
9. Tube 5 will remain sterile and used as a control.
10. Remove one molten NA tube from the water bath. Aseptically add 100 µL from dilution tube 1, mix well, and pour into plate A. Repeat the process using 100 µL from tubes 2 through 5 to plates B through E, respectively. Label the plates with the FDF (final dilution factors). Label plate E "control."
11. Allow the plates time to cool and solidify. Invert and incubate them at 25°C for 2 to 7 days.

Lab One: Groups 5 and 6

1. Obtain all materials including five Petri dishes. Label five plates SDA—A through E, respectively. Plate E will be the control.
2. Obtain four microtubes and label them 1 through 4. These are your dilution tubes; make sure they remain covered until needed.
3. Aseptically add 900 µL (0.9 mL) sterile water to dilution tubes 2 through 4. Cover when finished.
4. Group 1 will have added 10 g of the soil sample to the 100 mL water bottle. (This is DF 10^{-1}. If you don't remember how to calculate dilution factors, refer to Exercise 6-1.)
5. Obtain the bottle containing the sample, mix well and aseptically transfer 1500 µL of the suspended solution to dilution tube 1. Remember—this is DF 10^{-1}.
6. Aseptically transfer 100 µL from dilution tube 1 to dilution tube 2; mix well. This is DF 10^{-2}.
7. Aseptically transfer 100 µL from dilution tube 2 to dilution tube 3; mix well. This is DF 10^{-3}.
8. Tube 4 will remain sterile and used as a control.
9. Remove one molten SDA tube from the water bath. Aseptically add 1000 µL (1.0 mL) from dilution tube 1; mix well and pour into plate A.
10. Remove another molten SDA tube from the water bath. Aseptically add 100 µL from tube 1 to an SDA tube; mix well and pour into plate B. Repeat the process using 100 µL from tubes 2 through 4 to plates C through E, respectively. Label the plates with the FDF (the final dilution factor of plate A will remain the same as the source tube because it was inoculated with 1000 µL rather than 100 µL). Label plate E "control."

11. Allow the plates time to cool and solidify. Invert and incubate them at 25°C for 2 to 7 days.

Lab Two: All Groups

1. After incubation, set aside the countable plates (those with 30 to 300 colonies) and properly dispose of the uncountable plates. Only one plate should be countable.
2. Count the colonies on the plate and calculate the original cell density (OCD) of the water, using the formula provided.
3. Record the data on the chart in the Data Sheet.

References

Atlas, Ronald M., and Richard Bartha. 1998. Pages 367–374 in *Microbial Ecology —Fundamentals and Applications,* 4th ed. Benjamin/Cummings Science Publishing. Menlo Park, CA.

DIFCO Laboratories. 1984. Page 768 in *DIFCO Manual,* 10th ed. DIFCO Laboratories, Detroit.

Krieg, Noel R. and John G. Holt (Editor-in-Chief). 1984. Page 1383 in *Bergey's Manual of Systematic Bacteriology,* Vol. 1. Lippincott Williams and Wilkins, Baltimore.

Varnam, Alan H., and Malcolm G. Evans. 2000. Page 88 in *Environmental Microbiology*. ASM Press, Washington, DC.

Food Microbiology

Food microbiology is the study, utilization, and control of microorganisms in food. Many of our favorite foods—such as yogurt, wine, beer, sauerkraut, buttermilk, vinegar, bread, and cheeses—are produced by or with the help of microorganisms. Conversely, some extremely potent toxins can be found in under-processed canned foods. Many illnesses are caused by, and can be prevented by, avoiding improper handling of foods. In this section you will use beneficial organisms to produce a food; you will perform a test of milk quality; and you will test the effectiveness of a typical fermentation starter culture as a food preservative.

Exercise 8-5

Methylene Blue Reductase Test

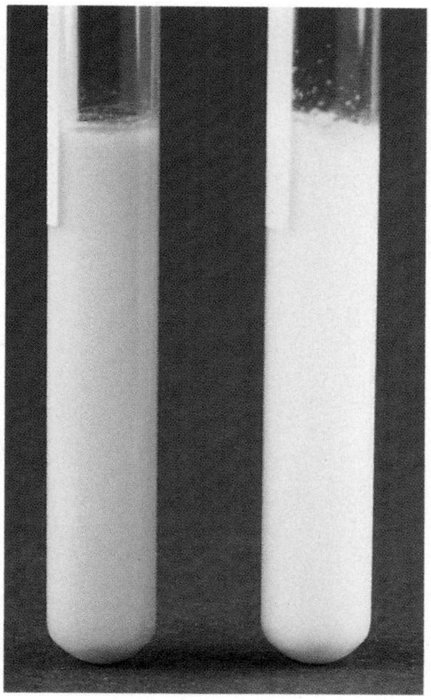

FIGURE 8-16 Methylene Blue Reductase Test The tube on the left is a control to illustrate the original color of the oxidized medium. The tube on the right indicates bacterial reduction of methylene blue after 20 hours. The speed of reduction is related to the concentration of microorganisms present in the milk.

Theory Methylene blue dye is blue when oxidized and colorless when reduced. It can be enzymatically reduced either aerobically or anaerobically. In the aerobic electron transport system, methylene blue is reduced by cytochromes, but immediately is returned to the oxidized state when it subsequently reduces oxygen. Anaerobically, the dye is in the reduced form, and in the absence of an oxidizing substance, remains colorless.

The reduction of methylene blue may be used as an indicator of milk quality. In the methylene blue reductase test, a small quantity of a dilute methylene blue solution is added to a sterilized test tube containing raw milk. The tube then is tightly sealed and incubated in a 35°C water bath. The time it takes the milk to turn from blue to white (because of methylene blue reduction) is a qualitative indicator of the number of microorganisms living in the milk (Figure 8-16). Good-quality milk takes longer than 6 hours to convert the methylene blue.

Application This test is helpful in differentiating the *enterococci* from other streptococci. It also tests for the presence of coliforms in raw milk.

In This Exercise...

You will test milk quality by measuring how long the indicator dye, methylene blue, takes to become oxidized by any bacterial contaminants present.

Materials

Per Student Group

- raw milk samples
- sterile screw-capped test tubes
- sterile 1 mL and 10 mL pipettes with mechanical pipettors
- hot water bath set at 35°C
- methylene blue reductase reagent
- overnight broth culture of *Escherichia coli*
- clock or wristwatch

Reagent

Methylene Blue Reductase Reagent

- Methylene blue dye 8.8 mg
- Distilled or deionized water 200.0 mL

Procedure

Lab One

1. Obtain sterile tubes for as many samples as you are testing, plus two more to be used as positive and negative controls. Label them appropriately.

2. Using a sterile 10 mL pipette, aseptically add 10 mL milk to each tube.
3. Inoculate the tube marked "positive control" with 1 mL of *E. coli* culture.
4. Aseptically add 1.0 mL methylene blue solution to each test tube. Cap the tubes tightly and invert them several times to mix thoroughly.
5. Place the tube marked "negative control" in the refrigerator to prevent it from changing color.
6. Place all other tubes in the hot water bath and note the time.
7. After 5 minutes, remove the tubes, invert them once to mix again, then return them to the water bath. Record the time in the chart in the Data Sheet under "Starting Time."
8. Using the control tubes for color comparison, check the tubes at 30-minute intervals and record the time when each becomes white. Poor-quality milk takes less than 2 hours; good-quality milk takes longer than 6 hours. If necessary, have someone check the tubes at 6 hours and record the results for you.
9. Using the chart provided, calculate the time it takes for each milk sample to become white.

References

Bailey, R. W., and E. G. Scott. 1966. Pages 114 and 306 in *Diagnostic Microbiology*, 2nd ed. Mosby, St. Louis.

Benathen, Isaiah. 1993. Page 132 in *Microbiology with Health Care Applications*. Star Publishing, Belmont, CA.

Power, David A. and Peggy J. McCuen. 1988. Page 62 in *Manual of BBL™ Products and Laboratory Procedures*, 6th ed. Becton Dickinson Microbiology Systems, Cockeysville, MD.

Richardson, first name or initials, editor. 1985. *Standard Methods for the Examination of Dairy Products*, 15th ed. American Public Health Association, Washington DC.

Exercise 8-6

Viable Cell Preservatives

Theory *Lactococcus lactis*, (formerly called *Streptococcus lactis*) is a naturally occurring organism in milk that produces the antibiotic **nisin.** It is believed that nisin from *L. lactis* (even when not actively metabolizing) slows growth of refrigerated **psychotrophic** organisms. Its presence, therefore, can slow milk spoilage.

Application This exercise examines the inhibitory effect of the *Lactococcus lactis* on refrigerated cultures.

In This Exercise...

Today you will begin a long-term exercise designed to compare the times required for *Enterococcus*, *Staphylococcus*, *Pseudomonas*, *Clostridium*,and *Salmonella* organisms to curdle treated and untreated refrigerated milk. Readings will be performed semi-weekly up to 4 weeks.

Materials

Per Student Group

- sterile skim milk medium
- sterile 1 ml pipettes
- Methylene blue
- pipette pump
- thermometer
- overnight broth cultures of:
  - *Pseudomonas fluorescens*
  - *Clostridium sporogenes*
  - *Staphylococcus aureus*
  - *Salmonella typhimurium*
  - *Enterococcus faecalis*
  - *Lactococcus lactis*

Procedure

Lab One

1. Obtain 12 tubes of skim milk and one of each culture. Organize the tubes into pairs and label them by name, two for each organism. Label one of each pair "L" to indicate that it is a tube to receive *L. lactis*. Label the two extra tubes "control" and "control L."
2. Mix the overnight cultures well.
3. Add three drops of each culture to its correspondingly labeled tubes.
4. Add one mL of *L. lactis* to each milk broth labeled with an "L," including "control L." Mix.
5. Immediately place all tubes in a refrigerator set at 5°C to 10°C.

Lab Two

1. Twice per week for up to 4 weeks, observe the skim milk medium for curdling.
2. Record your results in the chart on the Data Sheet.
3. On the graph paper provided, prepare a histogram of organism versus time. Be sure to include the control. It may be helpful to construct double bars comparing treated samples with untreated samples.

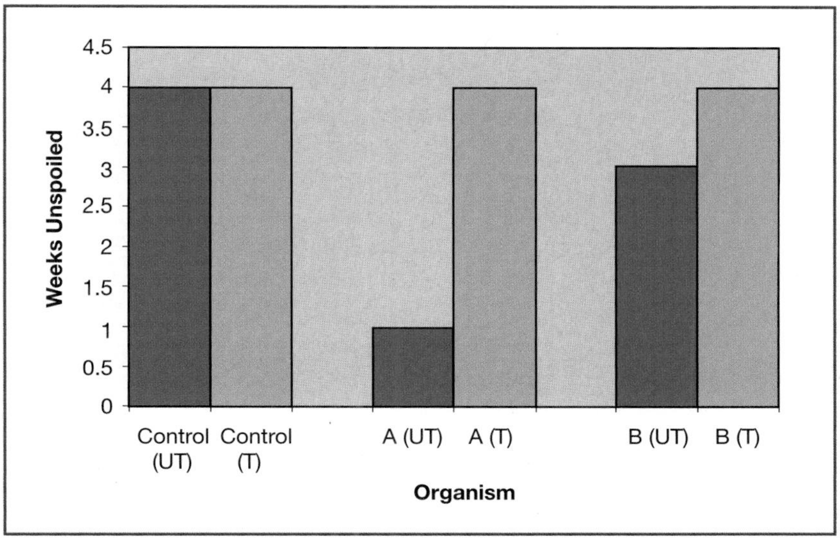

References

Bibek, Ray. 2001. Chapter 13 in *Fundamental Food Microbiology,* 2nd ed. CRC Press LLC, Boca Raton, FL.

Krieg, Noel R., and John G. Holt (Editor-in-Chief). 1984. Page 518 in *Bergey's Manual of Systematic Bacteriology,* Vol. 1. Lippincott Williams and Wilkins, Baltimore.

Exercise 8-7

Making Yogurt

Theory Several species of bacteria are used in the commercial production of yogurt. Most formulations include combinations of two or more species to synergistically enhance growth and to produce the optimum balance of flavor and acidity. One common pairing of organisms in commercial yogurt is that of *Lactobacillus delbrueckii* subsp *bulgaricus* and *Streptococcus thermophilus*.

Yogurt gets its unique flavor from acetaldehyde, diacetyl, and acetate produced from fermentation of the milk sugar lactose. The proportions of products, and ultimately the flavor, in the yogurt depend upon the types of enzyme systems possessed by the species used. Both species mentioned above contain **constitutive** β-galactosidase systems that break down lactose and convert the glucose to lactate, formate, and acetate via pyruvate in the Embden-Meyerhof-Parnas (glycolysis) pathway. (See Appendix A.)

As you may remember, lactose is a disaccharide composed of glucose and galactose. *S. thermophilus* does not possess enzymes needed to metabolize galactose, and *L. delbrueckii* preferentially metabolizes glucose. This results in an accumulation of galactose, which adds sweetness to the yogurt. Acetaldehyde is produced directly from pyruvate by *S. thermophilus* and through the conversion of proteolysis products threonine and glycine by *L. delbrueckii*. Some strains of *S. thermophilus* also produce glucose polymers that give the yogurt a viscous consistency.

Application This exercise is designed to keep you away from the TV for a few minutes.

In This Exercise…

You will produce yogurt with a simple home recipe using a commercial yogurt as a starter. Read the label to see which microorganisms are included. We hope you enjoy it.

Materials

Per Student Group

- whole, low-fat, or skim milk
- plain yogurt with active cultures (bring from home or supermarket)
- medium-size saucepan
- medium-size bowl
- wire whisk
- hotplate
- cooking thermometer
- measuring cup
- plastic wrap
- fresh fruit
- sugar (optional)
- pH meter or pH paper

Procedure

Lab One

1. Obtain all materials and set them up in a clean work area.
2. While stirring, slowly heat five cups milk in the saucepan to 185°F. Remove the milk from the heat, and let it cool to 110°F.
3. Place 1/4 cup starter yogurt in the bowl. Slowly, about 1/3 to 1/2 cup at a time, stir in cooled milk, mixing after each addition until smooth.
4. Cover the bowl with plastic wrap and puncture several times to allow gases and excess moisture to escape.
5. Label the bowl with your name, the date, and the cultures present in your yogurt starter.
6. Incubate 5–6 hours at 30–35°C. Remove the bowl from the incubator at the correct time and place it in the refrigerator.

Lab Two

1. Remove your yogurt from the refrigerator
2. Compare flavor, consistency, and starter cultures with other groups in the lab. Measure the pH of your yogurt with a pH meter or pH paper. Record your results in the chart provided on the Data Sheet.
3. Serve with fresh fruit and enjoy!

References

Downes, Frances Pouch, and Keith Ito. 2001. Chapter 47 in *Compendium of Methods for the Microbiological Examination of Foods,* 4th ed. American Public Health Association, Washington, DC.

Ray, Bibek. 2001. Chapter 13 in *Fundamental Food Microbiology,* 2nd ed. CRC Press LLC, Boca Raton, FL

Microbial Genetics

SECTION 9

In this unit you will be involved in four exercises dealing with DNA. First you will perform a simple extraction of *E. coli* DNA. In the next two exercises you will examine mutations—alterations of DNA. The first of these examines the effects of ultraviolet (UV) radiation on bacteria. It illustrates some characteristics of a particular *mutagen*—how it causes damage, factors affecting its impact on the cell, and how bacteria are able to repair that damage. The second mutation exercise—the "Ames Test"—illustrates a simple method of screening substances (commercial products) for mutagenicity, and in turn, carcinogenicity! As such, it is an important test in the fight against cancer. The fourth exercise incorporates into one exercise bacterial transformation, use of antibiotic selective media, genetic regulation, and genetic engineering. You will be transferring a specific gene into *E. coli* cells, selecting for only those cells that have been transformed successfully, then manipulating the environment so the organisms produce the gene product only when you want them to.

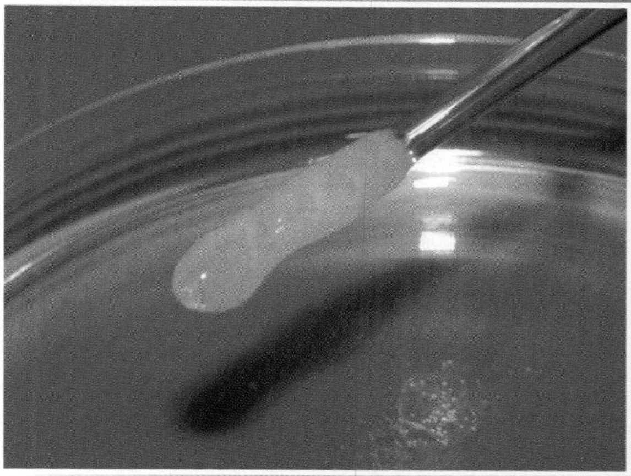

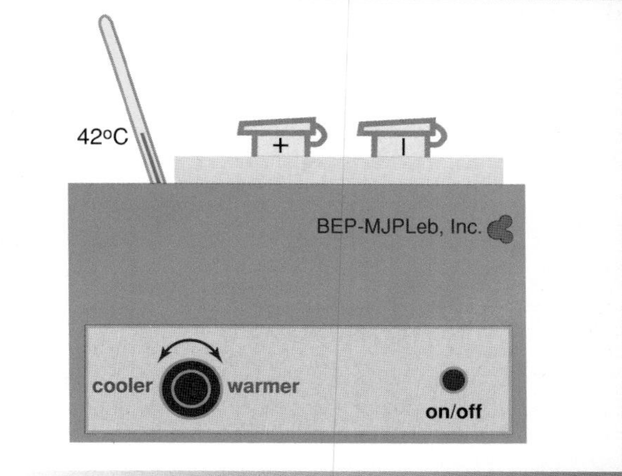

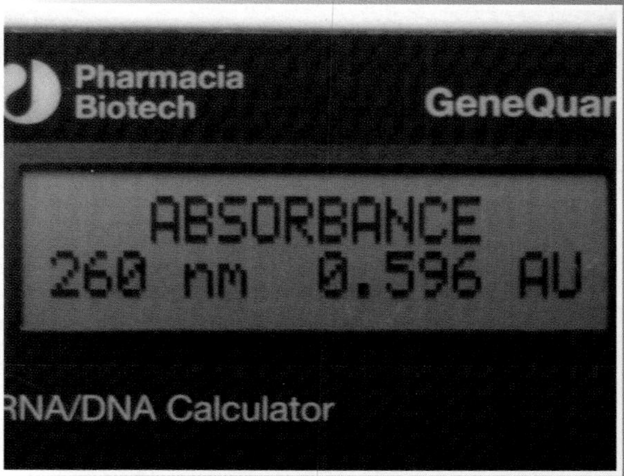

Exercise 9-1

Extraction of DNA from Bacterial Cells[1]

Theory DNA extraction from cells is surprisingly easy and occurs in three basic stages.

1. A detergent (*e.g.*, Sodium Dodecyl Sulfate—SDS) is used to lyse cells and release cellular contents, including DNA.
2. This is followed by a heating step (at approximately 65–70°C) that denatures protein (including DNases that would destroy the extracted DNA) and other cell components. Temperatures higher than 80°C will denature DNA, and this is undesirable. A protease also may be added to remove proteins. (Other techniques for purification may be used, but these will not be included in this exercise.)
3. Finally, the water-soluble DNA is precipitated in cold alcohol as a whitish, mucoid mass (Figure 9-1).

As an optional follow-up to extraction, an ultraviolet spectrophotometer (Figure 9-2) will be used to estimate DNA concentration in the sample by measuring absorbance at 260 nm, the optimum wavelength for absorption by DNA. An absorbance of A_{260nm} of 1 corresponds to 50 μg/mL of double-stranded DNA (dsDNA). Reading absorbance at 280 nm and calculating the following ratio can determine purity of the sample:

$$\frac{\text{Absorbance}_{260nm}}{\text{Absorbance}_{280nm}}$$

[1] *Note:* Thanks to the following individuals who offered helpful suggestions for this protocol: Dr. Melissa Scott and Allison Shearer of San Diego City College, Donna Mapston and Dr. Ellen Potter of Scripps Institute for Biological Studies, and Dr. Sandra Slivka of Miramar College.

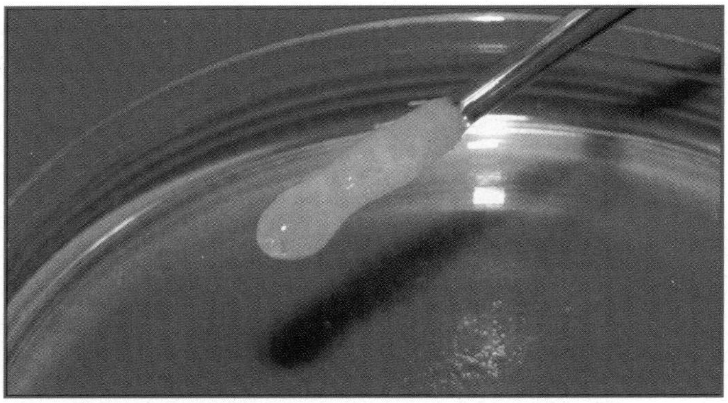

FIGURE 9-1 PRECIPITATED DNA
This onion DNA has been spooled onto a glass rod.

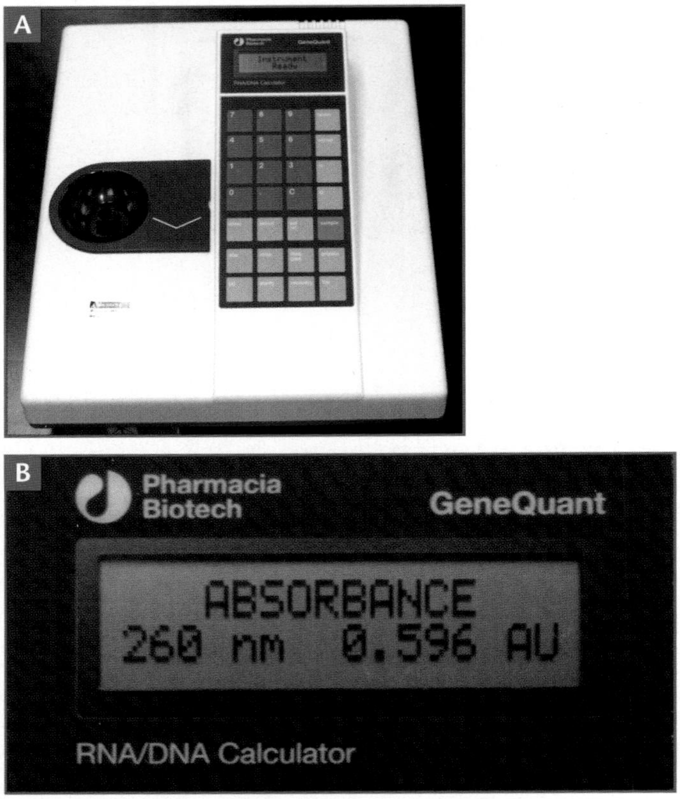

FIGURE 9-2 ULTRAVIOLET SPECTROPHOTOMETER
(A) A UV spectrophotometer can be used to determine DNA concentration. A quartz cuvette is shown in the sample port. (B) This specimen has an A_{260nm} of 0.596. Because an A_{260nm} of 1.0 is equal to 50 μg/mL of dsDNA, this specimen has a concentration of 29.8 μg/mL. Absorbance also can be used to determine purity of the sample. A relatively pure DNA sample will have an A_{260nm}/A_{280nm} value of 1.8.

If the sample is reasonably pure nucleic acid, the ratio will be about 1.8 (between 1.65 and 1.85). Because protein absorbs maximally at 280 nm, a ratio of less than 1.6 is likely because of protein contamination. If purity is crucial, the DNA extraction is repeated. If the ratio is greater than 2.0, the sample is diluted and read again.

Application Extraction of DNA is a starting point for many lab procedures, including DNA sequencing and cloning.

In This Exercise...

You will extract DNA from *E. coli*. To improve yield, you first will concentrate an *E. coli* broth culture. The actual extraction involves cell lysis, denaturation of protein, and precipitation of the DNA in alcohol. Following extraction, an optional procedure may be used to determine DNA yield and the purity of your extract.

Materials

Per Pair of Students

- overnight culture of *Escherichia coli* in Luria-Bertani broth (young cultures work best)
- water bath set to 65°C
- floats
- ice bath (small cups with crushed ice work well)
- 500 µL 10% Sodium Dodecyl Sulfate (SDS)
- 50 µL 20 mg/mL Proteinase K solution (stored in freezer between uses) (***Note:*** Meat tenderizer is an inexpensive substitute, though it may result in DNA hydrolysis if too much is used.)
- 500 µL 1.0 M Sodium Acetate solution (pH = 5.2)
- 3 mL 1X M Tris-Acetate-EDTA (TAE) buffer (dilute 10X TAE 9+1)
- 2 mL 90% Isopropanol (stored in a freezer or an ice bath)
- two calibrated disposable transfer pipettes
- 100–1000 µL digital pipettor and tips
- 25 mL centrifuge tube
- tabletop centrifuge
- minicentrifuge
- straight glass rod or disposable inoculating loop
- two microtubes (at least 1.5 mL in volume)
- vortex mixer (optional)
- ultraviolet spectrophotometer (optional)
- two quartz cuvettes (optional)

Medium Recipe

Luria-Bertani Broth

• Tryptone	10.0 g
• Yeast Extract	5.0 g
• NaCl	10.0
• Distilled or deionized water	1.0 L

pH 7.4 at 25°C

Procedure

Refer to the procedural diagram in Figure 9-3 as you read and follow this protocol.

1. Obtain the *E. coli* culture. Mix the suspension until uniform turbidity is seen, then transfer 5 mL to a clean, nonsterile centrifuge tube.
2. Spin the 5 mL sample slowly for 10 minutes to produce a cell pellet.
3. After spinning, remove 4.5 mL of the supernatant with a transfer pipette. Dispose of the supernatant in the original culture tube.
4. Transfer the remaining 0.5 mL to a nonsterile microtube.
5. Add 200 µL of 10% SDS to the *E. coli.*
6. Add 30 µL of 20 mg/mL Proteinase K solution.
7. Gently mix the tube for 5 minutes by tipping it upside down every few seconds.
8. Place the tube in a float and incubate in a 65°C water bath for 5 minutes.
9. Add 200 µL mL 1 M Sodium Acetate solution and mix gently.
10. Place the tube in the ice bath until it is at or below room temperature.
11. When cooled, squirt 400 µL mL of cold 95% isopropanol into the preparation. (If there are bubbles on the surface, remove them with a nonsterile transfer pipette prior to adding the isopropanol.)
12. Use a disposable loop to mix the preparation, moving in and out and turning occasionally. A glob of mucoid DNA will begin to appear as you mix and will adhere to the loop.

Optional Procedure (if your lab has an ultraviolet spectrophotometer)

Refer to the procedural diagram in Figure 9-3 as you read and complete the following protocol.

1. Remove the DNA from the original microtube and transfer it to a second microtube, using the loop.
2. Resuspend the DNA in 1000 µL isopropanol using a nonsterile transfer pipette, then spin it in a minicentrifuge for a few seconds. The DNA should be at the bottom of the tube.
3. With a nonsterile transfer pipette, remove the supernatant and allow the DNA pellet to air-dry.
4. Resuspend the dried DNA in 1000 µL of TAE buffer. Mix vigorously to dissolve the DNA in the TAE. This mixing may be done by hand, or a vortex mixer may be used.
5. Transfer the suspended DNA solution into a quartz cuvette.
6. Prepare a second cuvette containing 1000 µL of TAE as a blank.
7. Set the spectrophotometer to 260 nm wavelength. Follow the instructions for your UV spectrophotometer to check the absorbance of the extracted DNA, and record on the Data Sheet.
8. Set the spectrophotometer to 280 nm wavelength. Follow the instructions for your UV spectrophotometer to check the absorbance of the extracted DNA, and record on the Data Sheet.

9. Calculate the probable purity of your extracted DNA sample using the formula provided, and record on the Data Sheet.
10. If desired, absorbencies at other wavelengths may be taken to produce an absorption spectrum for DNA. Suggested wavelengths are: 220 nm, 240 nm, 300 nm, and 320 nm, in addition to the measurements for 260 and 280 nm taken above. Record these on the Data Sheet.
11. If step 10 was done, plot the absorption spectrum (Absorption versus Wavelength) of the DNA sample on the Data Sheet.

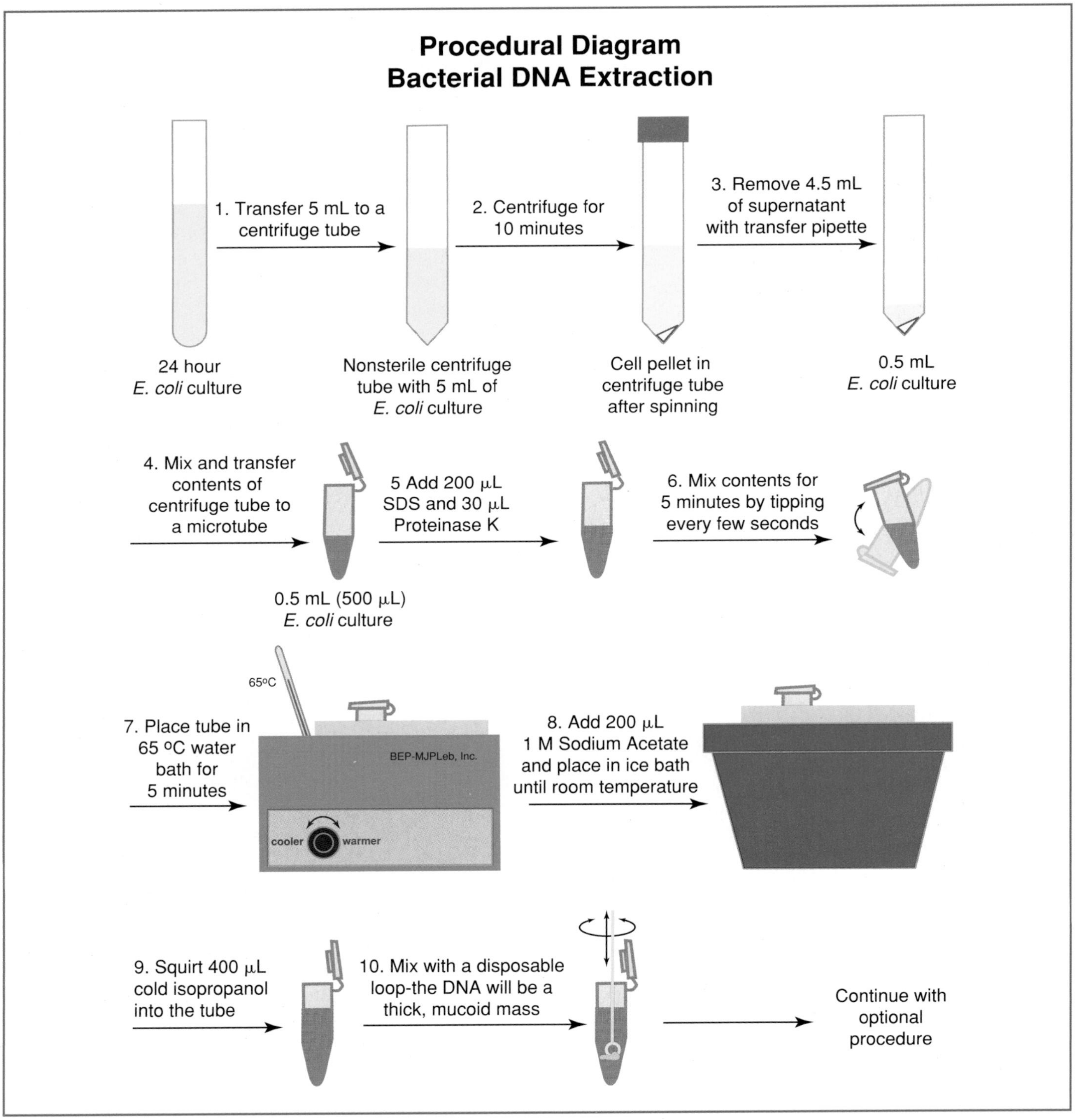
Procedural Diagram
Bacterial DNA Extraction
1. Transfer 5 mL to a centrifuge tube
2. Centrifuge for 10 minutes
3. Remove 4.5 mL of supernatant with transfer pipette
24 hour
E. coli culture
Nonsterile centrifuge tube with 5 mL of E. coli culture
Cell pellet in centrifuge tube after spinning
0.5 mL
E. coli culture
4. Mix and transfer contents of centrifuge tube to a microtube
5 Add 200 µL SDS and 30 µL Proteinase K
6. Mix contents for 5 minutes by tipping every few seconds
0.5 mL (500 µL)
E. coli culture
65°C
7. Place tube in 65 °C water bath for 5 minutes
BEP-MJPLeb, Inc.
cooler
warmer
8. Add 200 µL 1 M Sodium Acetate and place in ice bath until room temperature
9. Squirt 400 µL cold isopropanol into the tube
10. Mix with a disposable loop-the DNA will be a thick, mucoid mass
Continue with optional procedure

FIGURE 9-3 PROCEDURAL DIAGRAM FOR BACTERIAL DNA EXTRACTION

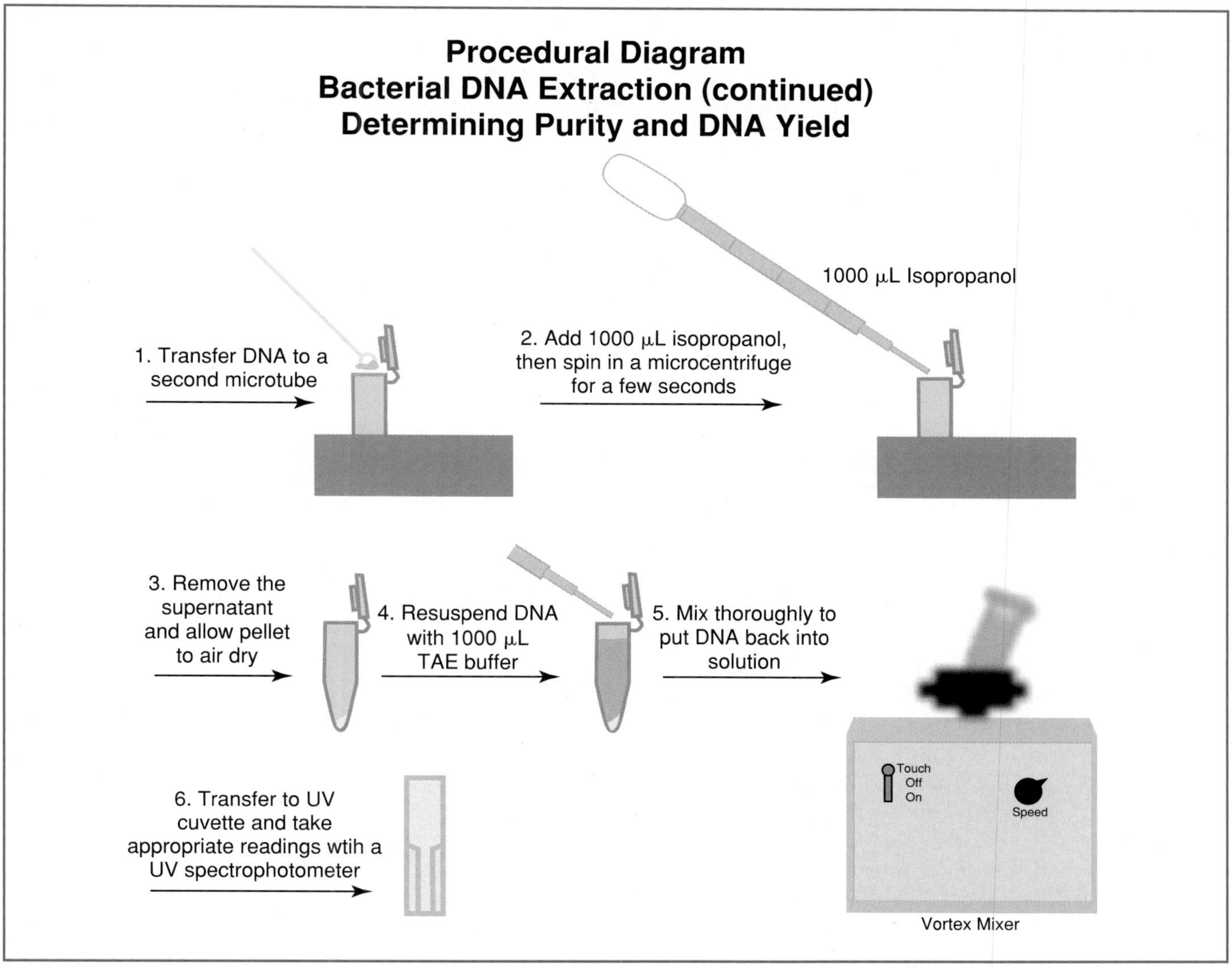
Procedural Diagram
Bacterial DNA Extraction (continued)
Determining Purity and DNA Yield
1000 μL Isopropanol
1. Transfer DNA to a second microtube
2. Add 1000 μL isopropanol, then spin in a microcentrifuge for a few seconds
3. Remove the supernatant and allow pellet to air dry
4. Resuspend DNA with 1000 μL TAE buffer
5. Mix thoroughly to put DNA back into solution
Touch
Off
On
Speed
6. Transfer to UV cuvette and take appropriate readings wtih a UV spectrophotometer
Vortex Mixer

FIGURE 9-3 PROCEDURAL DIAGRAM FOR BACTERIAL DNA EXTRACTION *(continued)*

References

Bost, Rod. 1989. *Down and Dirty DNA Extraction*. Carolina Genes. North Carolina Biotechnology, Research Triangle Park, NC.

Davis, Leonard G., Mark D. Dibner, and James F. Battey. (1986). *Basic Methods in Molecular Biology*. Elsevier Science Publishing, New York.

Freifelder, David. (1982). Pages 504–505 in *Physical Biochemistry*, 2nd ed. W. H. Freeman and Co, New York.

Kreuzer, Helen, and Adrianne Massey. 2001. *Recombinant DNA and Biotechnology—A Guide For Teachers*, 2nd ed. ASM Press, Washington, DC.

Zyskind, Judith W., and Sanford I. Bernstein. 1992. *Recombinant DNA Laboratory Manual*. Academic Press, San Diego.

Exercise 9-2

Ultraviolet Radiation Damage and Repair

Theory

Ultraviolet radiation is part of the electromagnetic spectrum, but with shorter, higher energy wavelengths than visible light. Prolonged exposure can be lethal to cells because when DNA absorbs UV radiation at 254 nm, the energy is used to form new covalent bonds between adjacent pyrimidines: cytosine-cytosine, cytosine-thymine, or thymine-thymine. Collectively, these are known as **pyrimidine dimers**, with thymine dimers being the most common. These dimers distort the DNA molecule and interfere with DNA replication and transcription (Figure 9-4).

Many bacteria have mechanisms to repair such DNA damage. *Escherichia coli* performs **light repair** or **photoreactivation**, in which the repair enzyme, **DNA photolyase**, is activated by light (340–400nm) and simply monomerizes the dimer by reversing the original reaction.

A second *E. coli* repair mechanism, **excision repair** or **dark repair**, involves a number of enzymes (Figure 9-5). The thymine dimer distorts the sugar-phosphate backbone of the strand. This is detected by an **endonuclease** (UvrABC) that breaks two bonds—eight nucleotides in the 5' direction from the dimer, and the other four nucleotides in the 3' direction. A **helicase** (UvrD) removes the 13-nucleotide fragment (including the dimer), leaving single-stranded DNA. **DNA polymerase I** inserts the appropriate complementary nucleotides in a 5' to 3' direction to make the molecule double-stranded again. Finally, **DNA ligase** closes the gap between the last nucleotide of the new segment and the first nucleotide of the old DNA, and the repair is complete. Both mechanisms are capable of repairing a small amount of damage, but long and/or intense exposures to UV produce more damage than the cell can repair, making UV radiation lethal.

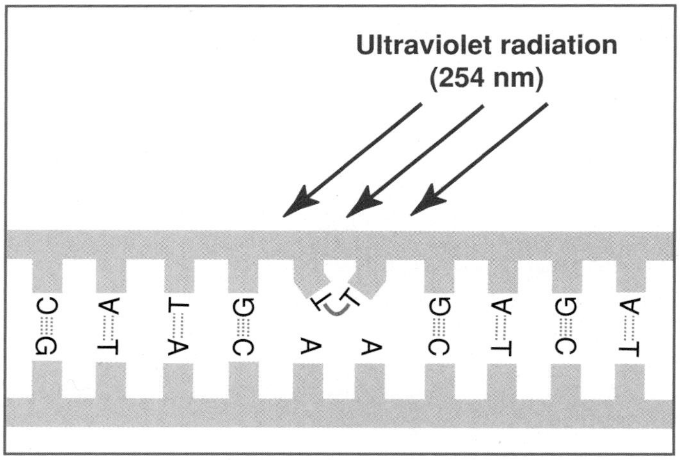

FIGURE 9-4 A THYMINE DIMER IN ONE STRAND OF DNA
Thymine dimers form when DNA absorbs UV radiation with a wavelength of 254 nm. The energy is used to form two new covalent bonds between the thymines, resulting in distortion of the DNA strand. The enzyme DNA photolyase can break this bond to return the DNA to its normal shape and function. If it doesn't, the distortion interferes with DNA replication and transcription.

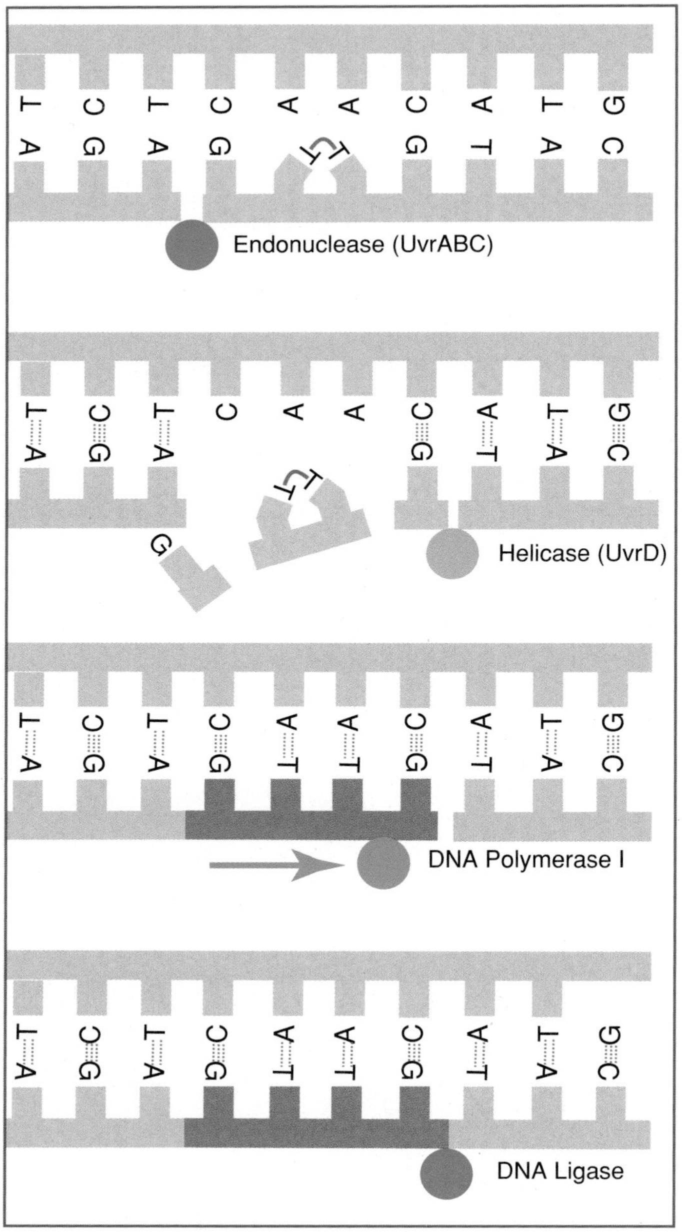

FIGURE 9-5 EXCISION OR DARK REPAIR IN *E. COLI*
In the sequence, four enzymes are used: (1) An endonuclease (UvrABC) to break two covalent bonds in the sugar-phosphate backbone of the damaged strand, (2) a helicase (UvrD) to remove the nucleotides in the damaged segment, (3) DNA polymerase I to synthesize a new strand, and (4) DNA ligase to form a covalent bond between the new and the original strands. In reality, the segment excised is 13 nucleotides long.

Application

Because ultraviolet radiation has a lethal effect on bacterial cells, it can be used in disinfection. Its use is limited, however, because it penetrates materials such as glass and plastic poorly. In addition, bacterial cells have mechanisms to repair UV damage. This exercise illustrates the effects of ultraviolet radiation on bacterial cells and allows observation of cellular repair in response to the exposure. It also demonstrates the ability of UV to penetrate other matter.

In This Exercise...

The lethal effects of ultraviolet radiation, its ability to penetrate various objects, and the cells' ability to repair UV damage will all be demonstrated. Do not be misled by the apparent simplicity of the experimental design—there's a lot going on!

Materials

Per Student Group

- seven Nutrient Agar plates
- disinfectant
- posterboard masks with 1" to 2" cutouts (Figure 9-6)
- UV light (shielded for eye protection)
- sterile cotton applicators
- 24-hour broth culture of *Serratia marcescens*

Procedure

Lab One

Refer to the Procedural Diagram in Figure 9-7 as you read and follow this procedure:[1]

1. Using a sterile cotton applicator, streak a Nutrient Agar plate to form a bacterial lawn over the entire surface by using a tight pattern of streaks. Rotate the plate one-third of a turn and repeat, then rotate it another one-third of a turn and streak one last time. Repeat this process for all remaining plates.
2. Number the plates 1, 2, 3, 4, 5, 6, and 7.
3. Remove the lid from plate 1 and set it on a disinfectant-soaked towel. Place the plate under the UV light and cover it with a mask.

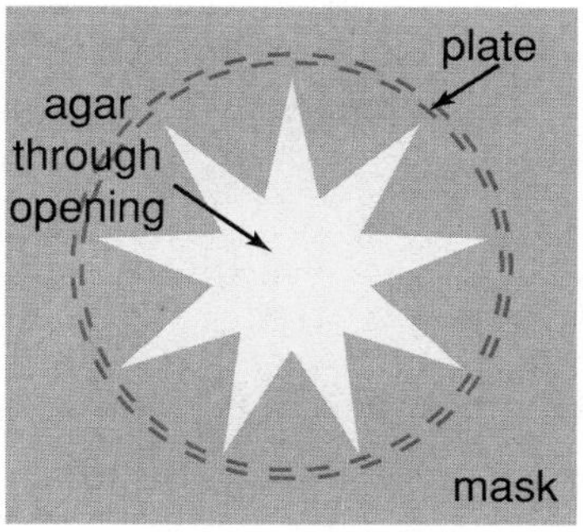

FIGURE 9-6
Posterboard Mask
This is an example of a mask placed over a Petri dish (shown as a dotted line). The cutout may be any shape but should leave the outer 25% of the plate masked.

4. Turn the UV light on, but do not look at it. After 30 seconds, turn off the UV light, remove the mask and plate, and replace its lid.
5. Repeat the process for plate 2. (If space allows, you may combine this step with step 4.)
6. Repeat the process for plate 3, but leave the UV light on for 3 minutes.
7. Irradiate plate 4 for 3 minutes, but leave the lid on and cover with the mask.
8. Repeat step 7 for plate 5. (If space allows, you may combine this step with step 7.)
9. Do not irradiate plates 6 and 7.
10. Incubate plates 1, 3, 4, and 6 for 24 to 48 hours at room temperature in an inverted position where they can receive natural light (*e.g.*, a windowsill) for 24–48 hours.
11. Wrap plates 2, 5, and 7 in aluminum foil, invert them, and place with the others for 24–48 hours.

Lab Two

1. Remove the plates and examine the growth patterns.
2. Record your results on the Data Sheet.

References

Moat, Albert G., John W. Foster, and Michael P. Spector. 2002. Chapter 3 in *Microbial Physiology*, 4th ed. Wiley-Liss, New York.

Nelson, David L., and Michael M. Cox. 2005. Chapter 25 in *Lehninger's Principles of Biochemistry*. W. H. Freeman and Co., New York.

White, David. 2000. Chapter 19 in *The Physiology and Biochemistry of Prokaryotes*. Oxford University Press, Inc. New York.

[1] Thanks to Roberta Pettriess of Wichita State University for her helpful suggestions to improve this exercise.

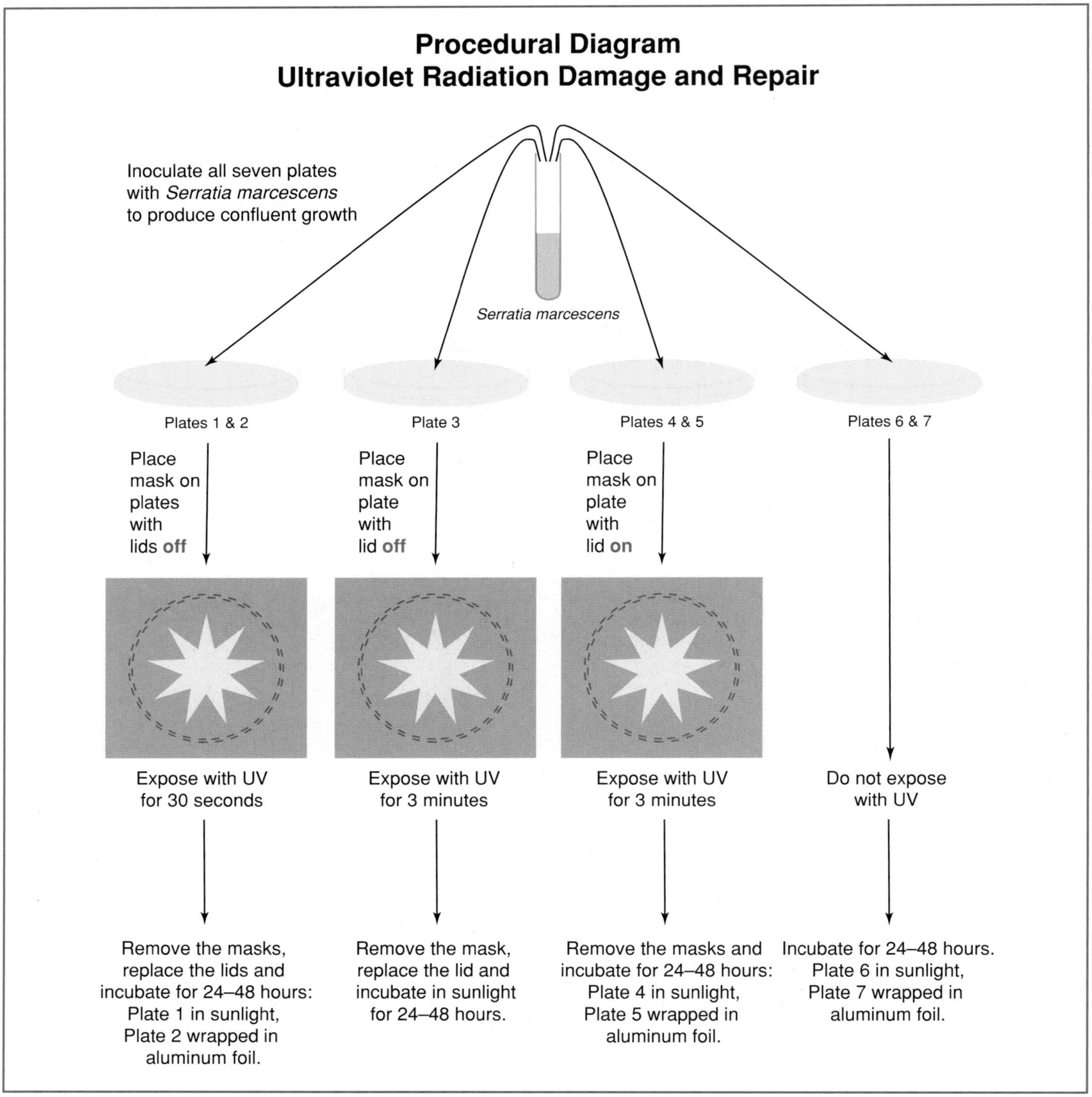

FIGURE 9-7 PROCEDURAL DIAGRAM
Inoculate and expose the plates as directed. Be sure to shield the UV light source adequately. Do not look at the light.

Exercise 9-3

Ames Test

Theory

Bacteria that are able to enzymatically synthesize a particular necessary biochemical (such as an amino acid) are called **prototrophs** for that biochemical. If the bacterial strain can make the amino acid histidine, for instance, it is classified as a histidine prototroph. Bacteria that must be supplied with that biochemical (*e.g.,* histidine) are called (histidine) **auxotrophs.** Auxotrophs typically are created from prototrophs when a mutation occurs in the gene coding for an enzyme used in the pathway of the biochemical's synthesis.

The Ames Test employs mutant strains of *Salmonella typhimurium* that have lost their ability to synthesize histidine. One strain of histidine auxotrophs possesses a **frameshift mutation**; they are missing one or have an extra nucleotide in the DNA sequence that otherwise would code for an enzyme necessary for histidine production. Other strains are **substitution mutants**, in which one nucleotide in the histidine gene has been replaced, resulting in a faulty gene product.

The Ames Test determines the ability of chemical agents to cause a reversal—a **back mutation**—of these auxotrophs to the original prototrophic state. In this test, histidine auxotrophs are spread onto **minimal agar plates** that contain all nutrients for growth but only a trace of histidine.

When a filter paper disk saturated with a suspected mutagen is placed in the middle of the minimal agar plate, the substance will diffuse outward into the medium. If it is mutagenic, it will cause back mutation in some cells (converting them to histidine prototrophs) that freely grow into full-size colonies. (***Note:*** Histidine initially is included in the medium to allow the auxotrophs to grow for several generations and expose them to the effects of the mutagen. The unmutated auxotrophs typically exhibit faint growth but do not develop into full-size colonies, because of the rapid exhaustion of the histidine.) A sample plate is shown in Figure 9-8.

Several variations of the Ames test are possible. This exercise uses two minimal agar plates and two **complete agar plates**. Complete agar contains all of the nutrients necessary for growth of *Salmonella*. All four plates are inoculated with histidine auxotrophs. One minimal agar plate and one complete agar plate each receive a filter paper disk saturated with the test substance. The second minimal agar plate and complete agar plate each receive a filter paper disk saturated with Dimethyl Sulfoxide, DMSO (a substance known to be nonmutagenic).

The minimal agar plate containing the test substance determines mutagenicity of the test substance. Only

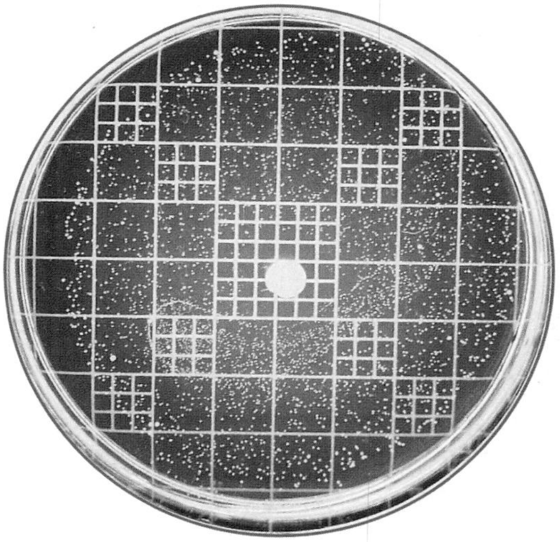

FIGURE 9-8 TESTING FOR MUTAGENICITY
In this example, a *Salmonella* histidine auxotroph is grown on histidine minimal agar and is exposed to a suspected mutagen. The colonies are derived from cells that have undergone back mutations. It is not known from this plate alone, however, whether they are a result of the effects of the test substance, so it must be compared to a control minimal agar plate in which a nonmutagenic substance (DMSO) is substituted for the test substance.

prototrophs will grow on minimal agar, so recovery of any colonies on the minimal agar inoculated with auxotrophs is indicative of back mutation. Depending on the strain used, the type of mutation (either frameshift or base substitution) also can be determined. The minimal agar plate with DMSO serves as a control for the minimal agar/test substance plate by measuring **spontaneous back mutations** (natural mutations that occur without the presence of a mutagen).

The purpose of the complete agar plate containing the test substance is a control to evaluate toxicity of the test substance. Creation of a **zone of inhibition** around the disk indicates toxicity (Figure 9-9). The more toxic the substance is, the larger the zone will be. If the substance is toxic, there may be no indication of mutagenicity because the cells are killed before they can back-mutate. Finally, the complete agar plate containing DMSO serves as a control for comparison to the growth and zone of inhibition on the complete agar/test substance plate.

Application

Many substances that are mutagenic to bacteria are also carcinogenic to higher animals. The Ames Test is a rapid, inexpensive means of using specific bacteria to evaluate the mutagenic properties of potential carcinogens. Many variations of the basic Ames Test are used in specific applications.

In This Exercise...

You will test a household substance for its mutagenic properties. If you want the best chance of a positive

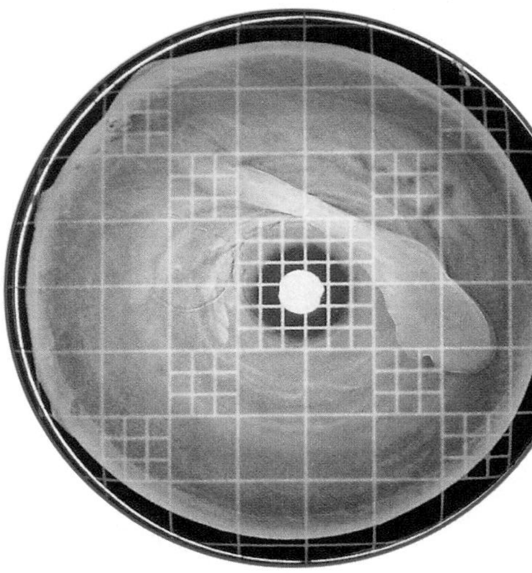

FIGURE 9-9 Testing for Toxicity
Shown is a *Salmonella* histidine auxotroph grown on complete agar (containing histidine) and exposed to the test substance. The zone of inhibition around the paper disk indicates toxicity of the substance. (Note that a zone of inhibition is visible on the plate shown in Figure 9-8, but it is not as well defined as on complete agar because the growth is not as dense.)

result, check below your sink and in the garage for products that have many organic chemicals in the ingredient list. When handling these materials, be sure to wear gloves. Use caution in the laboratory if the material is flammable.

Materials

Per Group

- four Minimal Medium (MM) plates
- four Complete Medium (CM) plates
  (*Note:* This exercise can be run with two plates of each medium if replicates are not desired.)
- centrifuge
- two sterile centrifuge tubes
- small beaker containing alcohol and forceps
- bottle of 1x Vogel-Bonner salts (To make 1x Vogel-Bonner salts, add 1.0 mL 50x Vogel-Bonner solution to 49 mL water.) (Appendix A)
- eight sterile filter disks made with a paper punch
- two sterile Petri dishes (for soaking filter paper disks)
- two sterile transfer pipettes
- 100 µL–1000 µL digital pipettors and tips
- container for disposal of supernatant (to be autoclaved)
- DMSO
- test substance (any substance that has possible mutagenic properties and does not contain histidine or protein)
- hand counter
- broth culture[1] of *Salmonella typhimurium TA 1535*
- broth culture[1] of *Salmonella typhimurium TA 1538*

Procedure

Day One

Follow the procedural diagram in Figure 9-10.

1. Soak four filter paper disks in DMSO and four filter paper disks in the test substance.
2. Pipette 10.0 mL TA 1535 into a sterile centrifuge tube. Do the same with TA 1538, then label the tubes.
3. Centrifuge the tubes on high speed for 10 minutes. Be sure the centrifuge is balanced.
4. Being careful not to disturb the cell pellet at the bottom, decant the supernatant from each centrifuge tube using a transfer pipette.
5. Resuspend the cell pellets by adding 1.0 mL sterile 1x Vogel-Bonner salts to each tube and mixing well.
6. Using the spread plate technique (see Exercise 1-4), inoculate two MM plates and two CM plates with 100 µL of the resuspended TA 1535. Do the same with TA 1538.
7. Flame the forceps by passing them through the Busen burner flame, and allowing the alcohol to burn off.
8. Using the flamed forceps, place the DMSO and test disks in the centers of the plates. Gently tap down the disks with the forceps to prevent them falling off when the plates are inverted.
9. Incubate the plates aerobically at 35°C for 48 hours.

Day Two

1. Measure and compare the zones of inhibition measured in millimeters on the CM plates.
2. Count the colonies on the MM plates and compare. Count only the large colonies, not the "hazy" background growth. (These are the colonies produced by auxotrophs that did not back-mutate to prototrophs and only grew until the histidine in the minimal medium was exhausted.)
3. Record your observations for the Ames Test plates on the Data Sheet.

[1] The cultures used for this exercise must be prepared as follows:

1. 24 hours before the test, inoculate two 10.0 mL broth tubes with TA 1535 and TA 1538. Incubate at 35 ± 2°C together with two sterile 90.0 mL broths (in a 100mL diluent bottle).
2. 5½ hours before the test, pour the TA 1538 culture into one of the sterile 90.0 mL broths and return it to the incubator until time for the exercise.
3. 4 hours before the test, pour the TA 1535 culture into the other 90.0 mL sterile broth and return it to the incubator until time for the exercise.

References

Eisenstadt, Bruce, C. Carlton, and Barbara J. Brown. 1994. Page 311 in *Methods for General and Molecular Bacteriology*, edited by Philipp Gerhardt, R. G. E. Murray, Willis A. Wood, and Noel R. Krieg, American Society for Microbiology, Washington, DC.

Maron, D. M. and B. N. Ames. 1983. *Mutation Research*, 113:173–215.

Nelson, David L., and Michael M. Cox. 2005. Chapter 25 in *Lehninger's Principles of Biochemistry*. W. H. Freeman and Co., New York.

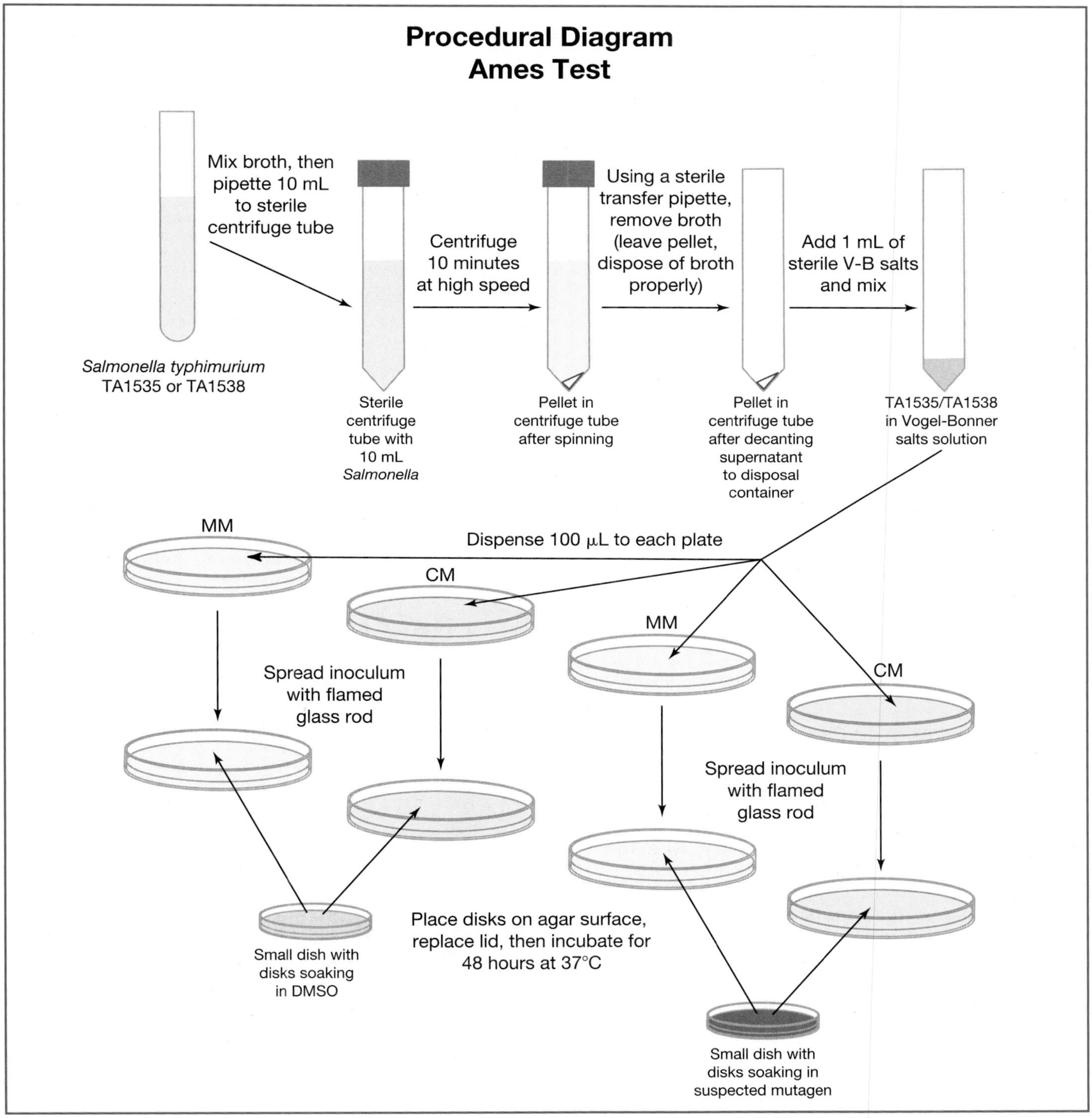

FIGURE 9-10 PROCEDURAL DIAGRAM

Be sure to dispose of all pipettes, broth, and tubes properly as you perform the experiment. After 48 hours of incubation, use a metric ruler to measure the diameter of the cleared zone on all CM plates. Use a colony counter to count the back-mutant colonies on all the MM plates. (***Note:*** These will be fairly large. Do not count the tiny "hazy" growth over the plate's surface.) Mark each colony with a toothpick (to avoid counting it more than once) as you keep track of the number using a hand tally counter.

Exercise 9-4

Bacterial Transformation: the pGLO™ System

Theory

This exercise utilizes a kit produced by Bio-Rad Laboratories[1] that efficiently illustrates the following principles of microbial genetics:

- bacterial transformation,
- use of an antibiotic selective medium to identify transformed cells, and
- the operon as a mechanism of microbial genetic regulation.

Bacterial **transformation** is the process by which **competent** bacterial cells pick up DNA from the environment and make use of the genes it carries. It was first demonstrated in a strain of pneumococcus in 1928 by Fred Griffith and since has been found to occur naturally in only certain genera. Modern techniques, however, have allowed biologists to make most cells (including eukaryotic cells) artificially competent, and this has made transformation a useful tool in genetic engineering. You will be using a $CaCl_2$ Transforming Solution and heat shock to make your *Escherichia coli* cells competent.

Green fluorescent protein (GFP) is responsible for **bioluminescence** in the jellyfish *Aequorea victoria*. The GFP gene was isolated and altered so the GFP in this experiment fluoresces more than the natural version. You will be introducing the GFP gene into competent *E. coli* cells and will be using the ability of the cell to fluoresce as visual evidence of successful transformation and subsequent gene expression.

Operons are structural and functional genetic units of prokaryotes. Each operon minimally includes a **promoter site** (for binding **RNA polymerase**) and two or more **structural genes** coding for enzymes in the same metabolic pathway. In this exercise you will be using part of the arabinose operon.

The complete arabinose operon (Figure 9-11) consists of the promoter (P_{BAD}) and three structural genes (*ara*B, *ara*A, and *ara*D) that code for enzymes used in arabinose digestion. *In vivo,* the arabinose enzymes are needed only when arabinose is present. After all, there is no point in the cell expending a lot of energy making the enzymes if the substrate is not there. A DNA binding protein called *ara*C attaches to the promoter of the arabinose operon and acts like a switch.

When arabinose is not present, *ara*C prevents RNA polymerase from binding to the promoter, so transcription can't occur—the switch is "off." When arabinose is

[1] Catalog Number 166-0003-EDU. Bio-Rad Laboratories Main Office 2000 Alfred Nobel Drive, Hercules, CA 94547; www.bio-rad.com, 1-800-424-6723.

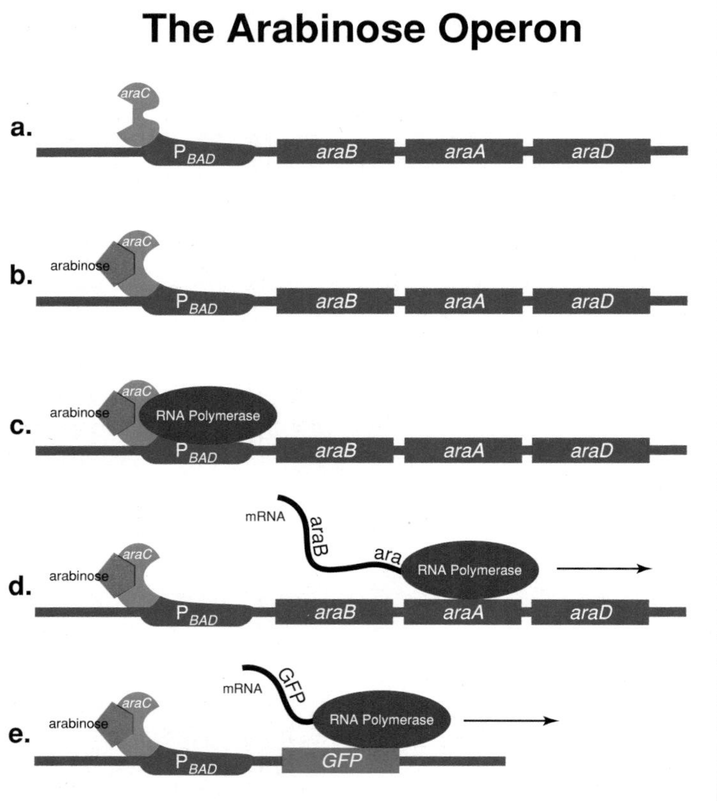

FIGURE 9-11 The Arabinose Operon—Normal and Genetically Engineered

Operons are prokaryotic structural and functional genetic units: They carry genes for enzymes in the same metabolic pathway, and they are regulated together. Operons with genes for catabolic enzymes are transcribed only when the specific substrate is present. In this case, the substrate is the sugar arabinose. (A) The arabinose operon consists of a promoter (P_{bad}) and three structural genes (*ara*B, *ara*A, and *ara*D). The DNA binding protein (*ara*C) attaches to the promoter and acts like a switch. Without arabinose present, RNA Polymerase is unable to bind to the promoter and begin transcription of the genes. (B) and (C) Arabinose binds to a receptor on *ara*C and causes it to change to a shape that allows binding of RNA Polymerase to the promoter. (D) Transcription of the structural genes occurs, the enzymes are produced, and arabinose is catabolized. (E) The pGLO™ plasmid has been engineered to contain the arabinose promoter and the gene for Green Fluorescent Protein (GFP) instead of the genes for arabinose catabolism. If arabinose is present, the GFP gene will be transcribed. In addition, the pGLO™ plasmid has a replication origin and genes for antibiotic resistance and the DNA binding protein.

present, it binds to *ara*C and changes its shape so RNA polymerase *can* bind to the promoter and transcribe the genes—the switch is "on." Then this sequence of events occurs: The genes are transcribed, the enzymes are synthesized, they catalyze their reactions, and eventually the arabinose is consumed. Now, without arabinose, *ara*C returns to its "off" shape and the genes are no longer transcribed. In this exercise you will be using the regulatory portion of the arabinose operon (the arabinose promoter and the *ara*C gene), but the structural genes have been replaced with the GFP gene.

All that remains is a means of carrying the GFP gene into the cell and replicating it—a **vector**. In this exercise you will be using a **plasmid** as a vector: the pGLO™ plasmid. Plasmids are small, naturally occurring, circular DNA molecules that possess only a few genes and replicate independent of the chromosome (because they have their own replication origin). Although they are nonessential, they often carry genes that are beneficial to the bacterium, such as antibiotic resistance. The antibiotic resistance gene (*bla*) used in this experiment produces an enzyme called β-lactamase, which hydrolyzes certain antibiotics, including penicillin and ampicillin.

The bottom line is this: The pGLO™ plasmid used in this experiment has been genetically engineered to contain the arabinose promoter (P_{BAD}), the gene for *ara*C, an antibiotic-resistance gene (*bla)*, the gene for green fluorescent protein (GFP), and a replication origin (*ori*).

As you interpret the results of your experiment, you will see how these components come together and provide you with information about what is happening at the molecular level in your *E. coli* culture. Their functions are summarized in Table 9-1.

Application

Introduction of foreign DNA into a cell, identification of transformed cells, and regulation of expression of the introduced genes are skills used in genetic engineering.

In This Exercise…

You will transform a competent *E. coli* cell with a plasmid containing the gene for Green fluorescent protein. This technique of introducing a plasmid into cells is done routinely in genetic engineering protocols. You also will use arabinose as a genetic switch to regulate expression of the GFP gene.

Materials

Per Student Group

One kit contains enough material for eight student workstations. Each workstation requires the following:

- one Luria-Bertani (LB) agar plate of *E. coli*
- one sterile LB plate

Name	Symbol	Function in This Experiment
Green fluorescent protein	GFP	Used an indicator of gene transcription in this experiment.
Plasmid	pGLO™	Used as a vector to introduce the GFP gene into recipient *E. coli* cells.
Arabinose promoter	P_{BAD}	It is the attachment site for RNA polymerase during transcription of the GFP gene on the pGLO™ plasmid. Normally, it is the attachment site for RNA polymerase during transcription of the *ara*B, *ara*A, and *ara*D genes.
DNA binding protein	*ara*C	It is a regulatory molecule for the arabinose promoter. In the presence of arabinose, *ara*C has a shape that allows RNA polymerase to bind to the promoter. Without arabinose, *ara*C's shape prevents RNA polymerase from binding to the promoter.
Antibiotic resistance gene	*bla*	Produces β-lactamase, an enzyme that hydrolyzes certain antibiotics, including ampicillin. The gene provides a means of differentiating cells that were transformed and those that were not.
Replication origin	*ori*	Necessary for DNA replication. The pGLO™ plasmid is capable of replicating inside the cell because it has this.

TABLE 9-1 Cast of Characters and a Legend of Abbreviations

- two sterile LB + ampicillin plates
- one sterile LB + ampicillin + arabinose plate
- one vial of transformation solution
- LB Broth
- seven disposable inoculating loops
- five disposable calibrated transfer pipettes
- one foam microtube holder/float
- one container of crushed ice
- one marking pen

Per Class

In addition, a class supply of the following is required:

- hydrated pGLO™ plasmid
- 42°C water bath and thermometer
- long-wave UV lamp

Procedure

Lab One

Refer to the procedural diagram in Figure 9-12 as you read and perform the following procedure.

1. Obtain two closed microtubes. Label them with your group's name, then label one "+DNA" and the other "−DNA." Put both tubes in the microtube holder/float.
2. Using a sterile calibrated transfer pipette, dispense 250 µL of Transformation Solution into each tube. Close the caps.
3. Return the two tubes to the microtube holder/float and place them in the ice bath.
4. With a sterile loop, transfer one entire *E. coli* colony into the +DNA tube. Agitate the loop until all the growth is off of it and the cells are dispersed uniformly in the Transformation Solution. Close the lid and properly dispose of the loop.
5. Repeat Step 4 with a sterile loop and the −DNA tube.
6. Hold the UV light next to the vial of pGLO plasmid solution. Record your observation on the Data Sheet.
7. Using a sterile loop, remove a loopful of pGLO plasmid DNA solution. Be sure there is a film across the loop. Then transfer the loopful of plasmid solution to the +DNA tube and mix.
8. Leave the tubes on ice. Make sure they are far enough down in the microtube holder/float that they make good contact with the ice. Leave them on ice for 10 minutes.
9. As the tubes are cooling on ice for 10 minutes, label the four LB agar plates.
   - Label one LB/amp plate: +DNA
   - Label the LB/amp/ara plate: +DNA
   - Label the other LB/amp plate: −DNA
   - Label the LB plate: −DNA
10. After 10 minutes on ice, transfer the microtube holder/float (still containing both tubes) to the 42°C water bath for exactly 50 seconds. This transfer from ice to warm water must be done rapidly. Also, make sure the tubes make good contact with the water.
11. After 50 seconds, quickly place both tubes back in the ice for 2 minutes. This process of heat shock makes the cell membranes more permeable to DNA. (Timing is critical. According to the kit's manufacturer, 50 seconds is optimal. No heat shock results in a 90% reduction in transformants, whereas a 90 second heat shock yields about half the transformants.)
12. After 2 minutes, remove the microtube holder/float from the ice and place it on the table.
13. Using a sterile pipette, add 250 µL of LB broth to the +DNA tube. Repeat with another sterile pipette and the −DNA tube. Properly dispose of both pipettes.
14. Incubate both tubes for 10 minutes at room temperature. Then mix the tubes by tapping them with your fingers.
15. Using a different sterile pipette for each transfer, inoculate the "LB/amp/+DNA" plate and the "LB/amp/ara/+DNA" plate with 100 µL from the +DNA tube.
16. Using a different sterile pipette for each transfer, inoculate the "LB/amp/−DNA" plate and the "LB/−DNA" plate with 100 µL from the −DNA tube.
17. Using a different sterile loop for each, spread the inoculum over the surface of all four plates to get confluent growth. Properly dispose of the loops.
18. Tape the plates together in a stack in which they face the same direction. Then label them with your group name and the date, and incubate them for 24 hours at 37°C in an inverted position.

Lab Two

1. Retrieve your plates. Observe them in ambient room light, then in the dark with UV illumination. Record your observations on the Data Sheet, and answer the questions.
2. When finished, properly dispose of all plates.

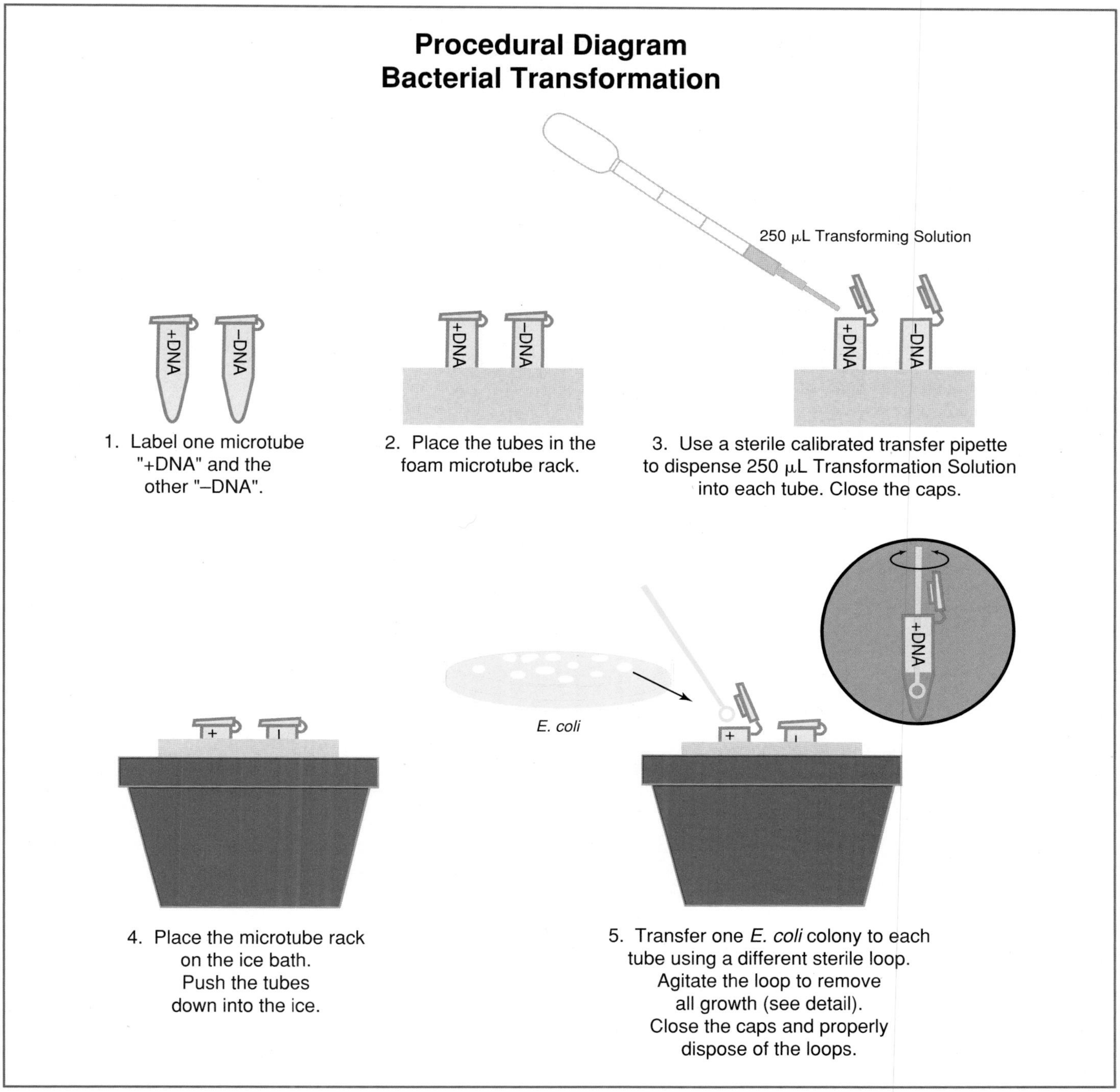
Procedural Diagram
Bacterial Transformation
250 μL Transforming Solution
+DNA
–DNA
1. Label one microtube "+DNA" and the other "–DNA".
2. Place the tubes in the foam microtube rack.
3. Use a sterile calibrated transfer pipette to dispense 250 μL Transformation Solution into each tube. Close the caps.
E. coli
4. Place the microtube rack on the ice bath. Push the tubes down into the ice.
5. Transfer one E. coli colony to each tube using a different sterile loop. Agitate the loop to remove all growth (see detail). Close the caps and properly dispose of the loops.

FIGURE 9-12 **PROCEDURAL DIAGRAM**
Be sure to dispose of all pipettes and loops properly. (Continue with next page.)

Reference

Bio-Rad Laboratories. Instruction pamphlet for the *Bacterial Transformation–The pGLO™ System* kit (Catalog Number 166-0003-EDU). Bio-Rad Laboratories, 2000 Alfred Nobel Drive, Hercules, CA 94547.

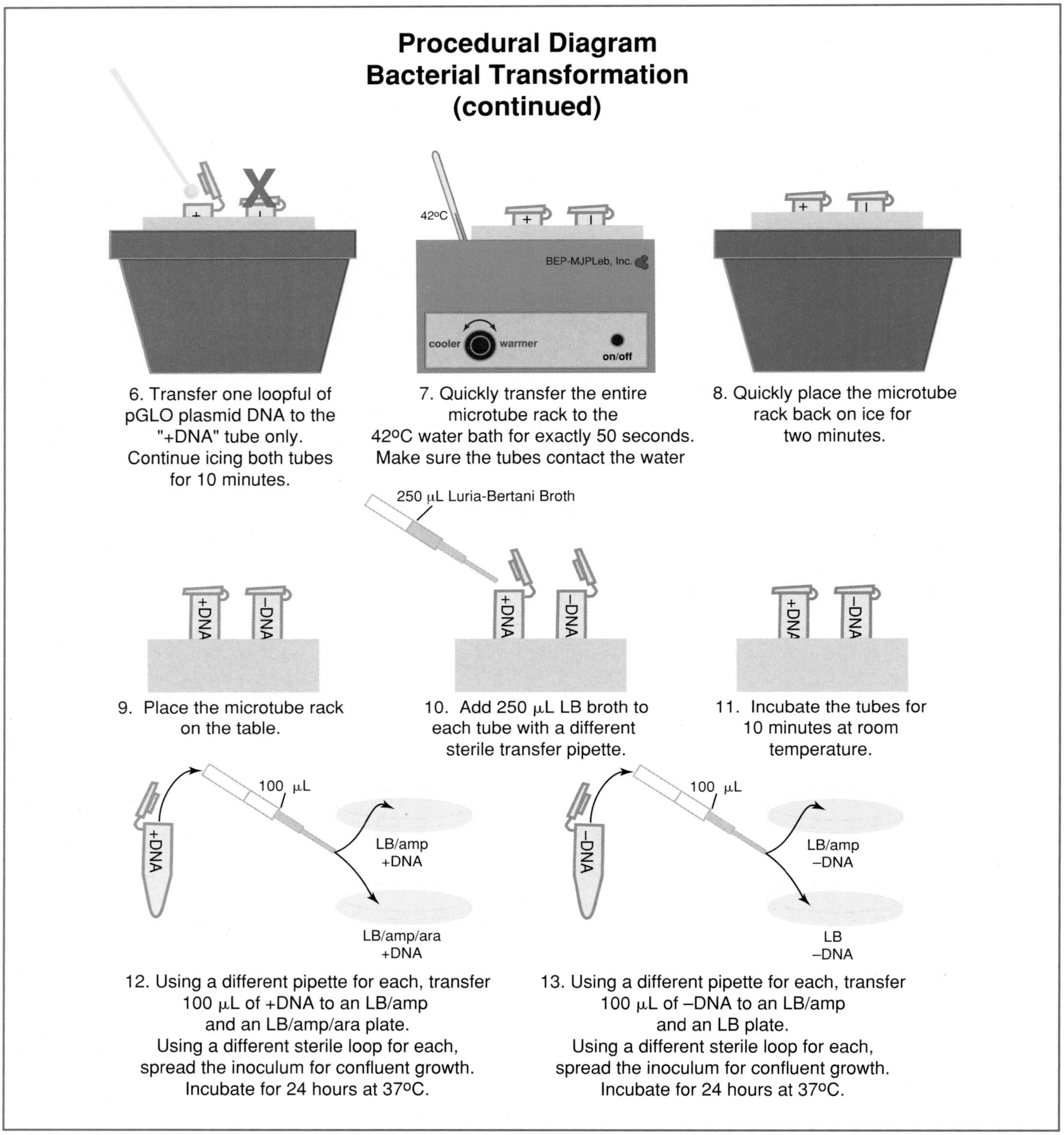
Procedural Diagram
Bacterial Transformation
(continued)
42°C
BEP-MJPLeb, Inc.
cooler
warmer
on/off
6. Transfer one loopful of pGLO plasmid DNA to the "+DNA" tube only. Continue icing both tubes for 10 minutes.
7. Quickly transfer the entire microtube rack to the 42°C water bath for exactly 50 seconds. Make sure the tubes contact the water
8. Quickly place the microtube rack back on ice for two minutes.
250 µL Luria-Bertani Broth
+DNA
–DNA
9. Place the microtube rack on the table.
10. Add 250 µL LB broth to each tube with a different sterile transfer pipette.
11. Incubate the tubes for 10 minutes at room temperature.
100 µL
LB/amp
+DNA
LB/amp/ara
+DNA
LB/amp
–DNA
LB
–DNA
12. Using a different pipette for each, transfer 100 µL of +DNA to an LB/amp and an LB/amp/ara plate. Using a different sterile loop for each, spread the inoculum for confluent growth. Incubate for 24 hours at 37°C.
13. Using a different pipette for each, transfer 100 µL of –DNA to an LB/amp and an LB plate. Using a different sterile loop for each, spread the inoculum for confluent growth. Incubate for 24 hours at 37°C.

FIGURE 9-12 PROCEDURAL DIAGRAM *(continued)*
Be sure to dispose of all pipettes and loops properly.

SECTION 10

Hematology and Serology

This section deals with blood cells and other aspects of the body's defenses. In Exercise 10-1 you will have the opportunity to perform a differential blood cell count. Exercises 10-2 through 10-5 allow you to perform serological tests that are used to detect the presence of specific antigens or antibodies in a sample.

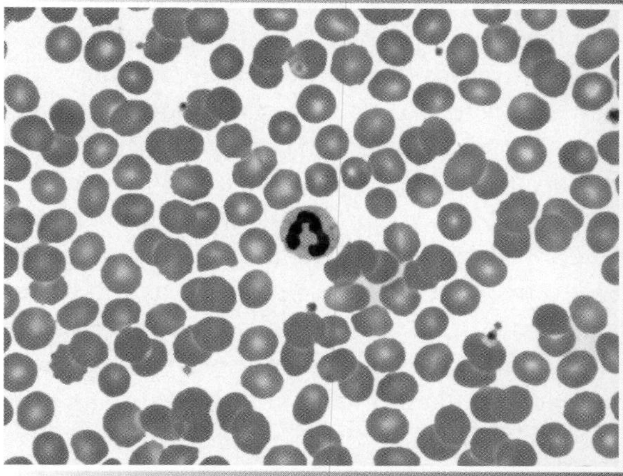

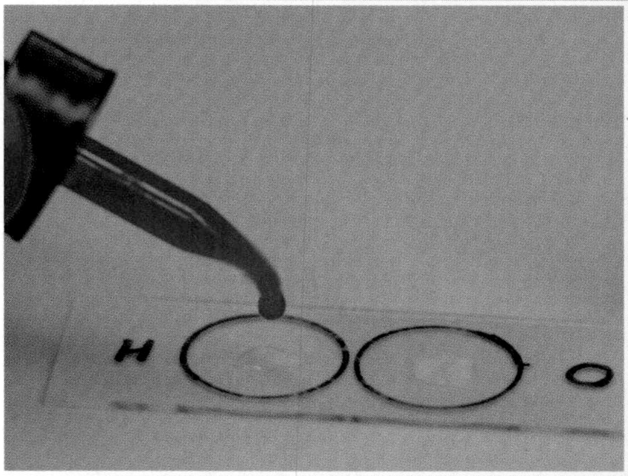

Exercise 10-1

Differential Blood Cell Count

Theory

Leukocytes (white blood cells, or WBCs) are divided into two groups: **granulocytes** (which have prominent cytoplasmic granules) and **agranulocytes** (which lack these granules). There are three basic types of granulocytes: **neutrophils, basophils,** and **eosinophils**. The two types of agranulocytes are **monocytes** and **lymphocytes**.

Neutrophils (Figure 10-1A) are the most abundant WBCs in blood. They leave the blood and enter tissues to phagocytize foreign material. An increase in neutrophils in the blood is indicative of a systemic bacterial infection. Mature neutrophils sometimes are referred to as **segs** because their nucleus usually is segmented into two to five lobes. Because of the variation in nuclear appearance, they also are called **polymorphonuclear neutrophils (PMNs)**. Immature neutrophils lack this nuclear segmentation and are referred to as **bands** (Figure 10-1B).

This distinction is useful because a patient with an active infection increases neutrophil production, which creates a higher percentage of the band (immature) type. Neutrophils are 12–15 µm in diameter—about twice the size of an erythrocyte (RBC).[1] Their cytoplasmic granules are neutral-staining and thus do not have the intense color of other granulocytes when prepared with Wright's or Giemsa stain.

Eosinophils are phagocytic, and their numbers increase during allergic reactions and parasitic infections (Figure 10-2). They are 12–15 µm in diameter (about twice the size of an RBC) and generally have two lobes in their nucleus. Their cytoplasmic granules stain red in typical preparations.

Basophils (Figure 10-3) are the least abundant WBCs in normal blood. They are structurally similar to tissue mast cells and produce some of the same chemicals (histamine and heparin) but are derived from different stem cells in bone marrow. They are 12–15 µm in diameter. The nucleus usually is obscured by the dark-staining cytoplasmic granules, but it either has two lobes or is unlobed.

Agranulocytes include monocytes and lymphocytes. Monocytes (Figure 10-4) are the blood form of **macrophages**. They are the largest of the leukocytes, being two to three times the size of RBCs (12–20 µm). Their nucleus is horseshoe-shaped, and the cytoplasm lacks prominent granules (but may appear finely granular).

Lymphocytes (Figure 10-5A) are cells of the immune system. Two functional types of lymphocytes are the **T-cell**, involved in cell-mediated immunity, and the **B-cell**, which converts to a **plasma cell** when activated, and produces antibodies. The nucleus usually is spherical and takes up most of the cell. Lymphocytes are approximately the same size as RBCs or up to twice their size. The larger ones form a third functional group of lymphocytes, the **null cell**, many of which are **natural killer (NK) cells** that kill foreign or infected cells without antigen–antibody interaction (Figure 10-5B).

In a differential white cell count, a sample of blood is observed under the microscope, and at least 100 WBCs are counted and tallied (this task is automated now). Approximate normal percentages for each leukocyte are as follows and as summarized in Table 10-1.

neutrophils (mostly segs) 55%–65%
lymphocytes 25%–33%
monocytes 3%–7%
eosinophils 1%–3%
basophils 0.5–1%.

[1] It is convenient to discuss leukocyte size in terms of erythrocyte size because RBCs are so uniform in diameter. In an isotonic solution, erythrocytes are 7.5 µm in diameter.

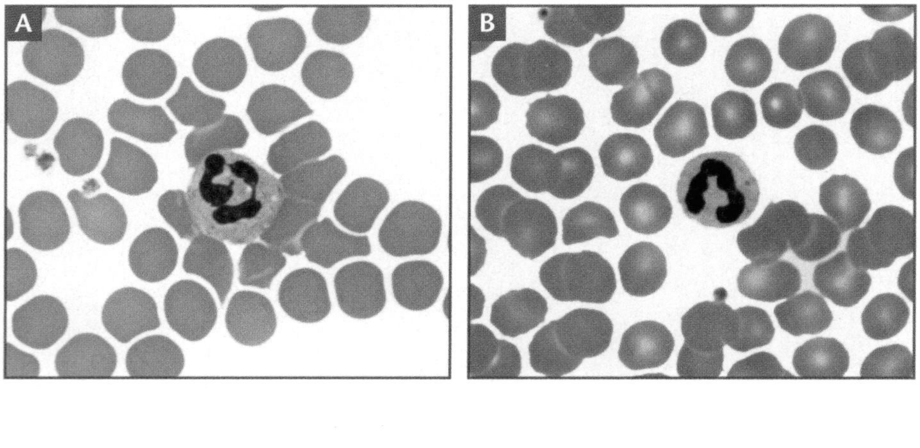

FIGURE 10-1 NEUTROPHIL
(A) The segmented nucleus of this cell identifies it as a mature neutrophil (segs). About 30% of neutrophils in blood samples from females demonstrate a "drumstick" protruding from the nucleus, as in this specimen. This is the region of the inactive X chromosome. (B) This is an immature band neutrophil with an unsegmented nucleus. Both specimens were prepared with Wright's stain and were magnified X1000.

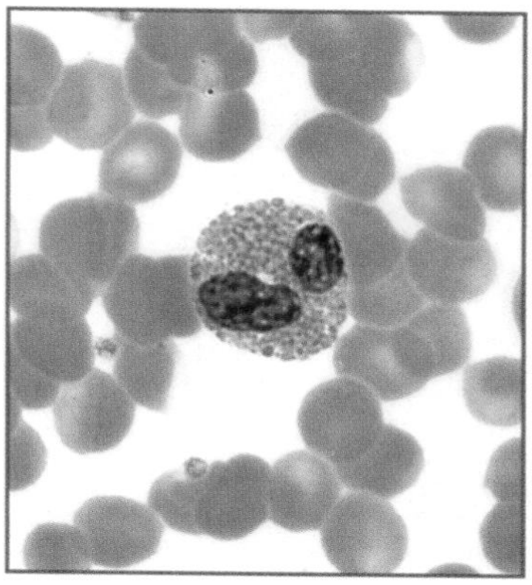

FIGURE 10-2 Eosinophil These granulocytes are relatively rare and are about twice the size of red blood cells. Their cytoplasmic granules stain red, and their nucleus usually has two lobes. This specimen was prepared with Wright's stain and was magnified X1000.

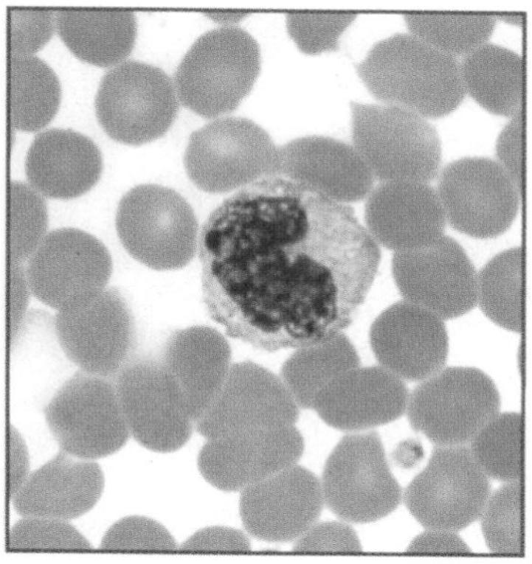

FIGURE 10-4 Monocyte Monocytes are the blood form of macrophages. They are about twice the size of red blood cells and have a round or indented nucleus. (Wright's stain, X1000.)

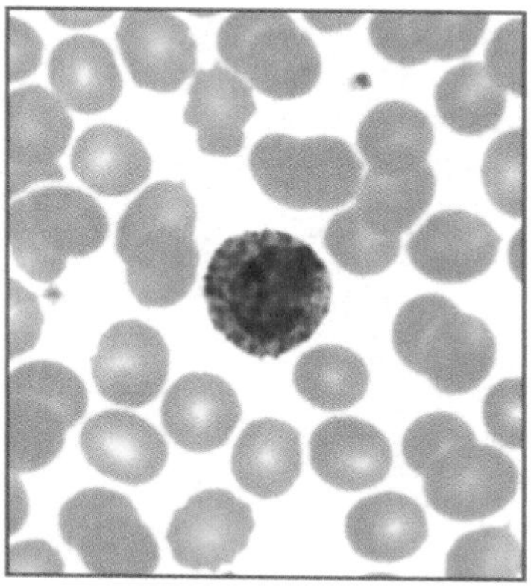

FIGURE 10-3 Basophil Basophils comprise only about 1% of all white blood cells. They are slightly larger than red blood cells and have dark purple cytoplasmic granules that obscure the nucleus. This micrograph was prepared with Wright's stain and magnified X1000.

Application A differential blood cell count is done to determine approximate numbers of the various leukocytes in blood. Excess or deficiency of all or a specific group is indicative of certain disease states. Even though differential counts are automated now, it is good training to perform one "the old-fashioned way" using a blood smear and a microscope to get an idea of the principle behind the technique.

TABLE OF RESULTS

Cell	Abundance in Blood (%)	Diameter (µm)	Nucleus	Cytoplasmic Granules (Wright's or Giemsa Stain)	Functions
Granulocyte					
Neutrophil	55–65	12–15	2–5 lobes	Present, but stain poorly; contain antimicrobial chemicals	Phagocytosis and digestion of (usually) bacteria
Eosinophil	1–3	12–15	2 lobes	Present and stain red; contain antimicrobial chemicals and histaminase	Present in inflammatory reactions and immune response against some multicellular parasites (worms)
Basophil	0.5–1	12–15	Unlobed or 2 lobes	Present and stain dark purple; contain histamine and other chemicals	Participate in inflammatory response
Agranulocyte					
Lymphocyte	25–33	7–18 (rare)	Spherical (leaving little visible cytoplasm)	Absent	Active in specific acquired immunity (as T and B cells)
Monocyte	3–7	12–20	Horseshoe-shaped (cytoplasm is prominent)	Absent	Phagocytosis (as macrophages)

TABLE 10-1 Typical Features of Human Leukocytes in Blood

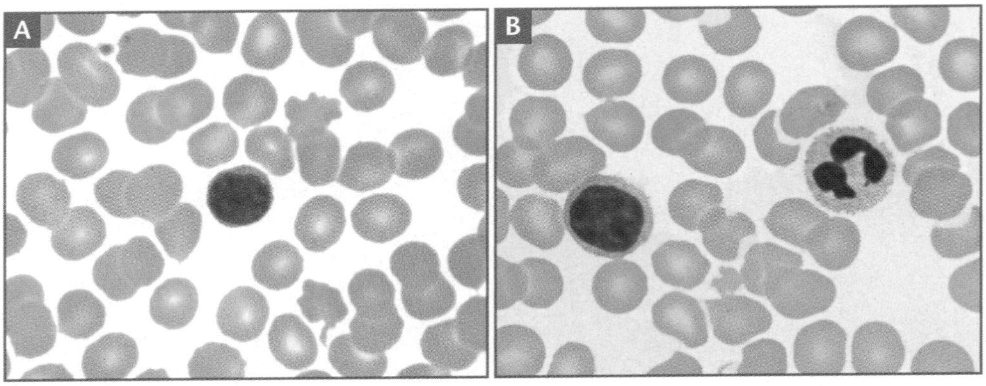

FIGURE 10-5 Lymphocyte
Lymphocytes are common in the blood, comprising up to 33% of all WBCs. Most are about the size of red blood cells and have only a thin halo of cytoplasm encircling their round nucleus. They belong to functional groups called B cells and T cells (which are morphologically indistinguishable). Micrograph (A) is a small lymphocyte and was prepared with Wright's stain. Some lymphocytes are larger, as in micrograph (B). These are natural killer (NK) cells or some other type of null cell. Also visible is a neutrophil. All micrographs are X1000.

In This Exercise...

You will be examining prepared blood smears and doing a differential count of white blood cells. As an optional activity, you may look at smears of abnormal blood and compare the differential count to normal blood.

Materials

Per Student Group

- commercially prepared human blood smear slides (Wright's or Giemsa stain)
- (optional) commercially prepared abnormal human blood smear slides (*e.g.*, infectious mononucleosis, eosinophilia, or neutrophilia)

Procedure

1. Obtain a blood smear slide and locate a field where the cells are spaced far enough apart to allow easy counting. (The cells should be fairly dense on the slide, but not overlapping.)
2. Using the oil immersion lens, scan the slide using the pattern shown in Figure 10-6. Be careful not to overlap fields when scanning the specimen. Choose a "landmark" blood cell at the right side of the field, and move the slide horizontally until that cell disappears off the left side. Avoid diagonal movement of the slide. As you scan, use the mechanical stage knobs separately to move the slide up and back or to the right and left in straight lines.

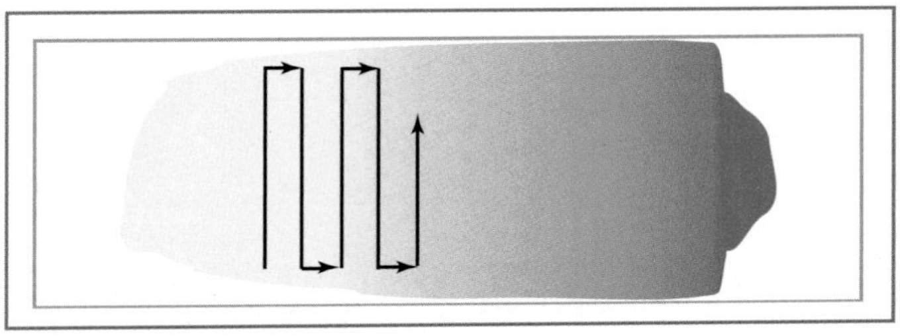

FIGURE 10-6 Following a Systematic Path
A systematic scanning path is used to avoid wandering around the slide and perhaps counting some cells more than once. Remember that a microscope image is inverted. If you want the image to move left, you must move the slide to the right.

3. Make a tally mark in the appropriate box in the chart on the Data Sheet for the first 100 leukocytes you see.
4. Calculate percentages and compare your results with the accepted normal values.
5. Repeat with a pathological blood smear (if available).

References

Brown, Barbara A. 1993. *Hematology—Principles and Procedures,* 6th ed. Lea and Febiger, Philadelphia.

Diggs, L. W., Dorothy Sturm, and Ann Bell. 1978. *The Morphology of Human Blood Cells,* 4th ed. Abbott Laboratories, North Chicago, IL.

Junqueira, L. Carlos, and Jose Carneiro. 2003. *Basic Histology, Text and Atlas,* 10th ed. Lange Medical Books, McGraw Hill, New York.

Leboffe, Michael J. 2003. *A Photographic Atlas of Histology.* Morton Publishing, Englewood, CO.

Simple Serological Reactions

Antigen–antibody reactions are highly specific and occur *in vitro* as well as *in vivo*. Serology is the discipline that exploits this specificity as an *in vitro* diagnostic tool. Two simple serological reactions—agglutination and precipitation—are used in the following exercises because they result in the formation of complexes that can be viewed with the naked eye and without sophisticated equipment.

The precipitin ring test (Exercise 10-2) and the radial immunodiffusion test (Exercise 10-3) illustrate precipitation. Both may be used to identify antigens or antibodies in a sample; the latter also is used to compare antigens in more than one sample. The slide agglutination test (Exercise 10-4) can be an important diagnostic (and highly specific) tool for the identification of organisms. It is especially useful for serotyping large genera such as *Salmonella*. Hemagglutination, which detects specific antigens on red blood cells, is the standard test for determining blood type (Exercise 10-5). Other hemagglutination tests are used for diagnosing infections.

Exercise 10-2

Precipitin Ring Test

Theory **Soluble antigens** may combine with **homologous antibodies** to produce a visible **precipitate**. Precipitate formation thus serves as evidence of antigen–antibody reaction and is considered to be a positive result.

Precipitation is produced because each antibody has (at least) two **antigen binding sites** and many antigens have multiple **epitopes** (sites for antibody binding). This results in the formation of a complex lattice of antibodies and antigens and produces the visible precipitate—a positive result. As shown in Figure 10-7, if either antibody or antigen is found in a concentration that is too high relative to the other, no visible precipitate will be formed even though both are present. **Optimum proportions** of antibody and antigen are necessary to form precipitate, and they occur in the **zone of equivalence.**

Several styles of precipitation tests are used. The **precipitin ring test** is performed in a small test tube or a capillary tube. Antiserum (containing antibodies homologous to the antigen being looked for) is placed in the bottom of the tube. The sample with the suspected antigen is layered on the surface of the antiserum in such a way that the two solutions have a sharp interface. As the two fluids diffuse into each other, precipitation occurs where optimum proportions of antibody and antigen are found (Figure 10-8). This test also may be run to test for antibody in a sample.

Application Precipitation reactions can be used to detect the presence of either antigen or antibody in a sample. They have mostly been replaced by more sensitive serological techniques for diagnosis but are still useful to simply demonstrate serological reactions.

In This Exercise...

You will perform a simple serological test to illustrate homology between equine albumin and equine albumen antiserum.

Materials

Per Student Group

- two clean 6 × 50 mm Durham tubes
- equine serum (containing equine albumin)
- equine albumin antiserum (containing equine albumin antibodies)
- 0.9% saline solution
- Pasteur pipettes

Procedure

1. Carefully add equine antiserum to both Durham tubes. Fill from the bottom of the tube until it is about 1/3 full.
2 Mark one tube "A." Add the equine serum in such a way that a sharp and distinct second layer is formed without any mixing of the two solutions. It is critical not to allow any mixing. Success usually can be achieved by allowing the serum to trickle slowly down the inside of the glass (Figure 10-9).
3. Mark the second tube "B." Add the 0.9% saline in the same way you added equine serum to tube A.
4. Incubate both at 35 ± 2°C undisturbed for 1 hour.

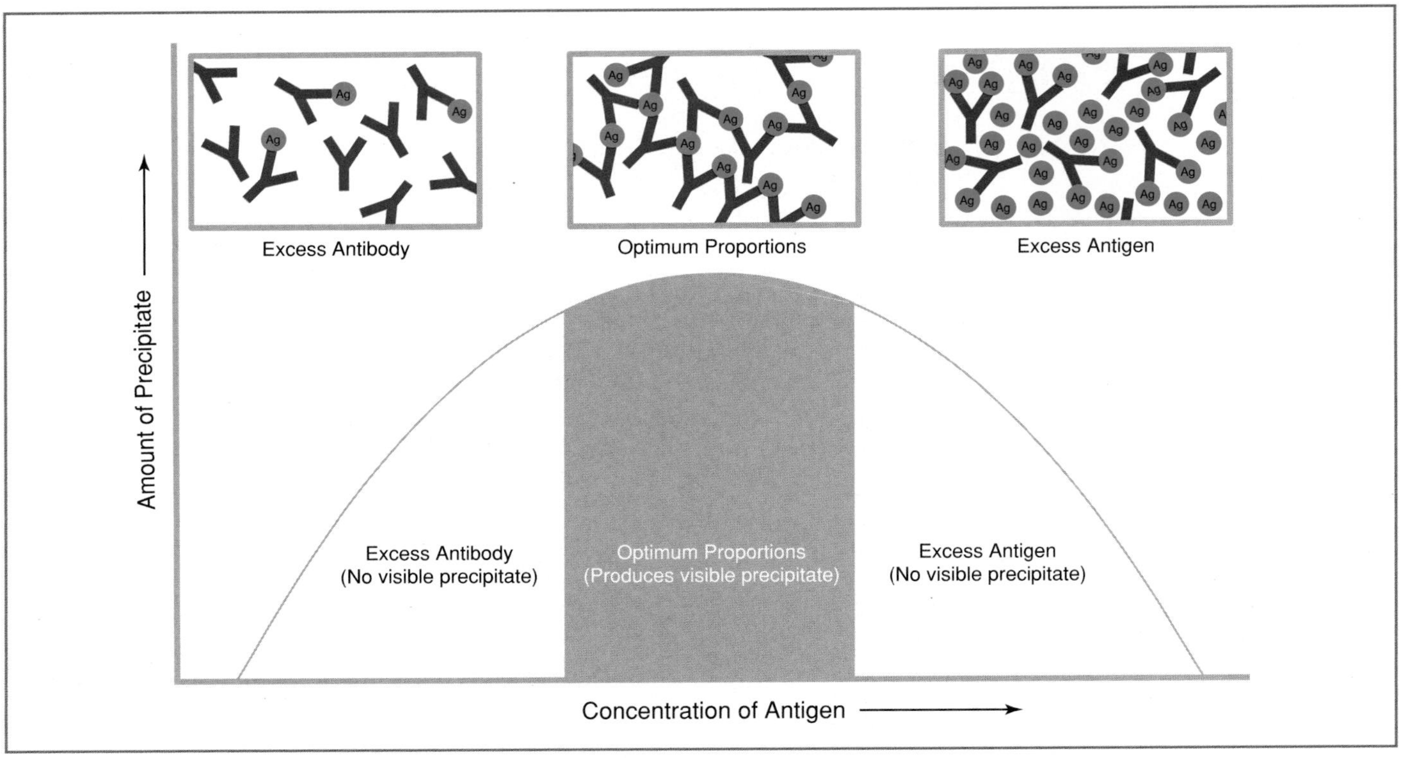

FIGURE 10-7 Precipitation Reactions
Precipitation occurs between soluble antigens and homologous antibodies where they are found in optimal proportions to produce a cross-linked lattice. Excess antigen or excess antibody prevents substantial cross-linking, so no lattice is formed and no visible precipitate is seen—even though both antigen and antibody are present. In this graph, antibody concentration is kept constant as antigen concentration is adjusted.

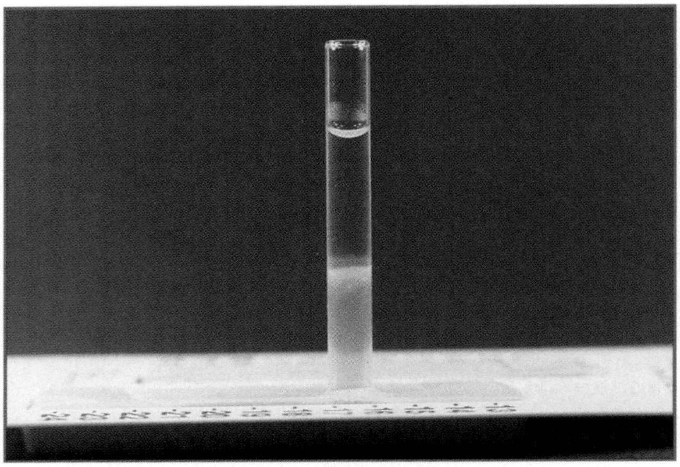

FIGURE 10-8 Positive Precipitin Ring Test
A sample of antigen has been layered over an antiserum. The white precipitation ring formed at the site of optimum proportions of antibodies and antigens.

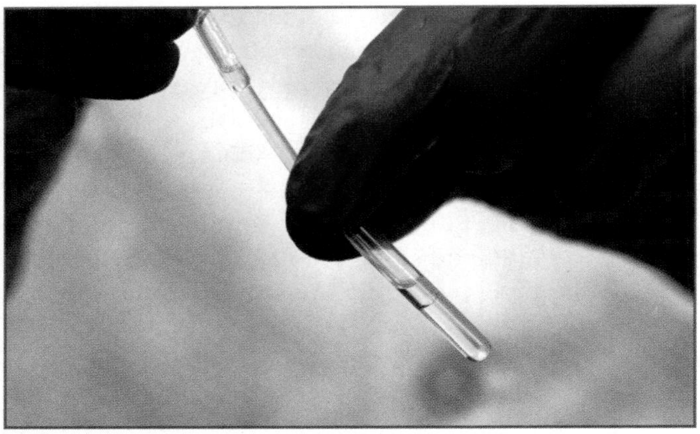

FIGURE 10-9 Layering the Antigen on Top of the Antibody
When adding the antigen layer to the tube containing antiserum, it is essential that there is no mixing of the two solutions. Place the pipette containing the antigen into the tube about 5 mm from the antiserum and let it slowly trickle down the inside of the glass. Allow the tube to stand for 1 hour undisturbed.

5. Observe the tubes for the characteristic ring formed at the interface of the two solutions (zone of optimum proportions). If after 1 hour there is no ring in either tube, place both in the refrigerator for 12 to 24 hours and then recheck.

Reference

Lam, Joseph S., and Lucy M. Mutharia. 1994. Page 120 in *Methods for General and Molecular Bacteriology*, edited by Philipp Gerhardt, R. G. E. Murray, Willis A. Wood, and Noel R. Krieg, American Society for Microbiology, Washington, DC.

Exercise 10-3

Gel Immunodiffusion

Theory

Precipitation reactions occur between soluble antigens and homologous antibodies. The precipitate forms when the two are at concentrations that allow maximum cross-linking in the antigen–antibody complex that forms (Figure 10-7).

In its simplest form, **gel immunodiffusion** involves two wells formed in a gel (Saline Agar). In one well is the antiserum; in the other is the sample with unknown antigen composition. The two diffuse out of their respective wells toward each other. If the sample has an antigen that will react with the antibody, then a precipitation line will form at the region of optimal proportions as their diffusion paths pass. If no precipitation line forms, the test is considered negative. That is, the sample has no antigen that will react with the antibodies. This procedure may also be used to identify the presence of antibody in a sample if a known antigen is used.

The diffusion path out of the well is not linear. Rather, it is radial in all directions. In a more complex form of immunodiffusion test, several wells are placed around a center well (Figure 10-10). Antiserum (which rarely contains only one kind of antibody) is placed in the center well. Samples of known or unknown antigen composition are put in the surrounding wells. Because diffusion out of the center well is radial, precipitation lines may form between the center well and any or all of the surrounding wells, depending on their antigen composition. The precipitation line pattern between neighboring wells is indicative of antigen relatedness in those wells (Figure 10-11). A single, smooth, curved precipitation line indicates the two antigens in neighboring wells are identical (**identity**). Two spurs indicate unrelated antigens (**nonidentity**) because two completely separate precipitation lines formed. A line with a spur

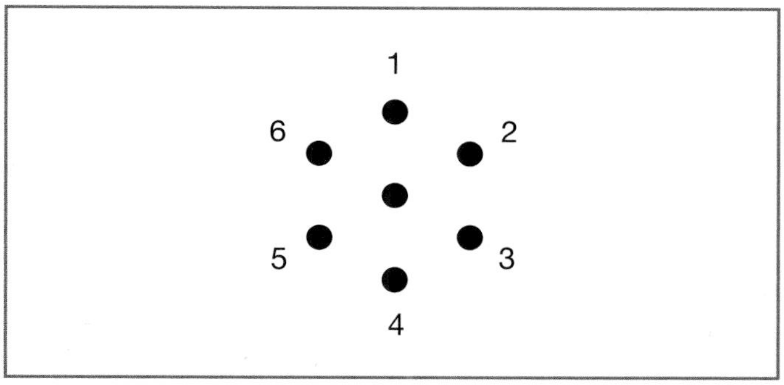

FIGURE 10-10 Double-Gel Immunodiffusion Well Template
Use this template as a guide when boring the wells in the saline agar plate.

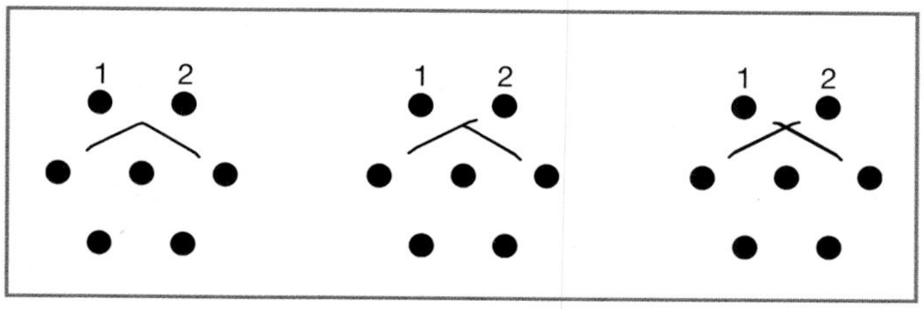

FIGURE 10-11 Precipitation Patterns
This diagram shows three possible precipitation patterns formed between the antibody in the center well and antigens in wells #1 and #2. The pattern on the left demonstrates identity between the antigens. This indicates that the antigens are identical. The pattern in the middle demonstrates partial identity, which means that the antigens are related but not identical (they share some epitopes). The pattern on the right shows nonidentity; the antigens are not related.

indicates that the antigens are related but not identical (**partial identity**), because two lines formed—a line of identity and a line of nonidentity.

In this exercise we will simulate degrees of antigen relatedness by using mixtures of antibodies and antigens.

Application

Gel immunodiffusion tests can be used to identify the presence of antigen or antibody in a sample. They also are used to check samples for identical, related, or unrelated antigens.

In This Exercise...

You will be performing gel radial immunodiffusion, a technique to compare antigen relatedness. Precipitation line patterns that illustrate antigen identity, nonidentity, and partial identity will be seen. In a real life situation, two antigens would be compared to see if they share one or more epitopes, but we will produce the various patterns artificially by mixing different combinations of horse and cow albumin.

Materials

Per Student Group

- one Saline Agar plate
- 3 mm punch (a glass dropper with a 3 mm diameter tip will work)
- template for cutting wells (Figure 10-10)
- bovine antiserum (containing bovine albumin antibodies)
- equine antiserum (containing equine albumin antibodies)
- 10% bovine serum (Prepared by adding 9 drops physiological saline to 1 drop 100% serum)
- 10% equine serum (Prepared by adding 9 drops physiological saline to 1 drop 100% serum)

- 0.9% saline solution
- disposable micropipettes
- two small test tubes

Procedure

Lab One

1. Center the saline agar plate over the template in Figure 10-10.
2. Using the punch or glass dropper, cut the seven wells in the agar as shown in Figure 10-12. If the small agar disks don't come out with the dropper, use suction with the dropper and bulb to dislodge and remove them. Avoid lifting the agar when removing the punched-out disks.
3. Number the peripheral wells 1 through 6 as shown in Figure 10-10.
4. In a small test tube or mixing cup prepare a 1:1 mixture of equine albumin antiserum and bovine albumin antiserum. In a second test tube, make a 1:1 mixture of equine and bovine serum.
5. Using a *different* micropipette for each transfer and the technique illustrated in Figure 10-13, carefully fill the wells. Be careful when adding the solutions not to overfill the wells.
   a. Fill the center well with the equine/bovine antiserum mixture.

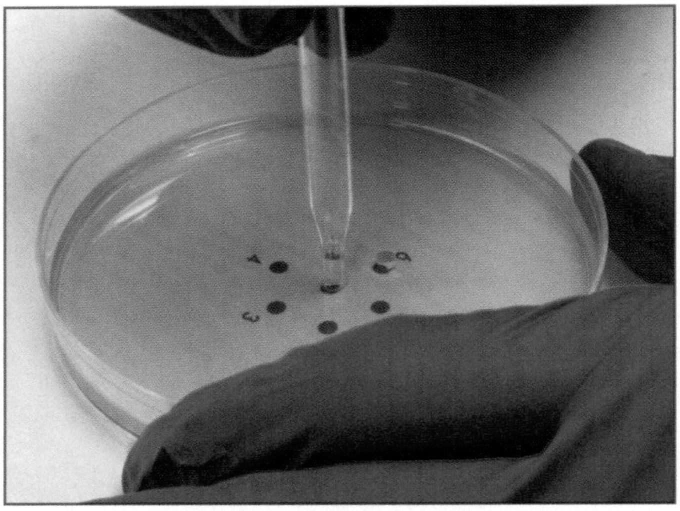

FIGURE 10-12 Boring Wells in the Saline Agar Plate
When cutting wells in the agar, press straight down; *do not twist.* Twisting the cutter or dropper may create fissures in the agar, which could disrupt the diffusion of the solutions.

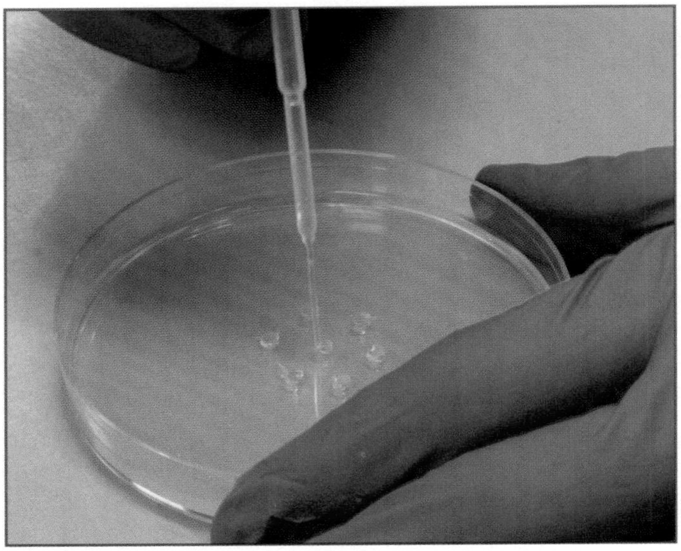

FIGURE 10-13 Filling the Wells
Place the tip of the pipette in the bottom of the well. Fill slowly to prevent creating air bubbles and to minimize spilling serum over the sides.

   b. Fill wells 1 and 2 with 10% equine serum.
   c. Fill wells 3 and 4 with 10% bovine serum.
   d. Fill well 5 with 0.9% saline.
   e. Fill well 6 with the equine/bovine serum mixture.
6. Cover the plate and allow it to sit undisturbed for 30 minutes.
7. Incubate at room temperature for up to 72 hours or until precipitation lines appear.

Lab Two

1. Examine the plate for precipitation lines.
2. Record your results on the Data Sheet. For each pair of wells, interpret the precipitation patterns.

References

Lam, Joseph S., and Lucy M. Mutharia. 1994. Page 120 in *Methods for General and Molecular Bacteriology*, edited by Philipp Gerhardt, R. G. E. Murray, Willis A. Wood, and Noel R. Krieg, American Society for Microbiology, Washington, DC.

Ouchterlony, O. 1968. Page 20 in *Handbook of Immunodiffusion and Immunoelectrophoresis.* Ann Arbor Science Publishers, Ann Arbor, MI.

Exercise 10-4

Slide Agglutination

Theory

Particulate antigens (such as whole cells) may combine with homologous antibodies to form visible clumps called **agglutinates. Agglutination** thus serves as evidence of antigen–antibody reaction and is considered a positive result. Agglutination reactions are highly sensitive and may be used to detect either the presence of antigen or antibody in a sample.

There are many variations of agglutination tests (Figure 10-14). **Direct agglutination** relies on the combination of antibodies and naturally particulate antigens. **Indirect agglutination** relies on artificially constructed systems in which agglutination will occur. These involve coating particles (such as RBCs or latex microspheres) with either antibody or antigen, depending on what is being looked for in the sample. Addition of the homologous antigen or antibody then will result in clumping of the artificially constructed particles.

Slide agglutination is an example of a direct agglutination test. Samples of antigen and antiserum are mixed on a microscope slide and allowed to react. Visible aggregates indicate a positive result.

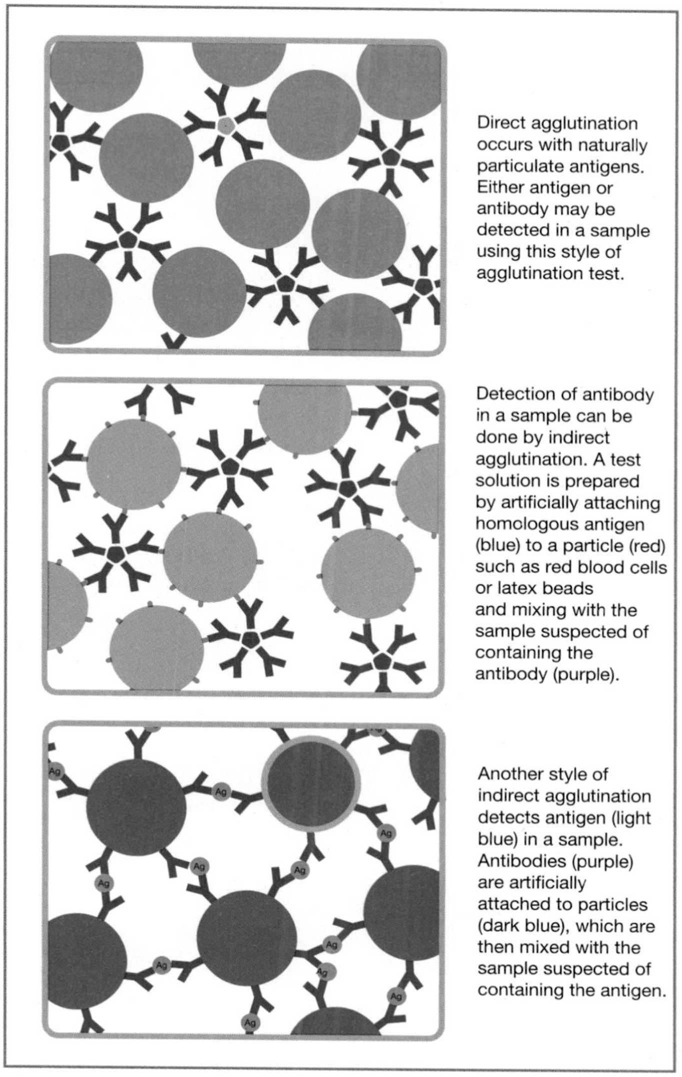

Direct agglutination occurs with naturally particulate antigens. Either antigen or antibody may be detected in a sample using this style of agglutination test.
Detection of antibody in a sample can be done by indirect agglutination. A test solution is prepared by artificially attaching homologous antigen (blue) to a particle (red) such as red blood cells or latex beads and mixing with the sample suspected of containing the antibody (purple).
Another style of indirect agglutination detects antigen (light blue) in a sample. Antibodies (purple) are artificially attached to particles (dark blue), which are then mixed with the sample suspected of containing the antigen.
Ag

FIGURE 10-14 DIRECT AND INDIRECT AGGLUTINATION TESTS Direct agglutinations involve naturally particulate antigens. Indirect agglutination relies on attaching either the antigen or antibody to a particle, such as a latex bead or red blood cell.

Application

Agglutination reactions may be used to detect the presence of either antigen or antibody in a sample. Direct agglutination reactions are used to diagnose some diseases, determine if a patient has been exposed to a certain pathogen, and are involved in blood typing. Indirect agglutination is used in some pregnancy tests as well as in diagnosing disease.

In This Exercise...

You will perform a simple agglutination test to identify the presence of *Salmonella* H antigen in a sample. You will compare this reaction to a reaction between the nonhomologous *Salmonella* anti-H antiserum and the *Salmonella* O antigen.

Materials

Per Student Group

- one clean microscope slide
- toothpicks
- marking pen
- *Salmonella* H antigen
- *Salmonella* O antigen
- *Salmonella* anti-H antiserum

Procedure

1. Using a marking pen, draw two circles approximately the size of a dime on a microscope slide (Figure 10-15). Label one "O" and the other "H."
2. Place a drop of *Salmonella* anti-H antiserum in each circle.
3. Place a drop of *Salmonella* O antigen in the "O" circle and a drop of *Salmonella* H antigen in the "H" circle (Figure 10-16). Be careful not to touch the dropper to the antiserum already on the slide.
4. Using a *different* toothpick for each circle, mix until each of the antigens is completely emulsified with the antiserum. Do not over-mix. Discard the toothpicks in a biohazard container.

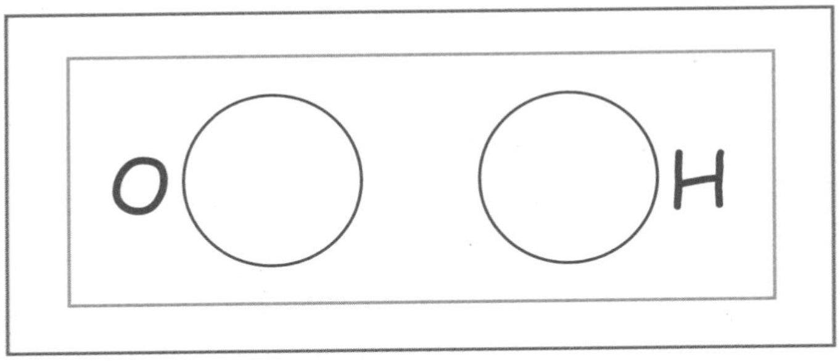

FIGURE 10-15 PREPARE THE SLIDE
With your marking pen, draw two dime-sized circles on the slide. Label one circle "O" and the other "H." The circles will be where you check for the presence of *Salmonella* O and H antigens, respectively.

5. Allow the slide to sit for a few minutes and observe for agglutination. Record your observations on the Data Sheet.

Alternative Test Procedure

Your instructor will cover the labels of the two *Salmonella* antigen bottles. You will use this procedure to identify which one contains the *Salmonella* H antigen.

1. Using a marking pen, draw two circles approximately the size of a dime on a microscope slide (Figure 10-15).
2. Place a drop of one *Salmonella* unknown antigen in one circle and a drop of the other *Salmonella* unknown antigen in the other circle (Figure 10-16).
3. Place a drop of *Salmonella* anti-H antiserum in each circle. Be careful not to touch the dropper to the antigen solutions already on the slide.
4. Using a *different* toothpick for each circle, mix until each of the antigens is completely emulsified with the antiserum. Do not *over*-mix. Discard the toothpicks in a biohazard container.
5. Allow the slide to sit for a few minutes, then observe for agglutination. Record your observations on the Data Sheet.

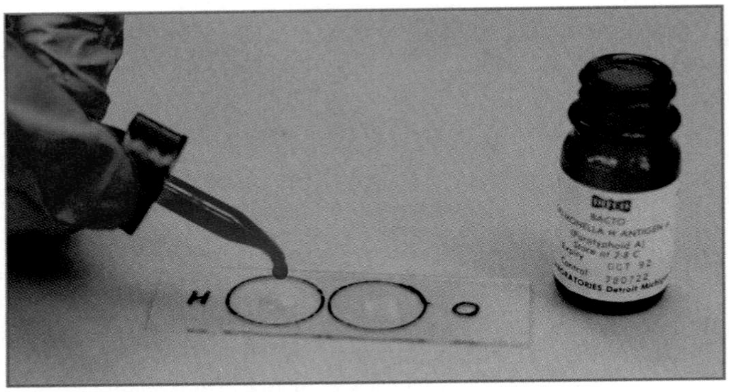

FIGURE 10-16 ADDING THE ANTIGEN
Carefully add the H and O antigens to the drops of Anti-H antiserum already on the slide. Do not touch the dropper to the antiserum.

References

Bopp, Cheryl A., Frances W. Brenner, Joy G. Wells, and Nancy A. Strockbine. 1999. Pages 467–471 in *Manual of Clinical Microbiology*, 7th ed., edited by Patrick R. Murray, Ellen Jo Baron, Michael A. Pfaller, Fred C. Tenover, and Robert H. Yolken. American Society for Microbiology, Washington, DC.

Collins, C. H., Patricia M. Lyne, J. M. Grange. 1995. Page 118 in *Collins and Lyne's Microbiological Methods*, 7th ed. Butterworth-Heinemann, Oxford, United Kingdom.

Constantine, Niel T., and Dolores P. Lana. 2003. Pages 222–223 in *Manual of Clinical Microbiology*, 8th ed., edited by Patrick R. Murray, Ellen Jo Baron, James H. Jorgensen, Michael A. Pfaller, and Robert H. Yolken. American Society for Microbiology, Washington, DC.

Forbes, Betty A., Daniel F. Sahm, and Alice S. Weissfeld. 2002. Pages 206–207 in *Bailey & Scott's Diagnostic Microbiology*, 11th ed. Mosby, St. Louis.

Lam, Joseph S., and Lucy M. Mutharia. 1994. Page 120 in *Methods for General and Molecular Bacteriology*, edited by Philipp Gerhardt, R. G. E. Murray, Willis A. Wood, and Noel R. Krieg. American Society for Microbiology, Washington, DC.

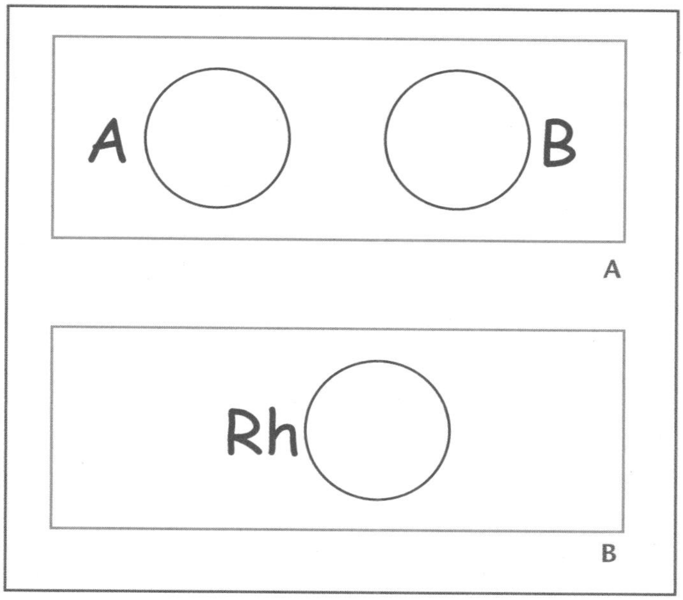

FIGURE 10-18 PREPARE THE SLIDES
(A) Draw two circles on a slide and label them "A" and "B" for the type of antiserum each will receive. (B) Draw one circle on a second slide and label it "Rh".

have someone else prick your finger, but make sure he or she wears protective gloves.

10. Discard the lancet in the sharps container.
11. Put an adhesive bandage on your wound.
12. Using a circular motion, mix each set of drops with a toothpick. Be sure to *use a different toothpick* for each antiserum.
13. Gently rock the slides back and forth for a few minutes or until agglutination occurs.
14. After the agglutination reaction is complete, record the results in the table provided on the Data Sheet. Compare your results with the possible results in Table 10-2 to determine your blood type.
15. Collect class data and record these in the table provided on the Data Sheet. Compare class data with U. S. population data.

Reference

American Association of Blood Banks Website. 2005. http://www.aabb.org/All_About_Blood/FAQs/aabb_faqs.htm#Facts

Eukaryotic Microbes

SECTION 11

Microbiology in college courses is usually dominated by bacteriology, but the discipline also includes eukaryotic microscopic organisms. You have had the opportunity to look at some eukaryotic microorganisms in Exercise 3-3. In this section, you will examine some representative protozoans and some microscopic fungi that are of medical or commercial importance. These exercises are followed by one dealing with parasitic helminths (worms) because their identification is often the responsibility of microbiologists who examine patient samples. Thus, these parasites have entered the domain of microbiology.

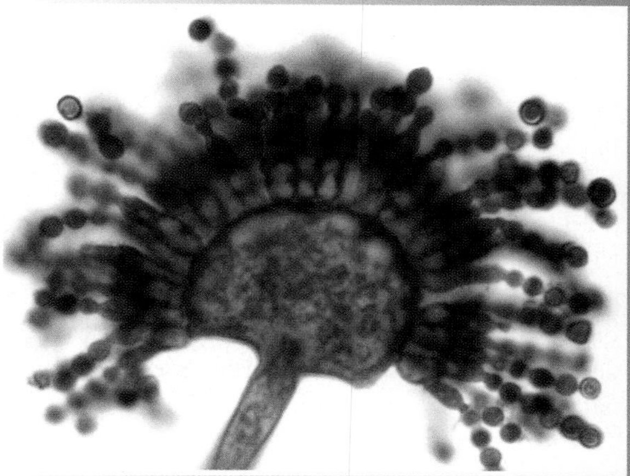

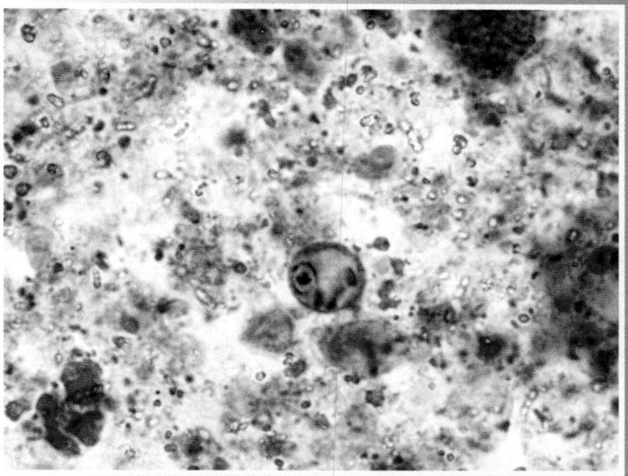

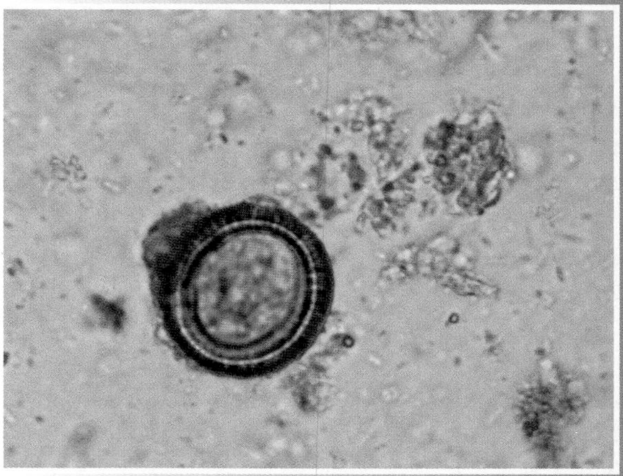

Exercise 11-1

The Fungi—Common Yeasts and Molds

Members of the Kingdom Fungi are nonmotile eukaryotes. Their cell wall is usually made of the polysaccharide **chitin**, not cellulose as in plants. Unlike animals (that ingest, then digest their food), fungi are **absorptive heterotrophs:** They secrete exoenzymes into the environment, then absorb the digested nutrients. Most are **saprophytes** that decompose dead organic matter, but some are **parasites** of plants, animals, or humans.

Fungi are informally divided into unicellular **yeasts** (Figure 11-1) and filamentous **molds** (Figure 11-2) based on their overall appearance. **Dimorphic fungi** have both mold and yeast life-cycle stages. Filamentous fungi that produce fleshy reproductive structures—mushrooms, puffballs, and shelf fungi—are referred to as **macrofungi** (Figure 11-3), even though the majority of the fungus is filamentous and hidden underground or within decaying matter.

Individual fungal filaments are called **hyphae**, and collectively they form a **mycelium**. The hyphae are darkly pigmented in **dematiaceous fungi** and unpigmented in **hyaline** or **moniliaceous fungi** (Figure 11-4). Hyphae may be **septate**, in which walls separate adjacent cells, or **nonseptate** if walls are absent (Figure 11-5).

Fungal life cycles usually are complex, involving both sexual and asexual forms of reproduction. Gametes are produced by **gametangia**, and spores are produced by a variety of **sporangia**. Typically, the only diploid cell in the fungal life cycle is the zygote, which undergoes meiosis to produce haploid spores that are characteristic of the fungal group (see the following paragraph). Various asexual spores also may be produced during the life cycle of many fungi. If they form at the ends of hyphae, they are called **conidia**. Other asexual spores are **blastospores**, which are produced by budding, and **arthrospores**, which are produced when a hypha breaks. **Chlamydospores** (**chlamydoconidia**) are formed at the end of some hyphae and are a resting stage.

Formal taxonomic categories are based primarily on the pattern of sexual spore production and the presence of crosswalls in the hyphae. Members of the **Class Zygomycetes** are terrestrial, have nonseptate hyphae, and produce nonmotile **sporangiospores** and **zygospores**. Members of the **Class Ascomycetes** produce a sac (an **ascus**) in which the zygote undergoes meiosis to produce haploid **ascospores**. Ascomycete hyphae are septate. Members of the **Class Basidiomycetes** have septate hyphae and during sexual reproduction produce a **basidium** that undergoes meiosis to produce four **basidiospores** attached to its surface. The **Class Deuteromycetes** is an unnatural assemblage of fungi in which sexual stages are either unknown or are not used in classification. Most deuteromycetes resemble ascomycetes.

Following is a brief survey of fungi that are likely to be encountered in an introductory microbiology class and selected medically important fungi.

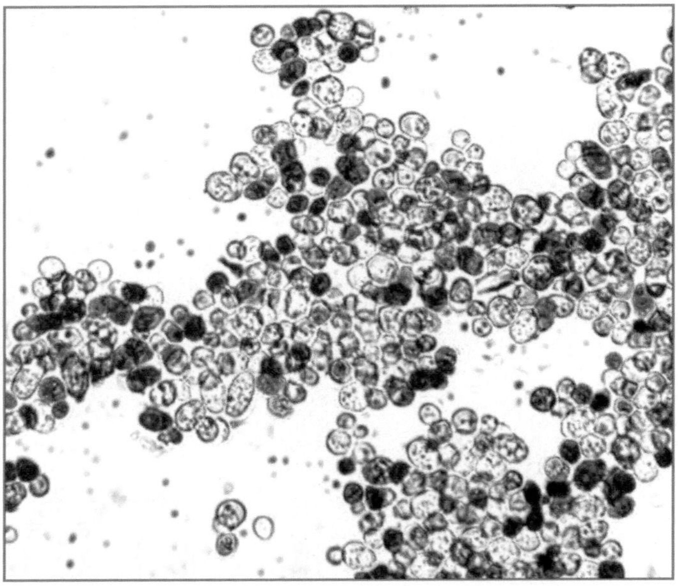

FIGURE 11-1 YEAST
This is a wet mount of the Brewer's yeast, *Saccharomyces cerevisiae* stained with Methylene Blue. (X600)

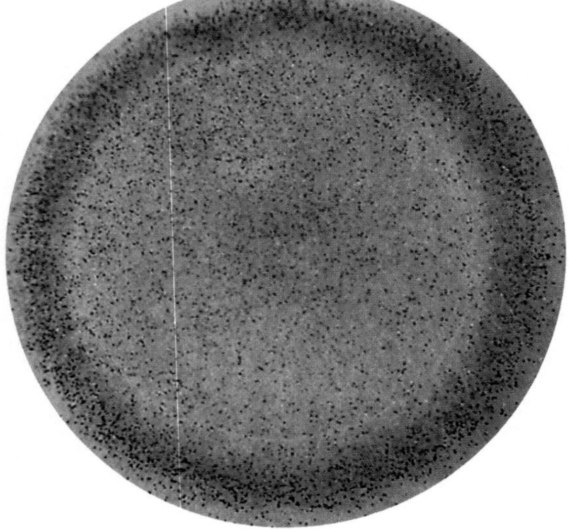

FIGURE 11-2 MOLD
Molds grow as fuzzy colonies. Their spores are abundant in the environment and frequently show up as contaminants on agar plates.

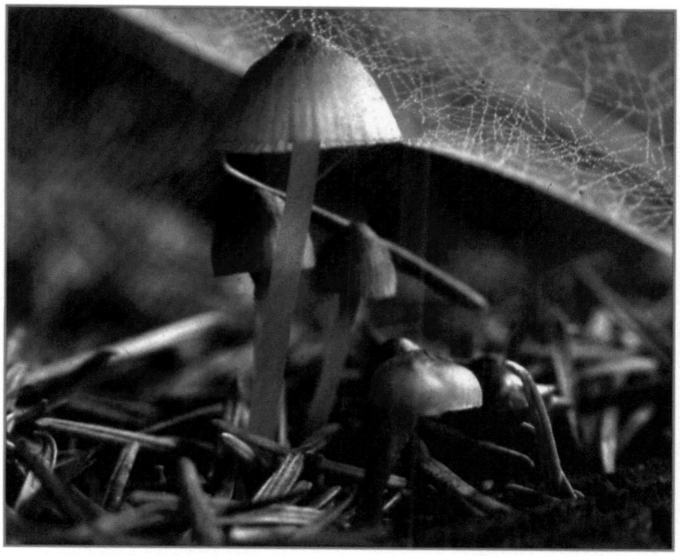

FIGURE 11-3 **MACROFUNGUS**
This mushroom is the fruiting body from a mycelium that is busily engaged in decomposing the redwood log below it.

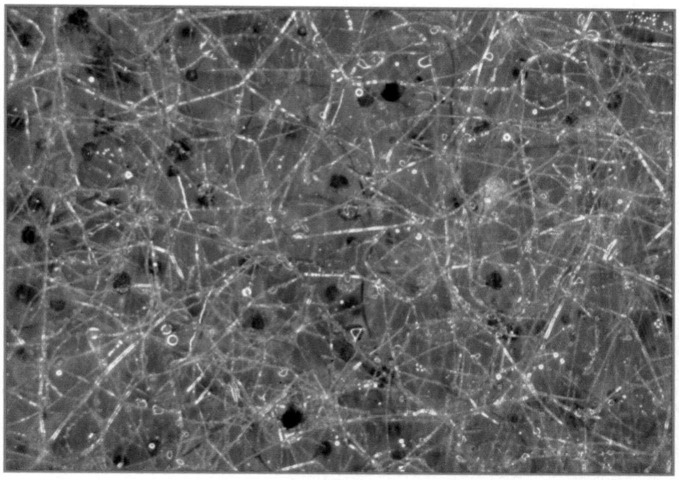

FIGURE 11-4 **HYALINE HYPHAE**
This tangled mass of hyphae belongs to the bread mold, *Rhizopus*. They are hyaline because they lack pigment.

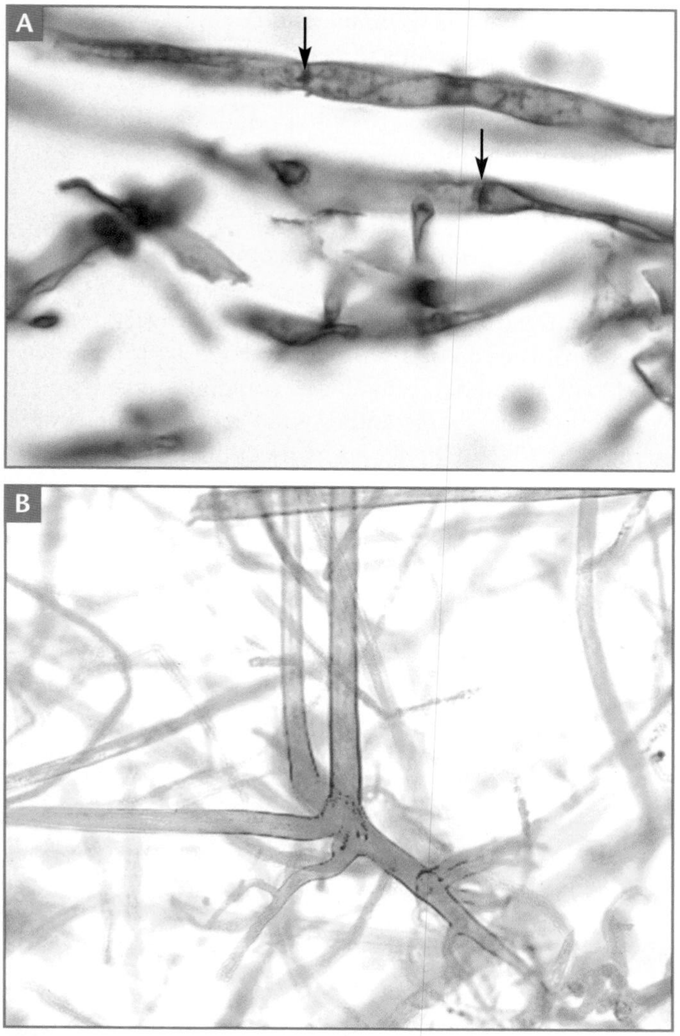

FIGURE 11-5 **SEPTATE AND NONSEPTATE HYPHAE**
(A) The hyphae of *Aspergillus* are septate. Note the crosswalls (designated by arrows) dividing cells. (X1000). (B) *Rhizopus* provides an example of nonseptate hyphae. (X200)

Yeasts of Medical or Economic Importance

Candida albicans

Candida albicans (Figure 11-6) is part of the normal respiratory, gastrointestinal and female urogenital tract floras. Under the proper circumstances, it may flourish and produce pathological conditions, such as thrush in the oral cavity, vulvovaginitis of the female genitals, and cutaneous candidiasis of the skin. Systemic candidiasis may follow infection of the lungs, bronchi, or kidneys. Entry into the blood may result in endocarditis. Individuals most susceptible to *Candida* infections are diabetics, those with immunodeficiency (*e.g.,* AIDS), catheterized

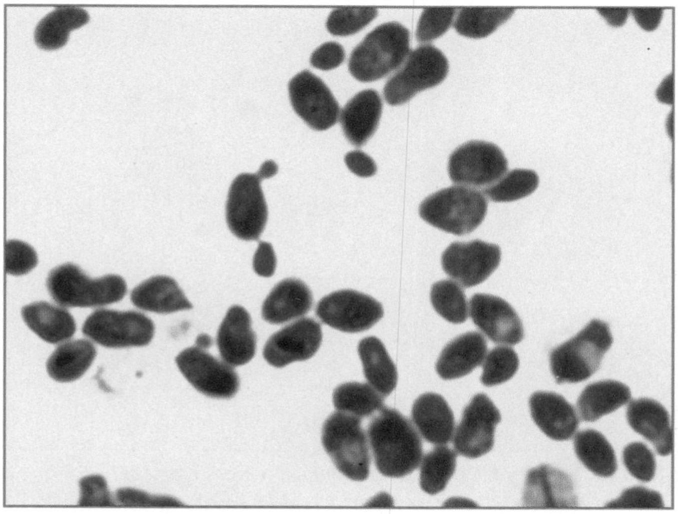

FIGURE 11-6 ***CANDIDA ALBICANS*** **VEGETATIVE CELLS (X2640)**
Note the oval shape and nuclei.

patients, and individuals taking antimicrobial medications. Budding results in chains of cells called **pseudohyphae,** which produce clusters of round, asexual blastoconidia at the cell junctions. Large, round, thick-walled **chlamydospores** form at the ends of pseudohyphae.

Saccharomyces cerevisiae

Saccharomyces cerevisiae is an ascomycete used in the production of bread, wine, and beer but is not an important human pathogen. It does not form a mycelium but, rather, produces a colony similar to bacteria (Figure 11-7). The vegetative cells (**blastoconidia**) are generally oval to round in shape, and asexual reproduction occurs by budding (Figure 11-8). Short **pseudohyphae** are sometimes produced when the budding cells fail to separate. Meiosis produces one to four ascospores within the vegetative cell, which acts as the ascus. Ascospores may fuse to form another generation of diploid vegetative cells or they may be released to produce a population of haploid cells that are indistinguishable from diploid cells. Haploid cells of opposite mating types may also combine to create a diploid cell.

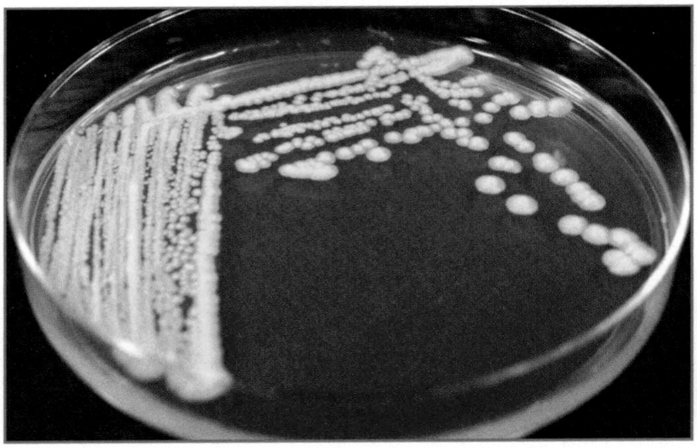

FIGURE 11-7 *SACCHAROMYCES CEREVISIAE* COLONY
Note that the appearance is similar to a typical bacterial colony, not "fuzzy" like mold colonies.

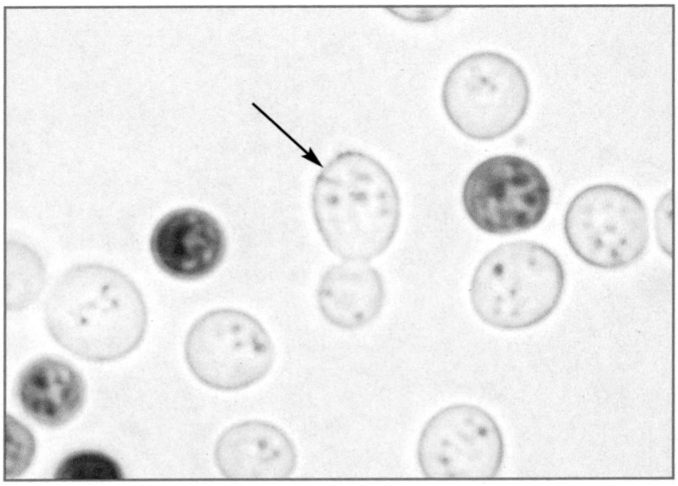

FIGURE 11-8 WET MOUNT OF *SACCHAROMYCES CEREVISIAE* VEGETATIVE CELLS (CRYSTAL VIOLET STAIN, X1320)
Note the budding cell (blastoconidium) in the center of field (arrow). This wet mount was stained with Methylene Blue.

Molds of Medical or Economic Importance

Rhizopus

Rhizopus species are fast-growing zygomycetes that produce white or grayish, cottony growth. The mycelium becomes darker with age as sporangia are produced, giving it a "salt and pepper" appearance (Figure 11-9). Microscopically, *Rhizopus* species produce broad (10 µm), hyaline, and usually nonseptate surface and aerial hyphae. Anchoring **rhizoids** (Figure 11-10) are produced where the surface hyphae (**stolons**) join the bases of the long, unbranched **sporangiophores.**

The *Rhizopus* life cycle (Figure 11-11) has both sexual and asexual phases. Asexual **sporangiospores** are produced by large, circular sporangia (Figure 11-12) borne at the ends of long, nonseptate, elevated sporangiophores. A hemispherical **columella** supports the sporangium. The spores develop into hyphae that are identical to those that produced them. On occasion, sexual

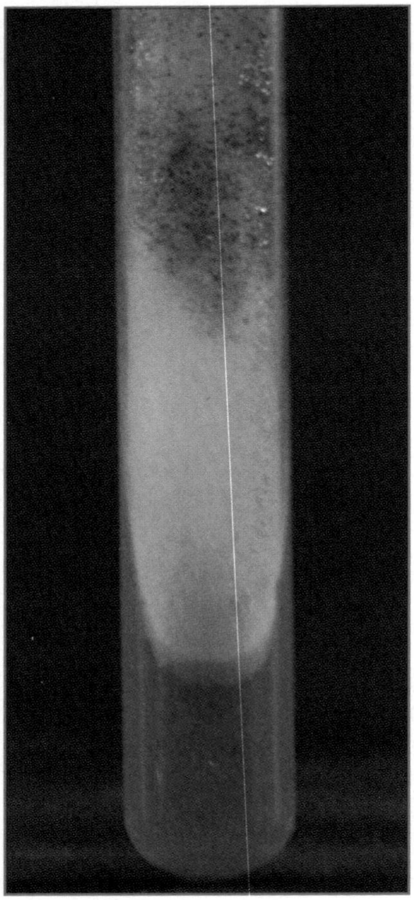

FIGURE 11-9 *RHIZOPUS STOLONIFER.*
Black asexual sporangia of this bread mold have begun to form, giving the growth a "salt and pepper" appearance.

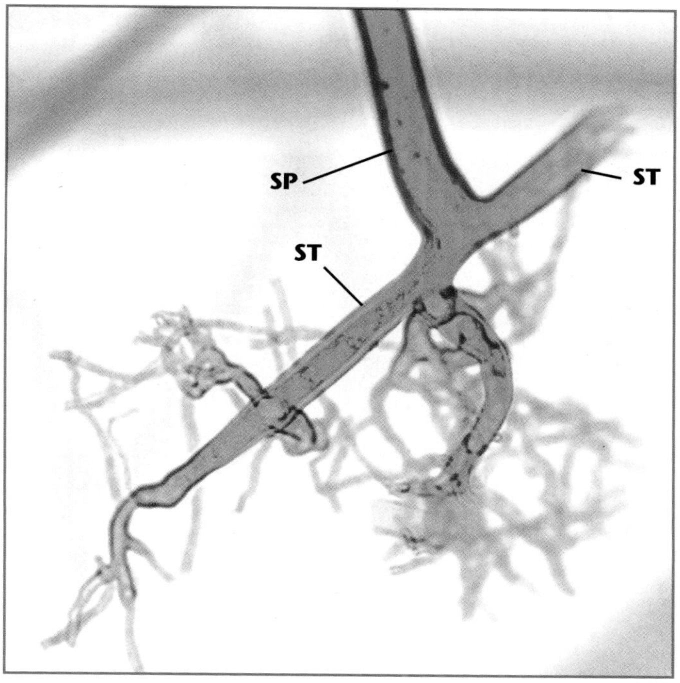

FIGURE 11-10 *RHIZOPUS* **RHIZOIDS (X200)**
Anchoring rhizoids form at the junction of each sporangiophore (SP) and the stolon (ST). Note the absence of the septa.

reproduction occurs when hyphae of different mating types (designated + and − strains) make contact.

Initially, **progametangia** (Figure 11-13) extend from each hypha. Upon contact, a septum separates the end of each progametangium into a gamete (Figure 11-14). The walls between the two gametangia dissolve, and a thick-walled **zygospore** develops (Figures 11-15 and 11-16). Fusion of nuclei occurs within the zygospore and produces one or more diploid nuclei, or **zygotes**. After a dormant period, meiosis of the zygotes occurs. The zygospore then germinates and produces a sporangium similar to asexual sporangia. Haploid spores are released, develop into new hyphae, and the life cycle is completed.

Rhizopus species are common contaminants. *R. stolonifer* is the common bread mold. *R. oryzae* and *R. arrhizus* are responsible for producing zygomycosis, a condition found most frequently in diabetics and immunocompromised patients. Inhalation of spores may lead to hypersensitivity reactions in the respiratory system. Entry into the blood leads to rapid spreading of the organism, occlusion of blood vessels, and necrosis of tissues.

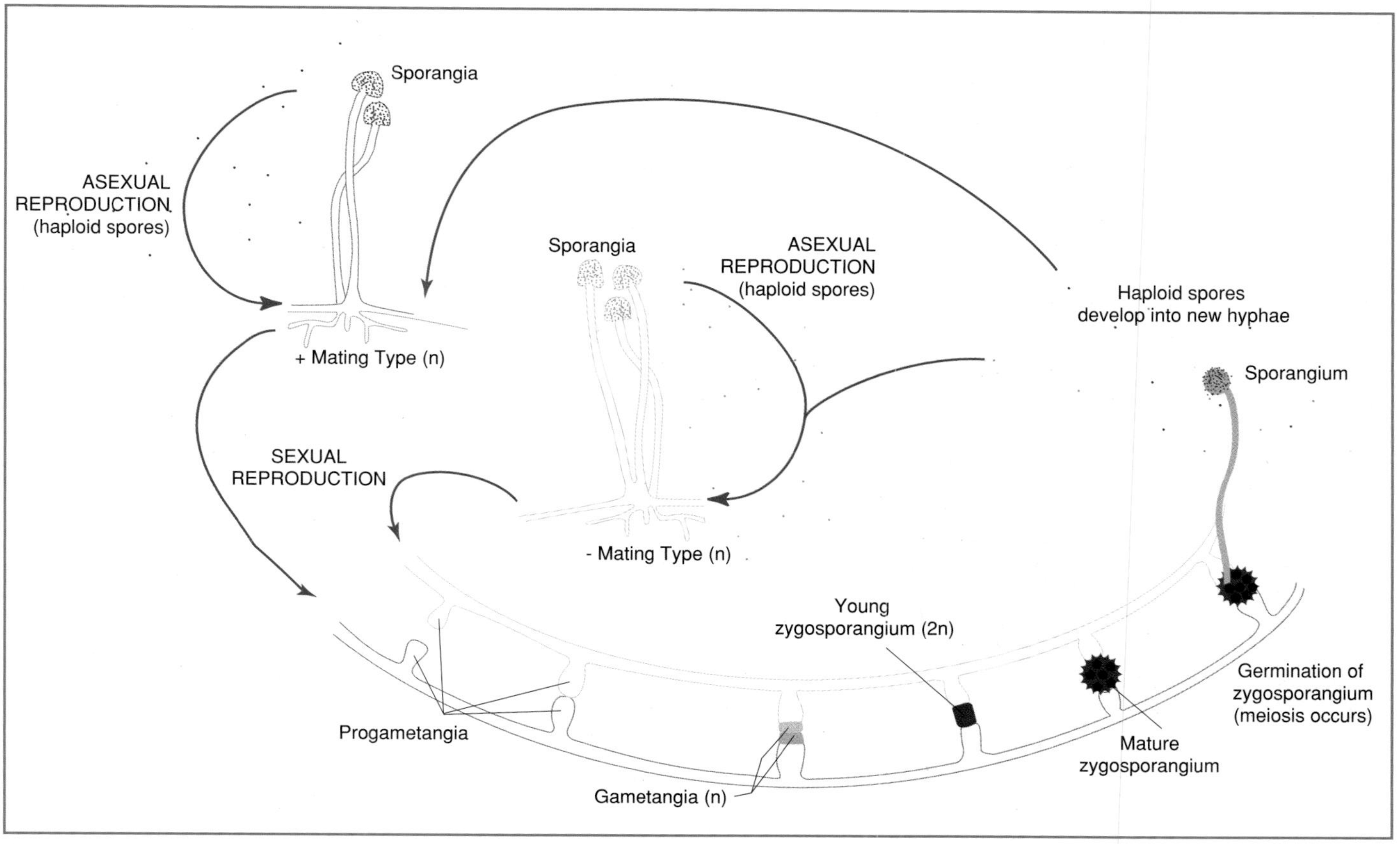

FIGURE 11-11 *RHIZOPUS* **LIFE CYCLE**
Please refer to the text for details.

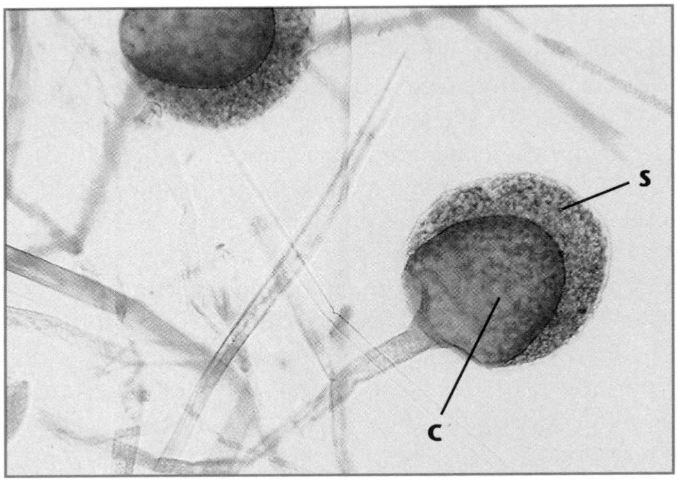

FIGURE 11-12 *RHIZOPUS* SPORANGIOPHORES (X264)
The sporangium is found at the end of a long, unbranched, and nonseptate sporangiophore. The haploid asexual sporangiospores (S) cover the surface of the columella (C), which has a flattened base.

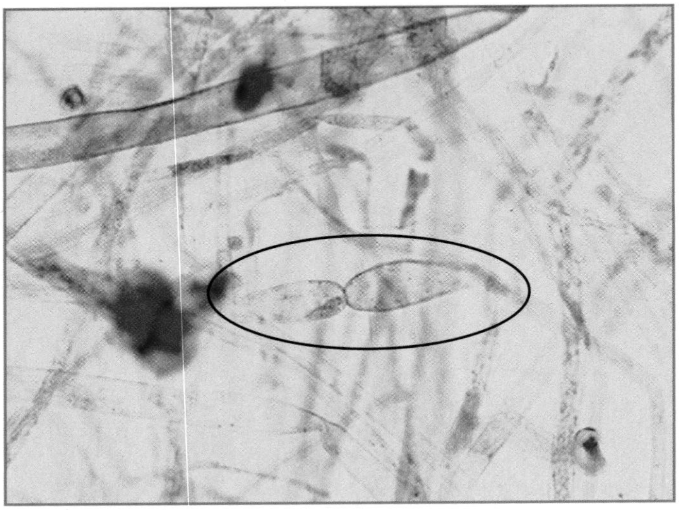

FIGURE 11-13 *RHIZOPUS* PROGAMETANGIA (X264)
Progametangia from different hyphae are shown in the center of the field. Contact between the progametangia results in each forming a gamete.

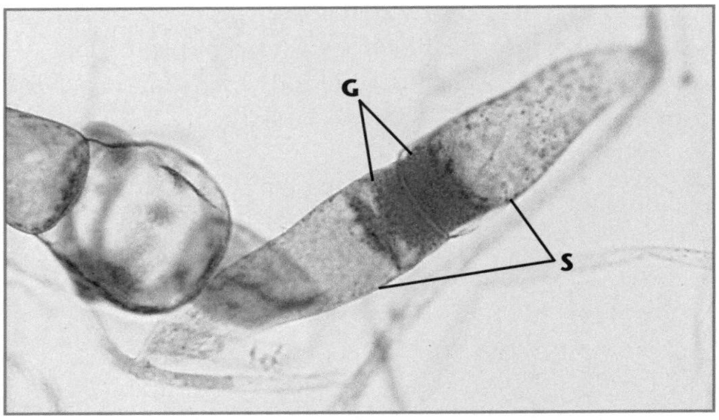

FIGURE 11-14 *RHIZOPUS* GAMETANGIA AND SUSPENSORS (X264)
Gametangia (G) and suspensors (S) are shown in the center of the field. Gametangia contain haploid nuclei from each mating type.

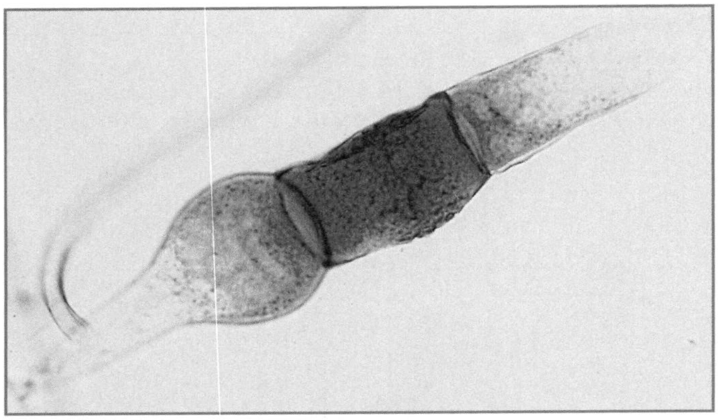

FIGURE 11-15 YOUNG *RHIZOPUS* ZYGOSPORE (X264)
The zygospore forms when the cytoplasm from the two mating strains fuse (plasmogamy).

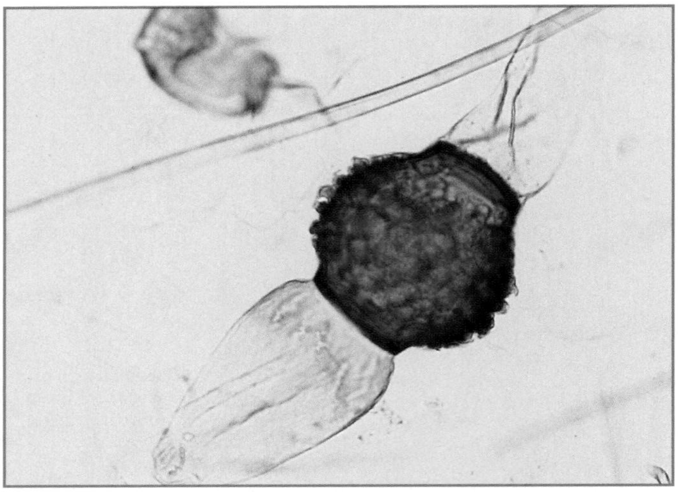

FIGURE 11-16 MATURE *RHIZOPUS* ZYGOSPORE (X264)
Haploid nuclei from each strain fuse within the zygospore (**karyogamy**) to produce many diploid nuclei. Meiosis occurs to produce numerous haploid spores.

Aspergillus

The genus *Aspergillus* is characterized by green to yellow or brown granular colonies with a white edge (Figure 11-17). One species, *A. niger*, produces distinctive black colonies. Vegetative hyphae are hyaline (unpigmented) and septate. The *Aspergillus* fruiting body is distinctive, with chains of conidia arising from one (uniseriate) or two (biseriate) rows of **phialides** attached to a swollen **vesicle** at the end of an unbranched conidiophore (Figure 11-18). The conidiophore grows from a foot cell in the vegetative hypha (Figure 11-19). Fruiting body structure and size, and conidia color are useful in species identification.

A. fumigatus and other species are opportunistic pathogens that cause aspergillosis, an umbrella term covering many diseases. One form of pulmonary aspergillosis (referred to as fungus ball) involves colonization of the bronchial tree or tissues damaged by tuberculosis.

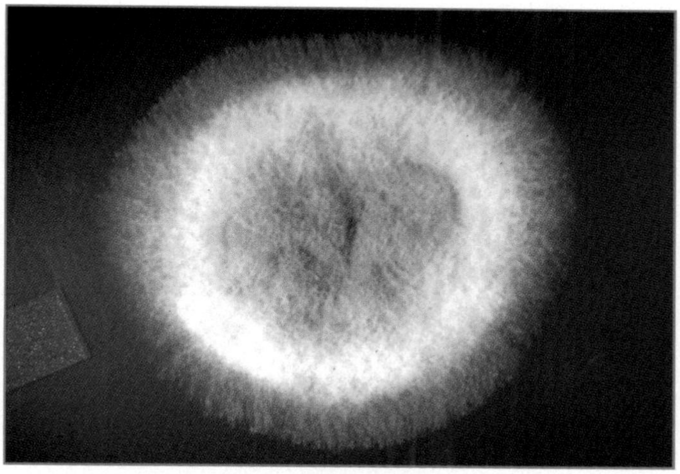

FIGURE 11-17 *ASPERGILLUS FUMIGATUS* **COLONY ON SABOURAUD DEXTROSE AGAR**
Note the rugose topography and green, granular appearance with a white margin. The reverse is white. Although a common cause of aspergillosis, normally healthy people are not at great risk from *A. fumigatus* infection.

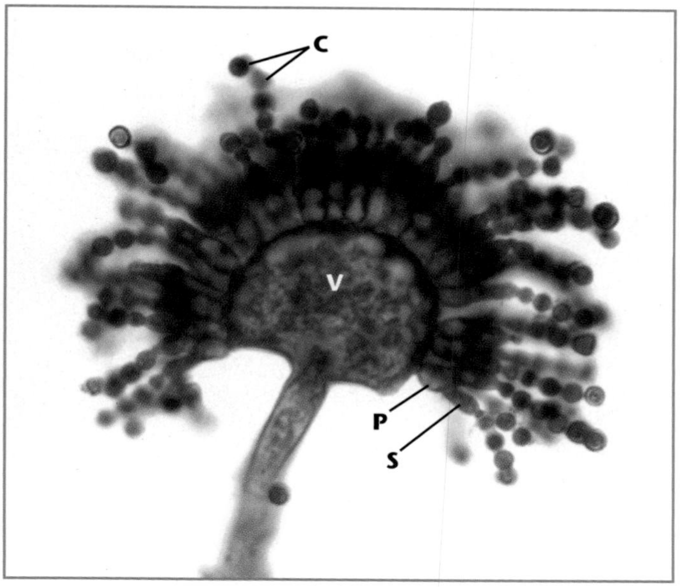

FIGURE 11-18 *ASPERGILLUS* **CONIDIAL HEADS**
(A) Shown is a section of an *Aspergillus niger* conidiophore (X1000). The conidia (C), primary (P) and secondary (S) phialides, and vesicle (V) are visible. This is a biseriate conidium.

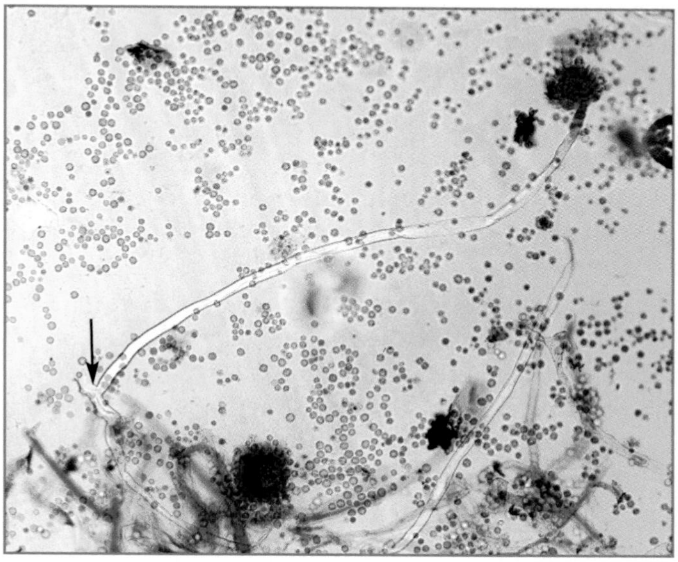

FIGURE 11-19 *ASPERGILLUS* **FOOT CELL (LACTOPHENOL COTTON BLUE STAIN, X100)**
Sporangiophores emerge from a foot cell with the shape of an inverted "T" (arrow). This is a wet mount of *A. flavus*. Notice all the conidia in the field.

Allergic aspergillosis may occur in individuals who are in frequent contact with the spores and become sensitized to them. Subsequent contact produces symptoms similar to asthma. Invasive aspergillosis is the most severe form. It results in necrotizing pneumonia and may spread to other organs.

Some species of *Aspergillus* are of commercial importance. Fermentation of soybeans by *A. oryzae* produces soy paste. Soy sauce is produced by fermenting soybeans with a mixture of *A. oryzae* and *A. soyae*. *Aspergillus* is also used in commercial production of citric acid.

Penicillium

Members of the genus *Penicillium* produce distinctive green, powdery, radially furrowed colonies with a white apron (Figure 11-20) and light colored reverse surface. The hyphae are septate and thin. Distinctive *Penicillium* fruiting bodies, consisting of **metulae, phialides** and chains of spherical conidia, are located at the ends of branched or unbranched **conidiophores** (Figure 11-21). Although not an important feature in laboratory identification, sexual reproduction results in the formation of ascospores within an ascus.

Penicillium is best known for its production of the antibiotic penicillin, but it is also a common contaminant. One pathogen, *P. marneffei*, is endemic to Asia and is responsible for disseminated opportunistic infections of the lungs, liver, and skin in immunosuppressed and immunocompromised patients. It is thermally dimorphic, producing a typical velvety colony with a distinctive red pigment at 25°C but converting to a yeast form at 35°C. Other species of *Penicillium* are of commercial importance for fermentations used in cheese production. Examples include *P. roquefortii* (Roquefort cheese) and *P. camembertii* (Camembert and Brie cheeses).

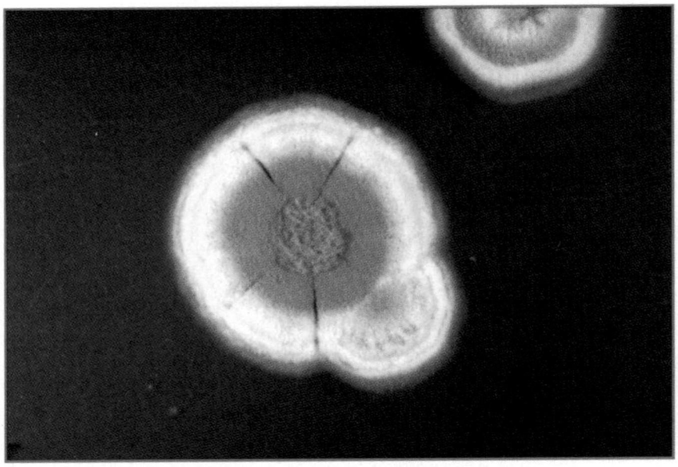

FIGURE 11-20 *PENICILLIUM NOTATUM* COLONY ON SABOURAUD DEXTROSE AGAR
The green, granular surface with radial furrows and a white apron are typical of the genus.

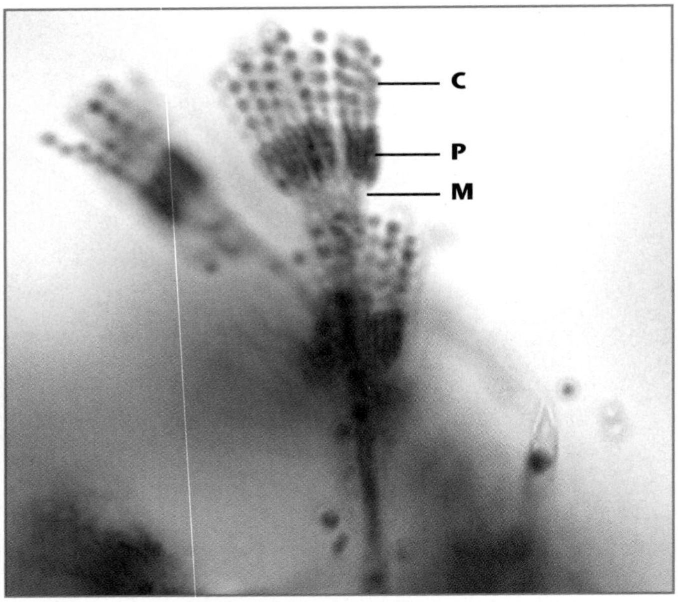

FIGURE 11-21 *PENICILLIUM* CONIDIOPHORE (X1000)
Penicillium species produce a characteristic brush-shaped conidiophore (penicillus). Metulae (M), phialides (P), and chains of spherical conidia (C) are visible.

Materials

Per Student Group

- Agar slant of *Saccharomyces cerevisiae*
- Potato Dextrose Agar or Sabouraud Dextrose Agar plate culture of *Aspergillus spp.* (with lid taped on)
- Potato Dextrose Agar or Sabouraud Dextrose Agar plate culture of *Penicillium spp.* (with lid taped on)
- Potato Dextrose Agar or Sabouraud Dextrose Agar plate culture of *Rhizopus spp.* (with lid taped on)
- Gram's iodine stain or Methylene Blue
- dissecting microscope
- prepared slides of:
  - *Aspergillus* spp. conidiophore
  - *Candida albicans*
  - *Penicillium spp.* conidiophore
  - *Rhizopus spp.* sporangia
  - *Rhizopus spp.* gametangia

Procedure

Yeasts

1. Using an inoculating loop and aseptic technique, make a wet mount slide of *Saccharomyces cerevisiae* as illustrated in Figure 3-18, and stain with iodine or Methylene Blue. Observe under high-dry and oil immersion. Identify vegetative cells and budding cells. Sketch representative cells in the space provided on the Data Sheet.
2. Observe prepared slides of *Candida albicans*. Identify vegetative cells and budding cells. Sketch representative cells in the space provided on the Data Sheet.

Molds

1. Obtain the plate culture of *Rhizopus. Do not remove the lid. Uncovering the organism will spread spores and contaminate the laboratory.*
   a. Examine the colony morphology, and sketch a representative colony in the space provided on the Data Sheet. Record the ***color*** on both the front (obverse) and reverse surfaces. Also record the ***colony texture*** as glabrous (leathery), velvety, yeast-like, cottony, or granular (powdery), and the ***colony topography*** as flat, rugose (with radial grooves), folded, crateriform, verrucose (warty, rough) or cerebriform (brain-like).
   b. Examine the colony under the dissecting microscope and identify hyphae, rhizoids, and sporangia. Sketch and label representative structures in the space provided on the Data Sheet.
2. Examine prepared slides of *Rhizopus* sporangia using medium and high-dry powers. Identify the following: sporangiophores, sporangia, and spores. Sketch and label representative structures in the space provided on the Data Sheet.
3. Examine prepared slides of *Rhizopus* gametangia using medium and high dry power. Identify the following: progametangia, gametangia, young zygosporangia, mature zygosporangia. Sketch and label representative structures in the space provided on the Data Sheet.

4. Obtain the plate culture of *Penicillium. Do not remove the lid from the plate or you will spread spores and contaminate the laboratory.*
   a. Examine the colony morphology and sketch a representative colony in the space provided on the Data Sheet. Record the ***color*** on both the front (obverse) and reverse surfaces. Also record the ***colony texture*** as glabrous (leathery), velvety, yeast-like, cottony, or granular (powdery), and the ***colony topography*** as flat, rugose (with radial grooves), folded, crateriform, verrucose (warty, rough) or cerebriform (brain-like).
   b. Examine the colony under the dissecting microscope. Identify hyphae and conidia. Sketch and label representative structures in the space provided on the Data Sheet.
5. Observe prepared slides of *Penicillium* conidiophores. Identify the following: hyphae, conidiophores, and chains of conidia. Sketch and label representative structures in the space provided on the Data Sheet.
6. Obtain the plate culture of *Aspergillus. Do not remove the lid from the plate or you will spread spores and contaminate the laboratory.*
   a. Examine the colony morphology and sketch a representative colony in the space provided on the Data Sheet. Record the ***color*** on both the front (obverse) and reverse surfaces. Also record the ***colony texture*** as glabrous (leathery), velvety, yeast-like, cottony, or granular (powdery), and the ***colony topography*** as flat, rugose (with radial grooves), folded, crateriform, verrucose (warty, rough), or cerebriform (brainlike).
   b. Examine the colony under the dissecting microscope. Identify hyphae and conidia. Sketch and label representative structures in the space provided on the Data Sheet.
7. Observe prepared slides of *Aspergillus* conidiophores. Identify hyphae, conidiophores, and conidia. Sketch and label representative structures in the space provided on the Data Sheet.

References

Collins, C. H., Patricia M. Lyne, and J. M. Grange. 1995. Chapter 51 in *Collins and Lyne's Microbiological Methods,* 7th ed. Butterworth-Heineman, Oxford, United Kingdom.

Fisher, Fran, and Norma B. Cook. 1998. Chapter 2 in *Fundamentals of Diagnostic Mycology.* W. B. Saunders, Philadelphia.

Forbes, Betty A., Daniel F. Sahm, and Alice. S. Weissfeld. 2002. Chapter 53 in *Bailey and Scott's Diagnostic Microbiology,* 11th ed. Mosby-Year Book, St. Louis.

Koneman, Elmer W., Stephen D. Allen, William M. Janda, Paul C. Schreckenberger, and Washington C. Winn, Jr. 1997. Chapter 19 in *Color Atlas and Textbook of Diagnostic Microbiology,* 5th ed. J. B. Lippincott, Philadelphia.

Exercise 11-2

Examination of Common Protozoans of Clinical Importance

Protozoans are unicellular eukaryotic heterotrophic microorganisms. A typical life cycle includes a vegetative **trophozoite** stage and a resting **cyst** stage. Some have additional stages, making their life cycles more complex.

As discussed in Exercise 3-3, protozoans are classified as follows:

Phylum Sarcomastigophora (including Subphylum Mastigophora [the flagellates] and Subphylum Sarcodina [the amoebas]),
Phylum Ciliophora (the ciliates), and
Phylum Apicomplexa (sporozoans and others).

Following is a survey of some commonly encountered prototzoans of clinical importance. You will be examining these on prepared slides because most are pathogens that are not handled appropriately in a beginning microbiology laboratory.

Amoeboid Protozoans Found in Clinical Specimens

Entamoeba histolytica

Entamoeba histolytica is the causative agent of amoebic dysentery (amebiasis), a disease most common in areas with poor sanitation. Identification is made by finding either trophozoites (Figure 11-22) or cysts (Figure 11-23) in a stool sample. The diagnostic features of each are described in the captions.

Infection occurs when a human host ingests cysts, either through fecal–oral contact or, more typically, contaminated food or water. Cysts (but not trophozoites) are able to withstand the acidic environment of the stomach. Upon entering the less acidic small intestine, the cysts undergo **excystation**. Mitosis produces eight small trophozoites from each cyst.

The trophozoites parasitize the mucosa and submucosa of the colon, causing ulcerations. They feed on red blood cells and bacteria. The extent of damage determines whether the disease is acute, chronic, or asymptomatic. In the most severe cases, infection may extend to other organs, especially the liver, lungs, or brain. Among the symptoms of amoebic dysentery are abdominal pain, diarrhea, blood and mucus in feces, nausea, vomiting, and hepatitis.

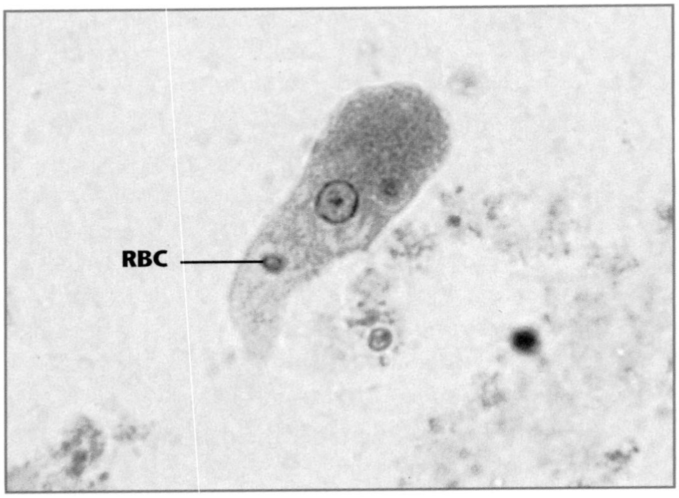

FIGURE 11-22 *Entamoeba histolytica* Trophozoite (X800, Iron Hematoxylin Stain)
Trophozoites range in size from 12 to 60 μm. Notice the small, central karyosome, the beaded chromatin at the margin of the nucleus, the ingested red blood cells, and the finely granular cytoplasm. Compare with an *Entamoeba coli* trophozoite in Figure 11-24.

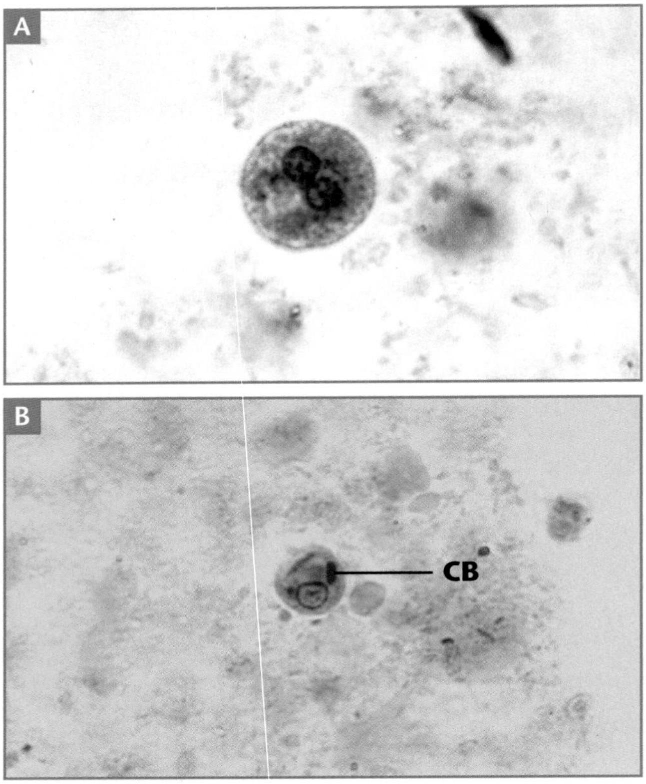

FIGURE 11-23 *Entamoeba histolytica* Cysts
(A) Cysts are spherical with a diameter of 10 to 20 μm. Two of the four nuclei are visible; other nuclear characteristics are as in the trophozoite. Compare with an *Entamoeba coli* cyst in Figure 11-25 (X1320, Iron Hematoxylin Stain). (B) *E. histolytica* cyst (X1200, Trichrome Stain) with cytoplasmic chromatoidal bars (CB). These are found in approximately 10% of the cysts, have blunt ends, and are composed of ribonucleoprotein.

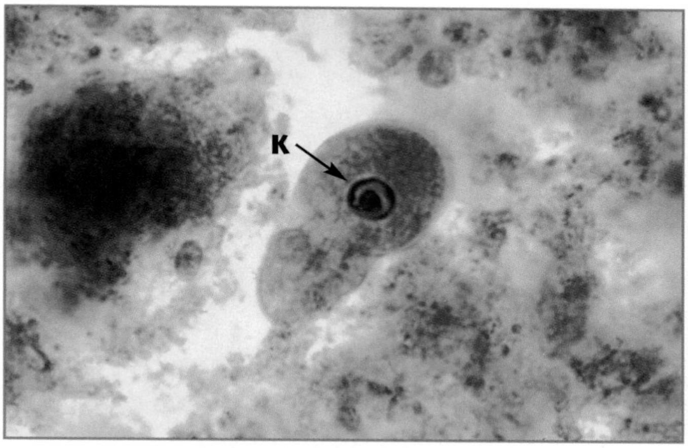

FIGURE 11-24 *ENTAMOEBA COLI* TROPHOZOITE (X1000, TRICHROME STAIN)
Trophozoites range in size from 15 to 50 μm. Notice the relatively large and eccentrically positioned karyosome (K), the unclumped chromatin at the periphery of the nucleus, and the vacuolated cytoplasm lacking ingested red blood cells. The usually nonpathogenic *E. coli* must be distinguished from the potentially pathogenic *E. histolytica*, so compare with Figure 11-22.

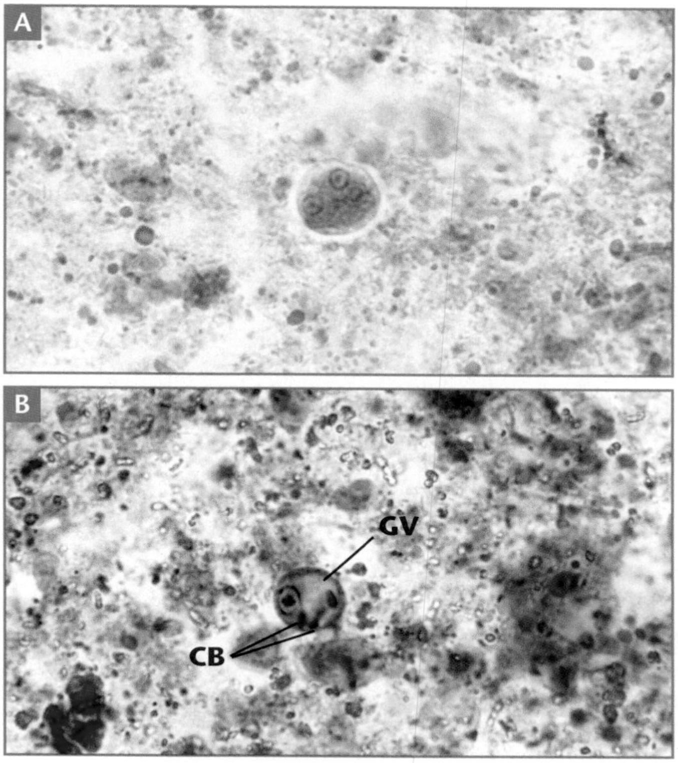

FIGURE 11-25 *ENTAMOEBA COLI* CYSTS
Cysts typically are spherical and are between 10 and 35 μm in diameter. They contain 8, or sometimes 16 nuclei. This makes differentiation from *E. histolytica* cysts simpler, as they never have more than 4 nuclei. (A) Five nuclei are visible in this specimen (X1000, trichrome stain). (B) Chromatoidal bars (CB) and a large glycogen vacuole (GV) characteristic of immature cysts are visible in this specimen (X1000, Trichrome Stain).

Developing cysts undergo mitosis to produce mature quadranucleate cysts, which are shed in the feces and are infective. They also may persist in the original host, resulting in an **asymptomatic carrier**—a major source of contamination and infection.

Another member of the genus *Entamoeba* deserves mention here. *Entamoeba coli* is a fairly common, nonpathogenic intestinal commensal that must be differentiated from *E. histolytica* in stool samples. Its characteristic features are given in the captions to Figures 11-24 and 11-25.

Ciliate Protozoan Found in Clinical Specimens

Balantidium coli

Balantidium coli (Figures 11-26 and 11-27) is the causative agent of balantidiasis and exists in two forms: a vegetative trophozoite and a cyst. Laboratory diagnosis is made by identifying either the cyst or the trophozoite, with the latter being more commonly found.

The trophozoite is highly motile because of the cilia and has a macronucleus and a micronucleus. Cysts in sewage-contaminated water are the infective form. Trophozoites may cause ulcerations of the colon mucosa, but not to the extent produced by *Entamoeba histolytica*. Symptoms of acute infection are bloody and mucoid feces. Diarrhea alternating with constipation may occur in chronic infections. Most infections probably are asymptomatic.

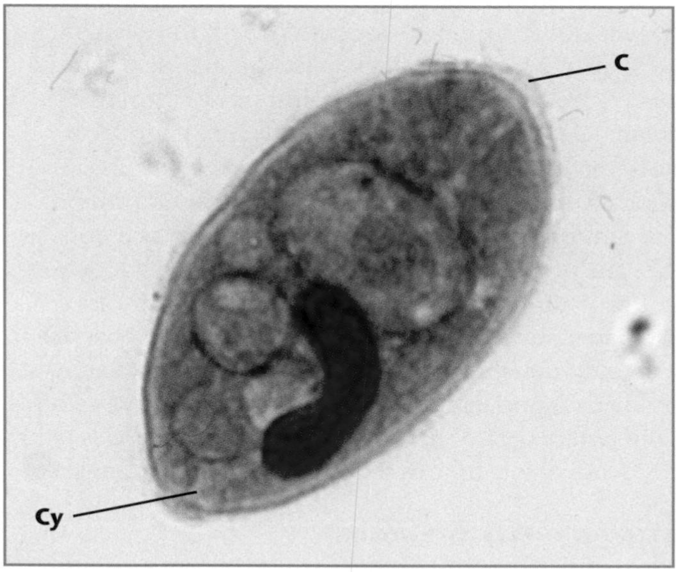

FIGURE 11-26 *BALANTIDIUM COLI* TROPHOZOITE (X800)
Trophozoites are oval in shape with dimensions of 50 to 100 μm long by 40 to 70 μm wide. Cilia (C) cover the cell surface. Internally, the macronucleus is prominent; the adjacent micronucleus is not. An anterior cytostome (Cy) is usually visible.

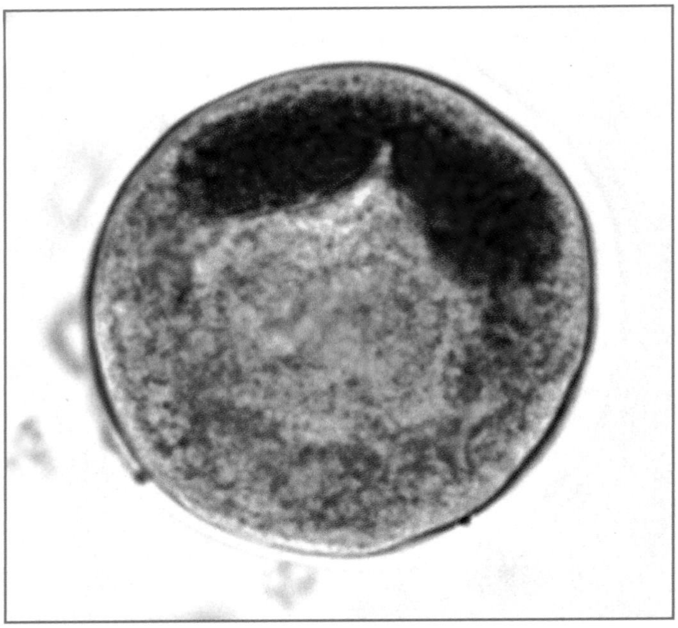

FIGURE 11-27 ***Balantidium coli* Cyst (X1000)**
Cysts usually are spherical and have a diameter in the range of 50 to 75 μm. There is a cyst wall, and the cilia are absent. As in the trophozoite, the macronucleus is prominent, but the micronucleus may not be.

Flagellate Protozoans Found in Clinical Specimens

Giardia lamblia

Giardiasis is caused by *Giardia lamblia* (also known as *Giardia intestinalis)*, a flagellate protozoan. It is seen most frequently in the duodenum as a heart-shaped vegetative trophozoite (Figure 11-28) with four pairs of flagella and a sucking disc that allows it to resist gut peristalsis. Multinucleate cysts lacking flagella (Figure 11-29) are formed as the organism passes through the colon. Cysts are shed in the feces and may produce infection of a new host upon ingestion. Transmission typically involves fecally contaminated water or food, but direct fecal–oral contact transmission is also possible.

The organism attaches to epithelial cells but does not penetrate to deeper tissues. Most infections are asymptomatic. Chronic diarrhea, dehydration, abdominal pain, and other symptoms may occur if the infection produces a population large enough to involve a significant surface area of the small intestine. Diagnosis is made by identifying trophozoites or cysts in stool specimens.

Trichomonas vaginalis

Trichomonas vaginalis (Figure 11-30) is the causative agent of trichomoniasis (vulvovaginitis) in humans. It has four anterior flagella and an **undulating membrane.**

Trichomoniasis may affect both sexes but is more common in females. *T. vaginalis* causes inflammation

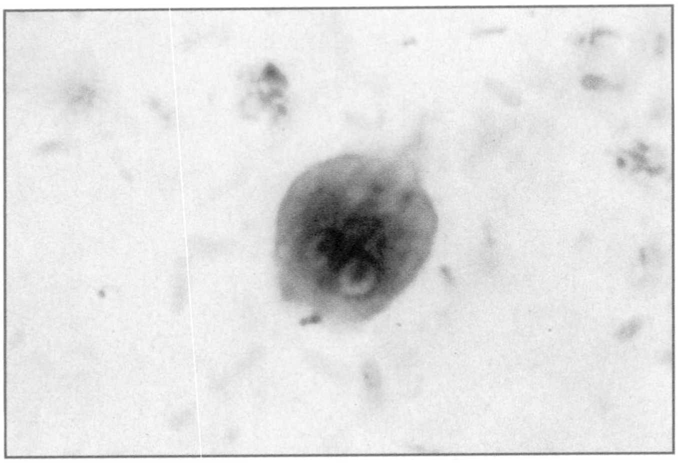

FIGURE 11-28 ***Giardia lamblia* Trophozoite (X1320, Iron Hematoxylin Stain)**
Trophozoites have a long, tapering posterior end and range in size from 9 to 21 μm by 5 to 15 μm. There are two nuclei with small karyosomes. Two median bodies are visible, but the four pairs of flagella are not.

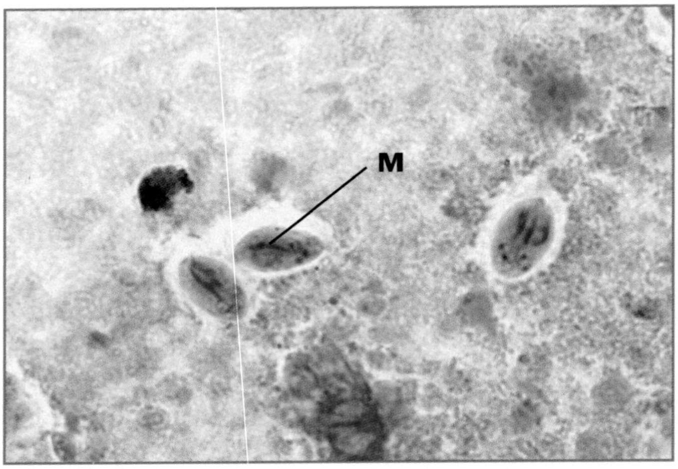

FIGURE 11-29 ***Giardia lamblia* Cysts (X1000, Trichrome Stain).**
Giardia cysts are smaller than trophozoites (8 to 12 μm by 7 to 10 μm), but the four nuclei with eccentric karyosomes and the median bodies (M) are still visible.

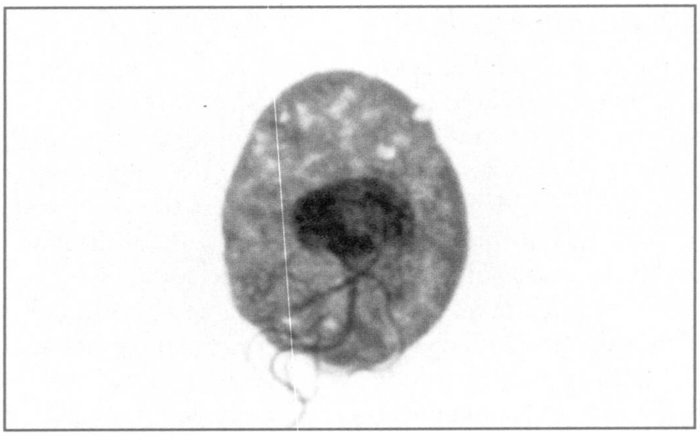

FIGURE 11-30 ***Trichomonas vaginalis* (X2027)**
The trophozoite is the only stage of the *Trichomonas* life cycle. Several flagella are visible.

of genitourinary mucosal surfaces—typically the vagina, vulva, and cervix in females and the urethra, prostate, and seminal vesicles in males. Most infections are asymptomatic or mild. Some erosion of surface tissues and a discharge may be associated with infection. The degree of infection is affected by host factors, especially the bacterial flora present and the pH of the mucosal surfaces. Transmission typically is by sexual intercourse.

The morphologically similar nonpathogenic *Trichomonas tenax* and *T. hominis* are residents of the oral cavity and intestines, respectively.

Trypanosoma brucei

Trypanosoma brucei (Figure 11-31) is a species of flagellated protozoans divided into subspecies: *T. brucei brucei* (which is nonpathogenic), and *T. brucei gambiense* and *T. brucei rhodesiense*, which produce African trypanosomiasis, also known as African sleeping sickness. The organisms are very similar morphologically but differ in geographic range and disease progress. West African trypanosomiasis (caused by *T. brucei gambiense*) is generally a mild, chronic disease that may last for years, whereas East African trypanosomiasis (caused by *T. brucei rhodesiense*) is more acute and results in death within a year. Modern molecular methods that compare proteins, RNA, and DNA are used to differentiate between them.

Trypanosomes have a complex life cycle. One stage of the life cycle, the **epimastigote**, multiplies in an intermediate host, the tsetse fly (genus *Glossina*).

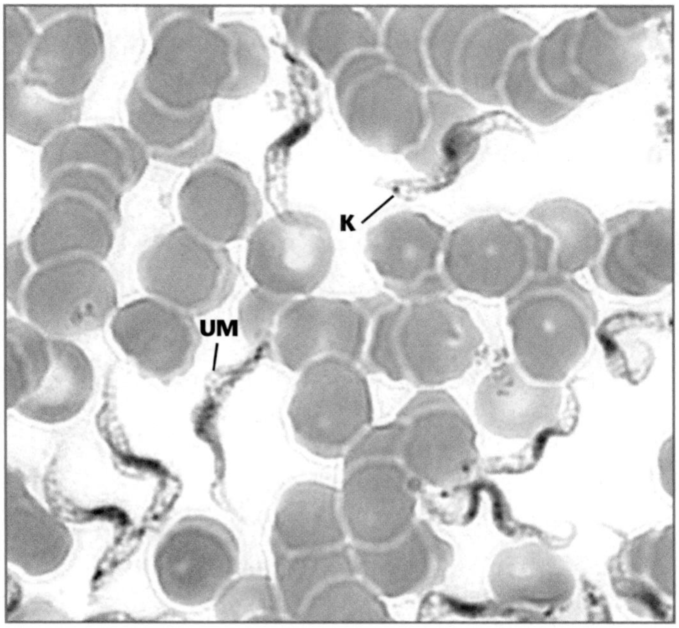

FIGURE 11-31 *TRYPANOSOMA BRUCEI* **TRYPOMASTIGOTES IN A BLOOD SMEAR (X1500)**
The central nucleus, posterior kinetoplast (K), and undulating membrane (UM) are visible.

The infective **trypomastigote** stage then is transmitted to the human host through tsetse fly bites. Once introduced, trypomastigotes multiply and produce a chancre at the site of the bite. They enter the lymphatic system and spread through the blood, and ultimately to the heart and brain.

Immune response to the pathogen is hampered by the trypanosome's ability to change surface antigens faster than the immune system can produce appropriate antibodies. This antigenic variation also makes development of a vaccine unlikely. Diagnosis is made from clinical symptoms and identification of the trypomastigote in patient specimens (*e.g.*, blood, CSF, and chancre aspirate). An ELISA and an indirect agglutination test also have been developed to detect trypanosome antigens in patient samples.

Progressive symptoms include headache, fever, and anemia, followed by symptoms characteristic of the infected sites. The symptoms of sleeping sickness—sleepiness, emaciation, and unconsciousness—begin when the central nervous system becomes infected. Depending on the infecting strain, the disease may last for months or years, but the mortality rate is high. Death results from heart failure, meningitis, or severe debility of some other organ(s).

The infective cycle is complete when an infected individual (humans, cattle, and some wild animals are reservoirs) is bitten by a tsetse fly, which ingests the organism during its blood meal. It becomes infective for its lifespan.

Sporozoan Protozoans Found in Clinical Specimens

Plasmodium spp.

Plasmodia are sporozoan parasites with a complex life cycle, part of which is in various vertebrate tissues while the other part involves an insect. In humans, the tissues are the liver and red blood cells, and the insect vector is the female *Anopheles* mosquito. A generalized life cycle is shown in Figure 11-32. Representative life cycle stages for the various species are shown in Figures 11-33 to 11-35.

Four species of *Plasmodium* cause malaria in humans:

P. vivax (benign tertian malaria),
P. malariae (quartan malaria),
P. falciparum (malignant tertian malaria), and
P. ovale (ovale malaria).

The life cycles are similar for each species, as is the progress of the disease, so *P. falciparum* will be discussed as an example, with unique aspects compared

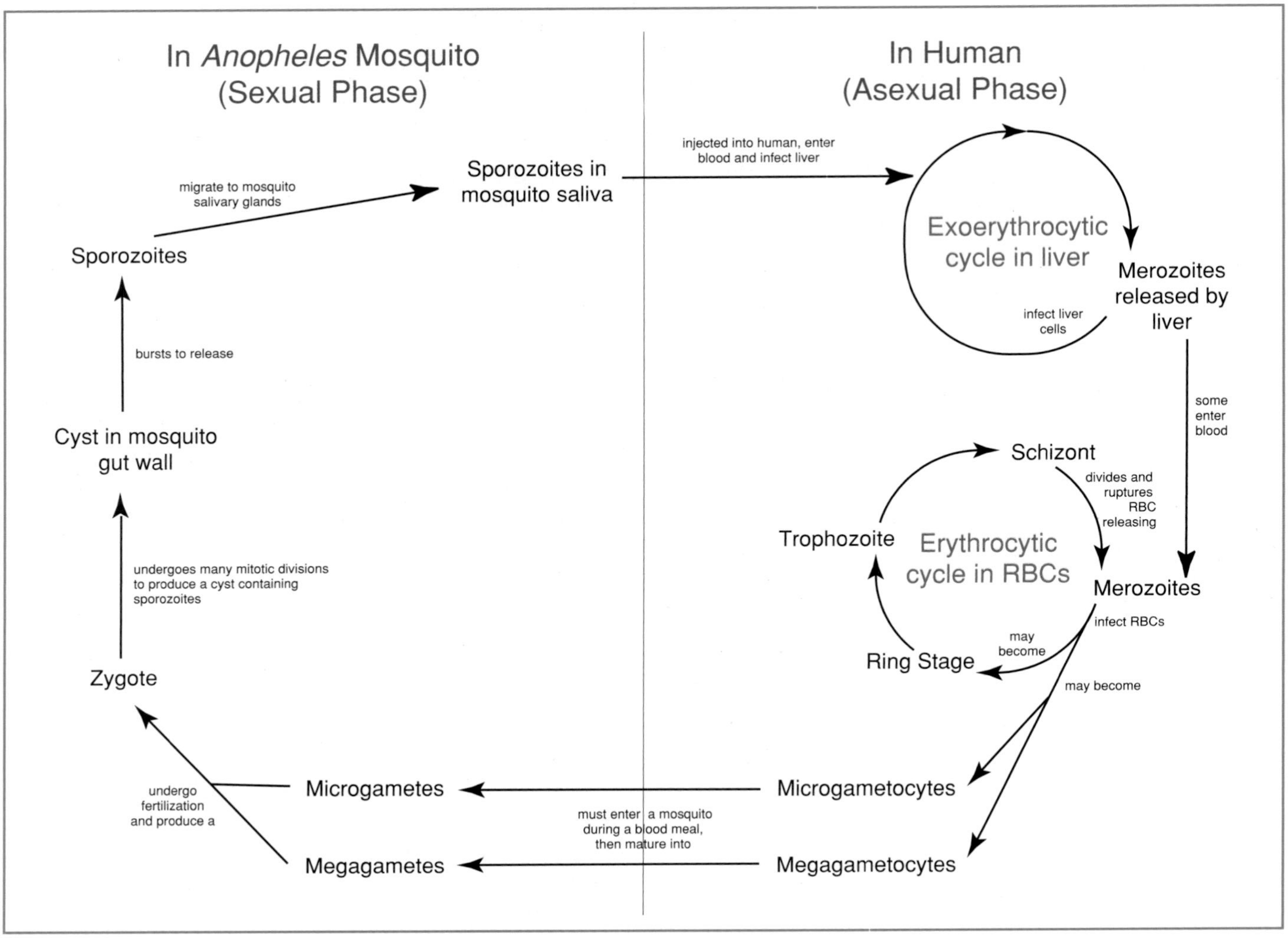

FIGURE 11-32 ***PLASMODIUM* LIFE CYCLE**

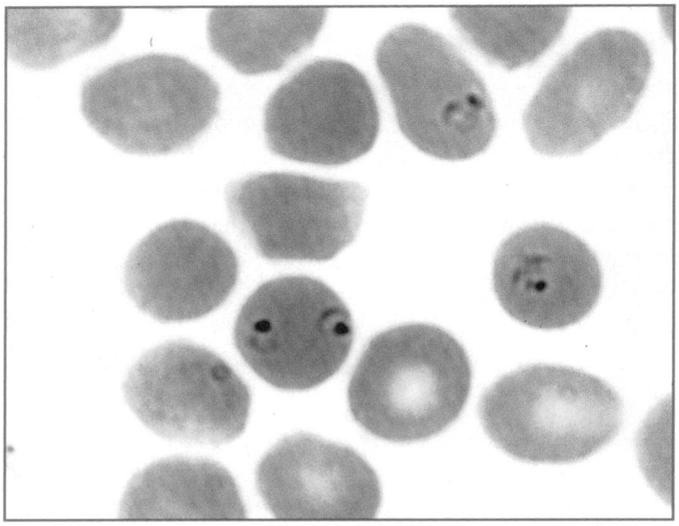

FIGURE 11-33 ***PLASMODIUM FALCIPARUM* DOUBLE INFECTION OF A RED BLOOD CELL (X2640)**
Double infections are commonly seen in *P. falciparum* infections. A single infection is seen at the right. Young trophozoites are said to be in the "ring stage."

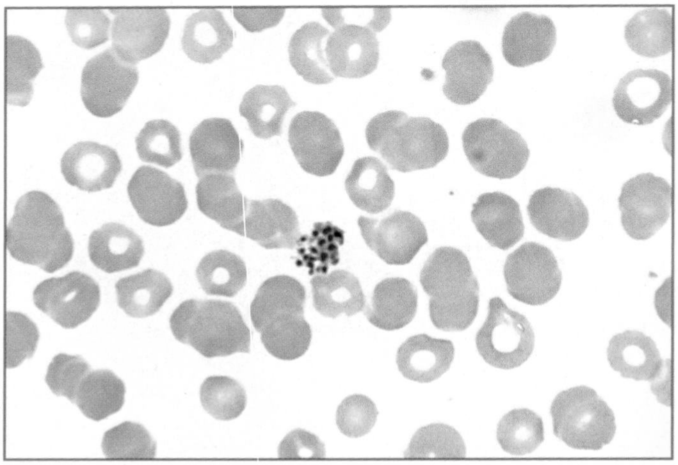

FIGURE 11-34 A MATURE *PLASMODIUM VIVAX* SCHIZONT COMPOSED OF APPROXIMATELY 16 MEROZOITES (X1200)
More than 12 merozoites distinguish *P. vivax* from *P. malariae* and *P. ovale*, both of which typically have eight, but up to 12 merozoites. *P. falciparum* may have up to 24 merozoites, but they typically are not seen in peripheral blood smears and so are not confused with *P. vivax*.

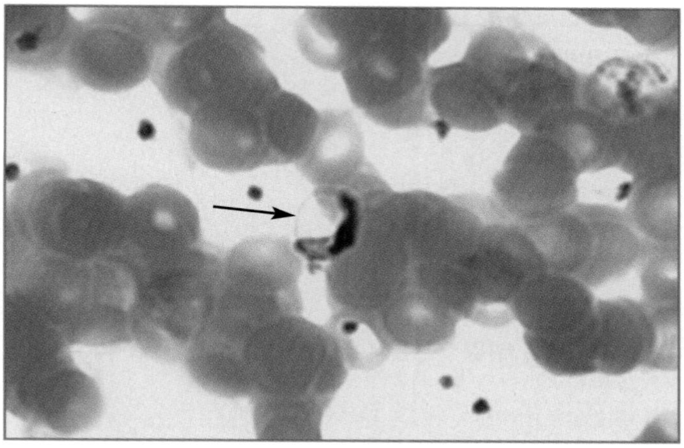

FIGURE 11-35 ***Plasmodium falciparum*** **Gametocyte in an Erythrocyte (X1000)**
Differentiation between microgametocytes and megagametocytes is difficult in this species. The erythrocyte membrane is visible around the gametocyte (arrow).

to the others.

The **sporozoite** stage of the pathogen is introduced into a human host during a bite from an infected female *Anopheles* mosquito. Sporozoites then infect liver cells and produce the asexual **merozoite** stage. Merozoites are released from lysed liver cells, enter the blood, and infect erythrocytes. (Reinfection of the liver occurs at this stage in all except *P. falciparum* infections.) Once in RBCs, merozoites enter a cyclic pattern of reproduction in which more merozoites are released from the red cells synchronously every 48 hours (hence *tertian*—every third day—malaria).

These events are tied to the symptoms of malaria. A chill, nausea, vomiting, and headache are symptoms that correspond to rupture of the erythrocytes. A spiking fever ensues and is followed by a period of sweating, after which the exhausted patient falls asleep. During this latter phase, the parasites reinfect the red cells and the cycle repeats.

The sexual phase of the life cycle begins when certain merozoites enter erythrocytes and differentiate into male or female **gametocytes**. The sexual phase of the life cycle continues when ingested by a female *Anopheles* mosquito during a blood meal. Fertilization occurs, and the zygote eventually develops into a cyst within the gut wall of the mosquito. After many divisions, the cyst releases sporozoites, some of which enter the mosquito's salivary glands ready to be transmitted back to the human host.

Most malarial infections are cleared eventually, but not before the patient has developed anemia and has suffered permanent damage to the spleen and liver. The most severe infections involve *P. falciparum*. Erythrocytes infected by *P. falciparum* develop abnormal projections that cause them to adhere to the lining of small blood vessels. This can lead to obstruction of the vessels, thrombosis, or local ischemia, which account for many of the fatal complications of this type of malaria—including liver, kidney, and brain damage.

Toxoplasma gondii

Like other sporozoans, the *Toxoplasma gondii* (Figure 11-36) life cycle has sexual and asexual phases. The sexual phase occurs in the lining of cat intestines where **oocysts** are produced and shed in the feces. Each oocyst undergoes division and contains eight **sporozoites.** If ingested by another cat, the sexual cycle may be repeated as the sporozoites produce **gametocytes,** which in turn produce gametes. If ingested by another animal host (including humans) the oocyst germinates in the duodenum and releases the sporozoites. Sporozoites enter the blood and infect other tissues, where they become trophozoites, which continue to divide and spread the infection to lymph nodes and other parts of the reticuloendothelial system. Trophozoites ingested by a cat eating an infected animal develop into gametocytes in the cat's intestines. Gametes are formed, fertilization produces an oocyst, and the life cycle is completed.

Infection via ingestion of the oocyst typically is not serious. The infected person may notice fatigue or muscle aches. The more serious form of the disease involves infection of a fetus across the placenta from an infected mother. This type of infection may result in stillbirth, or liver damage and brain damage. AIDS patients may incur fatal complications from infection.

Materials

Per Student Group

- Prepared slides of:
  - *Entamoeba histolytica* trophozoite and cyst
  - *Entamoeba coli* trophozoite and cyst
  - *Balantidium coli* trophozoite and cyst

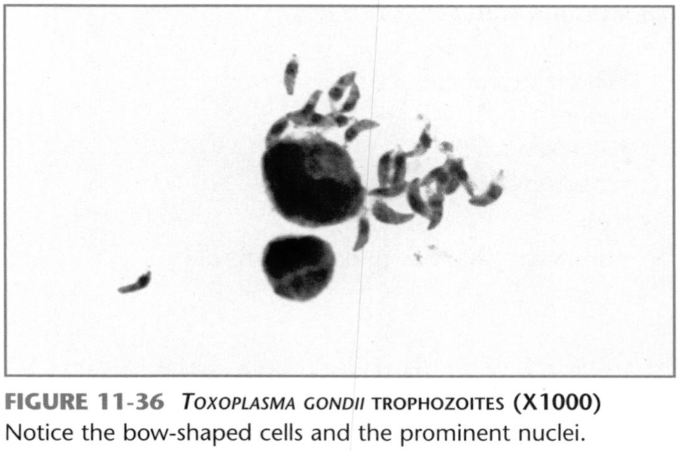

FIGURE 11-36 ***Toxoplasma gondii*** **trophozoites (X1000)**
Notice the bow-shaped cells and the prominent nuclei.

- *Giardia lamblia* trophozoite and cyst
- *Trichomonas vaginalis* trophozoite
- *Trypanosoma spp.*
- *Plasmodium spp.*
- *Toxoplasma gondii* trophozoite

Procedure

1. Obtain prepared slides of the protozoan pathogens and observe them under appropriate magnification. You should observe the assigned structures on each organism. (Many of these slides are made from patient samples, so there will be a lot of other material on the slide besides the desired organism. You must search carefully and with patience.)
2. Sketch and label the assigned components on the Data Sheet.

Entamoeba histolytica
- Trophozoite
  - pseudopods
  - nucleus with small, central karyosome and beaded nucleus
  - ingested erythrocytes
- Cyst
  - multiple nuclei (up to four) with karyosomes and chromatin as in the trophozoite
  - cytoplasmic chromatoidal bars (maybe)

Entamoeba coli
- Trophozoite
  - same as *E. histolytica* except with eccentric karyosome and unclumped chromatin
- Cyst
  - up to eight nuclei (more than four is enough to distinguish it from *E. histolytica*) that are the same as in the trophozoite
  - cytoplasmic chromatoidal bars (maybe)

Balantidium coli
- Trophozoite
  - elongated shape
  - cilia
  - macronucleus
  - micronucleus (maybe)
- Cyst
  - spherical shape with multiple nuclei

Giardia lamblia
- Trophozoite
  - oval shape
  - flagella (four pairs)
  - nuclei (two)
  - median bodies (two)
- Cyst
  - multiple nuclei (four)
  - median bodies (four)

***Trichomonas vaginalis* trophozoite**
- nucleus
- flagella (four)

Trypanosoma spp.
- nucleus
- flagellum
- undulating membrane
- kinetoplasm

Plasmodium spp.
- ring stage
- mature trophozoite
- schizont
- male gametocyte
- female gametocyte

Toxoplasma gondii
- trophozoite
- bow-shaped cells
- nucleus

References

Ash, Lawrence R., and Thomas C. Orihel. 1991. Chapter 14 in *Parasites: A Guide to Laboratory Procedures and Identification.* American Society for Clinical Pathology (ASCP) Press, Chicago.

Forbes, Betty A., Daniel F. Sahm, and Alice. S. Weissfeld. 2002. Chapter 52 (pages 650–687) in *Bailey and Scott's Diagnostic Microbiology,* 11th ed. Mosby-Year Book, St. Louis.

Garcia, Lynne Shore. 2001. Chapters 2, 3, 5, 7, and 9 in *Diagnostic Medical Parasitology*, 4th ed. ASM Press, Washington, DC.

Koneman, Elmer W., Stephen D. Allen, William M. Janda, Paul C. Schreckenberger, and Washington C. Winn, Jr. 1997. Chapter 20 in *Color Atlas and Textbook of Diagnostic Microbiology,* 5th ed. J. B. Lippincott, Philadelphia.

Lee, John J., Seymour H. Hutner, and Eugene C. Bovee. 1985. *Illustrated Guide to the Protozoa.* Society of Protozoologists, Lawrence, KS.

Markell, Edward K., Marietta Voge and David T. John. 1992. *Medical Parasitology,* 7th ed. W.B. Saunders, Philadelphia.

Exercise 11-3

Parasitic Helminths

A study of helminths is appropriate to the microbiology lab because clinical specimens may contain microscopic evidence of helminth infection. The three major groups of parasitic worms encountered in lab situations are trematodes (flukes), cestodes (tapeworms), and nematodes (round worms). Life cycles of the parasitic worms are often complex, sometimes involving several hosts, and are beyond the scope of this book. Emphasis here is on a brief background and clinically important diagnostic features of selected worms.

Trematode Parasites Found in Clinical Specimens

Clonorchis (Opisthorchis) sinensis

Clonorchis sinensis is the Oriental liver fluke (Figure 11-37) and causes clonorchiasis, a liver disease. It is a common parasite of people living in Japan, Korea, Vietnam, China, and Taiwan and is becoming more common in the United States with the influx of Southeast Asian immigrants. Infection typically occurs when undercooked infected fish is ingested. The adults migrate to the liver bile ducts and begin laying eggs in approximately one month. Degree of damage to the bile duct epithelium and surrounding liver tissue is due to the number of worms and the duration of infection. Diagnosis is made by identifying the characteristic eggs in feces (Figure 11-38).

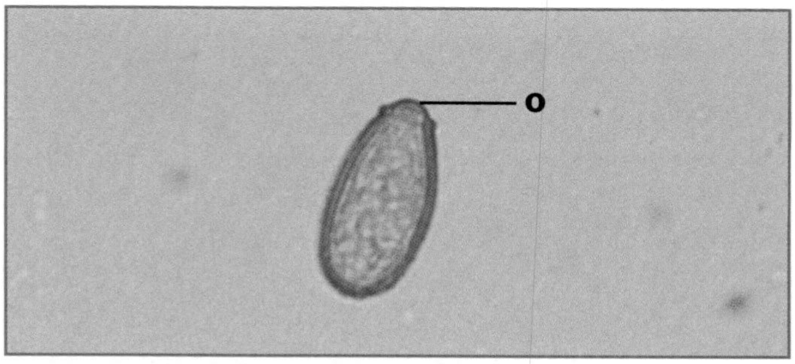

FIGURE 11-38 *Clonorchis sinensis* Egg in a Fecal Specimen (X1000, D'Antoni's Iodine Stain)
The eggs are thick shelled and between 27 and 35 µm long. There is a distinctive operculum (O) (positioned to give the appearance of shoulders) and often a knob at the aboperular end (not visible in this specimen).

Paragonimus westermani

Paragonimus westermani (Figure 11-39) is a lung fluke and one of several species to cause paragonimiasis, a disease mostly found in Asia, Africa, and South America. *P. westermani* is primarily a parasite of carnivores, but humans (omnivores) may get infected when eating undercooked crabs or crayfish infected with the cysts. Ingested juveniles excyst in the duodenum and travel to the abdominal wall. After several days, they resume their journey and find their way to the bronchioles where the adults mature. Eggs (Figure 11-40) are released in approximately two to three months and are diagnostic of infection. They may be recovered in sputum, lung fluids, or feces. Consequences of lung infection are a local inflammatory response followed by possible ulceration. Symptoms include cough with discolored or bloody sputum and difficulty breathing. These cases are rarely fatal, but may last a couple of decades. Occasionally, the wandering juveniles end up in other tissues, such as the brain or spinal cord, which can cause paralysis or death.

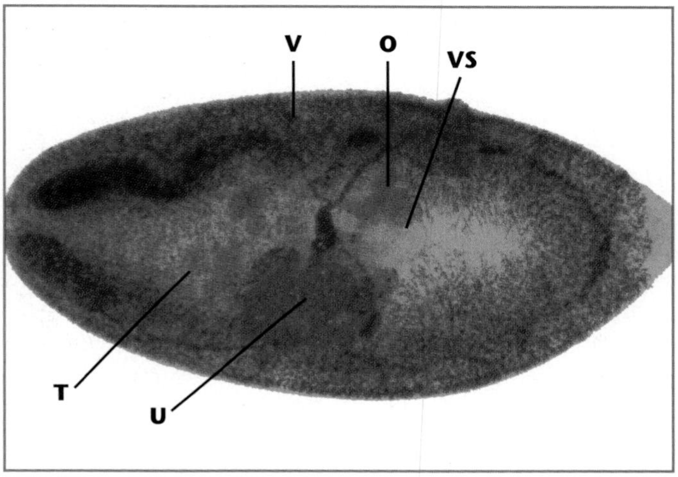

FIGURE 11-39 *Paragonimus westermani* Adult
Adults are up to 1.2 cm long, 0.6 cm wide and 0.5 cm thick. Visible in this specimen are the ventral sucker (VS), ovary (O), uterus (U), testis (T), and vitellaria (V).

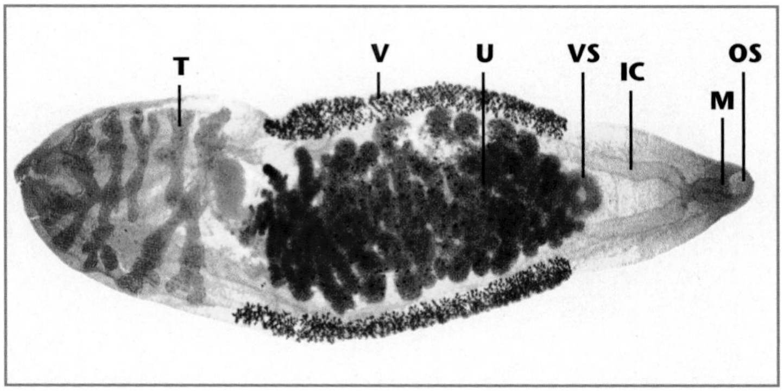

FIGURE 11-37 *Clonorchis sinensis* Adult (X5)
Adults range in size from 1 cm to 2.5 cm. Visible in this specimen are the oral sucker (OS), mouth (M), intestinal ceca (IC), ventral sucker (VS), uterus (U), vitellaria (V), and testis (T).

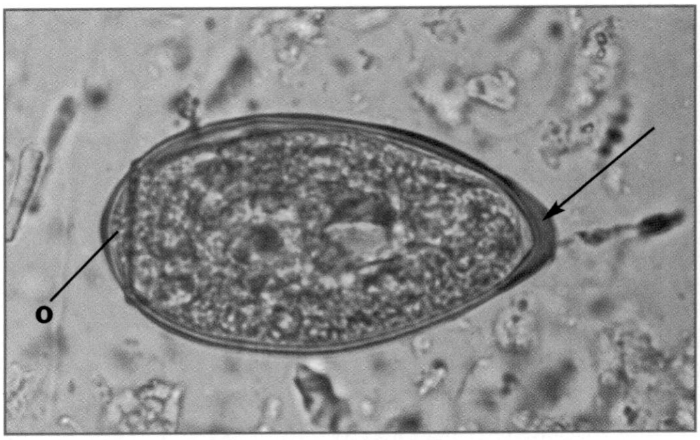

FIGURE 11-40 *PARAGONIMUS WESTERMANI* **EGG IN A FECAL SPECIMEN (X1000, D'ANTONI'S IODINE STAIN)**
Paragonimus westermani eggs are ovoid and range in size from 80 to 120 μm long by 45 to 70 μm wide. They have an operculum (O) and the shell is especially thick at the abopercular end (arrow). They are unembryonated when seen in feces.

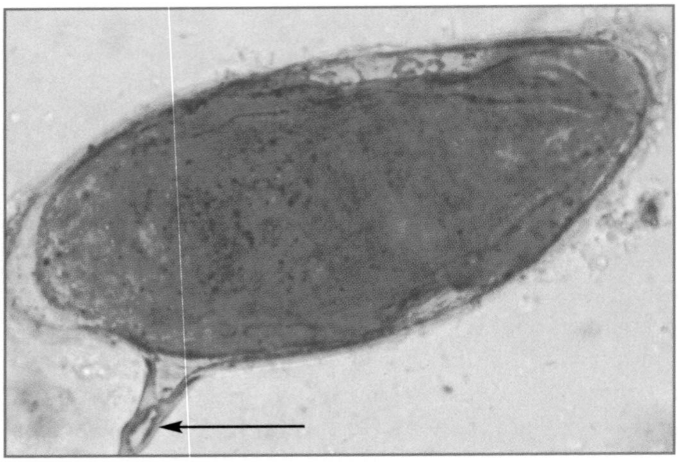

FIGURE 11-42 *SCHISTOSOMA MANSONI* **EGG IN A FECAL SPECIMEN (X1000, D'ANTONI'S IODINE STAIN)**
Schistosoma mansoni eggs are large (114 to 175 μm long by 45 to 70 μm wide) and contain a larva called a miracidium. They are thin-shelled, lack an operculum, and have a distinctive lateral spine (arrow).

Schistosoma mansoni

Schistosoma mansoni (Figure 11-41) is found in Brazil, some Caribbean islands, Africa, and parts of the Middle East. Infection occurs via contact with fecally contaminated water containing juveniles of the species. The juveniles penetrate the skin, enter circulation and continue development in the intestinal veins. Eggs (Figure 11-42) from the adults penetrate the intestinal wall and are passed out with the feces. Presence of eggs in the feces indicates infection. Some patients are asymptomatic, whereas others have bloody diarrhea, abdominal pain, and lethargy.

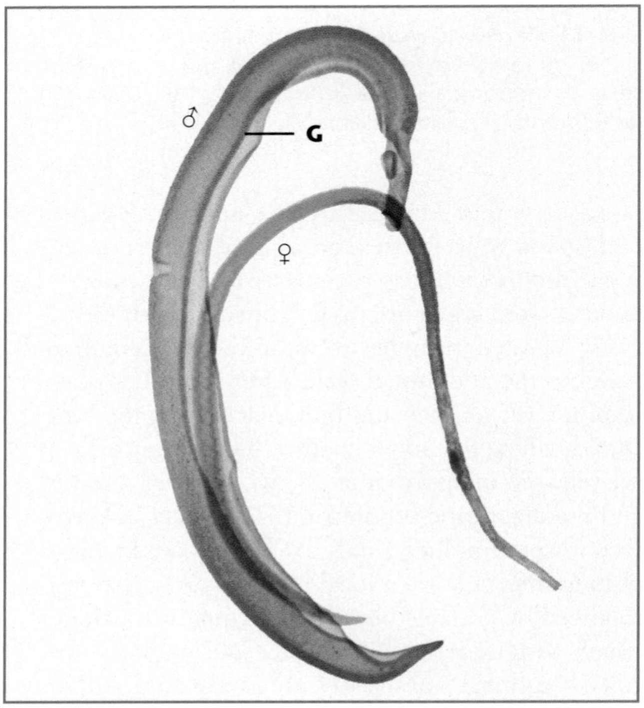

FIGURE 11-41 *SCHISTOSOMA MANSONI* **ADULTS (X20)**
The male is larger and has a gynecophoric groove (G) in which the slender female resides during mating.

Cestode Parasites Found in Clinical Specimens

Dipylidium caninum

Dipylidium caninum (Figure 11-43) is a common parasite of dogs and cats. Human infection usually occurs in children. The adult worms reside in dog or cat intestines and release proglottids (Figure 11-44) containing egg packets that migrate out of the anus. When these dry, they look like rice grains. Larval fleas may eat the eggs and become infected. If the dog, cat, or child ingests one of these fleas, the life cycle is completed in the new host. Infection may be asymptomatic or produce mild

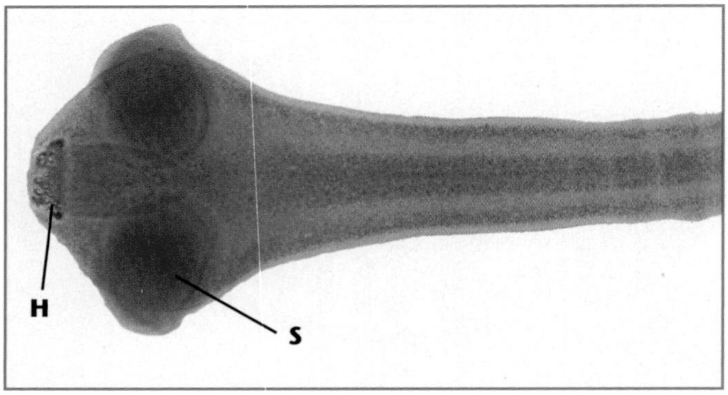

FIGURE 11-43 *DIPYLIDIUM CANINUM* **SCOLEX (X4)**
Adult worms may reach a length of 50 cm with a width of 3 mm. The scolex has four suckers (S) and a retractable rostellum with rows of hooks (H).

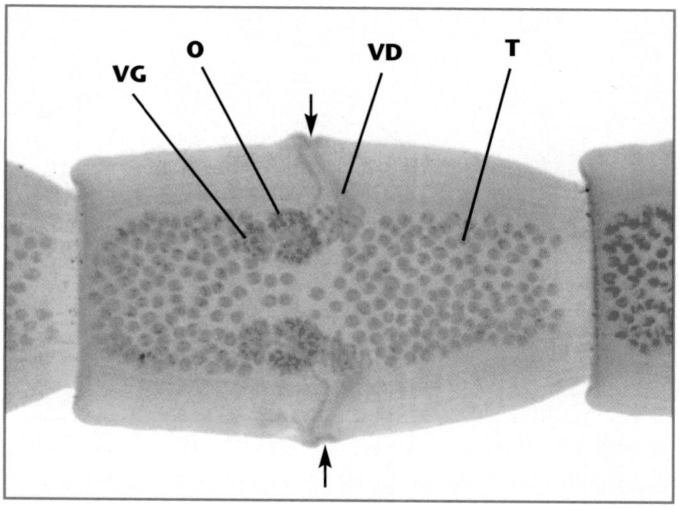

FIGURE 11-44 *DIPYLIDIUM CANINUM* **PROGLOTTIDS (X40)**
Visible are the testes (T), vasa deferentia (VD), ovaries (O), and vitelline glands (VG). The reproductive openings (arrows) on each side of the proglottid give this worm its common name—the "double-pored tapeworm."

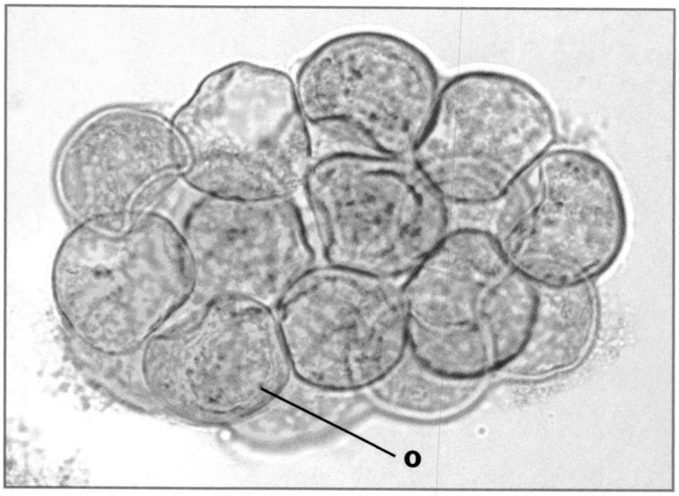

FIGURE 11-45 *DIPYLIDIUM CANINUM* **EGG PACKET (X1000)**
Each *Dipylidium caninum* egg packet is composed of 5 to 15 eggs each with an onchosphere (O). The onchosphere contains six hooklets.

abdominal discomfort, loss of appetite and indigestion. Diagnosis is made by identifying the egg packets (Figure 11-45).

Echinococcus granulosus

The definitive host of *Echinococcus granulosus* (Figure 11-46) is a carnivore, but the life cycle requires an intermediate host, usually an herbivorous mammal. Humans involved in raising domesticated herbivores (*e.g.*, sheep with their associated dogs) are most susceptible as intermediate hosts and develop hydatid disease. Ingestion of a juvenile *E. granulosus* leads to development of a hydatid cyst in the lung, liver, or other organ, a process that may take many years. The cyst (Figure 11-47) has a thick wall and develops many protoscolices within (Figure 11-48). The protoscolices, if ingested, are infective to the definitive host. Symptoms depend on the location and size of the hydatid cyst, which interferes with normal organ function. Due to sensitization by the parasite's antigens, release of fluid from the cyst can result in anaphylactic shock of the host. Diagnosis is made by detection of the cyst by ultrasound or X-ray.

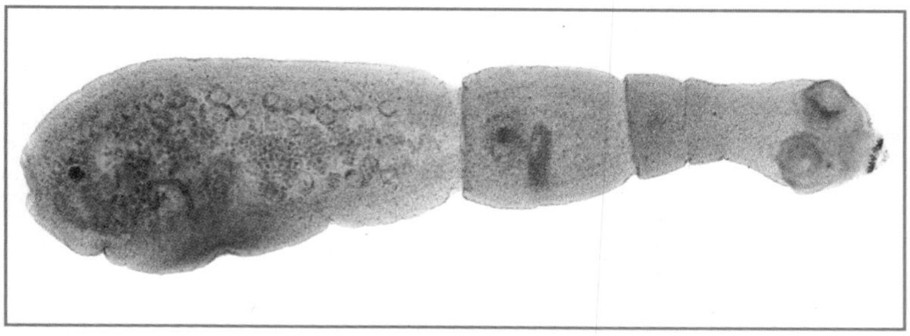

FIGURE 11-46 *ECHINOCOCCUS GRANULOSUS* **ADULT (X20)**
Adult worms are about 0.5 cm in length. There is a scolex with a ring of hooks, a neck, and one proglottid that contains up to 1500 eggs.

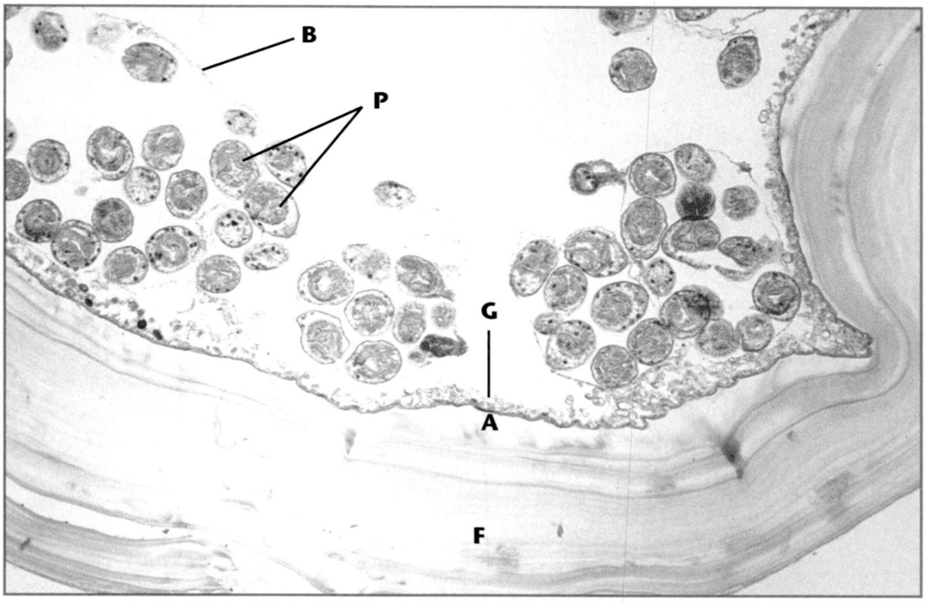

FIGURE 11-47 *ECHINOCOCCUS GRANULOSUS* **CYST IN LUNG TISSUE (SEC., X96)**
The cyst wall consists of a fibrous layer of host tissue (F), an acellular layer (A), and a germinal epithelium (G) that gives rise to stalked brood capsules (B). Brood capsules produce many *E. granulosus* protoscolices (P).

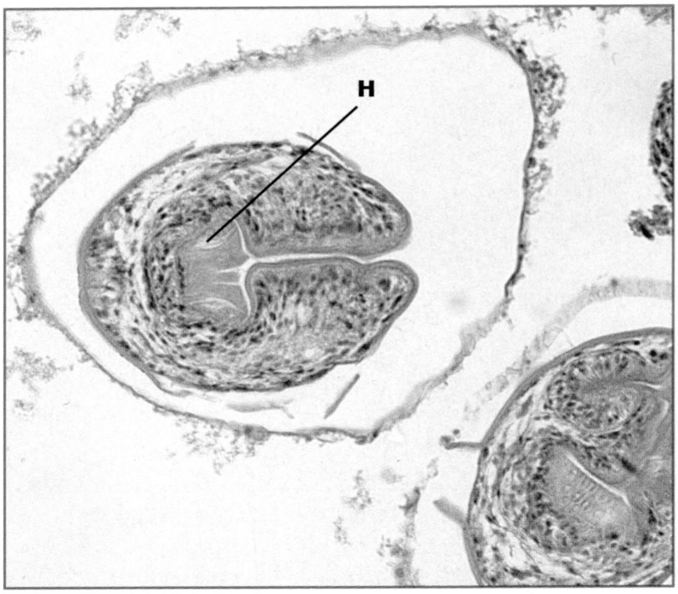

FIGURE 11-48 *ECHINOCOCCUS GRANULOSUS* PROTOSCOLEX (L.S., X320)
The protoscolex contains an invaginated scolex with hooks (H). Upon ingestion by the host, the protoscolex evaginates and produces an infectious scolex that attaches to the intestinal wall, matures and produces eggs.

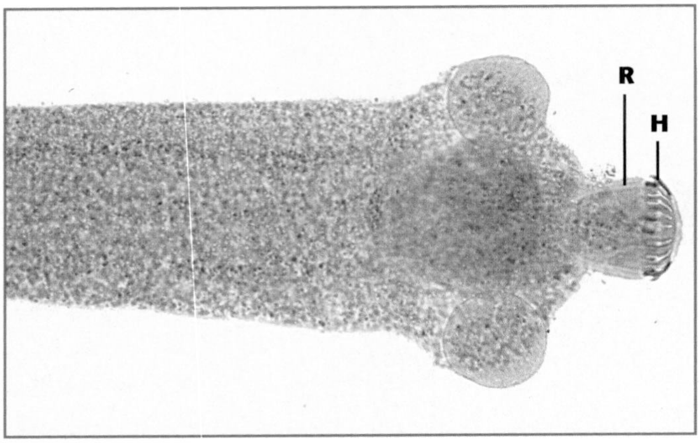

FIGURE 11-49 *HYMENOLEPIS NANA* SCOLEX (X40)
Adults gain a length of up to 10 cm, but are only 1 mm in width. The rostellum (R) is armed with up to 30 hooks (H).

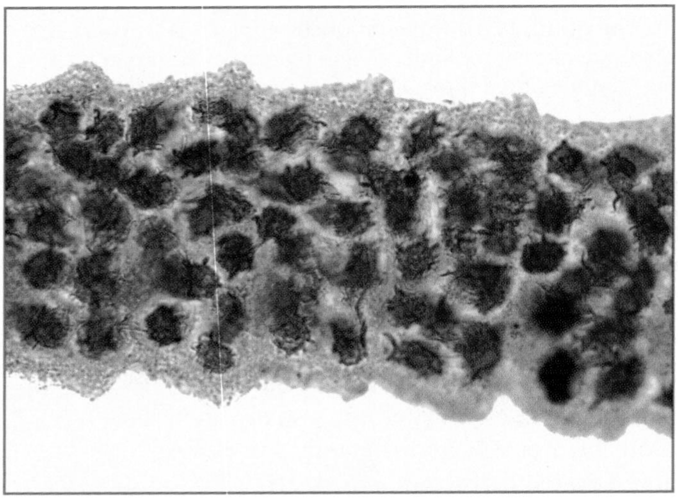

FIGURE 11-50 *HYMENOLEPIS NANA* PROGLOTTIDS (X40)
H. nana is hermaphroditic, but in this specimen, the numerous testes obscure most other structures.

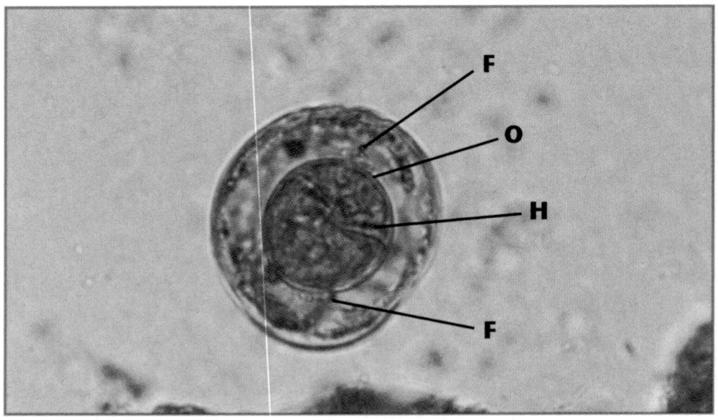

FIGURE 11-51 *HYMENOLEPIS NANA* EGG IN FECES (X1000, D'ANTONI'S IODINE STAIN)
The *Hymenolepis nana* egg is 30 to 47 µm in diameter and has a thin shell. The oncosphere (O) is separated from the shell and contains six hooks (H). Another distinguishing feature is the presence of between four and eight filaments (F) arising from either end of the oncosphere.

Hymenolepis (Vampirolepis) nana

Hymenolepis nana (Figures 11-49 and 11-50) is the dwarf tapeworm and is the most common cestode parasite of humans in the world. When eggs (Figure 11-51) are ingested, the oncospheres develop into juveniles in the lymphatics of intestinal villi. These juveniles are then released into the lumen within a week and attach to the mucosa to mature into adults. Infection may involve hundreds of worms, yet symptoms are usually mild: diarrhea, nausea, loss of appetite, or abdominal pain. Eggs may reinfect the same host or pass out with the feces to infect a new host. Eggs in the feces are used for identification, but proglottids are not as they are rarely passed.

Taenia spp.

Two taeniid worms are important human pathogens. These are *Taenia saginata* (*Taeniarhynchus saginatus*)—the beef tapeworm—and *Taenia solium*—the pork tapeworm.

T. saginata infects humans who eat undercooked beef containing juvenile worms. In the presence of bile salts, the juveniles develop into adults and begin producing gravid proglottids within a few weeks. Symptoms of infection are usually mild nausea, diarrhea, abdominal pain, and headache. Diagnosis to species is impossible with only the eggs (Figure 11-52); specific identification requires a scolex or gravid proglottid (Figure 11-53).

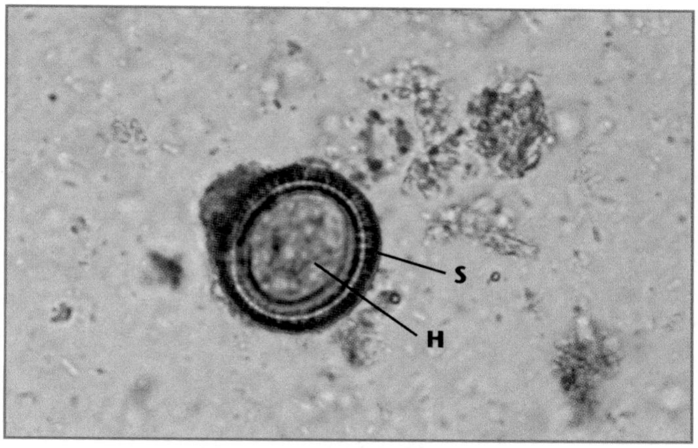

FIGURE 11-52 *TAENIA* **EGG IN FECES (X1000)**
Taeniid eggs are distinctive looking enough to identify to genus, but not distinctive enough to speciate. Eggs are spherical and approximately 40 μm in diameter with a striated shell (S). The oncosphere contains six hooks (H).

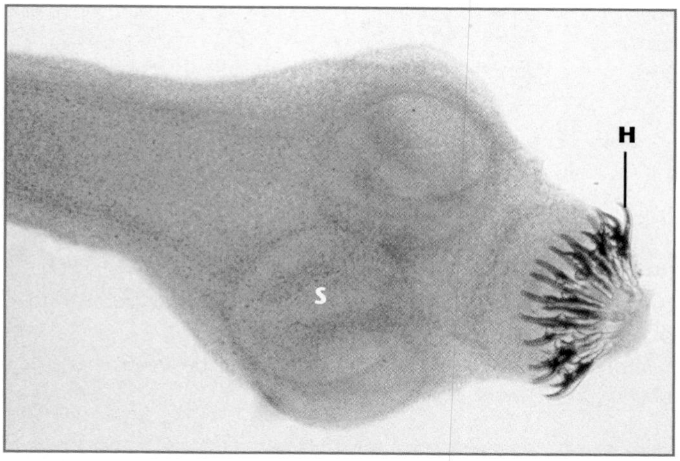

FIGURE 11-54 *TAENIA SOLIUM* **SCOLEX (X64)**
The *Taenia solium* scolex has two rings of hooks (H) and four suckers (S).

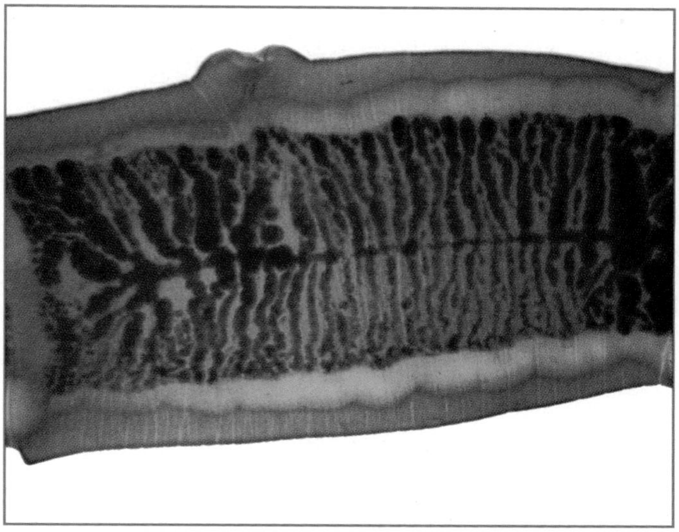

FIGURE 11-53 *TAENIA SAGINATA* **PROGLOTTID (X12)**
The uterus of *Taenia saginata* proglottids consists of a central portion with 15 to 20 lateral branches (compare to the *T. solium* uterus in Figure 11-55).

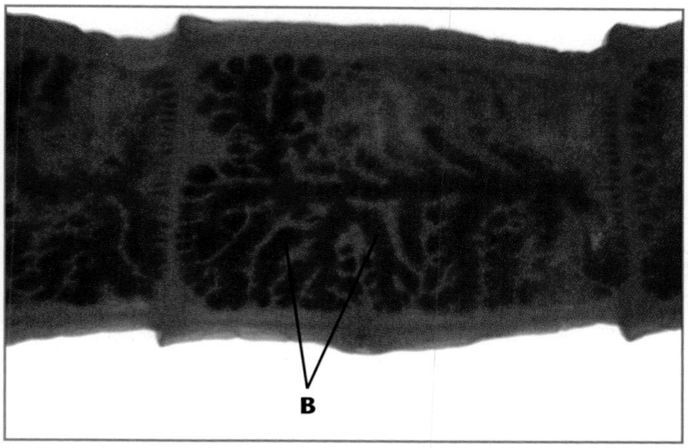

FIGURE 11-55 *TAENIA SOLIUM* **PROGLOTTID (X33)**
The uterus of *Taenia solium* proglottids consists of a central portion with 7 to 13 lateral branches (compare to the *T. saginata* uterus in Figure 11-53).

The *T. solium* life cycle is similar to *T. saginata*, but the host is pork, not beef, so human infection occurs when undercooked pork is eaten. If eggs are ingested, a juvenile form called a cysticercus develops. Cysticerci may be found in any tissue, especially subcutaneous connective tissues, eyes, brain, heart, liver, lungs, and coelom. Symptoms of cysticercosis depend on the tissue infected, but mostly they are not severe. However, death of a cysticercus can produce a rapidly fatal inflammatory response. As with *T. saginata*, diagnosis to species is impossible with only the eggs; specific identification requires a scolex or gravid proglottid (Figures 11-54 and 11-55).

Nematode Parasites Found in Clinical Specimens

Ascaris lumbricoides

Ascaris lumbricoides (Figures 11-56 and 11-57) is a large nematode—females may reach a length of 49 cm! Human infection occurs when eggs in fecally contaminated soil or food are ingested. Juveniles emerge in the intestine, penetrate its wall, and then migrate to the lungs and other tissues. After a period of development in the lungs, the juveniles move up the respiratory tree to the esophagus and are swallowed again. Adults then reside in the small intestine and produce eggs (Figures 11-58 and 11-59). Infection may result in inflammation in organs other than the lungs where juvenile worms settled incorrectly. *Ascaris* pneumonia occurs in heavy infections due to the lung damage caused by the juveniles.

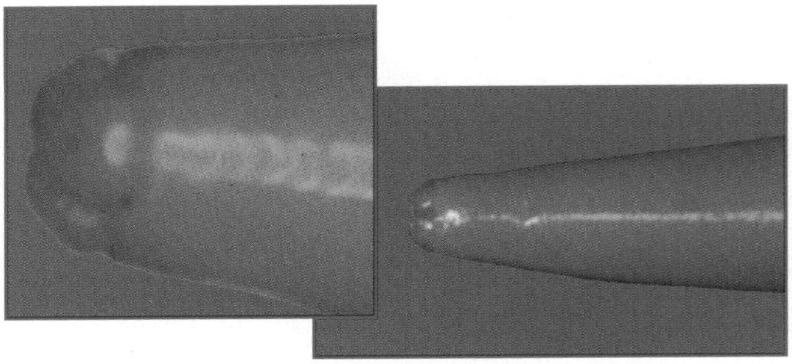

FIGURE 11-56 *ASCARIS LUMBRICOIDES* **ANTERIOR**
A. lumbricoides has a cylindrical shape with three prominent mouth parts (see inset).

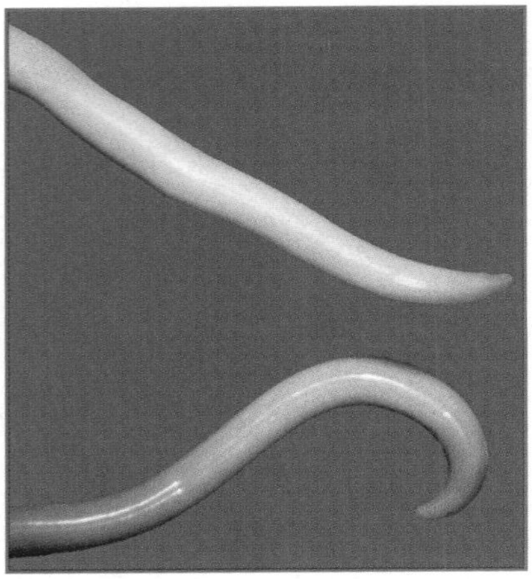

FIGURE 11-57
ASCARIS LUMBRICOIDES
ADULT WORMS
A. lumbricoides males (bottom) are shorter than females (up to 31 cm vs. 35 cm) and have a curved posterior.

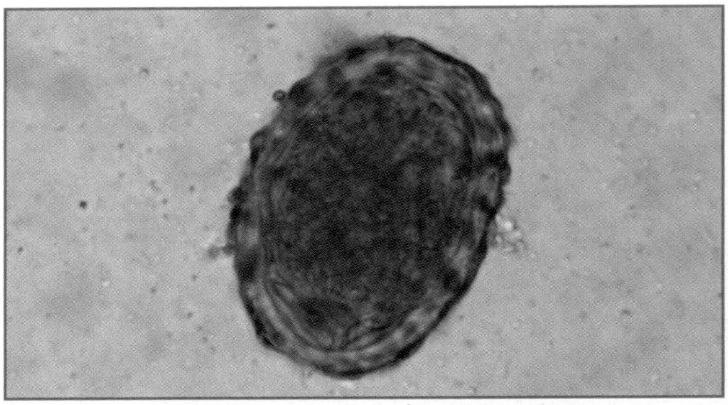

FIGURE 11-58 FERTILE *ASCARIS LUMBRICOIDES* FERTILE EGG IN FECES (X1000, D'ANTONI'S IODINE STAIN)
Fertile *Ascaris lumbricoides* eggs are 55 to 75 μm long and 35 to 50 μm wide and are embryonated. Their surface is covered by small bumps called mammillations.

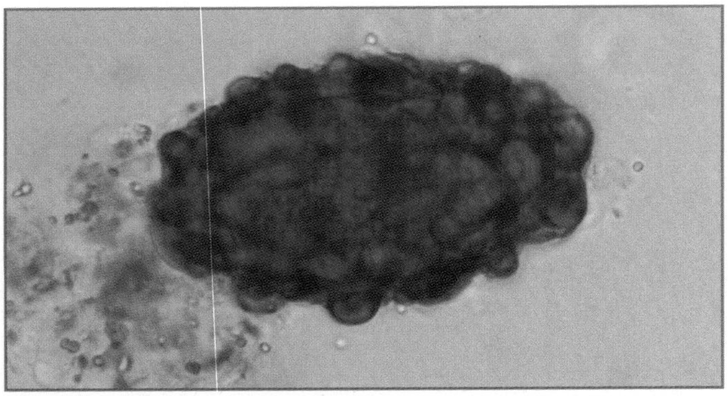

FIGURE 11-59 INFERTILE *ASCARIS LUMBRICOIDES* EGG IN FECES (X1000, D'ANTONI'S IODINE STAIN)
Infertile eggs are longer (up to 90 μm) than fertile eggs. There is no embryo inside.

If secondary bacterial infections occur, the pneumonia can be fatal. Blockage of the intestines and malnutrition also are possible in heavy infections. Lastly, under certain conditions, worms can wander to other body locations and cause damage or blockage. Identification of an *Ascaris* infection is made by observing the eggs in feces.

Enterobius vermicularis

Enterobius vermicularis (Figure 11-60) is the human pinworm. It is found worldwide and is especially prevalent among people in institutions (such as orphanages and mental hospitals) because conditions favor fecal-oral transmission of the parasite. Bedding, clothing, and the fingers (from scratching) become contaminated and may be involved in transmission. Poor sanitary habits of children make them especially prone to infecting others. Transmission may also involve eggs (Figure 11-61) being carried on air currents and then inhaled by a susceptible host. After ingestion, eggs hatch in the duodenum and mature in the large intestine where the adults reside. Adult females emerge from the anus at night to lay between 4,600 and 16,000 eggs in the perianal region. About one-third of pinworm infections are asymptomatic.

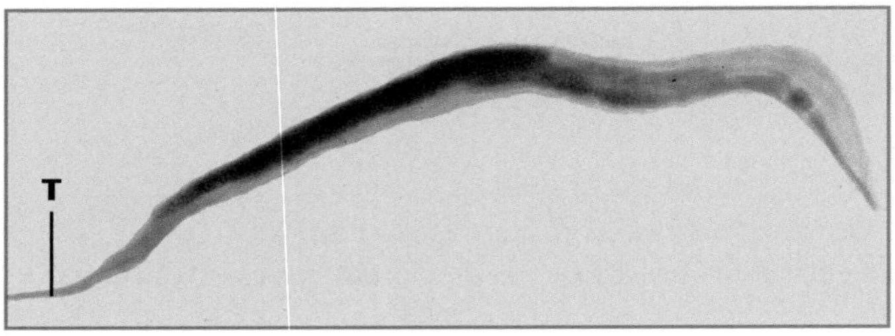

FIGURE 11-60 *ENTEROBIUS VERMICULARIS* **ADULT FEMALE**
Female pinworms are about 1 cm long and have a pointed tail (T) from which this group derives its common name—pinworm. Males are about half that size and have a hooked tail.

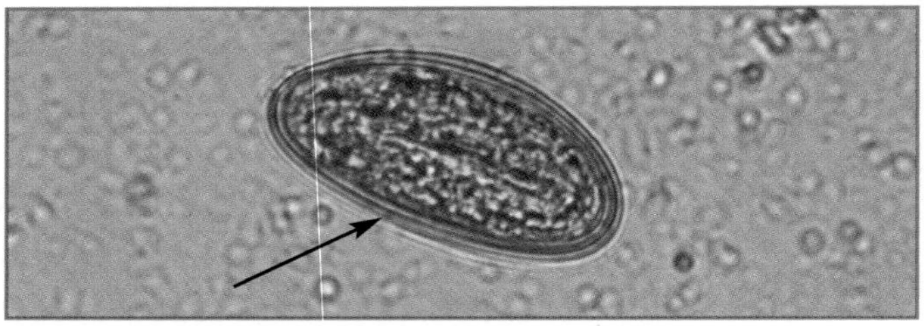

FIGURE 11-61 *ENTEROBIUS VERMICULARIS* **EGG (X1000, D'ANTONI'S IODINE STAIN)**
The eggs of *Enterobius vermicularis* are 50 to 60 μm long and 20 to 40 μm wide with one side flattened (arrow). They are usually embryonated in typical preparations.

The other two-thirds usually do not produce serious symptoms. Diagnosis is made by identifying the eggs. Since the eggs are laid externally, they are rarely found in feces. Instead, they are collected on cellophane tape from the perianal region and examined microscopically.

Hookworms (*Ancylostoma duodenale* and *Necator americanus*)

The hookworm *Ancylostoma duodenale* (Figure 11-62) and *Necator americanus* (Figure 11-63) have very similar morphologies and life cycles, and the eggs are indistinguishable, so they are considered together here. Infection occurs when juveniles penetrate the skin, enter the blood and travel to the lungs. They penetrate the respiratory membrane and are carried up and out of the lungs by ciliary action to the pharynx, where they are swallowed. When they reach the small intestine, they attach and mature into adults that feed on blood and tissues of the host. Adults are rarely seen as they remain attached to the intestinal mucosa. Eggs (Figures 11-64) are passed in the feces and are diagnostic of infection. The severity of hookworm disease symptoms is related to the parasite load, and most infections are asymptomatic. As a rule, severe symptoms of bloody diarrhea and iron deficiency anemia are only seen in acute heavy or chronic infections.

Strongyloides stercoralis

Strongyloides stercoralis is the intestinal threadworm. Infection occurs by penetration of the skin by infective juveniles from fecally contaminated soil. The juveniles then migrate to the lungs and develop into parthenogenetic females that migrate to the pharynx, are swallowed, and then burrow into the intestinal mucosa. Each day they release a few dozen eggs that develop

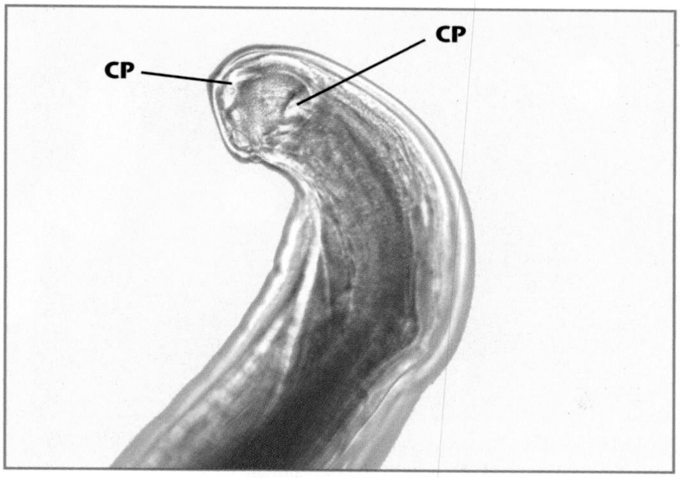

FIGURE 11-63 *NECATOR AMERICANUS* HEAD (X160)
This detail of *N. americanus* shows the cutting plates (CP) that help to differentiate it from *A. duodenale* (compare with Figure 11-62). Also notice the hooked head.

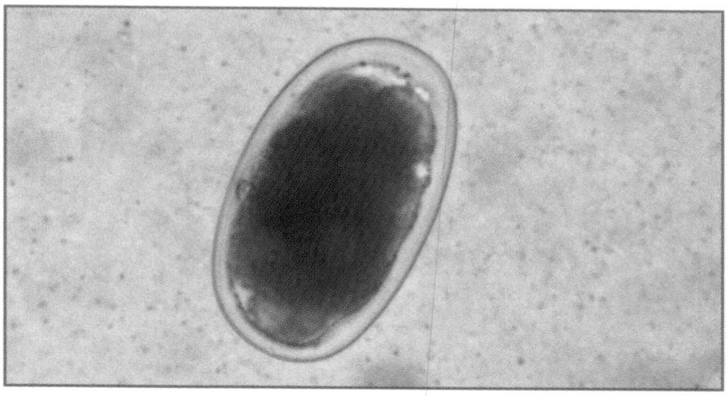

FIGURE 11-64 HOOKWORM EGG IN FECES (X1000, D'ANTONI'S IODINE STAIN)
Hookworm eggs are 55 to 75 µm long and 36 to 40 µm wide. They have a thin shell and contain a developing embryo (seen here at about the 16 cell stage) that is separated from the shell when seen in fecal samples.

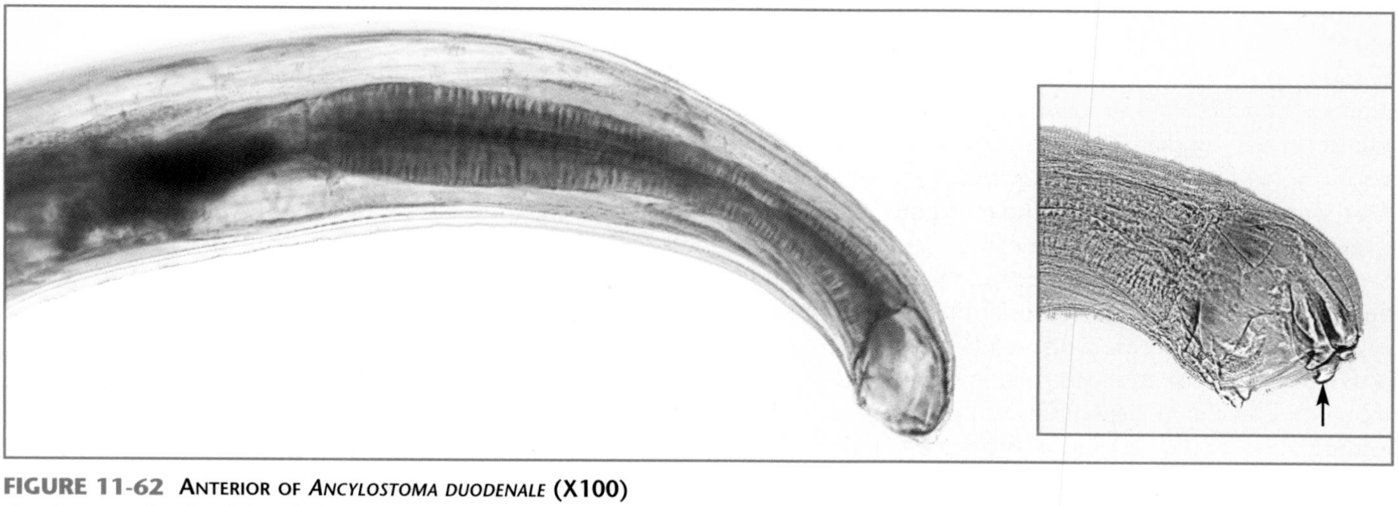

FIGURE 11-62 ANTERIOR OF *ANCYLOSTOMA DUODENALE* (X100)
Ancylostoma duodenale head showing the mouth and thick-walled esophagus. The bend in the head gives this group its common name—hookworm. In the inset, the chitinous teeth (arrow) are visible (compare with Figure 11-63).

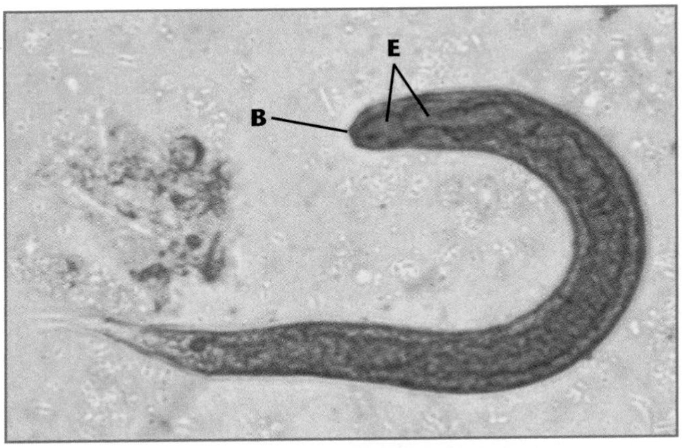

FIGURE 11-65 *STRONGYLOIDES STERCORALIS* **RHABDITIFORM LARVA IN FECES (X1000, D'ANTONI'S IODINE STAIN)**
These larvae may be distinguished from hookworm larvae (which are rarely in feces) by their short buccal cavity (B). The name "rhabditiform" refers to the esophagus (E) shape, which has a constriction within it.

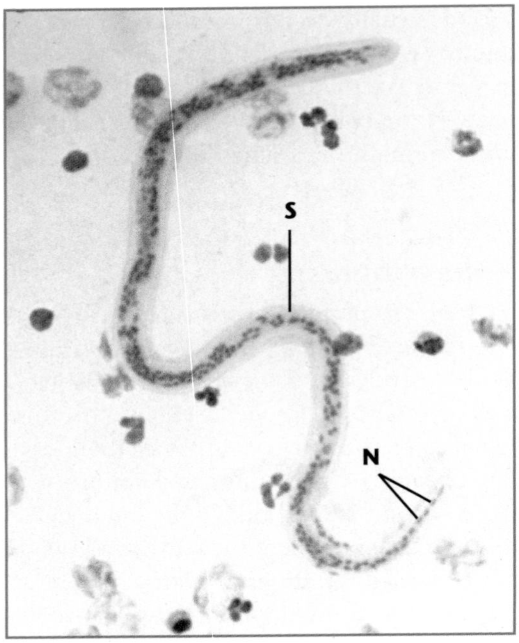

FIGURE 11-66 *WUCHERERIA BANCROFTI* **MICROFILARIA IN A BLOOD SAMPLE (X480)**
The microfilariae of *Wuchereria bancrofti* can be distinguished from others in the blood by the sheath (S) and the single column of nuclei (N) not extending to the tip of the tail.

into juveniles (Figure 11-65) before they are passed in the feces. These juveniles may become infective or may follow a developmental path that produces free-living adults. These adults eventually produce more infective juveniles and the cycle is completed. Symptoms of infection may be itching or secondary bacterial infection at the site of entry by the infective juveniles, a cough and burning of the chest during the pulmonary phase, and abdominal pain and perhaps septicemia during the intestinal phase. Diagnosis is by finding rhabditiform larvae in fresh fecal samples.

Wuchereria bancrofti

Wuchereria bancrofti is a filarial worm that causes lymphatic filariasis. Infection occurs from the bite of a mosquito harboring infective juveniles. Upon injection into the host, the worms migrate into the large lymphatic vessels of the lower body and mature. Adults are found in coiled bunches and the females release microfilariae (Figure 11-66) by the thousands. Microfilariae enter the blood and circulate there, often with a daily periodicity—most abundant at night when the mosquito vector is active and hidden away in lung capillaries during the day when it is hot. Some infections are asymptomatic, whereas others result in acute inflammation of lymphatics associated with fever, chills, tenderness, and toxemia. In the most serious cases, obstruction of lymphatic vessels occurs and results in elephantiasis, a disease caused by accumulation of lymph fluid in the tissues, an accumulation of fibrous connective tissue, and a thickening of the skin. Diagnosis of infection is made by identifying microfilariae in blood smears.

Materials

- Prepared slides of:
  - *Ascaris lumbricoides* eggs in a fecal smear
  - *Clonorchis sinensis* in a fecal smear
  - *Dipylidium caninum* eggs in a fecal smear
  - *Echinococcus granulosus* hydatid cyst in section
  - *Enterobius vermicularis* eggs in a fecal smear
  - Hookworm (*Ancylostoma duodenale* or *Necator americanus*) eggs in a fecal smear
  - *Hymenolepis nana* eggs in a fecal smear
  - *Paragonimus westermani* eggs in a fecal smear
  - *Schistosoma mansoni* eggs in a fecal smear
  - *Strongyloides stercoralis* rhabditiform larva in a fecal smear
  - *Taenia solium* proglottid (whole mount)
  - *Taenia solium* scolex (whole mount)
  - *Taenia spp.* eggs in a fecal smear
  - *Wuchereria bancrofti* microfilariae in a blood smear

Procedure

Observe the prepared slides provided of the helminth specimens in fecal smears and other tissues. Scanning on low power (10x objective) is best for most preparations, then move to high-dry or oil immersion to see detail. Most egg specimens are in fecal smears, so there will be a lot of other material on the slide besides the eggs. You must search carefully and with patience. Sketch what you see in the spaces provided on the Data Sheet and measure dimensions. Be able to identify each to species if shown an unlabeled specimen.

References

Ash, Lawrence R. and Thomas C. Orihel. 1991. Chapters 15 and 16 in *Parasites: A Guide to Laboratory Procedures and Identification.* American Society for Clinical Pathology (ASCP) Press, Chicago, IL.

Forbes, Betty A., Daniel F. Sahm, and Alice. S. Weissfeld. 2002. Chapter 52 in (Pages 687–698) *Bailey and Scott's Diagnostic Microbiology,* 11th Ed. Mosby-Year Book, Inc. St. Louis.

Garcia, Lynne Shore. 2001. Chapters 10, 12, 13, 14, 16, and 17 in *Diagnostic Medical Parasitology*, 4th Ed. ASM Press, Washington, DC.

Koneman, Elmer W., Stephen D. Allen, William M. Janda, Paul C. Schreckenberger and Washington C. Winn, Jr. 1997. Chapter 20 in *Color Atlas and Textbook of Diagnostic Microbiology,* 5th Ed. J. B. Lippincott Company, Philadelphia.

Orihel, Thomas C., and Lawrence R. Ash. 1999. Chapter 112 in *Manual of Clinical Microbiology,* 7th Ed., edited by Patrick R. Murray, Ellen Jo Baron, Michael A. Pfaller, Fred C. Tenover, and Robert H. Yolken. American Society for Microbiology, Washington, DC.

Roberts, Larry S. and John Janovy, Jr. *Foundations of Parasitology,* 5th Ed. Wm. C. Brown Publishers, Dubuque, IA.

Biochemical Pathways

APPENDIX A

So much of what is done in microbiology relies on an understanding of basic biochemical pathways. It's not as important to memorize them (although, with exposure they will become second nature) as it is to understand their importance in metabolism and to interpret diagrams of them when available. The following discussion is provided so you can see how the various biochemical tests presented in this manual fit into the overall scheme of cellular chemistry.

Oxidation of Glucose: Glycolysis, Entner-Doudoroff, and Pentose-Phosphate Pathways

Most organisms use **glycolysis** (also known as the "Embden-Meyerhof-Parnas pathway, Figure A-1) in energy metabolism. It performs the stepwise disassembly of glucose into two pyruvates, releasing some of its energy and electrons in the process. The exergonic (energy-releasing) reactions are associated with ATP synthesis by a process called **substrate phosphorylation**. Although a total of four ATPs are produced per glucose in glycolysis, two ATPs are hydrolyzed early in the pathway, leaving a net production of two ATPs per glucose. In one glycolytic reaction, the loss of an electron pair (oxidation) from a three-carbon intermediate occurs simultaneously with the reduction of NAD^+ to $NADH+H^+$. The $NADH+H^+$ then may be oxidized in an electron transport chain or a fermentation pathway, depending on the organism and the environmental conditions. The former yields ATP, and the latter generally does not. In summary, each glucose oxidized in glycolysis yields two pyruvates, $2\ NADH+2\ H^+$, and a net of 2 ATPs (Table A-1).

Although the intermediates of glycolysis are carbohydrates, many are entry points for amino acid, lipid, and nucleotide catabolism. Many glycolytic intermediates also are a source of carbon skeletons for the synthesis of these other biochemicals. Some of these are shown in Figure A-1. ***Note***: For clarity, many details have been omitted from these other pathways in Figure A-1. Single arrows may represent several reactions, and other carbon compounds not illustrated may be required to complete a particular reaction.

The **Entner-Doudoroff pathway** (Figure A-2) is an alternative means of degrading glucose into two pyruvates. This pathway is found exclusively among prokaryotes (*e.g.*, *Pseudomonas* and *E. coli*, as well as other Gram-negatives and certain archaebacteria). It allows utilization of a different category of sugars (aldonic acids) than glycolysis and therefore improves the range of resources available to the organism. It is less efficient than glycolysis because only one ATP is phosphorylated but more reducing power is produced (one NADH and one NADPH per glucose). Table A-2 summarizes this pathway.

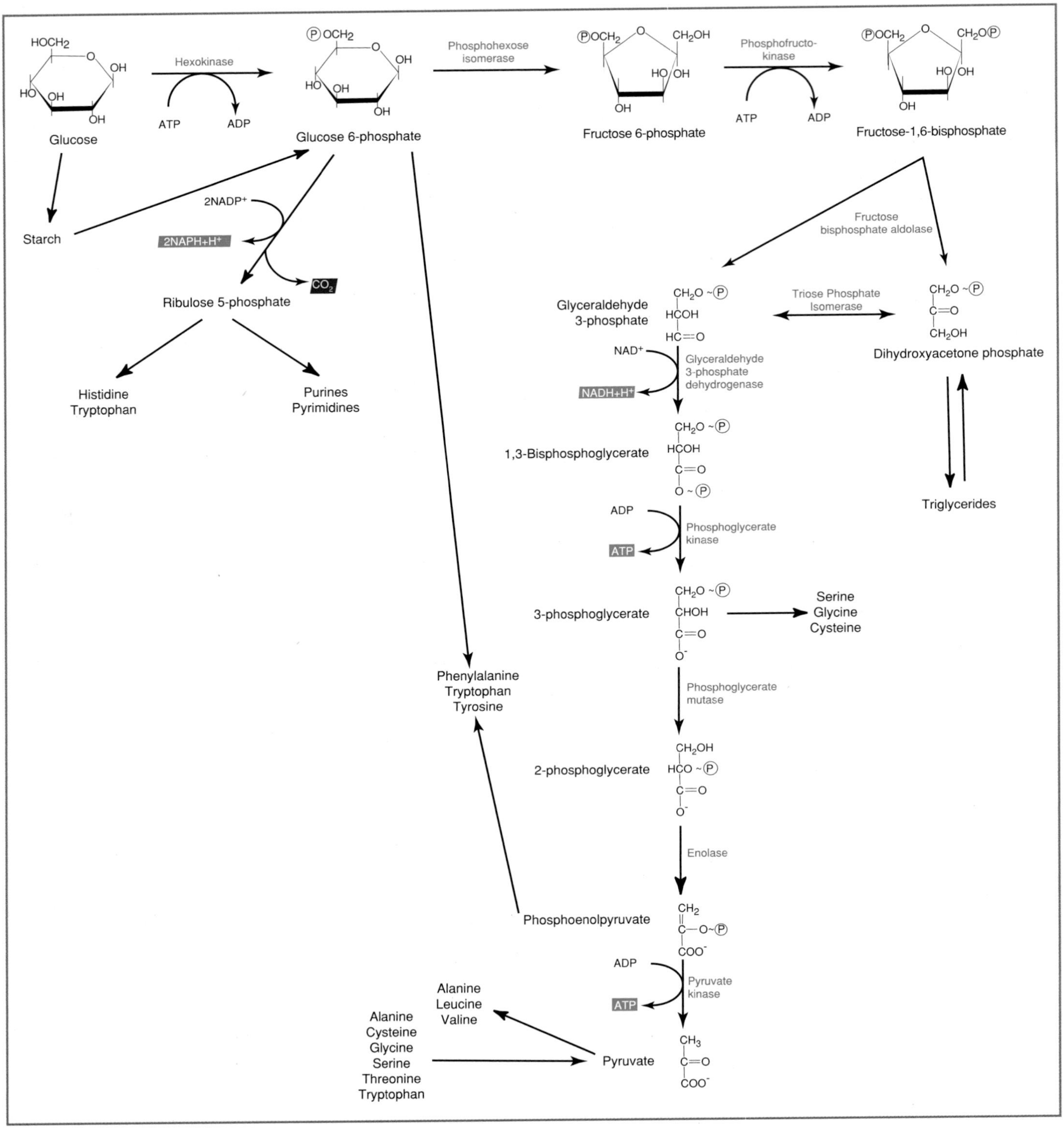

FIGURE A-1 Glycolysis and Associated Pathways

The names of glycolytic intermediates are printed in black ink; the enzyme names are in red. Reducing power (in the form of NADH+H+) and ATP are highlighted in blue. The major key to getting product yields correct is to recognize that *both* C_3 compounds (Glyceraldehyde 3-phosphate and Dihydroxyacetone phosphate) produced from splitting Fructose 1,6-bisphosphate can pass through the remainder of the pathway because of the isomerase reaction. The conversion of each into pyruvate results in the formation of 2 ATPs and 1 NADH+H+ (Table A-1).

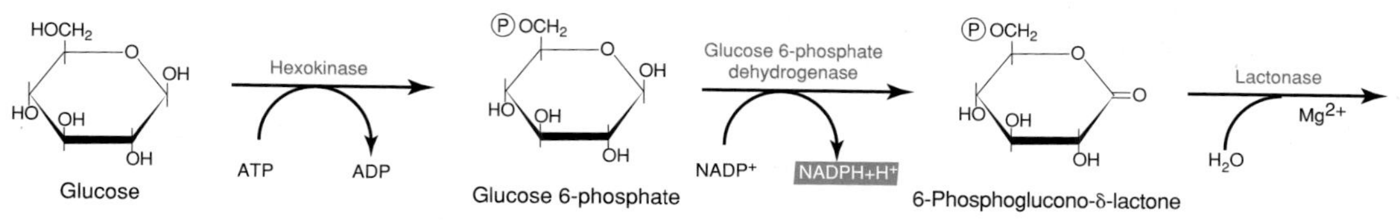

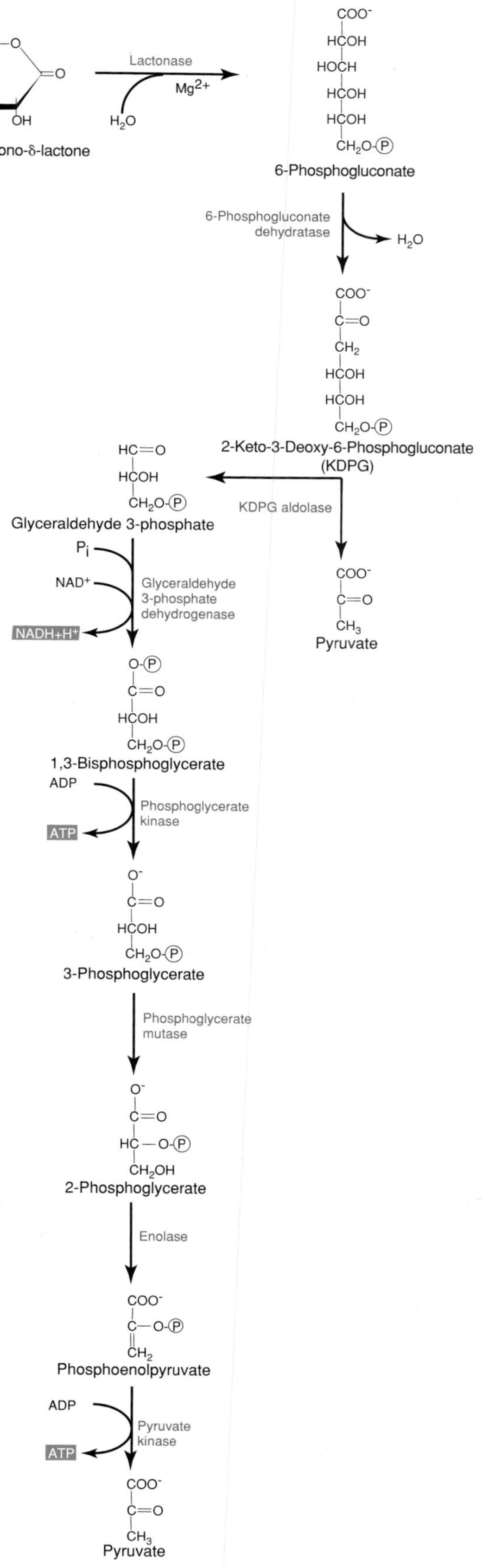

FIGURE A-2 ENTNER-DOUDOROFF PATHWAY

Notice the similarities of this pathway with glycolysis. The main difference is in the six-carbon compound that is split into two three-carbon compounds. The result of this split is pyruvate and glyceraldehydes-3-phosphate, which is oxidized as in glycolysis to pyruvate. Because only one three-carbon compound goes through the sequence of reactions leading to pyruvate, the ATP and NADH yield is one-half that of glycolysis. But one NADPH is produced that is not made in glycolysis.

Reactant	Product
Glucose ($C_6H_{12}O_6$)	2 Pyruvates ($C_3H_3O_3$)
2 ATP	2 ADP
4 ADP	4 ATP
NET: 2 ADP	NET: 2 ATP
2 NAD^+	2 NADH + 2 H^+

TABLE A-1 SUMMARY OF GLYCOLYTIC REACTANTS AND PRODUCTS PER GLUCOSE

Reactant	Product
Glucose ($C_6H_{12}O_6$)	2 Pyruvates ($C_3H_3O_3$)
1 ATP	1 ADP
2 ADP	2 ATP
NET: 1 ADP	NET: 1 ATP
1 NAD^+	1 NADH + 1H^+
1 $NADP^+$	1 NADPH + 1H^+

TABLE A-2 SUMMARY OF ENTNER-DOUDOROFF REACTANTS AND PRODUCTS PER GLUCOSE

Reactant	Product
Glucose-6-phosphate (C_6)	6 CO_2 + 1 P_i
12 $NADP^+$	12 NADPH + 12 H^+

TABLE A-3 Summary of Pentose Phosphate Reactants and Products per Glucose-6-phosphate

The **pentose phosphate pathway** is a complex set of cyclic reactions that provides a mechanism for producing five-carbon sugars (**pentoses**) from six-carbon sugars (**hexoses**). Pentose sugars are used in ribonucleotides and deoxyribonucleotides, as well as being precursors to aromatic amino acids. Further, this pathway produces NADPH, which is used as an electron donor in anabolic pathways. Unlike NADH, produced in glycolysis and Entner-Doudoroff, NADPH is not used as an electron donor in an electron transport chain for oxidative phosphorylation of ADP.

The pentose phosphate reactants and products are listed in Table A-3, and the overall path is shown in Figure A-3. To completely oxidize one hexose to $6CO_2$, a total of six hexoses must enter the cycle as glucose-6-phosphate and follow one of three different routes (notice the symmetry of pathways as drawn). Notice in Figure A-3 that each hexose loses a CO_2 upon entry into the cycle, but at the end, five hexoses are produced. Thus, the net reaction is one hexose being oxidized to 6 CO_2. Notice also the reactions that transfer two-carbon and three-carbon fragments between the five-carbon intermediates. **Transketolase** catalyzes the two-carbon transfer, whereas **transaldolase** catalyzes the three-carbon transfer. Alternatively, the five-carbon intermediates can be redirected into pathways for synthesis of aromatic amino acids and nucleotides (not shown).

Oxidation of Pyruvate: The Krebs Cycle and Fermentation

Pyruvate represents a major crossroads in metabolism. Some organisms are able to further disassemble the pyruvates produced in glycolysis and Entner-Doudoroff and make more ATP and NADH + H^+ in the **Krebs cycle.** Other organisms simply reduce the pyruvates with electrons from NADH + H^+ without further energy production in **fermentation.**

The Krebs cycle is a major metabolic pathway used in energy production by organisms that respire aerobically or anaerobically (Figure A-4). Pyruvate produced in glycolysis or other pathways is first converted to acetyl-coenzyme A during the **entry step** (also known as the **intermediate** or **gateway step**). Acetyl-CoA enters the Krebs cycle through a condensation reaction with oxaloacetate. Products for each pyruvate that enters the cycle via the entry step are: 3 CO_2, 4 NADH + H^+, 1 $FADH_2$, and 1 GTP. (Because two pyruvates are made per glucose, these numbers are doubled in Table A-4). The energy released from oxidation of reduced coenzymes (NADH + H^+ and $FADH_2$) in an electron transport chain is then used to make ATP. ATP yields are summarized in Table A-5.

Like glycolysis, many of the Krebs cycle's intermediates are entry points for amino acid, nucleotide and lipid catabolism, as well as a source of carbon skeletons for synthesis of the same compounds. These pathways are shown, but details have been omitted. Single arrows may represent several reactions, and other carbon compounds, not illustrated, may be required to complete a given reaction.

Figure A-5 illustrates some major fermentation pathways exhibited by microbes (though no single organism is capable of all of them). Pyruvate (shown in the blue box) is typically the starting point for each. End products of fermentation are shown in red. Fermentation allows a cell living under anaerobic conditions to oxidize reduced coenzymes (such as NADH + H^+ and shown in blue) generated during glycolysis or other pathways. Some bacteria (aerotolerant anaerobes) rely solely on fermentation and do not use oxygen even if it is available. Table A-6 summarizes major fermentations and some representative organisms that perform each.

Notice that fermentation end products typically fall into three categories: acid, gas, or an organic solvent (an alcohol or a ketone). The specific fermentation performed is the result of the enzymes present in a species and often is used as a basis of classification.

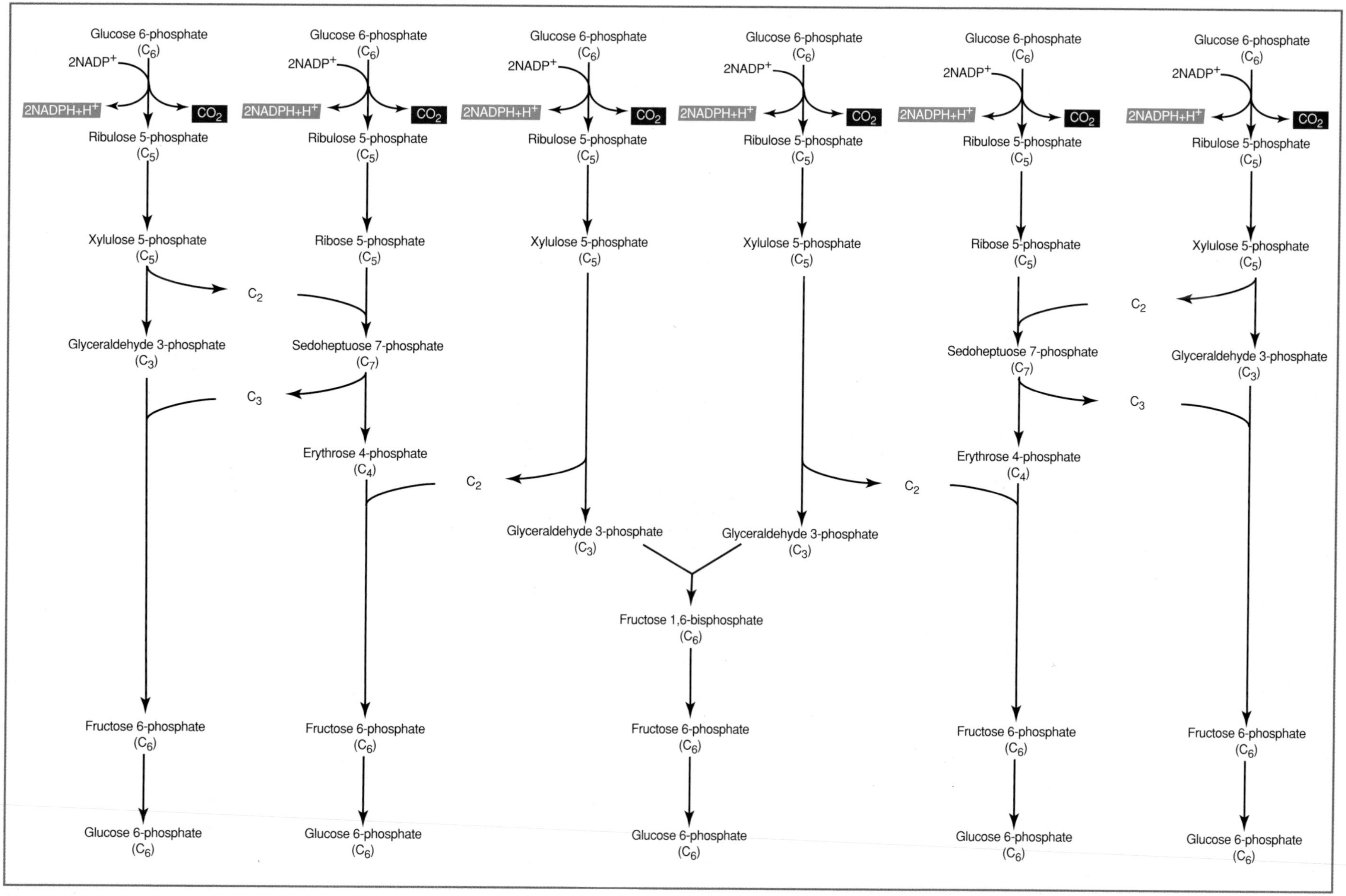

FIGURE A-3 Pentose Phosphate Cycle

For every six glucose-6-phosphates that enter and complete the cycle, 6 CO_2 and 12 NADPH are produced. Some of the five-carbon intermediates, however, may be redirected into synthesis of aromatic amino acids and nucleotides.

Entry Step		Krebs Cycle	
Reactant	**Product**	**Reactant**	**Product**
2 Pyruvates	2 Acetyl CoA + 2 CO_2	2 Acetyl CoA	4 CO_2
2 Coenzyme A			2 Coenzyme A
2 NAD^+	2 NADH + $2H^+$	6 NAD^+	6 NADH + $6H^+$
		2 GDP + 2 P_i (= 2 ADP + 2 P_i)	2 GTP (= 2 ATP)

TABLE A-4 Summary of Reactants and Products per Glucose in the Entry Step and the Krebs Cycle

Compound	Number Produced	ATP Value	Total ATPs per Glucose
NADH + H^+	10	3	30
$FADH_2$	2	2	4
ATP (by substrate phosphorylation)	4		4

TABLE A-5 ATP Yields from Complete Oxidation of Glucose to CO_2 by a Prokaryote Using Glycolysis and with Oxygen as the Final Electron Acceptor

Fermentation	Major End Products	Representative Organisms
Alcoholic fermentation	Ethanol and CO_2	*Saccharomyces cerevisiae*
Homofermentation	Lactate	*Streptococcus* and some *Lactobacillus*
Heterofermentation	Lactate, ethanol, and acetate	*Streptococcus, Leuconostoc,* and *Lactobacillus*
Mixed acid fermentation	Acetate, formate, succinate, CO_2, H_2, and ethanol	*Escherichia, Salmonella, Klebsiella,* and *Shigella*
2,3-Butanediol fermentation	2,3-Butanediol	*Enterobacter, Serratia,* and *Erwinia*
Butyrate/butanol fermentation	Butanol, butyrate, acetone, and isopropanol	*Clostridium, Butyrivibrio,* and some *Bacillus*
Propionic acid fermentation	Propionate, acetate and CO_2	*Propionibacterium, Veillonella,* and some Clostridium

TABLE A-6 Major Fermentations, Their End-Products, and Some Organisms That Perform Them

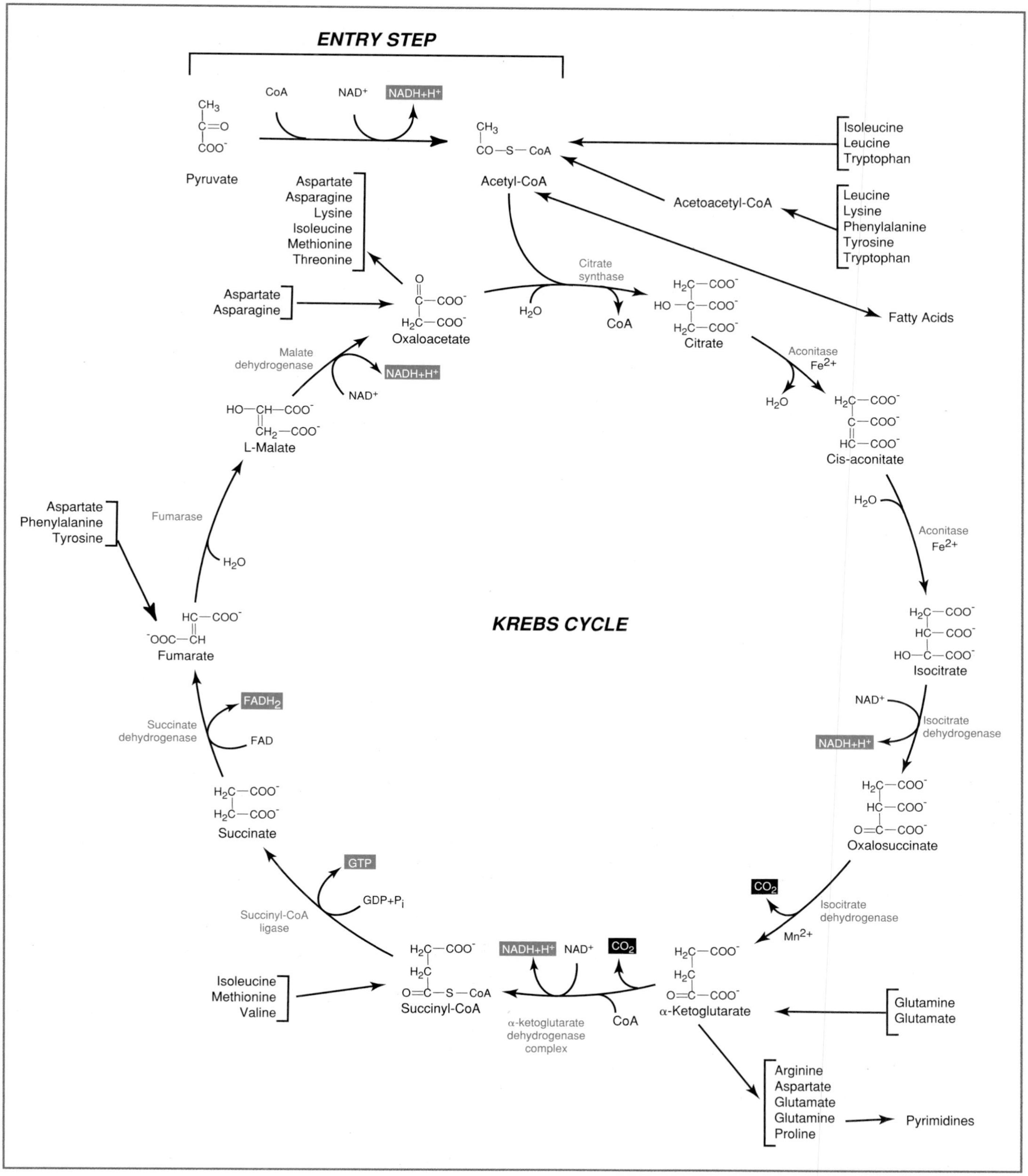

FIGURE A-4 THE ENTRY STEP AND KREBS CYCLE

The names of intermediates are printed in black ink; enzymes are in red. Reducing power (in the form of NADH + H$^+$ and $FADH_2$) and GTP are highlighted in blue. CO_2 produced from the oxidation of carbon is highlighted in black.

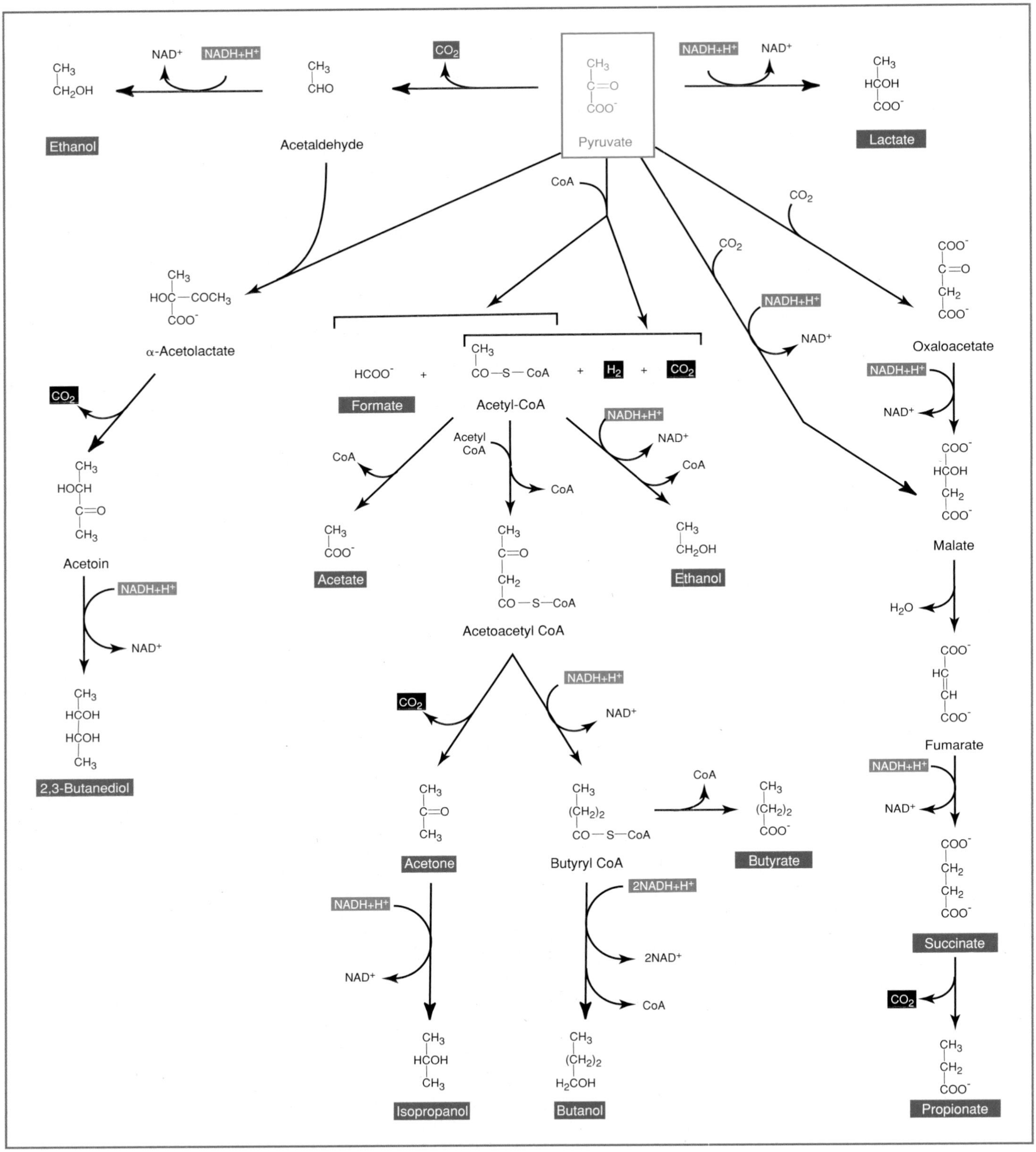

FIGURE A-5 A SAMPLING OF FERMENTATION PATHWAYS

Note that all pathways start with pyruvate, have a step(s) where NADH + H+ (in blue) is oxidized to NAD+, and produce end-products falling into one of three categories: acid, gas, or alcohol.

Miscellaneous Transfer Methods

APPENDIX B

Following are instructions for transfer methods that are performed less routinely than those in Exercise 1-2. As in Exercise 1-2, new skills in each process are printed in blue type.

Transfers Using a Sterile Cotton Swab

A sterile cotton swab generally is used to obtain a sample from a primary source —either a patient or an environmental site. Occasionally, swabs are used to transfer pure cultures. Sterile swabs may be dry, or they may be in sterile water, depending on the source of the sample. In either case, care must be taken not to contaminate the swab by touching unintended surfaces with it. Your instructor may provide specific instructions on collection of samples from sources other than the ones below.

Obtaining a Sample from a Patient's Throat with a Cotton Swab

1. Use a sterile tongue depressor and swab prepared in sterile water to obtain a sample from the throat. Have the patient open his/her mouth, then gently press down on the tongue (Figure B-1).
2. With the swab in the dominant hand, carefully sample the patient's throat with a swirling motion. Touching other parts of the oral cavity is likely to cause contamination. Also, avoid touching the soft palate or it may initiate a gag reflex!
3. Transfer the sample to an appropriate plated medium (see Exercise 1-3) as quickly as possible.
4. If plating is to be done at a later time, place the swab in an appropriate sterile container (such as a sterile, capped test tube).

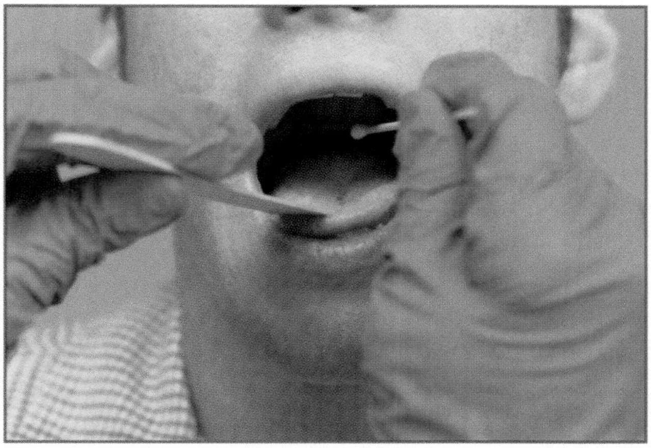

FIGURE B-1
TAKING A THROAT SAMPLE
Use a sterile tongue depressor and cotton swab to obtain a sample from a patient's throat. Be careful not to touch other parts of the oral cavity or the sample will get contaminated. Transfer the sample to a sterile medium as soon as possible.

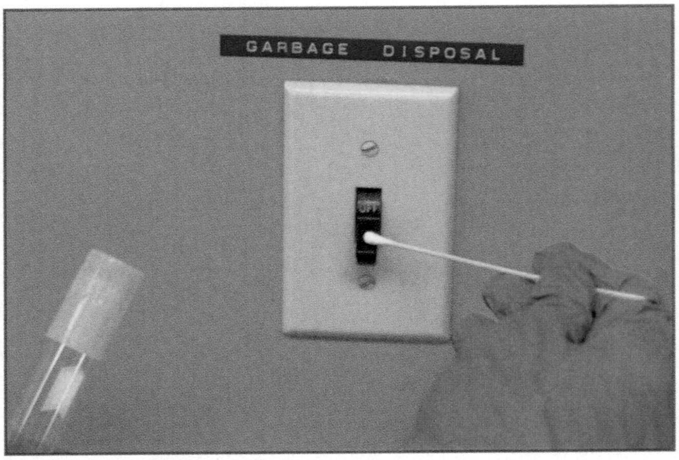

FIGURE B-2 Taking an Environmental Sample
Use a spinning motion of a sterile swab to sample inanimate objects in the environment. The swab may be placed in a sterile test tube until it is convenient to transfer the sample to a growth medium.

Obtaining an Environmental Sample with a Cotton Swab

1. Use a sterile cotton swab prepared in sterile water to obtain a sample from an environmental source.
2. Rotate the swab to collect from the area to be sampled (Figure B-2).
3. Transfer the sample to an appropriate plated medium (see Exercise 1-3) as quickly as possible.
4. If plating is to be done at a later time, place the swab in an appropriate sterile container (such as a sterile, capped test tube).

Stab Inoculation of Agar Tubes Using an Inoculating Needle

Stab inoculations of agar tubes are used for several types of differential media (usually to examine growth under anaerobic conditions or to observe motility). A stab is *not* used to produce a culture of microbes for transfer to another medium.

1. Remove the cap of the sterile medium with the little finger of your inoculating needle hand, and hold it there.
2. Flame the tube by quickly passing it through the Bunsen burner flame a couple of times. Keep your needle-hand still.
3. Hold the open tube on an angle to minimize airborne contamination. Keep your needle hand still.
4. Carefully move the agar tube over the needle wire (Figure B-3). Insert the needle into the agar to about 1 cm from the bottom.

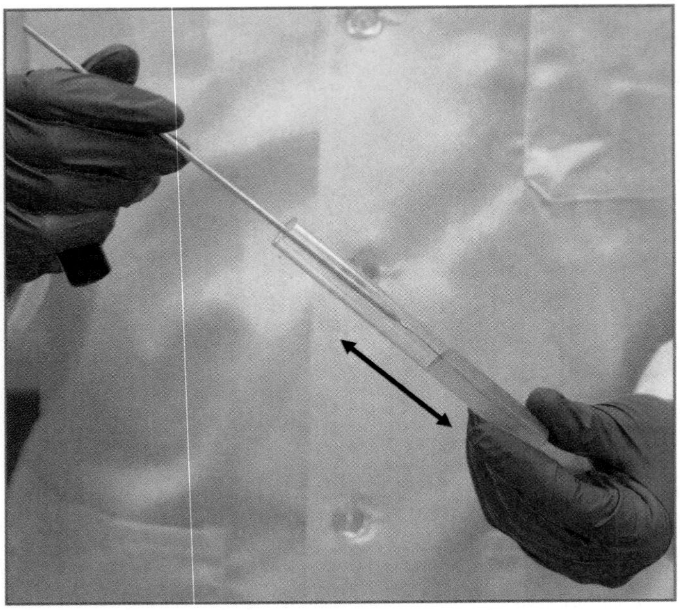

FIGURE B-3 Agar Deep Stab
Use the inoculating needle to stab the agar to a depth about 1 cm from the bottom. It is generally desirable to remove the needle along the original stab line and not create a new one. Upon completion, sterilize the needle.

5. Withdraw the tube carefully so the needle leaves along the same path it entered. (When removing the tube, be especially careful not to catch the needle tip on the tube lip. This springing action of the needle creates bacterial aerosols.)
6. Flame the tube lip as before. Keep your needle hand still.
7. Keeping the needle hand still (remember, it has growth on it), move the tube to replace its cap.
8. Sterilize the needle as before by incinerating it in the Bunsen burner flame. It is especially important to flame it from base to tip now because the needle has lots of bacteria on it.
9. Label the tube with your name, date, and organism. Incubate at the appropriate temperature for the assigned time.

Spot Inoculation of an Agar Plate

Sometimes an agar plate may be used to grow several different specimens at once. This is a typical practice with plated *differential* media (*i.e.*, media designed to differentiate organisms based on growth characteristics). Prior to beginning the transfer, divide the plate into as many as four sectors, using a marking pen. (Some plates already have marks on the base for this purpose.) Then each may be inoculated with a different organism. Inoculation involves touching the loop to the agar surface once so growth is restricted to a single spot—hence the name "spot inoculation."

1. Lift the lid of the sterile agar plate and use it as a shield to prevent airborne contamination.
2. Touch the agar surface toward the periphery of the sector (Figure B-4).
3. Remove the loop and replace the lid.
4. Sterilize your loop as before. It is especially important to flame it from base to tip now because the loop has lots of bacteria on it.
5. Label the base of the plate with your name, date, and organism(s) inoculated.
6. Incubate the plate in an inverted position for the assigned time at the appropriate temperature.

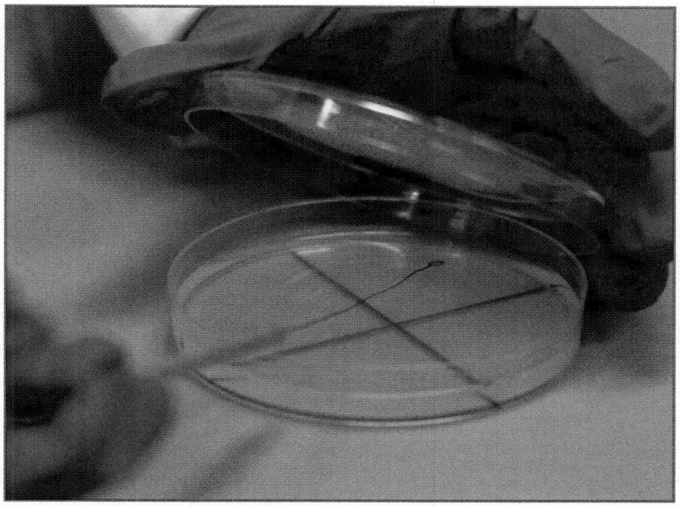

FIGURE B-4 Spot Inoculation of a Plate
Each of four sectors is spot-inoculated with a different organism by touching the loop to the surface and making a mark about 1 cm in length. Generally, spot inoculations are done toward the edge (rather than the crowded middle) to prevent overlapping growth and/or test results.

APPENDIX C

Using Glass Pipettes to Transfer from a Broth Culture

Glass pipettes are used to transfer a known volume of liquid diluent, media, or culture. Originally pipettes were filled by sucking on them like a drinking straw, but mouth pipetting is dangerous and has been replaced by mechanical pipettors. Three examples are shown in Figure C-1, each with its own method of operation. Your instructor will show you how to properly use the type of pipettor available in your lab.

To use a pipette correctly, you must be able to correctly read the calibration. Examine Figure C-2. The numbers indicate the pipette's *total volume* and its *smallest calibrated increments*. This is a 10.0 mL pipette divided into 0.1 mL increments.

When reading volumes, use the base of the meniscus (Figure C-3). The volume in the center pipette is read at exactly 3.0 mL because the meniscus is resting on the line. The left pipette is read as 2.9 mL and the right pipette is read as 3.1 mL (0 is always at the top of the pipette). Although the difference in volume between these three pipettes may seem negligible (1 part in 70, about a 1.5% error), it may be enough to introduce significant error into your work.

Two pipette styles are used in microbiology (Figure C-4): the **serological pipette**

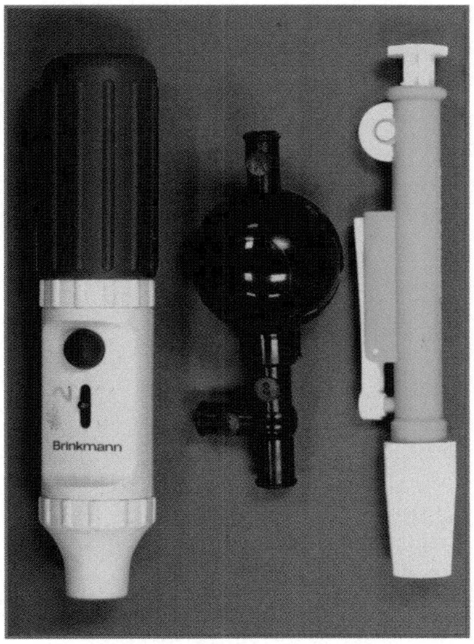

FIGURE C-1 MECHANICAL PIPETTORS
Three examples of mechanical pipettors are shown here, each with its own method of operation. Your instructor will show you how to properly use the style of pipettor available in your lab. From left to right: A pipette filler/dispenser, a pipette bulb, and a plastic pump.

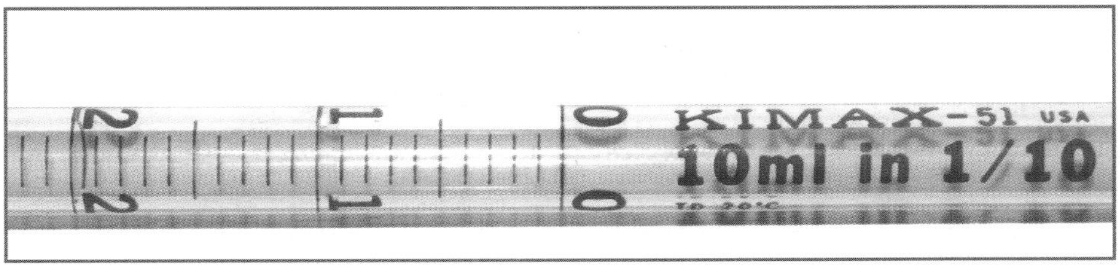

FIGURE C-2 PIPETTE CALIBRATION
Prior to using a pipette, read the pipette calibration. The numbers indicate the pipette's *total volume* and its *smallest calibrated increments*. This is a 10.0 mL pipette divided into 0.1 mL increments.

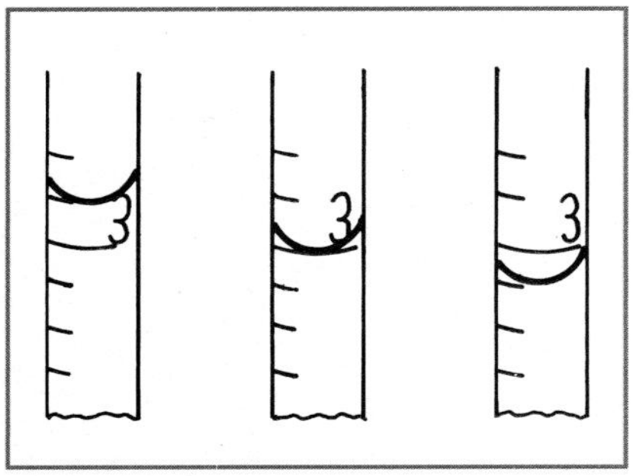

FIGURE C-3
Read the Base of the Meniscus
When reading volumes, use the base of the meniscus. The volume in the center pipette is read at exactly 3.0 mL because the meniscus is resting on the line. The left pipette is read as 2.9 mL and the right pipette is read as 3.1 mL (0 is always at the top of the pipette).

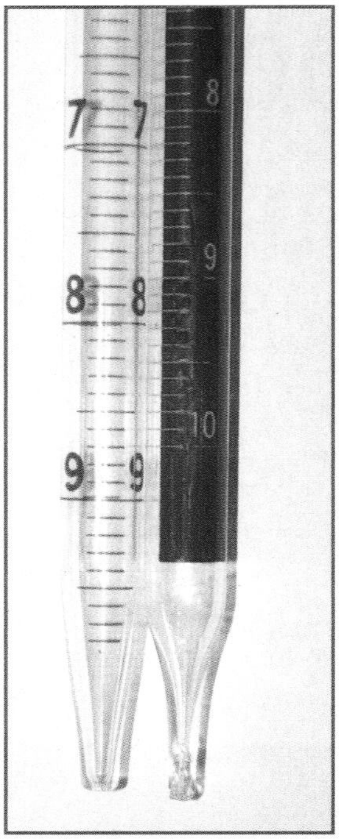

FIGURE C-4
Two Types of Pipettes
Two pipette styles—the *serological pipette* (left) and the *Mohr pipette* (right—are used in microbiology. A serological pipette is calibrated *to deliver* (TD) its volume by completely draining it and blowing out the last drop. The tip of a Mohr pipette is not graduated, so fluid flow must be stopped at a calibration line. Stopping the fluid beyond the last line on a Mohr pipette results in an unknown volume being dispensed.

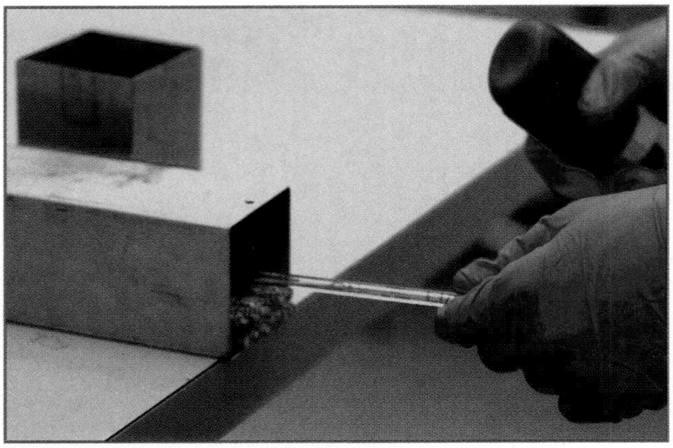

FIGURE C-5 Getting the Sterile Pipette
Pipettes of the same size are autoclaved in canisters, which then are opened and placed flat on the table. Pipettes are removed as needed.

and the **Mohr pipette**. A serological pipette is calibrated *to deliver* (TD) its volume by completely draining it and blowing out the last drop. The tip of a Mohr pipette is not graduated, so fluid flow must be stopped at a calibration line. Stopping the fluid beyond the last line on a Mohr pipette results in an unknown volume being dispensed. In either case, volumes are read at the bottom of the meniscus of fluid.

Important: If pipetting a bacterial culture, be careful not to allow any to drop from the pipette before disposing of it in the autoclave container. Clean up any spills.

Following are instructions for using a glass pipette. As in Exercise 1-2, new skills in each process are printed in blue type.

Filling a Glass Pipette

1. Bacteria should be suspended in the broth with a vortex mixer (Figure 1-14) or by agitating with your fingers (Figure 1-15). Be careful not to splash the broth into the cap or lose control of the tube.
2. Pipettes are sterilized in metal canisters, individually in sleeves, or as multiples in packages (if disposable). They typically are stored in groups of a single size (Figure C-5). *Be sure you know what volume your pipette will deliver.* Set the canister at the table edge and remove its lid. (Canisters should not be stored in an upright position, as they may fall over and break the pipettes or become contaminated.) If using pipettes in a package, open the end *opposite the tips*. Grasp *one pipette only* and remove it.
3. Carefully insert the pipette into the mechanical pipettor (Figure C-6). It's best to grasp the pipette near the end with your fingertips. This gives you more control and reduces the chance that you will break the pipette and cut your hand. *Do not touch* any part of the pipette that will contact the specimen or the medium or you will risk introducing a contaminant. Also, do not lay the pipette on the tabletop while you continue.
4. While keeping the pipette hand still, bring the culture tube toward it. Use your little finger to remove and hold its cap.
5. Flame the open end of the tube by passing it through a Bunsen burner flame two or three times.
6. Hold the tube at an angle to prevent contamination from above.
7. Insert the pipette and withdraw the appropriate volume (Figure C-7). Bring the pipette to a vertical position briefly to read the meniscus accurately. (Remember: The volumes in the pipette are correct only if the meniscus of the fluid inside is resting *on* the line, not below it.) Then carefully remove the pipette from the tube.
8. Flame the tube lip as before. Keep your pipette hand still.
9. Keeping the pipette hand still (remember, it contains fluid with microbes in it), move the tube to replace its cap.

10. What you do next depends on the medium to which you are transferring the growth. Please continue with the appropriate inoculation section.

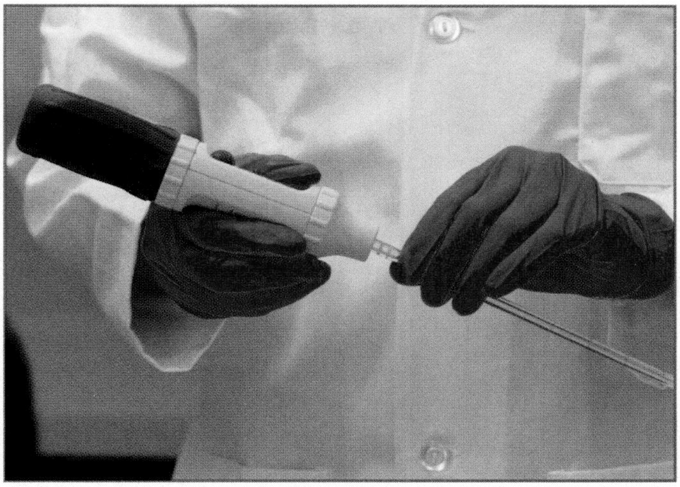

FIGURE C-6 **Assembling the Pipette**
Carefully insert the pipette into a mechanical pipettor. Notice that the pipette is held near the end with the fingertips. For safety, the hand is out of the way in case the pipette breaks.

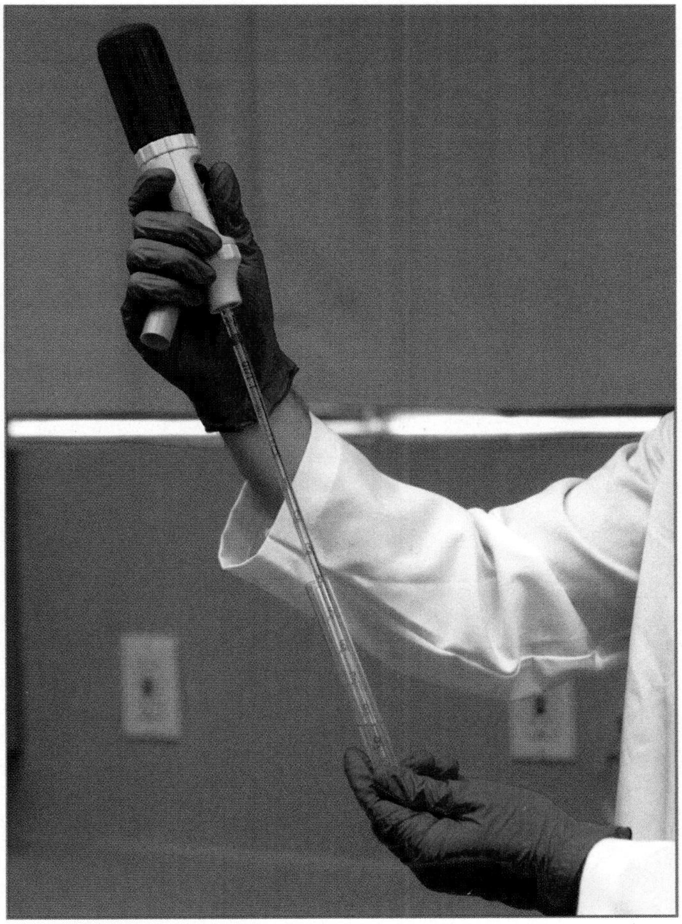

FIGURE C-7 **Filling the Pipette**
Carefully draw the fluid into the pipette. Briefly bring it to a vertical position and read the volume.

Inoculation of Broth Tubes with a Pipette

Pipettes are often used to inoculate a known volume of culture into a known volume of diluent during serial dilutions or to dispense a known volume to an agar plate.

1. While keeping the pipette hand still, bring the broth tube toward it. Use your little finger to remove and hold its cap.
2. Flame the tube by quickly passing it through the Bunsen burner flame two or three times. Keep your pipette-hand still.
3. Hold the open tube on an angle to minimize airborne contamination. Keep your pipette hand still.
4. Insert the pipette tip and dispense the correct volume of inoculum.
5. Withdraw the tube from over the pipette. Before completely removing it, touch the pipette tip to the glass to remove any excess broth.
6. Completely remove the pipette, but avoid waving it around. This can create aerosols.
7. Flame the tube lip as before. Keep your pipette hand still.
8. Keeping the pipette hand still, move the tube to replace its cap.
9. The pipette is contaminated with microbes and must be disposed of correctly. Each lab has its own specific procedures, and your instructor will advise you what to do. A glass pipette typically is placed in a pipette disposal container containing a small amount of disinfectant until it is autoclaved and reused (Figure C-8). Disposable pipettes must be

FIGURE C-8
Dispose of the Pipette
The pipette is contaminated with microbes and must be disposed of correctly. Each lab has its own specific procedures, and your instructor will advise you what to do. Shown here is a canister used for glass pipettes. Disposable pipettes must be placed in an appropriate biohazard container. In either case, be careful when removing the pipette from the mechanical pipettor. There is danger of culture dripping from the pipette or of breaking the glass.

placed in an appropriate biohazard container. In either case, be careful when removing the pipette from the mechanical pipettor. There is danger of culture dripping from the pipette or of breaking the glass.

Inoculation of Agar Plates with a Pipette

1. Lift the lid of the plate and use it as a shield to protect it from airborne contamination.
2. Hold the pipette over the agar and dispense the correct volume (often 0.1 mL) onto the center of the agar surface (Figure C-9). From this point, the remainder of steps should be completed within about 15 seconds to prevent the inoculum from soaking into the agar.
3. The pipette is contaminated with microbes and must be disposed of correctly. Each lab has its own specific procedures, and your instructor will advise you what to do. Glass pipettes typically are placed in a pipette disposal container containing a small amount of disinfectant until they are autoclaved and reused. (Figure C-8) Disposable pipettes must be placed in an appropriate biohazard container. In either case, be careful when removing the pipette from the mechanical pipettor. There is danger of culture dripping from the pipette or of breaking the glass.
4. Continue with the spread plate technique (Exercise 1-4).

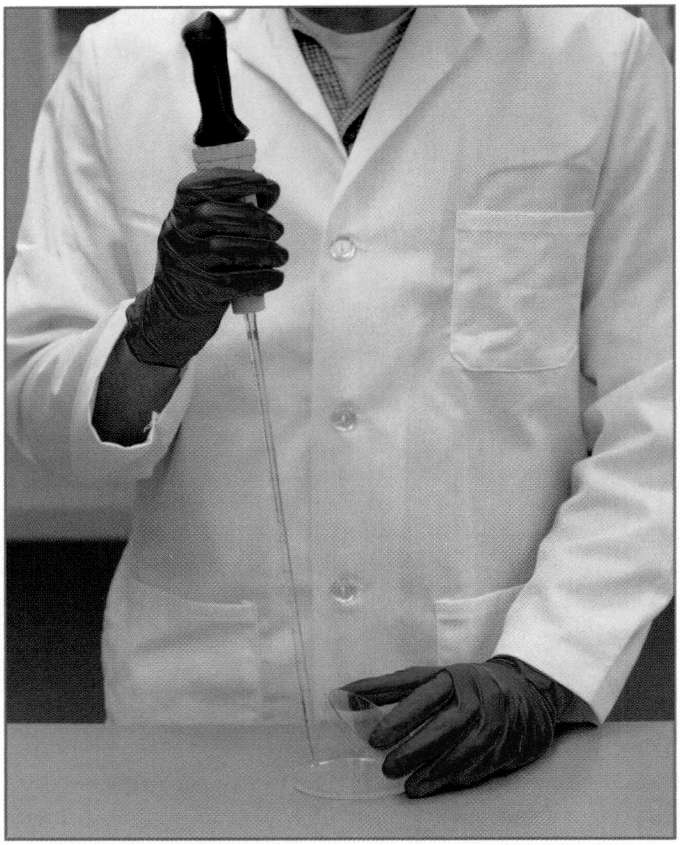

FIGURE C-9 INOCULATING THE PLATE WITH A PIPETTE
Open the lid and dispense the inoculum onto the surface of the agar near the middle.

APPENDIX D

Transfers from a Broth Culture Using a Digital Pipette

Modern molecular biology procedures often involve transferring extremely small volumes of liquid with great precision and accuracy. This has led to the development of digital pipettors (Figure D-1) that can be set up to dispense microliter volumes through milliliter volumes. Common digital pipettors are calibrated to dispense volumes of 1 to 10 µL, 10 to 100 µL, 100 to 1000 µL, or 1 to 5 mL, or more.

Filling a Digital Pipettor

Many manufacturers make digital pipettors, but they all work basically the same way. The following instructions are for Eppendorf Series 2100 models.

1. Growth may be suspended in the broth with a vortex mixer (Figure 1-14) or by agitating with your fingers (Figure 1-15). Be careful not to splash the broth into the cap or lose control of the tube.
2. Determine which digital pipettor should be used to dispense the desired volume.
3. Turn the setting ring to set the desired volume (Figure D-2). If you turn the dial past the volume range of the pipettor, you will damage it.
4. Hold the digital pipettor in your dominant hand.
5. Open the rack of appropriate pipette tips for your pipettor (these are color-coded and match the pipettor) and push the pipettor into a sterile tip (Figure D-3). Close the rack. Never touch the pipette tip with your hands or leave the rack open. *Never use a digital pipettor without a tip.*
6. Remove the cap from the culture tube with the little finger of your dominant hand, and flame the tube.
7. Press down the control button with your thumb to the first stop—the measuring stroke.

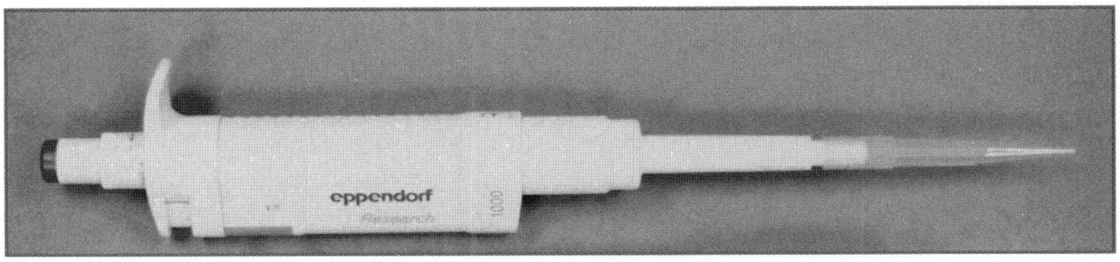

FIGURE D-1 Digital Pipettor
Shown is a digital pipettor to be used for dispensing volumes between 100 and 1000 µL. Always use a digital pipettor with the appropriate tip.

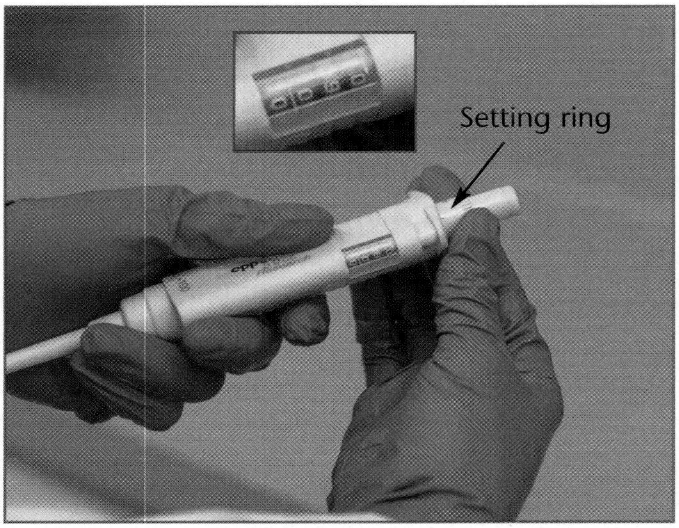

FIGURE D-2 Setting the Volume
Rotate the adjustment knob to set the volume on a digital pipettor. This pipettor has been set at 90.0 μL (see inset—the horizontal line between the third and fourth numerals is a decimal point). Never rotate the adjustment knob beyond the volume limits of the pipettor.

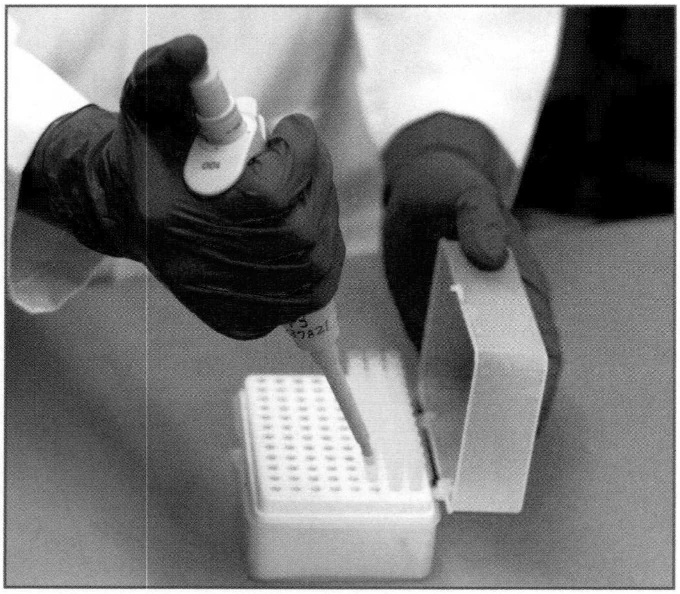

FIGURE D-3 Digital Pipettor Tips
Digital pipettors must be fitted with a sterile tip of appropriate size (color-coded to match the pipettor). Open the case and press the pipettor into a tip, then close the case to maintain sterility. Do not touch the pipettor tip.

8. Insert the tip into the broth approximately 3 mm while holding it vertically (Figure D-4).
9. Slowly release pressure with your thumb to draw fluid into the tip. Be careful not to pull any air into the tip.
10. Remove the pipettor and hold it still as you flame the tube as before.

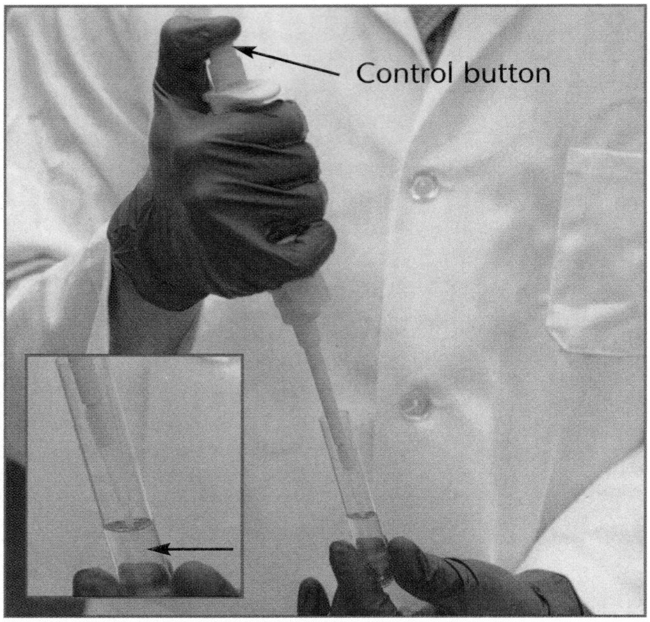

FIGURE D-4 Filling the Pipette
Depress the control button with your thumb to the first stop (measuring stroke). Holding the tube and pipettor in a vertical position, insert the tip into the fluid to a depth of 3 mm (see inset). Slowly release pressure with your thumb to fill the pipettor.

11. Keeping the pipettor hand still, move the tube to replace its cap.
12. What you do next depends on the medium to which you are transferring the growth. Please continue with the appropriate inoculation section.

Inoculation of Broth Tubes with a Digital Pipettor

1. Remove the cap of the sterile medium with the little finger of your pipettor hand, and hold it there.
2. Flame the tube by quickly passing it through the Bunsen burner flame a couple of times. Keep your pipettor hand still.
3. Insert the pipette tip into the tube. Hold it at an angle against the inside of the glass (Figure D-5).
4. Depress the control button slowly with your thumb to the first stop and pause until no more liquid is dispensed. Then continue pressing to the second stop (blow-out) to deliver the remaining volume.
5. While keeping pressure on the control button, carefully remove the pipettor from the tube by sliding it along the glass. Once it is out of the tube, slowly release pressure on the control button.
6. Flame the tube lip as before. Keep your pipette hand still.

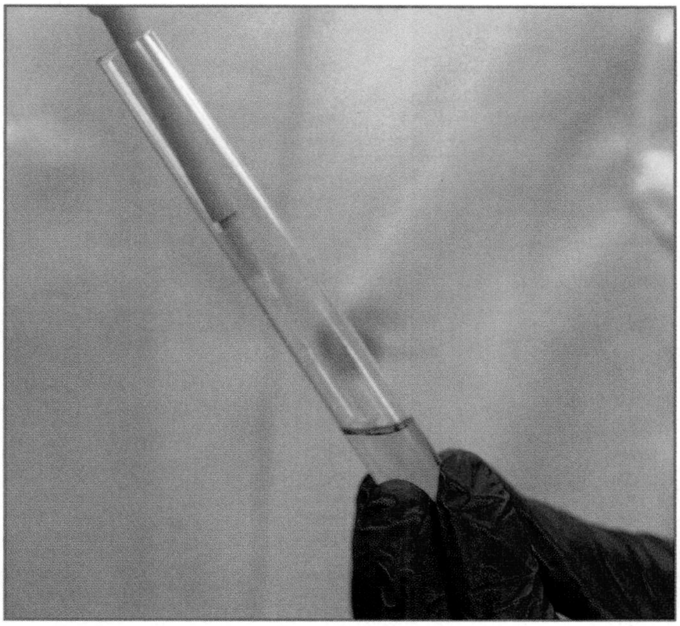

FIGURE D-5 DISPENSING TO A LIQUID
Hold the pipettor at an angle with the tip against the glass. Press the control button to the first stop. When no more fluid comes out, continue pressing to the second stop. Remove the pipettor, then slowly release pressure on the control button, and eject the tip into a biohazard container. Finally, replace the cap on the tube.

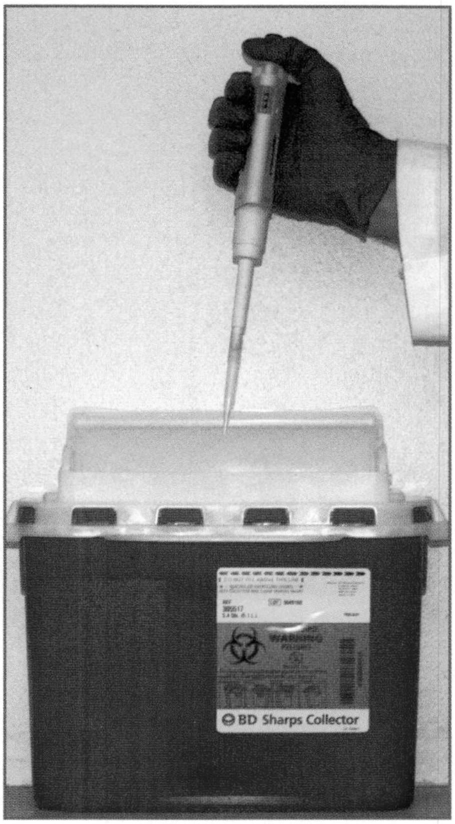

FIGURE D-6
EJECTING THE TIP
Using the tip ejector button, eject the contaminated tip into an appropriate biohazard container.

7. Keeping the pipette hand still, move the tube to replace its cap.
8. The pipettor tip is contaminated with microbes and must be disposed of correctly. Use the ejector button to remove the tip into an appropriate biohazard container (Figure D-6). Each lab has its own specific procedures, and your instructor will advise you what to do.

Inoculation of Agar Plates with a Pipettor

1. Lift the lid of the plate and use it as a shield to protect from airborne contamination.
2. Place the pipette tip over the agar surface. Be sure to hold the pipettor in a vertical position (Figure D-7).
3. Depress the control button slowly with your thumb to the first stop, and pause until no more liquid is dispensed. Then continue pressing to the second stop (blow-out) to deliver the appropriate volume. From this point, the remainder of steps should be completed within about 15 seconds to prevent the inoculum from soaking into the agar.
4. Because the pipettor tip is contaminated with microbes, you must dispose of it correctly. Use the ejector button to remove the tip into an appropriate

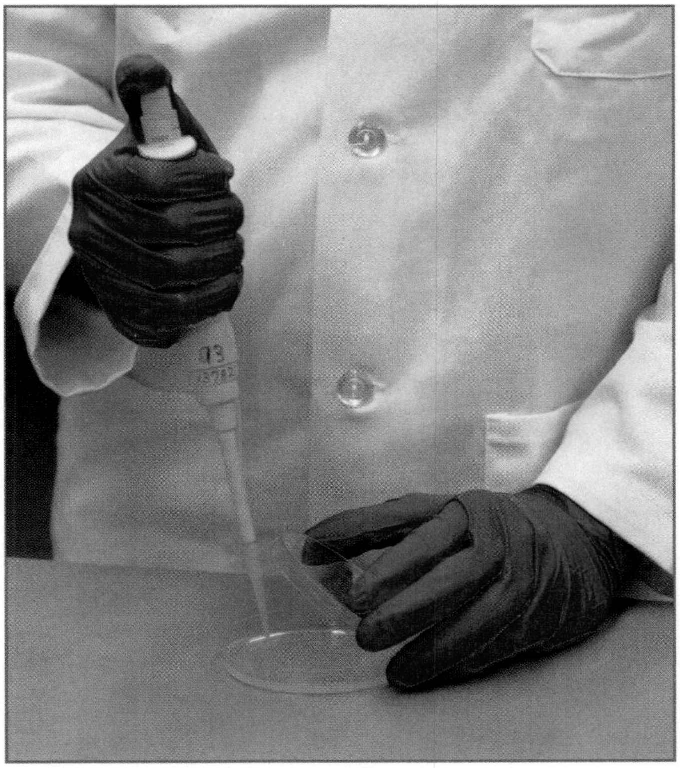

FIGURE D-7 DISPENSING TO A PLATE
Hold the pipette in a vertical position and press the control button to the first stop. When no more fluid comes out, continue pressing to the second stop. Slowly release pressure on the control button, then eject the tip into a biohazard container.

biohazard container (Figure D-6). Each lab has its own specific procedures, and your instructor will advise you what to do.

5. Continue with the Spread Plate Technique (Exercise 1-4).

Reference

Brinkman Instruments, *Eppendorf Series 2001 Pipette Instruction Manual*. One Cantiague Road, Westbury, New York, NY 11590-0207.

The Spectrophotometer

APPENDIX E

Theory The spectrophotometer (Figure E-1) is typically used in quantitative analysis; that is, to determine how much of something is in a sample in solution. In this manual, protocols in Exercises 2-10, 2-11, 2-12, 6-5, 7-2 and 9-1 use spectrophotometry to measure turbidity in liquid for a variety of purposes. Following is a brief introduction to the theory of spectrophotometry and instructions on using the spectrophotometer.

When light strikes a solution, any of several outcomes may occur. The light may be transmitted through it, it may be absorbed, it may be reflected, it may cause fluorescence, or it may be scattered. The spectrophotometer is designed to shine a beam of single wavelength light on a sample and measure **% Transmittance** (%T) or **absorbance** of that light. (Under the conditions in the spectrophotometer, reflection, fluorescence, and scatter are negligible.)

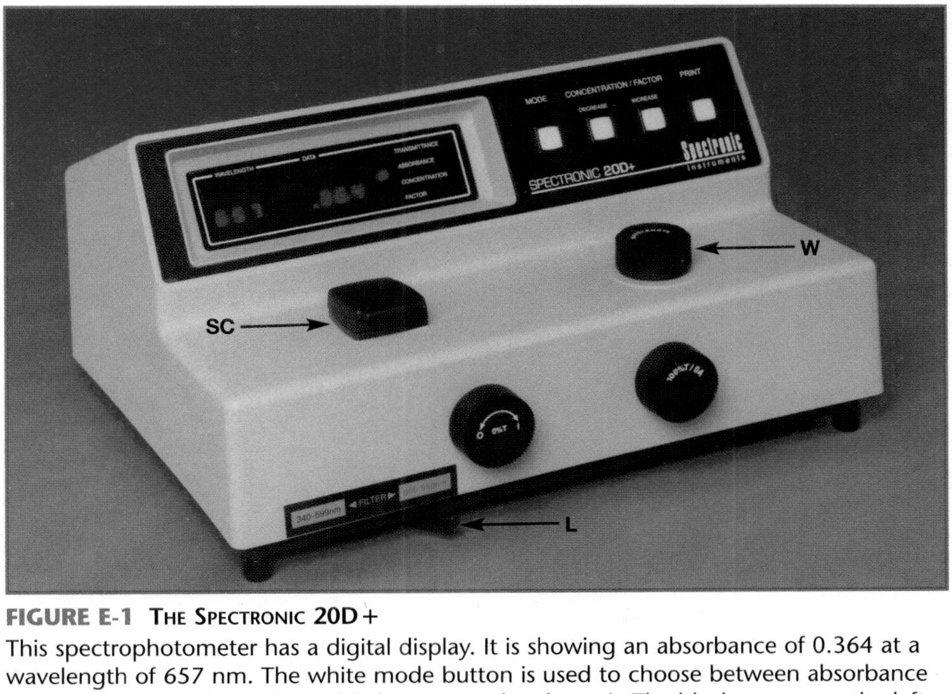

FIGURE E-1 The Spectronic 20D+

This spectrophotometer has a digital display. It is showing an absorbance of 0.364 at a wavelength of 657 nm. The white mode button is used to choose between absorbance and transmittance (note the red light next to absorbance). The black square on the left is the lid to the sample compartment (SC). The knob to its right is used to select wavelength (W). The lever (L) on the lower left positions and removes a filter for use with wavelengths of 600 to 950 nm. The knob on the front left is used to turn the machine on and off, as well as to set 0% Transmittance. The knob on the front right is used to set 100%T with a blank in the sample compartment.

(Photo courtesy of Thermo Spectronic, Rochester, NY)

When light is shined through an absorbing medium, each fraction of equal thickness absorbs an equal fraction of the light passing through it (Lambert's law). (For example, each 10% portion of the medium absorbs an equal amount.) Further, as long as the light doesn't change the physical or chemical state of the medium, the fraction of light transmitted by a given thickness is independent of the light's intensity. The fraction transmitted then may be expressed as

$$T = \frac{I}{I_o}$$

where I_o is the intensity of incident light and I is the intensity of the transmitted light. The percent transmittance is calculated by multiplying T by 100, as shown.

$$\%T = \frac{I}{I_o} \times 100$$

Plotting concentration of the absorbing medium against %T produces an exponential curve (Figure E-2A) that is cumbersome to use. Fortunately, concentration of a substance in a solution is directly proportional to the absorbance (A) of the solution (Beer-Lambert law), provided that the incident light is a single wavelength and that the light doesn't change the physical or chemical state of the medium. Figure E-2B illustrates the relationship between concentration and absorbance.

The mathematical relationship between absorbance and percent transmittance is as follows:

$$A = \log\frac{1}{T} = \log\frac{I_o}{I} = \log I_o - \log I$$

This equation may be modified to

$$A = 2 - \log\%T$$

because the spectrophotometer is calibrated such that the value of I_o is 100%. Most spectrophotometers allow reading of either %T or A rather than leaving the calculation up to the user.

A simplified schematic of how a spectrophotometer works is shown in Figure E-3. The beam of white light is broken up into its component colors by either a prism or a diffraction grating. The particular wavelength to be used is selected and allowed to pass through the exit slit. It then is directed through the sample contained in a glass or plastic cuvette with a 1 cm light path. The emerging transmitted light strikes a photodetector, which converts the light energy into an electrical signal proportional to the light intensity. This electrical signal is amplified and displayed by the spectrophotometer as either percent transmittance or absorbance.

In quantitative analysis, absorbance of the compound or particle being examined must be differentiated from absorbance because of the solvent in which it is dissolved and the cuvette itself. This is done by calibrating the spectrophotometer with an identical blank cuvette containing the solvent but lacking the compound or particle being measured.

Instructions for Use of the Spectronic D20+

1. Turn on the spectrophotometer and allow it to warm up for 15 minutes.
2. Choose the wavelength to be used by turning the wavelength control knob on the top of the machine.
3. Make sure the filter is set for the correct range of wavelength to be used.[1]

[1] On the older Spectronic 20 machines, a red filter will have to be manually inserted if wavelengths longer than 600 nm are used.

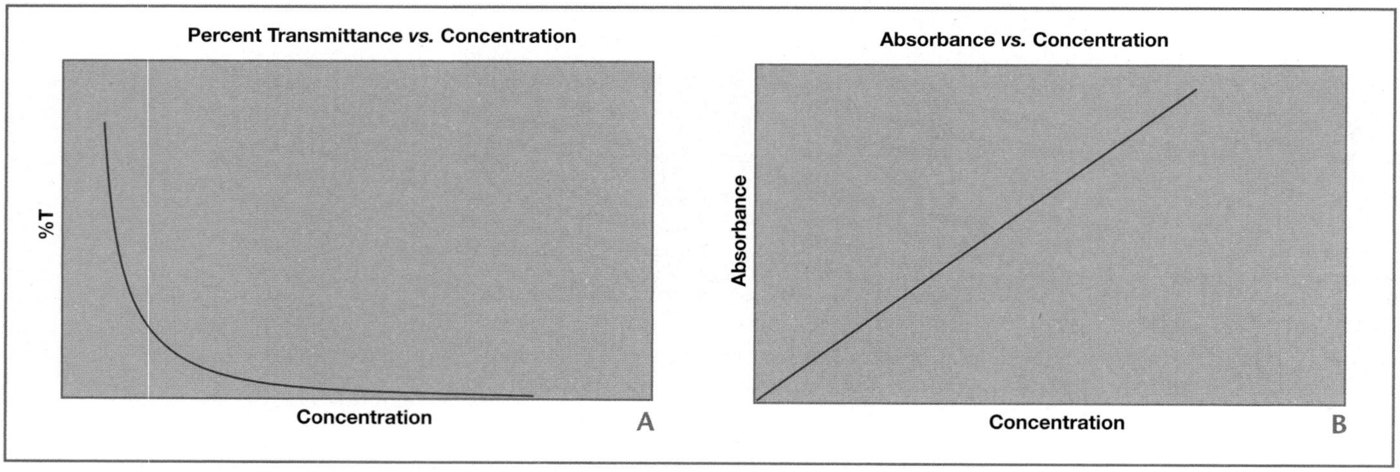

FIGURE E-2 TRANSMISSION AND ABSORBANCE AS A FUNCTION OF CONCENTRATION
(A) Transmittance varies exponentially with concentration. This makes using the graph for interpolation difficult, because many points are required to produce the standard curve. (B) Absorbance and concentration are linearly related, making production and interpretation of a standard curve easier.

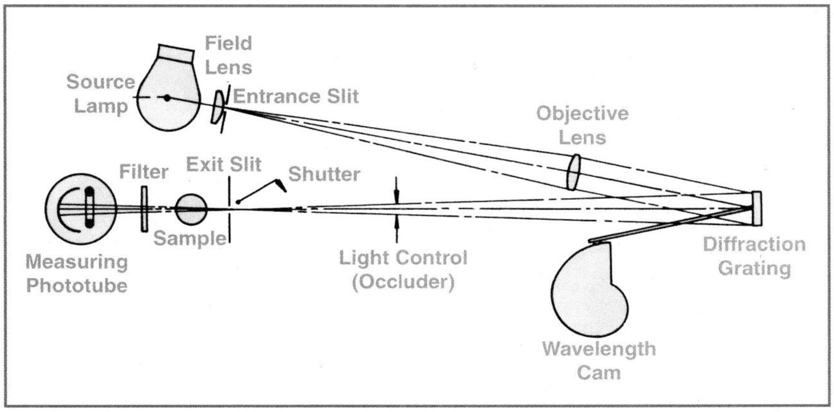

FIGURE E-3 A SCHEMATIC OF THE SPECTROPHOTOMETER
Light from the source lamp is focused on the diffraction grating by the field and objective lenses. The grating is responsible for breaking up the white light into its component wavelengths (colors). The desired wavelength is chosen by rotating the grating and aiming that wavelength through the exit slit. This incident light (I_o) then passes through the sample (with a light path of 1 cm) and into the phototube as transmitted light (I). The phototube then converts the transmitted light energy into an electrical signal that produces the readout on the display. Other components include the shutter, which closes when no sample is in place so the machine can be "zeroed," and the filter, which is used with wavelengths of 600 to 950 nm. (Diagram modified from Spectronic 20D+ literature, courtesy of Thermo Spectronic, Rochester, NY.)

4. Set the machine to read % Transmittance by pressing the mode button until the display indicates Transmittance.[2]
5. Prepare a "blank" cuvette identical to the experimental cuvette minus the particles you are measuring. (For instance, sterile medium if you are checking absorbance of bacterial cells.)
6. Use care to handle the cuvettes by the top only, as fingerprints influence results. Use tissue to remove fingerprints.
7. Use the power/zero control knob to set the machine to 0%T.
8. If measuring %T, go to step 9. If measuring A, go to step 10.
9. Be sure the blank is mixed uniformly, but do not introduce bubbles. Insert the blank into the sample compartment and close the cover. Use the Light Control knob to set 100 %T. Continue with step 11.
10. Select absorbance mode. Be sure the blank is mixed uniformly, but do not introduce bubbles. Then insert the blank into the sample compartment and close the cover. Use the Light Control knob to set zero A. Continue with step 11.
11. Remove the blank.
12. Be sure the experimental cuvette is mixed uniformly, but do not introduce bubbles. Insert it into the sample compartment and close the lid. Read the appropriate display, and record the value.
13. It is advisable to calibrate the spectrophotometer prior to each use if there is an interval of more than a few minutes between readings.
14. When finished with all readings, turn off the spectrophotometer.
15. *Note:* Your instructor must be notified immediately if you spill any culture in the spectrophotometer. You will be instructed on how to decontaminate the spill.

[2] On the Spectronic 20, the display has both %T and A scales.

References

Abramoff, Peter and Robert G. Thompson. 1982. Appendix A and Appendix B in *Laboratory Outlines in Biology—III.* W. H. Freeman and Co., San Francisco.

Thermo Spectronic. "Theory of UV-Visible Spectrophotometry," downloaded from http://www.thermo.com/eThermo/CDA/KnowledgeBase/Product_Knowledge_Base_Detail/0,1292,12100-119-10023,00.html. Thermo Spectronic, Rochester, NY.

Alernative Procedures for Section 6

APPENDIX F

Alternative Procedure for Exercise 6-1, Standard Plate Count

Materials

Per Student Group

- micropipettes (10–100 μL and 100–1000 μL) with sterile tips
- sterile microtubes
- flask of sterile water
- 8 Nutrient Agar plates
- beaker containing ethanol and a bent glass rod
- hand tally counter
- colony counter
- 24-hour broth culture of *Escherichia coli* (This culture will have between 10^7 and 10^{10} CFU/mL.)

Procedure

Refer to the procedural diagram in Figure 6-2 as needed. You will have to change the mL volumes to μL.

Lab One

1. Obtain eight plates, organize them into four pairs, and label them A_1, A_2, B_1, B_2, etc.
2. Obtain 5 microtubes and label them 1–5. These are your dilution tubes; make sure they remain covered until needed.
3. Aseptically add 990 μL sterile water to dilution tubes 1 and 2. Cover when finished. Aseptically add 900 μL sterile water to dilution tubes 3, 4, and 5. Cover when finished.
4. Mix the broth culture and aseptically transfer 10 μL to dilution tube 1; mix well. This is dilution factor 10^{-2} (DF 10^{-2}).
5. Aseptically transfer 10 μL from dilution tube 1 to dilution tube 2; mix well. This is DF 10^{-4}.
6. Aseptically transfer 100 μL from dilution tube 2 to dilution tube 3; mix well. This is DF 10^{-5}.

7. Aseptically transfer 100 µL from dilution tube 3 to dilution tube 4; mix well. This is DF 10^{-6}.
8. Aseptically transfer 100 µL from dilution tube 4 to dilution tube 5; mix well. This is DF 10^{-7}.
9. Aseptically transfer 100 µL from dilution tube 2 to plate A_1. Using the spread plate technique disperse the diluent evenly over the entire surface of the agar. Repeat the procedure with plate A_2 and label both plates "FDF 10^{-5}".
10. Following the same procedure, transfer 100 µL volumes from dilution tubes 3, 4, and 5 to plates B, C and D respectively. Label the plates with their appropriate FDF.
11. Invert the plates and incubate at 35 ± 2°C for 24 to 48 hours.

Lab Two

1. After incubation, examine the plates and determine the countable pair—plates with 30 to 300 colonies. Only one pair of plates should be countable.
2. Count the colonies on both plates and calculate the average (Figure 6-1). Record it in the chart provided on the Data Sheet. (***Note:*** Because of error, you may have more than one pair that is countable. Count all plates that have between 30 and 300 colonies for the practice, and try to identify which plate(s) you have the most confidence in. If no plates are in the 30–300 range, count the pair that is closest just for the practice and for purposes of the calculations.
3. Using the formula provided in Data and Calculations on the Data Sheet, calculate the density of the original sample and record it in the space provided.

Alternative Procedure for Exercise 6-4, Plaque Assay

Materials

Per Student Group

- micropipettes (10–100 µL and 100–1000 µL) with sterile tips
- 14 sterile capped microtubes (1.5 mL or larger)
- 7 nutrient agar plates
- 7 tubes containing 2.5 mL soft agar
- small flask of sterile water
- hot water bath set at 45°C to keep the soft agar liquefied
- T4 coliphage
- 24-hour broth culture of *Escherichia coli B* (T-series phage host)

Procedure

Refer to the procedural diagram in Figure 6-9 as needed. You will have to change the mL volumes to µL.

Lab One

1. Obtain all materials except for the soft agar tubes. To keep the agar tubes liquefied, leave them in the water bath and take them out one at a time as needed.
2. Label seven microtubes 1 through 7. Label the other seven microtubes *E. coli* 1–7. Place all tubes in a rack, pairing like-numbered tubes.
3. Label the Nutrient Agar plates A through G. Place them in the 35°C incubator to warm them. Take them out one at a time as needed. This will keep the soft agar (added at step 14) from solidifying too quickly and result in a smoother agar surface.
4. Aseptically transfer 990 µL sterile water to dilution tube 1.
5. Aseptically transfer 900 µL sterile water to dilution tubes 2–7.
6. Mix the *E. coli* culture and aseptically transfer 300 µL into each of the *E. coli* microtubes.
7. Mix the T4 suspension and aseptically transfer 10 µL to dilution tube 1. Mix well. This is DF (dilution factor) 10^{-2}.
8. Aseptically transfer 100 µL from dilution tube 1 to dilution tube 2. Mix well. This is DF 10^{-3}.
9. Aseptically transfer 100 µL from dilution tube 2 to dilution tube 3. Mix well. This is DF 10^{-4}.
10. Continue in this manner through dilution tube 7. The dilution factor of tube 7 should be 10^{-8}.
11. Aseptically transfer 100 µL sterile water to *E. coli* tube 1. This will be used to inoculate a control plate.
12. Aseptically transfer 100 µL from dilution tube 2 to its companion *E. coli* tube. Repeat this procedure with the remaining five tubes.
13. This is the beginning of the preadsorption period. Let all seven tubes stand undisturbed for 15 minutes.
14. Remove one soft agar tube from the hot water bath and add the contents of *E. coli* tube 1. Mix well and immediately pour onto plate A. Gently tilt the plate back and forth until the soft agar mixture is spread evenly across the solid medium. Label the plate "Control".
15. Remove a second soft agar tube from the water bath and add the contents of *E. coli* tube 2. Mix well and immediately pour onto plate B. Tilt back and forth to cover the agar and label it FDF 10^{-4}.

16. Repeat this procedure with dilutions 10^{-4} thru 10^{-8} and plates C thru G. Label the plates with the appropriate FDF.
17. Allow the agar to solidify completely.
18. Invert the plates and incubate aerobically at $35 \pm 2°C$ for 24 to 48 hours.

Lab Two

1. After incubation, examine the control plate for growth and the absence of plaques.
2. Examine the remainder of your plates and determine which one is countable (30 to 300 plaques). Count the plaques and record the number in the table provided. Record all others as either TMTC (to many to count) or TFTC (to few to count).
3. Using the FDF on the countable plate and the formula provided on the Data Sheet, calculate the original phage density. (Be sure you have converted the µL volumes to mL before calculating FDF.) Record your results in the space provided on the Data Sheet.

Alternative Procedure for Exercise 6-6, Thermal Death Time Versus Decimal Reduction Value

Materials

Per Class

- stopwatch
- thermometers
- 2 flasks each containing 49.0 mL Nutrient Broth, maintained at 60°C in a water bath
- 24-hour broth cultures of:
  - *Escherichia coli*
  - *Staphylococcus aureus*

Per Group

- 7 Nutrient Agar plates
- 16 Nutrient Broth tubes
- micropipettes
- sterile 100 µL micropipette tips
- sterile 1000 µL micropipette tips
- sterile 1.5 mL or larger microtubes
- sterile deionized water in a small flask or beaker
- beaker of alcohol and bent glass rod for spreading organisms

Procedure

Note: This procedure uses the Spread-Plate Technique described in Exercise 1-4. Refer to it as needed. The Plate Subgroups will require at least two or three students for labeling microtubes and plates, performing the serial dilution, and spreading plates. The Broth Subgroups will require one or two students to label and inoculate the 15 Nutrient Broths, keep track of time, and mix the broth prior to all transfers.

Lab One: Group 1 (T_0)

Plate Subgroup

1. Obtain all plating materials. Place seven microtubes in a rack. Label them 1 through 7. Add 900 µL of sterile water to microtubes 2 through 7, and cover them until needed.
2. Label seven Nutrient Agar plates 1 through 7, respectively.
3. When your instructor adds 1.0 mL of the *E. coli* culture to the heated flask, he/she will start the stopwatch. This is T_0. Immediately swirl the flask to disperse the sample, and remove 1000 µL with your pipette. If possible, do this step without removing the flask from the water bath. Begin the following serial dilution:
   - Add the 1000 µL of broth from the flask to tube 1.
   - Without changing the pipette tip, transfer 100 µL from tube 1 to tube 2; mix. Discard the pipette tip. Use a clean pipette tip for each transfer from here forward.
   - Transfer 100 µL from tube 2 to tube 3; mix.
   - Transfer 100 µL from tube 3 to tube 4; mix.
   - Transfer 100 µL from tube 4 to tube 5; mix.
   - Transfer 100 µL from tube 5 to tube 6; mix.
   - Transfer 100 µL from tube 6 to tube 7; mix.
   - Transfer 100 µL from tube 1 to plate 1 and spread according to spread-plate procedure.
   - Transfer 100 µL from tube 2 to plate 2 and spread.
   - Transfer 100 µL from tube 3 to plate 3 and spread.
   - Transfer 100 µL from tube 4 to plate 4 and spread.
   - Transfer 100 µL from tube 5 to plate 5 and spread.
   - Transfer 100 µL from tube 6 to plate 6 and spread.
   - Transfer 100 µL from tube 7 to plate 7 and spread.
4. Calculate the final dilution factor* for each plate and enter it in the first chart on the Data Sheet.

*Use of final dilution factors and the formula given in the Lab Two Procedure require knowledge of dilutions and calculations (explained fully in Exercise 6-1 Standard Plate Count.) If you haven't completed a standard plate count, or you have any doubt about your ability to complete this exercise, we suggest that you read Exercise 6-1 carefully and do some or all of the practice problems.

5. Tape the plates together, invert and incubate them at $35 \pm 2°C$ for 48 hours.

Broth Subgroup

1. Obtain all broth transfer materials. Label 15 Nutrient Broth tubes T_1 through T_{15} respectively. Label the 16th "Control."
2. When your instructor adds 1.0 mL of the *E. coli* culture to the heated flask, he/she will start the stopwatch. This is T_0. Slightly before T_1, gently swirl the flask to mix the broth. At exactly T_1, transfer 100μL from the flask to the appropriately labeled Nutrient Broth tube. Discard the pipette tip.
3. Repeat this procedure at T_2 through T_{15}. Be sure to mix the broth before each transfer, and use a clean pipette tip.
4. Incubate all broths at $35 \pm 2°C$ for 48 hours.

Lab One: Group 2 (T_{10})

Plate Subgroup

1. Obtain all plating materials. Place six microtubes in a rack. Label them 1 through 6. Add 900 μL of sterile water to microtubes 2 through 6, and cover them until needed.
2. Label seven Nutrient Agar plates 1 through 7, respectively.
3. When your instructor adds 1.0 mL of the *E. coli* culture to the heated flask, he/she will start the stopwatch. This is T_0. At T_{10} remove 1500 μL with your pipette. Unless you have 5000 μL pipettes, this will take two 750 μL withdrawals. In this case, wait for the broth group to withdraw its sample, then make your transfers as quickly as possible.
4. Begin the following serial dilution:
   - Add the 1500 μL of broth from the flask to tube 1.
   - Without changing the pipette tip, transfer 100 μL from tube 1 to tube 2; mix. Discard the pipette tip. Use a clean pipette tip for each transfer from here forward.
   - Transfer 100 μL from tube 2 to tube 3; mix.
   - Transfer 100 μL from tube 3 to tube 4; mix.
   - Transfer 100 μL from tube 4 to tube 5; mix.
   - Transfer 100 μL from tube 5 to tube 6; mix.
   - Transfer 100 μL from tube 6 to tube 7; mix.
   - Transfer 1000 μL from tube 1 to plate 1 and spread according to the spread-plate procedure.
   - Transfer 100 μL from tube 1 to plate 2 and spread.
   - Transfer 100 μL from tube 2 to plate 3 and spread.
   - Transfer 100 μL from tube 3 to plate 4 and spread.
   - Transfer 100 μL from tube 4 to plate 5 and spread.
   - Transfer 100 μL from tube 5 to plate 6 and spread.
   - Transfer 100 μL from tube 6 to plate 7 and spread.
5. Calculate the final dilution factor for each plate and enter it in the first chart on the Data Sheet.
6. Tape the plates together, invert, and incubate them at $35 \pm 2°C$ for 48 hours.

Broth Subgroup

1. Obtain all broth transfer materials. Label 15 Nutrient Broth tubes T_{16} through T_{30} respectively. Label the 16th "Control." Enter these numbers in the appropriate boxes in the second chart on the Data Sheet.
2. When your instructor adds 1.0 mL of the *E. coli* culture to the heated flask, he/she will start the stopwatch. This is T_0. Slightly before T_{16}, gently swirl the broth flask, and at exactly T_{16}, transfer 100 μL from the flask to the appropriately labeled Nutrient Broth tube. Discard the pipette tip.
3. Repeat this procedure at T_{17} through T_{30}. Be sure to mix the broth before each transfer, and use a clean pipette tip.
4. Incubate all broths at $35 \pm 2°C$ for 48 hours.

Lab One: Group 3 (T_0)

Plate Subgroup

Using the *S. aureus* culture, follow the instructions for the Group 1 Plate Subgroup.

Broth Subgroup

Using the *S. aureus* culture, follow the instructions for the Group 1 Broth Subgroup.

Lab One: Group 4 (T_{10})

Plate Subgroup

Using the *S. aureus* culture, follow the instructions for the Group 2 Plate Subgroup.

Broth Subgroup

Using the *S. aureus* culture, follow the instructions for the Group 2 Broth Subgroup

Lab Two: All Groups

1. Remove all broths and plates from the incubator.
2. Pick the plate containing between 30 and 300 colonies. There should be only one countable plate. All others are either TFTC (too few to count) or TMTC (too many to count). Label them as such, and enter this information in Chart 1 on the Data Sheet.

3. If you haven't done so already, count the colonies on the countable plate and enter the number in Chart 1 on the Data Sheet.
4. Examine the broths for growth, comparing each one to the uninoculated control (tube #16). Any turbidity is read as positive for growth. Enter your results in Chart 2 on the Data Sheet. Circle the earliest time that no turbidity appears (i.e., growth did not occur).
5. Calculate the population density of your group's organism for your specified time (T_0 or T_{10}), using the following formula: (***Note:*** For best results, convert the volume plated to mL first, then calculate the Final Dilution Factor.)

$$\text{Cell density (cells/mL)} = \frac{\text{\# Colonies counted}}{\text{Final dilution factor}}$$

6. Enter your results in Chart 1 and again on Chart 3 on the Data Sheet. Also, be sure to convert your results to logarithmic form and enter that in the third chart as well.
7. Collect the data from other groups and enter it in Chart 3 on the Data Sheet. Your instructor will likely provide a transparency or chalkboard space for class data.
8. Follow the directions on the Data Sheet to plot *and* calculate the D values of both organisms.
9. Answer the questions on the Data Sheet.

Media, Reagent, and Stain Recipes

APPENDIX G

Media

Ames Test: Complete Medium

• Beef extract	3.0 g
• Peptone	5.0 g
• Sodium chloride	5.0 g
• Agar	20.0 g
• Distilled or deionized water	1.0 L

1. Suspend the dry ingredients in the water, mix well, and boil until completely dissolved.
2. Cover loosely and sterilize in the autoclave at 15 lbs. pressure (121°C) for 15 minutes.
3. Remove from the autoclave and cool slightly.
4. Aseptically pour into sterile Petri dishes (20 mL/plate) and cool to room temperature.

Ames Test: Minimal Medium

• Dextrose (glucose)	20.0 g
• 50x Vogel-Bonner salts	20.0 mL
• Histidine	0.00016 g
• Biotin	0.00025 g
• Agar	20.0 g
• Distilled or deionized water	1.0 L

1. Prior to preparation of the medium:
   a. Add 1.6 mg histidine to 10.0 mL distilled or deionized water and filter sterilize.
   b. Add 2.5 mg biotin to 10.0 mL distilled or deionized water and filter sterilize.
2. Prepare the 50x Vogel-Bonner salts solution by adding the ingredients to just enough water to dissolve them while heating and stirring. After the ingredients are dissolved, add enough water to bring the total volume up to exactly 1 liter.
3. Suspend, mix, and boil the agar in 500.0 mL of the water until completely dissolved.
4. Suspend and mix the dextrose in the remaining 500.0 mL of water until completely dissolved.

5. Cover the agar and dextrose containers loosely and sterilize in the autoclave at 121°C for 15 minutes.
6. Remove from the autoclave and allow to cool to 80°C.
7. Aseptically add 1.0 mL histidine solution, 1.0 mL biotin solution, and 20 mL 50x Vogel-Bonner salts to the glucose solution. Mix well.
8. Add the glucose solution to the agar solution, mix well and aseptically pour into sterile Petri dishes (20 mL/plate). Allow the medium to cool to room temperature.

Bile Esculin Agar

• Beef extract	3.0 g
• Peptone	5.0 g
• Oxgall	40.0 g
• Esculin	1.0 g
• Ferric citrate	0.5 g
• Agar	15.0 g
• Distilled or deionized water	1.0 L

pH 6.4–6.8 at 25°C

1. Suspend the dry ingredients in the water, mix well, and boil until completely dissolved.
2. Dispense 7.0 mL volumes into test tubes and cap loosely.
3. Sterilize in the autoclave at 15 lbs. pressure (121°C) for 15 minutes.
4. Remove from the autoclave, slant, and allow the medium to cool to room temperature.

Blood Agar

• Infusion from beef heart (solids)	2.0 g
• Pancreatic digest of casein	13.0 g
• Sodium chloride	5.0 g
• Yeast extract	5.0 g
• Agar	15.0 g
• Defibrinated sheep blood	50.0 mL
• Distilled or deionized water	1.0 L

pH 7.1–7.5 at 25°C

1. Suspend the dry ingredients in the water, mix well, and boil until completely dissolved. This is Blood Agar base.
2. Cover loosely and sterilize in the autoclave at 15 lbs. pressure (121°C) for 15 minutes.
3. Remove from the autoclave and cool to 45°C.
4. Aseptically add the sterile, room-temperature sheep blood to the blood agar base and mix well.
5. Pour into sterile Petri dishes and allow the medium to cool to room temperature.

Brilliant Green Lactose Bile Broth

• Peptone	10.0 g
• Lactose	10.0 g
• Oxgall	20.0 g
• Brilliant green dye	0.0133 g
• Distilled or deionized water	1.0 L

pH 7.0–7.4 at 25°C

1. Suspend the dry ingredients in the water, mix well, and boil until completely dissolved.
2. Dispense 10.0 mL portions into test tubes.
3. Place an inverted Durham tube in each broth and cap loosely.
4. Sterilize in the autoclave at 15 lbs. pressure (121°C) for 15 minutes.
5. Remove the medium from the autoclave and allow it to cool before inoculating.

Citrate Agar (Simmons)

• Ammonium dihydrogen phosphate	1.0 g
• Dipotassium phosphate	1.0 g
• Sodium chloride	5.0 g
• Sodium citrate	2.0 g
• Magnesium sulfate	0.2 g
• Agar	15.0 g
• Bromthymol blue	0.08 g
• Distilled or deionized water	1.0 L

pH 6.7–7.1 at 25°C

1. Suspend the dry ingredients in the water, mix well, and boil until completely dissolved.
2. Dispense 7.0 mL portions into test tubes and cap loosely.
3. Sterilize in the autoclave at 15 lbs. pressure (121°C) for 15 minutes.
4. Remove from the autoclave, slant, and cool to room temperature.

Complete Medium (*See* Ames Test)

Decarboxylase Medium (Møller)

• Peptone	5.0 g
• Beef extract	5.0 g
• Glucose (dextrose)	0.5 g
• Bromcresol purple	0.01 g
• Cresol red	0.005 g
• Pyridoxal	0.005 g
• L-Lysine, L-Ornithine, or L-Arginine	10.0 g
• Distilled or deionized water	1.0 L

pH 5.8–6.2 at 25°C

1. Suspend the dry ingredients in the water, mix well, and boil until completely dissolved. (Use only one of the listed L-amino acids.)

2. Adjust pH by adding NaOH if necessary.
3. Dispense 7.0 mL volumes into test tubes and cap.
4. Sterilize in the autoclave at 15 lbs. pressure (121°C) for 10 minutes.
5. Remove from the autoclave and cool to room temperature.

Desoxycholate Agar (Modified Leifson)

• Peptone	10.0 g
• Lactose	10.0 g
• Sodium desoxycholate	1.0 g
• Sodium chloride	5.0 g
• Dipotassium phosphate	2.0 g
• Ferric citrate	1.0 g
• Sodium citrate	1.0 g
• Agar	16.0 g
• Neutral red	0.033 g
• Distilled or deionized water	1.0 L

pH 7.1–7.5 at 25°C

1. Suspend the dry ingredients in the water, mix well, and boil until completely dissolved.
2. When cooled to 50°C, pour into sterile plates.
3. Allow the medium to cool to room temperature.

DNase Test Agar with Methyl Green

• Tryptose	20.0 g
• Deoxyribonucleic acid	2.0 g
• Sodium chloride	5.0 g
• Agar	15.0 g
• Methyl green	0.05 g
• Distilled or deionized water	1.0 L

pH 7.1–7.5 at 25°C

1. Suspend the dry ingredients in the water, mix well, and boil until completely dissolved.
2. Cover loosely and sterilize in the autoclave at 15 lbs. pressure (121°C) for 15 minutes.
3. Aseptically pour into sterile Petri dishes (20 mL/plate) and cool to room temperature.

EC Broth

• Tryptose	20.0 g
• Lactose	5.0 g
• Dipotassium phosphate	4.0 g
• Monopotassium phosphate	1.5 g
• Sodium chloride	5.0 g
• Distilled or deionized water	1.0 L

pH 6.7–7.1 at 25°C

1. Suspend the dry ingredients in the water, and mix well until completely dissolved.
2. Dispense 10.0 mL portions into test tubes.
3. Place an inverted Durham tube in each broth and cap loosely.
4. Sterilize in the autoclave at 15 lbs. pressure (121°C) for 15 minutes.
5. Remove the media from the autoclave and allow it to cool before inoculating.

Endo Agar

• Peptone	10.0 g
• Lactose	10.0 g
• Dipotassium phosphate	3.5 g
• Agar	15.0 g
• Basic Fuchsin	0.5 g
• Sodium Sulfite	2.5 g
• Distilled or deionized water	1.0 L

pH 7.3–7.7 at 25°C

1. Suspend the dry ingredients in the water, mix well, and boil until completely dissolved.
2. Autoclave for 15 minutes at 15 lbs. pressure (121°C).
3. When cooled to 50°C, pour into sterile plates.
4. Allow the medium to cool to room temperature.

Enriched TSA (*See* Tryptic Soy Agar)

Eosin Methylene Blue Agar (Levine)

• Peptone	10.0 g
• Lactose	10.0 g*
• Dipotassium phosphate	2.0 g
• Agar	15.0 g
• Eosin Y	0.4 g
• Methylene blue	0.065 g
• Distilled or deionized water	1.0 L

pH 6.9–7.3 at 25°C

1. Suspend the dry ingredients in the water, mix well, and boil until completely dissolved.
2. Autoclave for 15 minutes at 15 lbs. pressure (121°C).
3. When cooled to 50°C, pour into sterile plates.
4. Allow the medium to cool to room temperature.

Glucose Broth

• Peptone	10.0 g
• Glucose	5.0 g
• NaCl	5.0 g
• Distilled or deionized water	1.0 L

1. Suspend the dry ingredients the water. Agitate and heat slightly (if necessary) to dissolve completely.

*An alternative recipe replaces the 10.0 g of lactose with 5.0 g of lactose and 5.0 g of sucrose.

2. Dispense 7.0 mL portions into test tubes and cap loosely.
3. Autoclave for 15 minutes at 121°C to sterilize the medium.

Glucose Salts Medium

• Glucose	5.0 g
• NaCl	5.0 g
• $MgSO_4$	0.2 g
• $(NH_4)H_2PO_4$	1.0 g
• K_2HPO_4	1.0 g
• Distilled or deionized water	1.0 L

1. Suspend the dry ingredients in the water. Agitate and heat slightly (if necessary) to dissolve completely.
2. Dispense 7.0 mL portions into test tubes and cap loosely.
3. Autoclave for 15 minutes at 121°C to sterilize the medium.

Glycerol Yeast Extract Agar

• Glycerol	5.0 mL
• Yeast extract	2.0 g
• Dipotassium phosphate	1.0 g
• Agar	15.0 g
• Distilled or deionized water	1.0 L

1. Suspend the dry ingredients in the water, mix well, and boil until completely dissolved.
2. Autoclave for 15 minutes at 15 lbs. pressure (121°C).
3. When cooled to 50°C, pour into sterile plates.
4. Allow the medium to cool to room temperature.

Hektoen Enteric Agar

• Yeast extract	3.0 g
• Peptic digest of animal tissue	12.0 g
• Lactose	12.0 g
• Sucrose	12.0 g
• Salicin	2.0 g
• Bile salts	9.0 g
• Sodium chloride	5.0 g
• Sodium thiosulfate	5.0 g
• Ferric ammonium citrate	1.5 g
• Bromthymol blue	0.064 g
• Acid fuchsin	0.1 g
• Agar	13.5 g
• Distilled or deionized water	1.0 L

pH 7.4–7.8 at 25°C

1. Suspend the dry ingredients in the water, mix well, and boil until completely dissolved.
2. Do not autoclave.
3. When cooled to 50°C, pour into sterile plates.
4. Cool to room temperature with lids slightly open.

Kligler's Iron Agar

• Beef extract	3.0 g
• Yeast extract	3.0 g
• Peptone	15.0 g
• Proteose peptone	5.0 g
• Lactose	10.0 g
• Dextrose (glucose)	1.0 g
• Ferrous sulfate	0.2 g
• Sodium chloride	5.0 g
• Sodium thiosulfate	0.3 g
• Agar	12.0 g
• Phenol red	0.024 g
• Distilled or deionized water	1.0 L

pH 7.2–7.6 at 25°C

1. Suspend the dry ingredients in the water, mix well, and boil until completely dissolved.
2. Transfer 7.0 mL portions to test tubes and cap loosely.
3. Sterilize in the autoclave at 15 lbs. pressure (121°C) for 15 minutes.
4. Remove from the autoclave and slant in such a way as to form a deep butt.
5. Allow the medium to cool to room temperature.

Lauryl Tryptose Broth

• Tryptose	20.0 g
• Lactose	5.0 g
• Dipotassium phosphate	2.75 g
• Monopotassium phosphate	2.75 g
• Sodium chloride	5.0 g
• Sodium lauryl sulfate	0.1 g
• Distilled or deionized water	1.0 L

pH 6.6–7.0 at 25°C

1. Suspend the dry ingredients in the water until completely dissolved. Heat slightly if necessary.
2. Dispense 10.0 mL portions into test tubes.
3. Place an inverted Durham tube in each broth and cap loosely.
4. Sterilize in the autoclave at 15 lbs. pressure (121°C) for 15 minutes.
5. Remove the medium from the autoclave and allow it to cool before inoculating.

Litmus Milk Medium

• Skim milk	100.0 g
• Azolitmin	0.5 g

• Sodium sulfite	0.5 g
• Distilled or deionized water	1.0 L

pH 6.3–6.7 at 25°C

1. Suspend and mix the ingredients in the water and heat to approximately 50°C to dissolve completely.
2. Transfer 7.0 mL portions to test tubes and cap loosely.
3. Sterilize in the autoclave at 113–115°C for 20 minutes.
4. Remove from the autoclave and allow the medium to cool to room temperature.

Luria-Bertani Agar

• Tryptone	10.0 g
• Yeast extract	5.0 g
• NaCl	10.0 g
• Agar	15.0 g
• Distilled or deionized water	1.0 L

pH 7.4 at 25°C

1. Suspend the dry ingredients in the water, mix well, and boil until completely dissolved.
2. Cover loosely and sterilize in the autoclave at 15 lbs. pressure (121°C) for 15 minutes.
3. Aseptically pour into sterile Petri dishes (20 mL/plate) and cool to room temperature.

Luria-Bertani Broth

• Tryptone	10.0 g
• Yeast extract	5.0 g
• NaCl	10.0 g
• Distilled or deionized water	1.0 L

pH 7.4 at 25°C

1. Suspend the ingredients in one liter of distilled or deionized water. Agitate and heat slightly (if necessary) to dissolve completely.
2. Dispense 7.0 mL portions into test tubes and cap loosely.
3. Autoclave for 15 minutes at 121°C to sterilize the medium.

Lysine Iron Agar

• Peptone	5.0 g
• Yeast extract	3.0 g
• Dextrose	1.0 g
• L-Lysine hydrochloride	10.0 g
• Ferric ammonium citrate	0.5 g
• Sodium thiosulfate	0.04 g
• Bromcresol purple	0.02 g
• Agar	15.0 g

pH 6.5–6.9 at 25°C

1. Suspend the dry ingredients in the water, mix well, and boil until completely dissolved.
2. Transfer 8.0 mL portions to test tubes and cap loosely.
3. Sterilize in the autoclave at 121°C for 15 minutes.
4. Remove from the autoclave and slant in such a way as to produce a deep butt. Allow the medium to cool to room temperature.

MacConkey Agar

• Pancreatic digest of gelatin	17.0 g
• Pancreatic digest of casein	1.5 g
• Peptic digest of animal tissue	1.5 g
• Lactose	10.0 g
• Bile salts	1.5 g
• Sodium chloride	5.0 g
• Neutral red	0.03 g
• Crystal violet	0.001 g
• Agar	13.5 g
• Distilled or deionized water	1.0 L

pH 6.9–7.3 at 25°C

1. Suspend the dry ingredients in the water, mix well, and boil until completely dissolved.
2. Autoclave for 15 minutes at 15 lbs. pressure (121°C).
3. When cooled to 50°C, pour into sterile plates.
4. Allow the medium to cool to room temperature.

Malonate Broth

• Yeast extract	1.0 g
• Ammonium sulfate	2.0 g
• Dipotassium phosphate	0.6 g
• Monopotassium phosphate	0.4 g
• Sodium chloride	2.0 g
• Sodium malonate	3.0 g
• Dextrose	0.25 g
• Bromthymol blue	0.025 g
• Distilled or deionized water	1.0 L

pH 6.5–6.9 at 25°C

1. Suspend the dry ingredients in the water. Agitate and heat slightly (if necessary) to dissolve completely.
2. Dispense 7.0 mL portions into test tubes and cap loosely.
3. Autoclave for 15 minutes at 121°C to sterilize the medium.

Mannitol Salt Agar

• Beef extract	1.0 g
• Peptone	10.0 g
• Sodium chloride	75.0 g
• D-Mannitol	10.0 g

- Phenol red 0.025 g
- Agar 15.0 g
- Distilled or deionized water 1.0 L

pH 7.2–7.6 at 25°C

1. Suspend the dry ingredients in the water, mix well, and boil until completely dissolved.
2. Autoclave for 15 minutes at 15 lbs. pressure (121°C).
3. When cooled to 50°C, pour into sterile plates.
4. Allow the medium to cool to room temperature.

Milk Agar

- Beef extract 3.0 g
- Peptone 5.0 g
- Agar 15.0 g
- Powdered nonfat milk 100.0 g
- Distilled or deionized water 1.0 L

pH 7.0–7.4 at 25°C

1. Suspend the powdered milk in 500.0 mL of water in a 1-liter flask, mix well, and cover loosely.
2. Suspend the remainder of the ingredients in 500.0 mL of water in a 1-liter flask, mix well, boil to dissolve completely, and cover loosely.
3. Sterilize in the autoclave at 113–115°C for 20 minutes.
4. Remove from the autoclave and allow the mixtures to cool slightly.
5. Aseptically pour the agar solution into the milk solution. Mix *gently* (to prevent foaming).
6. Aseptically pour into sterile Petri dishes (15 mL/plate).
7. Allow the medium to cool to room temperature.

Minimal Medium (*See* Ames Test)

Motility Test Medium

- Beef extract 3.0 g
- Pancreatic digest of gelatin 10.0 g
- Sodium chloride 5.0 g
- Agar 4.0 g
- Triphenyltetrazolium chloride (TTC) 0.05 g
- Distilled or deionized water 1.0 L

pH 7.1–7.5 at 25°C

1. Suspend the dry ingredients in the water, mix well, and boil until completely dissolved.
2. Dispense 7.0 mL portions into test tubes and cap loosely.
3. Sterilize in the autoclave at 15 lbs. pressure (121°C) for 15 minutes.
4. Remove from the autoclave and allow it to cool in the upright position.

MRVP Broth

- Buffered peptone 7.0 g
- Dipotassium phosphate 5.0 g
- Dextrose (glucose) 5.0 g
- Distilled or deionized water 1.0 L

pH 6.7–7.1 at 25°C

1. Suspend the dry ingredients in the water, mix well, and warm until completely dissolved.
2. Transfer 7.0 mL portions to test tubes and cap loosely.
3. Sterilize in the autoclave at 15 lbs. pressure (121°C) for 15 minutes.
4. Remove from the autoclave and allow the medium to cool to room temperature.

Mueller-Hinton II Agar

- Beef extract 2.0 g
- Acid hydrolysate of casein 17.5 g
- Starch 1.5 g
- Agar 17.0 g
- Distilled or deionized water 1.0 L

pH 7.1–7.5 at 25°C

1. Suspend the dry ingredients in the water, mix well, and boil until completely dissolved.
2. Cover loosely and sterilize in the autoclave at 121°C (15 lbs.) for 15 minutes.
3. Remove from the autoclave, allow to cool slightly.
4. Aseptically pour into sterile Petri dishes to a depth of 4 mm.
5. Allow the medium to cool to room temperature.

Nitrate Broth

- Beef extract 3.0 g
- Peptone 5.0 g
- Potassium nitrate 1.0 g
- Distilled or deionized water 1.0 L

pH 6.6–7.0 at 25°C

1. Suspend the ingredients in the water; mix well, and warm until completely dissolved.
2. Transfer 7.0 mL portions to test tubes, and cap loosely. (Add inverted Durham tubes before capping, if desired.)
3. Sterilize in the autoclave at 15 lbs. pressure (121°C) for 15 minutes.
4. Remove from the autoclave and allow the medium to cool to room temperature.

Nutrient Agar

• Beef extract	3.0 g
• Peptone	5.0 g
• Agar	15.0 g
• Distilled or deionized water	1.0 L

pH 6.6–7.0 at 25°C

Plates

1. Suspend the dry ingredients in the water, mix well, and boil until completely dissolved.
2. Cover loosely and sterilize in the autoclave at 15 lbs. pressure (121°C) for 15 minutes.
3. Remove from the autoclave, allow to cool slightly, and aseptically pour into sterile Petri dishes (20 mL/plate).
4. Allow the medium to cool to room temperature.

Tubes

1. Suspend the dry ingredients in the water, mix well, and boil until completely dissolved.
2. Dispense 7 mL or 10 mL portions into test tubes, and cap loosely.
3. Autoclave for 15 minutes at 121°C to sterilize the medium.
4. Cool to room temperature with the tubes in an upright position (10 mL) for agar deep tubes. Cool with the tubes on an angle (7 mL) for agar slants.

Nutrient Broth

• Beef extract	3.0 g
• Peptone	5.0 g
• Distilled or deionized water	1.0 L

pH 6.6–7.0 at 25°C

1. Suspend the dry ingredients in the water. Agitate and heat slightly (if necessary) to dissolve completely.
2. Dispense 7.0 mL portions into test tubes, and cap loosely.
3. Autoclave for 15 minutes at 121°C to sterilize the medium.

Nutrient Gelatin

• Beef extract	3.0 g
• Peptone	5.0 g
• Gelatin	120.0 g
• Distilled or deionized water	1.0 L

pH 6.6–7.0 at 25°C

1. *Slowly* add the dry ingredients to the water while stirring.
2. Warm to >50°C and maintain temperature until completely dissolved.
3. Dispense 7.0 mL volumes into test tubes and cap loosely.
4. Sterilize in the autoclave at 15 lbs. pressure (121°C) for 15 minutes.
5. Remove from the autoclave immediately and allow the medium to cool to room temperature in the upright position.

O-F Basal Medium (*see* O-F Carbohydrate Solution)

• Pancreatic digest of casein	2.0 g
• Sodium chloride	5.0 g
• Dipotassium phosphate	0.3 g
• Agar	2.5 g
• Bromthymol blue	0.03 g
• Distilled or deionized water	1.0 L

pH 6.6–7.0 at 25°C

O-F Carbohydrate Solution (*see* O-F Basal Medium)

• Carbohydrate (glucose, lactose, sucrose)	1.0 g
• Distilled or deionized water to total	10.0 mL

1. Suspend the dry ingredients, *without the carbohydrate*, in the water, mix well, and boil to dissolve completely. This is basal medium.
2. Divide the medium into ten aliquots of 100.0 mL each.
3. Cover loosely and sterilize in the autoclave at 121°C for 15 minutes.
4. Prepare carbohydrate solution, cover loosely, and autoclave at 118°C for 10 minutes.
5. Allow both solutions to cool to 50°C.
6. Aseptically add 10.0 mL sterile carbohydrate solution to a basal medium aliquot, and mix well.
7. Aseptically transfer 7.0 mL volumes to sterile test tubes, and allow the medium to cool to room temperature.

Phenol Red (Carbohydrate) Broth

• Pancreatic digest of casein	10.0 g
• Sodium chloride	5.0 g
• Carbohydrate (glucose, lactose, sucrose)	5.0 g
• Phenol red	0.018 g
• Distilled or deionized water	1.0 L

pH 7.1 ± 7.5 at 25°C

1. Suspend the dry ingredients in the water; mix well, and warm slightly to dissolve completely.
2. Dispense 7.0 mL volumes into test tubes.

3. Insert inverted Durham tubes into the test tubes and cap loosely.
4. Sterilize in the autoclave at 116–118°C for 15 minutes.
5. Remove from the autoclave and allow the medium to cool to room temperature.

Phenylalanine Deaminase Agar

• DL-Phenylalanine	2.0 g
• Yeast extract	3.0 g
• Sodium chloride	5.0 g
• Sodium phosphate	1.0 g
• Agar	12.0 g
• Distilled or deionized water	1.0 L

pH 7.1–7.5 at 25°C

1. Suspend the dry ingredients in the water, mix well, and boil until completely dissolved.
2. Dispense 7.0 mL volumes into test tubes and cap loosely.
3. Sterilize in the autoclave at 15 lbs. pressure (121°C) for 10 minutes.
4. Remove from the autoclave, slant, and allow the medium to cool to room temperature.

Phenylethyl Alcohol Agar

• Tryptose	10.0 g
• Beef extract	3.0 g
• Sodium chloride	5.0 g
• Phenylethyl alcohol	2.5 g
• Agar	15.0 g
• Distilled or deionized water	1.0 L

pH 7.1–7.5 at 25°C

1. Suspend the dry ingredients in the water, mix well, and boil until completely dissolved.
2. Autoclave for 15 minutes at 15 lbs. pressure (121°C).
3. When cooled to 50°C, pour into sterile plates.
4. Allow the medium to cool to room temperature.

Photobacterium Broth

• Tryptone	5.0 g
• Yeast extract	2.5 g
• Ammonium chloride	0.3 g
• Magnesium sulfate	0.3 g
• Ferric chloride	0.01 g
• Calcium carbonate	1.0 g
• Monopotassium phosphate	3.0 g
• Sodium glycerol phosphate	23.5 g
• Sodium chloride	30.0 g
• Distilled or deionized water	1.0 L

pH 6.8–7.2 at 25°C

1. Suspend the dry ingredients in the water, mix well, and boil until completely dissolved.
2. Dispense into flasks to form a shallow layer.
3. Autoclave for 15 minutes at 15 lbs. pressure (121°C).
4. Allow the medium to cool to room temperature.

Potato Dextrose Agar

• Potato flakes	20.0 g
• Dextrose	10.0 g
• Agar	15.0 g
• Distilled or deionized water	1.0 L

1. Suspend the dry ingredients in the water, mix well, and boil until completely dissolved.
2. Autoclave for 15 minutes at 15 lbs. pressure (121°C).
3. Remove the agar mixture from the autoclave and cool to 50°C.
4. Mix and pour into sterile Petri dishes. ***Note:*** Swirl the flask frequently to keep the flakes uniformly distributed.

Purple Broth

• Peptone	10.0 g
• Beef extract	1.0 g
• Sodium chloride	5.0 g
• Bromcresol purple	0.02 g
• Carbohydrate (glucose, lactose, or sucrose)	10.0 g
• Distilled or deionized water	1.0 L

pH 6.6–7.0 at 25°C

1. Suspend the dry ingredients in the water, mix well, and warm slightly to dissolve completely.
2. Dispense 9.0 mL volumes into test tubes.
3. Insert inverted Durham tubes into the test tubes and cap loosely.
4. Sterilize in the autoclave at 118°C for 15 minutes.
5. Remove from the autoclave and allow the medium to cool to room temperature.

Sabouraud Dextrose Agar

(with antibiotics added to inhibit bacterial growth)

• Peptone	10.0 g
• Dextrose	40.0 g
• Agar	15.0 g
• Penicillin*	20000.0 units
• Streptomycin*	0.00004 g

*To obtain the desired proportions of antibiotics in the medium, prepare as follows:

1. Dissolve 100,000 units penicillin in 10 mL sterile distilled or deionized water. Add 2 mL to 1.0 liter of agar medium.
2. Dissolve 1.0 g streptomycin in 10 mL sterile distilled or deionized water. Add 1.0 mL of this mixture to 9.0 mL sterile distilled or deionized water. Add 4 mL of this diluted mixture to 1.0 liter of agar medium.

- Distilled or deionized water 1.0 L

pH 5.2–5.6 at 25°C

1. Suspend the peptone, dextrosen, and agar in the water, mix well, and boil until completely dissolved.
2. Autoclave for 15 minutes at 15 lbs. pressure (121°C).
3. Remove the agar mixture from the autoclave and cool to 50°C.
4. Aseptically add antibiotics. Mix and pour into sterile Petri dishes.
5. Allow the medium to cool to room temperature.

Saline Agar: Double Gel Immunodiffusion

- Sodium chloride 10.0 g
- Agar 20.0 g
- Distilled or deionized water 1.0 L

1. Suspend the dry ingredients in the water, mix well, and boil until completely dissolved.
2. Pour into Petri dishes to a depth of 3 mm. Do not replace the lids until the agar has solidified and cooled to room temperature.

Seawater Agar

- Peptone 5.0 g
- Yeast extract 5.0 g
- Beef extract 3.0 g
- Agar 15.0 g
- Seawater (synthetic) 1.0 L

1. Suspend the ingredients in the seawater, mix well, and boil to dissolve completely.
2. Sterilize in the autoclave at 15 lbs. pressure (121°C) for 15 minutes.
3. When cooled to 50°C, pour into sterile plates.
4. Allow the medium to cool to room temperature.

SIM (Sulfur-Indole-Motility) Medium

- Pancreatic digest of casein 20.0 g
- Peptic digest of animal tissue 6.1 g
- Ferrous ammonium sulfate 0.2 g
- Sodium thiosulfate 0.2 g
- Agar 3.5 g
- Distilled or deionized water 1.0 L

pH 7.1–7.5 at 25°C

1. Suspend the dry ingredients in the water, mix well, and boil until completely dissolved.
2. Dispense 7.0 mL volumes into test tubes and cap loosely.
3. Sterilize in the autoclave at 15 lbs. pressure (121°C) for 15 minutes.
4. Remove from the autoclave and allow the medium to cool to room temperature.

Snyder Test Medium

- Pancreatic digest of casein 13.5 g
- Yeast extract 6.5 g
- Dextrose 20.0 g
- Sodium chloride 5.0 g
- Agar 16.0 g
- Bromcresol green 0.02 g
- Distilled or deionized water 1.0 L

pH 4.6–5.0 at 25°C

1. Suspend the dry ingredients in the water, mix well, and boil until completely dissolved.
2. Transfer 7.0 mL portions to test tubes and cap loosely.
3. Sterilize in the autoclave at 118–121°C for 15 minutes.
4. Remove from the autoclave and place in a hot water bath set at 45–50°C. Allow at least 30 minutes for the agar temperature to equilibrate before beginning the exercise.

Soft Agar

- Beef extract 3.0 g
- Peptone 5.0 g
- Sodium chloride 5.0 g
- Tryptone 2.5 g
- Yeast extract 2.5 g
- Agar 7.0 g
- Distilled or deionized water 1.0 L

1. Suspend the dry ingredients in the water, mix well, and boil until completely dissolved.
2. Transfer 2.5 mL portions to test tubes and cap loosely.
3. Sterilize in the autoclave at 15 lbs. pressure (121°C) for 15 minutes.
4. Remove from the autoclave and place in a hot water bath set at 45°C. Allow 30 minutes for the agar temperature to equilibrate.

Starch Agar

- Beef extract 3.0 g
- Soluble starch 10.0 g
- Agar 12.0 g
- Distilled or deionized water 1.0 L

pH 7.3–7.7 at 25°C

1. Suspend the dry ingredients in the water, mix well, and boil until completely dissolved.

2. Sterilize in the autoclave at 15 lbs. pressure (121°C) for 15 minutes.
3. Remove from the autoclave and allow to cool slightly.
4. Aseptically pour into sterile Petri dishes (20 mL per plate). Allow the medium to cool to room temperature.

Thioglycollate Medium (Fluid)

• Yeast extract	5.0 g
• Casitone	15.0 g
• Dextrose (glucose)	5.5 g
• Sodium chloride	2.5 g
• Sodium thioglycolate	0.5 g
• L-Cystine	0.5 g
• Agar	0.75 g
• Resazurin	0.001 g
• Distilled or deionized water	1.0 L

pH 7.1–7.5 at 25°C

1. Suspend the dry ingredients in the water, mix well, and boil until completely dissolved.
2. Dispense 10.0 mL into sterile test tubes.
3. Autoclave for 15 minutes at 15 lbs. pressure (121°C) to sterilize. Allow the medium to cool to room temperature before inoculating.

Tributyrin Agar

• Beef extract	3.0 g
• Peptone	5.0 g
• Agar	15.0 g
• Tributyrin oil	10.0 mL
• Distilled or deionized water	1.0 L

pH 5.8–6.2 at 25°C

1. Suspend the dry ingredients in the water, mix well, and boil until completely dissolved.
2. Cover loosely, and sterilize together with the tube of tributyrin oil in the autoclave at 15 lbs. pressure (121°C) for 15 minutes.
3. Remove from the autoclave, and aseptically pour agar mixture into a sterile glass blender.
4. Aseptically add the tributyrin oil to the agar mixture and blend on "High" for 1 minute.
5. Aseptically pour into sterile Petri dishes (20 mL/plate). Allow the medium to cool to room temperature.

Triple Sugar Iron Agar

• Beef extract	3.0 g
• Yeast extract	3.0 g
• Peptone	15.0 g
• Proteose peptone	5.0 g
• Dextrose (glucose)	1.0 g
• Lactose	10.0 g
• Sucrose	10.0 g
• Ferrous sulfate	0.2 g
• Sodium chloride	5.0 g
• Sodium thiosulfate	0.3 g
• Agar	12.0 g
• Phenol red	0.024 g
• Distilled or deionized water	1.0 L

pH 7.2–7.6 at 25°C

1. Suspend the dry ingredients in the water, mix well, and boil until completely dissolved.
2. Transfer 7.0 mL portions to test tubes and cap loosely.
3. Sterilize in the autoclave at 121°C for 15 minutes.
4. Slant in such a way as to form a deep butt.
5. Allow the medium to cool to room temperature.

Tryptic Nitrate Medium

• Tryptose	20.0 g
• Dextrose	1.0 g
• Disodium phosphate	2.0 g
• Potassium nitrate	1.0 g
• Agar	1.0 g
• Distilled or deionized water	1.0 L

pH 7.0–7.4 at 25°C

1. Suspend the dry ingredients in the water, mix well, and boil until completely dissolved.
2. Transfer 10.0 mL portions to test tubes, and cap loosely.
3. Sterilize in the autoclave at 15 lbs. pressure (121°C) for 15 minutes.
4. Remove from the autoclave and allow the medium to cool to room temperature.

Tryptic Soy Agar

• Tryptone	15.0 g
• Soytone	5.0 g
• Sodium Chloride	5.0 g
• Agar	15.0 g
• Distilled or deionized water	1.0 L

pH 7.1–7.5 at 25°C

1. Suspend the dry ingredients in the water, mix well, and boil until completely dissolved.
2. Transfer 7.0 mL portions to test tubes and cap loosely.
3. Sterilize in the autoclave at 121°C for 15 minutes.
4. Slant the tubes and allow the medium to cool to room temperature.

Tryptic Soy Agar (Enriched with Yeast Extract)

- Tryptone 15.0 g
- Soytone 5.0 g
- Sodium Chloride 5.0 g
- Yeast Extract 5.0 g
- Agar 15.0 g
- Distilled or deionized water 1.0 L

pH 7.1–7.5 at 25°C

1. Suspend the dry ingredients in the water, mix well, and boil until completely dissolved.
2. Transfer 10.0 mL portions to test tubes and cap loosely.
3. Sterilize in the autoclave at 121°C for 15 minutes.
4. Allow the medium to cool to room temperature with the tubes in an upright position (to be used for agar stabs).

Tryptic Soy Broth

- Tryptone 17.0 g
- Soytone 3.0 g
- Sodium chloride 5.0 g
- Dipotassium phosphate 2.5 g
- Distilled or deionized water 1.0 L

pH 7.1–7.5 at 25°C

1. Suspend the dry ingredients in the water. Agitate and heat slightly (if necessary) to dissolve completely.
2. Dispense 7.0 mL portions into test tubes and cap loosely.
3. Autoclave for 15 minutes at 121°C to sterilize the medium.

Urease Agar

- Peptone 1.0 g
- Dextrose (glucose) 1.0 g
- Sodium chloride 5.0 g
- Potassium phosphate, monobasic 2.0 g
- Agar 15.0 g
- Phenol red 0.012 g
- Distilled or deionized water 1.0 L

pH 6.6–7.0 at 25°C

1. Suspend the agar in 900 mL of the water, mix welln and boil to dissolve completely.
2. Cover loosely and sterilize by autoclaving at 15 lbs. pressure (121°C) for 15 minutes.
3. Remove from the autoclave and allow to cool to 55°C.
4. Suspend the remaining ingredients in 100 mL of the water, mix well, and filter sterilize. *Do not autoclave.* This is urease agar base.
5. Aseptically add the urease agar base to the agar solution, and mix well.
6. Aseptically transfer 7.0 mL portions to sterile test tubes, and cap loosely.
7. Slant in such a way that the agar butt is approximately twice as long as the slant.
8. Allow to cool to room temperature.

Urease Broth

- Yeast extract 0.1 g
- Potassium phosphate, monobasic 9.1 g
- Potassium phosphate, dibasic 9.5 g
- Urea 20.0 g
- Phenol red 0.01 g
- Distilled or deionized water 1.0 L

pH 6.6–7.0 at 25°C

1. Suspend the dry ingredients in the water and mix well.
2. Filter sterilize the solution. *Do not autoclave.*
3. Aseptically transfer 1.0 mL volumes to small sterile test tubes and cap loosely.

Xylose Lysine Desoxycholate Agar

- Xylose 3.5 g
- L-Lysine 5.0 g
- Lactose 7.5 g
- Sucrose 7.5 g
- Sodium chloride 5.0 g
- Yeast extract 3.0 g
- Phenol red 0.08 g
- Sodium desoxycholate 2.5 g
- Sodium thiosulfate 6.8 g
- Ferric ammonium citrate 0.8 g
- Agar 13.5 g
- Distilled or deionized water 1.0 L

pH 7.3–7.7 at 25°C

1. Suspend the dry ingredients in water and mix. Heat only until the medium boils.
2. Cool in a water bath at 50°C.
3. When cooled, pour into sterile plates.
4. Allow the medium to cool to room temperature.

Yeast Extract Broth

- Beef extract 3.0 g
- Peptone 5.0 g
- NaCl 5.0 g
- Yeast extract 5.0 g

- Distilled or deionized water 1.0 L

pH 6.6–7.0 at 25°C

1. Suspend the dry ingredients in the water. Agitate and heat slightly (if necessary) to dissolve completely.
2. Dispense 7.0 mL portions into test tubes and cap loosely.
3. Autoclave for 15 minutes at 121°C to sterilize the medium.

Reagents

Direct Count—Staining/Diluting Agents (*See* Stains)

Lysozyme Buffer

- NaCl 22.67 g
- Na_2HPO4 3.56 g
- KH_2PO4 6.65 g
- Distilled or deionized water 1.0 L

1. Dissolve all ingredients in approximately 900 mL water.
2. Add water to bring the total volume up to 1000 mL.

Lysozyme Substrate

- Lysozyme Buffer 200.0 mL
- *Micrococcus lysodeikticus* (freeze-dried) 0.1 g

1. Add the freeze-dried *Micrococcus lysodeikticus* to the lysozyme buffer and mix well.
2. Using a spectrophotometer with the wavelength set at 540 nm, adjust the solution's light transmittance to 10% by adding water or *M. lysodeikticus* as needed.

MR-VP (Methyl Red–Voges-Proskauer Test Reagents

Methyl Red

- Methyl red dye 0.1 g
- Ethanol 300.0 mL
- Distilled water to bring volume to 500.0 mL

1. Dissolve the dye in the ethanol.
2. Add water to bring the total volume up to 500 mL.

VP Reagent A (Barritt's)

- β-naphthol 5.0 g
- Absolute Ethanol to bring volume to 100.0 mL

1. Dissolve the β-naphthol in approximately 95 mL of water.
2. Add water to bring the total volume up to 100 mL.

VP Reagent B (Barritt's)

- Potassium hydroxide 40.0 g
- Distilled water to bring volume to 100.0 mL

1. Dissolve the potassium hydroxide in approximately 60 mL of water. (***Caution:*** This solution is highly concentrated and will become hot as the KOH dissolves. It should be prepared in appropriate glassware on a stirring hot plate.) Allow it to cool to room temperature.
2. Add water to bring the volume up to 100 mL.

McFarland Turbidity Standard (0.5)

- Barium chloride ($BaCl_2 \bullet 2\ H_2O$) 1.175 g
- Sulfuric acid, concentrated (H_2SO_4) 1.0 mL
- Distilled or deionized water ≅200.0 mL

1. Pour approximately 90 mL of water into a small Erlenmeyer flask.
2. Add the $BaCl_2$ and mix well.
3. Remeasure and add water to bring the total volume up to 100 mL.
4. Add the H_2SO_4 to approximately 90 mL of water.
5. Remeasure and add water to bring the total volume up to 100 mL.
6. Add 0.5 mL of the $BaCl_2$ solution to 99.5 mL of H_2SO_4 and mix well.
7. While keeping the solution well mixed (the barium sulfate will precipitate and settle out) distribute 7 to 10 mL volumes into very clean screw cap test tubes.

Methyl Red (*See* MR-VP)

Methylene Blue Reductase Reagent

- Methylene blue dye 8.8 mg
- Distilled or deionized water 200.0 mL

Nitrate Test Reagents

Reagent A

- Sulfanilic acid 1.0 g
- 5N Acetic acid 125.0 mL

Reagent B

- Dimethyl-β-naphthylamine 1.0 g
- 5N Acetic acid 200.0

Oxidase Test Reagent

- Tetramethyl-*p*-phenylenediamine dihydrochloride 1.0 g
- Distilled or deionized water 100.0 mL

Phenylalanine Deaminase Test Reagent

- Ferric chloride 10.0 g
- Distilled or deionized water ≅90.0 mL

1. Dissolve the ferric chloride in approximately 90 mL of distilled or deionized water.
2. Add water to bring the total volume up to 100 mL.

10X Tris-Acetate-EDTA buffer

- Trisma base 48.4 g
- Glacial acetic acid 11.42 mL
- 0.5 *M* EDTA, pH 8.0 20.0 mL
- Distilled or deionized water 1.0 L

Voges-Proskauer Reagents A and B (*See* MR-VP)

Stains

Acid-Fast, Cold Stain Reagents (Modified Kinyoun)

Carbolfuchsin

- Basic fuchsin 1.5g
- Phenol 4.5 g
- Ethanol (95%) 5.0 mL
- Isopropanol 20.0 mL
- Distilled or deionized water 75.0 mL

1. Dissolve the basic fuchsin in the ethanol and add the isopropanol.
2. Mix the phenol in the water.
3. Mix the solutions together and let stand for several days.
4. Filter before use.

Decolorizer

- H_2SO_4 1.0 mL
- Ethanol (95%) 70.0 mL
- Distilled or deionized water 29.0 mL

Brilliant Green

- Brilliant green dye 1.0 g
- Sodium azide 0.01g
- Distilled or deionized water 100.0 mL

Acid-Fast, Hot Stain Reagents (Ziehl-Neelson)

Carbolfuchsin

- Basic fuchsin 0.3 g
- Ethanol 10.0 mL
- Distilled or deionized water 95.0 mL
- Phenol 5.0 mL

1. Dissolve the basic fuchsin in the ethanol.
2. Dissolve the phenol in the water
3. Combine the solutions and let stand for a few days.
4. Filter before use.

Decolorizer

- Ethanol 97.0 mL
- HCl (concentrated) 3.0 mL

Methylene Blue Counterstain

- Methylene blue chloride 0.3 g
- Distilled or deionized water 100.0 mL

Capsule Stain

Congo Red

- Congo red dye 5.0 g
- Distilled or deionized water 100.0 mL

Maneval's Stain

- Phenol (5% aqueous solution) 30.0 mL
- Acetic acid, glacial (20% aqueous solution) 10.0 mL
- Ferric chloride (30% aqueous solution) 4.0 mL
- Acid fuchsin (1% aqueous solution) 2.0 mL

Direct Count: Staining/Diluting Agents

Agent A

- 100% saturated crystal violet-ethanol solution 1.0 mL
- Ethanol 40.0 mL
- NaCl 0.9 g
- Distilled or deionized water ≅58.1 mL

1. Dissolve the crystal violet in ethanol and filter.
2. Mix 0.9 g NaCl in approximately 55 mL of distilled or deionized water.
3. Add the Ethanol and crystal violet solutions.
4. Add water to bring the total volume up to 100 mL.

Agent B

- Ethanol 40.0 mL
- NaCl 0.9 g
- Distilled or deionized water ≅60.0 mL

1. Dissolve 0.9 g NaCl in approximately 55 mL of distilled or deionized water.
2. Add the mixture to 40 mL of ethanol.
3. Add water to bring the total volume up to 100 mL.

Flagella Stain (Ryu Method)

Solution I (Mordant)

• 5% Phenol (aqueous)	10.0 mL
• Tannic acid	2.0 g
• Aluminum potassium sulfate • 12 H_2O (saturated aqueous solution)	10.0 mL

Solution II (Stain)

• Crystal violet (saturated ethanolic solution—12 g / 100 mL 95% ethanol)	1.0 mL

1. Mix 10 parts of solution I with one part solution II.
2. Filter through Whatman No. 4 filter paper.
3. Keep the final stain at room temperature in a syringe fitted with a 0.22 μm pore-size membrane filter. No need to use a needle.
4. Cap the syringe to prevent drying of the stain. This preparation will keep for several weeks.

Gram Stain Reagents

Gram Crystal Violet (Modified Hucker's)

Solution A

• Crystal violet dye (90%)	2.0 g
• Ethanol (95%)	20.0 mL

Solution B

• Ammonium oxalate	0.8 g
• Distilled or deionized water	80.0 mL

1. Combine solutions A and B. Store for 24 hours.
2. Filter before use.

Gram Decolorizer

- Ethanol (95%)
- Can use 75% ethanol and 25% acetone, but decolorization time must be reduced

Gram Iodine

• Potassium iodide	2.0 g
• Iodine crystals	1.0 g
• Distilled or deionized water	300.0 mL

1. Dissolve the potassium iodide in the water *first*.
2. Dissolve the iodine crystals in the solution.
3. Store in an amber bottle.

Gram Safranin

• Safranin O	0.25 g
• Ethanol (95%)	10.0 mL
• Distilled or deionized water	100.0 mL

1. Dissolve the safranin O in the ethanol.
2. Add the water.

Negative Stain Nigrosin

• Nigrosin	10.0 g
• Distilled or deionized water	100.0 mL

Simple Stains

Crystal Violet (*See* Gram Stain)

Methylene Blue (*See* Acid Fast, Hot)

Safranin (*See* Gram Stain)

Carbolfuchsin (*See* Acid Fast, Hot)

Spore Stain

Malachite Green

• Malachite green dye	5.0 g
• Distilled or deionized water	100.0 mL

Safranin (*See* Gram Stain)

Vogel-Bonner Salts (50x)

• Magnesium sulfate	10.0 g
• Citric acid	100.0 g
• Dipotassium phosphate	500.0 g
• Monosodium ammonium phosphate	175.0 g
• Distilled or deionized water to bring volume to	1.0 liter*

References

Baron, Ellen Jo, Lance R. Peterson, and Sydney M. Finegold. 1994. *Bailey & Scott's Diagnostic Microbiology,* 9th ed. Mosby–Year Book, St. Louis.

Eisenstadt, Bruce, C. Carlton, and Barbara J. Brown. 1994. *Methods for General and Molecular Bacteriology*, edited by Philipp Gerhardt, R. G. E. Murray, Willis A. Wood, and Noel R. Krieg. American Society for Microbiology, Washington, DC.

Farmer, J. J. III. 2003. Chapter 41 in *Manual of Clinical Microbiology,* 8th ed. Edited by Patrick R. Murray, Ellen Jo Baron, James H. Jorgensen, Michael A. Pfaller, and Robert H. Yolken. ASM Press, American Society for Microbiology, Washington, DC.

Forbes, Betty A., Daniel F. Sahm, and Alice S. Weissfeld. 2002. Chapter 19 in *Bailey and Scott's Diagnostic Microbiology,* 11th ed. Mosby, St. Louis.

Koneman, Elmer W., et al. 1997. *Color Atlas and Textbook of Diagnostic Microbiology,* 5th ed. Lippincott-Raven Publishers, Philadelphia.

MacFaddin, Jean F. 1980. *Biochemical Tests for Identification of Medical Bacteria*, 2nd ed. Williams & Wilkins, Baltimore.

Zimbro, Mary Jo, and David A. Power, editors. 2003. *Difco™ and BBL™ Manual —Manual of Microbiological Culture Media.* Becton Dickinson and Co., Sparks, MD.

*The solid ingredients raise the solvent level significantly in this preparation. Use only enough water initially to dissolve the ingredients, then add water to bring the volume to 1 liter.

Data Sheets

EXERCISE 1-1

NUTRIENT BROTH AND NUTRIENT AGAR PREPARATION

Data Sheet

Name .... Date..............................

Lab Section .... I was present and performed this exercise (initials)............

Observations and Interpretations

Record the number of each medium type you prepared, then record the number of apparently sterile ones. Calculate your percentage of successful preparations for each. In the last column, speculate as to probable/possible sources of contamination.

Medium	Total Number Prepared	Number of Sterile Preparations	Percentage of Successful Preparations	Probable Sources of Contamination (if any)
Nutrient Agar Tubes (Slant or Deep)				
Nutrient Agar Plates				
Nutrient Broths				

Questions

1. Which medium was most difficult to prepare without contamination? Why do think this might be so?

2. For each of the following types of contamination, suggest the most likely point in preparation (or later) at which the contaminant was introduced.

a. Growth in all broth tubes.

b. Growth in one broth tube.

c. Growth only on the surface of a plate.

d. Growth throughout the agar's thickness on a plate.

e. Growth only in the upper 1 cm of agar in an agar deep tube.

f. All plates in a batch have the same type and density of contaminants.

g. Only a few plates in a batch are contaminated, and each looks different.

EXERCISE 1-2

COMMON ASEPTIC TRANSFERS AND INOCULATION METHODS

Data Sheet

Name .... Date...............................

Lab Section .... I was present and performed this exercise (initials)...........

Observations and Interpretations

Describe the appearance of growth on/in each medium. Draw representative samples of each growth type.

Organism	Medium Inoculated	
	NA Slant	NA Broth
B. subtilis (NA slant)		
E. coli (NA slant)		
M. luteus (NA slant)		
S. epidermidis (NB culture)		
Kocuria rosea (*M. roseus*) (NA plate)		

Questions

1. Considering the cultures used to inoculate each medium in this exercise, how many different microbial types should you expect to see in/on each medium?

2. You were asked to describe different growth types in each culture, if present. In which medium was this the most difficult to determine? What made this difficult?

3. Which culture was most difficult to transfer from? Which medium was most difficult to inoculate?

EXERCISE 1-3

STREAK PLATE METHODS OF ISOLATION

Data Sheet

Name .... Date..............................

Lab Section .... I was present and performed this exercise (initials)...........

Results and Interpretations

1. Using your pencil, perform a quadrant streak on the "practice plate" below. Relax your wrist and use the whole surface.

2. Examine the streak plate made from the mixture of two organisms. Have your lab partner write a critique of your isolation technique in the space below. The following should be addressed: Do you obtain isolation? Were the first three streaks near the edges of the plate? Did any streaks intersect streaks they shouldn't have? Was the whole surface of the agar used? Was the agar cut by the loop?

3. Did you achieve isolation of both species from each mixture? If so, in which streak (1, 2, 3, or 4) did it occur? If you did not achieve isolation, what might you do differently next time to improve your results?

4. Most colonies on streak plates grow from isolated colony-forming units (CFUs). On rare occasions, however, a colony can be a mixture of two different organisms. If a culture is started from this colony (thinking it is pure), correct identification will be next to impossible because the extra organism will confound the test results. How could you verify the purity of a colony? (The answers may vary depending on what experience you have had prior to performing this exercise.) If you found the colony to be a mixture of organisms, what could you do to purify it?

5. Consider the plates made from mixtures of pure cultures. Which colonies in this exercise likely started as more than single cells? That is, which colonies make use of the CFU designation appropriate? (If you haven't covered cell morphology and arrangement yet, omit this question.)

6. Examine the environmental sample and the gumline/cheek swab plates. Which had greater density and diversity of growth? Were the different streak methods appropriate to the cell densities recovered?

EXERCISE 1-4

SPREAD PLATE METHOD OF ISOLATION

Data Sheet

Name .... Date..............................

Lab Section .... I was present and performed this exercise (initials)............

Observations and Interpretations

Record your observations in the table below.

Organism	Plate(s) With Isolation	Comments
E. coli		
S. marcescens		

Questions

1. On which plate did you obtain isolation with *E. coli*? How about *S. marcescens*? Do you have reason to suspect that they should become isolated on the same dilution plate? Why or why not?

2. Once you obtained isolation at a particular dilution, did you continue to have isolation on subsequent dilution plates? Is this what you would expect? Why or why not?

3. What is the consequence of not spreading the inoculum adequately over the agar surface?

4. To get isolated colonies on a plate, only about 300 cells can be in the inoculum. What will happen if the cell density of the inoculum exceeds this number?

5. Suppose you have two organisms in a mixture and Organism A is 1000 times more abundant than Organism B. Will you (without counting on good luck!) be able to isolate Organism B using the spread plate technique? Explain your answer.

Exercise 2-1

UBIQUITY OF MICROORGANISMS

Data Sheet

Name .... Date.............................

Lab Section .... I was present and performed this exercise (initials)...........

Observations and Interpretations

1. Using the diagrams below as Petri dishes, draw the patterns of growth from each of your plates. (Use a representative colony of each type; you do not have to draw the entire plate.) Be sure to label the plates according to incubation time, temperature, and source of inoculum. Also include other useful colony information, such as color and relative abundance.
2. Save the plates for Exercise 2-2.

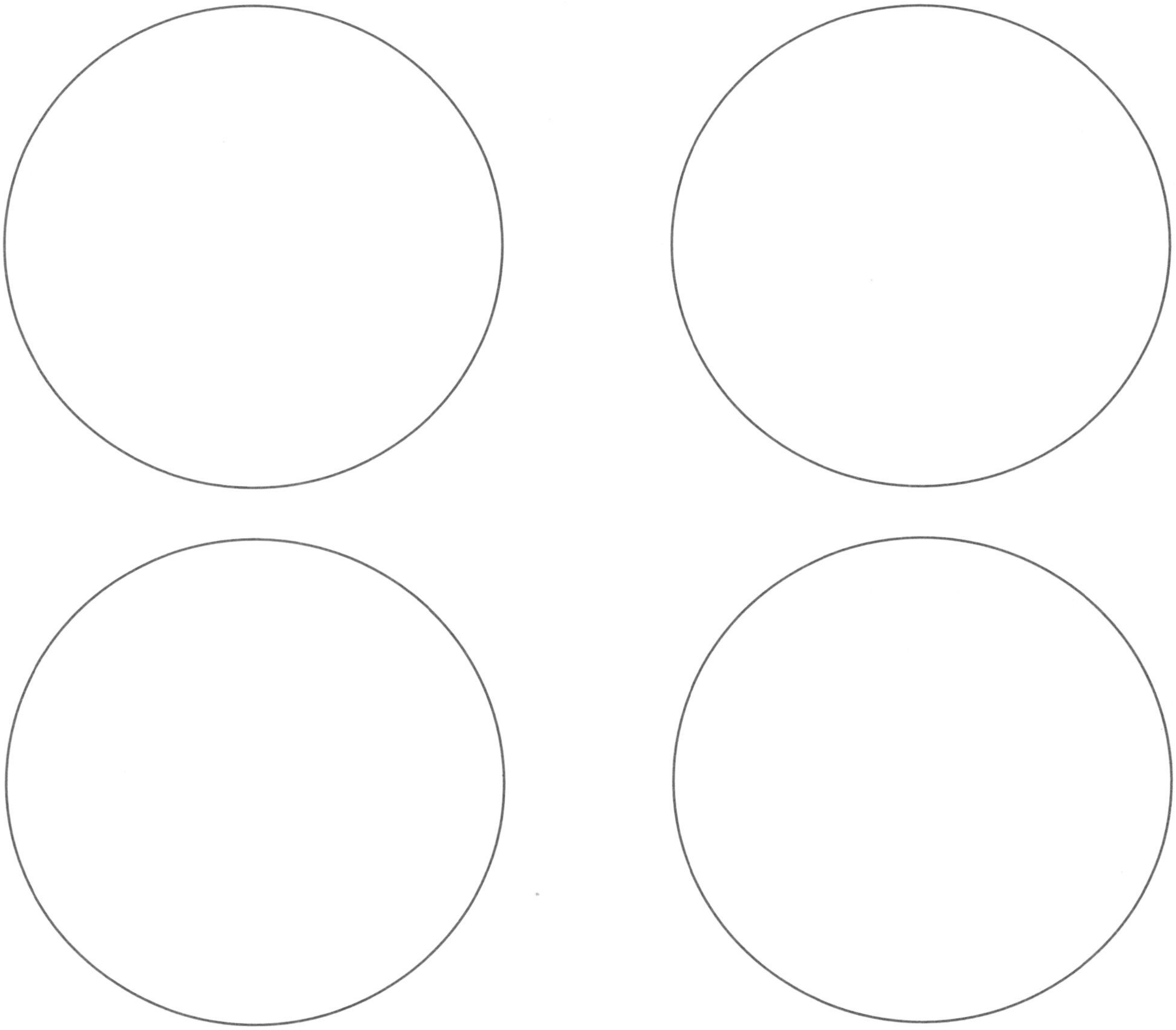

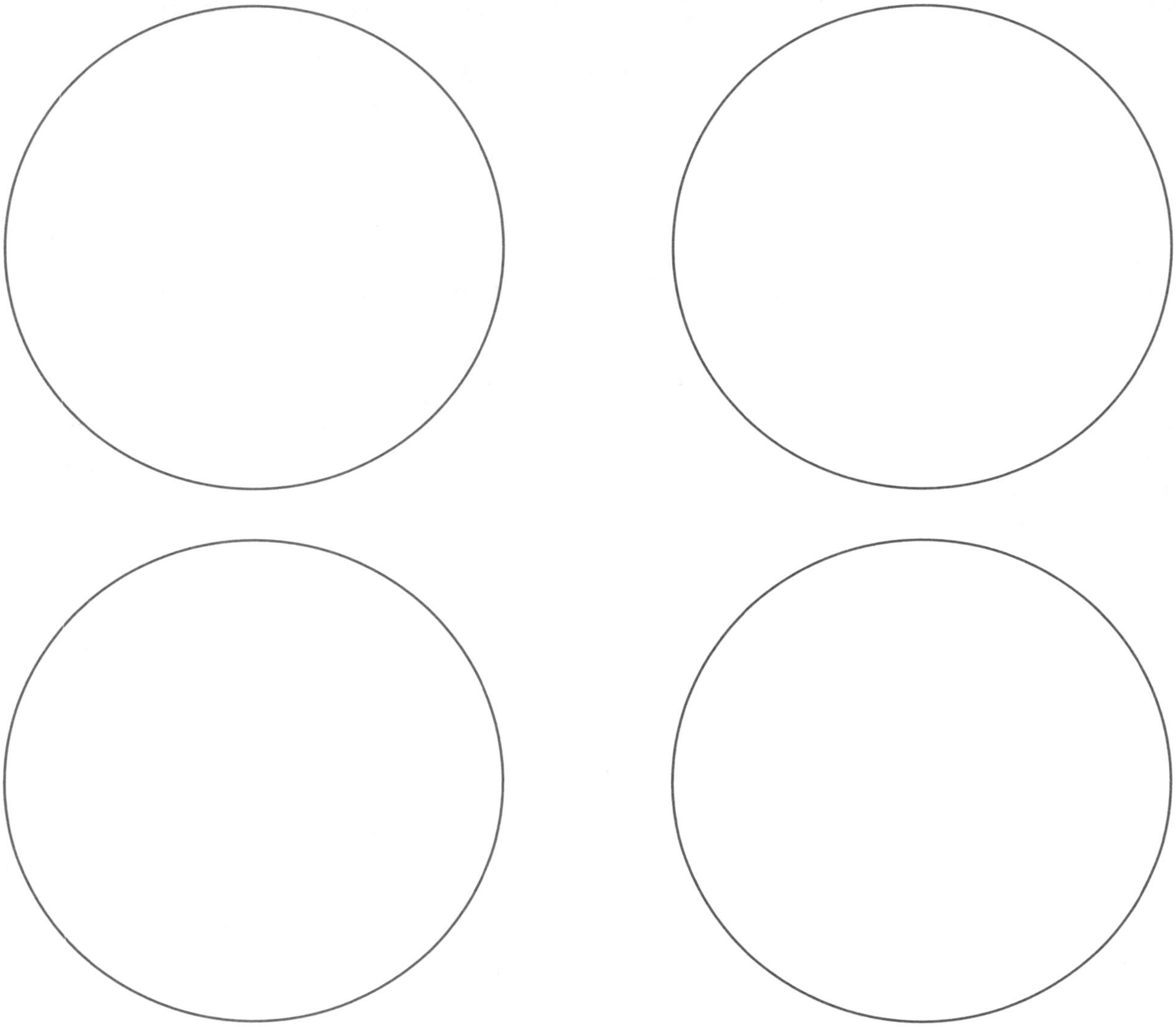

Questions

1. What was the purpose of incubating the unopened plates? Be specific. What is an appropriate label for these plates?

2. If growth appears on both unopened plates, what are some likely explanations? What if growth appears on only one plate? How does growth on the unopened plates affect the reliability (your interpretation) of the other plates?

3. Why were the specific types of exposure (air, hair, tabletop, *etc.*) chosen for this exercise?

4. Why were you asked to incubate the plates at two different temperatures? Be specific. What is the likely source (reservoir) of organisms that grew best at 37°C, and how do they survive at room temperature without nutrients?

5. Explain why you might have gotten different appearing colonies on plates 2 and 3.

6. The plates you are using for this lab will be autoclaved eventually to completely sterilize them. The measures taken to disinfect the tabletops (the source of the organisms on plates 2 and 3) are not as extreme. Why?

EXERCISE 2-2

COLONY MORPHOLOGY

Data Sheet

Name .... Date..............................

Lab Section .... I was present and performed this exercise (initials)...........

Observations and Interpretations

Using the terms in Figure 2-3, describe and sketch the colonies on your plates from today's exercise and the plates from Exercise 2-1. Use a colony counter if necessary. Measure colony diameters and include them with your descriptions.

Organism/Plate	Colony Description and Sketch

Organism/Plate	Colony Description and Sketch

Questions

1. A description of colony morphology provides important information about an organism. What other information should you include when describing physical growth characteristics?

2. Colony size, color, and shape are three critical aspects of a description of bacterial growth. At least three other important factors—not physical descriptions—typically are included when describing bacterial growth. Can you guess what they are and why they are important? *Hint*: Look in *Bergey's Manual of Systematic Bacteriology*.

Exercise 2-3

GROWTH PATTERNS ON SLANTS

Data Sheet

Name .... Date..............................

Lab Section .... I was present and performed this exercise (initials)...........

Observations and Interpretations

In the chart below, describe the growth on your slants including shape, margin, texture, and color.

Organism	Growth Description
Uninoculated control	

Questions

1. List some reasons why growth characteristics are more useful on agar plates than on agar slants.

2. Why are agar slants better suited to maintain stock cultures than agar plates?

3. Match the following:

_____ Filiform	1. Produces colored growth
_____ Spreading edge	2. Smooth texture with solid edge
_____ Transparent	3. Solid growth seeming to radiate outward
_____ Friable	4. Almost invisible or easy to see light through
_____ Pigmented	5. Rough texture with a crusty appearance

EXERCISE 2-4

GROWTH PATTERNS IN BROTH

Data Sheet

Name .... Date..............................

Lab Section .... I was present and performed this exercise (initials)...........

Observations and Interpretations

Examine the tubes and enter the descriptions in the chart below. Draw a picture if necessary, and include other information about the conditions of incubation such as time, temperature and the medium used.

Organism	Description of Growth in Broth
Uninoculated Control	

Questions

1. What factors besides physical growth characteristics are important when recording data about an organism? Why?

2. Match the following:

_____ Flocculent	a. Evenly cloudy throughout
_____ Sediment	b. Growth at top around the edge
_____ Ring	c. Growth on the bottom
_____ Pellicle	d. Membrane at the top
_____ Uniform fine turbidity	e. Suspended chunks or pieces

EXERCISE 2-5

EVALUATION OF MEDIA

Data Sheet

Name .... Date.............................

Lab Section .... I was present and performed this exercise (initials)...........

Observations and Interpretations

1. On the top line of the data table, write "defined" or "undefined" for each medium.

2. Score the relative amount of growth in each tube. Compare the five tubes for each medium with each other, as well as the amount of growth for each organism in the various media. Use "0" for no growth, "3" for abundant growth, and "1" and "2" for degrees of growth in between.

Organism	Nutrient Broth	Glucose Broth	Yeast Extract Broth	Glucose Salts Medium	Interpretation (relative fastidiousness)
Defined/Undefined					
Uninoculated Broth					
Ability of medium to support a wide range of microorganisms (good, fair, poor)					

Questions

1. What does comparing growth of the four organisms in a particular medium tell you? (That is, are you gaining information primarily about the organism or about the medium?)

2. What does comparing growth of a particular organism in the four media tell you? (Again, are you gaining information primarily about the organism or about the medium?)

3. Evaluate the media.

   a. Which medium supports growth of the widest range of organisms?

   b. Which medium supports the fewest organisms?

   c. Is there a correlation between your answers to Questions 3a and 3b, and the terms "defined medium" and "undefined medium"? If so, what is it? If not, attempt to explain the relationship you observe.

4. Evaluate the organisms.

   a. Which organism appears to be most fastidious? How can you tell?

   b. Which organism appears to be least fastidious? How can you tell?

5. What is the biochemical basis for the spectrum of fastidiousness seen in the microbial world? (That is, why are some organisms fastidious and others are nonfastidious?)

Exercise 2-6

AGAR DEEP STABS

Data Sheet

Name .... Date..............................

Lab Section .... I was present and performed this exercise (initials)...........

Observations and Interpretations

1. Examine all tubes and enter a description of each in the chart below.

2. Categorize each organism with respect to its aerotolerance group based on the location of growth in the media.

Organism	Region of Growth	Aerotolerance Category
Sterile stab		

Questions

1. Why is it important to use this medium soon after preparation?

2. If you have a tube that indicates uniform growth throughout the agar, can the aerotolerance category of the organism be determined? Explain.

3. When inoculating agar deep stabs, why do you have to insert and remove the needle along the same stab line?

4. If you inoculate an organism known to chemically reduce sulfur, where would you expect to see its growth in the test medium used today? Could it be seen in more than one zone? Explain.

5. If you have no growth in the stab after 24 hours' incubation, what are some possible explanations?

EXERCISE 2-7

AGAR SHAKES

Data Sheet

Name .... Date..............................

Lab Section .... I was present and performed this exercise (initials)............

Observations and Interpretations

1. Examine all tubes and enter a description of each in the chart below.

2. Categorize each organism with respect to aerotolerance category based on the location of growth in the medium.

Organism	Region of Growth	Aerotolerance Category
Uninoculated Control		

Questions

1. How might your results differ if you were to shake this medium vigorously before incubating it? Why?

2. How would you explain why an organism might grow throughout an Agar Stab, but grow only on the bottom of an Agar Shake?

3. If you have no growth in the agar after 24 hours' incubation, what are some possible explanations?

EXERCISE 2-8

FLUID THIOGLYCOLLATE MEDIUM

Data Sheet

Name .... Date..............................

Lab Section .... I was present and performed this exercise (initials)...........

Observations and Interpretations

Draw a diagram of each broth showing the location of growth. Indicate the amount of growth in the broth as: Good + + +, Fair + +, Poor +, No Growth 0

Organism	Location of Growth in Medium	Aerotolerance Category
Uninoculated Control		

Questions

1. Why is there a colored band at the surface of Fluid Thioglycollate Medium? Which is more desirable: a thick- or a thin-colored band?

2. Where would you expect to see growth of a strict aerobe? Anaerobe? Microaerophile? Facultative anaerobe?

3. Why is it important that this medium be fresh? Which type of organism (aerobe, anaerobe, microaerophile, facultative anaerobe) would most likely be affected negatively by use of old media? Which would most likely be affected positively?

EXERCISE 2-9

ANAEROBIC JAR

Data Sheet

Name .... Date..............................

Lab Section .... I was present and performed this exercise (initials)...........

Observations and Interpretations

Indicate the amount of growth on each plate as: Good + + +, Fair + +, Poor +, No Growth 0

Organism	Growth on Aerobic Plate	Growth on Anaerobic Plate	Aerotolerance Category

Questions

1. If, after incubation, you observed that the methylene blue indicator strip inside the jar was blue, what would you guess the internal environment to be—aerobic or anaerobic? How would you expect the growth on the plate inside the jar to differ from the plate incubated outside the jar?

2. Which of the three organisms would be most affected by the conditions described in question 1?

3. An alternative to the anaerobic jar is a candle jar, in which a candle is placed in the jar, lit, and the lid closed to enable the flame to use the available oxygen. Typically, not all of the oxygen is used in this system. Which types of organisms would most likely benefit from this environment?

EXERCISE 2-10

THE EFFECT OF TEMPERATURE ON MICROBIAL GROWTH

Data Sheet

Name .... Date..............................

Lab Section .... I was present and performed this exercise (initials)...........

Observations and Interpretations

1. Record your absorbance values or numeric values for each organism at each temperature.

Broth Data						
Organism	10°C	20°C	30°C	40°C	50°C	Classification
Uninoculated Control						

If using a spectrophotometer, enter the absorbance. Enter visual readings with the numeric values 0, 1, 2, or 3 (0 is clear and 3 is very turbid).

2. Record the cultural characteristics of the *Serratia marcescens* incubated at two temperatures.

Plate Data	
Incubation Temperature	Description of Growth
20°C	
35°C	

Questions

1. Using the data from the chart, determine the cardinal temperatures for each of the four organisms. Circle the optimum temperature for each organism. Use brackets to designate the range for each.

2. Plot the data for each organism on the graph paper provided. If using a spectrophotometer, plot absorbance versus temperature. If not using a spectrophotometer, plot the numeric values versus temperature.

3. Why is it not advisable to connect the data points for each organism in your graph?

4. In what way(s) could you adjust incubation temperature to grow an organism at less than its optimal growth rate?

5. Why do different temperatures produce different growth rates?

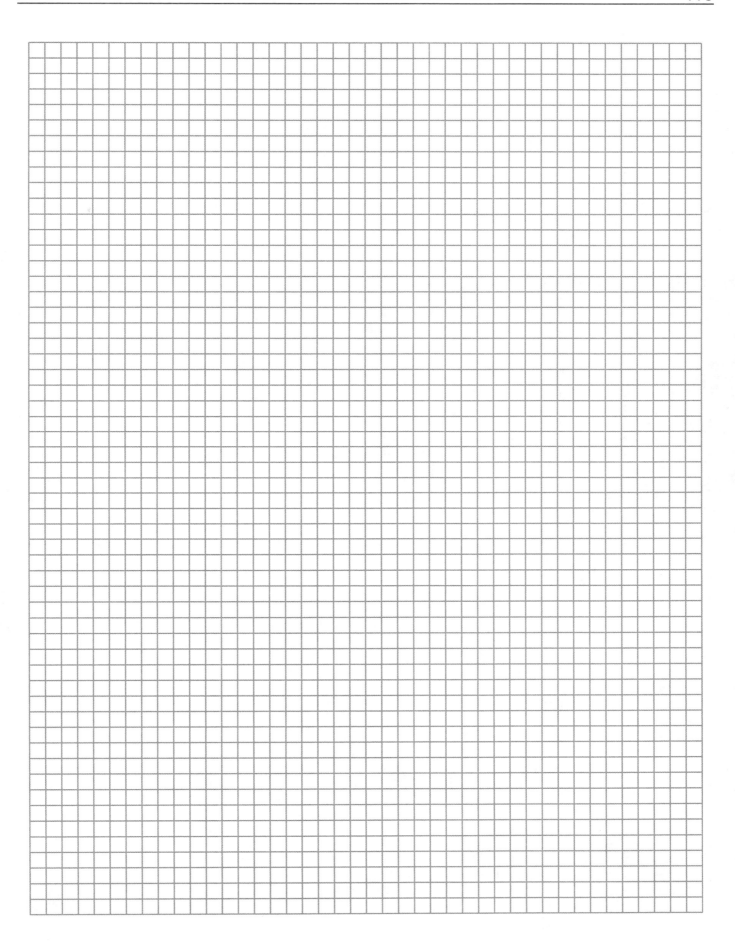

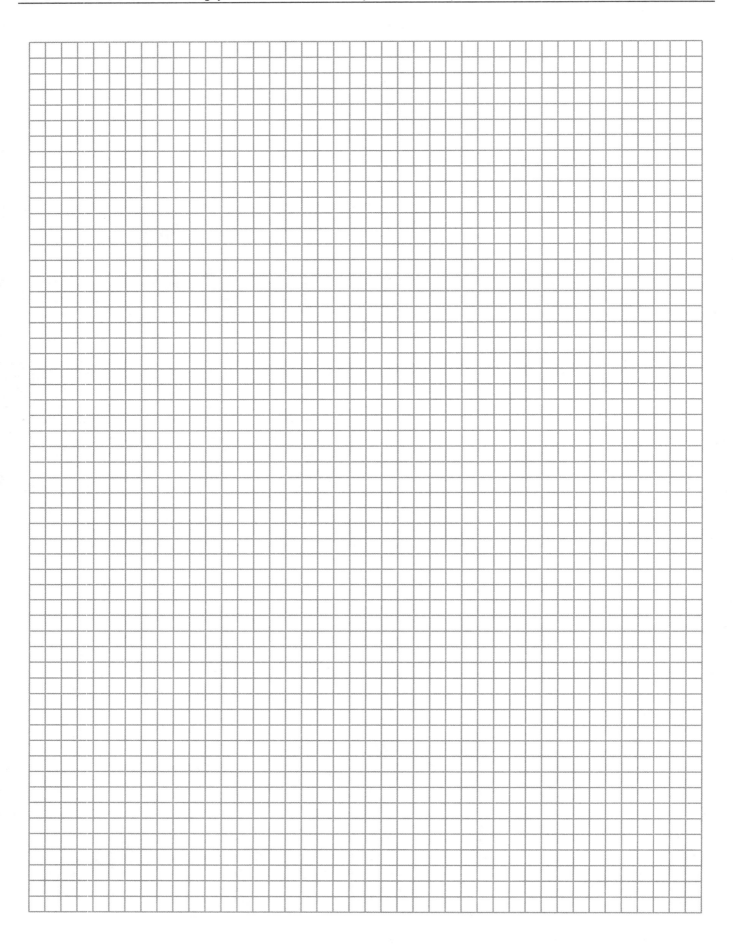

EXERCISE 2-11

THE EFFECT OF pH ON BACTERIAL GROWTH

Data Sheet

Name .... Date..............................

Lab Section .... I was present and performed this exercise (initials)...........

Observations and Interpretations

Record your absorbance values or numeric values for each organism at each pH.

Organism	pH 2	pH 4	pH 6	pH 8	pH 10	Classification
Uninoculated Control						

If using a spectrophotometer, enter the absorbance. Enter visual readings as 0, 1, 2, or 3.

Questions

1. Circle the pH optimum for each organism. Place brackets around the range. Is there any overlap between species?

2. Account for the inability of organisms to grow outside their pH ranges. Why, for instance, are alkaliphiles able to survive at high pHs when neutrophiles cannot?

3. Where is the pH optimum relative to the pH range for each organism? Do you see any parallels between these data and the data produced in Exercise 2-10? Explain.

4. Plot the data (absorbance or numeric values versus pH) for each organism on the graph paper provided.

5. Why is it not advisable to connect the data points for each organism in your graph?

6. Why do we organize the tubes into groups by pH when using the spectrophotometer?

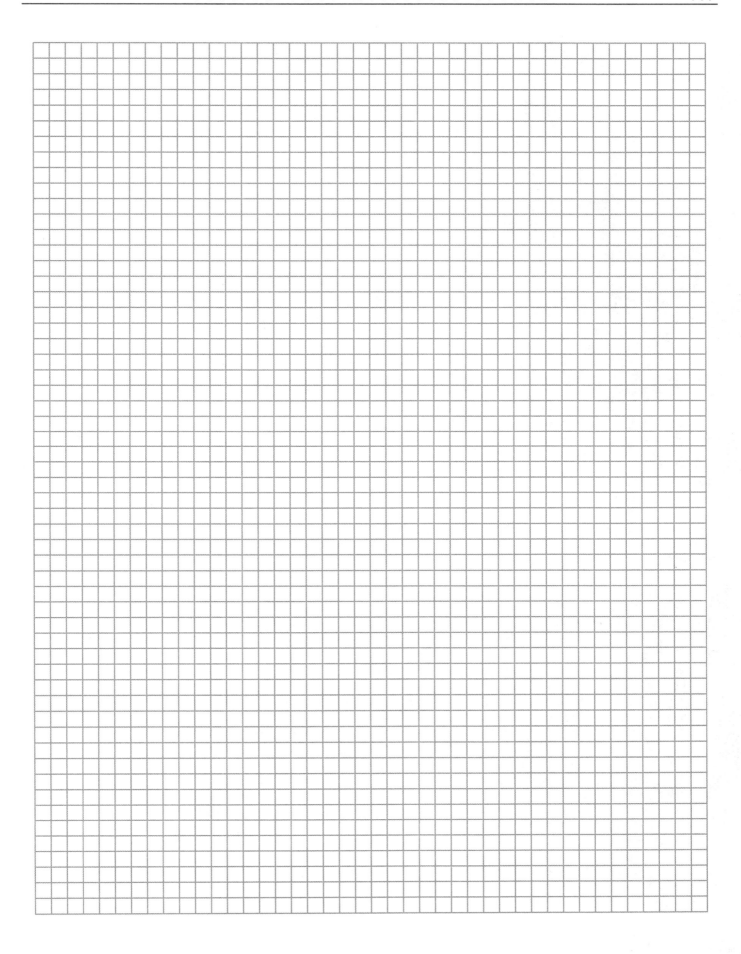

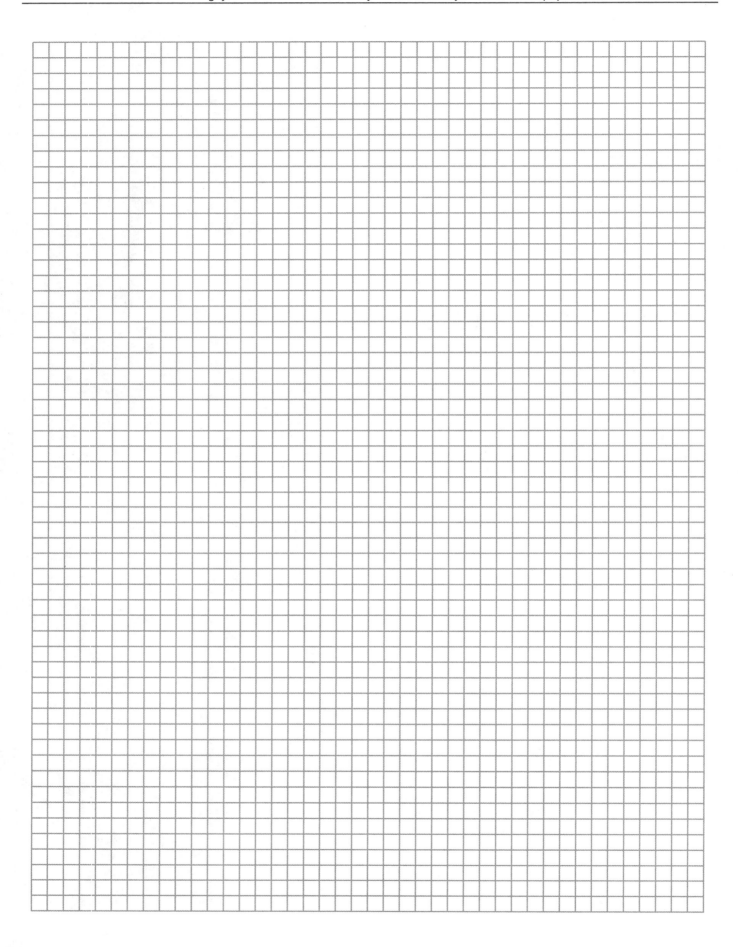

EXERCISE 2-12

THE EFFECT OF OSMOTIC PRESSURE ON MICROBIAL GROWTH

Data Sheet

Name .... Date..............................

Lab Section .... I was present and performed this exercise (initials)...........

Observations and Interpretations

Record your absorbance values or numeric values for each organism at each NaCl concentration.

Organism	NaCl Concentration			
	2%	5%	8%	11%

If using a spectrophotometer, enter the absorbance. Enter visual readings as 0, 1, 2, or 3.

Questions

1. Circle the optimum salinity for each organism. Place brackets around the range.
2. Which organism demonstrates the greatest tolerance range? Does this agree with your prediction? Explain.
3. Using the graph paper provided, plot the data (absorbance or numeric values versus % NaCl) of both organisms.
4. Why is it not advisable to connect the data points for each organism in your graph?
5. Why do we organize the tubes into groups by NaCl concentration when using the spectrophotometer?

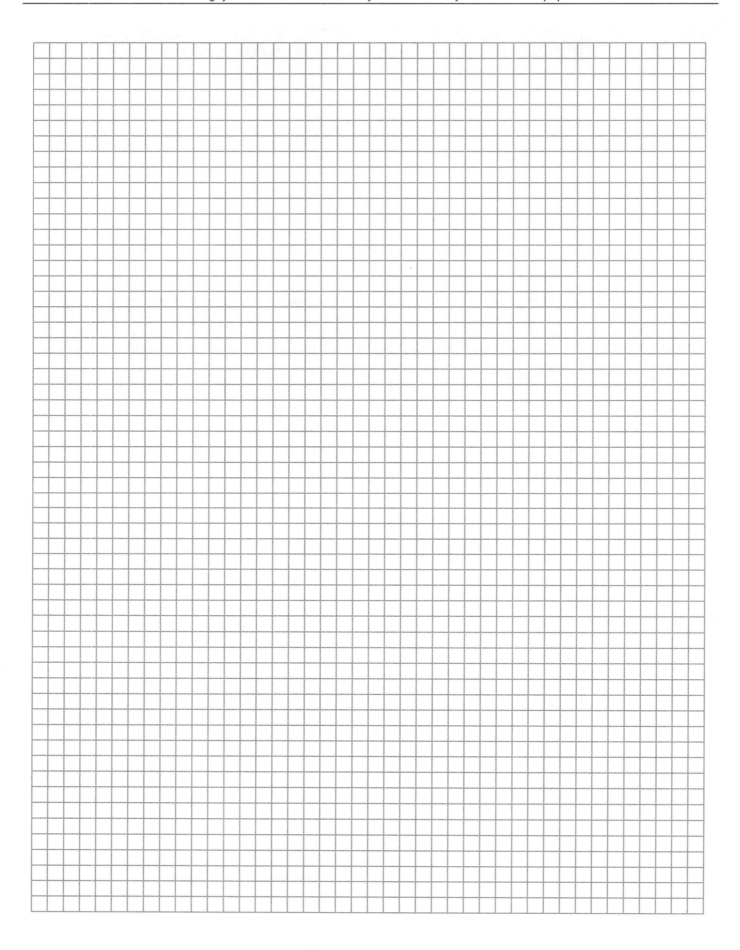

EXERCISE 2-13

THE LETHAL EFFECT OF ULTRAVIOLET LIGHT ON MICROBIAL GROWTH

Data Sheet

Name .... Date..............................

Lab Section .... I was present and performed this exercise (initials)...........

Observations and Interpretations

Group # ________________ Exposure time ______________________

1. Enter your class data in the chart below. Score the relative amount of growth on each plate. Use the symbols "0" for no growth, "3" for abundant growth, and "1" and "2" for degrees of growth in between.

Organism	No UV	5 min.	10 min.	15 min.	20 min.	25 min.	30 min.

Questions

1. The purpose of this exercise is to demonstrate the comparative effect of UV on three bacterial populations. This could have been accomplished without the cardboard cover. Why was the cover used?

2. This is not a quantitative exercise. Keeping this in mind, can you see a general trend between bacterial death and UV exposure time?

3. Did the spore-forming organisms survive longer? If so, which of the two cultures—24-hour or 7-day—lasted the longest? Why?

4. Why were you told to remove the plate covers prior to exposing them to UV?

5. What might account for the differences in survival of the various bacterial species?

6. Using the graph paper provided, construct a single graph of growth versus UV exposure time for the three organisms.

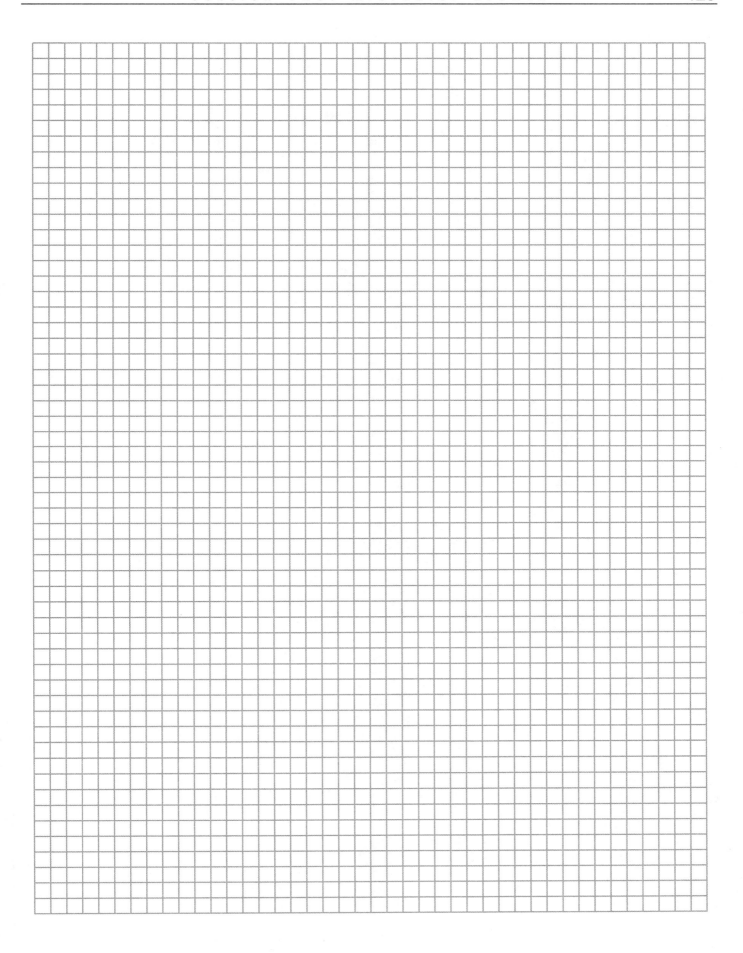

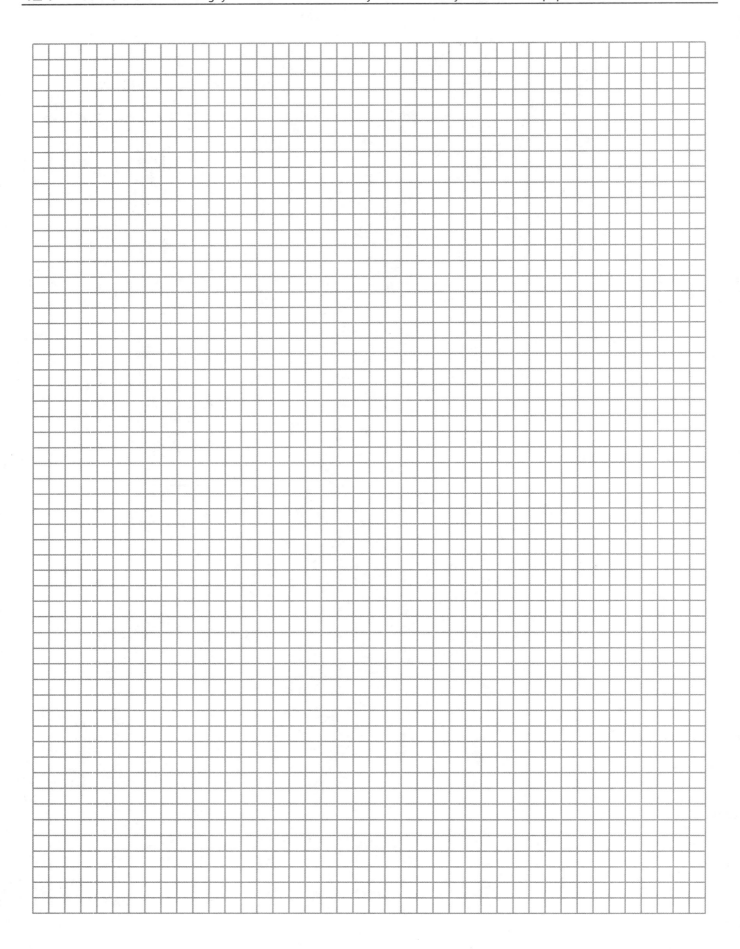

EXERCISE 2-14

DISINFECTANTS: USE-DILUTION METHOD

Data Sheet

Name .... Date..............................

Lab Section .... I was present and performed this exercise (initials)...........

Observations and Interpretations

1. Enter your individual data below.

Controls	Growth
Inoculated bead (must have growth)	
Uninoculated bead (must have no growth)	

G = Growth, NG = No Growth

Disinfectant Solution	Growth

G = Growth, NG = No Growth

2. Enter the class data in the chart below. Highlight the box with the minimum *effective* concentration for each organism.

Organism	Household Bleach			Hydrogen Peroxide			Ethyl Alcohol			Isopropyl Alcohol		
	0.01%	0.1%	1.0%	0.03%	0.3%	3%	10%	30%	50%	10%	30%	50%
E. coli												
S. aureus												

G = Growth, NG = No Growth

Questions

1. What is the general relationship between concentration of the antimicrobial agent and bactericidal effect? Would this trend continue indefinitely as the concentration is increased? Why would this be important to know?

2. Compare your results with the class data. Which disinfectant was most effective? Which was least effective? Defend your choices.

3. Were both organisms affected equally by the disinfectants? Which species, if any, seemed to be most resistant to the disinfectant? Why do you suppose it had greater resistance?

4. In a real-life situation, you probably would be unaware of what organisms are present in an area to be disinfected. Based on the information in question #2, what are the minimum concentrations of each disinfectant that could be reliably used to produce satisfactory results?

Exercise 2-15

EFFECTIVENESS OF HAND SCRUBBING

Data Sheet

Name .... Date..............................

Lab Section .... I was present and performed this exercise (initials)...........

Observations and Interpretations

1. Score and record relative growth in each of the quadrants below, using "0" for no growth, "3" for maximum growth, and "1" and "2" for degrees of growth in between.
2. Count the number of different colony morphologies you see, and record these numbers in the chart.

Quadrant	Soap or detergent used (if applicable)	Relative Amount of Growth		Number of Different Colony Morphologies	
		Soap	No Soap	Soap	No Soap
1					
2					
3					
4					

Questions

1. What is the purpose of drying one index finger with a paper towel and the other with sterile cotton?

2. Did you see a significant difference between the hand using the towel and the hand using the sterile cotton?

3. Was there a significant reduction in the number and diversity of colonies after washing?

4. This is a simple experiment. What are some of its obvious limitations?

Exercise 3-1

INTRODUCTION TO THE LIGHT MICROSCOPE

Data Sheet

Name .... Date...............................

Lab Section .... I was present and performed this exercise (initials)............

Data and Calculations

Record the relevant values off your microscope and perform the calculations of total magnification for each lens.

Lens System	Magnification of Objective Lens	Magnification of Ocular Lens	Total Magnification	Numerical Aperture
Scanning				
Low Power				
High-Dry				
Oil Immersion				
Condenser Lens				

Questions

1. Why aren't the magnifications of both ocular lenses of a binocular microscope used to calculate total magnification?

2. What is the total magnification for each lens setting on a microscope with 15X oculars and 4X, 10X, 45X, and 97X objectives lenses?

3. Assuming that all other variables remain constant, explain why light of shorter wavelength will produce a clearer image than light of longer wavelengths.

4. Why is wavelength the main limiting factor on limit of resolution in light microscopy?

5. On a given microscope, the numerical apertures of the condenser and low power objective lenses are 1.25 and 0.25, respectively. You are supplied with a filter that selects a wavelength of 520nm.

a. What is the limit of resolution on this microscope?

b. Will you be able to distinguish two points that are 330nm apart as being separate, or will they blur into one?

6. On the same microscope as in Question #5, the high dry objective lens has a numerical aperture of 0.85.

a. What is the limit of resolution on this microscope?

b. Will you be able to distinguish two points that are 250nm apart as being separate, or will they blur into one?

7. Calculate the limit of resolution for the oil lens of your microscope. Assume an average wavelength of 500 nm.

Exercise 3-2

CALIBRATION OF THE OCULAR MICROMETER

Data Sheet

Name .... Date..............................

Lab Section .... I was present and performed this exercise (initials)............

Data and Calculations

Record two or three values where the ocular micrometer and the stage micrometer line up for the scanning, low, high dry, and oil immersion objective lenses. Then calculate the calibration for each.

Scanning Objective Lens

Stage Micrometer (μm)	Ocular Micrometer (OU)	Calibration (μm/OU)

Low-Power Objective Lens

Stage Micrometer (μm)	Ocular Micrometer (OU)	Calibration (μm/OU)

High Dry Objective Lens

Stage Micrometer (µm)	Ocular Micrometer (OU)	Calibration (µm/OU)

Oil Immersion Objective Lens

Stage Micrometer (µm)	Ocular Micrometer (OU)	Calibration (µm/OU)

Calculate the average value for each calibration and record them in the chart below. If necessary, use one of the values to calculate the calibration of the oil lens.

Average Calibrations for My Microscope

Objective Lens	Average Calibration (µm/OU)
Scanning	
Lower Power	
High Dry Power	
Oil Immersion	

EXERCISE 3-3

EXAMINATION OF EUKARYOTIC MICROBES

Data Sheet

Name .... Date..............................

Lab Section .... I was present and performed this exercise (initials)...........

Observations and Interpretations

Fill in the chart for each eukaryotic microbe you observe.

Organism (Include wet mount or prepared slide)	Sketch (Include magnification and stain)	Dimensions	Identifying Characteristics (List only those that you observed)

Questions

1. What features did the cells you observed have in common? How were they different?

2. If you observed the same organism on a prepared slide and a wet-mount, how did the images compare?

EXERCISE 3-4

BACTERIAL STRUCTURE AND SIMPLE STAINS

Data Sheet

Name .... Date..............................

Lab Section .... I was present and performed this exercise (initials)...........

Observations and Interpretations

Record your observations in the table chart below.

Organism	Stain and Duration	Cellular Morphology and Arrangement (Include a sketch)	Cell Dimensions

Questions

1. What is the consequence of leaving a stain on the bacterial smear too long (overstaining)?

2. What is the consequence of not leaving a stain on the smear long enough (understaining)?

3. Choose a coccus and a bacillus from the organisms you observed, and calculate their surface-to-volume ratios. Consider the coccus to be a perfect sphere and the bacillus to be a rectangular block in which height and width are the same dimension. Use the equations supplied.

Cell Morphology	Surface Area	Volume
Coccus	$SA = 4\pi r^2$	$V = \frac{4}{3}\pi r^3$
Bacillus	$SA = 2(H \times W) + 4(L \times H)$	$V = H \times W \times L$

r = radius, H = height, L = length, W = width, $\pi = 3.14$

Surface-to-Volume Ratio of Sample Cells

Genus	Cell Morphology	Surface Area (μm^2)	Volume (μM^3)	Surface-to-Volume Ratio

4. Consider a coccus and a rod of equal volume.

   a. Which is more likely to survive in a dry environment? Explain your answer.

   b. Which would be better adapted to a moist environment? Explain your answer.

EXERCISE 3-5

NEGATIVE STAIN

Data Sheet

Name .... Date..............................

Lab Section .... I was present and performed this exercise (initials)...........

Observations and Interpretations

Record your observations in the chart below.

Organism	Stain	Cellular Morphology and Arrangement (Include a sketch)	Cell Dimensions

Questions

1. Why doesn't a negative stain colorize the cells in the smear?

2. Eosin is a red stain and methylene blue is blue. What should be the result of staining a preparation with a mixture of eosin and methylene blue?

3. Compare the diameter of *M. luteus* cells as measured using a basic stain (Exercise 3-4) and an acidic stain. What might account for any differences?

EXERCISE 3-6

GRAM STAIN

Data Sheet

Name .... Date..............................

Lab Section .... I was present and performed this exercise (initials)...........

Observations and Interpretations

Record your observations in the chart below.

Organism or Source	Cellular Morphology and Arrangement (Include a sketch)	Cell Dimensions	Color	Gram Reaction (+/−)

Questions

1. Predict the effect of the following "mistakes" made when performing a Gram stain. Consider each mistake independently.

   a. Failure to add the iodine.

   b. Failure to apply the decolorizer.

   c. Failure to apply the safranin.

   d. Reversal of crystal violet and safranin stains.

2. Both crystal violet and safranin are basic stains and may be used to do simple stains on Gram-positive and Gram-negative cells. This being the case, what makes the Gram stain a differential test?

3. If you saw large, eukaryotic cells in the preparation made from your gumline, they were most likely your own epithelial cells. Are you Gram-positive or Gram-negative? (You can make a good guess about this even if you didn't see your cells.)

EXERCISE 3-7

ACID-FAST STAINS

Data Sheet

Name .... Date..............................

Lab Section .... I was present and performed this exercise (initials)...........

Observations and Interpretations

Record your observations in the chart below.

Organism	Staining Method (ZN or K)	Cellular Morphology and Arrangement (Include a sketch)	Cell Dimensions	Color	Acid-Fast Reaction (+/−)

Questions

1. How does heating the bacterial smear during a ZN stain promote entry of carbolfuchsin into the acid-fast cell wall?

2. Are acid-fast negative cells stained by carbolfuchsin? If so, how can this be a differential stain?

3. Why do you suppose the acid-fast stain is not as widely used as the Gram stain? When is it more useful than the Gram stain?

EXERCISE 3-8

CAPSULE STAIN

Data Sheet

Name .... Date..............................

Lab Section .... I was present and performed this exercise (initials)............

Observations and Interpretations

Record your observations in the chart.

Organism	Cellular Morphology and Arrangement (Include a sketch)	Cell Dimensions	Capsule (+/−)	Width of Capsule, if Present

Questions

1. What is the purpose of emulsifying the bacteria in serum in this staining procedure?

2. Some oral bacteria produce an extracellular "capsule." Of what benefit is a capsule to these cells?

3. Sketch any cells from your mouth sample that display an unusual morphology or arrangement.

EXERCISE 3-9

SPORE STAIN

Data Sheet

Name .... Date..............................

Lab Section .... I was present and performed this exercise (initials)...........

Observations and Interpretations

Record your observations in the chart.

Organism (Include culture age)	**Cellular Morphology and Arrangement** (Include a sketch)	**Cell Dimensions**	**Spores** (Present or absent)	**Spore Shape** (If present)	**Spore Position** (If present)

Questions

1. Why does this exercise call for an older (5-day) culture of *Bacillus*?

2. What does a positive result for the spore stain indicate about the organism? What does a negative result for the spore stain indicate about the organism?

3. Why is it not necessary to include a negative control for this stain procedure?

4. Spores do not stain easily. Perhaps you have seen them as unstained white objects inside *Bacillus* species in other staining procedures. If they are visible as unstained objects in other stains, of what use is the endospore stain?

Exercise 3-10

WET MOUNT AND HANGING DROP PREPARATIONS

Data Sheet

Name .... Date..............................

Lab Section .... I was present and performed this exercise (initials)...........

Observations and Interpretations

Record your observations in the chart.

Organism	Procedure (Wet Mount or Hanging Drop)	Cellular Morphology and Arrangement (Include a sketch)	Cell Dimensions	Motility (+/−)

Questions

1. In a wet mount, each of the following complications could lead to a false result for motility. For each, write "false positive" or "false negative," depending on how it could interfere with your reading of a motile organism and a nonmotile organism.

   a. over-inoculation of the slide with organisms

   b. cells attaching to the glass slide or cover glass

   c. receding water line

   d. using an old culture

2. You are told that viewing is best done with as little illumination as possible. Why will transparent cells be easier to view with less light?

3. Why would you expect Brownian motion to increase the longer you observe a hanging drop preparation?

EXERCISE 3-11

FLAGELLA STAIN

Data Sheet

Name .... Date..............................

Lab Section .... I was present and performed this exercise (initials)...........

Observations and Interpretations

Record your observations in the table.

Organism	Cellular Morphology and Arrangement (Include a sketch)	Cell Dimensions	Flagella (Present or absent)	Flagellar Arrangement (Include a sketch)

Questions

1. Why can't flagella be observed in action?

2. Flagella have a diameter of about 1 nm. A hypothetical question: To resolve flagella, what is the maximum wavelength of the electromagnetic spectrum that would have to be used to create the image (given numerical apertures of 1.25 for both the condenser and the oil lens)?

Exercise 3-12

MORPHOLOGICAL UNKNOWN

Data Sheet

Name .... Date..............................

Lab Section .... I was present and performed this exercise (initials)...........

Observations and Interpretations

1. Complete the flowchart using the information in Table 3-4. Each path should end with a single organism.

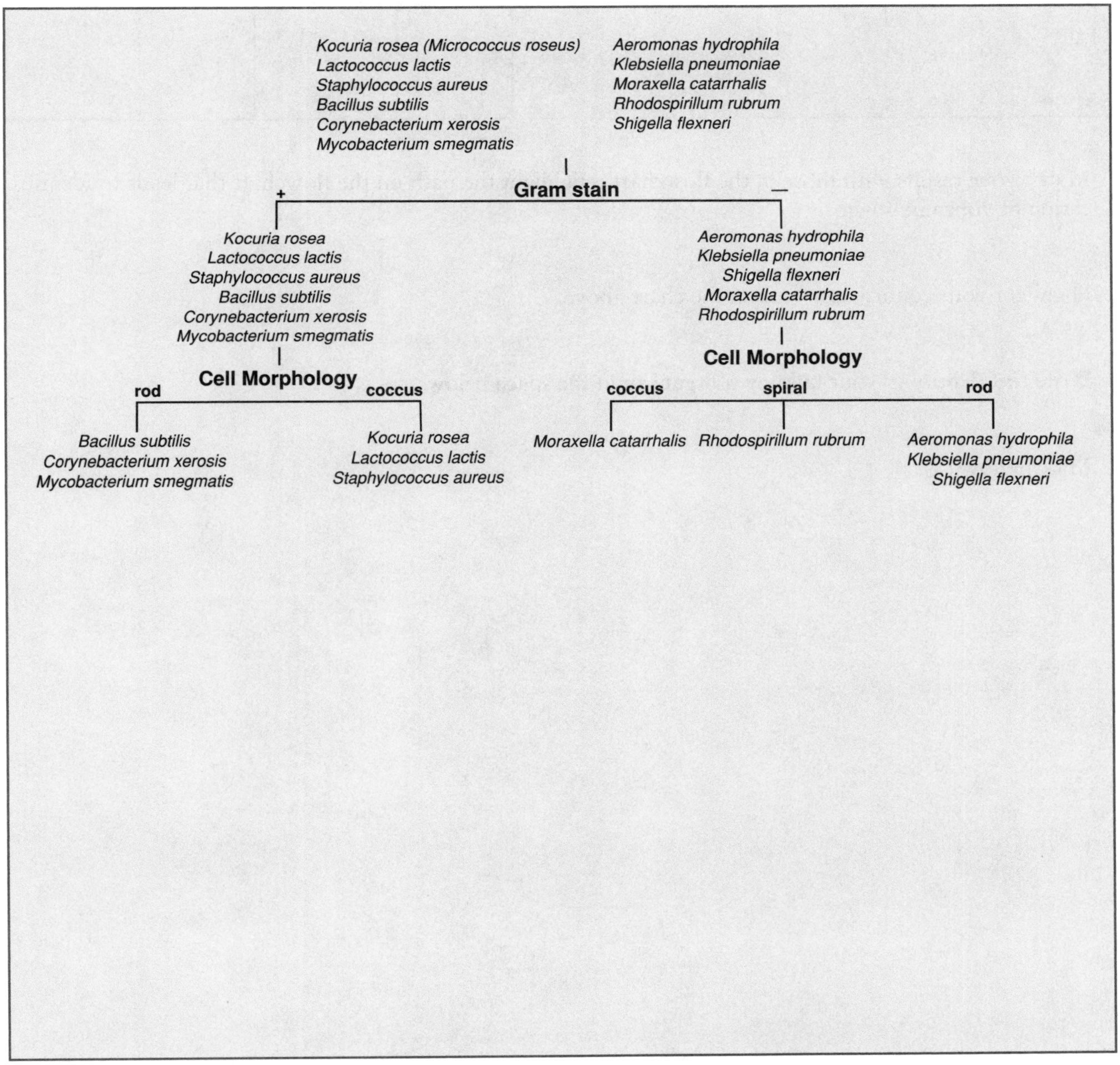

2. Record your stain results and the date each was run in the chart below.

Unknown number: ________________

	Gram Stain	Cell Morphology	Cell Arrangement (in Broth)	Cell Dimensions (μm)	Acid-Fast Stain	Motility (Wet Mount)	Capsule Stain	Spore Stain (shape and location)
Date Run								
Result								

3. Match your results with those in the flowchart. Highlight the path on the flowchart that leads to identification of your unknown.

4. Highlight your confirmatory test in the chart above.

5. Write the identity of your unknown organism in the space below.

My Unknown is:

Exercise 4-1

MANNITOL SALT AGAR

Data Sheet

Name .... Date..............................

Lab Section .... I was present and performed this exercise (initials)...........

Observations and Interpretations

Refer to Table 4-1 when recording your results and interpretations in the chart below. Use abbreviations or symbols as needed.

Organism	Growth (P/G)		MSA Growth Color (Y/R)	Interpretation
	MSA	NA		

Questions

1. What purpose does the Nutrient Agar plate serve? In what way does it increase the validity of the test result?

2. What would be the likely consequences of omitting the NaCl in Mannitol Salt Agar? Why?

3. Would omitting the NaCl alter the medium's specificity or sensitivity? Explain.

4. Which ingredient(s) supply(ies)

   a. Carbon?

   b. Nitrogen?

5. With the diversity of microorganisms in the world, how can a single test such as MSA be used to confidently identify *Staphylococcus aureus*?

EXERCISE 4-2

PHENYLETHYL ALCOHOL AGAR

Data Sheet

Name .... Date..............................

Lab Section .... I was present and performed this exercise (initials)...........

Observations and Interpretations

Refer to Table 4-2 when recording your results and interpretations in the chart below.

Organism	Growth (P/G)		Interpretation
	PEA	NA	

Questions

1. You were instructed to compare the growth on the NA and PEA plates. Because the two media used in this exercise were completely different and likely to produce differing amounts of growth, why not simply compare the organisms to each other on the PEA plate? In your explanation, include the information provided by the growth on the NA plate.

2. PEA contains only 0.25% phenylethyl alcohol because high concentrations inhibit both Gram-negative organisms and Gram-positive organisms. List some possible reasons why this is true.

3. If you observed growth of Gram-negative organisms on your PEA plate, does this negate the usefulness of PEA as a selective medium? Why or why not?

4. Is PEA a defined or an undefined medium? Why is this formulation desirable?

5. Which ingredient(s) in PEA supply(ies)

   a. Carbon?

   b. Nitrogen?

EXERCISE 4-3

DESOXYCHOLATE AGAR

Data Sheet

Name .... Date..............................

Lab Section .... I was present and performed this exercise (initials)...........

Observations and Interpretations

Refer to Table 4-3 when recording your results and interpretations in the chart below. Use abbreviations or symbols as needed.

Organism	Growth (P/G)		DOC Growth Color (R/C)	Interpretation
	DOC	NA		

Questions

1. What purpose does the Nutrient Agar plate serve? In what way does it increase the validity of the test results?

2. With respect to DOC medium, what would be the likely consequence of:
   a. adding glucose to the medium?

   b. not adding desoxycholate?

   c. not adding neutral red?

3. How would the changes in question #1 affect the sensitivity and specificity of the medium?

4. If you saw two DOC plates streaked with water samples suspected of fecal contamination—one with a mixture of pink and colorless colonies and one with all colorless colonies—which would you be more concerned about? Explain.

5. Which ingredient(s) in DOC supply(ies)
   a. Carbon?

   b. Nitrogen?

6. Compare the recipes of Nutrient Agar and Desoxycholate Agar. If an organism can grow on both media, on which would you expect it to grow better? Why?

7. DOC is a selective medium. Is it also a defined or an undefined medium? Why is that formulation desirable?

EXERCISE 4-4

ENDO AGAR

Data Sheet

Name .... Date..............................

Lab Section .... I was present and performed this exercise (initials)...........

Observations and Interpretations

Refer to Table 4-4 when recording and interpreting your results in the chart below. Use abbreviations or symbols as needed.

Organism	Growth (G/P)		Endo Growth Color (R/RD/C)	Interpretation
	ENDO	NA		

Questions

1. What purpose does the Nutrient Agar plate serve? In what way does it increase the validity of the test result?

2. With respect to Endo Agar, what would be the likely consequence of:
   a. adding glucose to the medium?

b. removing the lactose?

c. not adding sodium sulfite?

d. not adding basic fuchsin?

3. How would the changes in question #2 affect the sensitivity and specificity of the medium?

4. Which ingredient(s) in Endo Agar supply(ies)

a. Carbon?

b. Nitrogen?

Exercise 4-5

EOSIN METHYLENE BLUE AGAR

Data Sheet

Name .... Date..............................

Lab Section .... I was present and performed this exercise (initials)...........

Observations and Interpretations

Refer to Table 4-5 when recording and interpreting your results in the chart below. Use abbreviations or symbols as needed.

Organism	Growth (P/G)		EMB Growth Color (Pi/D/C)	Interpretation
	EMB	NA		

Questions

1. What purpose does the Nutrient Agar plate serve? In what way does it increase the validity of the test result?

2. Dipotassium phosphate is a buffer added to EMB that adjusts the pH to the proper starting level. What would be a possible consequence of adding buffers to raise the starting pH to 7.8?

3. Would the change in starting pH suggested in question #1 alter the medium's sensitivity or specificity?

4. You are becoming aware of the diversity and abundance of bacteria, and that we always have some degree of uncertainty about the identity of an isolate. With this in mind, how can we be so certain that growth with a green metallic sheen is truly a coliform bacterium?

5. Compare the recipes of Nutrient Agar and EMB Agar. If an organism can grow on both media, on which would you expect it to grow better? Why?

6. EMB is a selective medium. Is it also a defined or an undefined medium? Why is that formulation desirable?

7. Which ingredient(s) in EMB supply(ies)

   a. Carbon?

   b. Nitrogen?

EXERCISE 4-6

HEKTOEN ENTERIC AGAR

Data Sheet

Name .... Date.............................

Lab Section .... I was present and performed this exercise (initials)...........

Observations and Interpretations

Refer to Table 4-6 when recording your results and interpretations in the chart below. Use abbreviations or symbols as needed.

Organism	Growth (P/G)		HE Growth Color (Pi/Bppt/B)	Interpretation
	HE	NA		

Questions

1. What purpose does the Nutrient Agar plate serve? In what way does it increase the validity of the test result?

2. Which ingredient(s) in this medium supply(ies)

   a. Carbon?

   b. Nitrogen?

3. Compare the recipes of Nutrient Agar and HE agar. If an organism can grow on both media, on which would you expect it to grow better? Why?

4. HE Agar is a selective medium. Is it also a defined or an undefined medium? Why is that formulation desirable?

5. This medium was designed to differentiate *Salmonella* and *Shigella* from other enterics. *Salmonella* species sometimes produce a black precipitate in their growth; *Shigella* species do not. If you were designing a medium to differentiate these two genera, which ingredients from this medium would you include? Explain.

6. All enterics ferment glucose. What would be some consequences of replacing the sugars in this medium with glucose? What color combinations would you expect to see?

7. List all the things you know about a viable organism that produces green colonies with black centers on the medium in question #6.

Exercise 4-7

MacCONKEY AGAR

Data Sheet

Name .... Date..............................

Lab Section .... I was present and performed this exercise (initials)...........

Observations and Interpretations

Refer to Table 4-7 when recording your results and interpretations in the chart below. Use abbreviations or symbols as needed.

Organism	Growth (P/G)		MAC Growth Color (R/C)	Interpretation
	MAC	NA		

Questions

1. What purpose does the Nutrient Agar plate serve? In what way does it increase the validity of the test results?

2. With respect to the MacConkey Agar, what would be the possible consequence of:

   a. replacing the lactose with glucose?

   b. replacing the neutral red with phenol red (yellow when acidic; red or pink when alkaline)?

3. How would removing crystal violet from MacConkey Agar alter the sensitivity and specificity of the medium?

4. Compare the recipes of Nutrient Agar and MacConkey Agar. If an organism can grow on both media, on which would you expect it to grow better? Why?

5. MacConkey Agar is a selective medium. Is it also a defined or an undefined medium? Why is that formulation desirable?

EXERCISE 4-8

XYLOSE LYSINE DESOXYCHOLATE AGAR

Data Sheet

Name .... Date..............................

Lab Section .... I was present and performed this exercise (initials)...........

Observations and Interpretations

Refer to Table 4-8 when recording your results and interpretations in the chart below. Use abbreviations or symbols as needed.

Organism	Growth (P/G)		XLD Growth Color (Y/RB/R)	Interpretation
	XLD	NA		

Questions

1. What purpose does the Nutrient Agar plate serve? In what way does it increase the validity of the test result?

2. Which ingredient(s) in this medium supply(ies)
   a. Carbon?
   b. Nitrogen?

3. Compare the recipes of Nutrient Agar and XLD Agar. If an organism can grow on both media, on which would you expect it to grow better? Why?

4. XLD is a selective medium. Is it also a defined or an undefined medium? Why is that formulation desirable?

5. What would be the likely consequence of:

   a. incubating the medium for 48 hours?

   b. not adding desoxycholate?

   c. not adding sucrose and lactose?

   d. not adding lysine?

   e. not adding ferric ammonium citrate?

6. How would each of the changes in question #5 affect the sensitivity and specificity of the medium?

7. *Shigella* and *Providencia* species virtually never ferment xylose, sucrose or lactose. Since the medium was designed to directly identify these organisms, why were three unusable carbohydrates included when only one would seem to accomplish the goal?

EXERCISE 5-1

O-F MEDIUM

Data Sheet

Name .... Date..............................

Lab Section .... I was present and performed this exercise (initials)...........

Observations and Interpretations

Refer to Table 5-1 when recording your results and interpretations in the chart below.

Organism	Color Results		Symbol	Interpretation
	Sealed	Unsealed		
Uninoculated Control				

Questions

1. What is the purpose of the uninoculated control tubes used in this test?

2. Some microbiologists recommend inoculating a pair of O–F basal media (without carbohydrate) along with the carbohydrate media. Why do you think this is done?

3. All enterics are facultative anaerobes; that is, they have both respiratory and fermentative enzymes. What color results would you expect organisms in O–F glucose media inoculated with an enteric? Remember to describe both sealed and unsealed tubes.

4. Considering question #3, would you be more concerned about an infection isolate producing yellow in both tubes or one producing yellow in only the unsealed tube? Why?

5. If, upon examining your tubes, you were to discover that all tubes except the *sealed* control were yellow, what conclusions could you safely make? What if all tubes except the *unsealed* control were yellow?

EXERCISE 5-2

PHENOL RED BROTH

Data Sheet

Name .... Date...............................

Lab Section .... I was present and performed this exercise (initials)............

Observations and Interpretations

Enter your results in the chart below, using the symbols shown in Table 5-2.

Organism	Results				Interpretation
	PR Base	PR Glucose	PR Lactose	PR Sucrose	
Uninoculated Control					GLU
					LAC
					SUC
					GLU
					LAC
					SUC
					GLU
					LAC
					SUC
					GLU
					LAC
					SUC
					GLU
					LAC
					SUC

Questions

1. This test produces many color combinations. Yellow is easy to distinguish from pink or red. Why use both base broth controls and carbohydrate controls, and what specific purpose do they serve in this exercise?

2. Early formulations of this medium used a smaller amount of carbohydrate and occasionally produced false alkaline (pink) results after 48 hours. Why do you think this happened? List at least two steps, as a microbiologist, you could take to prevent the problem.

3. Suppose you inoculate a PR broth with a slow-growing fermenter. After 48 hours, you see slight turbidity but score it as (–/–). Is this result a false positive or a false negative? Is the failure a result of a lack of specificity or sensitivity of the test system?

4. Interpret each of the combinations of results below. Explain how the results might be obtained and whether they are reliable or not.

PR Glucose	PR Lactose	PR Sucrose	Interpretation	Reliable? Y/N
A / G	A / G	A / G		
A / –	A / –	– / –		
A / G	A / G	K		
– / –	A / G	A / G		
K	A / –	A / –		
K	A / –	– / –		
A / –	A / G	– / –		
A / –	A / G	A / G		

EXERCISE 5-3

PURPLE BROTH

Data Sheet

Name .... Date..............................

Lab Section .... I was present and performed this exercise (initials)...........

Observations and Interpretations

Enter your results in the chart below using the symbols shown in Table 5-3.

Organism	Results				Interpretation
	Purple Base Broth	Purple Glucose Broth	Purple Lactose Broth	Purple Sucrose Broth	
Uninoculated Control					GLU
					LAC
					SUC
					GLU
					LAC
					SUC
					GLU
					LAC
					SUC
					GLU
					LAC
					SUC
					GLU
					LAC
					SUC

Questions

1. When you examine these tubes after 48 hours, what color do you expect the Purple Base Broths to be? How will you proceed if they are yellow? Purple?

2. If this exercise produces a variety of results, why not just compare the tubes to each other instead of using a control of each carbohydrate broth? What purpose do the controls serve in this exercise?

3. Suppose after 48 hours incubation, you examined all of the broths inoculated with one of the organisms and found that they were turbid (*i.e.,* the organism grew well in all of the tubes). Upon further investigation, you discovered that the color of the glucose broth and lactose broth were unchanged while the sucrose broth was yellow. Are these results reliable? Explain.

4. Interpret each of the combinations of results below. Explain how the results might be obtained and whether they are reliable or not.

Purple Broth	Purple Lactose Broth	Purple Sucrose Broth	Interpretation	Reliable? Y/N
A/G	A/G	A/G		
A/–	A/–	–/–		
A/G	A/G	K		
–/–	A/G	A/G		
K	A/–	A/–		
K	A/–	–/–		
A/–	A/G	–/–		
A/–	A/G	A/G		

Exercise 5-4

METHYL RED AND VOGES-PROSKAUER TESTS

Data Sheet

Name .... Date..............................

Lab Section .... I was present and performed this exercise (initials)...........

Observations and Interpretations

Refer to Tables 5-4 and 5-5 as you record your results and interpretations in the chart below.

Organism	MR Result	VP Result	Interpretation
Uninoculated Control			

Questions

1. Some protocols call for a shorter incubation time for the MR and VP tests. Other protocols allow for up to 10 days incubation with virtually no risk of producing a false positive.

   a. Which of the two tests would likely produce more false negatives with a shorter incubation time. Why?

   b. Which test would likely benefit most from a longer incubation time? Why?

2. Would a false negative result for the VP test more likely be attributable to a lack of sensitivity or to specificity of the test system?

3. Why were you told to shake the VP tubes after the reagents were added?

4. Why is the Methyl Red Test read immediately and the Voges-Proskauer read after 60 minutes?

5. Why are organisms producing Methyl Red-negative results re-incubated for an additional 2 to 3 days?

EXERCISE 5-5

CATALASE TEST

Data Sheet

Name .... Date..............................

Lab Section .... I was present and performed this exercise (initials)...........

Observations and Interpretations

Using Table 5-6 as a guide, record your results and interpretations in the chart below.

Organism	Bubbles? Y / N	+ / −	Interpretation
Uninoculated control			

Questions

1. When flavoprotein transfers electrons directly to the final electron acceptor, hydrogen peroxide is produced. What other consequences might result from electron carriers in the ETC being bypassed?

2. Reduction often is referred to as the addition of hydrogens to a compound. It is more chemically correct to refer to reduction as the addition of electrons. Provide an example of a reduction reaction performed by cells in which hydrogen is *not* added to the reduced compound.

3. Would a false positive from the reaction between the inoculating loop and hydrogen peroxide be caused by a lack of specificity or by a lack of sensitivity of the test system? Explain.

4. If a weakly catalase-positive organism initially gives a weak positive result (recorded as negative) and is not checked under the microscope, would the false negative result be attributable to a failure of specificity or to sensitivity of the test system? Explain.

5. Why is it advisable to perform this test on a known catalase-positive organism along with the organism you are testing?

6. What is the purpose of adding hydrogen peroxide to the uninoculated tube?

Exercise 5-6

OXIDASE TEST

Data Sheet

Name .... Date..............................

Lab Section .... I was present and performed this exercise (initials)...........

Observations and Interpretations

Using Table 5-7 as a guide, enter your results in the chart below.

Organism	Color Result	+ / −	Interpretation
Positive Control			

Questions

1. Why is it advisable to test a known oxidase-positive organism along with the organism you are checking?

2. Is blue color formation after 45 seconds in the presence of oxidized cytochrome c oxidase an example of a false positive or a false negative result? Is this a result of lack of sensitivity or of specificity of the test system? Explain.

3. Cytochrome c oxidase is the last component of the electron transport chain prior to oxygen. Provide a possible explanation as to why this test, and not the other electron carriers, identifies the presence of cytochrome c oxidase.

EXERCISE 5-7

NITRATE REDUCTION TEST

Data Sheet

Name .... Date..............................

Lab Section .... I was present and performed this exercise (initials)...........

Observations and Interpretations

Using Table 5-8 as a guide, enter your results and interpretations in the chart below.

	Results			
	Gas	Color		
Organism	Y / N	After Reagents	After Zinc	Interpretation
Uninoculated Control				

Questions

1. Why is gas production not recognized as nitrate reduction when the organism is a known fermenter?

2. Suppose you remove your test cultures from the incubator and notice that one of them—a known fermenter—has a gas bubble in the Durham tube. Knowing that fermenters frequently produce hydrogen gas, you ignore the bubble and proceed to the next step. Adding reagents produces no change, nor does adding zinc. Is this occurrence consistent with what you have learned about this test? Why?

3. Would you change your answer to #2 above if the control broth did not change color after the addition of reagents? What if the control broth changed color only after the addition of reagents and zinc?

4. When testing microaerophiles, some microbiologists prefer to use a semi-solid nitrate medium that contains a small amount of agar. Why do you think this is done?

EXERCISE 5-8

CITRATE TEST

Data Sheet

Name .... Date..............................

Lab Section .... I was present and performed this exercise (initials)...........

Observations and Interpretations

Using Table 5-9 as a guide, enter your results and interpretations in the chart below.

Organism	Color Result	+ / −	Interpretation
Uninoculated Control			

Questions

1. Many bacteria that are able to metabolize citrate are not citrate-positive. Why? Be specific. Refer to Appendix A for help.

2. If an organism is able to convert the sodium citrate in Simmons Citrate Agar to pyruvate, list some possible reasons why an organism might ferment it rather than metabolize it oxidatively in the Krebs cycle. Refer to Appendix A for help.

3. Explain how a citrate-positive organism might not produce a color change in the medium. Is this a false positive or a false negative result? Is the false result attributable to a lack of sensitivity or lack of specificity in the test system?

EXERCISE 5-9

MALONATE TEST
Data Sheet

Name .... Date...............................

Lab Section .... I was present and performed this exercise (initials)...........

Observations and Interpretations

Using Table 5-10 as a guide, enter your results and interpretations in the chart below.

Organism	Color Result	+ / −	Interpretation
Uninoculated Control			

Questions

1. Examine the structural formulas below. You have learned that there is specificity between an enzyme and its substrate. How is it possible for succinate dehydrogenase to bind to two different substrates?

Malonic Acid Succinic Acid

2. The cell has hundreds of succinate dehydrogenase enzymes. How do you suppose malonate concentration affects cell growth?

3. How can a yellow color and no color change both be considered negative results for this test?

4. What is the purpose of the uninoculated control?

EXERCISE 5-10

DECARBOXYLATION TEST

Data Sheet

Name .... Date..............................

Lab Section .... I was present and performed this exercise (initials)...........

Observations and Interpretations

Using Table 5-11 as a guide, record your results and interpretations in the chart below..

Organism	Results								Interpretation
	Lysine		Ornithine		Arginine		Base		
	Color	+/−	Color	+/−	Color	+/−	Color	+/−	
Uninoculated Control									

Questions

1. What was the purpose of inoculating the base broths? Would a positive result in any of them affect the rest of the exercise? Why?

2. How can *no change* and a *conversion to a yellow color* of the broth both be considered negative results?

3. Incubation time for this medium is 1 week. Under what circumstances would an early reading be allowable? Not allowable? Explain.

4. Would not adding the sterile mineral oil likely reduce the specificity or the sensitivity of the test? Explain.

EXERCISE 5-11

PHENYLALANINE DEAMINASE TEST

Data Sheet

Name .... Date..............................

Lab Section .... I was present and performed this exercise (initials)...........

Observations and Interpretations

Using Table 5-12 as a guide, record your results and interpretations in the chart below.

Organism	Color Result	+ / −	Interpretation
Uninoculated Control			

Questions

1. What is the purpose of the uninoculated control?

2. If you are performing this test on an unknown organism, why is it a good idea to run simultaneous tests on known phenylalanine-positive and phenylalanine-negative organisms?

3. What do deamination and decarboxylation reactions have in common?

4. Phenylalanine medium is sometimes prepared in broth form with a pH indicator similar to that used in Decarboxylase Medium increases in pH can then be detected by development of purple color. When testing an organism for the ability to deaminate phenylalanine, would you expect to overlay this medium with mineral oil? Why or why not?

EXERCISE 5-12

BILE ESCULIN TEST

Data Sheet

Name .... Date..............................

Lab Section .... I was present and performed this exercise (initials)...........

Observations and Interpretations

Refer to Table 5-13 when recording your results and interpretations in the chart below.

Organism	Color Result	+ / −	Interpretation
Uninoculated Control			

Questions

1. Some organisms are able to grow on the medium without hydrolyzing the esculin. Other organisms are able to hydrolyze esculin but darken only a portion of the medium. Are either of these types of organisms considered positive for this test? If yes, why? If no, under what circumstances could they be considered positive? Explain.

2. In terms of the selectivity and differential capabilities of the medium, how would its utility be affected by not including sodium azide? Ferric citrate? Explain.

3. This medium sometimes is prepared in plates. Under those circumstances, even slight blackening of the agar (in contrast to more than half of the slanted version) is considered positive. Why do you think this is so?

4. Would it be acceptable to read an early negative result for this test? Why or why not?

EXERCISE 5-13

STARCH HYDROLYSIS

Data Sheet

Name .... Date..............................

Lab Section .... I was present and performed this exercise (initials)...........

Observations and Interpretations

Using Table 5-14 as a guide, enter your results and interpretations in the chart below.

Organism	Result	+ / −	Interpretation
Uninoculated Control			

Questions

1. Suppose you poured iodine on your plate and noticed clearings in the uninoculated area, as well as around both of your transferred cultures. What are some possible explanations for this occurrence? Has the integrity of the exercise been compromised? What kinds of things might be done to avoid this problem in future exercises?

2. How would you expect the results of this exercise to change if you were to add glucose to the medium?

3. In many tests it is acceptable to read a positive result before the incubation time is completed. Why is this not the case with Starch Agar?

4. Suppose you could selectively prevent production of α-amylase or oligo-1,6-glucosidase in an organism that normally hydrolyzes starch. Which enzyme would the organism miss the most?

EXERCISE 5-14

ONPG TEST

Data Sheet

Name .... Date...............................

Lab Section .... I was present and performed this exercise (initials)............

Observations and Interpretations

Using Table 5-15 as a guide, enter your results and interpretations in the chart below.

Organism	Result	+ / −	Interpretation
Uninoculated Tube			

Questions

1. β-galactosidase is an inducible enzyme.

   a. What does "inducible" mean?

   b. What is the purpose of inoculating the ONPG tubes with growth from Kligler's Iron Agar cultures?

2. What would be a possible consequence of inoculating the ONPG medium with an ONPG-positive organism grown on Nutrient Agar instead of Kligler's Iron Agar or Triple Sugar Iron Agar?

3. Examine the phenotypes of the organisms listed below. For each organism, predict the ONPG result.

Organism	β-galactoside permease	β-galactosidase	Predicted ONPG Result (+ or −)
A	Present	Present	
B	Present	Absent	
C	Absent	Present	
D	Absent	Absent	

4. When running this test on an unknown organism, why should you run a simultaneous test on a known ONPG-positive organism?

EXERCISE 5-15

UREASE TESTS

Data Sheet

Name .... Date.............................

Lab Section .... I was present and performed this exercise (initials)...........

Observations and Interpretations

1. Using Table 5-16 as a guide, enter your results for the agar test each day for 6 days or until pink color appears.

Urease Agar								
	Color							
Organism	24 Hrs.	2 Days	3 Days	4 Days	5 Days	6 Days	+/−	Interpretation
Uninoculated Control								

2. Using Table 5-17 as a guide, enter your results for the broth test after 24 hours.

Urease Broth			
Organism	Color Result	+ / −	Interpretation
Uninoculated Control			

Questions

1. What is the purpose of the uninoculated control?

2. Explain why it is acceptable to record positive tests before the suggested incubation time is completed but it is not acceptable to record a negative result early.

3. How can two urease-positive organisms take different amounts of time to turn the medium pink?

4. Did your results from the broth and solid media agree for each organism? Did you expect them to? Why or why not? How could you explain any differences?

EXERCISE 5-16

CASEASE TEST

Data Sheet

Name .... Date..............................

Lab Section .... I was present and performed this exercise (initials)...........

Observations and Interpretations

Using Table 5-18 as a guide, record your results in the chart below.

Organism	Result	+ / −	Interpretation
Uninoculated Control			

Questions

1. What does the enzyme casease have in common with amylase?

2. How do we know that casease is an exoenzyme and not a cytoplasmic enzyme?

3. Is it acceptable to read a positive casease test before the incubation time is completed? How about an early negative result?

4. Why is the uninoculated control relatively unnecessary in this test?

5. Why is it advisable to use a positive control along with organisms that you are testing?

EXERCISE 5-17

GELATINASE TEST

Data Sheet

Name .... Date..............................

Lab Section .... I was present and performed this exercise (initials)...........

Observations and Interpretations

Using Table 5-19 as a guide, record your liquefaction results in the chart below.

Organism	Result	+ / −	Interpretation
Uninoculated Tube			

Questions

1. What is the purpose of the uninoculated control in this test?

2. If the control is solid and an inoculated tube is liquid, is it acceptable to read the result before the complete incubation time has elapsed?

3. If the control is solid and an uninoculated tube is also solid, is it acceptable to read the result before the complete incubation time has elapsed?

4. Suggest some ways by which an organism could be a slow gelatin liquefier.

5. Suppose that after 7 days there is no evidence of liquefaction in a tube inoculated with a slow liquefier. Is this occurrence a result of failure of the test system's sensitivity or specificity?

Exercise 5-18

DNase Test

Data Sheet

Name .... Date..............................

Lab Section .... I was present and performed this exercise (initials)............

Observations and Interpretations

Using Table 5-20 as a guide, record your DNase results in the chart below.

Organism	Result	+ / −	Interpretation
Uninoculated Tube			

Questions

1. On page 173 the disassembly of DNA was described as a "depolymerization." What other term applies to the process?

2. A positive result for the DNase test does not distinguish between *Staphylococcus* DNase and the DNase produced by *Serratia*. Is this a failure of the system's specificity or sensitivity?

3. Suggest a reason why this test is read after only 24 hours, while other tests (*e.g.*, gelatinase test) may take a week.

4. Why is the uninoculated control relatively unnecessary in this test?

5. Why is it advisable to use a positive control along with organisms that you are testing?

EXERCISE 5-19

LIPASE TEST

Data Sheet

Name .... Date..............................

Lab Section .... I was present and performed this exercise (initials)...........

Observations and Interpretations

Using Table 5-21 as a guide, record your results and interpretations in the table below.

Organism	Result	+ / −	Interpretation
Uninoculated Tube			

Questions

1. Is the inability of Tributyrin Agar to differentiate between positive results produced by a variety of different lipases a failure of the sensitivity or the specificity of the test system?

2. Tributyrin agar has a shelf life of only a few days before it loses its opacity.

   a. With this in mind, explain the importance of positive and negative controls in this test.

   b. How would expired Tributyrin Agar affect the results of lipase (+) and lipase (−) organisms?

3. Imagine this situation: Species #1 and Species #2 are unrelated, but each produces an enzyme capable of hydrolyzing soybean oil. How can this be reconciled with the fact that enzymes are very specific to their substrates.

4. Why is Tributyrin Agar prepared as an emulsion rather than a solution?

Exercise 5-20

SIM Medium

Data Sheet

Name .... Date..............................

Lab Section .... I was present and performed this exercise (initials)...........

Observations and Interpretations

1. Using Table 5-22 as a guide, record your results and interpretations in the chart below.

Sulfur Reduction			
Organism	Black PPT? Y / N	+ / −	Interpretation
Uninoculated Control			

2. Using Table 5-23 as a guide, record your results and interpretations in the chart below.

Indole Production			
Organism	Red Color? Y / N	+ / −	Interpretation
Uninoculated Control			

3. Using Table 5-24 as a guide, record your results and interpretations in the chart below.

Motility			
Organism	Growth Pattern	+ / −	Interpretation
Uninoculated Control			

Questions

1. The sulfur reduction test is not able to differentiate between H_2S produced by anaerobic respiration and H_2S produced by putrefaction. Is this a failure of sensitivity or specificity of the test system?

2. What factors dictate the choice of tests included in a combination medium?

3. Which ingredient could be eliminated if this medium were used strictly for testing motility among sulfur reducers?

4. Which ingredient could be eliminated if this medium were used strictly for testing motility in indole producers?

EXERCISE 5-21

TRIPLE SUGAR IRON AGAR

Data Sheet

Name .... Date..............................

Lab Section .... I was present and performed this exercise (initials)...........

Observations and Interpretations

Refer to Table 5-25 when recording and interpreting your results.

Organism	Result	+ / −	Interpretation
Uninoculated Control			

Questions

1. TSI is a combination medium. Which other test(s) does it replace?

2. Why is TSI medium inoculated with both a stab and a streak?

3. TSI is a complex medium with many ingredients. What would be the consequences of the following mistakes in preparing this medium? Consider each independently.

   a. 1% glucose is added rather than the amount specified in the recipe.

   b. Ferrous sulfate is omitted.

   c. Beef extract is omitted.

   d. Sodium thiosulfate is omitted.

   e. Phenol red is omitted.

   f. The initial pH is 8.2.

   g. The agar butt is shallow rather than deep.

EXERCISE 5-22

LYSINE IRON AGAR

Data Sheet

Name .... Date..............................

Lab Section .... I was present and performed this exercise (initials)...........

Observations and Interpretations

Refer to Table 5-26 when recording and interpreting your results.

Organism	Result	Symbol	Interpretation
Uninoculated Control			

Questions

1. LIA is a combination medium. Which test(s) does it replace?

2. Lysine Iron Agar is not designed to identify ability of organisms to ferment. Why, then, is glucose added to the medium (beyond being just another nutrient)?

3. LIA is a complex medium with many ingredients. What would be the consequences of the following mistakes in preparing this medium? Consider each independently.

   a. Inclusion of 1% dextrose rather than the amount called for in the recipe.

   b. Omission of L-Lysine hydrochloride.

   c. Omission of Ferric ammonium citrate.

   d. Omission of Sodium thiosulfate.

   e. Omission of Bromcresol purple.

   f. Preparation of the medium with 1.5% agar rather than the amount called for in the recipe.

EXERCISE 5-23

LITMUS MILK

Data Sheet

Name .... Date...............................

Lab Section .... I was present and performed this exercise (initials)...........

Observations and Interpretations

Refer to Table 5-27 when recording and interpreting your results below.

Organism	Results	Interpretation
Uninoculated Control		

Questions

1. An acid clot can appear as pink or white with a pink band at the top. Explain the different conditions that would produce each of these occurrences. If the reaction to produce a white color occurs, why does the surface remain pink?

2. What do you think is the principal difference between a bacterial species that produces a curd in litmus milk and a species that does not?

3. What reaction would you predict from an organism growing in litmus milk that has the following results in other media?

   a. a clear zone around the growth on a milk agar plate

   b. A/– in PR lactose broth

   c. A/G in purple lactose broth

   d. K in PR glucose broth

EXERCISE 5-24

BACITRACIN SUSCEPTIBILITY TEST DATA SHEET

Data Sheet

Name .... Date..............................

Lab Section .... I was present and performed this exercise (initials)...........

Observations and Interpretations

Refer to Table 5-28 when recording and interpreting your results in the chart below.

Organism	Zone Diameter (in MM)	S / R	Interpretation

Questions

1. The Bacitracin Test is typically used to differentiate between Gram-positive cocci. Would you predict it to be an effective differential test for Gram-negative organisms? Why or why not?

2. Why is it important to get a bacterial lawn rather than isolated colonies on the plate?

3. Does the zone of inhibition's edge indicate the limit of bacitracin diffusion into the agar? If not, why doesn't the zone extend as far as the bacitracin diffuses?

Exercise 5-25

β-LACTAMASE TEST

Data Sheet

Name .... Date..............................

Lab Section .... I was present and performed this exercise (initials)...........

Observations and Interpretations

Refer to Table 5-29 when recording and interpreting your results in the chart below.

Organism	Result	+ / −	Interpretation

Questions

1. Why is it advisable to perform this test on a β-lactamase–positive organism while checking an unknown isolate?

2. Why are fresh cultures required for this test?

3. Why is Nitrocefin instead of a different cephalosporin used in this test? Does this address an issue of sensitivity or specificity of the test system?

4. How might the rapid increase of penicillin resistance in bacterial populations be explained?

EXERCISE 5-26

BLOOD AGAR

Data Sheet

Name .... Date..............................

Lab Section .... I was present and performed this exercise (initials)...........

Observations and Interpretations

Refer to Table 5-30 when recording and interpreting your results in the chart below.

Source of Culture	Colony Morphology	Hemolysis Result	Interpretation

Questions

1. The streak–stab technique, used to promote streptolysin activity, is preferred over incubating the plates anaerobically. Why do you think this is so? Compare and contrast what you see as the advantages and disadvantages of each procedure.

2. Assuming that all of the organisms cultivated in this exercise came from healthy students' throats, why is it important to cover and tape the plates?

3. Why is the streak plate preferred over the spot inoculations in this procedure?

EXERCISE 5-27

COAGULASE TESTS

Data Sheet

Name .... Date..............................

Lab Section .... I was present and performed this exercise (initials)............

Observations and Interpretations

Refer to Tables 5-31 and 5-32 when recording and interpreting your results in the charts below.

Slide Test Results			
Organism	**Slide**	**Result**	**Interpretation**
	A		
	B		
	A		
	B		

Slide Test Results		
Organism	**Result**	**Interpretation**
Uninoculated Control		

Questions

1. Why is it more important to use fresh cultures in the Coagulase Test than in a test medium such as Milk Agar? *Hint*: Examine the medium recipes.

2. How would you interpret a negative Slide Test and a positive Tube Test using the same organism?

3. Consider the Slide Test.

   a. What is the role of sterile saline plus organism on the Slide Test?

   b. Why is it advisable to run a known coagulase-positive organism along with your unknown organism?

   c. How will the validity of the test be affected if clumping occurs on both smears of the known coagulase-positive organism?

4. List possible reasons why the Slide Test is not appropriate for detecting free coagulase.

EXERCISE 5-28

MOTILITY TEST

Data Sheet

Name .... Date..............................

Lab Section .... I was present and performed this exercise (initials)............

Observations and Interpretations

Refer to Table 5-33 when recording and interpreting your results in the chart below.

Organism	Result	+ / −	Interpretation

Questions

1. Why is it important to carefully insert and remove the needle along the same stab line?

2. List possible ways you might obtain false positive and negative results using motility test medium.

3. You were advised to use extreme caution when transferring motile bacteria to a microscope slide for the flagella stain. Why is a similar degree of caution not necessary when inoculating motility agar?

4. Why is it essential that the reduced TTC be insoluble? Why is there less concern about the solubility of the oxidized form of TTC?

Exercise 5-29

API 20 E

Data Sheet

Name .... Date.................................

Lab Section .... I was present and performed this exercise (initials)............

Observations and Interpretations

Tape your API 20 E Result Sheet here.

Questions

1. Why is it important to perform the reagent tests last?

2. In clinical applications of this test system, reagents are added only if the glucose (oxidation/fermentation) test result is yellow or at least three other tests are positive. If these conditions are not met, a MacConkey Agar plate is streaked and additional tests are performed confirming glucose metabolism, nitrate reduction, and motility. Why do you think this is so? Be specific.

3. Suppose, after 24 hours incubation, you notice no growth in the tubes containing mineral oil. Assuming that it is behaving properly under these conditions, what do you know about the organism and what predictions can you safely make about its performance in the decarboxylase tests, fermentation tests, and nitrate reduction test? Is it a member of *Enterobacteriaceae*?

EXERCISE 5-30

ENTEROTUBE® II

Data Sheet

Name .... Date..............................

Lab Section .... I was present and performed this exercise (initials)...........

Observations and Interpretations

1. Tape your BBL® Enterotube® II Result Sheet here.

2. Enter the results from the CCIS booklet below.

CCIS Five-Digit Code	Possible Organisms	VP Result + / −	Identified Organism

Questions

1. Fecal coliforms such as *Escherichia coli, Enterobacter aerogenes,* and *Klebsiella pneumoniae* are enterics that ferment lactose to acid and gas at 35°C within 48 hours. The chart below summarizes most of these organisms' reactions in the Enterotube® II. Fill in the missing information and, using colored pencils, fill in the appropriate colors for each positive test.

	Glu	Gas	Lys	Orn	Ind	H_2S	Adon	Lact	Arab	Sorb	VP	Dulc	PA	Urea	Citrate
E. coli			+	−	+	−	−		+	+	−	−	−	−	−
E. aerogenes			+	+	−	−	+		+	+	+	−	−	−	+
K. pneumoniae			+	−	−	−	+		+	+	+	−	−	+	+

2. Based on the information above, enter the five-digit codes for the three organisms.

E. coli ________________, *E. aerogenes* ________________, *K. pneumoniae* ________________

3. Examine the following chart. Fill in the color reactions for each test based on the ID value given.

ID Value	Glucose/ Gas	Lysine	Ornithine	Indole/ H_2S	Adonitol	Lactose	Arabinose	Sorbitol	VP	Dulcitol/ PA	Urea	Citrate
31122												
27345												
04004												
05511												

4. Which of the four organisms are not members of *Enterobacteriaceae*? Why?

5. Which of the four ID values is questionable? Why?

EXERCISE 5-31

GRAM-POSITIVE UNKNOWN

Data Sheet

Name .... Date..............................

Lab Section .... I was present and performed this exercise (initials)...........

Unknown Number _______

Isolation Procedure (Please record all activities associated with isolation of your organisms—from mixed culture to pure culture. Always include the date, source of inoculum, destination, incubation temperature, and any other relevant information. Also, make note of transfers made to keep your pure culture fresh. This log must be kept current.)

Preliminary Observations

Colony Morphology (include medium) ________________________________

Gram Stain ______________ Cell Dimensions ______________ Optimum Temperature______________

Cellular Morphology and Arrangement ________________________________

Differential Tests (Include all information through the confirmatory test. This log must be kept current.)

Test #1:____________________ Date Begun:___________ Date Read:___________ Result:_____________

Comments: ____

Test #2:____________________ Date Begun:___________ Date Read:___________ Result:_____________

Comments: ____

Test #3:____________________ Date Begun:___________ Date Read:___________ Result:_____________

Comments: ____

Test #4:____________________ Date Begun:___________ Date Read:___________ Result:_____________

Comments: ____

Test #5:____________________ Date Begun:___________ Date Read:___________ Result:_____________

Comments: ____

Test #6:____________________ Date Begun:___________ Date Read:___________ Result:_____________

Comments: ____

Test #7:____________________ Date Begun:___________ Date Read:___________ Result:_____________

Comments: ____

Test #8:____________________ Date Begun:___________ Date Read:___________ Result:_____________

Comments: ____

Test #9:____________________ Date Begun:___________ Date Read:___________ Result:_____________

Comments: ____

Test #10:____________________ Date Begun:___________ Date Read:___________ Result:_____________

Comments: ____

Test #11:____________________ Date Begun:___________ Date Read:___________ Result:_____________

Comments: ____

Test #12:____________________ Date Begun:___________ Date Read:___________ Result:_____________

Comments: ____

Exercise 5-31

GRAM-NEGATIVE UNKNOWN

Data Sheet

Name .... Date..............................

Lab Section .... I was present and performed this exercise (initials)...........

Unknown Number _______

Isolation Procedure (Please record all activities associated with isolation of your organisms—from mixed culture to pure culture. Always include the date, source of inoculum, destination, incubation temperature, and any other relevant information. Also, make note of transfers made to keep your pure culture fresh. This log must be kept current.)

Preliminary Observations

Colony Morphology (include medium) ____________________

Gram Stain ____________ Cell Dimensions ____________ Optimum Temperature____________

Cellular Morphology and Arrangement ____________________

[1] Your instructor will write what tests to rerun in this space if you misidentify your unknown.

Differential Tests (Include all information through the confirmatory test. This log must be kept current.)

Test #1:________________ Date Begun:__________ Date Read:__________ Result:__________

Comments: ____

Test #2:________________ Date Begun:__________ Date Read:__________ Result:__________

Comments: ____

Test #3:________________ Date Begun:__________ Date Read:__________ Result:__________

Comments: ____

Test #4:________________ Date Begun:__________ Date Read:__________ Result:__________

Comments: ____

Test #5:________________ Date Begun:__________ Date Read:__________ Result:__________

Comments: ____

Test #6:________________ Date Begun:__________ Date Read:__________ Result:__________

Comments: ____

Test #7:________________ Date Begun:__________ Date Read:__________ Result:__________

Comments: ____

Test #8:________________ Date Begun:__________ Date Read:__________ Result:__________

Comments: ____

Test #9:________________ Date Begun:__________ Date Read:__________ Result:__________

Comments: ____

Test #10:________________ Date Begun:__________ Date Read:__________ Result:__________

Comments: ____

Test #11:________________ Date Begun:__________ Date Read:__________ Result:__________

Comments: ____

Test #12:________________ Date Begun:__________ Date Read:__________ Result:__________

Comments: ____

My Unknown is: ________________________ Rerun:[2]________________________

[2] Your instructor will write what tests to rerun in this space if you misidentify your unknown.

EXERCISE 6-1

STANDARD PLATE COUNT

Data Sheet

Name .... Date..............................

Lab Section .... I was present and performed this exercise (initials)...........

Data and Calculations

1. Enter the number of colonies counted on each countable plate. Only one pair of plates should be countable, but for practice, record all countable plates anyway. For all plates containing more than 300 colonies, enter TMTC ("too many to count"). For plates containing fewer than 30, enter TFTC ("too few to count").
2. Take the average number of colonies from the two (or more) countable plates and record it below.

Plate	A_1	A_2	B_1	B_2	C_1	C_2	D_1	D_2
Colonies Counted								
Average # Colonies								

3. Calculate the original density in CFU/mL using the following formula.

$$OCD = \frac{CFU}{FDF}$$

Original density of *E. coli* in the broth (CFU/mL)	

Questions

1. How would you produce a 10^{-1} dilution of 2.0 mL of concentrated cells using the full 2 mL volume?

2. How would you produce a 10^{-2} dilution of 2.0 mL of concentrated cells using the full 2 mL volume?

3. How would you produce a 10^{-2} dilution of a 5 mL bacterial sample using the full 5 mL volume?

4. You have 0.05 mL of an undiluted culture at a concentration of 3.6×10^6 CFU/mL. You then add 4.95 mL sterile diluent. What is the dilution factor, and what is the final concentration of cells?

5. What would be the dilution factor if 96 mL of diluent is added to 4 mL of a bacterial suspension?

6. You were instructed to add 1.0 mL out of 5.0 mL of an undiluted sample to 99 mL of sterile diluent. Instead, you add all 5.0 mL to the 99 mL. What was the intended dilution factor and what was the actual dilution factor?

7. Suppose you were instructed to add 0.2 mL of sample to 9.8 mL of diluent, but instead added 2.0 mL of sample. What was the intended dilution factor and what was the actual dilution factor?

8. Plating 1.0 mL of a sample diluted by a factor of 10^{-3} produced 43 colonies. What was the original concentration in the sample?

9. Plating 0.1 mL of a sample diluted by a factor of 10^{-3} produced 43 colonies. What was the original concentration in the sample?

10. There are 72 colonies on the plate with a final dilution factor of 10^{-7}. What was the original concentration in the sample?

11. There are 259 colonies on the plate with a final dilution factor of 10^{-6}. What was the original concentration in the sample?

12. How many colonies should be on the plate with a final dilution factor of 10^{-7} using the same sample as in Question #11?

13. How many colonies should be on the plate with a final dilution factor of 10^{-5} using the same sample as in Question #11?

14. A plate with a final dilution factor of 10^{-7} produced 170 colonies. What was the original concentration in the sample?

15. After incubation, how many colonies should be on the FDF = 10^{-8} plate from the dilution series in Question #14?

16. After incubation, how many colonies should be on the FDF = 10^{-6} plate from the dilution series in Question #14?

17. You have inoculated 100 µL of a sample diluted by a factor of 10^{-3} on a nutrient agar plate. After incubation, you count 58 colonies. What was the original cell density?

18. A plate that received 1000 µL of a bacterial sample diluted by a factor of 10^{-6} had 298 colonies on it after incubation. What was the original cell density?

19. A nutrient agar plate with an FDF of 10^{-5} had 154 colonies after incubation. What was the cell density in the original sample? What volume was used to inoculate this plate?

20. The original concentration in a sample is 2.79×10^6 CFU/mL. Which final dilution factor should yield a countable plate?

21. The original concentration in a sample is 5.1×10^9 CFU/mL. Which final dilution factor should yield a countable plate?

22. A sample has a density of 1.37×10^5 CFU/mL. What FDF should yield a countable plate? Which two dilution tubes could be used to produce this FDF? How?

23. A sample has a density of 7.9×10^9 CFU/mL. What FDF should yield a countable plate? Which two dilution tubes could be used to produce this FDF? How?

24. You are told that a sample has between 2.5×10^6 and 2.5×10^9 cells/mL. Devise a complete but efficient (that is, no extra plates!) dilution scheme that will ensure getting a countable plate.

25. A sample has between 3.3×10^4 and 3.3×10^8 CFU/mL. Devise a complete but efficient (that is, no extra plates!) dilution scheme that will ensure getting a countable plate.

26. Two plates received 100 µL from the same dilution tube. The first plate had 293 colonies, whereas the second had 158 colonies. Suggest reasonable sources of error.

27. Two parallel dilution series were made from the same original sample. The plates with an FDF of 10^{-5} from each dilution series yielded 144 and 93 colonies. Suggest reasonable sources of error.

28. What are the *only* circumstances that would *correctly* produce countable plates from two different dilutions?

EXERCISE 6-2

URINE CULTURE

Data Sheet

Name .... Date..............................

Lab Section .... I was present and performed this exercise (initials)...........

Data and Calculations

Enter your colony count and loop volume data in the chart below. Then calculate the original cell density using the following formula.

$$\text{OCD} = \frac{\text{CFU}}{\text{loop volume}}$$

Urine Sample	Colonies Counted	Loop volume (0.01 mL or 0.001 mL)	Original Cell Density (DFU/mL)

Questions

1. The plate pictured in Figure 6-5 was inoculated with a 0.01 mL volumetric loop and contains approximately 75 colonies. What was the original cell density?

2. The equation shown in the Theory portion of this exercise is used for calculating cell density in urine when using a 0.001 mL calibrated loop. The urine transferred in the loop is not literally diluted, yet its volume is equivalent to a dilution factor. What is the dilution factor (based on loop volume) expressed as a fraction? What is the dilution factor in scientific notation?

3. Calculation of original density in this exercise differs slightly from that offered in Exercise 6-1. Compare and contrast the formula used today with that used in Exercise 6-1. How do they differ? Could you have used the formula in Exercise 6-1 for today's calculations? Explain.

4. Using a volumetric loop is a semiquantitative technique. Why do you think it is not quantitative? Design a procedure that would make it quantitative. (*Hint:* Refer to Exercise 6-1 if necessary.)

EXERCISE 6-3

DIRECT COUNT

Data Sheet

Name .... Date..............................

Lab Section .... I was present and performed this exercise (initials)...........

Data and Calculations

1. Calculate your dilution factor using the following formula. For an explanation of dilution factors, refer to the footnote on page 224.

$$D_2 = \frac{V_1 D_1}{V_2}$$

2. Enter your data below.

Total Cells Counted	Squares Counted	Dilution Factor	Original Cell Density (calculated)

3. Calculate the original cell density of the *P. vulgaris* culture using the following equation. Record your answer in mL in the table.

$$\text{Original cell density} = \frac{\text{Total cells counted}}{(\text{Squares counted})(\text{Dilution factor})(5 \times 10^{-8}\ \text{mL})}$$

Questions

1. What are the advantages and disadvantages of the direct count over the plate count technique?

2. What would be the original cell density of a sample having a DF of 10^{-1} and a count of 75 cells in five small squares? Record your answer in cells/mL.

3. Suppose you had to count a culture that was very turbid and found that the recommended dilution of 0.4 mL stain A and 0.5 mL stain B per 0.1 mL culture (10^{-1} DF) was not enough. How would you adjust the volumes to make a 10^{-2} DF? Show your work.

4. Suppose you were given a diluted broth to count that already had a dilution factor of 10^{-3}. Assuming that you used the staining procedure described in the Procedure, what would the dilution factor of the solution be when it reached the counting chamber? What would the original cell density be if you counted a total of 116 cells in 16 small squares? Record your answer in cells/mL.

EXERCISE 6-4

PLAQUE ASSAY

Data Sheet

Name .... Date..............................

Lab Section .... I was present and performed this exercise (initials)...........

Data and Calculations

1. Enter the number of plaques counted on the countable plate. Only one plate should be countable. If there are more than 300 plaques, enter TMTC. If there are fewer than 30, enter TFTC.

Plate	A	B	C	D	E	F	G
Plaques Counted							
FDF							

2. Calculate the original density using the following formula, and enter the result below. Record your answer in PFU/mL.

$$\text{Original phage density} = \frac{\text{PFU}}{\text{FDF}}$$

Original density of the Bacteriophage (PFU/mL)	

Questions

1. In this exercise, there must be enough bacteria inoculated to produce a lawn of growth. Why is that important?

2. How might the results be altered if you skipped the preadsorption phase?

3. Suppose you followed all the necessary steps outlined in the Procedure and found no plaques on any of your plates after incubation. Suppose further that you knew with certainty that the bacteriophage was viable and had worked in other labs prior to yours. What possible explanations could there be for this occurrence? Explain.

4. Why was the water bath set at 45°C? What might be some consequences of changing the temperature?

5. Why was Soft Agar used for the agar overlay? What would you expect to see happen if standard Nutrient Agar had been used instead?

EXERCISE 6-5

CLOSED SYSTEM GROWTH
Data Sheet

Name .... Date..............................

Lab Section .... I was present and performed this exercise (initials)...........

Data and Calculations

1. Enter the absorbance values for all groups in Chart A below.
2. On a computer or graph paper, plot your absorbance values versus time. This will produce a growth curve including growth phases as far as the stationary phase. (***Note:*** Some cultures may not even reach the stationary phase. Record what you get!). Also plot the growth curves of the other four samples on the same set of axes.
3. From your graph, determine the approximate length of time spent in the various growth stages for all samples, and enter those values in Chart B.
4. From your graph, determine the absorbance for the lag phase and the stationary phase of each culture, and enter those in Chart B.
5. Calculate the mean growth rate constant for each sample. Do this by choosing two points clearly on the linear part of exponential growth. These are A_1 and A_2. Determine the absorbance values of each, and the time in minutes (t) between the two points. Substitute your values in the equation and solve for (k). Enter your results in Chart C.

$$k = \frac{A_2 - A_1}{0.301t}$$

6. On a computer or the graph paper provided, plot the mean growth rate versus temperature of the different samples.
7. Calculate generation times of the different samples. Enter your results in Chart C.

Chart A — Absorbance Readings

Temp (°C)	T_0	T_{15}	T_{30}	T_{45}	T_{60}	T_{75}	T_{90}	T_{105}	T_{120}	T_{135}	T_{150}	T_{165}	T_{180}	T_{195}	T_{210}	T_{225}	T_{240}
20																	
25																	
30																	
35																	
40																	

Chart B — Growth Phase Duration					
Temp (°C)	Lag Phase (Minutes)	Exponential Phase (Minutes)	Stationary Phase (Minutes)	Lag Phase (Absorbance)	Stationary Phase (Absorbance)
20					
25					
30					
35					
40					

Chart C — Calculations					
Temp (°C)	A_1	A_2	k	g	t
20					
25					
30					
35					
40					

Questions

1. Examine the microbial growth curve in Figure 6-12. In what ways would you expect it to be different from a growth curve of an organism living in a natural environment?

2. Aside from temperature, what factors in today's experiment likely had an effect on the growth? What factors likely kept the organism from realizing true exponential growth?

3. If an organism has a mean generation time of 22 minutes, what is its mean growth rate?

4. If an organism has a mean generation time of 47 minutes, what is its mean growth rate?

5. If an organism has a mean growth rate constant of 1.7 generations per hour, what is its generation time?

6. If an organism has a mean growth rate constant of 0.012 generations per minute, what is its generation time?

7. How would you calculate mean growth rate of an organism that quadrupled every generation?

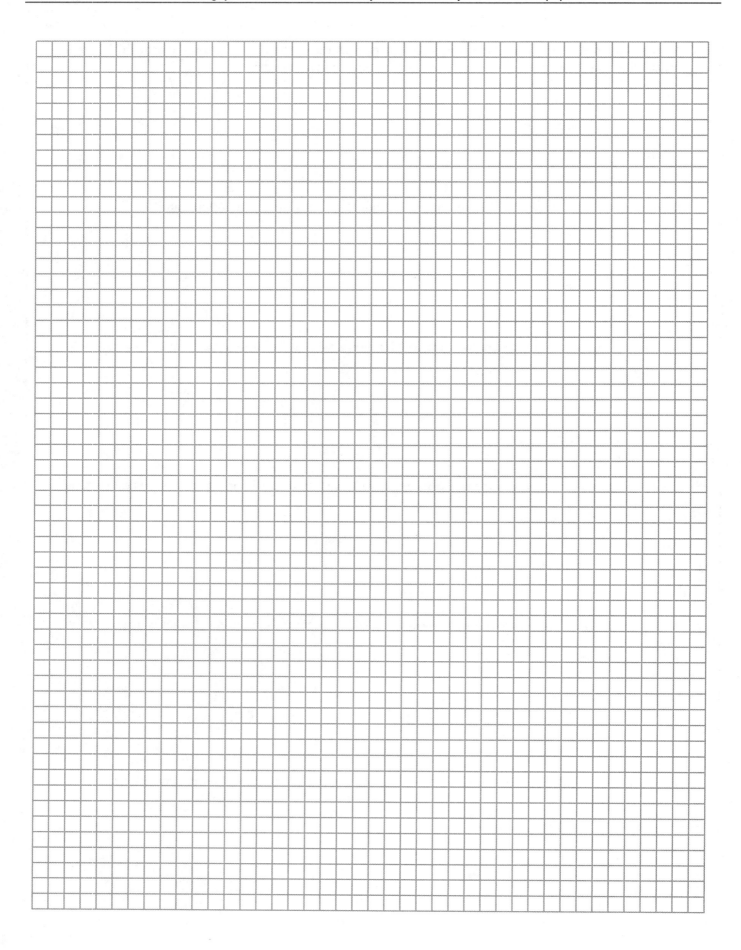

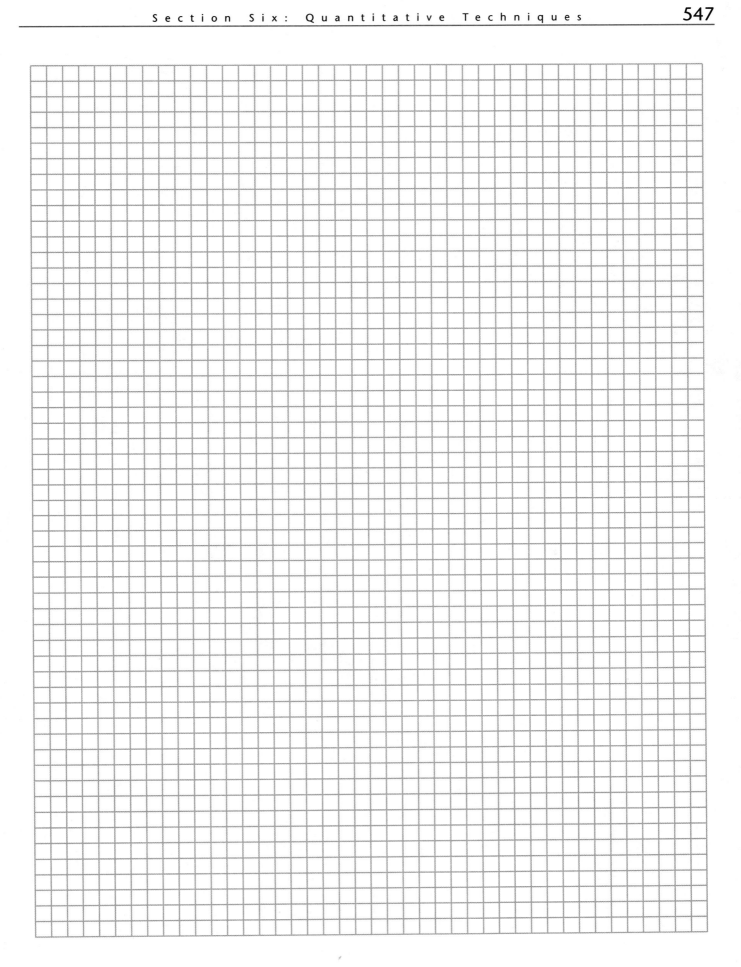

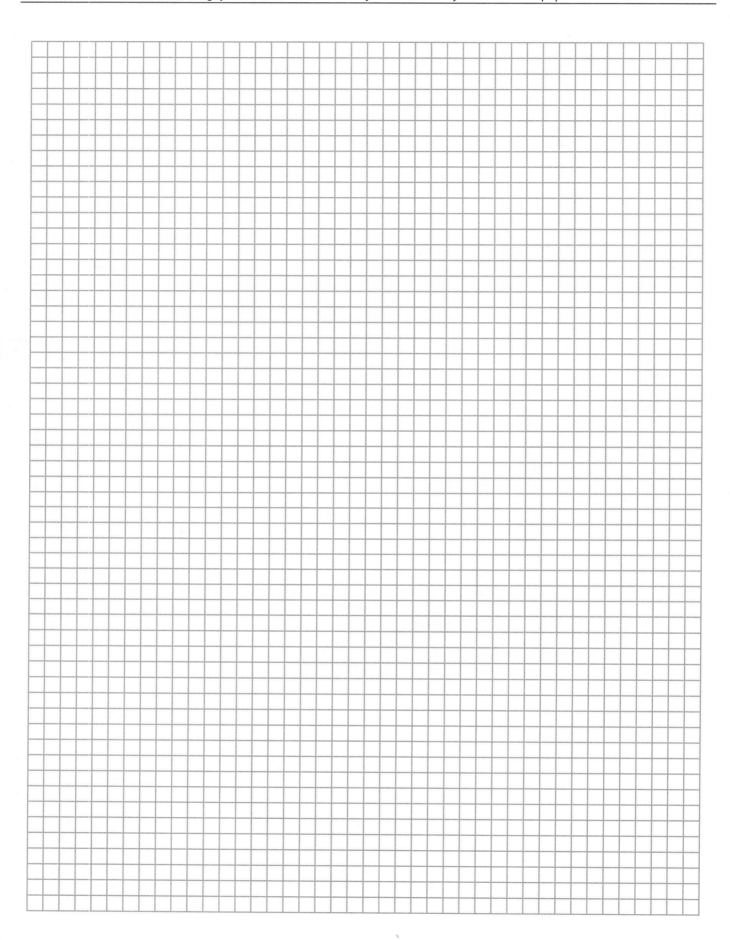

Exercise 6-6

THERMAL DEATH TIME VERSUS DECIMAL REDUCTION VALUE

Data Sheet

Name .... Date...............................

Lab Section .... I was present and performed this exercise (initials)...........

Data and Calculations

It works best to reproduce the Data Sheet charts on the blackboard or a transparency so that the entire class can enter and share data.

Chart 1 — Group Plate Data

Organism:				T$_?$:			
Plate Number	**1**	**2**	**3**	**4**	**5**	**6**	**7**
Plate Number							
Final Dilution Factor							
Colonies Counted							
Cell Density Cells/ML							

Fill in all boxes. For plates containing fewer than 30 or more than 300 colonies, enter TFTC or TMTC respectively.

Chart 2 — Group Broth Data

Organism:					
Time	**G/NG**	**Time**	**G/NG**	**Time**	**G/NG**
T__		T__		T__	
T__		T__		T__	
T__		T__		T__	
T__		T__		T__	
T__		T__		T__	

Circle the first time at which no growth appears in the broth.

Chart 3 — Class Data					
	T_0		T_{10}		
Organism	Cell Density (Cells/ML)	Log G_{10} of Cell Density	Cell Density (Cells/ML)	Log G_{10} of Cell Density	First Broth Without Growth (T_x)
E. coli					
S. aureus					

Chart 4 — Thermal Death Time Versus Decimal Reduction Value			
Organism	Thermal Death Time (min.)	Plotted D_{60} Value (min.)	Calculated D_{60} Value (min.)
E. coli			
S. aureus			

Perform the following for *each* organism.

Constructing the TDT curve

1. On the graph paper provided, construct a graph of Population size (log) versus Time at 60°C, as shown in Figure 6-14. To do this:
   a. Enter a data point on the *y*-axis representing the log of the original cell density of the broth culture at T_0 (from Chart 3).
   b. Enter a second data point on the *x*-axis at the time when growth in the Nutrient Broth tube stopped—thermal death time (from Chart 4).
   c. Draw a straight line between the two points. This is the TDT curve for this culture at 60°C.
2. Use this curve for plotting the D value.

Plotting D value from TDT curve

3. Locate two successive log values on the *y*-axis below the highest population point for the organism, representing one complete log cycle (*e.g.*, 10^6 and 10^5, 10^5 and 10^4, *etc.*).
4. Draw a horizontal line from each of these points on the *y*-axis to where they intersect the TDT curve.
5. Draw a vertical line from these intersection points downward to where they intersect with the *x*-axis.
6. The horizontal distance between these two points reveals the time required to kill 90% of the population. This is the D value of the organism. Enter it in Chart 4 in the box labeled "Plotted D_{60} value."

Calculating D value

1. From Chart 3, obtain the logs of the microbial cell densities before and after heating.
2. Using the formula below, calculate the D_{60} value of each organism.

$$D_T = \frac{t}{\log_{10} x - \log_{10} y}$$

3. Enter your results in Chart 4 in the box labeled "Calculated D_{60} Value."

Questions

1. Is the D value of either organism affected by the size of the population? Explain.

2. How much did your plotted and calculated values differ? How do you explain any differences?

3. Using the information from this exercise, how long would it take to kill a population of *E. coli* with a density of 10^3 cells/mL? How long to kill a population of *S. aureus* with the same density?

4. Because this is a quantitative procedure, why is it not necessary to know the volume of broth in the 30 Nutrient Broth tubes?

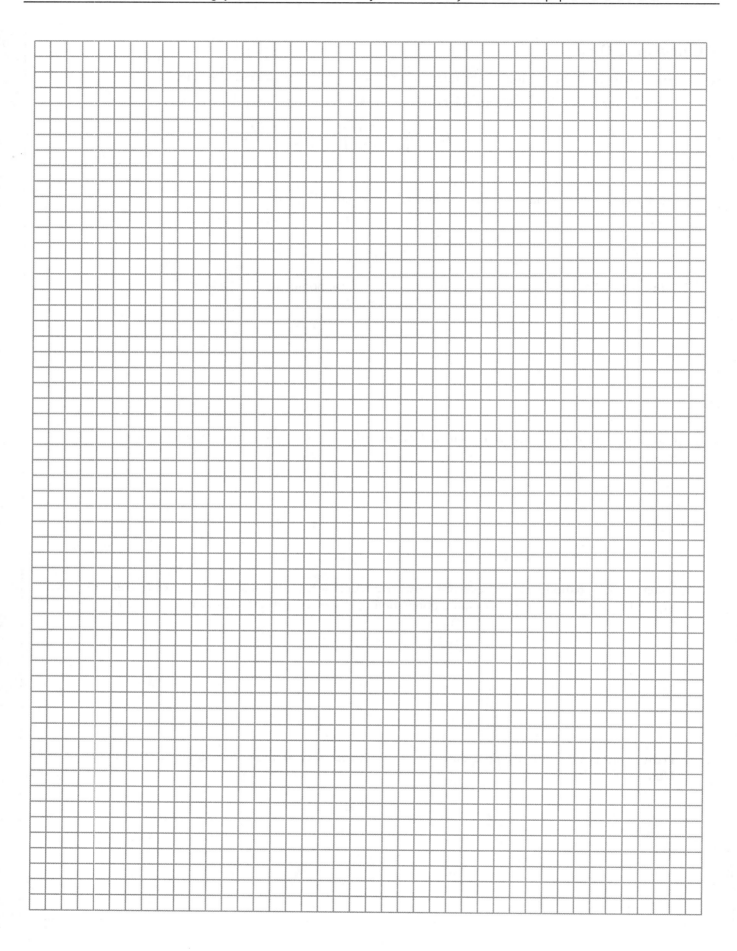

EXERCISE 7-1

SNYDER TEST

Data Sheet

Name .... Date..............................

Lab Section .... I was present and performed this exercise (initials)...........

Observations and Interpretations

Refer to Table 7-1 when recording and interpreting your results below.

	TIME		
	24 Hrs.	48 Hrs.	72 Hrs.
Color			

Based on the results of this test, I have a ______________________ susceptibility to dental caries.

Questions

1. *Lactobacillus* species ferment glucose, as demonstrated by this exercise. What other metabolic functions might they perform, judging by the list of ingredients?

2. In your estimation, what kinds of dietary items would likely increase the number of lactobacilli in saliva?

3. Why isn't the molten Snyder Test Agar allowed to solidify as a slant?

Exercise 7-2

LYSOZYME ASSAY

Data Sheet

Name .... Date...............................

Lab Section .... I was present and performed this exercise (initials)...........

Data and Calculations

1. Enter the standard curve data from Group 1.

Lysome Concentration	Actual Time of Transfer (T_n)	Light Absorbance at T_{20}
0.15625 mg/100 mL		
0.3125 mg/100 mL		
0.625 mg/100 mL		
0.125 mg/100 mL		
0.25 mg/100 mL		
0.5 mg/100 mL		

2. Enter class data for all samples.

Body Fluid	Actual Time of Transfer (T_0)		Light Absorbance at T_{20}	
	10^{-1}	10^{-2}	10^{-1}	10^{-2}

3. Using the information in the tables and the graph paper provided, construct a standard curve of lysozyme concentration (x-axis) versus absorbance (y-axis). Don't forget to label the axes and title the graph.

4. Using the standard curve, estimate the concentration of lysozyme in each diluted sample.

5. Calculate the original concentration (before dilution) of lysozyme in all samples, using the following formula (where OC is original concentration, LC is the lysozyme concentration obtained from the standard curve, and DF is the dilution factor):

$$OC = \frac{LC}{DF}$$

Questions

1. Lysozyme alone is not enough to kill bacterial cells. What other condition must be met in order for cell death to occur?

2. Why do you suppose Lysozyme is less effective against Gram-negative bacteria than Gram-positives?

3. Assuming the conditions are present for Lysozyme to be effective, would it work on actively-growing cells? Non-growing cells?

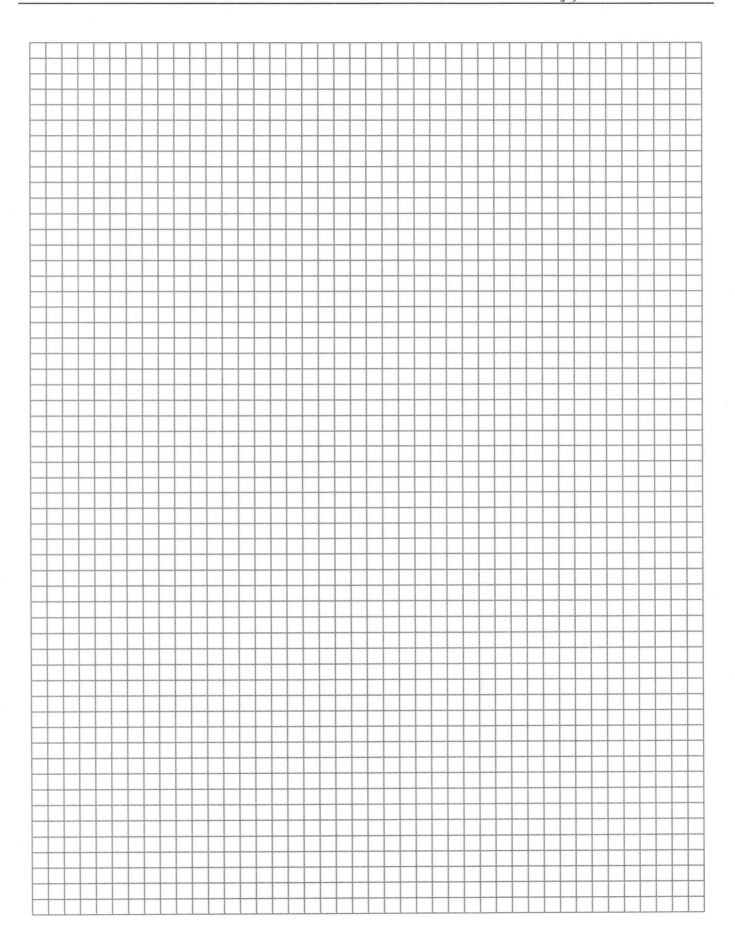

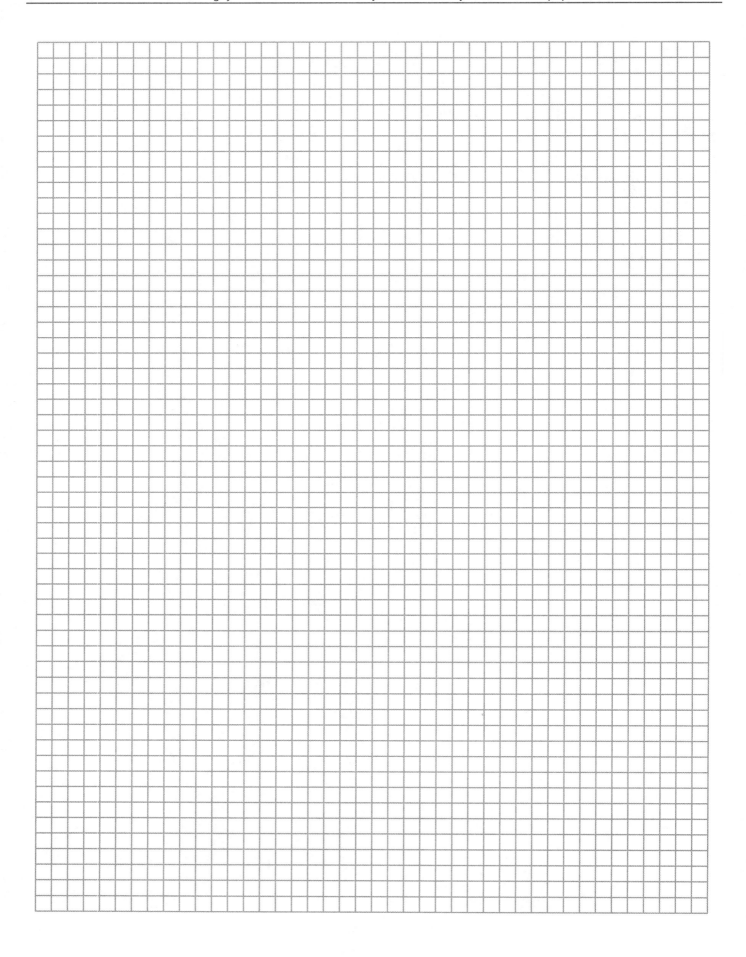

EXERCISE 7-3

ANTIMICROBIAL SUSCEPTIBILITY TEST

Data Sheet

Name .... Date..............................

Lab Section .... I was present and performed this exercise (initials)............

Observations and Interpretations

Record the zone diameters in mm. Using Table 7-3, enter "S" if the organism is susceptible, "R" if it is resistant to the antibiotic. (Not all combinations of antibiotics and organisms are available.)

	Streptomycin		Tetracycline		Penicillin		Chloramphenicol	
Organism	Zone Diameter	S/R	Zone Diameter	S/R	Zone Diameter	S/R	Zone Diameter	S/R

Questions

1. All aspects of the Kirby-Bauer test are standardized to assure reliability.

   a. What might the consequence be of pouring the plates 2 mm deep instead of 4 mm?

   b. The plates are supposed to be used within a specific time after their preparation and should be free of visible moisture. Why do you think this is so?

2. In clinical applications of the Kirby-Bauer test, diluted cultures (for the McFarland standard comparison) must be used within 30 minutes. Why is this important?

3. *E. coli* and *S. aureus* were chosen to represent Gram-negative and Gram-positive bacteria, respectively. For a given antibiotic, is there a difference in susceptibility between the Gram-positive and Gram-negative bacteria?

4. Suppose you do this test on a hypothetical *Staphylococcus* species with the antibiotics penicillin (P-10) and tetracycline (TE-30). You record zone diameters of 20 mm for the tetracycline disk and 25 mm for the penicillin disk. Which antibiotic would be most effective against this organism? What does this tell you about comparing zone diameters to each other and the importance of the interpretive chart?

EXERCISE 7-4

MMWR REPORT

Data Sheet

Name .... Date..............................

Lab Section .... I was present and performed this exercise (initials)............

Disease ______________________________________

Data and Calculations

1. Go to the MMWR Morbidity Tables at the CDC Web site. In the chart on pages 562–563, under the column labeled "Weekly Totals," record the cumulative number of cases for your chosen disease for the last two complete years. No calculation is necessary regarding these numbers at this point. Because they are cumulative totals, they should be greater than or equal to the preceding week. Nevertheless, because the reported numbers are provisional and subject to change, this may not always be the case. Simply record the numbers from the CDC Web site as they are reported.
2. Calculate the number of *new* cases during each 4-week period (weeks 1–4, weeks 5–8, *etc.*) by subtracting the total at the end of a 4-week period from the total at the end of the next 4-week period. That is, for the second four weeks, subtract the total at the end of week 4 from the total at the end of week 8. Record these in the chart on pages 562–563 in the column labeled "4-Week Totals." Show a sample calculation in the space below.

3. Calculate national incidence values of each 4-week period for both years, using the calculated 4-week totals (weeks 1–4, weeks 5–8, *etc.*) in the numerator. Use the U. S. population size on July 1 of each respective year in the denominator. This value can be obtained from the U. S. Census Bureau Web site at http://www.census.gov/popest/states/NST-ann-est.html. Be sure to choose an appropriate value for K, and don't abuse the significant figures! Show a sample calculation in the space below.

   Estimated population size for the year 200__:

   Estimated population size for the year 200__:

$$\text{Incidence Rate} = \frac{\text{number of new cases in a time period}}{\text{size of at-risk population at midpoint of time period}} \times K$$

Morbidity Data for ____ during the years 200__ and 200__.

Week	Cumulative Totals by Week Year (200■)	4-Week Totals Year (200■)	Incidence Values Year (200■)	Cumulative Totals by Week Year (200■)	4-Week Totals Year (200■)	Incidence Values Year (200■)
1						
2						
3						
4						
5						
6						
7						
8						
9						
10						
11						
12						
13						
14						
15						
16						
17						
18						
19						
20						
21						
22						
23						
24						

Week	Cumulative Totals by Week Year (200■)	4-Week Totals Year (200■)	Incidence Values Year (200■)	Cumulative Totals by Week Year (200■)	4-Week Totals Year (200■)	Incidence Values Year (200■)
25						
26						
27						
28						
29						
30						
31						
32						
33						
34						
35						
36						
37						
38						
39						
40						
41						
42						
43						
44						
45						
46						
47						
48						
49						
50						
51						
52						

Questions

1. Characterize the disease you have chosen. Be sure to include causative agent (by scientific name), symptoms, mode of transmission, and treatment.

2. Prepare a graph that illustrates the cumulative data for each week over the two years. Be sure to include a title, axis labels with appropriate units, and legends in your graph.

3. Graph the calculated national incidence values over the two years studied. In your graph, be sure to include a title, axis labels with appropriate units, and legends.

4. Briefly compare national trends of your chosen disease for the two years. Are the number of cases similar? Does the incidence appear to be seasonal?

EXERCISE 7-5

EPIDEMIC SIMULATION

Data Sheet

Name .... Date...............................

Lab Section .... I was present and performed this exercise (initials)............

Data and Calculations

Compile data for your class population (each case is the student's number in the exercise). The organism spread during contact between student pairs was *Serratia marcescens,* which produces reddish-orange colonies. You will likely see growth on both sides of the plate, but red colonies are the only sign of disease. If there is reddish-orange growth on side B of the plate, consider the person to have the disease and record a " + " in the appropriate box. If there is no red growth on the plate or there is no difference between the two sides, consider the person to be healthy and record a " – " in the appropriate box.

Case	Week	Disease (+/–)
1	1	
2	1	
3	1	
4	1	
5	1	
6	2	
7	2	
8	2	
9	2	
10	2	
11	2	
12	3	
13	3	
14	3	
15	3	
16	3	
17	4	
18	4	
19	4	
20	4	

Case	Week	Disease (+/–)
21	4	
22	4	
23	4	
24	4	
25	4	
26	4	
27	5	
28	5	
29	5	
30	5	
31	5	
32	5	
33	5	
34	5	
35	6	
36	6	
37	6	
38	6	
39	6	
40	6	

Identify the case number of the index patient.

Use the data obtained to calculate incidence and prevalence rates in your class population. For this purpose, assume that the cases have been identified over a 6-week period (or a shorter period, depending on how many students you have). Also assume that the duration of the disease is 7 days, so new cases in one week are still diseased in the next week but are healthy by the following week.

Record the population size: ____________________

Record incidence and point prevalence rates for each week in the chart below.

Week	Incidence	Prevalence
1		
2		
3		
4		
5		
6		

Questions

1. Explain how you determined the index case.

2. What was the purpose of side A on each plate?

3. Suggest possible reasons (based on the execution of the simulation) for cases showing no growth between cases that exhibited growth.

4. Suggest how the gaps could represent "subclinical infections" or "carriers" in this simulation.

5. Design a simulation that would be an example of a common source epidemic.

EXERCISE 7-6

IDENTIFICATION OF *ENTEROBACTERIACIAE*

Data Sheet

Name .... Date..............................

Lab Section .... I was present and performed this exercise (initials)...........

Unknown Number ______________________________

Isolation Procedure

Please record all activities associated with isolation of your organisms—from mixed culture to pure culture. Always include the date, source of inoculum, destination, type of inoculation, incubation temperature, and any other relevant information. Also make note of transfers made to keep your pure culture fresh. (This log must be kept current.)

____
____
____
____
____
____
____
____
____
____
____
____
____
____
____
____
____

Preliminary Observations

Colony Morphology (include medium) ____
____
____

Gram Stain ______________ Cell Dimensions ______________ Optimum Temperature______________

Cellular Morphology and Arrangement ____

Differential Tests

Please begin recording with your oxidase test, followed by the IMViC results. This log must be kept current. Record the Identification Chart

you are using: ________________

Test #1:____________________ Date Begun:___________ Date Read:___________ Result:_____________

Comments: ____

Test #2:____________________ Date Begun:___________ Date Read:___________ Result:_____________

Comments: ____

Test #3:____________________ Date Begun:___________ Date Read:___________ Result:_____________

Comments: ____

Test #4:____________________ Date Begun:___________ Date Read:___________ Result:_____________

Comments: ____

Test #5:____________________ Date Begun:___________ Date Read:___________ Result:_____________

Comments: ____

Test #6:____________________ Date Begun:___________ Date Read:___________ Result:_____________

Comments: ____

Test #7:____________________ Date Begun:___________ Date Read:___________ Result:_____________

Comments: ____

Test #8:____________________ Date Begun:___________ Date Read:___________ Result:_____________

Comments: ____

Test #9:____________________ Date Begun:___________ Date Read:___________ Result:_____________

Comments: ____

Test #10:____________________ Date Begun:___________ Date Read:___________ Result:_____________

Comments: ____

Test #11:____________________ Date Begun:___________ Date Read:___________ Result:_____________

Comments: ____

Test #12:____________________ Date Begun:___________ Date Read:___________ Result:_____________

Comments: ____

My Unknown is: ______________________________________ Rerun:[2] ______________________________

______________________________________ Rerun: ______________________________

[2] Your instructor will write what tests to rerun in this space if you misidentify your unknown.

EXERCISE 7-7

IDENTIFICATION OF GRAM-POSITIVE COCCI

Data Sheet

Name .... Date..............................

Lab Section .... I was present and performed this exercise (initials)...........

Unknown Number ________________________________

Isolation Procedure

Please record all activities associated with isolation of your organisms—from mixed culture to pure culture. Always include the date, source of inoculum, destination, type of inoculation, incubation temperature, and any other relevant information. Also make note of transfers made to keep your pure culture fresh. (This log must be kept current.)

Preliminary Observations

Colony Morphology (include medium) ________________________________

Gram Stain ________________ Cell Dimensions ________________ Optimum Temperature________________

Cellular Morphology and Arrangement ________________________________

Differential Tests

Please begin recording with your catalase results. This log must be kept current. Record the Identification Chart you are using: ______________________________

Test #1:______________ Date Begun:__________ Date Read:__________ Result:__________

Comments: ______________________________

Test #2:______________ Date Begun:__________ Date Read:__________ Result:__________

Comments: ______________________________

Test #3:______________ Date Begun:__________ Date Read:__________ Result:__________

Comments: ______________________________

Test #4:______________ Date Begun:__________ Date Read:__________ Result:__________

Comments: ______________________________

Test #5:______________ Date Begun:__________ Date Read:__________ Result:__________

Comments: ______________________________

Test #6:______________ Date Begun:__________ Date Read:__________ Result:__________

Comments: ______________________________

Test #7:______________ Date Begun:__________ Date Read:__________ Result:__________

Comments: ______________________________

Test #8:______________ Date Begun:__________ Date Read:__________ Result:__________

Comments: ______________________________

Test #9:______________ Date Begun:__________ Date Read:__________ Result:__________

Comments: ______________________________

Test #10:______________ Date Begun:__________ Date Read:__________ Result:__________

Comments: ______________________________

Test #11:______________ Date Begun:__________ Date Read:__________ Result:__________

Comments: ______________________________

Test #12:______________ Date Begun:__________ Date Read:__________ Result:__________

Comments: ______________________________

My Unknown is: ______________________________ Rerun:[2] ______________________

______________________________ Rerun: ______________________

[2] Your instructor will write what tests to rerun in this space if you misidentify your unknown.

EXERCISE 8-1

MEMBRANE FILTER METHOD

Data Sheet

Name .... Date..............................

Lab Section .... I was present and performed this exercise (initials)...........

Data and Calculations

1. Enter the number of colonies counted in each sample and calculate the total coliforms per 100 milliliters of water using the following formula:

$$\frac{\text{coliform colonies}}{100\text{ mL}} = \frac{\text{coliform colonies} \times 100\text{ mL}}{\text{volume of original sample in mL}}$$

2. Enter the results below.

Sample	Number of Colonies	Colonies/100 mL	Potable? Y/N

Questions

1. For this test, why is Endo Agar used instead of Nutrient Agar?

2. How would adding glucose to this medium affect the results? Would it affect its sensitivity or specificity? Explain.

3. Suppose you were to count one coliform colony produced from a 10 mL water sample. What is the coliform density of the water in cells per 100 mL. Is the water potable? Explain.

Exercise 8-2

MULTIPLE TUBE FERMENTATION METHOD

Data Sheet

Name .... Date.............................

Lab Section .... I was present and performed this exercise (initials)...........

Data and Calculations

1. Enter your data here.

BGLB Data				
Group	A	B	C	Totals (A + B + C)
Dilution Factor (DF)	10^{0}	10^{-1}	10^{-2}	NA
Portion of dilution added to LTB tubes that is original sample (1.0 mL × DF)	1.0 mL	0.1 mL	0.01 mL	NA
# Tubes in group	5	5	5	NA
# Positive results (Gas)				
# Negative results (No gas)				NA
Volume of original sample in negative LTB tubes (DF × 1.0 mL × # negative tubes)				
Volume of original sample in all LTB tubes (DF × 1.0 mL × # tubes)	5.0 mL	0.5 mL	0.05 mL	5.55 mL

EC Data				
Group	A	B	C	Totals (A + B + C)
Dilution Factor (DF)	10^0	10^{-1}	10^{-2}	NA
Portion of dilution added to LTB tubes that is original sample (1.0 mL × DF)	1.0 mL	0.1 mL	0.01 mL	NA
# Tubes in group	5	5	5	NA
# Positive results (Gas)				
# Negative results (No gas)				NA
Volume of original sample in negative LTB tubes (DF × 1.0 mL × # negative tubes)				
Volume of original sample in all LTB tubes (DF × 1.0 mL × # tubes)	5.0 mL	0.5 mL	0.05 mL	5.55 mL

2. Enter your final results here.

Total coliform MPN/100 mL (from BGLB data)	
E. coli MPN/100 mL (from EC data)	

Questions

1. When using the formula on page 272, why is it important to use the volume of original sample in the denominator instead of the total volume of the solution added to the broths?

2. Suppose you were to run this test on a water sample and after 48 hours incubation of the LTB tubes found no gas bubbles. What should you do next?

3. What if you found gas in the LTB tubes but none in the BGLB? Should you go on to the EC test?

4. All coliforms ferment glucose, but none of the media used for this test includes glucose. Why is glucose not used?

5. Would adding glucose increase or decrease the sensitivity? Specificity?

EXERCISE 8-3

BIOLUMINESCENSE

Data Sheet

Name .... Date..............................

Lab Section .... I was present and performed this exercise (initials)...........

Questions

1. Considering the relationship between bioluminescent bacteria and the Flashlight Fish, what evolutionary advantage do you think quorum sensing gives the microorganism? Explain.

2. Why does the medium used for this exercise contain seawater?

3. If after 48 hours incubation, your organism has grown but does not emit light, what is the likely explanation? What should you do to correct the problem?

EXERCISE 8-4

SOIL MICROBIAL COUNT

Data Sheet

Name .... Date..............................

Lab Section .... I was present and performed this exercise (initials)...........

Data and Calculations

1. Count the colonies on the countable plates and enter the data in the chart below.

Group	Colonies Counted	Final Dilution Factor (FDF) of Plated Sample	Cell Density per mL of H_2O	Cell Density per Gram of Soil
1				
2				
3				
4				
5				
6				

2. Calculate the original cell density of the water using the following formula, where OCD is original cell density in cells per milliliter, CFU is colony forming units (colonies counted on the plate), and FDF is the final dilution factor:

$$OCD = \frac{CFU}{FDF}$$

3. Because your task is to determine the original density of soil, you must convert cell density in milliliters to grams as follows (remembering that the original 100 milliliters of solution contained 10 grams of soil):

$$\frac{CFU}{mL} \times \frac{100\text{ mL}}{10\text{ g}} = \frac{CFU}{g}$$

Questions

1. Which type of organism was most abundant? Least abundant?

2. Calculations for this exercise could have been done in grams and milligrams because 1 milliliter of water weighs 1 gram, but the volume of water in the dilution bottle would have to have been adjusted to do this. To what should the volume be changed for calculations in grams to be correct?

3. Using only the dilution schemes for each type of organism, which would you predict to be the most abundant? Least abundant?

EXERCISE 8-5

METHYLENE BLUE REDUCTASE TEST

Data Sheet

Name .... Date.............................

Lab Section .... I was present and performed this exercise (initials)...........

Data and Calculations

Enter your data below and determine the quality of the milk sample. Under "Milk Quality, record "G" (good) if the milk takes longer than 6 hours to turn white, "P" (poor) if it turns white in 2 hours or less, and "M" (medium quality) if it turns white between 2 and 6 hours after inoculation.

Sample	Starting Time T_s (Milk is Blue)	Ending Time T_e (Milk is white)	Elapsed Time ($T_e - T_s$)	Milk Quality

Questions

1. Why were you told to cap the tubes tightly? Explain.

2. What results would you expect if the tubes were inoculated with a strict aerobe? A strict anaerobe?

EXERCISE 8-6

VIABLE CELL PRESERVATIVES

Data Sheet

Name .... Date...............................

Lab Section .... I was present and performed this exercise (initials)...........

Observations

1. Enter "C" for curdled; "NC" for not curdled.

Time (weeks)	Control		*P. fluorescens*		*C. sporogenes*		*E. faecalis*		*S. aureus*		*S. typhimurium*	
	Treated	Untreated	Treated	Untreated	Treated	Untreated	Treated	Untreated	Treated	Untreated	Treated	Untreated
0.5												
1												
1.5												
2												
2.5												
3												
3.5												
4												

2. In the data chart, circle the earliest time each culture showed curdling.
3. To illustrate the time necessary to curdle each sample, prepare a bar graph of organism versus time. Be sure to include the control. It may be helpful to construct double bars comparing treated samples with untreated samples, as in the following example that compares organisms A, B, and a control. (T = treated, UT = untreated):

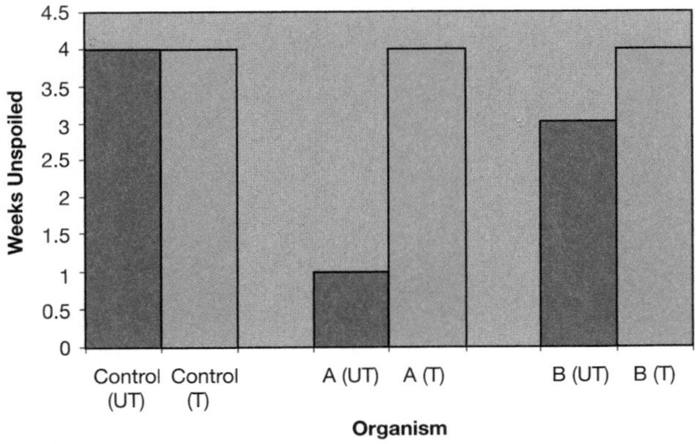

Questions

1. Would you expect the same results using raw or pasteurized milk? Milk from antibiotic-fed cows? Lactose-reduced milk?

2. Why is it important to use mesophilic organisms for this procedure?

3. What would be the optimum temperature of a refrigerator used for this procedure?

EXERCISE 8-7

MAKING YOGURT

Data Sheet

Name .... Date..............................

Lab Section .... I was present and performed this exercise (initials)...........

Observations and Interpretations

Culture Organisms	Flavor	Consistency	pH

Exercise 9-1

EXTRACTION OF DNA FROM BACTERIAL CELLS

Data Sheet

Name .... Date..............................

Lab Section .... I was present and performed this exercise (initials)............

Observations and Interpretations

In the chart below, record the absorbance values for the wavelengths used.

Wavelength (nm)	Absorbance
220	
240	
260	
280	
300	
320	

a. Calculate the DNA concentration in your sample. Be sure to take any dilutions into account.

b. Determine the probable purity of your sample using the equation:

$$\frac{\text{Absorbance}_{260\text{nm}}}{\text{Absorbance}_{280\text{nm}}}$$

c. Plot the absorption spectrum (Absorption versus Wavelength) of the DNA sample.

Questions

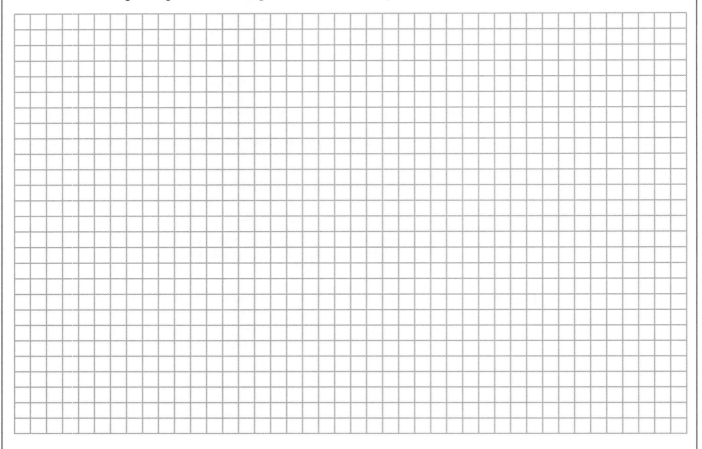

1. What is the importance of heating the cell lysate?

2. Why does the extraction work better with cold 95% ethanol than with room-temperature 95% ethanol?

3. Based on your results, what wavelength gives maximum absorption of DNA? Does this match the accepted wavelength for maximum absorption? If not, suggest reasons for the discrepancy.

Exercise 9-2

ULTRAVIOLET RADIATION DAMAGE AND REPAIR

Data Sheet

Name .... Date..............................

Lab Section .... I was present and performed this exercise (initials)...........

Observations and Interpretations

Using the terms "confluent–dense growth," "confluent–sparse growth," "individual colonies," and "none," describe the growth of *Serratia marcescens* in both parts of each plate (*i.e.,* where the mask shielded the growth from UV and where it did not). Record your observations in the chart below. Make a note as to the significance of each type of growth.

Plate	Mask/Lid	Exposure Time (min.)	Incubation	Growth on Agar Surface Covered by Mask	Growth on Agar Surface Beneath Opening of Mask
1	Mask, no lid	0.5	Sunlight		
2	Mask, no lid	0.5	Dark		
3	Mask, no lid	3.0	Sunlight		
4	Mask and lid	3.0	Sunlight		
5	Mask and lid	3.0	Dark		
6	Lid only	0	Sunlight		
7	Lid only	0	Dark		

Questions

1. Consider the exposed part of the plate.
   a. What does the relative sparseness of growth tell you about the effect of UV radiation?

   b. What do the colonies tell you about the cells from which they grew?

2. What does the relatively heavy growth covered by the mask and lid tell you about UV radiation?

3. What is the effect of longer UV exposure on *Serratia marcescens*? To answer this question, which plates should be compared?

4. What is the effect of the posterboard mask on UV radiation? To answer this question, which plates should be examined?

5. What is the effect of the plastic lid on UV radiation? To answer this question, which plates should be compared?

EXERCISE 9-3

AMES TEST

Data Sheet

Name .... Date..............................

Lab Section .... I was present and performed this exercise (initials)...........

Observations and Interpretations

Record your observations for the Ames Test plates below. On the CM plates, measure the zone of inhibition diameter in millimeters. On the MM plates, count only the large colonies, not the "hazy" background growth. (These are the colonies produced by auxotrophs that did not back-mutate to prototrophs and grew only until the histidine in the minimal medium was exhausted.)

	Zone Diameter		Colonies Counted	
	CM with TA 1535	CM with TA 1538	MM with TA 1535	MM with TA 1538
DMSO				
Test Substance ________________				

Questions

1. Which plates do you have to examine to determine if your test substance is toxic. What is your conclusion?

2. Which plates do you need to examine to determine if your test substance is mutagenic? What is your conclusion?

3. Why can't your test substance contain protein?

4. What is the purpose of including a small amount of histidine in minimal medium?

5. Why are the small colonies on the minimal medium plates not considered back-mutants? Why don't they grow to full size?

6. What are some advantages of the Ames Test over carcinogen tests using rats or mice? What are some disadvantages?

Exercise 9-4

BACTERIAL TRANSFORMATION: THE pGLO™ SYSTEM

Data Sheet

Name .... Date.............................

Lab Section .... I was present and performed this exercise (initials)...........

Observations and Interpretations

1. Record the appearance of the pGLO plasmid solution with and without UV illumination.

2. In the chart below, record your observations of the plates after incubation.

Plate	Inoculum	Number of Colonies	Appearance in Ambient Light	Appearance with UV Light
LB/amp	*E. coli* + DNA			
LB/amp/ara	*E. coli* + DNA			
LB/amp	*E. coli* − DNA			
LB	*E. coli* − DNA			

Questions

1. In the chart below, fill in the genotype of *E. coli* prior to transformation and after transformation.

E. coli	GFP Gene (+ or −)	*Bla* (+ or −)
Before transformation		
After transformation		

2. What was the purpose of examining the original pGLO™ solution with and without UV illumination?

3. What was the purpose of transferring the +DNA and −DNA tubes from ice, to hot water, to ice again?

4. Why were the vials incubated for 10 minutes in LB broth rather than transferring their contents directly to the plates?

5. What information is provided by the LB/−DNA plate?

6. Obviously, transformation could occur only if the pGLO plasmid was introduced into the solution. Which plate(s) exhibit transformation?

7. What information is provided by the LB/amp/−DNA plate?

8. Why does the LB/amp/ara/+DNA plate fluoresce when the LB/amp/+DNA plate does not?

9. Use the following information to calculate transformation efficiency.

   a. You put 10 μL (one loopful) of a 0.03 μg/μL pGLO solution into the +DNA tube. Calculate the μg of DNA you used.

   b. The +DNA tube contained 510 μL of solution prior to plating, but you did not use all of it. Calculate the fraction of the +DNA solution you used in each tube.

   c. Use your answers to questions 9a and 9b to calculate the micrograms of DNA you plated.

   d. Using the number of colonies on the LB/amp/ara/+DNA plate, calculate the transformation efficiency. (*Hint:* The units are transformants/μg of pGLO™ DNA).

   e. According to the manufacturer's (Bio-Rad Laboratories) manual, this protocol should yield a transformation efficiency between 8.0×10^2 and 7.0×10^3 transformed cells per microgram of pGLO™ DNA. How does your transformation efficiency compare? Account for any discrepancy.

Exercise 10-1

DIFFERENTIAL BLOOD CELL COUNT

Data Sheet

Name .... Date..............................

Lab Section .... I was present and performed this exercise (initials)............

Observations and Interpretations

Record your data from the differential blood cell count in the chart below. As you count the 100 white blood cells, make tally marks in the appropriate boxes. Then calculate the percentages of each type and compare them to the expected values.

Normal Blood						
	Monocytes	Lymphocytes	Segmented Neutrophils	Band Neutrophils	Eosinophils	Basophils
Number						
Percentage						
Expected Percentage	3–7%	25–33%	55–65% (all neutrophils)	—	1–3%	0.5–1%

Abnormal Blood (Condition:)						
	Monocytes	Lymphocytes	Segmented Neutrophils	Band Neutrophils	Eosinophils	Basophils
Number						
Percentage						
Expected Percentage	3–7%	25–33%	55–65% (all neutrophils)	—	1–3%	0.5–1%

Questions

1. How do the percentages of each WBC compare to the published values? What might account for any differences you noted?

2. If you did a differential count on abnormal blood, how did the percentages compare to normal blood? How can any differences you noted be explained in the context of the disease and/or defense process?

EXERCISE 10-2

PRECIPITIN RING TEST

Data Sheet

Name .... Date..............................

Lab Section .... I was present and performed this exercise (initials)...........

Observations and Interpretations

Draw the tubes after incubation. Label the solutions in each and any precipitation lines.

Questions

1. What was the purpose of tube B?

2. In general terms, describe how equine albumin antiserum could be obtained. What animal is least likely to be a source for it?

3. Suppose you were instructed to make two antiserum solutions: The first is identical to what you used in the lab exercise. The other is a 10^{-6} dilution of the antiserum. After incubation with the antigen, the full-strength antiserum produces a precipitin ring, but the diluted antiserum does not. Explain these results. Is this a failure of the test's specificity or sensitivity?

4. Suppose you were instructed to repeat this experiment, again using equine albumin antiserum in two tubes. You layer equine serum (containing equine albumin) over the antiserum in one tube and pig serum (containing pig albumin) over the antiserum in the other tube. After incubation, you see precipitin rings in both tubes. Explain these results. Is this a failure of the test's specificity or sensitivity?

EXERCISE 10-3

GEL IMMUNODIFFUSION

Data Sheet

Name .... Date..............................

Lab Section .... I was present and performed this exercise (initials)............

Observations and Interpretations

Draw and interpret the results of your immunodiffusion plate in the diagram below. (***Note:*** Do not use this as the template for the wells in your plate; use Figure 10-10.)

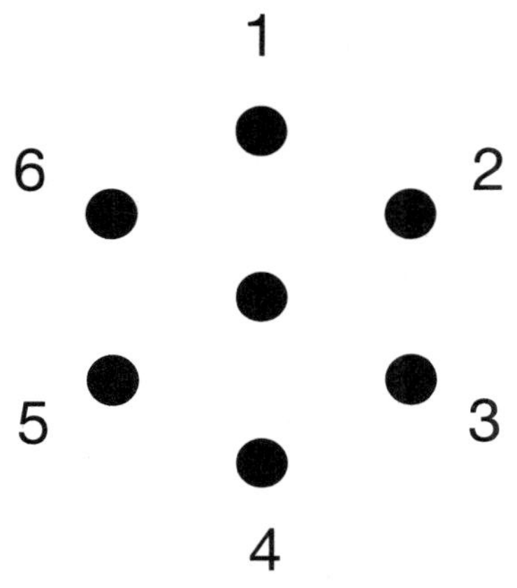

Questions

1. What was the purpose of well #5?

2. How many lines of identity did your plate demonstrate?

3. Between which two wells did you see nonidentity?

4. Between which two wells did you see partial identity?

5. Examine Figure 10-7. Suppose you performed gel immunodiffusion and all you got was the Zone of Antigen Excess. Is this a false positive or a false negative result? Is it a failure of the test's specificity or sensitivity?

6. Suppose you performed the gel immunodiffusion test again using the procedure in this exercise. How could you explain the presence of two precipitation lines between well #1 and the center well?

EXERCISE 10-4

SLIDE AGGLUTINATION

Data Sheet

Name .... Date..............................

Lab Section .... I was present and performed this exercise (initials)...........

Observations and Interpretations

Sketch and label your results in the diagram below. Indicate which sample contained *Salmonella* H antigen.

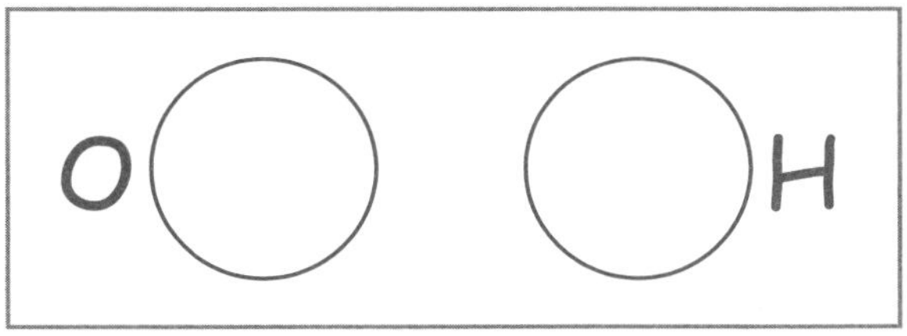

Questions

1. What is the purpose of performing this experiment with both H and O antigens?

2. Suppose you performed this test and got agglutination in *both* samples. Eliminating contamination of the antigen and antiserum samples as a possibility, provide an explanation of this hypothetical result. Is this attributable to a lack of sensitivity or specificity of the test system?

3. Suppose you performed this test and *neither* sample produced agglutination. Would this be attributable to a lack of sensitivity or specificity of the test system?

4. Higher vertebrates produce antibodies; *Salmonella* is a prokaryote. Given these facts, how can *Salmonella* antiserum be produced?

EXERCISE 10-5

BLOOD TYPING

Data Sheet

Name .... Date..............................

Lab Section .... I was present and performed this exercise (initials)...........

Observations and Interpretations

Record your results below.

Antiserum	Agglutination +/−
Anti-A	
Anti-B	
Anti-Rh	

My blood type is: ______________________

Record class data in the chart below.

Blood Type	Percentage in U. S. Population	Number in Class	Percentage in Class	Deviation from National Values
O+	38			
O−	7			
A+	34			
A−	6			
B+	9			
B−	2			
AB+	3			
AB−	1			

Source: American Association of Blood Banks;

Questions

1. Examine the blood type data obtained from your class. Attempt to explain any deviations from the national values.

2. At one time, people with Type "O" blood were said to be "Universal Donors" in blood transfusions. Explain the reasoning behind this. Which blood type was designated as the "Universal Recipient?"

3. What biological fact makes the concepts of "Universal Donor" and "Universal Recipient" misnomers?

4. Maternal-fetal Rh incompatibility (where mother's Anti-Rh antibodies destroy the fetus's Rh+ red blood cells) is a well-known phenomenon. Less well known are situations of maternal-fetal ABO incompatibility. Suggest combinations of maternal and fetal blood types that could lead to this situation.

5. Why are red blood cells used in many indirect agglutination tests?

EXERCISE 11-1

THE FUNGI—COMMON YEASTS AND MOLDS

Data Sheet

Name .... Date..............................

Lab Section .... I was present and performed this exercise (initials)...........

Observations and Interpretations

Sketch your observations of the microscopic structure of the assigned fungi in the spaces below.

Yeasts

Saccharomyces cerevisiae
vegetative cells

(X________)

Cell dimensions __________

Candida albicans
vegetative cells

(X________)

Cell dimensions __________

Molds

Rhizopus colony

(X________)

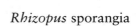

Rhizopus sporangia

(X________)

Sporangium dimensions __________

Rhizopus gametangia

(X________)

Gametangium dimensions __________

EXERCISE 11-1
THE FUNGI—COMMON YEASTS AND MOLDS *(continued)*

Molds *(continued)*

Penicilliums culture

(X________)

Penicillium conidiophores

(X________)

Conidia dimensions __________

Aspergillus culture

(X________)

Aspergillus conidiophores

(X________)

Conidia dimensions __________

____

(X_______)

Dimensions _________

____

(X_______)

Dimensions _________

____

(X_______)

Dimensions _________

____

(X_______)

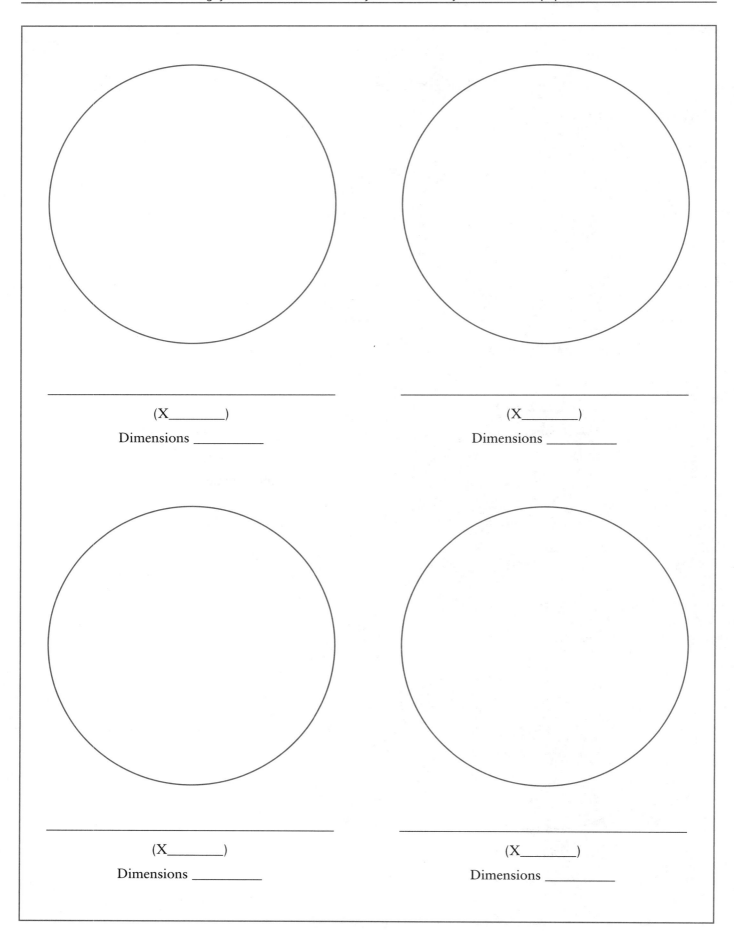

Dimensions _________

EXERCISE 11-2

EXAMINATION OF COMMON PROTOZOANS OF CLINICAL IMPORTANCE

Data Sheet

Name .... Date..............................

Lab Section .... I was present and performed this exercise (initials)...........

Observations and Interpretations

Sketch your observations of the assigned protozoans in the spaces below.

Entamoeba histolytica
trophozoite

(X________)

Dimensions __________

Entamoeba histolytica
cyst

(X________)

Dimensions __________

Entamoeba coli trophozoite

(X________)

Dimensions __________

Entamoeba coli cyst

(X________)

Dimensions __________

Balantidium coli trophozoite

(X________)

Dimensions __________

Balantidium coli cyst

(X________)

Dimensions __________

EXERCISE 11-2

EXAMINATION OF COMMON PROTOZOANS OF CLINICAL IMPORTANCE *(continued)*

Giardia lamblia
trophozoite

(X________)

Dimensions __________

Giardia lamblia
cyst

(X________)

Dimensions __________

Trichomonas vaginalis
trophozoite

(X________)

Dimensions __________

Trypanosoma spp.

(X________)

Dimensions __________

Plasmodium spp.
ring and mature trophozoites

(X________)

Dimensions __________

Plasmodium spp.
schizont

(X________)

Dimensions __________

Plasmodium spp.
gametocytes

(X________)

Dimensions __________

Toxoplasma gondii
trophozoite

(X________)

Dimensions __________

EXERCISE 11-3

PARASITIC HELMINTHS

Data Sheet

Name .... Date...............................

Lab Section .... I was present and performed this exercise (initials)...........

Observations and Interpretations

Sketch your observations of the assigned helminths in the spaces below.

Trematodes

Clonorchis sinensis
egg
(X________)
Dimensions __________

Paragonimus westermani
egg
(X________)
Dimensions __________

Trematodes *(continued)*

Schistosoma mansoni
egg

(X________)

Dimensions __________

Cestodes

Dipylidium caninum
egg

(X________)

Dimensions __________

Echinococcus granulosus
protoscolices in a hydatid cyst

(X________)

Protoscolex dimensions __________

EXERCISE 11-3
PARASITIC HELMINTHS *(continued)*

Cestodes *(continued)*

Hymenolepis nana
egg with oncosphere and filaments

(X________)

Dimensions __________

Taenia spp.
egg

(X________)

Dimensions __________

Taenia solium
scolex

(X________)

Taenia solium
proglottid

(X________)

Nematodes

Ascaris lumbricoides
eggs with mammillations

(X________)

Dimensions __________

Enterobius vermicularis
egg

(X________)

Dimensions __________

Hookworm (*Ancylostoma* or *Necator*)
egg

(X________)

Dimensions __________

Strongyloides stercoralis
rhabditiform larva

(X________)

Dimensions __________

EXERCISE 11-3
PARASITIC HELMINTHS *(continued)*

Nematodes *(continued)*

Wuchereria bancrofti
microfilariae in a blood smear

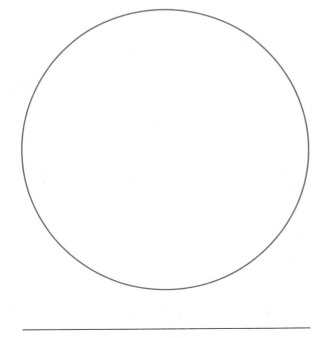

(X________)

Dimensions __________

____

(X________)

Dimensions __________

____

(X________)

Dimensions __________

____

(X________)

Dimensions __________

____

(X________)

Dimensions __________

____

(X________)

Dimensions __________

____

(X________)

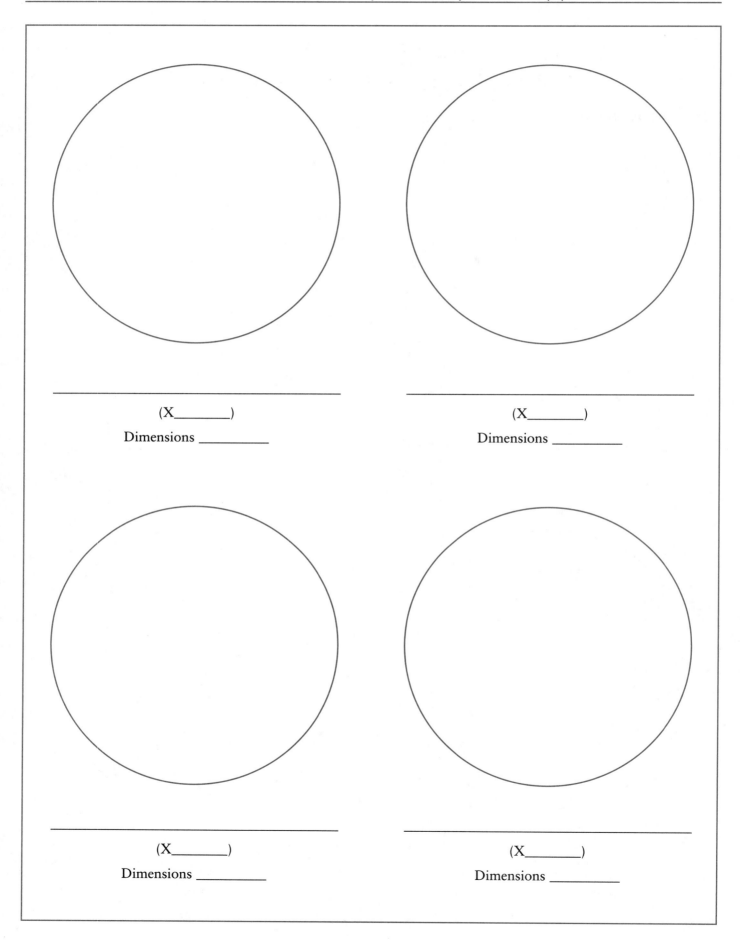

Dimensions __________

Glossary

A

absorbance A measurement, using a spectrophotometer, of how much light entering a substance is *not* transmitted.

acetoin A four-carbon intermediate in the conversion of pyruvic acid to 2,3-Butanediol; the compound detected in the Voges-Proskauer test.

acetyl-CoA Compound that enters the Krebs cycle by combining with oxaloacetate to form citrate; may be produced from pyruvate in carbohydrate metabolism or from -oxidation of fatty acids.

acid clot Pink clot formed in litmus milk from the precipitation of casein under acidic conditions; an indication of lactose fermentation.

acidic stain Staining solution with a negatively charged chromophore.

acidophil *See* eosinophil.

acidophile Microorganism adapted to a habitat below pH 5.5.

acid reaction Pink color reaction in litmus milk from acidic conditions produced by lactose fermentation.

adenosine triphosphate (ATP) In hydrolysis of ATP to adenosine diphosphate (ADP), -supplies the energy necessary to perform most work in the cell.

aerobe Microorganism that requires oxygen for growth.

aerobic respiration Metabolism in which oxygen is the final electron acceptor.

aerosol Droplet nuclei that remain suspended in the air for a long time.

aerotolerance Designation of an organism based on its ability to grow in the presence of oxygen.

aerotolerant anaerobe Anaerobe that grows in the presence of oxygen but does not use it metabolically.

agar overlay Used in plaque assays, the soft agar containing a mixture of bacteriophage and host poured over nutrient agar as an inoculum.

agglutination Visible clumping produced when antibodies and particulate antigens react.

agranulocyte Category of white blood cells characterized by the absence of prominent cytoplasmic granules; includes monocytes and lymphocytes.

alkaliphile Microorganism adapted to a habitat above pH 8.5.

alpha (α) hemolysis The greening around a colony on blood agar as a result of partial destruction of red blood cells; typical of certain members of *Streptococcus*.

amphitrichous Describes a flagellar arrangement with flagella at both ends of an elongated cell.

amylase Family of enzymes that hydrolyze starch.

anaerobe Microorganism that cannot tolerate oxygen.

anaerobic respiration Metabolism in which an inorganic substance other than oxygen is the final electron acceptor.

antibiotic Refers to an antimicrobial substance produced by a microorganism such as a bacterium or fungus.

antibody A glycoprotein produced by plasma cells, in response to an antigen, that reacts with the antigen specifically.

antigen A molecule of high molecular weight and complex three-dimensional shape that stimulates the production of antibodies and reacts specifically with them.

antimicrobic(al) Describes any substance that kills microorganisms, natural or synthetic.

antiporter A system that simultaneously transports substances in opposite directions across the cytoplasmic membrane.

antiserum General term applied to a serum that contains antibodies of a specific type.

Ascomycetes A division of fungi that produces sexual spores in a saclike structure called an ascus.

aseptic Describes the condition of being without contamination.

autoinducer Substance secreted by bioluminescent bacteria that, upon reaching sufficient concentration, triggers the bioluminescent reaction; believed to help conserve energy by synchronizing individual cellular reactions.

autotroph An organism that is able to make all of its organic molecules using CO_2 as the source of carbon. *See also* heterotroph.

auxotroph A nutritional mutant; a cell incapable of synthesizing the nutrient that is acting as the marker in a specific genetic experiment; will grow only on a complete medium. *See also* prototroph.

B

bacillus Rod-shaped cell.

back mutation Mutation in a gene that reverses the effect of an original mutation.

bacteriophage Virus that attacks bacteria; usually host-specific.

basic stain Staining solution with a positively charged chromophore.

Basidiomycetes Division of fungi characterized by club-like appendages (basidia) that produce haploid spores; include common mushrooms and rusts.

basophil Category of white blood cells; one of three types of granulocyte (along with eosinophils and neutrophils) characterized by a cytoplasm with dark-staining granules and a lobed nucleus (often difficult to see because of the dark cytoplasmic granules).

β-hemolysis The clearing around a colony on blood agar resulting from complete destruction of red blood cells; typical of certain members of *Streptococcus*.

β-lactamase Enzyme found in some bacteria that breaks a bond in the β-lactam ring of β-lactam antibiotics, rendering the antibiotic ineffective; the source of resistance against these antibiotics.

β-lactam antibiotic A group of structurally related antimicrobial chemicals that interfere with cross-linking of peptidoglycan subunits, which results in a defective wall structure and cell lysis; examples are penicillins and cephalosporins.

biofilm A layer of bacterial cells adhering to and reproducing on a surface; used commercially for acetic acid production and water purification.

bioluminescence Process by which a living organism emits light.

biosafety level (BSL) One of four sets of minimum standards for laboratory practices, facilities, and equipment to be used when handling organisms at each level; BSL-1 requires the least care, BSL-4 the most.

bound coagulase An enzyme (also called clumping factor) bound to the bacterial cell wall, responsible for causing the precipitation of fibrinogen and coagulation of bacterial cells.

C

capnophile Microaerophile that requires elevated CO_2 levels.

capsule Insoluble, mucoid, extracellular material surrounding some bacteria.

carbohydrate One of four families of biochemicals; characterized by containing carbon, hydrogen, and oxygen in the ratio 1:2:1.

carcinogen Any substance that causes cancer.

cardinal temperatures Minimum, optimum, and maximum temperatures for an organism.

casease Enzyme produced and secreted by some bacteria that catalyzes the breakdown of casein.

casein Milk protein that gives milk its white color.

catalase Enzyme produced by some bacteria that catalyzes the breakdown of metabolic H_2O_2.

cestode The class of parasitic worms commonly called tapeworms; characterized by having a head (scolex) with suckers and hooks and numerous segments (proglottids) that are little more than sex organs.

chromogenic reducing agent A substance that produces color when it gives up electrons (becomes oxidized).

chromophore The charged region of a dye molecule that gives it its color.

coagulase *See* bound coagulase; free coagulase.

coagulase-reacting factor Plasma component that reacts with free coagulase to trigger the clotting mechanism.

coccus Spherical cell.

coenzyme Nonprotein portion of an enzyme that aids the catalytic reaction (usually by accepting or donating electrons).

coliform Member of *Enterobacteriaceae* that ferments lactose with production of gas within 48 hours at 37°C.

colony Visible mass of cells produced on culture media from a single cell or single CFU (cell forming unit); a pure culture.

colony forming unit Term used to define the cell or group of cells (*e.g.*, staphylococci, streptococci) that produces a colony when transferred to plated media.

commensal Describes a synergistic relationship between two organisms in which one benefits from the relationship and the other is affected neither negatively nor positively.

common source epidemic An epidemic in which the disease is transmitted from a single source (such as water supply) and is not transmitted person-to-person.

compatible solutes Compounds such as amino acids that function both as metabolic constituents and solutes necessary to maintain osmotic balance between the internal and external environments.

competent cell Cell capable of picking up DNA from the environment; some cells are naturally competent, in other cases the cells are made to be competent artificially.

competitive inhibition Process whereby a substance attaches to an enzyme's active site, thereby preventing attachment of the normal substrate.

complete medium A medium used in genetic experiments that supplies all nutrients for growth required by prototrophs and auxotrophs.

confirmatory test One last test after identification of an unknown, to further support the identification.

constitutive Term used to describe an enzyme that is produced continuously, as opposed to an enzyme produced only when its substrate is present. *See also* induction.

counter-stain Stain applied after decolorization to provide contrast between cells that were decolorized and those that weren't.

culture A liquid or solid medium with microorganisms growing in or on it.

cyst "Resting stage" in life cycle of certain protozoans. *See also* trophozoite.

cytochrome oxidase An electron carrier in the electron transport chain of aerobes, facultative anaerobes, and microaerophiles that makes the final transfer of electrons to oxygen.

cytoplasm The semifluid component of cells inside the cellular membrane in which many chemical reactions take place.

D

deaminase Enzyme that catalyzes the removal of the amine group (NH_2) from an amino acid.

death phase Closed system microbial growth phase immediately following stationary phase; characterized by population decline usually resulting from nutrient deficiencies or accumulated toxins.

decarboxylase Enzyme that catalyzes the removal of the carboxyl group (COOH) from an amino acid.

decimal reduction value Amount of time at a specific temperature to reduce a microbial population by 90% (one log cycle).

defined medium Culture medium in which amounts and identities of all nutritional components are specified.

denitrification Conversion of nitrate or nitrite to gaseous nitrogen or nitrogen oxide.

deoxyribonuclease A class of enzymes that catalyze the depolymerization of deoxyribonucleic acid (DNA) by breaking (depending on the organism) either the 3'-carbon/phosphate bond or the 5'-carbon/phosphate bond, thereby producing nucleic acid fragments.

Deuteromycetes An unnatural grouping of fungi in which the sexual stages are either unknown or are not used in classification.

differential medium Growth medium that contains an indicator (usually color) to detect the presence or absence of a specific metabolic activity.

differential stain Staining procedure that allows distinction between cell types or parts of cells; often involves more than one stain, but not necessarily.

dilution factor Proportion of original sample present in a new mixture after it has been diluted; calculated by dividing the volume of the original sample by the total volume of the new mixture. (Subsequent dilutions in a serial dilution must be multiplied by the dilution factor of the previous dilution.)

diploid Defines a cell that has two complete sets of genetic information; one stage in the life cycle of all eukaryotic microorganisms (which also have a haploid [one set] stage).

direct agglutination Serological test in which the combination of antibodies and *naturally* particulate antigens, if positive, form a visible aggregate. *See also* indirect agglutination.

disaccharide A sugar made of two monosaccharide subunits; *e.g.*, the monosaccharides glucose and galactose can combine to make the disaccharide lactose.

DNA polymerase A group of enzymes that catalyze the addition of deoxyribonucleotides to the 3' end of an existing polynucleotide chain.

DNase *See* deoxyribonuclease

Durham tube A small, inverted test tube used in some liquid media to trap gas bubbles and indicate gas production.

E

electromagnetic energy The energy that exists in the form of waves (including x-rays, ultraviolet, visible light, and radio waves).

electron donor A compound that can be oxidized to transfer electrons to another compound, which in turn becomes reduced.

electron transport chain A series of membrane-bound electron carriers that participate in oxidation-reduction reactions whereby electrons are transferred from one carrier to another until given to the final electron acceptor (oxygen in aerobes, other inorganic substances in anaerobes) in respiration.

endospore Dormant, highly resistant form of bacterium; produced only by species of *Bacillus, Clostridium,* and a few others.

enteric bacteria Informal name given to bacteria that occupy the intestinal tract.

Enterobacteriaceae A group of Gram-negative rods that also are oxidase-negative, ferment glucose to acid, have polar flagella if motile, and usually are catalase-positive and reduce nitrate to nitrite.

enzyme A protein that catalyzes a metabolic reaction by interacting with the reactant(s) specifically; each metabolic reaction has its own enzyme.

eosinophil A category of white blood cells; one of three types of granulocyte (along with basophils and basophils) characterized by a cytoplasm with red (in typical stains) granules and lobed nucleus.

epitope That portion of an antigen that stimulates the immune system and reacts with antibodies.

eukaryote Type of cell with membranous organelles, including a nucleus, and having 80S ribosomes and many linear molecules of DNA.

exoenzyme Enzyme that operates in the external environment after being secreted from the cells.

exponential phase Closed-system microbial growth phase immediately following the lag phase; characterized by constant maximal growth during which population size increases logarithmically.

extracellular Pertains to the region outside a cell.

extreme halophile Organism that grows best at 15% or higher salinity.

extreme thermophile Organism that grows best at temperatures above 80°C.

F

facultative anaerobe Microorganism capable of both fermentation and respiration; grows in the presence or absence of oxygen.

facultative thermophile Microorganism that prefers temperatures above 40°C but will grow in lower temperatures.

false negative Test result that is negative when the sample is actually positive; usually a result of lack of sensitivity in the test system.

false positive Test result that is positive when the sample is actually negative; usually a result of lack of specificity of the test system.

fastidious (microorganism) Describes a microorganism with strict physiological requirements; difficult or impossible to grow unless specific conditions are provided.

fatty acid A long chain, organic molecule with a carboxylic acid at one end with the rest of the carbons being bonded to hydrogens.

fermentation Metabolic process in which an organic molecule acts as an electron donor and one or more of its organic products act as the final electron acceptor, marking the end of the metabolic sequence (differs from respiration in that an inorganic substance is not needed to act as final electron acceptor).

final electron acceptor (FEA) The molecule receiving electrons (it becomes reduced) at the end of a metabolic sequence of oxidation/reduction reactions.

flavin adenine dinucleotide (FAD) Coenzyme used in oxidation/reduction reactions that acts as an electron acceptor or donor, respectively.

flavoprotein A flavin-containing protein in the electron transport chain, capable of receiving and transferring electrons as well as entire hydrogen atoms; sometimes bypasses normal route of transfer and reduces oxygen directly, forming hydrogen peroxide and superoxide radicals.

free coagulase An enzyme produced and secreted by some microorganisms that initiates the clotting mechanism in plasma; seen as a solid mass in the coagulase tube test or clumps in the slide test.

free living A term used to describe non-parasitic organisms.

G

gametangium A structure that produces gametes; used in describing fungi.

gamete Reproductive cell that must undergo fusion with another gamete to continue the life cycle; usually are haploid. *See also* spore.

gelatinase Enzyme secreted by microorganisms, which catalyzes the hydrolysis of gelatin.

generation time The time required for a population to produce offspring (*i.e.,* in bacteria that reproduce by binary fission, the time needed for the population to double); inverse of mean growth rate.

germ theory Theory holding that diseases and infections are caused by microorganisms; first hypothesized in the 16th century by Girolamo Fracastoro of Verona.

glycolysis Metabolic process by which a glucose molecule is split into two 3-carbon pyruvic acid molecules, producing two ATP (net) and two $NADH_2$ molecules.

granulocyte A category of white blood cells characterized by prominent cytoplasmic granules; includes neutrophils, eosinophils, and basophils.

group A streptococci α-hemolytic members of the genus *Streptococcus,* characterized by possessing the Lancefield group A antigen; belong to the species *S. pyogenes*, an important human pathogen.

group B streptococci β-hemolytic members of the genus *Streptococcus,* characterized by possessing the Lancefield group B antigen; belong to the species *S. agalactiae*.

group D streptococci Streptococci possessing the Lancefield group D antigen; include *S. bovis* and species of the genus *Enterococcus*.

H

halophile Microorganism that grows best at 3% or higher salinity.

haploid Describes a cell that has one complete set of genetic information; all eukaryotic microorganisms have a haploid stage and a diploid (two sets) stage in their life cycle; all prokaryotes are haploid.

hemagglutination General term applied to any agglutination test in which clumping of red blood cells indicates a positive reaction.

hemoglobin Iron-containing protein of red blood cells that is responsible for binding oxygen.

hemolysin A class of chemicals produced by some bacteria that break down hemoglobin and produce hemolysis.

heterotroph An organism that requires carbon in the form of organic molecules. *See also* autotroph.

host An organism that serves as a habitat for another organism such as a parasite or a commensal.

hydrolysis The metabolic process of splitting a molecule into two parts, adding a hydrogen ion to one part and a hydroxyl ion to the other. (One water molecule is used in the process.)

hyperosmotic (solution) A term used to describe extracellular solute concentration relative to the cell; a solution that contains a higher concentration of solutes than the cell, such that water tends to move down its concentration gradient and diffuse out of the cell.

hypha A filament of fungal cells.

hyposmotic A term used to describe extracellular solute concentration relative to the cell; a solution that contains a lower concentration of solutes than the cell, such that water tends to move down its concentration gradient and diffuse into the cell.

I

IMViC Acronym representing the four tests used in identification of *Enterobacteriaceae:* Indole, Methyl Red, Voges-Proskauer, and Citrate.

incomplete medium *See* minimal medium.

incubation The process of growing a culture by supplying it with the necessary environmental conditions.

index case First occurrence of an infection or disease that results in an epidemic.

indirect agglutination Serological test in which artificially produced particulate antigens or antibodies are used to form a visible aggregate if positive. *See also* direct agglutination.

induction Process by which a substrate (inducer) causes the transcription of the genes used in its digestion.

infectious disease A transmissible illness or infection.

inoculum The organisms used to start a new culture or transferred to a new place.

inorganic molecule A molecule that does *not* contain carbon and hydrogen.

intracellular Within the cell; as an intracellular enzyme catalyzing reactions inside the cell.

isolate (*v.*) The process of separating individual cell types from a mixed culture; (*n.*) the group of cells resulting from isolation.

isosmotic Describe extracellular solute concentration relative to the cell; a solution that contains the same concentration of solutes as the cell, such that water tends to move equally into and out of the cell.

K

karyogamy The process of nuclear fusion that occurs after fertilization (plasmogamy) in sexual life cycles.

Krebs cycle A cyclic metabolic pathway found in organisms that respire aerobically or anaerobically.

L

lag phase Closed-system microbial growth phase immediately preceding the exponential phase; characterized by a period of adjustment in which no growth takes place.

limit of resolution The closest two points can be together for the microscope lens to make them appear separate; two points closer than the limit of resolution will blur together.

lipase A family of enzymes that hydrolyze lipids.

lipid A fat.

lophotrichous Describes a flagellar arrangement with a group of flagella at one end of an elongated cell.

lymphocyte A category of white blood cells; one of two types of agranulocyte (along with monocytes); characterized by a large nucleus and little visible cytoplasm; involved in specific acquired immunity as T-cells and B-cells.

lysozyme A naturally occurring bactericidal enzyme in saliva, tears, urine, and other body fluids; functions by breaking peptidoglycan bonds.

lytic cycle The viral life cycle from attachment to lysis of the host cell.

M

mean growth rate constant The number of generations produced per unit time; the inverse of generation time.

medium A substance used for growing microbes; may be liquid (usually a broth) or solid (usually agar).

meiosis The process in which the nucleus of a diploid eukaryotic cell nucleus divides to make four haploid nuclei.

mesophile A microorganism that grows best at temperatures between 15°C and 45°C.

microaerophile A microorganism that requires oxygen at less than atmospheric concentration.

microbial growth curve Graphic representation of microbial growth in a closed system, consisting of lag phase, exponential (log) phase, stationary phase, and death phase.

minimal medium A medium used in genetic experiments, supplying all nutrients for growth *except* the one required by auxotrophs, and thus supporting growth of only prototrophs.

minimum inhibitory concentration The lowest concentration of an antimicrobial substance required to inhibit growth of all microbial cells it contacts; on an agar plate, typically the outer edge of the zone of inhibition where the substance has diffused to the degree that it no longer inhibits growth.

mitosis The process in which a nucleus divides to produce two identical nuclei.

mixed acid fermentation Vigorous fermentation producing many acid products including lactic acid, acetic acid, succinic acid, and formic acid, and subsequently lowering the pH of the medium to pH 4.4 or below.

mold Informal grouping of filamentous fungi. *See also* yeast.

monocyte A category of white blood cells; one of two types of agranulocyte (along with lymphocytes); characterized by large size and lack of cytoplasmic granules; the blood form of macrophages.

monotrichous Describes a flagellar arrangement consisting of a single flagellum.

morbidity Epidemiological measurement of incidence of a disease; typically accompanied by "incidence rate," referring to the incidence of a disease over time.

morphology The shape of an organism.

mortality Epidemiological measurement of death caused by a disease; typically accompanied by "incidence rate," referring to the incidence of death from a disease over time.

mutagen A substance that causes mutation in DNA; most mutagens are carcinogens.

mutation Alteration in a cell's DNA.

mutualistic Describes a synergistic relationship between two organisms in which both benefit from the interaction.

mycelium A mass of fungal filaments (hyphae).

N

nematode A class of roundworms; environmentally abundant in some parasitic species.

neutrophil Category of white blood cells; one of three types of granulocyte (along with eosinophils and basophils) characterized by a granular cytoplasm and lobed nucleus; also known as "polymorphonuclear granulocytes" or "PMNs."

neutrophile Microorganism adapted to a habitat between pH 5.5 and 8.5.

nicotinamide adenine dinucleotide (NAD) A coenzyme used in oxidation/reduction reactions that acts as an electron acceptor or donor, respectively.

nisin Antibiotic produced by *Lactococcus lactis.*

nitrate A highly oxidized form of nitrogen; NO_3.

nitrate reductase An enzyme produced by all members of *Enterobacteriaceae* (and others) that catalyzes the reduction of nitrate (NO_3) to nitrite (NO_2).

nitrite NO_2, an oxidized form of nitrogen.

noninfectious diseases Conditions not caused by microorganisms; examples are stroke, heart disease, and emphysema.

O

objective lens The microscope lens that first produces magnification of the specimen in a compound microscope.

obligate (strict) aerobe Microorganism that requires oxygen to survive and grow.

obligate (strict) anaerobe Microorganism for which oxygen is lethal; requires the complete absence of oxygen.

obligate thermophile Microorganism that grows only at temperatures above 40°C.

ocular lens The lens the microscopist looks through; produces the virtual image by magnifying the real image.

ocular micrometer A uniformly graduated linear scale placed in the microscope ocular used for measuring microscopic specimens.

operon A prokaryotic structural and functional genetic unit consisting of two or more structural genes that code for enzymes in the same pathway and that are regulated together.

opportunistic pathogen A microorganism not ordinarily thought of as pathogenic (*i.e.*, most enterics) that will cause infection when out of its normal habitat.

organic molecule A molecule made of reduced carbon—that is, containing at least carbon and hydrogen.

osmosis Diffusion of water across a semipermeable membrane.

osmotolerant Microorganism that will grow outside of its preferred salinity range.

oxidase ADD def

oxidation/reduction Chemical reaction in which electrons are transferred. (The molecule losing the electrons becomes oxidized; the molecule gaining the electrons becomes reduced.)

oxidative phosphorylation The process by which an electron transport chain is used to add a phosphate to ADP to make ATP.

oxidizing agent A substance that removes electrons from (oxidizes) another. *See also* reducing agent.

P

parasite An organism that lives symbiotically with another, but to the detriment of the other organism (called a host).

parthenogenesis A process in which females produce offspring from an unfertilized egg.

peptidoglycan The insoluble, porous, cross-linked polymer comprising bacterial cell walls; generally thick in Gram-positive organisms and thin in Gram-negative organisms.

peptone A digest of protein used in formulating some bacteriological media.

peritrichous Describes a flagellar arrangement in which flagella arise from the entire surface of the cell.

pH The measure of a solution's alkalinity or acidity; the negative logarithm of the hydrogen ion concentration.

phage *See* bacteriophage.

phage host Bacteria attacked by a virus.

plaque The clearing produced in a bacterial lawn as a result of cell lysis by a bacteriophage; used to calculate phage titer (PFU/mL) when accompanied by a serial dilution.

plaque forming unit (PFU) Term that replaces "viral particle" or "single virus" when referring to phage titer; accounts for multiple particle arrangements in which more than one virus is responsible for initiating a plaque.

plasma The noncellular (fluid) portion of blood; consists of serum (including serum proteins) and clotting proteins.

plasmid Small, circular, extrachromosomal piece of DNA found in prokaryotic cells; often carries genes for antibiotic resistance.

plasmogamy The cytoplasmic fusion of gametes at the time of fertilization. *See also* karyogamy.

plasmolysis Shrinking of cell membrane (pulling away from the rigid cell wall) because of loss of water to the environment and reduced turgor pressure.

polar flagellum A single flagellum at one end of an elongated cell.

pour plate technique Method of plating bacteria in which the inoculum is added to the molten agar prior to pouring the plate.

preadsorption period When performing a plaque assay, the time given to allow a bacteriophage to attach to the host before adding the mixture to the molten agar being plated.

precipitate An insoluble material that comes out of a solution during a precipitation reaction.

presumptive identification Tentative identification of an isolate based on one or more key test results.

primary stain The first stain applied in many differential staining techniques; usually subjected to a decolorization step that forms the basis for the differential stain.

proglottid Tapeworm segments posterior to the scolex, used for absorption of nutrients and containing reproductive organs.

prokaryote Type of cell lacking internal compartmentalization (membranous organelles, including a nucleus) and having 70S ribosomes and a circular molecule of DNA; more primitive than eukaryotes.

promoter site The patch of DNA upstream from the structural gene(s), which binds RNA polymerase to begin transcription.

propagated transmission Conveying a disease person-to-person.

proteolytic Refers to catabolism of protein; *e.g.*, a *proteolytic* enzyme.

prototroph A strain that is capable of synthesizing the nutrient that is acting as the marker in a particular genetic experiment; prototrophs will grow on complete and minimal medium. *See also* auxotroph.

psychrophile A microorganism that grows only at temperatures below 20°C, with an optimum around 15°C.

psychrotroph A microorganism that grows optimally at temperatures between 20 and 30°C but will grow at temperatures as low as 0°C and as high as 35°C.

pure culture Microbial culture containing only a single species.

putrefaction The process of digesting dead organic material; decay.

pyruvic acid (pyruvate) A three-carbon compound produced at the end of glycolysis that may enter a respiration or a fermentation pathway; also serves as a starting point for synthesis of certain amino acids and an entry point for their digestion.

Q

quinoidal (compound) A color-producing compound containing quinone as its central structure.

quorum sensing The phenomenon in bioluminescing bacteria whereby the light-emitting reaction of all cells takes place simultaneously when a threshold concentration of secreted autoinducer is reached.

R

real image Magnified image of a specimen produced by the objective lens of a microscope; the real image is magnified again by the ocular lens to produce the virtual image.

reducing agent A substance that donates electrons to (reduces) another. *See also* oxidizing agent.

reductase An enzyme that catalyzes the transfer of electrons from donor molecule to acceptor molecule, thereby reducing the acceptor.

reduction *See* oxidation/reduction.

refraction The bending of light as it passes from a medium with one refractive index into another medium with a different refractive index.

reservoir A nonhuman host or other site in nature serving as a perpetual source of pathogenic organisms.

resolution The clarity of an image produced by a lens; the ability of a lens to distinguish between two points in a specimen; high resolution in a microscope is desirable.

resolving power *See* limit of resolution.

respiration Metabolic process by which an organic molecule acts as an electron donor and an inorganic substance—such as oxygen, sulfur, or nitrate—acts as the final electron acceptor in an electron transport chain, marking the end of the metabolic sequence (differs from fermentation, which uses one of its own organic products as the final electron acceptor).

reticuloendothelial system Combination of macrophages and associated cells located in the liver, spleen, bone marrow, and lymph nodes.

reversion In carbohydrate fermentation tests, the phenomenon of a microorganism fermentively depleting the carbohydrate and reverting to amino acid metabolism, thereby neutralizing acid products with alkaline products; produces a false negative.

rhizoid A root-like structure used for attachment of some fungi to the substrate.

RNA polymerase A group of enzymes that catalyze the addition of ribonucleotides to the 3' end of an existing polynucleotide chain.

S

saprophyte A heterotroph that digests dead organic matter; a decomposer.

scolex The "head" of a tapeworm, often with suckers and hooks for attachment.

selective medium Growth medium that favors growth of one group of microorganisms and inhibits or prevents growth of others.

sensitivity Ability of a test to identify true positive samples as positive.

serial dilution Series of dilutions used to reduce the concentration of a culture and thereby produce between 30 and 300 colonies when plated, providing a means of calculating the original concentration.

serology A discipline that utilizes a serum containing antibodies (antiserum) to detect the presence of antigens in a sample; also refers to identification of antibodies in a patient's serum.

serum Fluid portion of blood minus the clotting factors.

soft agar A semisolid growth medium containing a reduced concentration of agar; used in plaque assay to allow diffusion of bacteriophage while arresting movement of the bacteriophage host.

solute The dissolved substance in a solution.

solution The mixture of dissolved substance (solute) and solvent (liquid).

solvent The liquid portion of a solution in which solute is dissolved.

specificity Ability of a test to identify only true positives as positive.

spirillum Spiral-shaped cell.

sporangium Structure that produces spores.

spore In bacteria, a dormant form of a microbe protected by specialized coatings produced under conditions of, and resistant to, adverse conditions; also known as an endospore; in fungi and plants, spores are specialized reproductive cells; frequently a means of dissemination.

spread plate technique Method of plating bacteria in which the inoculum is transferred to an agar plate and spread with a sterile bent-glass rod or other spreading device.

stage micrometer A microscope ruler used to calibrate an ocular micrometer.

standard curve A graph constructed from data obtained using samples of known value for the independent variable; once made, can be used to experimentally determine the value of the independent variable when the dependent variable is measured on an unknown.

stationary phase Closed system microbial growth phase immediately following exponential phase; characterized by steady, level growth during which death rate equals reproductive rate.

stolons Surface hyphae of some molds (*e.g., Rhizopus*) that attach to the substrate with rhizoids.

stormy fermentation Vigorous fermentation produced in litmus milk by some species that produce an acid clot but subsequently break it up because of heavy gas production (members of *Clostridium*).

streptolysin Hemolysin (blood hemolyzing exotoxin) produced and secreted by members of *Streptococcus.*

superoxide dismutase Enzyme produced by some bacteria that catalyzes the conversion of superoxide radicals to hydrogen peroxide.

T

thermal death time The amount of time required to kill a population of a specific size at a specific temperature.

thermophile A microorganism that grows best at temperatures above 40°C.

thiosulfate reductase ADD DEF.

titer A measurement of concentration of a substance or particle in a solution; used in measurements of phage concentration.

transformant cells Cells that have undergone transformation by picking up foreign DNA in a genetic engineering experiment.

transformation A form of genetic recombination performed by some bacteria in which DNA is picked up from the environment and incorporated into its genome.

trematode A class of parasitic flatworms; also known as "flukes."

trend line A line drawn on a graph to show the general relationship between X and Y variables; also known as a regression line.

triacylglycerol *See* triglyceride.

tricarboxylic acid cycle *See* Krebs cycle

triglyceride A molecule composed of glycerol and three long chain fatty acids.

trophozoite "Feeding" stage in the life cycle of certain protozoans. *See also* cyst.

tryptophan An amino acid.

tryptophanase Enzyme that catalyzes the hydrolysis of tryptophan into indole and pyruvic acid; detected in the indole test.

turgor pressure The pressure inside a cell that is required to maintain its shape, tonicity, and necessary biochemical functions.

2,3-butanediol fermentation The end-product of a metabolic pathway leading from pyruvate through acetoin with the associated oxidation of NADH; detected in the Voges-Proskauer test.

U

undefined medium A growth medium in which the amount and/or identity of at least one ingredient is unknown.

urease Enzyme that catalyzes the hydrolysis of urea into two ammonias and one carbon dioxide.

utilization medium A differential medium that detects the ability or inability of an organism to metabolize a specific ingredient.

V

variable A factor in a scientific experiment that is changed; the control and experimental groups in a good experiment differ in only one variable, the one being tested.

vector In genetic engineering, a means, often a plasmid or a virus, of introducing DNA into a new host.

vegetative cell An actively metabolizing cell.

virtual image The image produced when the ocular lens of a microscope magnifies the real image; appears within or below the microscope.

Voges-Proskauer test A differential test used to identify organisms that are capable of performing a 2,3-butanediol fermentation.

W

wavelength Measurement of a wave from crest to crest, usually in nanometers—as in electromagnetic energy.

whey Watery portion of milk as seen upon coagulation of casein in the production of a curd.

X

X-Y scatter plot Graph presenting the relationship between two variables.

Y

yeast An informal grouping of unicellular fungi. *See also* mold.

Z

zone of inhibition On an agar plate, the area of nongrowth surrounding a paper disc containing an antimicrobial substance. (The zone typically ends at the point where the diffusing antimicrobial substance has reached its minimum inhibitory concentration, beyond which it is ineffective.)

zygospore The product of fertilization and the site of meiosis in some molds.

zygote The product of gamete fusion (plasmogamy) and nuclear fusion (karyogamy); a fertilized egg.

Index